Methods in Enzymology

Volume 327
APPLICATIONS OF CHIMERIC GENES
AND HYBRID PROTEINS
Part B
Cell Biology and Physiology

METHODS IN ENZYMOLOGY

EDITORS-IN-CHIEF

John N. Abelson   Melvin I. Simon

DIVISION OF BIOLOGY
CALIFORNIA INSTITUTE OF TECHNOLOGY
PASADENA, CALIFORNIA

FOUNDING EDITORS

Sidney P. Colowick and Nathan O. Kaplan

Methods in Enzymology

Volume 327

Applications of Chimeric Genes and Hybrid Proteins

Part B
Cell Biology and Physiology

EDITED BY

Jeremy Thorner

UNIVERSITY OF CALIFORNIA
BERKELEY, CALIFORNIA

Scott D. Emr

HOWARD HUGHES MEDICAL INSTITUTE
AND SCHOOL OF MEDICINE
UNIVERSITY OF CALIFORNIA, SAN DIEGO
LA JOLLA, CALIFORNIA

John N. Abelson

CALIFORNIA INSTITUTE OF TECHNOLOGY
PASADENA, CALIFORNIA

ACADEMIC PRESS

San Diego   London   Boston   New York   Sydney   Tokyo   Toronto

This book is printed on acid-free paper.

Academic Press
A Harcourt Science and Technology Company
525 B Street, Suite 1900, San Diego, California 92101-4495, USA

http://www.academicpress.com

Academic Press Limited
32 Jamestown Road, London NW1 7BY, UK

International Standard Book Number: 0-12-182228-1

PRINTED IN THE UNITED STATES OF AMERICA
00 01 02 03 04 05 06 MM 9 8 7 6 5 4 3 2 1

Table of Contents

Section I. Epitope Tags for Immunodetection

Section II. Markers for Cytology, Analysis of Protein Trafficking, and Lineage Tracing

Section III. Tools for Analysis of Membrane Proteins

Section IV. Signals for Addressing Proteins to Specific Subcellular Compartments

Section V. Application of Chimeras in Monitoring and Manipulating Cell Physiology

Contributors to Volume 327

Article numbers are in parentheses following the names of contributors.
Affiliations listed are current.

STEPHEN R. ADAMS (39, 40), *Department of Pharmacology and Howard Hughes Medical Institute, University of California, San Diego, La Jolla, California 92093*

THOMAS R. ANDERSON (1), *Covance Research Products, Inc., Richmond, California 94804*

V. ANDREEVA (28), *Engelhardt Institute of Molecular Biology, Russian Academy of Sciences, Moscow 117984, Russia*

BRIGITTE ANGRES (7), *Clontech Laboratories, Inc., Palo Alto, California 94303*

CHRISTOPHER AUSTIN (10), *Merck Research Laboratories, West Point, Pennsylvania 19486*

UDO BARON (30), *Zentrum für Molekulare Biologie, Universität Heidelberg, Heidelberg D-69120, Germany*

JON BECKWITH (12), *Department of Microbiology and Molecular Genetics, Harvard Medical School, Boston, Massachusetts 02115*

S. BELLUM (28), *Center for Molecular Medicine, Maine Medical Center Research Institute, South Portland, Maine 04106*

CAROLYN R. BERTOZZI (20), *Departments of Chemistry, and Molecular and Cell Biology, University of California at Berkeley, Berkeley, California 94720*

ANASTASIYA D. BLAGOVESHCHENSKAYA (4), *Medical Research Council Laboratory for Molecular Cell Biology and Department of Biochemistry and Molecular Biology, University College London, London WC1E 6BT, England, United Kingdom*

HERMANN BUJARD (30), *Zentrum für Molekulare Biologie, Universität Heidelberg, Heidelberg D-69120, Germany*

CHRISTOPHER G. BURD (5), *Department of Cell and Developmental Biology and Institute for Human Gene Therapy, University of Pennsylvania School of Medicine, Philadelphia, Pennsylvania 19104-6160*

SHAWN BURGESS (11), *Center for Cancer Research, Massachusetts Institute of Technology, Cambridge, Massachusetts 02139*

JANICE E. BUSS (26), *Department of Biochemistry and Biophysics, Iowa State University, Ames, Iowa 50011*

CONSTANCE L. CEPKO (10), *Department of Genetics, Harvard Medical School and Howard Hughes Medical Institute, Boston, Massachusetts 02115*

RAY CHANG (34), *Affymax Research Institute, Palo Alto, California 94304-1218*

NEIL W. CHARTERS (20), *Department of Molecular and Cell Biology, University of California at Berkeley, Berkeley, California 94720*

HWAI-JONG CHENG (2, 15), *Howard Hughes Medical Institute and Department of Anatomy, University of California, San Francisco, San Francisco, California 94143*

GEOFFREY J. CLARK (26), *Department of Cell and Cancer Biology, Division of Clinical Science, Medical Branch, National Cancer Institute, Rockville, Maryland 20850-3300*

DANIEL F. CUTLER (4), *Medical Research Council Laboratory for Molecular Cell Biology and Department of Biochemistry and Molecular Biology, University College London, London WC1E 6BT, England, United Kingdom*

TAMARA DARSOW (8), *Department of Biology, University of California, San Diego, La Jolla, California 92093-0668*

CHANNING J. DER (26), *Department of Pharmacology, Lineberger Comprehensive Cancer Center, University of North Carolina at Chapel Hill, Chapel Hill, North Carolina 27599*

Scott D. Emr (8), *Howard Hughes Medical Institute and School of Medicine, University of California, San Diego, La Jolla, California 92093-0668*

Michael A. Farrar (31), *Merck Research Laboratories, Rahway, New Jersey 07065-0900*

John D. Fayen (27), *Department of Pathology, Case Western Reserve University, Cleveland, Ohio 44106*

David A. Feldheim (2), *Department of Cell Biology, Harvard Medical School, Boston, Massachusetts 02115*

Shawn Fields-Berry (10), *Department of Genetics, Harvard Medical School and Howard Hughes Medical Institute, Boston, Massachusetts 02115*

John G. Flanagan (2, 15), *Department of Cell Biology and Program in Neuroscience, Harvard Medical School, Boston, Massachusetts 02115*

Christian E. Fritze (1), *Covance Research Products, Inc., Richmond, California 94804-4609*

Clare Futter (3), *Medical Research Council Laboratory for Molecular Cell Biology, University College London, London WC1E 6BT, England, United Kingdom*

Adèle Gibson (3), *Medical Research Council Laboratory for Molecular Cell Biology, University College London, London WC1E 6BT, England, United Kingdom*

Jeffrey Golden (10), *Department of Pathology, Children's Hospital of Philadelphia, Philadelphia, Pennsylvania 19104*

Todd R. Graham (9), *Department of Molecular Biology, Vanderbilt University, Nashville, Tennessee 37235*

Gisele Green (7), *Clontech Laboratories, Inc., Palo Alto, California 94303*

B. Albert Griffin (40), *Aurora Biosciences Corporation, San Diego, California 92121*

Mitsuharu Hattori (2), *Department of Cell Biology, Harvard Medical School, Boston, Massachusetts 02115*

Koret Hirschberg (6), *Cell Biology and Metabolism Branch, National Institute of Child Health and Human Development, National Institutes of Health, Bethesda, Maryland 20892-5430*

Knut Holthoff (38), *Department of Biological Sciences, Columbia University, New York, New York 10027*

B. Diane Hopkins (9), *Department of Molecular Biology, Vanderbilt University, Nashville, Tennessee 37235*

Colin Hopkins (3), *Medical Research Council Laboratory for Molecular Cell Biology, University College London, London WC1E 6BT, England, United Kingdom*

Nancy Hopkins (11), *Biology Department and Center for Cancer Research, Massachusetts Institute of Technology, Cambridge, Massachusetts 02139*

Bryan A. Irving (16), *Department of Microbiology and Immunology, University of California, San Francisco, San Francisco, California 94143-0414*

Ehud Y. Isacoff (19), *Department of Molecular and Cell Biology, University of California at Berkeley, Berkeley, California 94720-3200*

Lara Izotova (42), *Department of Molecular Genetics and Microbiology, University of Medicine and Dentistry of New Jersey, Robert Wood Johnson Medical School, Piscataway, New Jersey 08854-5635*

Christina L. Jacobs (20), *Departments of Chemistry, and Molecular and Cell Biology, University of California at Berkeley, Berkeley, California 94720*

Jay Jones (40), *Aurora Biosciences Corporation, San Diego, California 92121*

Steven R. Kain (7, 37), *Cellomics, Inc., Palo Alto, California 94301*

Heike Krebber (22), *Institut für Molekularbiologie und Tumorforschung, Philipps-Universität Marburg, 35033 Marburg, Germany*

Markku S. Kulomaa (39), *Department of Biology, University of Jyvaskyla, FIN 40351, Jyvaskyla, Finland*

M. LANDRISCINA (28), *Center for Molecular Medicine, Maine Medical Center Research Institute, South Portland, Maine 04106*

JENNIFER A. LEEDS (12), *Department of Microbiology and Molecular Genetics, Harvard Medical School, Boston, Massachusetts 02115*

WARREN J. LEONARD (17), *Laboratory of Molecular Immunology, National Heart, Lung, and Blood Institute, National Institutes of Health, Bethesda, Maryland 20892-1674*

JOHN LIN (10), *Department of Genetics, Harvard Medical School and Howard Hughes Medical Institute, Boston, Massachusetts 02115*

LEI LIN (42), *Department of Molecular Genetics and Microbiology, University of Medicine and Dentistry of New Jersey, Robert Wood Johnson Medical School, Piscataway, New Jersey 08854-5635*

JENNIFER LIPPINCOTT-SCHWARTZ (6), *Cell Biology and Metabolism Branch, National Institute of Child Health and Human Development, National Institutes of Health, Bethesda, Maryland 20892-5430*

JUAN LLOPIS (39), *Facultad de Medicina de Albacete, Universidad de Castilla–La Mancha, 02071 Albacete, Spain*

QIANG LU (2), *Department of Cell Biology, Harvard Medical School, Boston, Massachusetts 02115*

TERRY E. MACHEN (39), *Department of Molecular and Cell Biology, University of California at Berkeley, Berkeley, California 94720*

THOMAS MACIAG (28), *Center for Molecular Medicine, Maine Medical Center Research Institute, South Portland, Maine 04106*

LARA K. MAHAL (20), *Departments of Chemistry, and Molecular and Cell Biology, University of California at Berkeley, Berkeley, California 94720*

YOSHIRO MARU (32), *Department of Genetics, Institute of Medical Science, University of Tokyo, Tokyo 108, Japan*

LARRY C. MATTHEAKIS (34), *Affymax Research Institute, Palo Alto, California 94304-1218*

J. MICHAEL MCCAFFERY (39), *Integrated Imaging Center, Department of Biology, Johns Hopkins University, Baltimore, Maryland 21218*

M. EDWARD MEDOF (27), *Departments of Pathology and Medicine, Case Western Reserve University, Cleveland, Ohio 44106*

TOBIAS MEYER (36), *Department of Pharmacology, Stanford University Medical School, Stanford, California 94305*

GERO MIESENBÖCK (38), *Cellular Biochemistry and Biophysics Program, Memorial Sloan-Kettering Cancer Center, New York, New York 10021*

REBECCA B. MILLER (38), *Cellular Biochemistry and Biophysics Program, Memorial Sloan-Kettering Cancer Center, New York, New York 10021*

ATSUSHI MIYAWAKI (35), *Brain Research Institute, RIKEN, Wako City, Saitama 351-0198, Japan*

HSIAO-PING H. MOORE (39), *Department of Molecular and Cell Biology, University of California at Berkeley, Berkeley, California 94720*

JOHN R. MURPHY (18), *Department of Medicine, Boston University School of Medicine, Boston, Massachusetts 02118*

AKIHIKO NAKUNO (9), *Molecular Membrane Biology Laboratory, RIKEN, Wako, Saitama 351-0198 Japan*

VALERIE NATALE (37), *Clontech Laboratories, Inc., Palo Alto, California 94303*

DAVID A. NAUMAN (20), *Departments of Chemistry, and Molecular and Cell Biology, University of California at Berkeley, Berkeley, California 94720*

ELENA OANCEA (36), *Department of Neurobiology, Childrens Hospital, Boston, Massachusetts 02115*

GREG ODORIZZI (8), *Division of Cellular and Molecular Medicine, University of California and Howard Hughes Medical Institute, San Diego, La Jolla, California 92093-0668*

STEVEN H. OLSON (31), *Merck Research Laboratories, Rahway, New Jersey 07065-0900*

HUGH R. B. PELHAM (21), *MRC Laboratory of Molecular Biology, Cambridge CB2 2QH, England, United Kingdom*

ROGER M. PERLMUTTER (31), *Merck Research Laboratories, Rahway, New Jersey 07065-0900*

SIDNEY PESTKA (42), *Department of Molecular Genetics and Microbiology, University of Medicine and Dentistry of New Jersey, Robert Wood Johnson Medical School, Piscataway, New Jersey 08854-5635*

ROBERT D. PHAIR (6), *BioInformatics Services, Rockville, Maryland 20854*

DIDIER PICARD (29), *Département de Biologie Cellulaire, Université de Genève, Sciences III, 1211 Genève 4, Switzerland*

PAOLO PINTON (33), *Department of Biomedical Sciences, CNR Centre of Biomembranes, University of Padova, 35121 Padova, Italy*

TULLIO POZZAN (33), *Department of Biomedical Sciences, CNR Centre of Biomembranes, University of Padova, 35121 Padova, Italy*

I. PRUDOVSKY (28), *Center for Molecular Medicine, Maine Medical Center Research Institute, South Portland, Maine 04106*

LAWRENCE A. QUILLIAM (26), *Department of Biochemistry and Molecular Biology, Indiana University School of Medicine, Indianapolis, Indiana 46202-5122*

STEPHEN REES (34), *Biological Chemistry Units, Glaxo Wellcome Research and Development, Stevenage, Hertfordshire SG1 2NY, England, United Kingdom*

MARILYN D. RESH (25), *Cell Biology Program, Memorial Sloan-Kettering Cancer Center, New York, New York 10021*

GARY W. REUTHER (26), *Department of Pharmacology, Lineberger Comprehensive Cancer Center, University of North Carolina at Chapel Hill, Chapel Hill, North Carolina 27599*

ROSARIO RIZZUTO (33), *Department of Experimental and Diagnostic Medicine, University of Ferrara, 44100 Ferrara, Italy*

VALERIE ROBERT (33), *Department of Biomedical Sciences, CNR Centre of Biomembranes, University of Padova, 35121 Padova, Italy*

ELIZABETH RYDER (10), *Department of Biology and Biotechnology, Worcester Polytechnic Institute, Worcester, Massachusetts 01609*

KEN SATO (9), *Molecular Membrane Biology Laboratory, RIKEN, Wako, Saitama 351-0198 Japan*

CHRISTIAN SENGSTAG (13), *ETH Zurich, Center for Teaching and Learning, Swiss Federal Institute of Technology, CH-8092 Zurich, Switzerland*

EVE SHINBROT (37), *Clontech Laboratories, Inc., Palo Alto, California 94303*

MICAH S. SIEGEL (19), *Computation and Neural Systems Graduate Program, California Institute of Technology, Pasadena, California 91125, and Department of Molecular and Cell Biology, University of California at Berkeley, Berkeley, California 94720-3200*

PAMELA A. SILVER (22), *Department of Biological Chemistry and Molecular Pharmacology, Harvard Medical School and The Dana Farber Cancer Institute, Boston, Massachusetts 02115*

D. SMALL (28), *Center for Molecular Medicine, Maine Medical Center Research Institute, South Portland, Maine 04106*

R. SOLDI (28), *Center for Molecular Medicine, Maine Medical Center Research Institute, South Portland, Maine 04106*

COLLIN SPENCER (37), *Rigel Corporation, South San Francisco, California 94080*

JENNY STABLES (34), *Lead Discovery, Glaxo Wellcome Research and Development, Stevenage, Hertfordshire SG1 2NY, England, United Kingdom*

IGOR STAGLJAR (14), *Institute of Veterinary Biochemistry, University of Zurich, 8057 Zurich, Switzerland*

JANE STINCHCOMBE (3), *Medical Research Council Laboratory for Molecular Cell Biology, University College London, London WC1E 6BT, England, United Kingdom*

STEPHAN TE HEESEN (14), *ETH Zurich, Microbiology Institute, CH-8093 Zurich, Switzerland*

KEN TETER (39), *Health Sciences Center, University of Colorado, Denver, Colorado 80262*

KOSTAS TOKATLIDIS (24), *School of Biological Sciences, University of Manchester, Manchester M13 9PT, England, United Kingdom*

VALERIA TOSELLO (33), *Department of Biomedical Sciences, CNR Centre of Biomembranes, University of Padova, 35121 Padova, Italy*

ROGER Y. TSIEN (35, 39, 40), *Department of Pharmacology and Howard Hughes Medical Institute, University of California, San Diego, La Jolla, California 92093*

MARK L. TYKOCINSKI (27), *Department of Pathology and Laboratory Medicine, University of Pennsylvania, Philadelphia, Pennsylvania 19104*

WOUTER VAN'T HOF (25), *Pulmonary Research Laboratories, Department of Medicine/Institute for Genetic Medicine, Weill Medical College of Cornell University, New York, New York 10021*

PIERRE VANDERHAEGHEN (2), *Department of Cell Biology, Harvard Medical School, Boston, Massachusetts 02115*

JOHANNA C. VANDERSPEK (18), *Department of Medicine, Boston University School of Medicine, Boston, Massachusetts 02118*

ALEXANDER VARSHAVSKY (41), *Division of Biology, 147-75, California Institute of Technology, Pasadena, California 91125*

KARSTEN WEIS (23), *Department of Molecular and Cell Biology, Division of Cell and Developmental Biology, University of California at Berkeley, Berkeley, California 94720-3200*

ARTHUR WEISS (16), *Howard Hughes Medical Institute and Departments of Medicine and of Microbiology and Immunology, University of California, San Francisco, San Francisco, California 94143*

MINNIE M. WU (39), *Department of Molecular and Cell Biology, University of California at Berkeley, Berkeley, California 94720*

WEI WU (42), *Department of Molecular Genetics and Microbiology, University of Medicine and Dentistry of New Jersey, Robert Wood Johnson Medical School, Piscataway, New Jersey 08854-5635*

KEVIN J. YAREMA (20), *Departments of Chemistry, and Molecular and Cell Biology, University of California at Berkeley, Berkeley, California 94720*

RAFAEL YUSTE (38), *Department of Biological Sciences, Columbia University, New York, New York 10027*

SHIFANG ZHANG (38), *Department of Biological Sciences, Columbia University, New York, New York 10027*

Preface

The modern biologist takes almost for granted the rich repertoire of tools currently available for manipulating virtually any gene or protein of interest. Paramount among these operations is the construction of fusions. The tactic of generating gene fusions to facilitate analysis of gene expression has its origins in the work of Jacob and Monod more than 35 years ago. The fact that gene fusions can create functional chimeric proteins was demonstrated shortly thereafter. Since that time, the number of tricks for splicing or inserting into a gene product various markers, tags, antigenic epitopes, structural probes, and other elements has increased explosively. Hence, when we undertook assembling a volume on the applications of chimeric genes and hybrid proteins in modern biological research, we considered the job a daunting task.

To assist us with producing a coherent work, we first enlisted the aid of an Advisory Committee, consisting of Joe Falke, Stan Fields, Brian Seed, Tom Silhavy, and Roger Tsien. We benefited enormously from their ideas, suggestions, and breadth of knowledge. We are grateful to them all for their willingness to participate at the planning stage and for contributing excellent and highly pertinent articles.

A large measure of the success of this project is due to the enthusiastic responses we received from nearly all of the prospective authors we approached. Many contributors made additional suggestions, and quite a number contributed more than one article. Hence, it became clear early on that given the huge number of applications of gene fusion and hybrid protein technology—for studies of the regulation of gene expression, for lineage tracing, for protein purification and detection, for analysis of protein localization and dynamic movement, and a plethora of other uses—it would not be possible for us to cover this subject comprehensively in a single volume, but in the resulting three volumes, 326, 327, and 328.

Volume 326 is devoted to methods useful for monitoring gene expression, for facilitating protein purification, and for generating novel antigens and antibodies. Also in this volume is an introductory article describing the genesis of the concept of gene fusions and the early foundations of this whole approach. We would like to express our special appreciation to Jon Beckwith for preparing this historical overview. Jon's description is particularly illuminating because he was among the first to exploit gene and protein fusions. Moreover, over the years, he and his colleagues have continued to develop the methodology that has propelled the use of fusion-based techniques from bacteria to eukaryotic organisms. Volume 327 is focused on procedures for tagging proteins for immunodetection, for using chimeric proteins for cytological purposes, especially the analysis of membrane proteins and intracellular protein trafficking, and for monitoring and manipulating various aspects of cell signaling and cell physiology. Included in this volume is a rather extensive section on the green fluorescent protein (GFP) that deals with applications not covered in Volume 302. Volume 328 describes protocols for using hybrid genes and proteins to identify and analyze protein–protein and protein–nucleic interactions, for mapping molecular recognition domains, for directed molecular evolution, and for functional genomics.

We want to take this opportunity to thank again all the authors who generously contributed and whose conscientious efforts to maintain the high standards of the *Methods in Enzymology* series will make these volumes of practical use to a broad spectrum of investigators for many years to come. We have to admit, however, that, despite our best efforts, we could not include each and every method that involves the use of a gene fusion or a hybrid protein. In part, our task was a bit like trying to bottle smoke because brilliant new methods that exploit the fundamental strategy of using a chimeric gene or protein are being devised and published daily. We hope, however, that we have been able to capture many of the most salient and generally applicable procedures. Nonetheless, we take full responsibility for any oversights or omissions, and apologize to any researcher whose method was overlooked.

Finally, we would especially like to acknowledge the expert assistance of Joyce Kato at Caltech, whose administrative skills were essential in organizing these books.

JEREMY THORNER
SCOTT D. EMR
JOHN N. ABELSON

METHODS IN ENZYMOLOGY

VOLUME XXXVI. Hormone Action (Part A: Steroid Hormones)
Edited by BERT W. O'MALLEY AND JOEL G. HARDMAN

VOLUME XXXVII. Hormone Action (Part B: Peptide Hormones)
Edited by BERT W. O'MALLEY AND JOEL G. HARDMAN

VOLUME XXXVIII. Hormone Action (Part C: Cyclic Nucleotides)
Edited by JOEL G. HARDMAN AND BERT W. O'MALLEY

VOLUME XXXIX. Hormone Action (Part D: Isolated Cells, Tissues, and Organ Systems)
Edited by JOEL G. HARDMAN AND BERT W. O'MALLEY

VOLUME XL. Hormone Action (Part E: Nuclear Structure and Function)
Edited by BERT W. O'MALLEY AND JOEL G. HARDMAN

VOLUME XLI. Carbohydrate Metabolism (Part B)
Edited by W. A. WOOD

VOLUME XLII. Carbohydrate Metabolism (Part C)
Edited by W. A. WOOD

VOLUME XLIII. Antibiotics
Edited by JOHN H. HASH

VOLUME XLIV. Immobilized Enzymes
Edited by KLAUS MOSBACH

VOLUME XLV. Proteolytic Enzymes (Part B)
Edited by LASZLO LORAND

VOLUME XLVI. Affinity Labeling
Edited by WILLIAM B. JAKOBY AND MEIR WILCHEK

VOLUME XLVII. Enzyme Structure (Part E)
Edited by C. H. W. HIRS AND SERGE N. TIMASHEFF

VOLUME XLVIII. Enzyme Structure (Part F)
Edited by C. H. W. HIRS AND SERGE N. TIMASHEFF

VOLUME XLIX. Enzyme Structure (Part G)
Edited by C. H. W. HIRS AND SERGE N. TIMASHEFF

VOLUME L. Complex Carbohydrates (Part C)
Edited by VICTOR GINSBURG

VOLUME LI. Purine and Pyrimidine Nucleotide Metabolism
Edited by PATRICIA A. HOFFEE AND MARY ELLEN JONES

VOLUME LII. Biomembranes (Part C: Biological Oxidations)
Edited by SIDNEY FLEISCHER AND LESTER PACKER

VOLUME LIII. Biomembranes (Part D: Biological Oxidations)
Edited by SIDNEY FLEISCHER AND LESTER PACKER

VOLUME LIV. Biomembranes (Part E: Biological Oxidations)
Edited by SIDNEY FLEISCHER AND LESTER PACKER

Section I

Epitope Tags for Immunodetection

[1] Epitope Tagging: General Method for Tracking Recombinant Proteins

By CHRISTIAN E. FRITZE and THOMAS R. ANDERSON

Introduction

Epitope tagging is a procedure whereby a short amino acid sequence recognized by a preexisting antibody is attached to a protein under study to allow its recognition by the antibody in a variety of *in vitro* or *in vivo* settings. Since its first use in 1987 by Munro and Pelham,[1] the epitope-tagging strategy has come to be widely utilized in molecular biology. As testimony to that fact, epitope tagging was employed in some manner in 30% (20 of 64) of the articles in a recent volume of the journal *Cell* (Volume 98, July–October, 1999).

Tagging a protein with an existing epitope is a simple procedure that allows researchers to readily purify or follow proteins through meaningful and revealing experiments quite promptly after expressing a cloned sequence. This stands in sharp contrast to the several months that would otherwise be spent generating and characterizing antisera against the protein itself. Highly specific antibodies and useful cloning vectors encoding epitope tags adjacent to cloning sites are readily available from commercial suppliers or erstwhile collaborators, adding to the ease of initiating such studies.

The most obvious advantage of epitope tagging is that the time and expense associated with generating and characterizing antibodies against multiple proteins are obviated. However, epitope tagging offers a number of additional advantages. For example, because the tag would be missing from extracts of cells that are not expressing a tagged protein, negative controls are unequivocal. Experiments using antibodies against epitopes found in the native molecule cannot provide a comparable negative control. Similarly, epitope tagging can allow for tracking closely related proteins without fear of spurious results resulting from cross-reactive antibodies. The intracellular location of epitope-tagged proteins can be identified in immunofluorescence experiments in a similarly well-controlled manner, without fear of cross-reactivity with the endogenous protein. Because the experimenter has a choice of the tag insertion site in a protein, a site can be selected that is not likely to result in antibody interference with functional

---

[1] S. Munro and H. R. Pelham, *Cell* **48**, 899 (1987).

sites in the molecule, for example, sites that might be the location of protein–protein interactions. Because the antigenic determinant of the epitope tag antibody is in each case defined by a specific peptide, that peptide can be used to elute fusion proteins in purification efforts, avoiding harsh conditions generally used in conventional affinity chromatography. Hence, tagging a protein immediately provides a straightforward purification strategy. Finally, the epitope-tagging approach may be particularly useful for discriminating among otherwise similar gene products that cannot be distinguished with conventional antibodies. For example, epitope tagging permits discrimination of individual members of closely related protein families or the identification of *in vitro*-mutagenized variants in the context of endogenous wild-type protein.

This chapter provides a brief summary of several common experimental procedures that make use of epitope tagging. An effort is made to suggest factors to be considered when designing or troubleshooting experiments involving epitope tagging. Interested readers are directed elsewhere for a description of the historical development of epitope tagging or for a more extensive listing of bibliographic citations,[2] or to past reviews on this topic.[3-5]

General Considerations

Choosing Tags

The most commonly used epitope tags are outlined in Table I. In each case, monoclonal and polyclonal antibodies as well as cloning vectors are widely available. As the use of epitope tagging has become more widespread, a number of observations have been made that can occasionally suggest the preferred use of one tag over another. Several "pros and cons" are noted to help guide the researcher in choosing a tag appropriate to the application. The reader is cautioned, however, that each disadvantage noted in Table I has its exceptions. For example, whereas Table I indicates that the 9E10 antibody is a poor choice for experiments that involve immunoprecipitation of tagged proteins, there are, of course, ample references in the literature to experiments in which immunoprecipitations were effectively accomplished with this antibody.

A number of less commonly used tags are presented in Table II. These

[2] *http://www.babco.com/etagging.html;* C. Fritze and T. Anderson, *Biotechniques,* in preparation.
[3] J. W. Jarvik and C. A. Telmer, *Annu. Rev. Genet.* **32,** 601 (1998).
[4] Y. Shiio, M. Itoh, and J. Inoue, *Methods Enzymol.* **254,** 497 (1995).
[5] P. A. Kolodziej and R. A. Young, *Methods Enzymol.* **194,** 508 (1991).

TABLE I

COMMONLY USED EPITOPE TAGS

Tag	Recognized sequence	Advantages	Disadvantages	Development of antibody[a]	First use as a tag[a]
HA	YPYDVPDYA	Highly specific second-generation antibodies available	Original 12CA5 antibody not optimized for use in epitope tagging	1, 2	3
Myc	EQKLISEEDL	Hybridoma line expressing the 9E10 monoclonal obtainable from the ATCC for use in large-scale projects	9E10 monoclonal may not immunoprecipitate reliably. Endogenous c-myc expression interferes with use as epitope tag	4	5
FLAG	DYKDDDK	Epitope easily cleaved off of tagged protein after purification	Detection by some antibodies requires placement at the protein termini	6–8	9
Polyhistidine	HHHHHH	Tagged proteins can be purified on Ni^{2+} affinity matrix. Rare sequence makes cross-reactivity with endogenous proteins unlikely	Some commercially available antibodies require additional amino acids to specify recognition		

[a] Key to references: (1) H. L. Niman, R. A. Houghten, L. E. Walker, R. A. Reisfeld, I. A. Wilson, J. M. Hogle, and R. A. Lerner, *Proc. Natl. Acad. Sci. U.S.A.* **80**, 4949 (1983); (2) I. A. Wilson, H. L. Niman, R. A. Houghten, A. R. Cherenson, M. L. Connolly, and R. A. Lerner, *Cell* **37**, 767 (1984); (3) J. Field, J. Nikawa, D. Broek, B. MacDonald, L. Rodgers, I. A. Wilson, R. A. Lerner, and M. Wigler, *Mol. Cell. Biol.* **8**, 2159 (1988); (4) G. I. Evan, G. K. Lewis, G. Ramsay, and J. M. Bishop, *Mol. Cell. Biol.* **5**, 3610 (1985); (5) S. Munro and H. R. Pelham, *Cell* **48**, 899 (1987); (6) K. S. Prickett, D. C. Amberg, and T. P. Hopp, *BioTechniques* **7**, 580 (1989); (7) B. L. Brizzard, R. G. Chubet, and D. L. Vizard, *BioTechniques* **16**, 730 (1994); (8) R. G. Chubet and B. L. Brizzard, *BioTechniques* **20**, 136 (1996); (9) T. P. Hopp, K. S. Prickett, V. L. Price, R. T. Libby, C. J. March, D. P. Cerretti, D. L. Urdal, and P. J. Conlon, *BioTechniques* **6**, 1204 (1988); (10) E. Hochuli, W. Bannwarth, H. Dobeli, and R. Gentz, *Bio/Technology* **6**, 1321 (1988).

TABLE II
OTHER EPITOPE TAGS

Tag	Recognized sequence	Comments	Ref.
AU1	DTYRYI	Optimized for immunostaining and immunohisto-chemistry, may be harder to detect via immunoblotting	a, b
AU5	TDFYLK	Optimized for immunostaining and immunohisto-chemistry, may be harder to detect via immunoblotting	a, c
IRS	RYIRS	Must be placed at protein C terminus. Small epitope; IRS is often sufficient to specify recognition by the antibody	d, e
B-tag	QYPALT	Epitope consists entirely of uncharged amino acids	f, g
Universal	HTTPHH	Ease of cloning: the DNA sequence encoding the epitope is translated as HTTPHH regardless of reading frame	
S-Tag	KETAAAKFERQHMDS	Tag binds S-protein for purification and detection	h
Protein C	EDQVDPRLIDGK	Available calcium-dependent antibodies facilitate purification	i, j
Glu-Glu	EYMPME or EFMPME		k, l
KT3	PPEPET		m, n
VSV	MNRLGK		o
T7	MASMTGGQQMG		p
HSV	QPELAPEDPED		q

[a] P. S. Lim, A. B. Jenson, L. Cowsert, Y. Nakai, L. Y. Lim, X. W. Jin, and J. P. Sundberg, *J. Infect. Dis.* **162,** 1263 (1990).

[b] D. J. Goldstein, R. Toyama, R. Dhar, and R. Schlegel, *Virology* **190,** 889 (1992).

[c] P. Crespo, K. E. Schuebel, A. A. Ostrom, J. S. Gutkind, and X. R. Bustelo, *Nature* (*London*) **385,** 169 (1997).

[d] B. Rubinfeld, S. Munemitsu, R. Clark, L. Conroy, K. Watt, W. J. Crosier, F. McCormick, and P. Polakis, *Cell* **65,** 1033 (1991).

[e] W. Luo, T. C. Liang, J. M. Li, J. T. Hsieh, and S. H. Lin, *Arch. Biochem. Biophys.* **329,** 215 (1996).

[f] D. H. Du Plessis, L. F. Wang, F. A. Jordaan, and B. T. Eaton, *Virology* **198,** 346 (1994).

[g] L. F. Wang, A. D. Hyatt, P. L. Whiteley, M. Andrew, J. K. Li, and B. T. Eaton, *Arch. Virol.* **141,** 111 (1996).

[h] N. C. Chi, E. J. H. Adam, and S. A. Adam, *J. Biol. Chem.* **272,** 6818 (1997).

[i] D. J. Stearns, S. Kurosawa, P. J. Sims, N. L. Esmon, and C. T. Esmon, *J. Biol. Chem.* **263,** 826 (1988).

[j] A. R. Rezaie, M. M. Fiore, P. F. Neuenschwander, C. T. Esmon, and J. H. Morrissey, *Protein Expr. Purif.* **3,** 453 (1992).

[k] T. Grussenmeyer, K. H. Scheidtmann, M. A. Hutchinson, W. Eckhart, and G. Walter, *Proc. Natl. Acad. Sci. U.S.A.* **82,** 7952 (1985).

[l] B. Rubinfeld, S. Munemitsu, R. Clark, L. Conroy, K. Watt, W. J. Crosier, F. McCormick, and P. Polakis, *Cell* **65,** 1033 (1991).

[m] H. MacArthur and G. Walter, *J. Virol.* **52,** 483 (1984).

[n] G. A. Martin, G. Viskochil, G. Bollag, P. C. McCabe, W. J. Crosier, H. Haubruck, L. Conroy, R. Clark, P. O'Connell, and R. M. Cawthon, *Cell* **63,** 843 (1990).

[o] T. E. Kreis, *EMBO J.* **5,** 931 (1986).

[p] G. Baier, D. Telford, L. Giampa, K. M. Coggeshall, G. Baier-Bitterlich, N. Isakov, and A. Altman, *J. Biol. Chem.* **268,** 4997 (1993).

[q] J. Sakai, E. A. Duncan, R. B. Rawson, X. Hua, M. S. Brown, and J. L. Goldstein, *Cell* **85,** 1037 (1996).

have found niches in the scientist's arsenal because of their suitability for specialized applications (e.g., the AU1 and AU5 tags are particularly useful for immunostaining), or because of experimental needs to use multiple different tags simultaneously.

Tag Placement

The driving motivation behind the epitope-tagging strategy is to attach a small "handle" onto a protein under study without disturbing native protein structure and function. The choice of tag location will be dictated primarily by whatever regions are not eliminated from consideration, based on the existence of known sequence motifs such as substrate-binding sites, extensive hydrophobic regions (which may be buried internally in the mature protein), sites of protein–protein interaction, and kinase recognition sites. It is advisable to compare the coding sequence to be tagged against PROSITE[6] or a similar motif database to identify probable sites of protein modification, interaction, or cleavage. The more that is known about such sites at the outset, the more dependable the educated guess about where insertion of a small epitope tag will be tolerated with little functional impact.

In most cases the ease of cloning leads to the choice of the N or C terminus of the protein for placement of the epitope tag, but this choice must be made cautiously. N-terminal myristoylation sites or signal sequences destined for removal, as well as C-terminal isoprenylation sites (CAAx) or PDZ domain-binding motifs (T/SxV/I), are among the sequences that may make terminal epitope tag placement ill advised. Although some of the common epitope tag antibodies recognize their epitopes *only* at one particular end of a molecule (see Tables I and II), most function well within the coding region as well. This is perhaps best exemplified by experiments designed to determine the topology of integral membrane proteins, in which multiple constructs were made with tags inserted at sites all along the length of the protein.[7–10]

There are times when attachment of a single epitope tag to a protein will give unacceptably low levels of recognition by the corresponding antibody. This is especially the case in experiments in which the protein must be recognized in its native conformation, such as immunostaining or immunoprecipitation. In such cases, addition of multiple copies of the tag may help to improve recognition of the tagged protein by the antibody, either

---

[6] K. Hofmann, P. Bucher, L. Falquet, and A. Bairoch, *Nucleic Acids Res.* **27,** 215 (1999).

[7] C. Kast, V. Canfield, R. Levenson, and P. Gros, *J. Biol. Chem.* **271,** 9240 (1996).

[8] V. A. Canfield L. Norbeck, and R. Levenson, *Biochemistry* **35,** 14165 (1996).

[9] J. Borjigin and J. Nathans, *J. Biol. Chem.* **269,** 14715 (1994).

[10] A. Charbit, J. Ronco, V. Michel, C. Werts, and M. Hofnung, *J. Bacteriol.* **173,** 262 (1991).

by providing additional sites for antibody binding or locally perturbing protein structure so that the tagged region is more exposed.[7] Alternatively, several groups have reported an increase in antibody sensitivity by adding a linker adjacent to the tag.[11] The addition of a short polyglycine motif, for instance, probably serves to distance the epitope from the rest of the protein structure, adds flexibility, and generally improves antibody accessibility.[9]

A further confounding circumstance may arise when the predominant full-length protein apparently does not contain the epitope tag. This may result from nonspecific degradation of the recombinantly expressed protein or outright cleavage by a specific protease. In such cases, it may be useful to pass the tagged protein sequence through the PROSITE database or other algorithm that identifies protease cleavage sites to eliminate the possibility that such a site has been inadvertently generated at the juncture of the epitope tag and native protein sequence. Nonspecific degradation may arise from a too rapid or robust induction of the expressed protein. Growth conditions and supplements that attenuate the onset of induction may be beneficial. The use of less inducer, induction at lower temperature, or induction in the presence of additives that slow isopropyl-β-D-thiogalactopyranoside (IPTG) induction can all be useful strategies.[12] Finally, if all else fails, use of a different tag location or choice of an alternative tag can be considered.

Attaching Tags to Proteins

Engineering an expression clone with the selected epitope tag fused to the open reading frame of interest is typically straightforward. If the tag is to be placed at the N or C terminus of the protein, the researcher can employ one of many publicly or commercially available vectors. Many modern cloning vectors contain one or more epitope tags flanking the polylinker, suitably coupled to bacterial and/or eukaryotic promoters and transcription terminators. Use of these sorts of vectors allows a single cloning step in which the coding sequence is ligated into the polylinker. DNA sequencing across the cloning junction and a quick Western analysis of extract from a transfected line are then used to confirm the integrity of the resulting construct. If an off-the-shelf vector is not appropriate, the small size of the tag coding sequence makes it possible to incorporate the entire tag sequence into a polymerase chain reaction (PCR) oligonucleotide primer homologous to the sequence of interest, so that PCR across the

[11] E. Grote, J. C. Hao, M. K. Bennett, and R. B. Kelly, *Cell* **81,** 581 (1995).
[12] K. Furukawa, C. E. Fritze, and L. Gerace, *J. Biol. Chem.* **273,** 4213 (1998).

sequence generates a tagged product. An epitope tag can also be introduced into a preexisting construct by ligating a double-stranded oligonucleotide into a suitable restriction site in the coding sequence. In all of these cases it is, of course, essential to be aware of the proper reading frame across cloning junctions, and to verify the fidelity of the final constructs by DNA sequencing.

Specific Methods and Considerations

This section presents a few of the more important protocols that most researchers would typically need to perform for the surveillance of epitope-tagged proteins. These methods are well represented elsewhere in the annals of immunology and molecular biology.[13-15] However, because epitope tagging sidesteps a major investment in antibody production and characterization, many researchers embark on their first forays into immunological studies by the way of this strategy. Hence, these basic protocols may be of use to the reader here. Concentration, dilutions, etc., indicated in the text are intended as suggested starting points. Specific parameters for each experiment cannot be defined *a priori,* and should be fine tuned in individual experiments. Emphasis is placed on common pitfalls and considerations to be made when less than optimal results are obtained in initial experiments. The experienced practitioner may wish to skim these protocols quickly, focusing instead on specific considerations.

Immunoblotting (Western Blots)

Protocol

1. Resolve sample proteins and controls via polyacrylamide gel electrophoresis. Transfer the proteins to nitrocellulose by standard methods.

2. Remove the blot from the transfer apparatus and soak in Tween–Tris-buffered saline [TTBS: 0.1% (v/v) Tween 20 in 100 mM Tris-HCl (pH 7.5), 0.9% (w/v) NaCl] for two rinses of 15 min each. The blot may be marked (with pencil or India ink) for identification at this stage if desired.

3. Block the blot with 10% (w/v) nonfat dried milk (NFDM) made

[13] J. Sambrook, E. F. Fritsch, and T. Maniatis, "Molecular Cloning: A Laboratory Manual." Cold Spring Harbor Laboratory Press, Cold Spring Harbor, New York, 1989.

[14] J. E. Coligan, A. M. Kruisbeek, D. H. Margulies, E. M. Shevach, and W. E. Strober, "Current Protocols in Immunology." John Wiley & Sons, New York, 1992.

[15] E. Harlow and D. Lane, "Antibodies: A Laboratory Manual." Cold Spring Harbor Laboratory Press, Cold Spring Harbor, New York, 1988.

fresh in TTBS; rock on a rotating shaker for 15 min at room temperature or overnight at 4°.

4. Rinse the blot three times in TTBS.

5. Probe with primary antibody in 1% (w/v) NFDM for 1 hr at room temperature. Primary antisera or ascites should be diluted 1:1000 to 1:10,000 in initial experiments; purified primary antibodies should be used at a concentration of 1 μg/ml.

6. Rinse the blot three times in TTBS.

7. Probe the blot with an enzyme-linked secondary antibody (typically horseradish peroxidase or alkaline phosphatase) in 1% (w/v) NFDM for 30 min at room temperature. Review instructions included with the secondary antibody to determine the appropriate dilution to use.

8. Rinse excess secondary antibody from the blot with three rinses in 20–50 ml of TTBS for 5 min each. The blot is now ready for use with standard colorimetric or chemiluminescent detection reagents.

Considerations. Perhaps the most common problem in Western blotting is the occurrence of high background. Fortunately, in the case of Western blots utilizing epitope tag antibodies, the antibodies are quite specific. Hence the best remedy for high background (in many cases) is simply to dilute the primary antibody further. Other solutions standard to Western blotting would include ensuring that detergent is used in the blocking reagent, using an alternative blocking reagent [casamino acids, bovine serum albumin (BSA), serum], and decreasing the amount of protein applied to the electrophoresis gel.

Lack of any signal at all is another frustrating result. In this instance it is vital to ensure that the protein is in fact being expressed, perhaps by using an antibody specific for the native sequence of the molecule as a test. Other strategies would include loading more protein or increasing the amount of primary antibody in developing the blot.

Occurrences of extra bands in the blot can sometimes be resolved by several strategies. Running a control blot omitting the primary antibody can determine if the secondary antibody is the source of the problem. Replacing the secondary antibody with a different lot or a similar reagent from a different source can provide resolution. Spurious bands below the targeted molecular weight suggest that the protein is being degraded in the experiment; inclusion of protease inhibitors can help. Although not commonly invoked as a strategy for immunoblotting, the signal-to-noise ratio in the experiment can also be enhanced by using a protein tagged with multiple copies of the tag, as discussed above.

Some antibodies will not bind in the presence of detergent; the data sheet for each antibody should be consulted prior to performing any procedure.

Immunoprecipitation

Protocol

1. Divide the preparation of antigen into two equally sized aliquots and place in microcentrifuge tubes. Adjust the volume of each aliquot to 0.5 ml with immunoprecipitation buffer [IP buffer: 50 mM Tris-HCl (pH 7.5), 150 mM NaCl, 0.1% (v/v) Tween 20 or 0.1% (v/v) Nonidet P-40 (NP-40), 1 mM EDTA (pH 8.0), 0.25% (w/v) gelatin, 0.02% (w/v) sodium azide].

2. To one aliquot, add antibody directed against the appropriate tag. To the other aliquot, add the same volume of a control antibody. Gently rock both aliquots for 1 hr at 4°.

3. Add protein G–Sepharose to the antigen–antibody mixtures, and incubate for 1 hr at 4° on a rocking platform.

4. Centrifuge the protein G–Sepharose antibody–antigen complex at 10,000g for 20 sec at 4° in a microcentrifuge tube. Remove the supernatant by gentle aspiration. Add 1 ml of IP buffer and resuspend the beads.

5. Incubate the resuspended beads for 20 min at 4° on a rocking platform.

6. Repeat steps 4 and 5 three times. Collect the final washed protein G–Sepharose antibody–antigen complex by centrifugation at 10,000g for 20 sec at 4° in a microcentrifuge. Take care to remove the last traces of the final wash.

7. Add reducing gel loading buffer [50 mM Tris-HCl (pH 6.8), 10% (v/v) glycerol, 2% (w/v) sodium dodecyl sulfate (SDS), 5% (v/v) 2-mercaptoethanol, 0.0025% (w/v) bromphenol blue] and boil for 3 min.

8. Remove the protein G–Sepharose from the complex by centrifugation at 10,000g for 20 sec at room temperature in a microcentrifuge. Transfer the supernatant to a fresh tube and separate the sample components by electrophoresis.

Considerations. Typically, complete immunoprecipitation of radiolabeled antigen from extracts of transfected mammalian cells requires between 1 and 5 μl of polyclonal antiserum, 5–100 μl of hybridoma tissue culture medium, or 1–3 μl of ascites. If more antibody is used than necessary, nonspecific background will increase. In fact, it is probably ideal to utilize an amount of antibody that does not have the capacity to capture all the antigen, to minimize nonspecific precipitation.

In some instances, antibodies are available already bound to Sepharose beads, eliminating some steps from the procedure.

As noted earlier, results from some epitope tag experiments have been enhanced by making a construct that has multiple copies of the tag expressed in tandem. This may be a particularly good strategy for immunoprecipitation efforts, inasmuch as it would allow a single copy of the protein to be bound by multiple antibodies, making for larger complexes and more efficient precipitation.

Note that in experiments in which the desire is to coimmunoprecipitate other proteins that interact with the tagged protein, it may be necessary to omit the detergent from the precipitation buffer or consider alternative detergents so that protein interactions are not disrupted.

Affinity Purifications with Epitope Tag Antibodies

Protocol

Two protocols are provided. The first is appropriate for small-scale immunopurification efforts performed in a microcentrifuge tube.

1. Combine affinity matrix and immunopurification buffer [200 mM 2-(N-morpholino)ethanesulfonic acid (MES, pH 6.2), 0.1 mM MgCl$_2$, 0.1 mM EDTA, 1 mM 2-mercaptoethanol (see Considerations), 1 mM phenylmethylsulfonyl fluoride, 0.05% (v/v) Tween, and 0.5 M NaCl] at a 1:5 dilution in a microcentrifuge tube. Add tagged protein to the mixture. Gently rock the aliquots for 1 hr at 4°.

2. Centrifuge the mixture at 10,000g for 20 sec at 4°. Remove the supernatant without disturbing the beads. Add half again as much immunopurification buffer to the beads and resuspend the matrix. Gently rock the aliquots for 20 min at 4°. Keep a portion of the supernatant from each rinse step to use in Western blot analysis.

3. Repeat step 2 four times.

4. Elute the bound protein with the appropriate epitope peptide at a 1-mg/ml concentration in immunopurification buffer. Resuspend the beads, incubate, centrifuge, and withdraw supernatant as in step 2, repeating for a total of four elutions. Recover as much of the eluate as possible at each stage.

5. For each elution sample, prepare at a 1:1 dilution with reducing gel loading buffer. Boil the tubes for 3 min.

6. Analyze the supernatant samples by Western blot. If using the first wash as a starting point, the tagged protein band should fade through the first four washes. After the addition of peptide, the eluted tagged protein band will appear again in the eluate. The elution in which the strongest band appears will have the greatest concentration of eluted protein.

The following protocol is appropriate for larger scale affinity purification efforts and is accomplished in a chromatography column. The starting material in this instance would be a crude extract from a 100-ml bacterial culture or equivalent.

1. Pass the material through a 1-ml Sepharose column to remove any proteins that nonspecifically bind to Sepharose.

2. Prepare the affinity matrix column by cross-linking 2 mg of purified monoclonal antibody to 1 ml of NHS-activated Sepharose, according to the manufacturer instructions, or purchase ready-made material. Resuspend the cross-linked beads in immunopurification buffer.

3. Pack the antibody-bound Sepharose resin into a column and wash with several column volumes of immunopurification buffer at 4°.

4. Load the sample onto the monoclonal antibody column, collecting the flowthrough, and then reload the material again.

5. Wash with 100 ml of buffer, and then close the column outflow.

6. Prepare elution buffer by resuspending the appropriate epitope peptide at 0.4 mg/ml in immunopurification buffer.

7. Add 2.5 ml of elution buffer to the column and incubate for 15 min at room temperature.

8. Open the column outflow and collect the eluate. Repeat the elution twice more.

9. Analyze fractions by gel electrophoresis, and concentrate if desired.

10. Strip the column with 100 mM glycine buffer, pH 2.9, followed by immunopurification buffer until the pH returns to neutral.

11. Store the column in phosphate-buffered saline (PBS) containing 0.03% (w/v) thimerosal at 4°.

Considerations. The ingredient 2-mercaptoethanol is not always necessary but is included in the immunopurification buffer to improve the solubility of the proteins in the lysate. At the concentration indicated, it should not reduce antibodies. Dithiothreitol (DTT) may be used as a substitute.

If recovery of purified protein is poor, the elution step can be carried out at 30° with prewarmed elution buffer.

Other elution buffers such as 0.1 *M* glycine, pH 2.8, or 40 mM diethylamine, pH 11.0, may be used to strip the column.

Immunofluorescence

Protocol

1. Rinse the cells attached to coverslips briefly twice with ice-cold PBS, removing liquid by gentle aspiration in this and subsequent steps.

2. Fix the cells with 4% (v/v) formaldehyde in PBS for 6 min at room temperature, and then rinse briefly twice with PBS.

3. Permeabilize the fixed cell with 0.2% (v/v) Triton X-100 in PBS for 6 min.

4. Wash the cells briefly three times with PBS, and then twice with PBS containing 1% (w/v) BSA (blocking reagent).

5. Dilute the primary antibody in 1% (w/v) BSA in PBS. Working quickly, aspirate area surrounding the coverslip to dryness, then gently add 100 μl of diluted primary antibody to the coverslip, so that the solution remains restricted to the coverslip by surface tension. Incubate for 1 hr at room temperature in a moist environment to prevent drying.

6. Wash the cells three times with PBS, and then twice with PBS–1% (w/v) BSA.

7. Dilute the fluorochrome-coupled secondary antibody in PBS–1% (w/v) BSA and apply as in step 5. Incubate for 1 hr at room temperature.

8. Wash the cells three times with PBS, then mount coverslips on the slides, using antifade mounting medium.

Considerations. Adherent cells may be grown directly on coverslips or chambered slides; suspension cells may be adhered to coverslips via poly-L-lysine treatment.

Care should be taken to use the highest quality primary and secondary antibodies in order to avoid nonspecific labeling. Ideally, the specificity of primary antibodies is confirmed via immunoblotting of cell extracts. A control immunofluorescence sample omitting the use of primary antibody will demonstrate nonspecific signal generated by the secondary antibody.

In case of high background, the use of less primary and/or secondary antibody as well as increased or alternative blocking reagent can be considered. If the assay involves localization of a tagged protein expressed from a heterologous promoter, then the researcher should keep in mind that overexpression of the protein may produce mislocalization and hence broader staining than expected from endogenous protein.

Several approaches can be considered in cases of an unacceptably low signal. The immunoflurescence protocol itself may be altered: Use increased amounts of primary antibody, extend the incubation of primary antibody to overnight at 4°, fix the cells with methanol or acetone, or consider the use of an epitope tag antibody from a different supplier. If these measures are not sufficient, it may be that the tag is not sufficiently exposed to the primary antibody in the context of the native protein structure. It may be necessary to move the tag to a different location within the protein or tag the protein with multiple tandem copies of the epitope.

Immunohistochemistry

Protocol

This protocol is for paraformaldehyde-fixed paraffin-embedded tissue sections, and a biotinylated primary antibody and a horseradish peroxidase–

avidin conjugate staining procedure. Other methods for tissue preparation are available (such as frozen sections) as are other protocols (such as unlabeled primary antibody detected with a secondary antibody) and other detection reagents (such as alkaline phosphatase). Many of the same considerations indicated for this protocol apply to those methods as well.

1. Prepare the tissue by standard means, such as by immersing the tissue fragment in 4% (w/v) paraformaldehyde in PBS for 6 hr.

2. Dehydrate the tissue by standard methods involving ethanol and xylene immersion followed by embedding in paraffin. Prepare 5- to 8-μm sections and affix the sections to slides.

3. Dry the slides and deparaffinize in Histoclear and ethanol.

4. Block endogenous peroxidase with 0.3% (v/v) H_2O_2 in methanol.

5. Wash the slides twice in PBS, 10 min each.

6. Block by incubating with 5% (v/v) goat serum or other blocking reagent.

7. Apply biotinylated primary antibody diluted 1 : 100 in PBS. Incubate for 1 hr at room temperature.

8. Wash the slides twice in PBS.

9. Apply a horseradish peroxidase–avidin conjugate, and incubate for 20 min.

10. Wash the slides twice in PBS.

11. Add the substrate solution, and incubate for 5 min.

12. Wash the slides.

13. Counterstain with hematoxylin, using standard techniques.

14. Wash the slide, apply Aquamount and a coverslip, and allow to dry.

Considerations. If this is the first foray of the reader into immunohistochemistry, enlistment of a collaborator in a dedicated histology laboratory would be well advised. Much of the equipment and many of the procedures for fixing, dehydrating, clearing, embedding, and sectioning tissues are routine in a histology laboratory but quite foreign to the molecular biologist.

Note that immunohistochemistry is notable for "variations on a theme." Primary antibodies can be applied unlabeled and be detected with a labeled secondary antibody, or can be labeled with biotin to form a link to an avidin-conjugated enzyme, or can be directly labeled with an enzyme. While use of secondary antibodies and/or use of a biotinylated epitope tag antibody leads to more steps in the procedure, both of those strategies also increase the signal generated by the antibody and thereby improve resolution in the experiment.

Double-staining experiments, in which two different primary antibodies directed against two different antigens are used, can be particulary revealing. In many cases this is accomplished with primary antibodies generated in two different species, along with species-specific secondary antibodies labeled with two different enzymes. Similar results can be achieved with directly labeled primary antibodies, although that strategy would quite likely result in diminished signal.

Immunohistochemistry, perhaps more than any of the other techniques presented in this chapter, will demand titration of reagent concentrations and incubation times for individual experiments. This is particularly true for more complicated "stacks" in the detection strategy (primary antibody detected by a biotinylated secondary antibody detected by an avidin-conjugated enzyme identified by a colorogenic substrate) or in experiments in which two antibodies are utilized to identify two antigens.

Note that secondary antibodies can cross-react with endogenous immunoglobulin, resulting in excess background in some experiments. For example, if a mouse-derived monoclonal antibody is used to detect an antigen in a rat tissue, the secondary antibody (directed against mouse immunoglobulin) might cross-react with endogenous rat immunoglobulin. That this is the source of the background can be confirmed with a control slide omitting the primary antibody. This sort of background can be prevented or diminished by using commercially available species-specific reagents, or by adsorbing the secondary antibody to serum derived from the species of the tissue being examined.

Note that a well-designed immunohistochemical analysis demands multiple controls. An isotype-matched antibody control for the primary antibody confirms that the signal is not due to background. Use of primary antibody preabsorbed to the antigen (in this case, adsorbed to the tag sequence) and controls omitting the primary antibody provide similar assurances. Controls are also necessary for endogenous peroxidase activity when horeseradish peroxidase is used as the detection system. This can be done with a substrate-only control. In double-staining procedures, the list of controls would expand to confirm that each stain is working independent of the other.

Summary

Epitope tagging has provided a useful experimental strategy with widespread applicability. The ample variety of epitope tag systems that have been put to use to date provide a collection of attributes relevant to virtually any experimental system. As a consequence, epitope tagging will continue to be a valuable tool for molecular biologists long into the future.

Acknowledgment

The authors thank Silvio Gutkind for helpful suggestions, Mendi Warren for providing useful protocols, and Feran Pete for editorial word-processing talents.

Section II

Markers for Cytology, Analysis of Protein Trafficking, and Lineage Tracing

[2] Alkaline Phosphatase Fusions of Ligands or Receptors as *in Situ* Probes for Staining of Cells, Tissues, and Embryos

By JOHN G. FLANAGAN, HWAI-JONG CHENG, DAVID A. FELDHEIM, MITSUHARU HATTORI, QIANG LU, and PIERRE VANDERHAEGHEN

Introduction

Polypeptide ligands and their cell surface receptors bind to one another with high affinity and specificity. These biological properties can be exploited to make affinity probes to detect their cognate ligands or receptors. This approach has been applied for decades, using radiolabeled ligands as probes to detect their receptors. More recently, it has also been found that receptor ectodomains can be used as soluble probes to detect their ligands.[1,2]

When producing soluble receptor or ligand affinity probes, it has been common to produce the probe as a fusion protein with a tag. This can make detection and purification procedures much easier. Two tags that are widely used for this purpose are alkaline phosphatase (AP)[1] or the immunoglobulin Fc region.[2] Both of these tags are dimeric, and both are expected to produce a fusion protein with a pair of ligand or receptor moieties facing away from the tag in the same direction. This dimeric structure is likely to be an important feature in many experiments, because it is likely to increase greatly the avidity of the fusion protein for ligands or receptors that are oligomeric, or are bound to cell surfaces or extracellular matrix. The principles of using either AP or Fc fusion proteins are similar; here we focus on procedures for the AP tag.

An advantage of the AP tag is that it possesses an intrinsic enzymatic marker activity. It is therefore generally not necessary to purify the fusion protein, chemically label it, or use secondary reagents such as antibodies. This simplifies probe production, and also helps make detection procedures simple and extremely sensitive. Fusions can be made at either the N or C terminus of AP. The human placental isozyme of AP[3] is used because it is highly stable, including a high heat stability that allows it to survive heat inactivation steps to destroy background phosphatase activities. The enzyme also has an exceptionally high turnover number (k_{cat}), allowing

[1] J. G. Flanagan and P. Leder, *Cell* **63,** 185 (1990).

[2] A. Aruffo, I. Stamenkoviv, M. Melnick, C. B. Underhill, and B. Seed, *Cell* **61,** 1301 (1990).

[3] J. Berger, A. D. Howard, L. Brink, L. Gerber, J. Hauber, B. R. Cullen, and S. Udenfriend, *J. Biol. Chem.* **263,** 10016 (1988).

sensitive detection. A wide variety of substrates for AP are available that allow either detection *in situ,* or quantitative assays in solution.

In many respects soluble ligand or receptor fusion probes resemble antibodies, and can be used in almost all the same types of procedure. They can, however, be produced far more quickly than antibodies. In our experience production of active fusion proteins has been reliable, although this will depend on the properties of the individual receptor or ligand. Detection procedures are quick and simple, usually taking only a few hours. Notably, because these fusion probes exploit natural receptor–ligand interactions, they can give information not available with antibody probes or other techniques. For example, they can be used to identify previously unknown ligands or receptors, they can allow quantitative characterization of ligand–receptor interactions, they can distinguish active from masked or degraded molecules, or they can allow the simultaneous detection of multiple cross-reacting ligands in an embryo.

Initial applications of receptor or ligand fusion protein probes focused on the identification and cloning of previously unknown ligands or receptors.[4] More recently it has been found that receptor and ligand fusion proteins can be used efficiently as *in situ* probes to detect the distribution of cognate ligands or receptors in embryos.[5] Increasingly, this approach is taking a place alongside other techniques to study the spatial distribution of biological molecules, such as immunolocalization or RNA hybridization, as a technique that can provide valuable and sometimes unique information in understanding biological systems (e.g., see Refs. 5–12). At the same time, it is important to remember that, because all the available techniques give different types of information, it can be valuable to obtain confirmatory information by using two or more of them.[13]

In this chapter we describe the production of AP fusion proteins. We

---

[4] J. G. Flanagan and H. J. Cheng, *Methods Enzymol.* **327,** Chap. 15, 2000 (this volume).

[5] H.-J. Cheng and J. G. Flanagan, *Cell* **79,** 157 (1994).

[6] H.-J. Cheng, M. Nakamoto, A. D. Bergemann, and J. G. Flanagan, *Cell* **82,** 371 (1995).

[7] R. Devos, J. G. Richards, L. A. Campfield, L. A. Tartaglia, Y. Guisez, J. Vanderheyden, J. Travernier, G. Plaetinck, and P. Burn, *Proc. Natl. Acad. Sci. U.S.A.* **93,** 5668 (1996).

[8] N. W. Gale, S. J. Holland, D. M. Valenzuela, A. Flenniken, L. Pan, T. E. Ryan, M. Henkemeyer, K. Strebhardt, H. Hirai, D. G. Wilkinson, T. Pawson, S. Davis, and G. D. Yancopoulos, *Neuron* **17,** 9 (1996).

[9] A. M. Koppel, L. Feiner, H. Kobayashi, and J. A. Raper, *Neuron* **19,** 531 (1997).

[10] U. Muller, D. N. Wang, S. Denda, J. J. Meneses, R. A. Pedersen, and L. F. Reichardt, *Cell* **88,** 603 (1997).

[11] T. Takahashi, F. Nakamura, and S. M. Strittmatter. *J. Neurosci.* **17,** 9183 (1997).

[12] Y. Yang, G. Drossopoulou, P. T. Chuang, D. Duprez, E. Marti, D. Bumcrot, N. Vargesson, J. Clarke, L. Niswander, A. McMahon, and C. Tickle, *Development* **124,** 4393 (1997).

[13] J. G. Flanagan, *Curr. Biol.* **10,** R52 (2000).

also describe *in situ* procedures in which these affinity probes are used to detect the distribution of cognate ligands or receptors in tissues or cells. In [15] in this volume[4] we describe other applications: molecular characterization of ligands and receptors, and the cloning of novel ligands and receptors. Although we focus on polypeptide ligands and their cell surface receptors, the same techniques could presumably be applied to other types of interacting biological molecules.

Designing Constructs Encoding Receptor– or Ligand–Alkaline Phosphatase Fusion Proteins

AP fusion proteins can be produced by inserting the cDNA for the molecule of interest—a ligand or a receptor ectodomain—into the APtag vectors (Figs. 1 and 2; vectors can be obtained from GenHunter, Nashville, TN).

For proteins that are membrane anchored in their native state, including receptors and many ligands, the protein is generally fused to the N terminus of AP. This allows the AP tag to be fused to the position where the native protein would enter the cell membrane, making it unlikely that the tag will interfere sterically with ligand binding. We generally position the fusion site immediately outside the hydrophobic transmembrane domain. The APtag-1, -2, and -5 vectors can be used for this purpose. The secretion signal sequence of the inserted protein is generally used, although with APtag-5 the signal sequence in the vector can be used instead.

For proteins that are not membrane anchored in their native state, such as soluble ligands, we generally suggest making both a fusion to the N terminus of AP (with APtag-1, -2, or -5) and a fusion to the C terminus of AP (with APtag-4 or -5). In the case of fusions to the C terminus of AP, secretion will be directed by the signal sequence of the AP, and so any secretion signal in the inserted sequence should be eliminated.

In addition to an AP tag, the APtag-5 vector includes a hexahistidine (His_6) tag that can be used for purification or concentration of the protein, and a Myc epitope tag. APtag-4 or -5 can be used to produce unfused AP as an important negative control that we use for most of our experiments.

Procedure to Insert Receptor or Ligand cDNA into APtag Vectors

1. Digest the APtag vector of choice with the appropriate restriction enzymes. When making fusions to the N terminus of AP (APtag-1, -2, or -5), we have generally used *Hin*dIII for the 5' end of the insert. At the 3' end, fusions at the *Bgl*II site will result in a four-amino acid linker, which should give plenty of conformational flexibility. Fusion proteins linked at

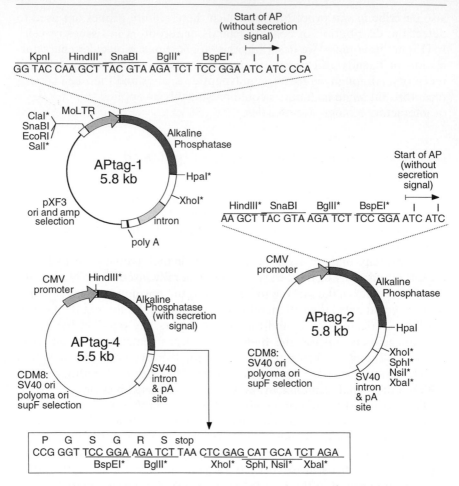

Start of AP
(without secretion
signal)

| KpnI | HindIII* | SnaBI | BglII* | BspEI* | I | I | P |

GG TAC CAA GCT TAC GTA AGA TCT TCC GGA ATC ATC CCA

ClaI*
SnaBI
EcoRI
SalI*

MoLTR

APtag-1
5.8 kb

Alkaline
Phosphatase

Hpal*

Xhol*

pXF3
ori and amp
selection

intron

poly A

Start of AP
(without
secretion
signal)

| HindIII* | SnaBI | BglII* | BspEI* | I | I |

AA GCT TAC GTA AGA TCT TCC GGA ATC ATC

CMV
promoter

CMV
promoter

HindIII*

Alkaline
Phosphatase
(with secretion
signal)

APtag-4
5.5 kb

SV40
intron
& pA
site

CDM8:
SV40 ori
polyoma ori
supF selection

Alkaline
Phosphatase

APtag-2
5.8 kb

Hpal

Xhol*
SphI*
NsiI*
Xbal*

CDM8:
SV40 ori
polyoma ori
supF selection

SV40
intron
& pA
site

| P | G | S | G | R | S | stop | | | |

CCG GGT TCC GGA AGA TCT TAA CTC GAG CAT GCA TCT AGA

BspEI*   BglII*   Xhol*   SphI, NsiI*   Xbal*

FIG. 1. Vectors to make AP fusion proteins. APtag-1,[1] APtag-2,[6] and APtag-4 (not previously published) vectors are diagrammed. APtag-2 and -4 are for transient transfection, whereas APtag-1 is for stable transfection. APtag-2 and -4 have a supF selection marker and must be grown in an appropriate bacterial strain such as MC1061/P3. APtag-1 and -2 are designed for fusions to the N terminus of AP, whereas APtag-4 is for fusions to the C terminus of AP. APtag-4 has its own secretion signal sequences and therefore, in addition to making fusion proteins, it can be used as a source of unfused AP as an important negative control. Asterisks indicate restriction sites that cut the vector only once.

the *Bsp*EI site have also worked well. Note that *Bgl*II and *Bsp*EI both produce sticky ends that are compatible with several other enzymes. To make fusions to the C terminus of AP (APtag-4 or -5), the 5′ end of the insert can be joined to any of the unique sites upstream of the stop codon

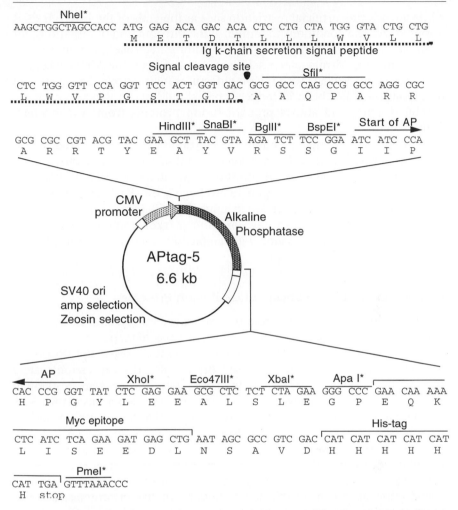

```
                 NheI*
AAGCTGGCTAGCCACC ATG GAG ACA GAC ACA CTC CTG CTA TGG GTA CTG CTG
                  M   E   T   D   T   L   L   L   W   V   L   L
                        Ig k-chain secretion signal peptide

              Signal cleavage site              SfiI*
CTC TGG GTT CCA GGT TCC ACT GGT GAC GCG GCC CAG CCG GCC AGG CGC
 L   W   V   P   G   S   T   G   D   A   A   Q   P   A   R   R

                  HindIII* SnaBI*   BglII*    BspEI*   Start of AP
GCG CGC CGT ACG TAC GAA GCT TAC GTA AGA TCT TCC GGA ATC ATC CCA
 A   R   R   T   Y   E   A   Y   V   R   S   S   G   I   I   P
```

CMV promoter

Alkaline Phosphatase

APtag-5
6.6 kb

SV40 ori
amp selection
Zeosin selection

```
     AP                 XhoI*      Eco47III*    XbaI*      Apa I*
CAC CCG GGT TAT CTC GAG GAA GCG CTC TCT CTA GAA GGG CCC GAA CAA AAA
 H   P   G   Y   L   E   E   A   L   S   L   E   G   P   E   Q   K
          Myc epitope                                      His-tag
CTC ATC TCA GAA GAT GAG CTG AAT AGC GCC GTC GAC CAT CAT CAT CAT CAT
 L   I   S   E   E   D   L   N   S   A   V   D   H   H   H   H   H
              Pmel*
CAT TGA GTTTAAACCC
 H  stop
```

FIG. 2. Additional vectors to make AP fusion proteins. The APtag-5 vector (not previously published) is diagrammed. The vector can be used for either transient or stable transfection. Proteins can be fused to either the N or C terminus of AP. The AP in this vector has its own secretion signal sequence, and thus it can be used as a source of unfused AP as a negative control. Asterisks indicate restriction sites that cut the vector only once.

(Figs. 1 and 2). In these cases, the C-terminal peptide sequence of AP is likely to act as a good linker.

2. Prepare a cDNA encoding a soluble form of the ligand or receptor, so that it has sticky ends compatible with the vector. We generally use the polymerase chain reaction (PCR) to amplify precisely the relevant region,

while introducing artificial restriction sites at the ends of the insert. If PCR is employed, use conditions to minimize the introduction of mutations; for example, use a polymerase with a 3'-5' editing nuclease function, such as *Pfu* polymerase (Stratagene, La Jolla, CA), and keep the NTP concentrations low in accordance with the manufacturer instructions. To ensure that mutations have not been introduced, one should preferably sequence the amplified gene, and always prepare fusion proteins from two independent clones.

3. Ligate the foreign gene into the restriction enzyme-digested vector, transform into competent *Escherichia coli,* and select recombinants. APtag-2 and -4 have a *supF* marker, and must be grown in the MC1061/P3 bacterial strain [available from InVitrogen (San Diego), Bio-Rad (Hercules, CA), and other suppliers] with selection in ampicillin (50 μg/ml) plus tetracycline (10 μg/ml). APtag-1 and -5 have an *amp* marker and can be grown in commonly used *E. coli* strains with ampicillin selection. Check plasmid structure by restriction mapping.

Production of Alkaline Phosphatase Fusion Proteins

AP fusion proteins are prepared by transfecting cultured cells. We generally use mammalian cells to minimize the risk of inappropriate protein modification or folding, and because the transient expression protocols are fast and reliable. We have also used the baculovirus expression system successfully. However, expression in bacterial or yeast systems is likely to be more risky, and we know of several examples where this has not worked.

Depending on the situation, it may be preferable to use either transient transfection (APtag-2, -4, or -5 vectors) or stable transfection (APtag-1 or -5 vectors). Transient transfection is much faster: it takes only about 1 week to obtain a fusion protein ready for experiments, whereas it takes at least 1 month in a stable expression system. Also, in our experience transient transfection has been reliable for expression of fusion proteins, whereas some proteins are expressed poorly after stable transfection. However, if a large amount of fusion protein is needed over a long period of time, a stable cell line may save time and money in the long term.

Transient transfection can be done with the APtag-2, -4, and -5 vectors, which have a simian virus 40 (SV40) origin so they will replicate in cell lines that express SV40 large T antigen, such as COS cells or 293T cells (American Type Culture Collection, Manassas, VA). We find that 293T cells give a severalfold higher yield of fusion protein than COS cells. We usually perform the transfection with LipofectAMINE (GIBCO-BRL, Gaithersburg, MD), according to the manufacturer instructions. Transiently transfected cells usually begin to express AP activity within 48 hr after the start of transfection. However, the production of transfected protein increases rapidly around this time, so we generally change to fresh medium at 48 hr after the start of transfection and condition it for approximately another 4 days, monitoring the AP activity daily.

Stable transfection can be done with the APtag-1 or -5 vector. For APtag-1 we transfect into NIH 3T3 cells by the calcium phosphate method, cotransfecting with a separate plasmid that provides a selectable marker such as the *neo* gene. For APtag-5 we transfect into 293T cells, using LipofectAMINE (GIBCO-BRL), and use the zeocin selection marker on the plasmid (500 μg/ml zeocin). As an efficient way to obtain clones of stably transfected cells, we split the cells at several dilutions into 96-well plates at the time when drug selection is initiated. Colonies should start to appear after about 5 to 10 days. Some supernatant can be taken to assay for AP activity a few days later, after the colonies have become confluent, and high-expression clones are then expanded. To collect AP supernatants in bulk, grow the cells to confluence, and then change the medium and condition it for a further 3 days.

Storage and Purification of Alkaline Phosphatase Fusion Proteins

After producing conditioned medium from transiently or stably transfected cells, spin out debris at maximum speed in a benchtop centrifuge, buffer the supernatant with 10 mM HEPES, pH 7.0, and then filter (0.45-μm pore size) the supernatant and store it at 4°. We usually also add sodium azide (0.05%, w/v) to prevent microbial growth (although this should be omitted if the conditioned medium is to be used in subsequent experiments that require actively metabolizing cells).

We usually find that supernatants containing AP fusion proteins are stable for months or even years at 4°. For most purposes, the supernatant should be ready to use as a reagent without further steps. If serum in the complete medium is a problem in subsequent experiments, the conditioned medium can be produced under serum-free conditions. For COS or 293T cells, Opti-MEM I (GIBCO-BRL) can be used, and for NIH 3T3 cells, Dulbecco's modified Eagle's medium (DMEM) with insulin, transferrin, and selenium (Redu-SER; Upstate Biotechnology, Lake Placid, NY). If necessary the protein can be concentrated by ultrafiltration with an Amicon (Danvers, MA) pressure cell and a PM30 or YM100 membrane (depending on the size of the fusion protein). We have also affinity purified fusion proteins with an anti-AP antibody, with elution by low pH[1] or 3 M MgCl$_2$ (M.-K. Chiang and J. G. Flanagan, unpublished). Alternatively, with APtag-5, the His$_6$ tag can be used to purify fusion proteins on a nickel column. However, these procedures are laborious and are rarely necessary.

We use untreated conditioned medium as the reagent for almost all our experiments.

Appropriate Working Concentrations of Alkaline Phosphatase Fusion Proteins

The concentration of receptor–AP fusion protein optimal for final use depends on the individual experiment. Most importantly, it depends on the affinity of the ligand–receptor interaction. For the interaction of receptor tyrosine kinases with their cognate ligands, the dissociation constant of binding, K_D, is generally in the range of approximately 10^{-8} to 10^{-12} M (K_D is equivalent to the concentration of the soluble fusion protein that will give half-maximal occupation of its binding sites). In general, increasing the concentration of receptor–AP fusion is expected to increase the signal (saturably) and also increase the background (nonsaturably). This means that there should be an optimal concentration that will maximize specific binding, while avoiding excessive nonspecific binding. The optimal concentration for any particular experiment may need to be determined empirically. For known ligands, we would typically use a receptor–AP concentration between 1 and 10 times the K_D. When testing for an unknown ligand, we might typically try concentrations of receptor–AP fusion in the range of 2 to 40 nM. If necessary, supernatants can be concentrated by ultrafiltration (see previous section) or can be diluted with HBAH buffer [Hanks' balanced salt solution, bovine serum albumin (0.5 mg/ml), 0.1% (w/v) NaN$_3$, 20 mM HEPES (pH 7.0)].

For all types of binding experiment, we use unfused AP, at the same concentration of AP activity as the fusion protein, as a negative control. For many types of experiment, such as staining of cells or tissues, the receptor–AP fusion protein can be saved after use and reused several times. The protein concentration remaining can be assessed by measuring the AP activity.

Measurement of Alkaline Phosphatase Activity

Because each fusion protein contains one AP tag, the concentration of fusion protein can be estimated from the AP activity. We measure the AP activity by adding the substrate p-nitrophenyl phosphate, which is converted into a yellow product that can be quantitated. The activity can be measured by the change in absorbance at 405 nm, either in a cuvette by spectrophotometer [optical density (OD) units per hour] or in a 96-well plate with a microplate reader (V_{max} in milli-OD units per minute). We perform all reactions at room temperature. To convert from V_{max} in a microplate to OD units per hour in a cuvette, divide by an approximate conversion factor of 59 (this assumes a volume of 200 μl and a light path length of 0.713 cm for the microplate, versus a reaction diluted to 1 ml and a path length of 1 cm for the cuvette). To convert from OD units per hour in a cuvette to picomoles of AP protein, divide by an approximate conversion factor of 36. However, please note that this is an approximate and empirically determined value. We find that different fusion proteins generally seem to have approximately the same specific activity of AP. However, to obtain accurate values for a particular protein, it would be necessary to measure the AP activity and compare it with the protein concentration.

Procedure for Quantitative Measurement of Alkaline Phosphatase Activity

1. Put 1 ml supernatant, or less, in an Eppendorf tube and heat inactivate the endogenous AP activity in a 65° water bath for 10 min.
2. Spin the tube in a microcentrifuge at maximum speed for 5 min. Collect the supernatant.
3. Take some of the supernatant and add an equal amount of 2× AP buffer to check the AP activity. The final volume would generally be approximately 1 ml for measurement in a spectrophotometer, or 200 μl for measurement in a plate reader. If the activity is reasonably high, it may be necessary to dilute the supernatant first, which can be done in HBAH or in another buffer containing carrier protein. However, avoid buffers containing phosphate, which is a competitive inhibitor of AP.

To prepare 2× AP substrate buffer, add 100 mg of *p*-nitrophenyl phosphate (Sigma, St. Louis, MO) and 15 μl of 1 M $MgCl_2$ to 15 ml of 2 M diethanolamine, pH 9.8. This stock should be kept on ice, and can be stored frozen at $-20°$ in aliquots. To make up 2 M diethanolamine, take 20 ml of liquid diethanolamine and bring to a final volume of 100 ml with water, adjusting the pH with HCl.

Always compare samples with negative controls, because cells can produce low levels of endogenous phosphatase activity, and because *p*-nitrophenyl phosphate has a low rate of spontaneous hydrolysis.

Immunoprecipitation of Alkaline Phosphatase Fusion Protein;
    Confirmation of Intact Polypeptide

To confirm that the AP fusion protein has the appropriate size it may be desirable to immunoprecipitate it and estimate the molecular weight by sodium dodecyl sulfate–polyacrylamide gel electrophoresis (SDS–PAGE). A protocol for this is given below. An equally good alternative is to assess the size by Western blotting. Unfused AP should migrate at an apparent molecular weight of approximately 67,000, and the ligand or receptor polypeptide would be added to this. It is worth noting that some receptors undergo a natural proteolytic cleavage into fragments that remain associated at the cell surface. Therefore, even if an AP fusion polypeptide is proteolytically cleaved, the fragments may remain associated in solution and may function perfectly well as a probe.

Procedure for Immunoprecipitation of Alkaline Phosphatase Fusion Protein

1. Couple a monocolonal antibody against AP to Sepharose beads.
   a. Weigh out about 3.5 g of CNBr-Sepharose powder (Pharmacia, Piscataway, NJ). Swell the gel for a few minutes by mixing it with 1 mM HCl in a 50-ml tube. Wash the mixture in a sintered glass funnel over vacuum with about 500 ml of 1 mM HCl over a period of 15 min.
   b. Resuspend the washed gel in a small amount of 1 mM HCl and pipette some into a 15-ml tube. Spin down at 2000 rpm for 5 min to estimate the packed volume of beads. Adjust the packed volume to 5 ml by suspending it again and removing the excess suspension. Spin down to pack the beads and remove the supernatant.
   c. Set up the coupling reaction. The final concentrations should be as follows: gel at 40–50%, v/v; 5 mg of antibody; 0.25 M sodium phosphate, pH 8.3. Incubate at 4° on a rotator overnight. Monoclonal antibodies to human placental AP can be bought from Genzyme Diagnostics (Cambridge, MA) (Genzyme MIA1801 has a relatively high affinity and is suitable for immunoprecipitations; MIA1802 has a lower affinity and can be used for affinity purification of AP fusions, although this is a procedure we rarely perform). Polyclonal antibodies can be purchased from several suppliers, including GenHunter, Zymed (South San Francisco, CA), and Dako (Carpinteria, CA).
   d. For a 10-ml coupling reaction, add 5 ml of 1 M ethanolamine HCl, pH 8.0, to stop the reaction. Incubate on a rotator at 4° for 4 hr.
   e. Wash the beads once with 0.5 M sodium phosphate, pH 8.3, and then three times with modified radioimmunoprecipitation assay (RIPA) buffer [0.5% Nonidet P-40 (NP-40), 0.5% (w/v) sodium deoxycholate, 0.1% (w/v) SDS, 0.1% (w/v) NaN_3, 144 mM NaCl, 50 mM Tris-HCl (pH 8.0)].
   f. Store the beads in modified RIPA buffer in a tightly closed tube at 4°.

2. For a six-well tissue culture plate, label the cells with 2 ml of labeling solution (DMEM without methionine, containing 10% [v/v] dialyzed serum, and 400 μCi of [^{35}S]methionine) at 37° for 3 to 6 hr.
3. Collect the supernatant and concentrate to about 200 μl on a Centricon-10 (Amicon).
4. Mix the supernatant with 20 μl of beads coupled with anti-AP antibody for 30 min or a rotator at room temperature.
5. Wash the beads twice in Tris-buffered saline (TBS)–0.1% (w/v) NP-40, three times in modified RIPA buffer, and once in TBS–0.1% (w/v) NP-40. Use ice-cold buffers and do this quickly. Spins are at 5000 rpm for 1 min in a microcentrifuge.
6. Add an equal volume of loading buffer and heat the sample for 2 min at 100°. The size of AP fusion protein can be analyzed on an SDS–polyacrylamide gel.

In Situ Binding with Alkaline Phosphatase Fusion Proteins on Whole-Mount Preparations

Whole-mount AP *in situ* can be done either on whole embryos, or on parts of embryos or adult tissues that have been dissected out carefully. Staining of whole mounts is often the most sensitive way to detect a binding partner. It is also relatively simple, and is generally the first approach to try, to form an initial impression of binding patterns. An example of *in situ* binding of a receptor–AP probe to an embryo whole mount is shown in Fig. 3.

Penetration of the AP fusion protein into a tissue may be limited, and this is always a factor to bear in mind. One factor limiting penetration can be the presence of skin, or outer membranes such as the pial membranes of the nervous system. Careful removal of these by dissection can greatly improve penetration, and may be essential. Because of this factor, as well as overall size, embryos should not be treated as a single whole-mount preparation at later stages: for mouse embryos, they should not be older than approximately day 10.5, and for chick embryos, day 4. For older embryos, one can dissect out the organ or tissue of interest, such as brain or other internal organs, and treat this as a whole mount.

The protocol here describes a procedure without prefixation. This protocol works especially well for molecules that are located near the superficial layers of an embryo or tissue. Penetration into deeper regions, beyond 1 mm or so from the surface, is likely to be limited. To detect molecules that might be deeper, and to visualize expression patterns most effectively, we recommend also trying to stain sections, as described below. Another approach to detect molecules that might be in deeper layers is to try pre-

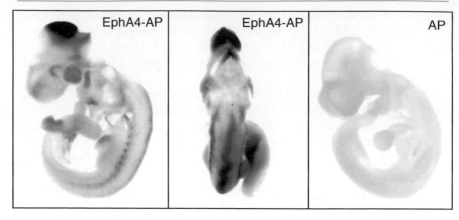

FIG. 3. *In situ* staining with a receptor–AP fusion protein on whole-mount embryos. Whole-mouse embryos at developmental day 9.5 were treated with supernatants containing an AP fusion of the EphA4 receptor ectodomain. Embryos were then washed and stained for bound AP activity; color development in this case was for 30 min. Unfused AP serves as a control. Embryos are viewed from the side (*left* and *right*) and dorsally (*middle*). Specific staining is seen in several areas, including the dorsal spinal cord, limb buds, somites, and branchial arches, and most strongly in the developing midbrain, which is seen here at the top of the photographs. This type of experiment has provided the first evidence of ligands for EphA4 expressed in the embryo.[5] Several ligands that can bind EphA4 have now been cloned, and their RNA expression patterns, as judged by *in situ* hybridization, appear consistent with the receptor–AP binding patterns.

fixing the embryo with either 4% (w/v) paraformaldehyde or 8% (v/v) formalin, and then to incubate embryos with AP fusion proteins in buffer containing a nonionic detergent such as NP-40. However, depending on the protein, signal detection may be reduced by these prefixation procedures.

Procedures for Staining Whole Mounts

1. Dissect the embryos and transfer them into 2-ml microcentrifuge tubes. We use tubes or vials with a round or flat but not conical base, to avoid embryos being trapped at the bottom. When individual tissues are dissected out, it may help greatly to carefully remove any surrounding membranes, such as the pial membranes surrounding neural tissue.

2. Rinse the embryos once with 1.5 ml of HBAH buffer.

3. Incubate the embryos with 1 ml of AP fusion protein, or enough to cover the tissue, for 90 min on a rotator or orbital shaker at room temperature. For some labile proteins incubation at 4° may work better.

4. Remove the AP fusion protein. Wash the embryos three times with 1.5 ml of HBAH buffer. For each wash, leave the tube on the rotator for

5 min. If the embryos later show a high background, it might be because this wash step was not sufficiently thorough. In our experience, washing 10 times or more, or even washing overnight in the cold, can still give a good signal and the background may be reduced significantly.

5. Fix the embryos with 1 ml of acetone–formalin fixative [65% (v/v) acetone, 8% (v/v) formalin in 20 mM HEPES, pH 7.0] for 2.5 min. A longer fixation time may reduce the signal. Formalin (8%, v/v) or 4% (w/v) paraformaldehyde for 5 min can also be used alone, if the acetone should cause any problem.

6. Wash out excess fixative with 1 ml of HBS (150 mM NaCl, 20 mM HEPES, pH 7.0), three times for 5 min each.

7. Incubate the tube containing the embryos and 1 ml of HBS in a 65° water bath for 1 hr, to heat inactivate endogenous phosphatases. Increase the incubation time if the background is high. We have had good results with heat inactivation for several hours, or even overnight, although there may be some risk of losing the specific signal.

8. Rinse the embryos once with 1 ml of AP staining buffer (100 mM NaCl, 5 mM MgCl$_2$, 100 mM Tris-HCl, pH 9.5). At this point the embryos can be transferred to a six-well plate to allow easier observation.

9. Add 1 ml of BCIP/NBT substrate solution. [This is bromochloro-indolyl phosphate (0.17 mg/ml) and nitroblue tetrazolium (0.33 mg/ml) in AP staining buffer; BCIP and NBT are stored as stock solutions at −20°, at 25 mg/ml in 70% (v/v) dimethylformamide and at 50 mg/ml in 50% (v/v) dimethylformamide, respectively.] Incubate at room temperature on a rotator under a shade of aluminum foil. Chromogenic AP substrates may darken significantly if exposed to light. During color development, samples should generally be kept in the shade. If they are to be viewed under a microscope, it should be done for only a short time. Staining can be monitored periodically. A strong signal may become visible in as little as 5 to 10 min. Weaker signals may take a few hours to develop, and the sample can even be incubated overnight, although background staining is then likely to become significant.

10. Stop the reaction by washing the embryos with 1 ml of phosphate-buffered saline (PBS)–10 mM EDTA. Fix the embryos in 8% (v/v) formalin or 4% (w/v) paraformaldehyde for 30 min. Wash and store in PBS–10 mM EDTA in the dark.

Embryonic tissues that are damaged or cut during dissection may show staining by ligand or receptor fusion probes nonspecifically; thus the dissection and interpretation must be done carefully. In addition, newly forming cartilage, bone, and nervous tissues can have endogenous AP activity, which can be hard to heat inactivate completely in a whole-mount preparation.

If there should be problems with background staining, try a longer time of heat inactivation and compare the staining carefully with negative controls.

In Situ Binding with Alkaline Phosphatase Fusion Proteins on Tissue Sections

The penetration of AP fusion proteins into whole mounts is likely to be limited, especially in the case of tissues that have not been fixed or permeabilized. Sectioning allows access to deeper layers. However, it tends to be less sensitive than whole-mount analysis. We generally feel that a combination of analyzing both whole mounts and sections is desirable, because each has technical advantages, and also because it can be much easier to visualize an expression pattern accurately and comprehensively by combining both types of information.

The procedure below describes treatment of lightly fixed frozen sections. An alternative is to cut thick (>100 μm) unfixed sections with a Vibratome and treat them essentially as whole mounts, using the protocol described above. This can prevent the loss of binding sites due to fixation procedures. For some ligands or receptors it may therefore be a much more sensitive approach, although technically more demanding. When sectioning with a Vibratome, it may help to keep the tissue chilled, and to embed it in agarose before sectioning. It may also help to fix the outside of the tissue with 4% (w/v) paraformaldehyde for a few minutes to keep it intact during Vibratome sectioning, although this might compromise subsequent binding efficiency.

Procedure to Prepare and Stain Frozen Sections

1. Prepare tissue sections:
   a. Dissect the embryos or tissues and fix them in 4% (w/v) paraform-aldehyde. Depending on the size of the tissue one can do this either at room temperature for 2 hr, or at 4° overnight. This protocol is good for tissues such as mouse embryos up to developmental day 11.5 to 14.5. If the tissue is older than that, it should be fixed longer or cut open to let the fixative penetrate deeper. To make 4% (w/v) paraformaldehyde, add 2 g of paraformaldehyde powder to 50 ml of PBS. Heat in a 55° water bath for 30 min, adding 5 to 50 μl of 10 N NaOH as necessary to dissolve the powder. Let it stand at room temperature to cool slowly, then filter (0.45-μm pore size). Paraformaldehyde should be prepared fresh, or can be stored at −20° and thawed before use.

  b. Rinse the tissues with PBS once to eliminate the fixative.
  c. Put the tissues in 30% (w/v) sucrose (in PBS) at 4° on a rotator to mix them, until they sink to the bottom of the tube when it is placed upright.
  d. Pour out the scrose solution until the surface is level with the upper part of tissue in the tube and add an equal amount of O.C.T. freezing solution. Mix the contents by placing the tube on a rotator at room temperature for 2 hr.
  e. Put the tissue in molds, add enough O.C.T. solution to cover the tissue, quick freeze the mold with tissue on dry ice, and transfer to a −70° freezer.
  f. Cryosection the tissue before the binding experiment and air dry the sections at room temperature overnight. The sections can be stored at −70° after they have been dried.
2. Wash the sections for 10 min in a jar containing HBS (10 mM HEPES, pH 7.0, 150 mM NaCl), to eliminate the O.C.T. solution on the slide.
3. Rinse twice in HBAH buffer.
4. Add AP fusion protein to cover all sections on the slide and incubate at room temperature for 90 min in a moist chamber.
5. Wash the sections six times in cold HBAH.
6. Add acetone–formalin fixative to the sections for 15 sec. Longer fixation may destroy some AP activity.
7. Wash the sections twice in HBS.
8. Incubate the sections in preheated HBS, in a 65° water bath, for 15 min. Increase the incubation time if the background is high. We have had good results with heat inactivation for several hours, or even overnight, although there may be some risk of losing the specific signal.
9. Wash the sections once in AP staining buffer.
10. Add BCIP/NBT substrate to cover the sections on the slide. Incubate at room temperature under a shade of aluminum foil in a moist chamber. Staining can be monitored periodically against a white background under a dissecting microscope. Color should become visible in about 30 min to 2 hr. Sometimes it takes a few hours, or even overnight, but background color is likely to appear after incubation of more than few hours.
11. Stop the reaction by putting the slides in PBS with 10 mM EDTA.
12. Fix the sections in 8% (v/v) formalin for 20 min.
13. Wash the sections in PBS with 10 mM EDTA.
14. Mount the sections and keep them in the dark at room temperature.

In Situ Binding with Alkaline Phosphatase Fusion Proteins
on Cultured Cells

In situ staining can also be done on cultured cells. This can give information on the cellular or subcellular distribution of ligands or receptors. It also provides a good method to identify individual positive cells when screening an expression library.[4] However, to screen cell lines for potential endogenous expression of a ligand, quantitative cell surface binding is much more sensitive and reliable than *in situ* staining.[4]

The procedure described below can be used to detect ligands or receptors at the cell surface. With a modification (step 2b, below) it can also be used to detect soluble ligands within the cell in the secretory pathway.[14]

Procedure to Stain Cultured Cells

1. Grow the cells to be tested on a 10-cm tissue culture plate. (The protocol can easily be adapted to use different sizes of plates, or cells grown on chamber slides or coverslips.) For library screening it is important to ensure a uniform density of cells over all parts of the plate, and to stain cells that are just under confluence, or just recently confluent. Overconfluent cells can pile up, trapping the fusion protein probe and sometimes causing unpredictable background staining. Also note that during washing procedures, cells can dry and fall off quickly if all the medium is siphoned out of the plate. The problem is mainly seen around the edges, and so is more severe in smaller plates. To minimize this effect one can pipette the medium out, leaving just enough to provide a thin covering. With experience this can be done quickly with a vacuum aspirator by withdrawing the tip of the pipette as soon as the liquid level reaches the bottom of the well at its center.

2a. To detect a cell surface ligand, wash the cells once with 10 ml of cold HBAH. Proceed to step 3.

2b. To detect a soluble ligand in the secretory pathway, wash the cells once with 10 ml of cold TBS, fix with TBS–4.5% (v/v) formalin for 15 min, and then incubate with HBAH containing 0.1% (v/v) Triton X-100 for 15 min to permeabilize the cells. Proceed to step 3.

3. Add 4 ml of AP fusion protein and incubate at room temperature for 90 min. The time is determined by the rate of binding, k_{on}. For the reaction of a cell surface receptor and a polypeptide ligand, k_{on} can be quite slow, but 60 to 90 min at room temperature should be enough to give

---

[14] S. Davis, T. H. Aldrich, P. F. Jones, A. Acheson, D. L. Compton, V. Jain, T. E. Ryan, J. Bruno, C. Radziejewski, P. C. Maisonpierre, and G. D. Yancopoulos, *Cell* **87,** 1161 (1996).

good binding. On ice the reaction is expected to be much slower. Swirl briefly to mix at approximately the 30- and 60-min time points.

4. Remove the AP fusion protein solution with a pipette. Wash the cells six times with 10 ml of cold HBAH. For each wash, incubate the cells with HBAH for 5 min and gently swirl the medium by hand or on a platform shaker.

5. Aspirate out the HBAH and add 10 ml of acetone–formalin fixative slowly and swirl for 15 sec. Longer fixation may destroy some AP activity.

6. Aspirate off the fixative and wash twice with 10 ml of HBS. Leave 10 ml of HBS in the plate.

7. Incubate the plate containing 10 ml of HBS on a flat shelf in a 65° preheated oven for 100 min.

8. Wash with 10 ml of AP staining buffer.

9. Add 4 ml of BCIP/NBT substrate in AP staining buffer. Incubate at room temperature under a shade of aluminum foil. Staining can be monitored periodically against a white background under a dissecting microscope. Color should become visible in about 30 min. Sometimes it takes a few hours, or can even be incubated overnight, although background color will begin to appear.

10. Stop the reaction by washing the plate with PBS and store the cells in 10 ml of PBS with 10 mM EDTA at 4° in the dark.

[3] Chimeric Molecules Employing Horseradish Peroxidase as Reporter Enzyme for Protein Localization in the Electron Microscope

By Colin Hopkins, Adèle Gibson, Jane Stinchcombe, and Clare Futter

Introduction

Cytochemical techniques for the detection of peroxidases by electron microscopy were pioneered in the 1960s by Werner Straus.[1] These studies showed that horseradish peroxidase (HRP), a heme enzyme with a molecular weight of 40,000, could be used as a macromolecular tracer because it remains soluble in blood and tissue fluid, diffuses freely through intercellular spaces, and, when taken up in the fluid phase, readily distributes along intracellular pathways leading to the lysosome. Subsequent experience in

[1] W. Straus, *J. Histochem. Cytochem.* **15**, 381 (1967).

a wide variety of cellular systems has shown HRP to be a relatively inert tracer with little tendency to absorb nonspecifically to membranes and to be tolerated by cells in culture at concentrations as high as 10 mg/ml. Smaller moieties with peroxidase activity have also been developed, the most widely used being microperoxidase, an 8- to 11-residue fragment derived from cytochrome c.[2]

Considerations in Using Horseradish Peroxidase as Reporter Enzyme

Peroxidase is detectable in the electron microscope because in the presence of hydrogen peroxide it reacts enzymatically with diaminobenzidine (DAB) to form an insoluble, electron-opaque product. Its enzymatic activity thus provides an amplification step that makes HRP an extremely sensitive tracer. An estimate[3] using liposomes suggested that a single HRP molecule enveloped within a 50-nm-diameter vesicle is sufficient to fill the vesicle lumen with DAB reaction product. This reaction product is believed to be a complex, cross-linked mixture of tarlike substances, which during its formation cross-links other macromolecules located within its immediate vacinity. The DAB reaction product does not cross cell membranes and remains limited within the intracellular compartments in which it forms. However, as shown originally by Stoorvogel et al.,[4] hydrogen peroxide and DAB readily cross living cell membranes so that HRP located in intracellular locations in cell growing in culture will also form cross-linked reaction products. As described below, opportunities to exploit the ability of HRP to cross-link intracellular components in living cells can be used in a variety of ways.

Endogenous peroxidase is present in peroxisomes and in other compartments, such the leukocyte granules, which contain myeloperoxidase.[5] Inhibitors that can selectively block this activity in living cells are available but they have not been widely used because localizations of endogenous peroxidases are readily demonstrable and are not a complicating factor for most tracer studies.

HRP is a heme enzyme with a compled three-dimensional structure. Its enzyme activity survives aldehyde fixatives, including the concentrations of glutaraldehyde used in routine electron microscopy. The activity of the enzyme is not significantly reduced when coupled to carrier proteins such as transferrin (Tf) and, as indicated by its ability to remain active within

[2] N. Feder, *J. Cell Biol.* **51,** 339 (1971).
[3] J. C. Stinchcombe, H. Nomoto, D. F. Cutler, and C. R. Hopkins, *J. Cell Biol.* **131,** 1387 (1995).
[4] W. Stoorvogel, V. Oorschot, and H. J. Geuze, *J. Cell Biol.* **132,** 21 (1996).
[5] Barnton and Farquhar (1967).

lysosomes for periods of more than 24 hr, it has a remarkable resistance to degradation by endogenous enzymes.

Because of the ease and sensitivity with which HRP can be detected, its inert and impermeant properties with regard to membrane boundaries, and its ready availability, it is currently the most widely used enzymatic tracer for electron microscopy. For double labeling in the electron microscope HRP can be used in parallel with immunogold labels. Also, because DAB reaction product has an appearance that is distinct from the discrete precipitates produced by inorganic salts (such as lead phosphate), it can be used in parallel with other reporter enzymes, such as alkaline phosphatase.

HRP has been widely used as a fluid-phase tracer because it is readily taken up into the lumena of budding endocytic vesicles and thereby becomes distributed throughout endocytic pathways. This uptake process is nonselective but sorting mechanisms located along the endocytic pathway concentrate fluid-phase materials so that free HRP is selectively routed along the length of this pathway to the lysosome. HRP coupled to transferrin (Tf) internalizes and recycles with the transferrin receptor and labels subcompartments of the endocytic system, such as the narrow (50-nm diameter) tubules involved in recycling. These tubules are penetrated by fluid-phase HRP only when high levels of exogenous tracer are internalized over prolonged periods.[6] Quantitative studies using radiolabeled Tf show that recycling Tf does not normally pass through the lysosome, and morphological studies using Tf–HRP show that the tracer does not penetrate further along the pathway than the multivesicular body state. Thus, although the reaction product of internalized fluid-phase HRP is usually widely distributed throughout the endosome compartment, suggesting that it is capable of accessing even minor pathways where traffic is relatively low, the reaction product of HRP attached to specific ligands is normally tightly restricted to the subcompartments within the pathway that processes them. As a result the distribution of HRP–ligand complexes such as Tf–HRP closely reflects the fidelity of the sorting mechanisms a pathway contains.

Biosynthetically Generated Horseradish Peroxidase Chimeras

HRP can be introduced into the biosynthetic pathway by transfecting cultured cells with cDNA constructs containing the sequence for HRP and a signal sequence.[7] HRP expressed without a signal sequence or as part of a chimera such that it is cytosolically located is inactive.

Vectors containing a variety of promoters, including those derived from viral DNA, making them capable of being introduced by infection rather

---

[6] J. Tooze and M. Hollinshead, *J. Cell Biol.* **118,** 813 (1991).
[7] Connolly, *et al.* (1994).

than transfection, have also been used successfully to introduce HRP.[8] The HRP chimera constructs prepared initially also included an eight-residue c-Myc sequence as a precaution against the possibility that the HRP may be translated efficiently but fail to fold properly or bind heme and thus provided an alternative means of detection by immunofluorescence. Recourse to this alternative method for identification was not found to be necessary even though some of our unpublished studies have suggested that a large fraction of the newly synthesized HRP is probably inactive. Nevertheless, when HRP is being synthesized as part of a chimera that includes a transmembrane protein we have usually included c-Myc as an additional sequence in the construct in order to increase the separation of the enzyme from the membrane.

Methods

General Approach to Manufacture of Horseradish Peroxidase Constructs

HRP (isoenzyme C^9), is generated from BBG10 (Rand D Systems, Oxford, UK). The signal sequence of human growth hormone[10] is added to generate ssHRP. In general, we have tried to separate the targeting signals from the HRP-coding region. Thus, in the construction of KDEL–HRP, the KDEL and ssHRP sequences are separated by the c-*Myc* epitope tag. To generate cDNAs encoding HRP-containing transmembrane proteins, the lumenal domain of the transmembrane protein is replaced with HRP, inserting a spacer of at least 10 amino acids between the HRP sequence and the predicted transmembrane domain.

Specific Examples of Manufacture of KDEL–HRP and Horseradish Peroxidase–Transferrin Receptor. HRP is generated by polymerase chain reaction (PCR) amplification with the oligonucleotides 5′-C GGA TTC CAG TTA ACG CCG ACT TTC-3′ and 5′-C CCC AAG CTT AGA GTT GAC 3′, so that the fragment produced is flanked by *Eco*RI and *Hin*dIII sites. The c-*myc* KDEL sequence is obtained by PCR amplification from the sequence 5′-TTG GAG CAA AAG CTC ATT TCT GAA GAG GAC TTG AAG GAC GAA CTT TAA GCT-3′, using the nucleotides 5′-CCC AAG CTT GAG CAA AAG CTC ATT TCT GAA GAG GAC-3′ and 5′-C CCG CTC GAG TTA AAG TTC GTC CTT CAA-3′, so that the fragment produced is flanked by *Hin*dIII and *Xho*I sites.

[8] G. Odorizzi, A. Pearse, D. Domingo, I. S. Trowbridge, and C. R. Hopkins, *J. Cell Biol.* **135,** 139 (1996).

[9] K. G. Welinder, *Eur. J. Biochem.* **96,** 483 (1979).

[10] J. Hall, G. P. Hazlewood, M. A. Surani, B. H. Hirst, and H. J. Gilbert, *J. Biol. Chem.* **265,** 19996 (1990).

The HRP and c-*myc* KDEL fragments are ligated together and the product separated by electrophoresis on low gelling temperature (LGT) agarose. This product is then ligated into PSR α.ss.

Sequences of the human transferrin receptor, which include only the transmembrane domain and the cytoplasmic domain or only the transmembrane domain plus residues on either side of the signal anchor (residues 60–91), are generated by PCR from cDNA provided by I. Trowbridge (Salk Institute, San Diego, CA). The primers used are 5′-C TCG GAA TTC GGA TCC TCG AGA TGA AAA GGT GTA GTG GAA GTA-3′ and 5′-C TAT GAA GCT TCA AGT CCT CTT CAG AAA TGA GCT TTT GCT CCC TTT AGA ATA GCC CAA-3′. They generate fragments of the receptor flanked by a new initiating methionine at the N terminus and by the c-*Myc* sequence at the C terminus. The cystein at position 89 is also changed to serine because this residue is thought to participate in dimerization of the receptor. The PCR product is digested by *Xho*I and *Hind*III, using the sites introduced by the primers.

Introduction of Horseradish Peroxidase Chimeric Proteins into Cells

For transient expression, cells are transfected with vectors containing cDNAs encoding HRP chimeric proteins, using electroporation as described in [4] in this volume.[10a] HRP chimeric proteins are also expressed in chick embryo fibroblasts by infection with recombinant Rous sarcoma virus. HRP constructs can also be expressed in mammalian cells by infection with this virus if the cells have been previously transfected with the receptor for Rous sarcoma virus subgroup A.[11] Full descriptions of this expression system can be found in Odorizzi and Trowbridge[12] and Odorizzi et al.[8]

The need for newly synthesized HRP to fold properly in order to gain enzymatic activity has been useful in studies designed to demonstrate that active HRP chimeras are exported efficiently from the rough endoplasmic reticulum (REF). In these studies cells expressing chimeras were incubated with dithiothreitol (DTT), a reagent that effectively penetrates living cells and inhibit disulfide bond formation.[13]

Chasing Horseradish Peroxidase Chimeras with Dithiothreitol

Transfected cells (1 day after transient transfection) are rinsed once in Dulbecco's modified Eagle's medium–fetal calf serum (DMEM–FCS)

[10a] A. Blagoveshchenskaya and D. F. Cutler, *Methods Enzymol.* **327,** Chap. 4, 2000 (this volume).

[11] P. Bates, J. A. T. Young, and H. E. Varmus, *Cell* **74,** 1043 (1993).

[12] G. Odorizzi and I. S. Trowbridge, *Methods Cell Biol.* **43,** 79 (1994).

[13] I. Braakman, J. Helenius, and A. Helenius, *EMBO J.* **11,** 1717 (1992).

containing 1 mM DTT, and chased at 37° in DMEM–FCS containing 1 mM DTT, a concentration shown to be sufficient for complete reduction of influenza hemagglutinin disulfide bonds.[13]

After 30 min in 1 mM DTT at 37°, detectable HRP activity can be chased from the RER without any apparent nonspecific side effects.[3] HRP chimeras in which the enzyme replaces the lumenal domains of the transferrin receptor (TR), sialyltransferase, fucosyltransferase, Lgp120, and TGN 38 have all been successfully expressed in a variety of cell types (including HEP-2, HeLa, and MDCK cells, and chick and human fibroblasts). When chased with DTT the localizations of these chimeras reflect the steady state concentrations of the native protein. Thus TR–HRP localizes to the plasma membrane and the endosome compartment, its distribution extending to the lysosome only in cells in which there is extremely high expression; the Golgi transferases localize to the *trans*-most cisternae of the Golgi stack and the *trans*-Golgi reticulum and are undetectable on the plasma membrane. TGN 38 and Lgp120 are distributed, as expected, in the *trans*-Golgi reticulum and lysosomes, respectively. In every instance studied thus far the HRP chimera has localized to the membrane boundary on which its endogenous counterpart has been shown to be localized in previous published studies.

The most useful HRP chimera developed for use as a ligand-coupled tracer in the biosynthetic pathway has been and HRP construct carrying the KDEL retrieval signal. This chimera, which is redistributed by the recycling KDEL receptor, strongly labels the entire RER compartment, including the nuclear envelope as well as the intermediate compartment between the RER and the Golgi stack and the *cis*-most Golgi stack cisternae (Fig. 1). It is not, however, detectable beyond the *cis*-most stack cisternae even in the most strongly expressing systems.

Cytochemistry and Embedding for Conventional Electron Microscopy

Cells are fixed in cacodylate buffer (0.1 M sodium cacodylate, pH 7.6) containing 2% (w/v) paraformaldehyde and 1.5% (v/v) glutaraldehyde for 15 min at room temperature. After being washed in cacodylate buffer, cells are washed three times in Tris buffer (0.5 M Tris-HCl, pH 7.6) and then incubated in DAB reaction mixture [DAB (0.75 mg/ml), 0.21% (v/v) hydrogen peroxide in Tris buffer] for 30 min at room temperature in the dark. Cells are then washed three times in Tris buffer and then in cacodylate buffer. Cells are then incubated in reduced osmium [1% (w/v) osmium tetroxide, 1.5% (w/v) potassium ferricyanide in water] for 1 hr at 4°. After washing in cacodylate buffer, cells are incubated in 0.05% (w/v) sodium cacodylate, pH 7.6, containing 1% (w/v) tannic acid, for 45 min at room

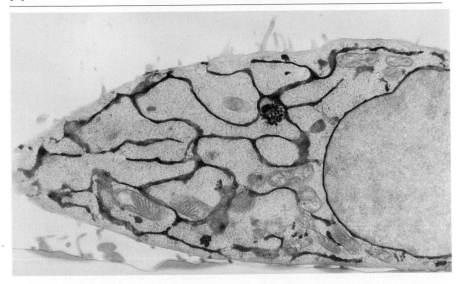

FIG. 1. Chick fibroblast expressing HRP–KDEL/DAB reaction product is distributed throughout the nuclear envelope, the rough endoplasmic reticulum, and vesicles of the intermediate compartment. The cell surface and mitrochondria show contrast but are negative for HRP. Original magnification: ×10,000.

temperature, rinsed in sodium sulfate buffer [0.05% (w/v) sodium cacodylate, pH 7.6, containing 1% (w/v) sodium sulfate], and then incubated for 5 min at room temperature in sodium sulfate buffer. Cells are then rinsed in water and dehydrated through graded ethanols (70, 90, and 100%) and embedded according to Hopkins and Trowbridge.[14] Sections are cut on a Reichert-Jung Ultracut E microtome set at either 50 to 70 nm (thin), or 200 to 400 nm (thick), and viewed either unstained, or stained with lead citrate, in a Philips (Mahwah, NJ) CM12 transmission electron microscope.

Exploiting Cross-Linking by Horseradish Peroxidase Chimeras in Intracellular Compartments

The enzyme activity of targeted HRP chimeras can be used to bring about DAB-incuded cross-linking in selected intracellular compartments. In one approach of wide utility the cross-linking has been used to preserve the compartments in living cells prior to detergent extraction. Thus, if living cells containing a chimera are treated with hydrogen peroxide–DAB and

[14] C. R. Hopkins and I. S. Trowbridge, *J. Cell Biol.* **97**, 508 (1983).

then extracted with Triton X-100, most cellular structures are removed.[15] The main features of these extracted cells as seen by electron microscopy are the filamentous elements of the cytoskeleton, the compartments containing the DAB reaction product, and nuclear chromatin. The membrane boundaries of the cross-linked compartments are removed but immunolabeling with gold shows that integral membrane proteins exposed to the DAB cross-linking in the lumen of a compartment are retained, as are many membrane-associated cytosolic complexes such as clathrin lattices. The accessibility of these proteins in the permeabilized cells allows them to be labeled with immunogold and it is evident from these studies that the density of labeling that can be achieved by this approach is significantly higher than that seen by postembedding methods, in which labeling is carried out on section surfaces.

Because cells can be so thoroughly extracted with detergent and aldehyde fixation can, if necessary, be delayed until after immunolabeling, the preservation by DAB cross-linking has significant advantages over other preembedding methods. Detergent extraction results, of course, in an arbitrary removal of membrane-associated components and for many purposes it is clear that digitonin is often to be preferred to Triton X-100 because there is less extraction of components from intracellular membranes. The choice of extraction procedure can, however, be readily assessed by preliminary comparisons using immunofluorescence. The major drawback with this extraction approach is that the plasma membrane is only poorly preserved. Also, on intracellular boundaries with membrane-associated complexes such as clathrin lattices, these complexes can prevent the penetration of probes such as antibodies to the membrane proteins that lie beneath them.

Horseradish Peroxidase–Diaminobenzidine Reaction on Living Cells

To preserve HRP-containing compartments before subsequent extraction with digitonin or Triton X-100, the same concentrations of DAB and hydrogen peroxide are used as described above for cytochemistry, but the buffer is modified as follows and the reaction is carried out at 4°. The allow DAB reaction to form where HRP is present 0.05 M Tris, pH 7.4, containing 0.9% (w/v) NaCl is used. Alternatively, to prevent DAB reaction product from forming on the cell surface while allowing stabilization of intracellular compartments, the membrane-impermeant substrate of HRP, ascorbic acid, is included in the buffer [0.88% (w/v) ascorbic acid, 0.47% (v/v) HEPES, 0.4% (w/v) NaCl, pH 7.4] according to Stoorvogel et al.[4]

---

[15] A. Gibson, C. E. Futter, S. Maxwell, E. H. Allchin, M. Shipman, J.-P. Kraehenbuhl, D. Domingo, G. Odorizzi, I. S. Trowbridge, and C. R. Hopkins, *J. Cell Biol.* **143,** 81 (1998).

Triton X-100 Extraction

After the DAB reaction, as described above, cells are washed three times at 4° in PBS+ [phosphate-buffered saline (PBS) containing 1 mM magnesium chloride, 0.1 mM calcium chloride] and then incubated in PBS+ containing 1% (v/v) Triton X-100, a protease inhibitor cocktail [chymostatin (1 mg/ml), and leupeptin, antipain, and pepstatin A in dimethyl sulfoxide (DMSO), diluted 1000-fold], and 0.2% (w/v) sodium azide for 10 min at 4°. Cells are then washed three times in PBS+ at 4° before fixation in PBS+ containing 2% (w/v) paraformaldehyde at 4° but are then allowed to warm to room temperature during a 15-min incubation period in fixative.

Digitonin Extraction

After the DAB reaction, as described above, cells are washed three times at 4° in permeabilization buffer (25 mM HEPES, 38 mM potassium aspartate, 38 mM potassium glutamate, 38 mM potassium gluconate, 2.5 mM magnesium chloride, 1.5 mM EGTA, pH 7.2) and then incubated in permeabilization buffer containing digitonin (40 μg/ml; diluted from a 100-mg/ml stock in DMSO) and protease inhibitor cocktail for 10–20 min (depending on the cell type) at 4°. Cells are then washed two times in permeabilization buffer, two times in permeabilization buffer containing 1% (w/v) bovine serum albumin (BSA; to quench the digitonin), and then a further two times in permeabilization buffer, all at 4°. Cells are then fixed in permeabilization buffer containing 2% (w/v) paraformaldehyde at 4° but are then allowed to warm to room temperature during a 15-min incubation period in fixative.

Immunogold Labeling of Extracted Cells

For digitonin-permeabilized cells reagents are made up in permeabilization buffer, and for Triton X-100-extracted cells reagents are made up in PBS+; all incubations are at room temperature unless otherwise indicated.

After extraction and paraformaldehyde fixation cells are quenched in 0.11% (w/v) glycine for 20 min with one change of solution. Cells are then blocked in 1% (w/v) BSA for 15 min and incubated in primary antibody in 1% (w/v) BSA for 1 hr at room temperature or for 15 hr at 4°. Cells are then washed twice in 1% (w/v) BSA and then three times (5 min each) in 2% (w/v) BSA–2% (v/v) FCS. Cells are then incubated in gold-conjugated secondary antibody in 2% (w/v) BSA–2% (v/v) FCS for 1 hr. Cells are washed twice in 1% (w/v) BSA and then three times 15 min each time) in 2% (w/v) BSA–2% (v/v) FCS. After washing twice in buffer containing no BSA or FCS cells are fixed, stained, and embedded as described above for cytochemistry and embedding, or are processed for whole mounts as described below.

For morphological analysis of pleiomorphic structures such as endosomes preservation by DAB cross-linking followed by detergent extraction produces preparations that can be examined in either whole-cell mounts or in very thick (>1.0-μm) plastic sections. In these preparations tubules and other attenuated continuities can be traced to their origins and are readily discriminated from free vesicles of similar diameter.

Preparation of Whole Mounts for Electron Microscopy

Cells grown on glass coverslips are extracted with Triton X-100 and immunogold labeled as described above, and are then fixed in cacodylate buffer, containing 4% (v/v) glutaraldehyde for 15 min at room temperature. Cells are then postfixed in 1% (w/v) reduced osmium and dehydrated through graded ethanols, as described for cytochemistry and embedding. Specimens, are then air dried and then platinum–carbon rotary shadowed and carbon coated with a Balzers-Pfeiffer (Nashua, NH) FDU coating unit. The coverslip is then removed by floating it on a solution of 8% (v/v) hydrofluoric acid. Specimens are finally washed by floating on water, mounted onto grids, dried, and examined in the electron microscope.

Light Microscope Studies. The ability of DAB cross-linking to destroy other signals contained within compartments containing targeted HRP constructs has also proved useful.[16] At the light microscope level it has been used to demonstrate that cross-linking by peroxidase in the presence of DAB will completely quench the fluoresence of probes carrying fluorescein isothiocyanate (FITC) contained within the same compartment. This approach has been particularly useful for studies of sorting mechanisms because it can discriminate between compartments containing both HRP and FITC probes from compartments to which the FITC probe has been selectively removed.

Fluorescence Quenching by Horseradish Peroxidase Activity

To quench FITC-labeled tracers with cointernalized HRP, tracer-loaded cells are fixed with 3% (v/v) formaldehyde and photographed, and then Tris buffer (0.05 M Tris-HCl, pH 7.6) containing 0.021% (v/v) hydrogen peroxide and DAB (0.75 mg/ml) is perfused across the coverslip.

The ability of HRP to cross-link in the presence of DAB and hydrogen peroxide can also be exploited in living cells. It has, for example, been possible to inhibit transfers to the lysosome by first chasing a pulse of fluid-

---

[16] Hopkins *et al.* (1994).

phase HRP down the endocytotic pathway to these target organelles and then cross-linking their content.[17] After this treatment with DAB and hydrogen peroxide other trafficking routes such as transferrin uptake and recycling continue to operate. However, further transfers into the lysosomes are prevented and earlier stages in the endocytic pathway such as those involving the multivesicular vacuole compartment become greatly expanded.[17]

Diaminobenzidine Cross-Linking for in Vivo Studies

Cells are washed at 4° with Tris buffer [0.05 M Tris, pH 7.4, containing 0.9% (w/v) NaCl] and then incubated in Tris buffer containing DAB (0.75 mg/ml) and 0.0021% (v/v) hydrogen peroxide for 30 min at 4° in the dark. Cells are then washed three times at 4° with Tris buffer and then with normal incubation medium.

Acknowledgments

The excellent assistance of Liz Allchin and Susan Maxwell is greatly appreciated. This work was supported by an MRC program grant to C. R. Hopkins.

[17] C. E. Futter, A. Pearse, L. Hewlett, and C. R. Hopkins, *J. Cell Biol.* **132**, 1011 (1996).

[4] Biochemical Analyses of Trafficking with Horseradish Peroxidase-Tagged Chimeras

By Anastasiya D. Blagoveshchenskaya and Daniel F. Cutler

Introduction

Quantitating the level of a protein within any intracellular compartment is rarely trivial, particularly when there is no organelle-specific posttranslational modification that can be monitored, as is often the case with post-Golgi trafficking. The determination of intracellular distribution in such cases usually relies on quantitating levels of the protein within subcellular fractions obtained by differential centrifugation of postnuclear supernatants. Using subcellular fractionation, we have been quantitating the steady state levels of chimeras comprising the transmembrane and cytoplasmic domains of proteins including P-selectin, sialyltransferase, tyrosinase, and the transferrin receptor (TrnR), in which their lumenal domains have been replaced by an enzymatic reporter—horseradish peroxidase (HRP)—in a variety of post-Golgi intracellular compartments. In particular, quantitative analyses of the targeting of mutant chimeras have been facilitated by this approach. For a brief discussion of the design of chimeras and their use for electron microscopy, see [1] in this volume.[1]

Typically, using any catalytically active enzymatic reporter will make it easier to obtain quantitative, reproducible data with the high sensitivity of detection allowing for increased resolution and ease of method optimization. First, quantitation of intracellular distribution of an enzyme activity is both simpler and much more accurate than analysis by Western blotting or immunoprecipitation of gradient fractions. Moreover, the ease of handling large numbers of samples allows for rapid characterization of many different constructs, which has been important in determining the sequence dependence of various trafficking steps.[2–4] Second, data obtained by enzymatic assay are reproducible between independent experiments. This is especially so when data are normalized by both the expression level of the chimera and by organelle recovery (see below). This combination even allows for quantitative analyses of subcellular localization after transient expression of the chimera of interest, generating data in which variation between independent experiments is acceptable: standard deviations do not usually exceed 10% of the mean. Third, the use of an enzymatically active reporter allows for sensitive detection, such that smaller amounts of material and lower levels of expression will be sufficient to perform an assay; analyzing the distribution of proteins at lower levels of expression reduces the possibility that overexpression will affect the pathway under examination. Fourth, both the sensitivity of detection and the simplicity of assay allow for analyzing a larger number of fractions from a gradient than would be practicable by other approaches, therefore increasing the resolution of the analysis. Fifth, the increased resolution, coupled with the speed of analysis, leads to an ease of development and fast optimization of protocols. Thus, altogether, we find that the use of an enzymatic reporter for biochemical analyses of the subcellular distribution of a protein confers considerable benefits.

Choosing HRP as a reporter also has specific advantages in the examination of the intracellular fate of proteins. It allows for an improved measurement of late endosomal/lysosomal targeting of HRP chimeras, carried out by exploiting the protease resistance of HRP, and for the determination

[1] C. Hopkins, A. Gibson, J. Stinchcombe, and C. Futter, *Methods Enzymol.* **327**, Chap. 3, 2000 (this volume).
[2] J. P. Norcott, R. Solari, and D. F. Cutler, *J. Cell Biol.* **134**, 1229 (1996).
[3] A. D. Blagoveshchenskaya, J. P. Norcott, and D. F. Cutler, *J. Biol. Chem.* **273**, 2729 (1998).
[4] A. D. Blagoveshchenskaya, E. W. Hewitt, and D. F. Cutler, *J. Biol. Chem.* **273**, 27896 (1998).

of colocalization of two or more proteins (where one is an HRP chimera), using a 3,3-diaminobenzidine (DAB) density shift procedure.

P-selectin can be used to exemplify the use of HRP chimeras in analyses of post-Golgi trafficking, because it has been shown to possess sequences within its cytoplasmic domain capable of targeting an HRP–P-selectin chimera to three major organelles: lysosomes, dense-core secretory granules, and synaptic vesicles. P-selectin belongs to a family of adhesion molecules and was originally identified in the secretory granules of platelets and endothelial cells. We have previously shown that when expressed in H.Ep.2 cells, HRP–P-selectin chimeras are efficiently endocytosed to lysosomes via the plasma membrane,[3,4] essentially as full-length P-selectin when expressed in other nonspecialized cell lines.[5,6] When heterologously expressed in neuroendocrine cells, P-selectin is also efficiently sorted to regulated secretory organelles: synaptic-like microvesicles (SLMVs) and/or dense core granules (DCGs).[2,7–10]

Table I summarizes the fractionation procedures we have successfully used for the purification of early and late endosomes, lysosomes, DCGs, SLMVs. After subcellular fractionation, the isolation of the organelle of interest is assessed by, on the one hand, a significant (30- to 150-fold) enrichment of the marker enzyme or protein (see Table I) over that found in the total homogenate and, on the other hand, by resolving this compartment from any other likely contaminant intracellular membranes, i.e., an absence of other pertinent markers.

Cell Lines and Transient Transfections

All experiments start with transfection, and we have found electroporation to give highly efficient transient expression of our cDNAs using a simple vector driven from the cytomegalovirus (CMV) promoter. Most of our work has been carried out with H.Ep.2 or PC12 cells, and so the protocols below have been developed for these cells. The human epithelioid cell line H.Ep.2 (CCL23; American Type Culture Collection, Rockville,

[5] S. A. Green, H. Setiadi, R. P. McEver, and R. B. Kelly, *J. Cell Biol.* **124,** 435 (1994).

[6] K. S. Straley, B. L. Daugherty, S. E. Aeder, A. L. Hockenson, K. Kim, and S. A. Green, *Mol. Biol. Cell.* **9,** 1683 (1998).

[7] J. A. Koedam, E. M. Cramer, E. Briend, B. Furie, B. C. Furie, and D. D. Wagner, *J. Cell Biol.* **116,** 617 (1992).

[8] J. C. Fleming, G. Berger, J. Guichard, E. M. Cramer, and D. D. Wagner, *Eur. J. Cell Biol.* **75,** 331 (1998).

[9] P. W. Modderman, E. A. Beuling, L. A. Govers, J. Calafat, H. Janssen, A. E. Von dem Borne, and A. Sonnenberg, *Biochem. J.* **336,** 153 (1998).

[10] A. D. Blagoveshchenskaya, E. W. Hewitt, and D. F. Cutler, *J. Cell Biol.* **145,** 1419 (1999).

TABLE I
ISOLATION AND IDENTIFICATION OF DIFFERENT INTRACELLULAR COMPARTMENTS

Subcellular compartment	Marker	Cell line	Gradient system
Early endosomes	[^{125}I]Trn for 1 hr at 37°, [^{125}I]EGF at 4° for 1 hr, followed by 5-min chase at 37°	PC12; H.Ep.2	1–16% Ficoll in HB for 45 min at 35,000 rpm followed by centrifugation on 3–16% Ficoll in HB for 45 min at 35,000 rpm
Late endosomes	[^{125}I]EGF at 4° for 1 hr followed by 10- or 20-min chase at 37°	H.Ep.2	1–16% Ficoll in HB for 45 min at 35,000 rpm
	[^{125}I]EGF at 4° for 1 hr followed by 10- or 20-min chase at 37°	PC12	1–16% Ficoll in HB for 45 min at 35,000 rpm followed by centrifugation on 0.9–1.85 M sucrose in 10 mM HEPES, pH 7.3, for 21 hr at 35,000 rpm
Lysosomes	NAGA; [^{125}I]EGF at 4° for 1 hr followed by 30-min chase at 37°	PC12; H.Ep.2	1–16% Ficoll in HB for 45 min at 35,000 rpm followed by centrifugation on 5–25% Ficoll in HB for 50 min at 35,000 rpm
SLMVs	Synaptophysin/p38	PC12	1–16% Ficoll in HB for 45 min at 35,000 rpm
SLMVs	Synaptophysin/p38	PC12	5–25% Glycerol in buffer A for 2 hr 50 min at 35,000 rpm
DCGs	[^{3}H]Dopamine at 37° for 2 hr	PC12	1–16% Ficoll in HB for 45 min at 35,000 rpm followed by 0.9–1.85 M sucrose in 10 mM HEPES, pH 7.3, for 21 hr at 35,000 rpm

Abbreviations: Tfn, transferrin; EGF, epidermal growth factor; NAGA, N-acetyl-β-D-glucosaminidase; SLMV, synaptic-like microvesicles; DCG, dense core granules.

MD) is maintained in Dulbecco's modified minimum Eagle's medium (DMEM) supplemented with 10% (v/v) fetal bovine serum (FBS) and gentamicin (100 μg/ml). Rat pheochromocytoma neuroendocrine PC12 cells [Cell Culture Facility, University of California at San Francisco (UCSF)] are cultivated in DMEM supplemented with 10% (v/v) heat-inactivated horse serum, 5% (v/v) FBS, and gentamicin (100 μg/ml). One day before transfection, cells are plated to ~75% confluency. On the day of transfection, a 9-cm dish is trypsinized, washed once with conditioned growth medium, and then twice in ice-cold Hebs [20 mM HEPES (pH 7.05), 137 mM NaCl, 5 mM KCl, 0.7 mM Na$_2$HPO$_4$, and 6 mM glucose]. The cells are finally resuspended in 250 μl of Hebs in a 0.4-cm electroporation cuvette chilled on ice. CsCl-purified DNA (10 μg) is added, and the DNA–cell mix is given two pulses in a GenePulser (400 V, 125 μF, and infinite resistance; Bio-Rad Laboratories, Hercules, CA). After 5 min at room temperature, the cells are reseeded into a 9-cm dish in fresh growth medium. H.Ep.2 cells expressing HRP chimeras are usually assayed 2 days posttransfection, whereas PC12 cells are analyzed 2–7 days posttransfection. Using this transfection procedure, we have been able to obtain expression of HRP chimeras in 40–70% of cells. Typically, if the transfected cells from one 9-cm dish are lysed in 1 ml of an HRP assay buffer (see below), 1 μl from this lysate is enough to detect the presence of the enzyme after the o-phenylenediamine (OPD) reaction performed in microplates (see below).

Subcellular Fractionation

Gradients and Ultracentrifugation

Ficoll (1–16%) Velocity Gradients. Most of our subcellular fractionation procedures leading to quantitation of targeting start with an initial 1–16% Ficoll velocity gradient, which provides a preliminary separation of most of the organelles discussed above (Fig. 1). To monitor the position of early recycling endosomes, the cells are fed with ^{125}I-labeled Trn (^{125}I-Trn) at 37° for 1 hr, while ^{125}I-labeled epidermal growth factor (^{125}I-EGF) allows for localization of a variety of early, late endosomal, and lysosomal compartments. Where necessary, the position of plasma membrane on the gradient can be determined by binding of ^{125}I-EGF or ^{125}I-Trn bound to the cells at 4° for 1 hr (data not shown). In addition, N-acetyl-β-D-glucosaminidase (NAGA) (see below for protocol) provides a marker for late endosomes (note shoulder of NAGA activity colocalizing with ^{125}I-EGF bound to the cells at 4° for 1 hr and internalized for 20 min at 37°) and lysosomes (Fig. 1). To monitor the position of DCG, PC12 cells are labeled with [^{3}H]dopamine (0.5 μCi/ml) for 2 hr at 37° in growth medium.

The fractionation procedure used is as follows: after two rinses with ice-cold HB (320 mM sucrose, 10 mM HEPES, pH 7.3), cells from one 9-cm dish are scraped into 1.5 ml of HB with a rubber policeman and passed 10 times through a ball-bearing homogenizer with a 0.009-mm clearance (made in the EMBL, Heidelberg, Germany). The cell homogenate is spun at 8500 g for 5 min and 1.3 ml of postnuclear supernatant (PNS) is layered on an 11-ml 1–16% preformed linear Ficoll gradient (Ficoll-400; Sigma, St. Louis, MO) made in HB. The gradients are centrifuged for 45 min at 35,000 rpm in an SW40Ti rotor (Beckman Instruments, Palo Alto, CA) and then fractionated in 0.5-ml fractions from the top of the tube, using an Autodensi-Flow IIC (Buchler Instruments, Kansas City, MO).

Glycerol Velocity Gradients for Synaptic-Like Microvesicle Isolation. The alternative gradient is that used for isolating SLMVs. This velocity

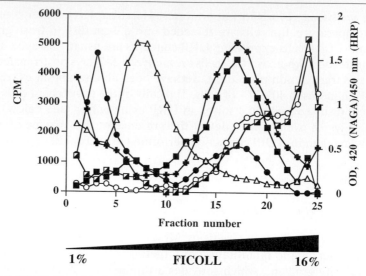

Fig. 1. Distribution of endocytic markers, [³H]dopamine, and expressed wild-type HRP–P-selectin in PC12 cells on an initial 1–16% Ficoll gradient. Cells were labeled with either ¹²⁵I-Trn or ¹²⁵I-EGF or [³H]dopamine as described in Table I. Cells were homogenized in HB, and the PNS was centrifuged on velocity 1–16% Ficoll gradients and fractionated. The position of early/recycling endosomes is revealed by the distribution of ¹²⁵I-Trn endocytosed for 60 min at 37° (△). Late stages of the endocytic pathway are shown by the distribution of ¹²⁵I-EGF internalized for 20 min at 37° (■) and for 30 min at 37° (◪). The distribution of HRP–P-selectin (●) is monitored by the amount of HRP activity (OD₄₅₀ ₙₘ) in each fraction across the gradient. Positions of DCG and lysosomes are monitored by distributions of [³H]dopamine (+) and NAGA activity (○) (OD₄₂₀ ₙₘ), respectively. (Adapted from the *Journal of Cell Biology*, 1999, Vol. 145, pp. 1419–1433 by copyright permission of The Rockefeller University Press.)

glycerol gradient is adapted for SW40 tubes from that developed by the Kelly laboratory.[11] Cells from one 9-cm dish are scraped into buffer A (150 mM NaCl, 0.1 mM MgCl₂, 1 mM EGTA, and 10 mM HEPES, pH 7.4) and homogenized by passage nine times through a ball-bearing homogenizer with a 0.009-mm clearance. The homogenate is then centrifuged for 15 min at 13,000 *g* in a microcentrifuge. The PNS is then layered on top of an 11-ml 5–25% preformed glycerol gradient made in buffer A and centrifuged in an SW40Ti rotor for 2 hr and 50 min. The distribution of HRP activity for wild-type HRP–P-selectin chimera relative to that for the SLMV marker protein, synaptophysin/p38, is shown in Fig. 2A and B.

Secondary Sucrose Equilibrium Gradients for Separation of Late Endosomes and Dense Core Granules. The 1–16% Ficoll gradients provide only

[11] L. Clift-O'Grady, C. Desnos, Y. Lichtenstein, V. Faundez, J.-T. Horng, and R. B. Kelly, *Methods Companion: Methods Enzymol.* **16,** 150 (1998).

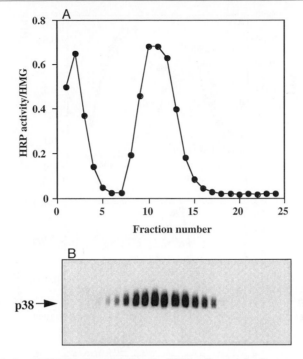

Fig. 2. Isolation of SLMVs from PC12 cells. PC12 cells, transiently expressing wild-type HRP–P-selectin, were homogenized in buffer A (see text) and PNS was centrifuged on the preformed 5–25% glycerol gradients as described. The distribution of HRP activity (●) for wild-type HRP–P-selectin relative to that of the endogenous marker protein synaptophysin/p38 immunoreactivity is shown in (A) and (B). (Adapted from the *Journal of Cell Biology,* 1999, Vol. 145, pp. 1419–1433 by copyright permission of The Rockefeller University Press.)

a preliminary isolation of many organelles. Further purification is sometimes needed for more accurate quantitation of levels of HRP chimera in any given organelle. In PC12 cells, it is possible to purify late endosomes from DCGs by recentrifuging the material obtained from the initial 1–16% Ficoll gradient (fractions 15–20; see Fig. 1) on 0.9–1.85 M sucrose equilibrium gradients. This can be achieved in the following manner: the fractions indicated above are diluted with HB and 4 ml of this material is then layered on top of a 9-ml preformed 0.9–1.85 M sucrose gradient. The gradients are then centrifuged at 35,000 rpm for 21 hr in an SW40Ti rotor and fractionated into 0.5-ml aliquots, as described above. The separation of late endosomes, and DCGs, as on this type of gradient, is shown in Fig. 3A. The lighter peak of HRP activity containing NAGA and internalized ^{125}I-EGF corresponds to late endosomes, while the denser peak, which overlaps with [^{3}H]dopamine distribution, corresponds to the DCGs.

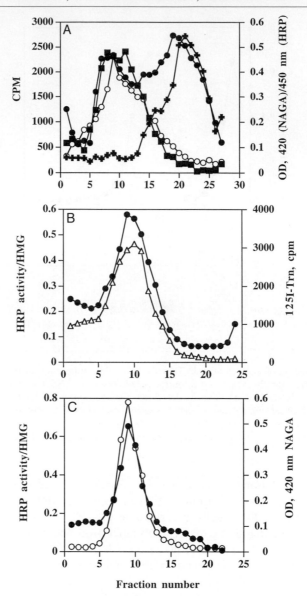

FIG. 3. Purification of DCGs, late endosomes, early endosomes, and lysosomes from PC12 cells on secondary gradients. (A) Separation of late endosomes and DCGs on equilibrium gradients. PC12 cells were transfected to transiently express wild-type HRP–P-selectin, labeled with ^{125}I-EGF or [3II]dopamine as described in Table I, and fractionated on initial 1–16% Ficoll gradients. Fractions 14–19 from these gradients (Fig. 1) were pooled, diluted with HB, and recentrifuged on 0.9–1.85 M sucrose gradients to equilibrium. The distribution of NAGA

Secondary 3–16% Ficoll Velocity Gradients for Early Endosome Isolation. As mentoned above, early endosomes on the initial 1–16% Ficoll gradients can easily be localized by distribution of ^{125}I-Trn radioactivity (Fig. 1). However, for HRP chimeras present within more than one endosomal compartment under steady state conditions, the quantitation of targeting to this particular endosomal population is hardly feasible using this type of gradient. We therefore designed a procedure for further purification of early Trn-positive endosomes using secondary velocity 3–16% Ficoll gradients. To do this, fractions from the initial 1–16% Ficoll gradient containing a peak of ^{125}I-Trn radioactivity (fractions 5–11) are pooled, diluted with HB up to a 4-ml final volume, and layered on top of a 9-ml 3–16% Ficoll gradient. The gradients are centrifuged in an SW40Ti rotor for 50 min at 35,000 rpm. Fractionation is performed as described above. The typical distribution of HRP activity for wild-type HRP–P-selectin chimera present within ^{125}I-Trn-containing early endosomes is shown in Fig. 3B.

Secondary 5–25% Ficoll Velocity Gradients for Lysosome Isolation. Another organelle that can be monitored on the initial Ficoll gradient but to which targeting is more accurately quantitated after further purification is the lysosome. To purify lysosomes from PC12 or H.Ep.2. cells, fractions 20–25 from the 1–16% Ficoll gradient are pooled, 2.5 ml of this material is diluted with HB to make 4 ml, and the material is then layered on a 9-ml 5–25% Ficoll gradient. Centrifugation and fractionation are carried out as described for the initial 1–16% Ficoll gradients. The distribution of HRP activity for wild-type HRP–P-selectin and NAGA after centrifugation on this type of gradient is shown in Fig. 3C.

Enzyme Assays

Horseradish Peroxidase Assay. After fractionation, the presence of HRP activity on any of these gradients is assayed by the following procedure. Subcellular fractions are aliquoted in triplicate, using nonsterile 96-well

---

activity (○), HRP activity (●), [3H]dopamine radioactivity (+), and ^{125}I-EGF internalized for 20 min at 37° (■) after centrifugation on this gradient are shown. (B) Purification of early endosomes. Cells transiently expressing wild-type HRP–P-selectin were fed with ^{125}I-Trn as described in Table I and PNS obtained from these cells was centrifuged on 1–16% Ficoll gradients. Fractions containing the peak of ^{125}I-Trn radioactivity (fractions 5–11) were pooled, diluted with HB, and recentrifuged on preformed 3–16% Ficoll gradients (see Table I). The distribution of HRP activity (●) relative to that of ^{125}I-Trn (△) is shown. (C) Purification of lysosomal compartment from PC12 cells. Cells expressing wild-type HRP–P-selectin were homogenized and the PNS was centrifuged on 1–16% Ficoll gradients. Fractions 20–25 were pooled and recentrifuged on 5–25% Ficoll velocity gradients (see Table I). Fractionation was followed by assaying for HRP (●) and NAGA (○) activities. (Adapted from the *Journal of Cell Biology,* 1999, Vol. 145, pp. 1419–1433 by copyright permission of The Rockefeller University Press.)

plates. Fifty to 100 μl of each fraction is added to an equal amount of the assay reagent [100 mM citrate buffer (pH 5.5), 0.1% (w/v) Triton X-100, 0.02% (v/v) H_2O_2, and 0.1% (w/v) o-phenylenediamine (OPD; Sigma)] made as a saturated solution in methanol. The reaction is carried out in the dark at room temperature for up to 1 hr and then stopped by adding 50–100 μl (depending on the original volume of the aliquots) of 2 M HCl. Plates are read at 490 nm in a microplate reader.

N-Acetyl-β-D-glucosaminidase Assay. To detect the presence of the lyso-somal enzyme, N-acetyl-β-D-glucosaminidase (NAGA), 100 μl of each sub-cellular fraction is transferred into 96-well plates and 10 μl of subcellular fraction buffer [1 mM NaHCO$_3$, 1 mM EDTA, 0.01% (v/v) Triton X-100] is added. One hundred microliters of 12 mM p-nitrophenyl-β-D-glucosami-nide (Sigma) made in ice-cold 0.1 M sodium citrate, pH 4.5, supplemented with 0.2% (w/v) Triton X-100 is then added to the samples. The NAGA substrate solution should be precleared by centrifugation at 2000g for 1 min and kept on ice before use. The reaction is performed in the dark at 37° for 30–60 min and then halted by adding 80 μl of stop buffer [0.25 M glycine, 0.2 M NaCl (pH 12.5), 4% (w/v) sodium dodecyl sulfate (SDS)] prewarmed to 37°. Plates are read at 415–420 nm in a microplate reader.

Quantitation of Targeting Data

To quantitate accurately the data from the transient transfection/subcel-lular fractionation experiments, a way of comparing the results is required. An example of when this might be needed is in analyzing the relative targeting efficiencies of different mutants to a particular organelle or series of organelles. Our approach, which uses the concept of a targeting index, is modeled on that used by Kelly and colleagues for examining the targeting of mutants of VAMP to SLMVs.[12] Thus, to determine the sequences needed for targeting an HRP–P-selectin chimera to various post-Golgi destinations, we evaluated differences in targeting efficiencies for different mutants ver-sus wild-type HRP–P-selectin. Targeting data are presented as targeting indexes (TIs), i.e., the amount of HRP activity present in the peak cor-responding to the appropriate subcellular compartment for each chimera normalized to that for wild-type HRP–P-selectin. In all experiments, the targeting indexes for wild-type HRP–P-selectin are set at 1. To take into account variations of expression level and organelle yield, the amount of HRP activity present in the peak (HRP peak) has been corrected for the amount of marker enzyme or protein (marker peak) (see Table I for specification) within the the same peak and for total HRP activity and amount of marker in the homogenate (HRP hmg and marker hmg). After

---

[12] E. Grote, J. C. Hao, M. K. Bennet, and R. B. Kelly, *Cell* **81,** 581 (1995).

simplifying the original equation, the targeting indexes are defined as follows:

$$TI = \frac{\dfrac{\text{mutant HRP peak}}{\text{mutant HRP hmg}} \cdot \dfrac{\text{mutant marker peak}}{\text{mutant marker hmg}}}{\dfrac{\text{wild-type HRP peak}}{\text{wild-type HRP hmg}} \cdot \dfrac{\text{wild-type marker peak}}{\text{wild-type marker hmg}}}$$

Typically, the TI for an HRP–P-selectin chimera in which the cytoplasmic tail has been deleted[2] is about 20% of that for wild type and is subtracted from those of the other chimeras in each experiment to provide a baseline; i.e., the TI for the tailless protein is set at 0. The TIs of the mutants are therefore described on a scale within a range set by wild-type HRP–P-selectin (1) and the tailless chimera (0). In each experiment, the chimeras setting the boundaries for the targeting indices, i.e., wild-type and tailless HRP–P-selectin, have always been analyzed in parallel with the mutants. In this way, cross-experimental variation can be reduced to levels such that the standard deviation about the mean is less than 10%, allowing for accurate determination of targeting efficiencies. Such quantitative analyses reveal subtleties of targeting that may not be apparent when other approaches are taken.

Determination of Colocalization of Proteins by Diaminobenzidine Density Shift

Many smooth membrane-bound structures, including vesicles, tubules, and vacuoles involved in membrane traffic, are of similar density, so that separation of these compartments by traditional subcellular fractionation procedures can be problematic. However, a particular advantage of HRP as a reporter is that the use of DAB as a substrate in the presence of H_2O_2 leads to an increase in the density of that organelle, containing HRP. Density-modification techniques for the isolation of endosomal compartments loaded with HRP or HRP-conjugated ligands in the presence of DAB and H_2O_2 were first pioneered by Courtoy and colleagues.[13,14] DAB, polymerized in vesicles containing HRP, causes a major shift in their density on gradients, as well as ablating immunoreactivity. This effect of the DAB reaction can therefore be exploited to determine whether a marker and HRP are present within the same vesicle. If this is so, then a shift in density

---

[13] P. J. Courtoy, J. Quintart, and P. Baudhuin, *J. Cell Biol.* **98,** 870 (1984).

[14] P. J. Courtoy, J. Quintart, J. N. Limet, C. DeRoe, and P. Baudhuin, Polymeric IgA and galactose-specific pathways in rat hepatocytes: evidence for intracellular ligand sorting. *In*: "Endocytosis" (I. Pastan and M. C. Willingham, eds.), pp. 163–194. Plenum Press, New York, 1985.

of the marker after the DAB reaction will occur. In previous studies, this approach was successfully applied to establishing the sorting kinetics of two distinct endocytic tracers at various time points after internalization.[15,16] In our studies, rather than using HRP-conjugated ligands, we have taken the technique a step further, using the HRP activity contained within chimeras in such analyses. This allows for detection of HRP activity within compartments that may not otherwise be accessible to analysis using HRP-conjugated ligands. Below, we show how we have used the density-shift approach to determine the presence of HRP–P-selectin within different endosomal compartments. To quantitate the degree of the density shift of radiolabeled endocytic markers, we divide the amount of radioactivity remaining within the peak after the DAB reaction by the original radioactivity in the endosomal peak in the controls not incubated with DAB and H_2O_2, expressed in percentages, and subtract this value from 100%.

Figure 4A shows a DAB-induced shift by the wild-type HRP–P-selectin chimera of late endosomes labeled with ^{125}I-EGF internalized for 20 min at 37° after the preliminary binding of ligand at 4° for 1 hr. In this experiment, H.Ep.2 cells, transiently expressing wild-type HRP–P-selectin, are fed with ^{125}I-EGF as described above to label the late endosomes, followed by centrifugation of PNS on 1–16% Ficoll gradients. Fractions 17–23, containing most of the ^{125}I-EGF radioactivity (see Fig. 1), are pooled. Seven hundred microliters of this material is incubated for 15 min at room temperature with 800 μl of DAB mixture [4.5 mM DAB in HB (pH 7.3), 0.06% (w/v) H_2O_2], which is prefiltered through a 0.22-μm pore size Millipore (Bedford, MA) filter. Another 700 μl is incubated with HB alone. These mixtures are subsequently layered onto secondary 5–25% Ficoll gradients supplemented with 1 mM imidazole to amplify the signal and centrifuged for 50 min in an SW40Ti rotor. Quantitation of DAB-shifted ^{125}I-EGF radioactivity indicates that a significant (71%) shift of ^{125}I-EGF has occurred. As a control, the DAB reaction is carried out for the same chimera when only plasma membrane vesicles are labeled with ^{125}I-EGF after the incubation of the cells with ligand at 4° for 1 hr (Fig. 4B). In this case, the HRP moiety is exposed outside the vesicular lumen and no shift of ^{125}I-EGF is observed.

We have also successfully exploited the same DAB shift procedure to examine the colocalization of HRP–P-selectin chimeras with internalized ^{125}I-Trn (Fig. 4C and D). H.Ep.2 cells transfected to transiently express wild-type HRP–P-selectin or a chimera that has had a portion of cytoplasmic domain deleted such that it cannot exit from Trn-positive early

[15] W. Stoorvogel, H. J. Geuze, and G. J. Strous, *J. Cell Biol.* **104,** 1261 (1987).
[16] R. S. Ajioka and J. Kaplan, *J. Cell Biol.* **104,** 77 (1987).

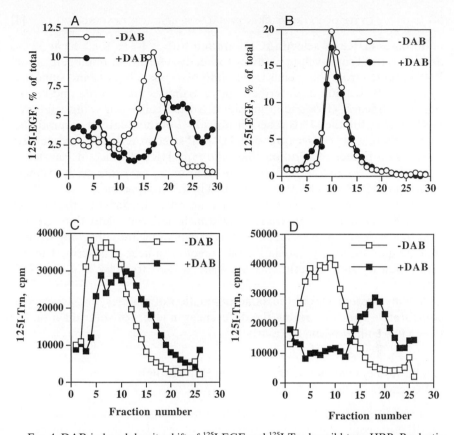

FIG. 4. DAB-induced density shift of ^{125}I-EGF and ^{125}I-Trn by wild-type HRP–P-selectin chimera or HRP–P-selectinΔC1 in H.Ep.2 cells. (A) Cells expressing wild-type HRP–P-selectin were allowed to bind ^{125}I-EGF at 4° for 1 hr. Cells were subsequently transferred to 37° for 20 min to induce internalization followed by removal of any ligand remaining on the cell surface by an acetic acid wash on ice twice for 3 min. Cells were then homogenized and a PNS was centrifuged on initial 1–16% Ficoll gradients. Dense fractions containing a peak of ^{125}I-EGF radioactivity were pooled and split into two parts. One part was incubated with DAB/H_2O_2 (●) and the another with HB alone (○). Material was recentrifuged on secondary 5–25% Ficoll gradients for lysosomal isolation and fractionated, and radioactivity in each fraction was counted. (B) Cells expressing wild-type HRP–P-selectin were incubated with ^{125}I-EGF at 4° for 1 hr, homogenized, and centrifuged on a 1–16% Ficoll gradient. The peak corresponding to the plasma membrane vesicles (fractions 6–9) was collected and split into two aliquots. One part was incubated with DAB/H_2O_2 (●) and the other with HB alone (○). After the DAB reaction, the samples were recentrifuged on secondary 3–16% Ficoll gradients in a SW40Ti rotor for 45 min. (C and D) Cells expressing either wild-type HRP–P-selectin (C) or HRP-P-selectinΔC1 (D) were fed with ^{125}I-Trn at 37° for 1 hr, washed, and homogenized. A PNS obtained from each dish was then split into two parts. One part was incubated with DAB/H_2O_2 (■), and the other with HB alone (□). The samples were centrifuged on 1–16% Ficoll gradients and fractionated, and radioactivity in each fraction was counted. (Panels A, B, and C are reproduced from A. D. Blagoveshchenskaya, J. P. Norcott, and D. F. Cutler, J. Biol. Chem. **273**, 2729 (1998).)

endosomes (HRP–P-selectinΔC1)[2] are fed with ^{125}I-Trn for 1 hr at 37° so as to load the Trn recycling pathway to steady state. After homogenization, PNS obtained from these cells is split into two equal aliquots and subjected to the DAB reaction as described above. Samples are layered and centrifuged on linear 1–16% Ficoll gradients. In the cells expressing wild-type HRP–P-selectin, the DAB reaction shifts 24% of the total ^{125}I-Trn radioactivity, whereas in cells expressing HRP–P-selectinΔC1 54% are shifted. These data reflect the finding that wild-type HRP–P-selectin transiently progresses through the Trn-positive endosomes *en route* to lysosomes, whereas HRP–P-selectinΔC1 is enriched in this compartment.

The values obtained in these experiments reflect the fact that the samples used were obtained from transiently transfected cells, and thus are not exact measures of the absolute amounts of chimera within the compartment that are quantitated by calculating the targeting indices as described above. These density-shift experiments are instead used to show that the HRP activity that colocalizes on a gradient with the marker, either EGF or Trn, is actually present within the same population of vesicles, and to provide comparative data, describing the differences in levels of colocalization between two chimeras and the marker.

Evaluation of Lysosomal Targeting by Horseradish Peroxidase-Clipping Assay

In addition to the utility of HRP in subcellular fractionation and in the DAB shift assay, the protease resistance of the HRP domain can be exploited in a novel and convenient analysis of targeting to a proteolytically active compartment, thereby providing a measure of lysosomal targeting of HRP chimeras. This assay uses Triton X-114 partitioning between the detergent-soluble, aqueous, and detergent-insoluble, hydrophobic phases of the HRP activity to determine the amount of hydrophilic HRP released (or "clipped") from its (hydrophobic) membrane anchor by proteolytic action. This assay therefore does not, as is usually the case, measure the disappearance of the protein but its transition between phases, and may provide a more accurate measure of exposure to proteases than other approaches. Obviously, when differential targeting to late endosomes and lysosomes is of interest, fractionation will be required: we have found that measuring exposure to protease action of HRP–P-selectin is capable of providing only an oversimple view of trafficking in the late endocytic pathway.[4] The protease resistance of HRP is also of benefit in such studies, because survival within late endocytic/lysosomal compartment(s) is clearly essential for detection.

To carry out the clipping assay, cells expressing a chimera, plated on

35-mm dishes, are placed on ice, rinsed twice with ice-cold PBS, and lysed in 1 ml of 1% (w/v) Triton X-114 (Sigma) made in PBS. Before use, commercially available TX-114 must be precondensed to remove any hydro-philic contaminants. To do this, TX-114 is dissolved in the ice-cold PBS to a final concentration of 3% (w/v). This solution is warmed to 37° and centrifuged at 2000 g for 5 min. The upper phase is discarded, while the detergent-enriched phase is redissolved in fresh ice-cold PBS. This proce-dure is then repeated three times. The detergent-enriched phase can be stored as a stock at 4 or −20°. When condensed at 37° in PBS, TX-114 forms a phase containing approximately 12% (w/v) detergent, which should be diluted to 1% (w/v) in ice-cold PBS and then used to lyse the cells. Lysates are then transferred into 1.5-ml microcentrifuge tubes and kept on ice for 10 min. The insoluble debris is pelleted in a refrigerated microcentri-fuge for 2 min at 13,000 g and the supernatants are then transferred into fresh tubes. The tubes are placed in a water bath at 30° for 2 min and then centrifuged at 13,000 g for 1 min at room temperature to separate the

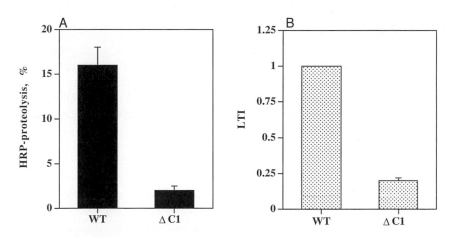

FIG. 5. A comparison of lysosomal targeting data as obtained by the HRP-clipping assay versus subcellular fractionation. (A) PC12 cells transiently expressing either wild-type HRP–P-selectin or HRP-P-selectinΔC1 were lysed with 1% Triton X-114 in PBS, followed by phase separation and assaying for HRP activity in each phase. The extent of proteolysis was expressed as a percentage of HRP activity in the soluble phase compared with the total activity in the lysate. Each bar represents the mean ± SE of four independent determinations. (B) Cells expressing either wild-type HRP–P-selectin or HRP–P-selectinΔC1 were fractionated to iso-late the lysosomal compartments as described in the legend to Fig. 3C. Targeting to lysosomes was quantitated by calculating lysosomal targeting indexes, allowing for HRP activity in the lysosomal peak to be normalized by the chimera expression level and the organelle recovery as judged by NAGA activity. Each bar represents the mean ± SE of three independent experiments. (Adapted from the *Journal of Cell Biology,* 1999, Vol. 145, pp. 1419–1433 by copyright permission of The Rockefeller University Press.)

phases. The upper, aqueous phase (~0.9 ml) is transferred to a fresh tube to which is added 0.1 ml of 12% (w/v) TX-114 stock. The lower detergent phase is mixed with 0.9 ml of ice-cold PBS. Tubes are placed on ice until a homogeneous solution is obtained (~5 min) followed by the standard HRP assay (see above) performed with both samples. Amounts of HRP activity present in the upper, hydrophilic phase, containing the clipped, soluble HRP domain, and the lower, hydrophobic phase containing the membrane-anchored HRP are determined and used to calculate the percentage of clipped chimera as a ratio of HRP activity in the soluble phase to the total activity in the lysate. To normalize for interexperimental variations, we subtract the percentage of HRP proteolysis for tailless HRP–P-selectin chimera (typically about 25%) as background from each data set. To relate this assay to a more traditional approach, results obtained by subcellular fractionation can be compared with those obtained by the clipping assay. The data obtained are shown in Fig. 5A and B, revealing that both assays give similar results.

Conclusion

In summary, we have described the subcellular fractionation procedures that we have established to quantitate targeting of HRP–P-selectin chimeras to a variety of post-Golgi destinations in different cell lines, and showed how the specific advantages of HRP can be exploited in the HRP-clipping assay of targeting to proteolytically active late endocytic/lysosomal compartments as well as in a DAB shift assay.

Acknowledgments

This work was supported by an international traveling postdoctoral fellowship to A.D.B. from the Wellcome Trust and by MRC program support to A.D.B. and D.F.C.

[5] Visualizing Protein Dynamics in Yeast with Green Fluorescent Protein

By CHRISTOPHER G. BURD

The use of the *Aequorea victoria* green fluorescent protein (GFP) as a real-time molecular tag to study protein localization and dynamics has had a major impact in yeast cell biology, as it has for so many other areas of biology. Investigators have only begun to explore the experimental possibilities to be manifested by integrating GFP-based technologies with the broad array of existing genetic and molecular biology yeast tools, but already the use of GFP has led to a greater appreciation of protein sorting and stability, phosphatidylinositol kinase signaling, nucleocytoplasmic shuttling, chromosomal dynamics, and RNA localization. In this chapter, some of the basic considerations and methodologies for using GFP in yeast are described and examples are provided of how GFP-based experiments can be used to reveal important aspects of protein dynamics.

One of the unique advantages of using yeast for GFP-based analyses of protein dynamics is the availability of mutant strains that allow one to examine the role(s) of individual cellular components in the behavior of the GFP-tagged protein. Two examples of GFP applications in yeast are shown in Figs. 1 and 2. In the first example, a region of the human early endosomal autoantigen (EEA1) protein containing its FYVE domain was expressed as a fusion to GFP. This domain binds phosphatidylinositol 3-phosphate [PtdIns(3)P], and this GFP fusion protein localizes to a small number of cytoplasmic puncta identified as endosomes.[1] In both yeast and human cells, localization of EEA1 to endosomes requires an intact FYVE domain, indicating that this FYVE domain recognizes a feature of endosomal compartments conserved between species[1,2] (Fig. 1). To investigate the basis for endosomal localization, the GFP–EEA1 (FYVE) fusion gene was introduced into mutant yeast strains defective in endosomal trafficking, and localization of GFP was examined.[1] The idea behind this screen was that any mutant lacking the feature of endosomes recognized by the GFP–EEA1(FYVE) protein would result in cytosolic localization, as occurs when mutant FYVE domains that do not fold properly are observed. One of the mutants that displayed this phenotype was *vps34Δ*. The *VPS34* gene encodes the only phosphatidylinositol 3-kinase (PI3K) in yeast, and so these

---

[1] C. G. Burd and S. D. Emr, *Mol. Cell.* **2**, 157 (1998).
[2] H. Stenmark, R. Aasland, B. H. Toh, and A. D'Arrigo, *J. Biol. Chem.* **271**, 24048 (1996).

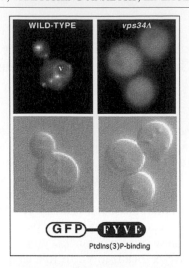

FIG. 1. Localization of a GFP–FYVE domain fusion protein in wild-type and mutant yeast strains. A vector encoding a fusion protein of GFP and the PtdIns(3)P-binding FYVE domain of human EEA1 was expressed in wild-type (*left*) and *vps34Δ* mutant yeast (*right*). The mutant strain lacks the enzyme that produces PtdIns(3)P, resulting in loss of GFP-FYVE-binding sites within the cell, and localization of GFP–FYVE throughout the cytosol. V, Vacuole. The vector used is pGO-GFP.[8]

results suggested that the GFP–EEA1 (FYVE) fusion protein may bind PI3K products. Another yeast mutant, *fab1*, was used to demonstrate that the EEA1 FYVE domain specifically binds PtdIns(3)P *in vivo*, but not other PI3K products. The *FAB1* gene encodes a PtdIns(3)P 5-kinase [which produces PtdIns(3,5)P_2], and when localization of the GFP–EEA1(FYVE) fusion protein was examined in *fab1* mutants that lack PtdIns(3,5)P_2, endo-somal localization was still observed. Taken together, the results of these experiments suggested that the Vps34 PI3K product, PtdIns(3)P, was essential for endosomal localization of the FYVE domain of EEA1 *in vivo*. Subsequent *in vitro* lipid-binding experiments indeed confirmed that the EEA1 FYVE domain, as well other FYVE domains, specifically bind PtdIns(3)P. In this example, the use of yeast mutants defective in phosphorylation of phosphoinositides allowed identification of PtdIns(3)P as the physiologically relevant ligand for FYVE domains *in vivo*.

In the second example, localization of a GFP-tagged mating pheromone receptor, Ste2p–GFP, was observed in wild-type cells. Ste2p is a heterotrimeric G protein-coupled receptor with seven transmembrane domains. It binds the secreted mating pheromone, α-factor, and initiates a cascade of events that regulates cell mating. Like many plasma membrane signal-

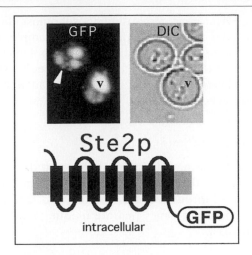

Fig. 2. Visualization of Ste2p-GFP in wild-type yeast cells. The *STE2* gene encodes a seven-transmembrane domain plasma membrane receptor for the mating pheromone, α-factor. The Ste2p–GFP fusion protein is localized to multiple compartments, including the plasma membrane, small endosomal compartments (white arrowhead), and the vacuole (V). Note that although the GFP moiety is on the cytosolic side of the membrane, the vacuolar GFP signal comes from the lumen of the vacuole, indicating that the entire Ste2p–GFP fusion protein had been internalized into the vacuole. The fusion gene was carried on a single-copy vector and expressed from the native *STE2* promoter.

transducing receptors, Ste2p is subject to downregulation; after binding ligand, it is rapidly internalized by endocytosis, and then delivered to the vacuole where it is degraded. A functional Ste2p–GFP fusion protein was constructed by cloning the coding sequence of GFP in frame immediately upstream of the *STE2* stop codon, resulting in fusion of GFP to the carboxy terminus of Ste2p.[3] Thus, the GFP moiety was oriented on the cytosolic side of the membrane (see Fig. 2). In wild-type cells, the Ste2p–GFP fusion protein is localized to the plasma membrane, endosomal vesicles, and the vacuole (Fig. 2). Surprisingly, the GFP signal associated with the vacuole was found to emanate from the lumen of the vacuole, rather than the limiting membrane of the vacuole, as had been expected. Because little staining of the vacuole membrane was observed, these results suggested that Ste2p–GFP had been delivered to the vacuole in internal vesicles of multivesicular bodies. Odorizzi *et al.*[4] used the Ste2p–GFP fusion protein to screen yeast mutants for defective delivery of the Ste2p–GFP fusion protein to the vacuole. Several genes were implicated in Ste2P–GFP traf-

[3] C. J. Stefan and K. J. Blumer, *J. Biol. Chem.* **274,** 1835 (1999).
[4] G. Odorizzi, M. Babst, and S. D. Emr, *Cell* **95,** 847 (1998).

ficking, including *FAB1*, thus implicating phosphoinositide signaling in the biogenesis of vacuolar internal vesicles. In this example, the use of GFP and fluorescence microscopy revealed important topological aspects of protein and membrane trafficking pathways to the vacuole, which had largely gone unappreciated in studies relying on Ste2p degradation as an indication of delivery to the vacuole.

Experimental Considerations

The use of GFP as a marker for localization studies requires many of the same considerations that apply for other types of molecular tags. Many of these are discussed throughout this volume, and therefore this chapter discusses several important issues that are unique or especially important for using GFP in yeast.

Which Vector to Use?

There are two ways to express GFP-tagged proteins in yeast: by plasmid-based expression vectors, and by integration of the GFP gene at a chromosomal locus to generate an in-frame fusion between GFP and the query protein. The decision as to which method to use is largely determined by experimental considerations. Plasmid-encoded GFP fusion genes are easily introduced into a large number of strains; however, expression of genes from plasmid vectors often does not recapitulate expression from the native, chromosomal locus. Integration of the GFP gene downstream of a locus may also alter the expression of the query protein. A drawback to this method is that each new strain needs to be confirmed for integration at the correct locus, and the GFP-tagged gene can be transferred to a different genetic background only by mating.

Several plasmid-based GFP fusion vectors have been described in the literature (Table I[5-8]). Use of these vectors entails cloning of the gene of interest in frame to the GFP gene, and then transformation of the plasmid into the appropriate strain for expression. Vectors are available that allow both amino- and carboxy-terminal fusions to GFP, and that allow constitutive or regulated expression of the fusion protein. If possible, it is best to include the native transcriptional regulation signals (e.g., the native pro-

[5] A. Wach, A. Brachat, C. Alberti-Segui, C. Rebischung, and P. Philippsen, *Yeast* **13,** 1065 (1997).
[6] M. S. Longtine, A. McKenzie III, D. J. Demarini, N. G. Shah, A. Wach, A. Brachat, P. Philippsen, and J. R. Pringle, *Yeast* **14,** 953 (1998).
[7] R. K. Niedenthal, L. Riles, M. Johnston, and J. H. Hegemann, *Yeast* **12,** 773 (1996).
[8] C. R. Cowles, G. Odorizzi, and S. D. Emr, *Cell* **91,** 109 (1997).

TABLE I
YEAST GFP VECTORS

Vector name	Selectable marker	GFP fusion	Expression	Ref.
pFA6a-GFP (series)	Kan^R; *HIS3*, *TRP1*	Amino and carboxy	Constitutive and regulated by galactose	5, 6
pGFP-N-FUS	*URA3*	Amino terminal	Regulated by methionine	7
pGFP-C-FUS	*URA3*	Carboxy terminal	Regulated by methionine	7
pGO-GFP	*TRP1, URA3*	Amino terminal	Constitutive	8

moter for carboxy-terminal fusions to GFP, and the native termination signals for amino-terminal fusions to GFP), which are generally (but not always) located within approximately 300 base pairs (bp) of the ends of the open reading frame.

Several vectors have been described that are useful for GFP tagging a chromosomal locus of interest. These vectors contain a GFP cassette linked to a selectable marker (e.g., *HIS3*) immediately downstream of the GFP gene.[5,6] Homologous recombination at the desired locus is targeted by sequences flanking the GFP cassette. Generally, the polymerase chain reaction (PCR) is used to amplify the GFP cassette with primers that append at least 40 bp of sequence from the locus to be analyzed on either side of the cassette. After transformation, the included selectable marker is used to select transformants that are then screened by PCR with a primer internal to the GFP cassette, and a primer outside the GFP cassette to determine if integration was at the correct locus. This method is particularly useful for appending GFP to the carboxy terminus of a protein, thus leaving the endogenous promoter intact and making it more likely that expression of the GFP-tagged gene is similar to that of the native gene. Vectors are also available for amino-terminal tagging, but these vectors also introduce relatively strong promoters to direct expression of the GFP fusion gene.

Does the Green Fluorescent Protein Tag Interfere with
Function/Localization of Query Protein?

Because expression of the GFP-tagged proteins in strains lacking the corresponding native wild-type protein is possible, yeast is well suited to determine whether the GFP tag interferes with the function/localization of the query protein. A relevant assay to determine the function of the query protein will determine if the GFP-tagged protein is functional. As with any cleavable molecular tag, it is important to determine if the GFP fusion protein is stable, so that complementation can be attributed to the tagged protein. This is easily accomplished by immunoblotting with antibodies to the query protein, to GFP, or best, to both.

Is Green Fluorescent Protein Fusion Protein Expressed at a Level Sufficient for Detection?

Early studies using wild-type GFP were not successful because of the low fluorescence signal when expressed in yeast. Since then, most studies have employed the S65T variant, which has proved sufficient for most localization studies. So far, only a few other GFP mutant variants have been used in yeast-based studies, and so the applicability of these is not yet known. No GFP gene that has been described in the literature is codon optimized for yeast, and so the effect of this variable on expression is also not yet known. For proteins that are localized to discrete regions of the cell (e.g., the membrane of a particular organelle), a smaller number of proteins can be detected (estimated to be approximately 300–3000[9]) compared with proteins localized to diffuse regions of the cell (e.g., the cytoplasm). Multicopy (2μ) vectors generally express genes at a 10- to 20-fold higher rate than single-copy (CEN) vectors, and can be useful for overcoming poor expression problems.

Do Experimental Conditions Affect Green Fluorescent Protein Signal or Detection?

Some experiments may require conditions that affect the GFP signal. For example, the use of temperature-conditional mutants requires temperature shifts that may affect folding of GFP and its fluorescence. Once folded, GFP is a stable molecule, and commonly used temperature shifts (e.g., 26 to 37°) do not pose any detection problems.

Required Equipment

Equipment required to visualize GFP-tagged proteins in yeast is essentially the same as required for immunofluorescence studies. A ×60 objective lens is sufficient for most studies; however, a ×100 lens is useful for high-resolution localization studies. For image capture and archiving, it is useful to have a digital camera to capture images directly from the microscope, and facilitate computer manipulation (e.g., false coloring). Currently, black-and-white CCD cameras are best because of the lower resolution of most color CCD cameras and the small size of yeast cells. If desired, black-and-

[9] R. Y. Tsien, *Annu. Rev. Biochem.* **67,** 509 (1998).

white images can be false-colored using appropriate software. For some applications, observation of single cells is not required or desired, and for these a relatively simple set-up can be constructed to observe GFP in cells *in situ* on a plate.[10]

Is Green Fluorescent Protein the Right Tag to Use?

There are several applications for which GFP may not be an appropriate tag for observing the behavior of a particular protein. Because folding of native GFP and formation of the chromophore can take between 30 min (for some mutant variants that fold more rapidly) and 2 hr (for wild-type GFP),[9] newly synthesized GFP-tagged proteins may not be visible during a relevant time course. This is important to keep in mind, for example, in studies of the secretory pathway because secretion from the cell can be rapid (less than 10 min) and it may not be possible to observe some aspects of protein dynamics. New GFP mutants that fold more efficiently, and/or generate the chromophore with more rapid kinetics, may diminish this concern in the future.

Little success has been reported for tagging soluble secretory proteins, or lumenal domains of membrane proteins. This is likely the result of the issues discussed above and protein-folding issues, but it also may reflect the sensitivity of GFP fluorescence to the differing environments of the secretory organelles. It had been anticipated that GFP-based localization studies of proteins to acidic organelles such as endosomes and the vacuole would be problematic because the fluorescence of GFP is quenched at acidic pH.[9] So far, however, this has not proved to be a problem (Fig. 2), probably because GFP is still highly fluorescent at the pH of the yeast vacuole (approximately pH 6 in standard media). Moreover, the highly compact GFP molecule is relatively resistant to proteolysis by vacuolar proteases.

General Protocol for Visualizing Green Fluorescent Protein-Tagged
  Protein in Yeast

Construction of Green Fluorescent Protein Fusion Gene

A plasmid vector is chosen and the query protein is cloned into the vector. Typically, this requires introducing appropriate restriction endonuclease sites to the ends of the open reading frame, usually by PCR. Alternatively, the chromosomal locus can be tagged by transformation with a DNA

[10] S. Cronin and R. Y. Hampton, *Trends Cell Biol.* **9,** 36 (1999).

that directs homologous recombination of GFP (along with a selectable marker) to the locus of interest. The decision concerning which approach to use (integrate GFP at the chromosomal locus or plasmid-based expression) is largely determined by the nature of the query protein and the nature of the question being asked. With plasmid vectors, it is possible to crudely regulate the level of expression by using low-copy (CEN) vectors (1–5 plasmids per cell), or multicopy (2μ) vectors (10–30 plasmids per cell).

Transformation

The appropriate yeast strain is transformed by the standard lithium acetate method,[11] or by electroporation. After colonies appear on the transformation plates, individual colonies are grown on selective plates for single colonies. For strains with GFP integrated at a chromosomal locus, it is necessary to confirm that the GFP gene has been integrated at the correct locus. This can be done easily by PCR to amplify a region of the chromosome with primers outside the region where GFP was integrated, and within the GFP gene. Integration is not always accurate, so it is best to test at least five colonies.

Visualization of Cells by Fluorescence Microscopy

Standard culture conditions for viewing cells need to be determined, because the physiological state of the cells can have a dramatic effect on the localization and function of the query protein. It has been observed, for example, that localization of some GFP-tagged proteins to endosomal compartments can differ dramatically between cells taken from exponentially growing cultures, and cells taken from cultures nearing saturation. Moreover, to compare results from different experiments, it is important to establish standard assay conditions. If possible, use conditions that allow results to be compared with those of other functional assays. For most applications, it is best to grow a preculture overnight, and to inoculate a fresh culture the following morning so that cells can be visualized from an exponentially growing culture (e.g., OD_{600} of approximately 0.5).

Microscope slides are prepared by spreading approximately 10 μl of concanavalin A (1 mg/ml) and allowing it to dry. This step facilitates adherence of the cells to the slide. A small amount of culture (approximately 3–5 μl) is then placed on a slide, and a coverslip is placed on top. Common yeast media can exhibit significant fluorescence, and therefore some investigators wash out the medium and resuspend the cells in distilled water. The

---

[11] H. Ito, Y. Fukuda, K. Murata, and A. Kimura, *J. Bacteriol.* **153,** 163 (1983).

effect of changing media should be tested. Alternatively, one may use a low fluorescence medium.[12]

It is often possible to view GFP in cells picked right off the original transformation plate, although caution should be used when interpreting results obtained by this method. There are significant physiological differences between cells grown on a plate and cells grown in liquid culture, and these differences can influence the localization and dynamics of many proteins. It should also be kept in mind that GFP will not be visible in all cells picked off a plate, and that for poorly expressed GFP fusion proteins, the signal from the autofluorescence of dead cells can be easily confused with a weak GFP signal.

Visualization of Green Fluorescent Protein in Colonies Growing on a Plate

It is possible to visualize GFP in cells comprising colonies growing on a plate. All that is required is a light source of appropriate wavelength to excite GFP, and an appropriate filter that permits observation of GFP emission. This method is particularly well suited for rapid genetic screening, especially for gene expression and protein stability experiments. Cronin and Hampton[10] found that a standard carousel slide projector fitted with an appropriate filter is sufficient to excite GFP in cells growing as a colony on a plate. GFP fluorescence is observed by viewing the plates through a filter that blocks the excitation light. This method is described in detail in an earlier volume of this series.[13]

---

[12] J. A. Waddle, T. S. Karpova, R. H. Waterston, and J. A. Cooper, *J. Cell Biol.* **132,** 861 (1996).
[13] S. Cronin and R. Y. Hampton, *Methods Enzymol.* [58] **302,** (1999).

[6] Kinetic Analysis of Intracellular Trafficking in Single Living Cells with Vesicular Stomatitis Virus Protein G–Green Fluorescent Protein Hybrids

By Koret Hirschberg, Robert D. Phair, and Jennifer Lippincott-Schwartz

Introduction

Intracellular trafficking in eukaryotic cells has been studied extensively with the tsO45 temperature-sensitive mutant vesicular stomatitis virus G protein (VSVG), using a variety of biochemical and morphological techniques.[1-5] VSVGtsO45 is a type I transmembrane protein that reversibly misfolds at nonpermissive temperature (39.5°) in the endoplasmic reticulum (ER) and is consequently accumulated. On temperature shift to 32°, VSVGtsO45 refolds and is efficiently exported from the ER through the Golgi to the plasma membrane (PM). Tagging VSVGtsO45 with the green fluorescent protein (GFP) generates a chimeric protein that maintains this temperature-sensitive phenotype,[6,7] making it a powerful tool with which to visualize and study protein transport in living cells. In this chapter we describe a method that employs digital time-lapse imaging to obtain virtually continuous sets of data on the transit of VSVG–GFP through the compartments of the secretory pathway in single cells after shift to permissive temperature. The VSVG–GFP distribution within the cell is monitored continuously by quantification of the fluorescence intensity in defined regions of interests (ROIs) containing the Golgi apparatus and the whole cell. These values are converted to the number of protein molecules and are used for the kinetic analysis. Commercially available software is employed to test the ability of a model hypothesis to account for the experimental data obtained. The modeling process yields kinetic parameters, which provide information on the rates of export, residence times, fluxes, peak fluxes, etc. The images are acquired at a rate of one every 30–120 sec, resulting in hundreds of data points for experiments run between 3 to 10 hr. These sets of data provide stringent and powerful means to test the model hypothesis and thus yield kinetic parameters of unprecedented accuracy. Ultimately, much more detailed kinetic models will be required for full analysis of the complete pathway.

The kinetic analysis of VSVG–GFP trafficking is particularly emphasized here because it can serve as a paradigm for diverse experimental strategies for studying the secretory pathway. The effect of miscellaneous perturbations, using pharmaceuticals or coexpressed proteins, on specific stages and kinetic parameters also can be carefully analyzed with VSVG–GFP. Taking into account the limitations and the principles addressed in this chapter, this method can be readily extended and applied to the use of other GFP-tagged proteins expressed in different cell types.

[1] J. Wehland, M. C. Willingham, M. G. Gallo, and I. Pastan, *Cell* **28,** 831 (1982).
[2] H. Arnheiter, M. Dubois-Dalcq, and R. A. Lazzarini, *Cell* **39,** 99 (1984).
[3] T. E. Kreis and H. F. Lodish, *Cell* **46,** 929 (1986).
[4] H. Plutner, H. W. Davidson, J. Saraste, and W. E. Balch, *J. Cell Biol.* **119,** 1097 (1992).
[5] W. E. Balch, J. M. McCaffery, H. Plutner, and M. G. Farquhar, *Cell* **76,** 841 (1994).
[6] J. F. Presley, N. B. Cole, T. A. Schroer, K. Hirschberg, K. J. Zaal, and J. Lippincott-Schwartz, *Nature (London)* **389,** 81 (1997).
[7] S. J. Scales, R. Pepperkok, and T. E. Kreis, *Cell* **90,** 1137 (1997).

Methods

The general protocol used for kinetic analysis of VSVG–GFP in living cells is outlined in Fig. 1. It utilizes mostly standard equipment and methods, but requires specific digital image capturing techniques that ensure proper imaging conditions appropriate for data collection that can be used in kinetic analysis.

Expression of VSVG–GFP in COS-7 Cells

For the kinetic analysis, COS-7 cells are used for the following reasons: COS-7 cells are large and flat (more than 50 μm across and ~5–10 μm

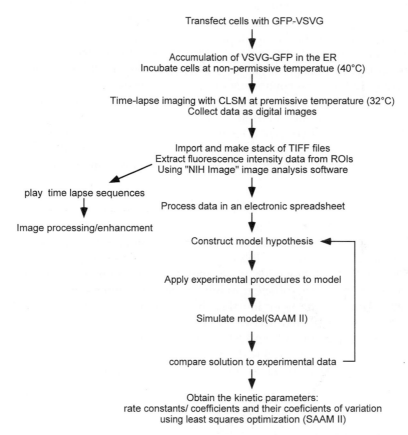

Fig. 1. A scheme depicting the stages in kinetic analysis of VSVG–GFP trafficking in living cells.

thick in the nuclear region) and they do not migrate significantly (thus enabling long time-lapse experiments). COS-7 cells are also readily transfected with high efficiency.

COS-7 cells are maintained in Dulbecco's modified Eagle's medium (DMEM) supplemented with 10% (v/v) fetal bovine serum (FBS), 2 mM glutamine, penicillin (100 U/ml), and streptomycin (100 μg/ml) at 37° in a 5% CO_2 incubator. Cells are grown in 15-cm-diameter tissue culture dishes or 176-cm^2 flasks. For electroporation, cells are replated 1 day before to obtain ~70% confluency on the day of transfection. Cells from a 15-cm-diameter dish are sufficient for replating in a 10-cm-diameter tissue culture dish containing glass coverslips or on chambered coverglasses (Lab Tek, Naperville, IL) that, combined, are equal in surface area. Cells are harvested, washed twice with electroporation buffer (RPMI–20 mM HEPES), incubated for 10 min on ice, and electroporated in 400 μl of electroporation buffer in electroporation cuvettes (200 V, 500 μF). After 20 min on ice cells are plated onto dishes containing medium prewarmed to 37°. COS-7 cells are also readily transfected with the Fugene-6 (Roche Molecular Biochemicals, Indianapolis, IN) transfection reagent; cells are plated 1 day before to obtain 50% confluency on the day of transfection. In both cases transfected cells are transferred the next day to a 39.5° incubator after replacing the medium. Cells should be kept for equivalent times in the 39.5° incubator, so that similar levels of VSVG–GFP are accumulated in the ER. An alternative to transient transfection is to use cells stably expressing VSVG–GFP, in which expression levels are roughly similar between cells.

Acquisition of Images for Kinetic Analysis

Cells are imaged in 2–3 ml of RPMI without phenol red, but containing HEPES buffer (20 mM, pH 7.4), 2 mM glutamine, penicillin (100 U/ml) streptomycin (100 μg/ml), 20% (v/v) fetal calf serum, and cycloheximide (150 μg/ml). Cycloheximide is an inhibitor of protein synthesis and the concentration used is sufficient to inhibit protein synthesis by 90% in COS-7 cells.[8] During time-lapse imaging the temperature is controlled with an air stream stage incubator (Nevtek, Burnsville, VA).

Cells are imaged with an LSM 410 (Carl Zeiss, Thornwood, NY), and GFP molecules are excited with the 488 line of a krypton–argon laser and imaged with a 512/27 bandpass filter and double dichroics (488/568). Time-

[8] N. B. Cole, J. Ellenberg, J. Song, D. DiEuliis, and J. Lippincott-Schwartz, *J. Cell Biol.* **140,** 1 (1998).

lapse images are captured at 30- to 120-sec intervals 30–50% laser power and 99% attenuation for as long as 3 to 10 hr.

It is imperative for quantitative imaging that the efficiency of fluorescence detection is constant during the capturing of a prolonged time-lapse sequence. This is particularly important when imaging living cells where the fluorescent label changes its localization and geometry within the cells. For example, on shift to 32°, VSVG–GFP redistributes from a widespread reticulum (the ER; Fig. 2A) to concentrated Golgi cisternae (Fig. 2B). The confocal laser scanning microscope (CLSM) was originally designed to sample a thin section from a given sample, using the pinhole apparatus that excludes out-of-focus fluorescence and determines the z axis boundaries of the focal plane. The thickness of the sampled slice is called the full-width half-maximum (FWHM). The FWHM is defined as the distance in the z direction between planes where the intensity is 50% of that at the plane of focus. Apart from the pinhole diameter, the FWHM is a function of the emission and excitation wavelengths, the magnification, and the numerical aperture (NA) as shown in Eq. (1) (obtained from the Zeiss confocal microscope manual):

$$\text{FWHM } (\mu\text{m}) = (1.41) \frac{\lambda_1}{\lambda_2} \left[\left(\frac{2n}{M \times \text{NA}} \right) \times 15.24P + \frac{n\lambda_2}{\text{NA}^2} \right.$$
$$\left. \times \exp \left(-\frac{2 \times \text{NA} \times 15.24P}{\lambda_2 \times M} \right) \right] \tag{1}$$

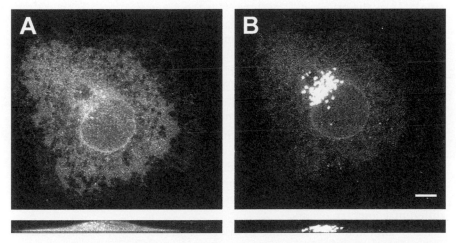

FIG. 2. Three-dimensional analysis of the distribution of VSVG–GFP in COS-7 cells. Projections at 180 and 90° of a three-dimensional reconstruction of cells expression VSVG–GFP before (A) and 40 min after shift to 32° (B). Bar: 5 μm.

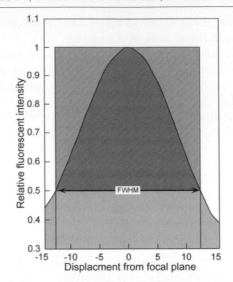

FIG. 3. Efficiency of fluorescence intensity detection within the FWHM. A series of sections. 1 μm apart in the z axis of COS-7 cells expressing VSVG–GFP, were taken with an ×25/ NA 0.8 Zeiss Plan-Neofluor oil immersion objective with pinhole open at 9.8 Airy disk units. The FWHM is indicated. The lightly shaded area in the box corresponds to the estimated total fluorescence intensity distribution of GFP standard solutions used to generate the standard curve for estimation of the number of VSVG–GFP molecules.

where λ_1 is the emission wavelength (nm); λ_2 is the excitation wavelength (nm); NA is the numerical aperture; M is the magnification; n is the refractive index of the immersion medium; P represents pinhole Airy units; and FWHM is the full-width half-maximum (μm).

High-magnification objectives with a numerical aperture value of about 1.4 will sample less than a 0.5-μm FHWM of a specimen. For the purpose of quantitative image analysis, a thicker FHWM is obtained with a wide open pinhole and low-magnification objectives. In the protocol described here a ×25/NA 0.8 Zeiss Plan-Neofluor oil immersion objective is used. The FWHM for this objective calculated on the basis of Eq. (1) is approximately 22 μm, suggesting that a relatively flat cell expressing a GFP-tagged molecule will be within the high-efficiency fluorescence detection zone. Thus, little or no change in efficiency of fluorescence detection in COS-7 cells during redistribution of GFP-tagged proteins to compartments with different geometry occurs.[9]

The combination of low-energy laser output (20–50%), high attenuation

[9] K. Hirschberg, C. M. Miller, J. Ellenberg, J. F. Presley, E. D. Siggia, R. D. Phair, and J. Lippincott-Schwartz, *J. Cell Biol.* **143**, 1485 (1998).

(99% of the laser power), and a less focused excitation-laser beam caused by the low NA objective results in negligible photobleaching during repetitive imaging for more than 3 hr.[9]

On reducing the temperature VSVG–GFP rapidly refolds and exits the ER. Therefore, the time after removing the cells from the 39.5° incubator and before image acquisition should be kept to a minimum and should be noted. We have found that VSVG–GFP is concentrated at least four- to fivefold when in the Golgi compared with when it is in ER or plasma membrane (PM). Because of this, care should be taken to avoid saturation of the fluorescent signal in the images captured during the experiment. To avoid saturation of the fluorescent signal, images can be captured in two separate channels, using two contrast settings for each time point. To incorporate the two data sets the effect of contrast setting on fluorescence intensity output should be determined. As a rough estimate VSVG–GFP can be four- to fivefold brighter in the Golgi than in the ER.

Autofocusing

In long time-lapse imaging experiments the object can end up moving from the plane-of-focus because of internal vibrations. An advantage of using a scanning LSM is that long time-lapse imaging sessions can be performed with recurrent autofocusings to maintain the specimen in the plane-of-focus. The autofocusing feature is embedded in the Zeiss 410 CLSM software. The autofocusing macro scans the z axis twice and finds the brightest plane. In the time-lapse experiments this macro is employed when the LSM is in the reflection mode using one of the visible emission laser lines such as the 568- or 647-nm line (Fig. 4). The autofocus in the reflection mode finds the surface of the coverslip, which reflects the laser beam to the reflection mode detector. The coverslip surface, the plane to which the microscope can automatically direct itself, is used as a fixed reference to find the plane-of-focus in which the imaged cells are to be found, usually 1–2 μm above the coverslip surface. Because a low-NA objective is used in these types of experiments, the autofocusing should be performed with the pinhole closed (~0.3 airy units or 5 Zeiss digital units) in order to facilitate accuracy of autofocusing.

Conversion of Fluorescence Intensity to Number of Green Fluorescent Protein Molecules

The number of VSVG–GFP molecules expressed in a single cell is estimated by comparing the total cellular pixel intensity value in digitized images with a standard curve generated with solutions of known concentrations of recombination purified GFP (Clontech, Palo Alto, CA) at identical

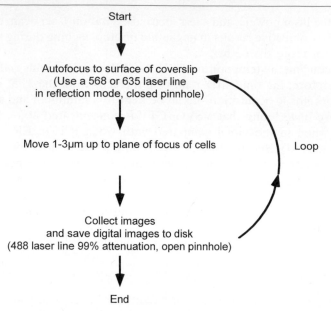

FIG. 4. A scheme of a time-lapse imaging macro with autofocusing used to control the LSM during a long experiment.

power, attenuation, contrast, and brightness settings on the confocal microscope.[10] The fluorescence in an ROI with fixed area is then plotted against the number of GFP molecules estimated to be within a corresponding volumetric ROI, determined by the product of the area and the FWHM. Plots of GFP molecules versus total fluorescence (the sum of the pixel values in an area) should fit precisely with a linear function.

This method is an estimation of the number of molecules because the FWHM is assumed to approximate the z-dimensional thickness of the sample (Fig. 3), and the efficiency of detection of fluorescence within that volume is considered constant (see shaded area in Fig. 3). The validity of these assumptions is confirmed by imaging in oil spherical droplets of GFP solution with diameters smaller then the FWHM within the range of analyzed organelles. This estimate also assumes that the fluorescence parameters (i.e., quantum yield, detection efficiency, and proportion of

[10] Lippincott-Schwartz, J. Presley, J. F. Zaal, K. Hirshberg, C. M. Miller, and J. Ellenberg, *in* "Methods in Cell Biology," Vol. 58 (K. Sullivan and S. Kay, eds.), p. 261. Academic Press, San Diego, 1998.

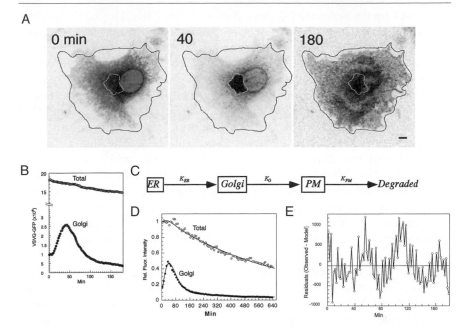

FIG. 5. Measurement and analysis of fluorescence intensity data in a single VSVG–GFP-expressing cell (A). ROIs used for quantitative analysis of transport. Black line represents total cell fluorescence. White line represents Golgi fluorescence. The images are inverted to better reveal the ROI. (B) Fluorescent intensities (converted to number of molecules on the basis of comparison of cellular VSVG–GFP fluorescence with known concentrations of GFP) associated with the Golgi ROI (Golgi, filled circles) and entire cell ROI (Total, open circles) for one representative cell after the shift to 32° are plotted at 30-sec intervals for 3 hr. (C) The three-compartment model for trafficking of VSVG–GFP. (D) VSVG–GFP-expressing cells incubated for 20 hr at 40° were shifted to 32° and imaged every 120 sec for 10 hr. The fluorescent intensities within the Golgi ROI (filled circles) and total cell ROI (open circles) arc plotted. Lines are the simulated data generated by testing the model in (C). (E) Plot of residuals used to check visually for randomness of deviation of simulated experimental data. Ideally, the residuals should be distributed randomly around zero.

properly folded and fluorescent GFP molecules) are similar for GFP chimeras in cells and for GFP in aqueous solution.[11]

Extracting Data from Images

Quantitation of digital images involves summing up the intensity values in defined ROIs as outlined in Fig. 5A. The data in Fig. 5A was analyzed

[11] D. Piston, G. H. Patterson, and S. M. Knobel, *in* "Methods in Cell Biology," Vol. 58 (K. Sullivan and S. Kay, eds.). Academic Press, San Diego, USA, 1998.

with NIH Image software (Wayne Rasband Analytics, Research Services Branch, NIH, Bethesda, MD), which is advantageous in being free of charge and extremely powerful because of the macro language that it supports. A limitation of NIH Image is that it operates only on 8-bit images with a range of 256 levels of fluorescence intensity. A myriad of image analysis software packages are available that can handle 12- and 16-bit images. Although the Zeiss 410 CLSM generates 8-bit images, CLSMs from other manufacturers are and will be available that produce 12-bit images, allowing a range up to 4096 levels of fluorescence intensity. The digital images captured with the Zeiss 410 LSM are TIFF images and can be readily imported to NIH Image, where they are visualized and analyzed as a stack. Fluorescence intensity in ROIs can be quantified in every image manually or automatically, using a macro that will sample the same ROI in all the images in a stack. For analysis of VSVG–GFP trafficking the total fluorescence in two ROIs is used. The first ROI contains the Golgi fluorescence intensity and its boundaries are determined from images ~40 min after the shift to permissive temperature (approximating the time when VSVG–GFP peaks at the Golgi). The second ROI includes the entire cell. The total cell fluorescence ROI is determined with the last image because the PM-labeled cell occupies more area than the ER-labeled cell at the beginning of the experiment. The fluorescence intensities in both ROIs are plotted against time in Fig. 5B. The Golgi ROI has a distinct peak and the total cell fluorescence is declining with time as a result of internalization and degradation of VSVG–GFP.

For further analysis the data obtained are transferred to an electronic worksheet. The extracellular background should be subtracted from the values collected. The Golgi ROI fluorescence intensity is contaminated with ER and PM fluorescence because there is always ER and PM present above or below the Golgi. This cross-contamination can be removed by subtracting the value of a nearby ER region, or it can be quantitatively accounted for in the modeling process.

Data Analysis and Hypothesis Testing

Translating Working Hypothesis to Computer Model

The principal benefit of collaborations between experimental and computational biologists is effective hypothesis testing. For this reason, the first step in data analysis is translation of the current working hypothesis to the language of mathematics. In the field of protein trafficking, the working hypothesis is most often a diagram representing the various cellular compartments along with arrows representing the known or hypothesized processes that constitute the secretory pathway for the protein being studied. This diagram will be different for different investigators, because each has his or her own working hypothesis, and each will necessarily focus on a subset of all the molecular interactions that define the full pathway. Because the VSVG–GFP data represent dynamics or kinetics, the mathematical form of the corresponding computer model must be capable of predicting dynamic behavior. This means the model will be made up of differential equations. A key decision in this early phase is whether to treat cellular compartments as well mixed. This decision should be made in joint discussions between experimentalists and theoreticians; it determines whether the model will consist of ordinary or partial differential equations.

In this chapter we treat the ER, Golgi, and PM as well-mixed compartments. Consequently, to perform this analysis the software need solve only ordinary differential equations. To write the appropriate differential equations, apply the principle of conservation of mass, successively, to each of the cellular compartments defined in the working hypothesis. Figure 5C is a diagram of the working hypothesis tested here. Taking the Golgi compartment as an example, the differential equation for the number of molecules of VSVG–GFP in the Golgi of a single analyzed cell is

$$dN_G/dt = \text{Flux}_{ER \to Golgi} - \text{Flux}_{Golgi \to PM} \tag{2}$$

In other words, the rate of change of the number of molecules of VSVG–GFP in the Golgi, N_G, is determined by the difference between the number of molecules per minute (flux) entering from the ER and the number of molecules per minute leaving for the PM. This is just a statement of conservation of mass for VSVG–GFP in the Golgi.

The next step is formulation of the appropriate rate laws for the fluxes in Eq. (2). This is a central feature of all of computational biology. Although linear rate laws are sufficient for the analysis described here, rate laws for nonlinear cell biological processes are readily formulated.[12] For many biologists, the most familiar rate law is that of Michaelis and Menten for an irreversible enzyme-catalyzed reaction in which the total amount of enzyme does not change with time. But an even simpler rate law, based on the principle of mass action, is all that is required to begin our analysis. This rate law says that the number of molecules per second (or per minute) moving along one of the arrows in our diagram is proportional to the number of molecules in the source compartment. In mathematical terms, and using the arrow from Golgi to PM as an example:

$$\text{Flux}_{Golgi \to PM} = k_G N_G \tag{3}$$

[12] R. D. Phair, *Metabolism* **46**, 1489 (1987).

Here, k_G is the rate constant that characterizes the intra-Golgi and Golgi-to-PM transport processes. Rate constants are fundamental cell biological parameters and we have more to say about them below.

Applying Experimental Protocol to Computer Model

The next step in the process of hypothesis testing is to carry out with the computer model the same experimental protocol that was used to collect the data to be analyzed. To test the hypothesis, we want to determine if it responds to this protocol in the same way as the living cells. Doing this properly again calls for careful discussion between the experimental and computational teams, because only when the modeled protocol is truly the same as that which produced the observations will the computed solutions truly be comparable to the experimental data. For the case of the temperature-sensitive mutant of VSVG, this process is straightforward. Because all of the fluorescent VSVG is retained in the ER at the nonpermissive temperature, the experimental protocol can be modeled with a single initial value (initial condition) in the ER compartment at the beginning of the experiment. An estimate of this initial condition plus estimates of the three rate constants, k_{ER}, k_G, and k_{PM}, are all that is required to calculate the model-predicated time courses of N_{ER}, N_G, and N_{PM} for this experimental protocol.

The next step is to discover whether there are values of the initial condition and the rate constants for which the computed solution is consistent with the experimental data. In all but the simplest cases, this is best done with one of many available software packages that solve ordinary differential equations numerically and permit easy comparison of solutions with data. We use SAAM II software (SAAM Institute, Seattle, WA), but the choice of software is largely a matter of personal preference. The strengths of SAAM II for analysis of GFP kinetics are (1) an intuitive graphical user interface, (2) flexible definition of both linear (mass-action) and nonlinear (saturated or allosteric) rate laws, (3) easy specification of experimental protocols in terms of the working model, (4) the ability to account for contributions from multiple cellular compartments to a single experimental measurement, and (5) modern, peer-reviewed algorithms for nonlinear least-squares parameter estimation.

If, during the transfection and imaging periods, the cells are exposed to some physiological or pharmacological perturbation, then the values of the model parameters can be usefully compared with those obtained by analysis of control cells. These comparisons yield direct information on the sites of action of the physiological or pharmacological agents as described below in Biological Interpretation of Kinetic Parameters (below).

Comparison of Model Solutions and Experimental Data

If experimental measurements were always pure observations of a single cellular compartment, we could compare the time courses directly with the experimental data and move on directly to parameter estimation. But in the real world of cell biology and fluorescence microscopy, the measured light will generally originate from more than one cellular structure. For example, total cellular fluorescence must be compared with the sum of the fluorescence in each model compartment. Also, because there is always some portion of the ER included in the region of interest (ROI) bounding the Golgi, the Golgi data must be compared with the time course of N_G with the addition of a contribution from N_{ER}. In mathematical terms, the Golgi fluorescence must be compared with a linear combination of the model solutions for N_G and N_{ER}:

$$F_{ROI}^G = c_1 N_G(t) + c_2 N_{ER}(t) + c_3 N_{PM}(t) \qquad (4)$$

Here, F_{ROI}^G is the calculated fluorescence that will be compared directly with the Golgi data: and c_1, c_2, and c_3 are constants (whose values can be derived from the experimental data) that account for several unavoidable experimental facts. First, they include the conversion from molecules to fluorescence as described above. Second, these constants include the fraction of the Golgi or ER or PM compartment that falls within the boundaries of the ROI. Third, they account for differences in the efficiency with which fluorescence is measured in different organelles.[9]

The only complication that arises from the requirement for Eq. (4) is that it is now necessary to estimate the quantitative contributions made, for example, by the ER to the Golgi fluorescence. For this purpose it is necessary to choose software that permits fitting the data against functions like the one in Eq. (4). SAAM II, for example, implements this requirement by allowing the user to define any function of N_G, N_{ER}, and N_{PM} as a "sample." Samples are then compared directly with the experimental data, and parameters are estimated by a new generalization of the classic least-squares procedure.[13]

A question that should be answered as early in a study as possible is whether the assumption of linear rate laws can be justified. This is relatively easy in the case of VSVG–GFP because the experimental data encompass a large dynamic range. Dynamic range can be thought of as the ratio of the largest to the smallest value of a variable, a group of variables, or a measurement. Intrinsically, fluorescence measurements are capable of enormous dynamic ranges (many orders of magnitude), but in practical

[13] B. M. Bell, J. V. Burke, and A. Schumitzky, *Comput. Stat. Data Anal.* **22,** 119 (1996).

situations we are usually limited by the background of cellular autofluorescence and by the size of the registers (8- or 12-bit) in the analog-to-digital image capture board associated with the microscope system. Nevertheless, the number of VSVG–GFP molecules in, say, the ER will traverse the full dynamic range in the course of the experiments described in this chapter. This follows from the fact that all the molecules are in the ER at the start, and essentially none are in the ER at the end of the experiment. Consider what this means for any saturable, rate-limiting process.

If, for example, the rate-limiting process in ER export of VSVG is saturated by the large number of molecules awaiting export at the beginning of the experiment (2×10^7 in our experiments), then the rate "constant" must increase as the experiment progresses. The increase occurs because, at some point, the inevitable decrease in "substrate" concentration will no longer saturate the process. Graphically, this is illustrated in Fig. 6. A rate "constant" that changes 25-fold during the trafficking experiment simply could not be approximated as constant. In fact, when we tested the hypothesis explicitly by allowing Michaelis–Menten rate laws, the least-squares optimizer consistently drove the value of the K_m so high that the entire experiment took place in the linear range. This is convincing evidence that the rate-limiting steps of VSVG–GFP transport to the PM are not saturated by the expression levels we attained.

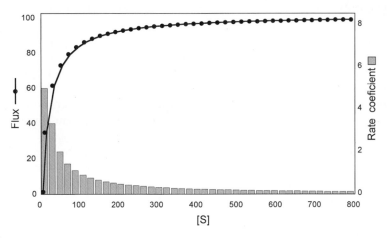

Fig. 6. Michaelis–Menten flux saturation curve plotted with the corresponding values of the rate "constant" (rate coefficient). The plot illustrates, for example, that as substrate concentration [S] decreases from 800 to 2 arbitrary units, the observed rate coefficient would increase ~25-fold, from ~0.2 to 5.0.

Why Golgi Data Alone Are Insufficient

A frequently asked question is: Why not measure just the Golgi fluorescence; doesn't it contain enough information to characterize both entry from the ER and exit to the PM? The short answer is, no. A longer answer requires some thought about rate-limiting steps. One of the most fundamental truths of kinetics is that it is not possible to know whether entry into a compartment or exit from the compartment is the slow step just by monitoring the time course in the compartment. The reason for this immutable rule is that the shape of the time course will be exactly the same even if the magnitudes of the two rate constants are interchanged.

In the context of VSVG trafficking, let us suppose that k_{ER} and k_G are 2 and 5%/min, respectively. Given only the data from the Golgi ROI, it is impossible to determine these two values. In fact, the data could be fitted just as well if k_{ER} were set to the fast rate, 5%/min, and k_G to the slow rate, 2%/min. If this problem is submitted to data analysis software, a message should appear, warning that the solution is not unique. In fact, no statistical information at all should appear concerning the final values of the parameters because, in effect, they are not resolvable from the data.

This problem can be solved, however, by recording fluorescence from the rest of the cell, that is, from the non-Golgi ROI, or even from the total cell. Simultaneous fitting of total cell fluorescence and Golgi fluorescence is sufficient to resolve k_{ER}, k_G, and k_{PM} uniquely. This is because the total cell data contain information on the source compartment, the ER. Two ROIs, which represent different linear combinations of the fluorescence from each cellular compartment, are generally sufficient to resolve all three parameters. Moreover, they permit estimation of the ER initial condition and the coefficients, c_i, in Eq. (4) and in the corresponding equations for other ROIs. This process is detailed in the next section.

Parameter Estimation and Statistical Evaluation

Most investigators are familiar with the concept of regression analysis, and while this is not the process used to fit time course data, there are similarities. The most familiar regression seeks to find the best straight line to describe data plotted on a $x-y$ graph. The word "best" has a formal mathematical definition; the best line is the one that minimizes the sum of the squares of the differences between the line and the data points. We minimize the sum of squares so that data points below the line will have the same effect on the solution as data points above the line. This process is commonly referred to at a "least-squares fit."

The difference between this common regression procedure and the one required for analysis of VSVG–GFP kinetics is that our "line" is not a simple straight line, but is instead the solution of a mechanistic system of differential equations. In other words, the "line" we want to fit to our data is a function of time such as Eq. (3). Moreover, it is not simply an arbitrary function of time such as a power series or a Fourier series or a sum of exponentials, but is instead the numerical solution of a system of differential equations—a system of equations that represents our working hypothesis about the secretory pathway. This is a critical point, often misunderstood. All mathematical models are not equal. Models consisting of arbitrary functions can be powerful in diagnosis, but they cannot make testable predictions about the consequences of a physiological or pharmacological perturbation. And testable predictions are the *sine qua non* of scientific hypothesis testing.

In the early days of numerical analysis, the procedure we follow would have required an enormous number of calculators, elaborate tabulation of intermediate results, and infinite patience. Today, the process of least-squares parameter estimation for the three rate constants, one initial condition, and three independent coefficients of the "sample" equations required about 15 sec on a desktop personal computer for each cell analyzed.

A step-by-step procedure is as follows: import the background-subtracted data (Fig. 5B) into the data analysis software. In SAAM II, for example, cut and paste the data from the experimental spreadsheet into the Data Window.

Assign statistical weight to the data points for the purposes of least-squares fitting. Choice of weighting schemes is a topic large enough for its own chapter, but a few rules of thumb can be enumerated. Ideally, each datum should be weighted with the reciprocal of its variance. If the variance of the measurements is known, it should definitely be used to weight the data, but in most practical situations this information is not available. So how is the variance estimated? Standard practice is to estimate the variance as a function of the data value (or the model prediction), N_i, at each time point. Weights, w_i, are then obtained as

$$w_i = \frac{1}{a + b|N_i|^c} \tag{5}$$

An excellent rule of thumb, when the error associated with the data is not known, is to use fractional standard deviations (FSDs), in which case it is assumed that $a = 0$, $c = 2$, and the square root of b is the fractional standard deviation. If, for example, FSD is set to 0.1, it is being asserted that the measured datum comes from a distribution whose standard deviation is 10% of the measured value. In our experience, this a good choice for fluorescence data collected from single cells.

Next, it is necessary to set upper and lower bounds on the initial estimates of the unknown parameters. It is good practice to initiate the least-squares fitting procedure with initial values that yield a reasonable fit, but it is often surprising how well a modern optimizer can do even with poor initial estimates. Set upper and lower bounds that are 10-fold greater than the initial estimate and 10-fold smaller than the initial estimate, respectively. A greater range may be needed if an approximate fit has not first been obtained by manually adjusting the parameters. And a smaller range can be advantageous if a fit is already nearly attained.

Now it is time to initiate the software optimizer. If the data are sufficient, the optimizer will converge (Fig. 5D) and a variety of output options will be presented depending on the software chosen. If convergence is not reached, it is sometimes helpful to add some *a priori* information about one or more parameters. This approach, sometimes referred to as a Bayesian term, says that some knowledge is available about what this parameter should be, perhaps based on previous studies. If so, it is possible to provide a mean and standard deviation for the distribution from which this parameter is thought to be drawn and thus provide some valuable assistance to the struggling optimizer.

If the option is available, plot the residuals (Fig. 5E; i.e., the differences between the optimal model solution and the experimental data) as a function of time through the experiment. The goal is that the residuals should be randomly distributed about zero. To test for this random distribution, do a standard runs test on the residuals. Even without this statistical confirmation, the plot of residuals can be highly informative. Even the appearance of randomness is a good sign, and a long sequence of uniformly positive or uniformly negative residuals is a bad sign suggesting that the model does not adequately account for the data. Now look at the statistical information provided by the optimizer. This is where the numbers to be reported for publication are found. First, of course, record the "point estimates" of the adjustable parameters. These are the optimized values of the three rates constants, the ER initial condition, and the coefficients of the "sample" equations.

Next, look at the coefficients of variation (CVs) for each of these adjusted parameters. The smaller, the better, of course, but CVs of 2–20% are excellent and CVs of 20–50% are reasonable. Coefficients of variation in the range of 100–1000% (or even greater) provide a strong hint that the data contain little, if any, information on the corresponding parameter. Alternatively, the software may provide 95% confidence limits for the parameter estimates. The tighter the limits, the better the estimate. Enter all of this information in a spreadsheet where the columns are the parameter values and the rows correspond to each cell analyzed. By calculating the means and standard deviations for the population of analyzed cells, useful information on cell-to-cell variability can be extracted. In our experience this variability is substantially greater than the variation associated with individual cell parameter estimates. This emphasizes the value of doing single-cell analysis. It demonstrates that values obtained by analyzing lumped data from thousands or millions of cells will have large variances, not because cells function imprecisely, but because cell-to-cell variation is quite significant even in established cell cultures.

Be sure to report not only the mean parameter values, but also their standard deviations. Also useful is a statement concerning the range of coefficients of variation obtained for each parameter during the fitting of individual cells. If a runs test to establish randomness of the residuals was performed, report this as well.

Biological Interpretation of Kinetic Parameters

Obtaining precise estimates of all those parameters would not be useful if they did not permit something interesting to be said about the biological system being studied. This section aims to explain the meaning or the biological significance of the parameters obtained in previous sections.

The most common question is, "Just what is a rate constant?" The answer is simple. A rate constant is the answer to the question, "What fraction of the material in this compartment leaves by this pathway every second or every minute?" Rate constants have units of inverse time. In the case of VSVG–GFP (in COS cells), rate constants for the rate-limiting process between ER and Golgi and for the rate-limiting process between Golgi and PM are both about 3%/min. In other words, about 3% of the VSVG–GFP in the ER is exported to the Golgi each minute, and about 3% of what is in the Golgi is exported to the PM per minute. Another way to effectively convey the meaning of a rate constant is to contrast it with V_{max} and K_m, the relatively well known Michaelis–Menten parameters. It is useful to think of a rate constant, especially for a protein-mediated process, as a ratio of V_{max} to K_m. In other words, the rate constant contains all the information on the current level of expression of the mediator protein and on all of its allosteric or pharmacologic regulators. This is one reason why knowledge of the rate constants in several physiological situations provides so much useful biological insight.

Knowing that the rate constant can be thought of as a ratio of V_{max} to K_m leads to a reasonable question. How can a single number, the rate constant, yield saturation behavior? The answer is, it cannot. This means, as emphasized earlier in this chapter, that our ability to fit VSVG–GFP data using a single rate constant over the entire 200-fold range of VSVG–GFP concentration implies that even with 2×10^7 molecules of VSVG expressed the rate-limiting steps in VSVG trafficking are not saturated. This observation may have profound implications for our mechanistic understanding of the secretory pathway. For example, it can be argued that the observed linearity supports the hypothesized existence of membrane microdomains or rafts, which play distinctive roles in protein sorting and export from the ER and the Golgi apparatus.[9]

Kineticists are often asked, "What is the rate-limiting step?" Moreover, we have used the phrase "rate-limiting steps" repeatedly in this chapter. The only useful definition of "rate-limiting step" is the process with the slowest (i.e., smallest) rate constant. From our analysis of VSVG–GFP trafficking in COS cells, it is apparent that neither ER export nor Golgi export is substantially slower than the other. But even if one were much slower, one would not want to conclude that it was the step that controls the flux (molecules per second) through the secretory pathway. A thoughtful group of theoreticians and experimentalists in the field of metabolic control analysis (MCA) has long since demonstrated that the control of flux is nearly always distributed over the entire pathway.[14]

Another useful number, commonly referred to as the residence time, can be calculated by summing all the rate constants that leave a compartment irreversibly and calculating the reciprocal of that sum. Because it can be obtained in this way, it is obvious that it contains no new information, but it does crystallize the meaning of a rate constant for some readers. The residence time is the average time spent by a molecule of VSVG–GFP in a particular compartment. For the analysis described here, where there is but a single exit route from each compartment, the ER residence time is just the reciprocal of k_{ER}. It should not be surprising to find that the average (over many cells studied) residence time is not equal to the reciprocal of the average k_{ER}; this is simply a consequence of the general truth that the reciprocal of the average is not equal to the average of the reciprocals. The residence time can be thought of as a measure of how long the average molecule spends traversing all of the processes that have been subsumed into this particular rate constant.

At any moment in time, the flux of VSVG–GFP in molecules per second can be calculated as the product of a rate constant and the number of molecules in the "source" compartment, that is, the compartment from which the rate constant exits. Estimating the number of molecules in, say, the Golgi at any time is relatively easy once the fluorescence has been calibrated and the contributions of quenching and ER overlap have been accounted for. Multiplying this number by k_G (in sec^{-1}) then gives the number of molecules per second, leaving the Golgi apparatus. As cell

[14] H. Kacser and J. A. Burns, *Biochem. Soc. Trans.* **23,** 341 (1995).

biology becomes more quantitative, such absolute abundances and absolute fluxes will become more essential and valuable as reported experimental results.

Concluding Remarks

Investigators who have worked in protein trafficking using both biochemical tools (amino acid pulse–chase, membrane fractionation, and immunoprecipitation) and fluorescence microscopy will recognize several parallels in the two approaches. Both, of course, are kinetic methods. Both yield time-course data whose interpretation is best done in the context of computer models, especially when the working hypothesis of the investigator consists of five or more interacting variables.

One of the principle complexities faced in data analysis for either pulse–chase or GFP experiments is in teasing apart the contributions made to any measurement by multiple cellular structures. This problem is seen in cell fractionation studies as contamination of a group of sucrose gradient fractions with membranes from several organelles. Microscopy has the entirely analogous problem that multiple organelles appear in every ROI. Fortunately, kinetic analysis is exceptionally good at recognizing and quantifying these secondary contributions. And the "sample" tools described earlier in this chapter can be used to extract optimal estimates of cellular parameter values, no matter which experimental approach is chosen. Interestingly, however, we find that the essentially continuous measurements made in living cells expressing GFP chimeras permit much greater certainty in parameter estimation. Rate constants estimated for VSVG–GFP have coefficients of variation that are frequently as small as 2%, while rate constants estimated from amino acid pulse–chase experiments often have coefficients of variation between 30 and 50%.

Applications of quantitative fluorescence microscopy to cell biology are rapidly developing. The ability to express and visualize more then one protein in living cells, using GFP fluorescence variants, makes it possible to imagine a synergistic combination of computational and experimental approaches that will permit analysis of complex experiments. The ability to image and record two proteins expressed in a single cell will provide a powerful system to monitor, localize, and quantitate interaction between protein components, for example, cargo and transport protein during movement through intracellular compartments. Together with monitoring protein intracellular redistribution during an experiment, biophysical phenomena such as fluorescence resonance energy transfer (FRET) can be quantified as well. Combining these computational and image analysis tools with experimental protocols involving physiological or pharmacological perturbations will be a powerful method for future work addressing key questions in intracellular membrane trafficking.

Acknowledgments

We thank Nihal Altan for critical review of the manuscript. We also thank J. Ellenberg, C. Miller, and J Presley for supplying macros for NIH Image and LSM 410. K. Hirschberg is funded by the Human Frontiers Science Program funds.

[7] Dual Color Detection of Cyan and Yellow Derivatives of Green Fluorescent Protein Using Conventional Fluorescence Microscopy and 35-mm Photography

By GISELE GREEN, STEVEN R. KAIN, and BRIGITTE ANGRES

Introduction

The green fluorescent protein (GFP) has been used successfully to monitor protein localization in living cells.[1] Since the early development of the reporter improvements have been made by mutating the sequence of the original gene to obtain brighter fluorescence and improved expression. Brighter versions of the green fluorescent protein have since been widely used in many experimental applications.

To monitor and distinguish multiple proteins in a cell or multiple cells in an organism, more mutants of the green fluorescent protein were generated that exhibit different excitation and emission wavelengths, allowing their differential detection. These shifts in wavelength, however, in many cases were too small to result in a different color visible to the human eye. Distinction between the variants required digital imaging or laser scanning confocal microscopy followed by pseudo-coloring of the images,[2] technologies that are not available in every laboratory or for routine use.

Among the mutants that were different in color were blue-shifted variants[3,4] that could be well distinguished from green and yellow-green variants

---

[1] M. Chalfie and S. Kain, "Green Fluorescent Protein: Properties, Applications, and Protocols." Wiley-Liss, New York, 1998.

[2] C. T. Baumann, C. S. Lim, and G. L. Hager, *J. Histochem. Cytochem.* **46,** 1073 (1998).

[3] R. Heim, D. C. Prasher, and R. Y. Tsien, *Proc. Natl. Acad. Sci. U.S.A.* **91** 12501 (1994).

[4] R. Heim and R. Y. Tsien, *Curr. Biol.* **6,** 178 (1996).

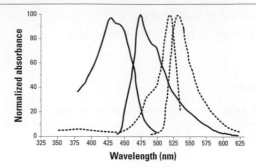

FIG. 1. Excitation and emission spectra of recombinant ECFP (solid lines) and EYFP (dashed lines). Spectra were obtained from equal concentrations of His_6-tagged recombinant proteins on a spectrofluorometer as described previously [T. Yang, P. Sinai, G. Green, P. A. Kitts, Y.-T. Chen, L. Lybarger, R. Chervenak, G. H. Patterson, D. W. Piston, and S. R. Kain, *J. Biol. Chem.* **273**, 8212 (1998)].

by conventional fluorescence microscopy equipment.[5,6] However, the blue variants were not as bright and photobleached quickly, which rendered microscopy more difficult. Again, the problem could be overcome to a certain extent by using sensitive detection systems such as cooled CCD cameras, for which exposure times for image documentation are in the millisecond range.[4] However, the weak, rapidly bleaching blue fluorescence did not satisfy the user of conventional fluorescence microscopy with 35-mm film for documentation, which requires longer exposure times.

The development of the W7 cyan variant,[4] another blue-shifted mutant of GFP, whose excitation and emission spectra are shifted toward longer wavelengths, provided considerable improvement. Its relatively high fluorescence intensity and vastly improved resistance to photobleaching make the cyan variant much more suitable for fluorescence microscopy. However, the cyan variant is difficult to pair with the green variant for dual labeling because the respective excitation and emission wavelengths are too close to each other for separate visualization. Yet spectra of the cyan variant are distant enough from those of the yellow-green variant EYFP[7] which makes the two the ideal pair for dual-labeling experiments (see Fig. 1).

To create an enhanced cyan fluorescent protein (ECFP) we incorporated

[5] R. Rizzuto, M. Brini, F. De Giorgi, R. Rossi, R. Heim, R. Y. Tsien, and T. Pozzan, *Curr. Biol.* **6**, 183 (1996).
[6] T. Yang, P. Sinai, G. Green, P. A. Kitts, Y.-T. Chen, L. Lybarger, R. Chervenak, G. H. Patterson, D. W. Piston, and S. R. Kain, *J. Biol. Chem.* **273**, 8212 (1998).
[7] T. Yang, S. R. Kain, P. Kitts, A. Kondepudi, M. M. Yang, and D. C. Youvan, *Gene* **173**, 19 (1996).

four mutations into the sequence of our enhanced GFP (EGFP)[8] that render the excitation and emission spectra similar to those of the W7 cyan mutant reported by Heim and Tsien.[4] The mutation Tyr-66 to Trp gives the ECFP the characteristic excitation and emission spectra and cyan color described by Heim and Tsien.[4] Mutations Ser-65 to Thr[9]; Asn-146 to Ile, Met-153 to Thr, and Val-163 to Ala[4]; Phe-64 to Leu[10]; and humanized codons[6,11] mediate the rapid folding, high solubility, increased brightness, and high expression of the variant in mammalian cells when compared with wild-type GFP.

Spectral Characteristics of Enhanced Cyan Fluorescent Protein and Enhanced Yellow Fluorescent Protein

Figure 1 shows the excitation and emission spectra of ECFP and EYFP. ECFP has two excitation peaks: a major peak at 434 nm and a minor peak at 453 nm. Two emission peaks are apparent, one at 477 nm and a minor peak at 496 nm. The extinction coefficient of ECFP is 26,000 M^{-1} cm^{-1} with a quantum yield of 40% (D. Piston, personal communication, 1998). When compared with the extinction coefficient and quantum yield of EBFP (E_m = 31,000 M^{-1} cm^{-1}, with a quantum yield of 18%; D. Piston, personal communication, 1998) the relative fluorescence intensity of ECFP is approximately twice as high.

Image Acquisition with a 35-mm Camera

We tested the utility of ECFP and EYFP for double labeling and image documentation on 35-mm color slide film, using a standard fluorescence microscope [Zeiss (Thornwood, NY) Axioskop] equipped with a 100-W mercury arc lamp and optical filter sets (Omega Optical, Brattleboro, VT) tailored to the specific excitation and emission spectra of ECFP (filter set XF114) and EYFP (filter set XF104). To be able to spatially distinguish ECFP from EYFP within one cell, we used plasmid constructs that encode GFP fusion proteins that are targeted to mitochondria (ECFP–Mito) and

---

[8] M. Ormö, A. B. Cubitt, K. Kallio, L. A. Gross, R. Y. Tsien, and S. J. Remington, *Science* **273**, 1392 (1996).

[9] R. Heim, A. B. Cubitt, and R. Y. Tsien, *Nature* (*London*) **373**, 663 (1995).

[10] B. P. Cormack, R. H. Valdivia, and S. Falkow, *Gene* **173**, 33 (1996).

[11] J. Haas, E. C. Park, and B. Seed, *Curr. Biol.* **6**, 315 (1996).

ECFP-Mito:

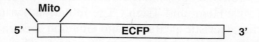

EYFP-Nuc:

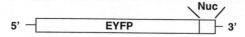

FIG. 2. Schematic drawing of ECFP–Mito and EYFP–Nuc fusion constructs. To generate ECFP–Mito the mitochondrial targeting sequence (Mito) from subunit VIII of human cytochrome *c* oxidase was fused to the 5' end of ECFP. To generate EYFP–Nuc three copies of the nuclear localization signal (Nuc) of the simian virus 40 large T-antigen were fused to the 3' end of EYFP.

the nucleus (EYFP–Nuc; see Fig. 2). HeLa cells were transiently transfected with these plasmids and imaged 48 hr after transfection. Images were taken with a ×40, 1.3 NA oil-immersion objective by double exposure on ISO 400 35-mm color slide film (Kodak, Rochester, NY). The exposure time for each variant was 60 sec. As can be seen in Fig. 3, the two colors were well separated. (See color insert.) No color bleed-through was observed and a good signal-to-noise ratio was obtained for both variants. However, ECFP still photobleached faster than EYFP and unequal fluorescence intensities became apparent after exposure of the fluorescent proteins for 60 sec (data not shown).

We found ECFP to be relatively photostable when observed in living cells or after fixation in formaldehyde and directly applied to microscopy. Mounting cells in mounting media, however, in our hands required the presence of antifade reagents to avoid rapid photobleaching of the cyan variant.

Conclusions

Our results demonstrate that the spectral characteristics and fluorescence intensities of ECFP and EYFP are well suited for dual-color labeling experiments using standard epifluorescence microscopy and image documentation with a 35-mm camera. The two fluorescent proteins fulfill two crucial requirements for documentation on film: they are distinguishable in color and are strong enough in fluorescence for the relatively long exposure times needed for conventional photography. Although EBFP and EGFP are well distinguishable in color, ECFP proved to be superior to

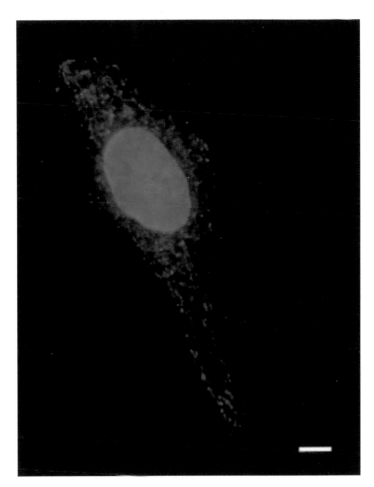

FIG. 3. Visualization of intracellular organelles by dual-color epifluorescence microscopy, using ECFP and EYFP in HeLa cells. HeLa cells were transiently transfected with a mixture of plasmids pEYFP-Nuc and pECFP-Mito (Clontech, Palo Alto, CA). Forty-eight hours posttransfection cells were fixed in a 1:10 dilution of formalin in PBS for 30 min, mounted in ProLong Antifade (Molecular Probes, Eugene, OR), and visualized with a Zeiss Axioskop (Carl Zeiss, Thornwood, NY) equipped with a ×40 1.3 NA Plan Neofluor oil-immersion objective. The cell was photographed by double exposing 35-mm color slide film (Kodak Elite, 400 ASA), using filter sets XF104 (EYFP) and XF114 (ECFP). Exposure time for each variant was 60 sec. Bar: 10 μm.

EBFP with respect to fluorescence intensity and photobleaching propertie~
The use of EYFP did not result in any disadvantage in comparison witl.
EGFP because its fluorescence intensity and photobleaching characteristics
are similar.

Materials and Methods

Targeting Green Fluorescent Protein Variants to Nucleus and Mitochondria

Plasmids pEYFP-Nuc and pECFP-Mito (Clontech, Palo Alto, CA) are
used for the targeting of GFP variants to the nucleus and mitochondria,
respectively. pEYFP-Nuc contains three copies of the nuclear localization
signal (NLS) of the simian virus 40 large T-antigen[12,13] fused to the 3' end
of EYFP. The reiteration of the NLS sequence significantly increases the
efficiency of translocation of EYFP into the nucleus of mammalian cells.[14]
pECFP-Mito contains the mitochondrial targeting sequence from subunit
VIII of human cytochrome c oxidase[5] fused to the 5' end of ECFP. The
mitochondrial targeting sequence localizes the fusion protein to the matrix
of mitochondria.

Transient Transfection of HeLa Cells by Electroporation

HeLa cells are cultured in growth medium consisting of Dulbecco's
modified Eagle's medium (DMEM; Sigma, St. Louis, MO) containing 10%
(v/v) fetal bovine serum (FBS) supplemented with antibiotics and glutamine
(Life Technologies, Gaithersburg, MD) at 37° and 5% CO_2. Cells growing
in log phase on 10-cm dishes are washed once with 5 ml of phosphate-
buffered saline (PBS) and subsequently removed from the culture dish by
trypsin treatment (trypsin–EDTA solution; Sigma) for 3 to 5 min. Trypsinization is stopped by adding 9 ml of growth medium. Cells are pelleted in
a tissue culture centrifuge and resuspended in 0.8 ml of growth medium.
The cell suspension is added to a 0.4-mm gap electroporation cuvette containing 4–8 μg of plasmid in a maximum volume of 30 μl. Cells and DNA
are incubated at room temperature for 5 to 10 min, briefly mixed, and
electroporated with a Bio-Rad (Hercules, CA) GenePulser II with a capacitance extender. Electroporation is performed at 0.24 kV and 950 μF. An
electroporation pulse of 5 to 30 msec is indicative of a successful electroporation. After electroporation cells are allowed to sit for 5 to 10 min at room

[12] D. Kalderon, B. L. Roberts, W. D. Richardson, and A. E. Smith, Cell 39, 499 (1984).
[13] R. E. Lanford, P. Kanda, and R. C. Kennedy, Cell 46, 575 (1986).
[14] L. Fischer-Fantuzzi and C. Vesco, Mol. Cell Biol. 8, 5495 (1988).

emperature. Cells are then seeded in tissue culture dishes containing 10 ml of growth medium and are incubated overnight. The next day cells are washed once in PBS, removed from the plate by trypsinization, and seeded on glass coverslips in a six-well tissue culture plate and incubated for an additional 1 day. Passaging the cells from one plate to another at this point is done to remove cell debris derived from dead cells generated during the electroporation procedure.

Cell Fixation

Cells are washed once with PBS and fixed in a 1:10 dilution of formalin (Sigma) in PBS for 30 min at room temperature. After fixation cells are washed three times in PBS and either inspected under the microscope directly by placing the coverslip face down on a microslide after a brief water rinse or by mounting them in mounting medium (ProLong antifade kit; Molecular Probes, Eugene, OR). We have found that mounting medium containing antifade reagents preserves the fluorescence of ECFP considerably better than mounting media without antifade reagents. In contrast, the preservation of EYFP fluorescence seems not to be dependent on the addition of antifade reagents.

Microscopy

Fluorescence is visualized with a Zeiss Axioskop equipped with a 100-W mercury arc lamp. ECFP and EYFP are detected by using filter sets tailored to the excitation and emission of each variant. The use of these specific filter sets allows the separate detection of each variant without any bleed-through of the respective other variant. ECFP is detected with filter set XF114 (exciter 440DF20, dichroic 455DRLP, emitter 480DF30) and EYFP is detected with filter set XF104 (exciter 500RDF25, dichroic 525DRLP, emitter 545RDF35), both obtained from Omega Optical (Brattleboro, VT). Appropriate filter sets are also available from Chroma Technology (Brattleboro, VT). Pictures showing double labeling of ECFP and EYFP are taken by double exposure of one frame of an ISO 400 35-mm color slide film (Kodak Elite Chrome 400) by using the respective filter set for each variant. Exposure times vary from 30 to 60 sec for each variant, depending on the intensity of the fluorescence.

Acknowledgments

The authors thank D. W. Piston and G. H. Patterson (Vanderbilt University Medical Center) for determining spectra for our GFP variants and providing us with data for Fig. 1.

[8] Invertase Fusion Proteins for Analysis of Protein Trafficking in Yeast

By TAMARA DARSOW, GREG ODORIZZI, and SCOTT D. EMR

Introduction

It is often convenient to examine the behavior of proteins through their fusion to enzyme reporters that can be assayed quickly and accurately. Invertase (β-D-fructofuranoside fructohydrolase; EC 3.2.1.26) of the yeast *Saccharomyces cerevisiae* hydrolyzes sucrose into the monosaccharides glucose and fructose. The invertase gene (*SUC2*) encodes two transcripts that differ in length by 60 nucleotides at the 5' end.[1] The longer transcript encodes a 532-amino acid protein containing an N-terminal hydrophobic signal sequence that directs its translocation into the endoplasmic reticulum (ER). During export to the cell surface, this secreted form of invertase is transported through the Golgi, where it receives extensive and heterogeneous modifications at up to 13 N-linked glycosylation sites. The shorter transcript encodes a 512-amino acid protein that does not contain the N-terminal prosequence and is consequently located entirely within the cytoplasm. In most strains of *S. cerevisiae,* secreted invertase represents the majority of total cellular invertase activity.[2]

As a reporter protein, invertase enzymatic activity can be measured by both quantitative and qualitative assays.[3,4] In addition, the invertase protein can be followed by methods that measure the extent and types of glycosylation it receives.[5] Like many other reporter proteins, invertase has been used in various applications, including quantitation of gene expression, identification of protein translocation into the ER, and determination of membrane protein topology.[6–9] However, because invertase is a secreted protein, it has been uniquely useful in the examination of protein trafficking

[1] M. Carlson, R. Taussig, S. Kustu, and D. Botstein, *Mol. Cell. Biol.* **3,** 439 (1983).
[2] A. Goldstein and J. O. Lampen, *Methods Enzymol.* **42,** 504 (1975).
[3] L. M. Johnson, V. A. Bankaitis, and S. D. Emr, *Cell* **48,** 875 (1987).
[4] G. Paravicini, B. F. Horazdovsky, and S. D. Emr, *Mol. Biol. Cell* **3,** 415 (1992).
[5] E. C. Gaynor, S. te Heesen, T. R. Graham, M. Aebi, and S. D. Emr, *J. Cell Biol.* **127,** 653 (1994).
[6] S. D. Emr, R. Schekman, M. C. Flessel, and J. Thorner, *Proc. Natl. Acad. Sci. U.S.A.* **80,** 7080 (1983).
[7] S. W. Van Arsdell, G. L. Stetler, and J. Thorner, *Mol. Cell. Biol.* **7,** 749 (1987).
[8] D. Feldheim, J. Rothblatt, and R. Schekman, *Mol. Cell. Biol.* **12,** 3288 (1992).
[9] W. Hoffmann, *Mol. Gen. Genet.* **210,** 277 (1987).

through the secretory pathway.[3,5,10] For example, invertase fusion proteins have been employed to define sequences that direct protein import into the ER,[11,12] to screen for mutants defective in retrograde transport from the Golgi to the ER,[13] and to identify secreted human proteins in large-scale screens.[14]

In this chapter, we describe assays to measure invertase enzymatic activity, focusing on the secretion of invertase fusion proteins as a reporter in studies of vacuolar protein trafficking. [Methods that measure invertase glycosylation as an indication of protein transport through the early part of the secretory pathway (i.e., ER to Golgi) are presented in [9] in this volume[14a].] Several important features of invertase have contributed to its utility in protein trafficking studies. Its enzymatic activity is not altered when fused to heterologous proteins by recombinant DNA techniques, and because invertase secretion is not essential under standard growth conditions, it can be easily manipulated without affecting cell viability. In addition, other than the N-terminal signal sequence that is cleaved after its translocation into the ER, invertase does not appear to contain any sequence information that is essential for its transport through the secretory pathway.[3,6] Thus, invertase enzymatic activity will be accurately localized within the cellular compartment(s) to which the native protein under study is transported.[15,16]

Below, we describe three assays for the detection of secreted invertase enzymatic activity, each of which has advantages and disadvantages to be considered depending on the purpose. In each assay, glucose oxidase is used to oxidize the glucose released by enzymatic hydrolysis of sucrose. A byproduct of this reaction is H_2O_2, which is used by peroxidase to oxidize a chromogen (o-dianisidine), resulting in a dark precipitate that can either be observed qualitatively on a plate or filter or measured quantitatively at 540 nm.

In addition to these analytical methods, we also describe a selection procedure to isolate yeast mutants that inappropriately secrete invertase

[10] J. H. Stack, D. B. DeWald, K. Takegawa, and S. D. Emr, *J. Cell Biol.* **129,** 321 (1995).
[11] C. A. Kaiser, D. Preuss, P. Grisafi, and D. Botstein, *Science* **235,** 312 (1987).
[12] C. A. Kaiser and D. Botstein, *Mol. Cell. Biol.* **10,** 3163 (1990).
[13] H. R. Pelham, K. G. Hardwick, and M. J. Lewis, *EMBO J.* **7,** 1757 (1988).
[14] K. A. Jacobs, L. A. Collins-Racie, M. Colbert, M. Duckett, M. Golden-Fleet, K. Kelleher, R. Kriz, E. R. LaVallie, D. Merberg, V. Spaulding, J. Stover, M. J. Williamson, and J. M. McCoy, *Gene* **198,** 289 (1997).
[14a] B. Diane Hopkins, Ken Sato, Akihiko Nakano, and Todd R. Graham, *Methods Enzymol.* **327,** Chap. 9, 2000 (this volume).
[15] D. J. Klionsky, L. M. Banta, and S. D. Emr, *Mol. Cell. Biol.* **8,** 2105 (1988).
[16] V. A. Bankaitis, L. M. Johnson, and S. D. Emr, *Proc. Natl. Acad. Sci. U.S.A.* **83,** 9075 (1986).

fusion proteins. This strategy has been highly successful in isolating a collection of more than 40 mutants that do not efficiently sort an invertase fusion protein from the Golgi to the vacuole.[16,17] Similar strategies can be applied to isolating mutants defective in other intracellular protein transport pathways. For example, a screen to detect mutants in retrieval of an invertase-HDEL fusion protein from the Golgi to the endoplasmic reticulum identified proteins required for retrieval of ER-resident HDEL motif-containing proteins from the Golgi compartment to the endoplasmic recticulum.[13]

Vectors for Construction and Expression of Invertase Fusion Proteins

A set of vectors has been designed for the construction and expression of invertase fusion proteins in yeast (Fig. 1). Both vectors contain genetic sequences for their selection and propagation in *Escherichia coli* and for their selection and maintenance at single-copy levels in yeast.[3,5] To generate a fusion protein in which amino acids 3–512 of mature invertase (lacking the N-terminal signal sequence) are located at the C terminus, a DNA fragment containing both the gene of interest and its 5' untranslated region can be cloned in the correct reading frame into one of several unique restriction enzyme sites located upstream of the *SUC2* coding region in pSEYC306 (Fig. 1A).[3] Because this construct does not contain any promotor region or ER signal sequence for invertase, these elements must be included in the heterologous sequences in order to drive expression and translocation of the fusion protein. To generate a fusion protein in which amino acids 3–512 of mature invertase are located at the N terminus, the coding region of the gene of interest can be cloned in the correct reading frame into one of three unique restriction enzyme sites in pSEYC350 located downstream of the *SUC2* coding region (Fig. 1B). Both the 5' untranslated region and the N-terminal signal sequence of *PRC1* (the CPY gene) is included in pSEYC350 to promote constitutive expression of the N-terminal invertase fusion protein.[5]

Plate Overlay and Filter Assays for Qualitative Analysis of Invertase Secretion

The plate overlay and filter assays provide a qualitative indication of invertase secreted directly from yeast cells growing on agar plates. Both procedures can be used for screening large numbers of individual colonies for mutations that result in mislocalization of invertase fusion proteins to the cell surface. These methods have been particularly useful for identifying

[17] J. S. Robinson, D. J. Klionsky, L. M. Banta, and S. D. Emr, *Mol. Cell. Biol.* **8,** 4936 (1988).

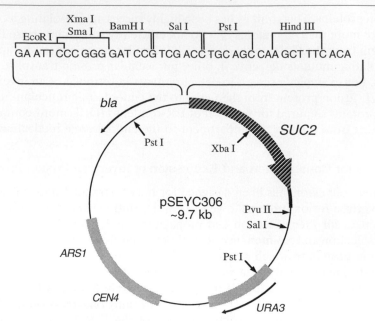

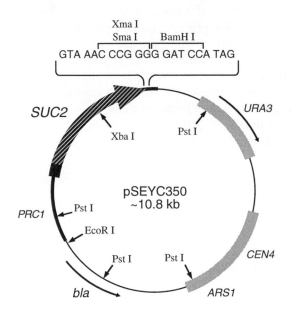

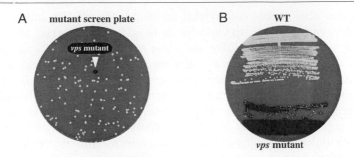

A   mutant screen plate   B   WT

vps mutant

vps **mutant**

FIG. 2. (A) A CPY–invertase fusion protein secretion screen is shown. A single mutant that secretes invertase is indicated by the arrow. (B) Wild-type and mutant from (A) are struck for single colonies and then tested for invertase secretion.

mutants in both positive and negative screens. For example, invertase secretion has been used to identify clones that have temperature-sensitive mutations resulting in the secretion of invertase activity.[10,18] These screening methods can also be used to identify complementing DNA fragments (such as library plasmids) that, when expressed in existing mutants, restore the correct intracellular retention of invertase activity.[19] The disadvantage of the plate overlay and filter assays is that both are extremely sensitive and will detect even small amounts of secreted invertase activity. Consequently, for an accurate indication of invertase secretion, it is imperative that little of the enzyme be secreted from wild-type cells and that appropriate controls be included in every experiment.

Overlay Assay for Invertase Activity

Cells grown on agar medium are overlaid with top agar containing a chromogenic solution, and secreted invertase activity is observed as a color change from white to reddish-brown within 10–15 min. An example in which the overlay assay is used to compare secretion of invertase activity from two different yeast strains is shown in Fig. 2. Total cellular invertase

[18] T. Darsow, S. E. Rieder, and S. D. Emr, *J. Cell Biol.* **138,** 517 (1997).
[19] P. K. Herman and S. D. Emr, *Mol. Cell. Biol.* **10,** 6742 (1990).

FIG. 1. Plasmid vectors for the construction of invertase fusion proteins. pSEY306 is used to make N-terminal invertase fusion protein constructs[3] and pSEYC350 is used for C-terminal invertase fusion protein constructs.[5] Nucleotides are listed in groups of three to indicate the phase of the reading frame for the *SUC2* gene. Multiple cloning sites are enlarged and restriction sites are indicated.

activity can also be assayed by this method by lysing the cells grown on agar plates with chloroform vapor prior to adding the overlay solution. To assay for temperature-sensitive invertase secretion, the cells to be tested are replica plated onto duplicate agar medium plates that are incubated at a permissive temperature. Prior to the assay, one set of plates is incubated for several hours at a nonpermissive temperature while the other set is maintained at the permissive temperature. Overlay solution is then added to both sets of plates, and invertase activity secreted only from colonies incubated at the nonpermissive temperature are identified by direct comparison of the permissive and nonpermissive temperature replica pairs.[10,18]

Reagents

Chloroform
YP-fructose agar medium:

Yeast extract	10 g
Peptone	20 g
Bacto Agar	25 g
H_2O	Up to 960 ml

SD-fructose agar medium:

Yeast nitrogen base without amino acids	6.7 g
Bacto Agar	25 g
Amino acid dropout mix*	0.7 g
H_2O	Up to 960 ml

* The amino acid dropout mix depends on the genetic auxotrophies of the strain to be examined. For example, the dropout mix used for the strain SEY6210 (*MATα leu2-3,112 ura3-52 his3Δ200 trpl-901 lys2-801 suc2-Δ9*) consists of 2.0 g of L-histidine, 4.0 g of L-leucine, 2.0 g of L-lysine, 2.0 g of methionine, 2.0 g of L-tryptophan, 2.0 g of L-tyrosine, plus 0.5 g of adenine and 2.0 g of uracil. One of these compounds can be eliminated to select for plasmids bearing the appropriate gene conferring prototrophy

1. Autoclave YP or SD agar medium.
2. Cool to ~55° and then add 40 ml of 50% (w/v) fructose solution.
3. Pour plates immediately (1 liter of agar medium is enough to pour about forty 10-cm plates).

Invertase overlay solution:

H_2O	60 ml
Sucrose, ultrapure	4.3 g
Sodium acetate buffer, 1.0 *M,* pH 5.5	10 ml
N-ethylmaleimide (NEM),* 20 m*M*	2 ml
Horseradish peroxidase (HRP), 1 mg/ml*	1 ml

Glucose oxidase, 1000 units/ml   0.8 ml
o-Dianisidine,* 10 mg/ml   6 ml
Agar solution, 3% (w/v) in H$_2$O   20 ml

* Stock solutions of NEM, HRP, and o-dianisidine can be made in
   advance and stored at 4°. It is important to maintain o-dianisidine
   in a light-proof bottle. One hundred milliliters of overlay solution
   is enough to assay ~20 plates

Procedure

Day 1. Replica plate, using a felt replicator, or patch cells to be tested
directly onto duplicate sets of YP-fructose or SD-fructose agar medium
plates and grow for 1–2 days at the desired temperature (usually 26–30°).
   Day 2 or 3. Perform the following steps:

1. Allow the plates to equilibrate to room temperature (~5 min). To
compare the amounts of secreted versus total invertase activity, lyse the
cells grown on one set of duplicate plates immediately prior to the assay
by inverting the plates for 3 min over Whatman (Clifton, NJ) filters soaked
with chloroform (can be applied to cut filters positioned into the lid of the
plate). If assaying for temperature-sensitive secretion of invertase activity,
incubate one set of duplicate plates at a nonpermissive temperature for
3–6 hr prior to the assay while maintaining the second set at permissive
temperature. (However, it is important to allow these plates to equilibrate
to room temperature prior to the assay as elevated temperatures can cause
the reaction to proceed much faster.)

2. Melt 3% (w/v) agar solution in a microwave and mix gently with
freshly prepared overlay solution.

3. Cover the entire surface of each plate with ~5 ml of overlay solution.
Avoid pouring directly onto patches of cells, as this may wash them off
the surface.

4. Positive colonies that turn from white to reddish-brown can be picked
with toothpicks and restruck onto fresh medium.

Filter Assay for Invertase Activity

Cells grown on agar medium are replica plated onto two filters, one
that is assayed for secreted invertase activity from intact cells and a second
that is assayed for total invertase activity from lysed cells.

Reagents

YP-fructose or SD-fructose agar medium (see above)
Whatman filters, 11 cm in diameter

Chloroform
Invertase indicator solution (15 ml is enough to assay seven filters):

H_2O	12 ml
Sucrose, ultrapure	0.64 g
Sodium acetate buffer, 1.0 M, pH 5.5	1.5 ml
NEM, 20 mM	0.3 ml
HRP, 1 mg/ml	0.15 ml
Glucose oxidase, 1000 units/ml	0.12 ml
o-Dianisidine, 10 mg/ml	0.9 ml

Procedure

Day 1. Replica plate or patch the cells to be tested onto YP-fructose or SD-fructose plates and grow for 1–2 days at the desired temperature (usually 26–30°).

Day 2 or 3. Perform the following steps:

1. Make fresh indicator solution and pour into a large petri dish.

2. Label Whatman filters with a pencil. Using forceps, dip the filters one at a time into the indicator solution, then drain off excess solution onto a clean paper towel.

3. Place a Kimwipe tissue onto a standard replica block. Place a filter containing indicator solution on top of the tissue, then replica plate cells from the master plate onto the filter. After replica plating the cells onto the first filter, lyse the cells remaining on the master plate by inverting the plates for 3 min over filters soaked with chloroform. Replica plate the lysed cells onto a second filter as described above.

4. Allow the reactions on each filter to proceed at room temperature for 10–20 min. Replicated cells secreting invertase activity appear red on the filter.

5. Stop the reactions by drying the filters in a microwave for 1–2 min. (The filters dry more evenly when they are sitting upright in a rack.) The filters must be thoroughly dry to stop the reaction and prevent the appearance of false positives.

Liquid Invertase Assay

The liquid invertase assay provides a quantitative estimate of the amount of glucose released from the enzymatic hydrolysis of sucrose into glucose and fructose. By comparing the amount of invertase activity in the extracellular fraction with the total activity present in cell lysates, the percentage of secreted invertase activity can be determined. In cases involving an invertase fusion protein that is targeted to a specific subcellular compartment, invertase activity can also be measured from individual organelle fractions generated by various subcellular fractionation procedures.[15,16]

Reagents

Sodium acetate buffer, 0.1 M, pH 4.9
Glucose, 1 mM, used as standard
Sucrose, 0.5 M, ultrapure
Potassium phosphate (K_2HPO_4) buffer, 0.2 M, pH 10.0
Hydrochloric acid, 6 N
Triton X-100, 20% (v/v) solution
Glucostat reagent (40 ml is enough to assay 20 samples):

Potassium phosphate (K_2HPO_4) buffer, 0.1 M, pH 7.0	39 ml
Glucose oxidase, 1000 units/ml	80 μl
HRP, 1 mg/ml	100 μl
NEM, 20 mM	200 μl
o-Dianisidine, 10 mg/ml	600 μl

Procedure

1. In a set of 12 × 100 mm test tubes, combine 0.1–2.0 OD_{600} units of yeast cells (or cell extracts) with sodium acetate buffer so that the final volume is 400 μl. Within each set of measurements, include a blank consisting only of sodium acetate buffer as well as three or four glucose standards (ranging from 20 to 100 μmol).

2. Divide the samples into two 200-μl aliquots. To one aliquot of each sample, lyse the cells by adding 5 μl of Triton X-100. This aliquot represents the amount of total cellular invertase activity.

3. Place the tubes in a 30° bath and start the reactions by adding 50 μl of sucrose solution. Stagger additions by 10–15 sec.

4. Incubate at 30° for 30 min.

5. Stop the reactions by adding to each tube 0.3 ml of 0.2 M K_2HPO_4, pH ~10.0, and immediately place the tubes in a boiling water bath for 3 min. The change to a basic pH takes the enzyme out of its active range and also renders the enzyme more sensitive to the heat treatment.

6. Chill the tubes on ice, and then add 2 ml of glucostat reagent and place the tubes in a 30° water bath for 30 min.

7. Stop the reactions by adding 2 ml of 6 N HCl.

8. Measure the absorbance of the contents of each tube at 540 nm (linear range is between 0.05 and 0.3 at OD_{540}).

Definition of Unit Activity. One unit of invertase activity is defined as the amount of enzyme at pH 4.9 that hydrolyzes sucrose to produce 1 μmol of glucose per minute at 30°.

A

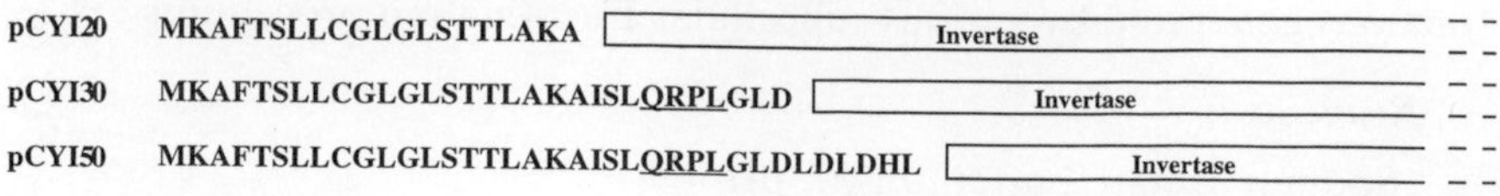

pCYI20	MKAFTSLLCGLGLSTTLAKA	Invertase
pCYI30	MKAFTSLLCGLGLSTTLAKAISL**QRPL**GLD	Invertase
pCYI50	MKAFTSLLCGLGLSTTLAKAISL**QRPL**GLDLDLDHL	Invertase

B

| | | Invertase Activity (Units) | | | |
Construct	Fermentation Phenotype	Total	Extracellular (Secreted)	Intracellular	% Secreted
pCYI20	suc+	300	280	20	98
pCYI30	suc+	253	101	152	37
pCYI50	suc-	300	15	285	<2

FIG. 3. CPY–invertase fusion proteins. (A) The amino acid sequence of CPY, which is fused to invertase, is shown for several CPY–invertase fusion constructs. The QPRL sequence of CPY, which is required for recognition by the Vps10 sorting receptor and efficient transport to the vacuole, is underlined. (B) The invertase secretion phenotype (Suc⁺ or Suc⁻) is indicated as well as the results of a qualitative liquid invertase assay. Invertase activity is shown in units of activity as defined in text.

An example of the utility of the liquid invertase activity assay was provided by the study of a hybrid protein in which invertase was fused to the C terminus of the soluble vacuolar hydrolase carboxypeptidase Y (CPY). Fusion of invertase to CPY facilitated the determination of the sequence requirements contained within CPY necessary for its recognition and sorting to the vacuole.[3] CPY is normally transported via the secretory pathway to a late Golgi compartment, where it is recognized by its trans-membrane receptor Vps10 and sorted to a prevacuolar endosome.[20,21] Within the endosome, CPY dissociates from its receptor (which recycles back to the Golgi), and CPY is subsequently delivered to the vacuole. Invertase was fused to regions of CPY containing progressive N-terminal truncations (Fig. 3), and the liquid invertase activity assay was used to measure the percentage of invertase secreted from yeast cells expressing these hybrid proteins. From these experiments, a 50-amino acid region of CPY was found to be necessary and sufficient to mediate sorting of the CPY–invertase fusion protein to the vacuole.[3] Contained within this region of CPY is the sequence Gln-Arg-Pro-Leu, which was later found to be the

[20] A. A. Cooper and T. H. Stevens, *J. Cell. Biol.* **133,** 529 (1996).
[21] J. H. Stack, B. Horazdovsky, and S. D. Emr, *Annu. Rev. Cell Dev. Biol.* **11,** 1 (1995).

recognition sequence required for interaction with the Vps10p transmembrane receptor.[22,23]

Selection of Yeast Mutants that Missort and Secrete Invertase Fusion Proteins

In addition to using invertase fusions to determine the *cis*-acting determinants necessary for CPY recognition and sorting,[3] the CPY–invertase fusion protein has also been used to identify *trans*-acting protein components required for transport of CPY and other vacuolar proteins.[21] As mentioned above, invertase secretion is not required under standard growth conditions. However, secreted invertase is essential for yeast cells to metabolize sucrose when provided as the sole carbon source. Under these conditions, wild-type cells (deleted for endogenous *SUC2*) expressing the CPY–invertase fusion protein are not viable, as all the invertase activity is retained within the cell and sorted to the vacuole and yeast cells lack a plasma membrane transport system to bring sucrose into the cytoplasm. However, cells harboring mutations in genes required for sorting of CPY to the vacuole missort and secrete CPY–invertase, thereby allowing survival through sucrose fermentation.[16] Using this approach, more than 40 genes have been identified that contain spontaneous mutations enabling growth on BCP-sucrose agar medium. Examination of these mutants has revealed the underlying cause for secretion of invertase activity to be defects in the vacuolar protein sorting pathway.[16,17]

A similar approach may be used to isolate mutants that exhibit defects in other protein trafficking pathways (e.g., see [9] in this volume[14a]). With such a selection procedure, however, the specificity of any mutant phenotypes observed must be confirmed by examining the trafficking of the native protein under study. For example, secretion of CPY can be directly determined by Western blotting with antibodies directed against CPY.[24] In addition, the efficiency with which CPY is transported to the vacuole can be monitored *in vivo* by pulse–chase immunoprecipitation assays that can detect the secretion of newly synthesized CPY.[10]

---

[22] L. A. Valls, J. R. Winther, and T. H. Stevens, *J. Cell Biol.* **111,** 361 (1990).

[23] E. G. Marcusson, B. F. Horazdovsky, J. L. Cereghino, E. Gharakhanian, and S. D. Emr, *Cell* **77,** 579 (1994).

[24] R. C. Piper, E. A. Whitters, and T. H. Stevens, *Eur. J. Cell Biol.* **65,** 305 (1994).

Reagents

BCP-sucrose agar medium:

Yeast extract	10 g
Peptone	20 g
Bromocresol purple (BCP), 0.4% (w/v)	8 ml
Bacto Agar	25 g
H_2O	Up to 960 ml

1. Autoclave the BCP-sucrose agar medium.
2. Cool to ~55° and then add 40 ml of 50% (w/v) sucrose solution.
3. Add antimycin A to a final concentration of 10 μg/ml. Addition of antimycin A, a potent inhibitor of the electron transport chain in mitochondria, eliminates respiratory metabolism due to any trace contamination of glucose.
4. Pour plates immediately (1 liter of agar medium is enough to pour about forty 10-cm plates).

Procedure

1. To select for spontaneous mutations, make a thick streak of cells expressing the invertase fusion protein of interest onto BCP-sucrose agar medium. To select for mutations induced by mutagenesis (e.g., with ethylmethane sulfonate[25]; replica plate colonies generated after mutagenesis and expressing the invertase gene fusion onto BCP-sucrose agar medium. It is imperative that the invertase fusion protein not be secreted from the parent strain, as this would result in all the colonies growing on BCP-sucrose medium.

2. The bromocresol purple (BCP) included in the agar medium is a chromogenic pH indicator. Thus, individual clones of cells harboring mutations that enable survival through sucrose fermentation are identified as bright yellow colonies, while cells that survive through respiratory metabolism will remain purple.

3. Restreak Suc+ colonies onto fresh agar medium and assay for secreted invertase activity, using the plate or liquid assays described above.

---

[25] C. W. Lawrence, *Methods Enzymol.* **194,** 273 (1991).

[9] Introduction of Kex2 Cleavage Sites in Fusion Proteins for Monitoring Localization and Transport in Yeast Secretory Pathway

By B. DIANE HOPKINS, KEN SATO, AKIHIKO NAKANO, and TODD R. GRAHAM

Introduction

The study of protein transport through the yeast secretory pathway has contributed substantially to our understanding of the molecular mechanisms underlying this process. An essential component of these studies has been the ability to follow the progression of a protein through the yeast secretory pathway by examining it for specific modifications that are catalyzed within the endoplasmic reticulum (ER) or Golgi complex. For example, the Kex2 protease (Kex2) is localized to a late Golgi compartment[1,2] comparable to the *trans*-Golgi network (TGN) of multicellular organisms; thus cleavage of a protein by Kex2 is a telltale sign that the protein has passed through the TGN. However, only a few natural substrates for Kex2 have been identified with M1 killer toxin and the mating pheromone precursor pro-α-factor being the best characterized.[1,3] Because most proteins are not substrates for Kex2, it is necessary to introduce a Kex2 cleavage site if one wishes to determine if a protein traffics through the TGN. We outline such a method below, which makes use of fusion constructs containing pro-α-factor sequences encoding a Kex2 cleavage site in order to analyze the transport of a fusion protein through the yeast secretory pathway. In addition, we describe how this procedure can also be used to identify mutants that fail to retain a fusion protein in the ER or early Golgi compartments.

Principle of Method

The biosynthesis and secretion of α-factor have been the subject of a few excellent reviews.[1,4] Briefly, this pheromone is encoded by two genes, *MFα1* and *MFα2,* although the *MFα1* gene is more highly expressed. The

[1] R. S. Fuller, R. E. Sterne, and J. Thorner, *Annu. Rev. Physiol.* **50,** 345 (1988).

[2] T. R. Graham and S. D. Emr, *J. Cell Biol.* **114,** 207 (1991).

[3] H. Bussey, D. Saville, D. Greene, D. J. Tipper, and K. A. Bostian, *Mol. Cell. Biol.* **3,** 1362 (1983).

[4] G. F. J. Sprague and J. W. Thorner, *in* "The Molecular and Cellular Biology of the Yeast *Saccharomyces*" (E. W. Jones, J. R. Pringle, and J. R. Broach, eds.), p. 657. Cold Spring Harbor Laboratory Press, Cold Spring Harbor, New York, 1992.

initial translation product of *MFα1* is a 165-amino acid, high molecular weight precursor called prepro-α-factor (Fig. 1A), which contains an N-terminal signal sequence (the "pre" sequence) that directs the translocation of this protein across the ER membrane. Within the lumen of the ER, the "pre" sequence is removed by signal peptidase and the "pro" region of the precursor receives three N-linked oligosaccharides. Pro-α-factor is then transported from the ER to the Golgi complex, where it encounters an α-1,6-mannosyltransferase in a *cis*-Golgi compartment, α-1,2- and α-1,3-mannosyltransferases in *medial* compartments, and finally the Kex2 protease within the last Golgi compartment or TGN (Fig. 1B). The pro-α-factor polypeptide contains four copies of the mature 13-amino acid peptide with a Kex2 cleavage site preceding each copy. Endoproteolytic cleavage of pro-α-factor by Kex2 produces four intermediate peptides that are further processed by the Kex1 carboxypeptidase and the Ste13 aminopeptidase within the TGN to produce the mature α-factor peptides. From the Kex2 compartment, mature α-factor is packaged into secretory vesicles for exocytosis. Only the fully processed mature form is capable of producing a

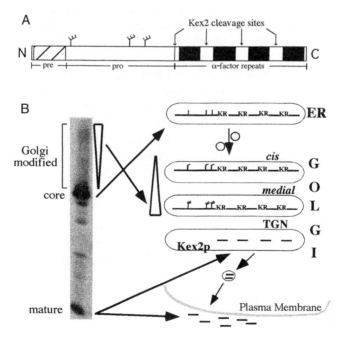

FIG. 1. Processing of prepro-α-factor precursor in yeast. (A) Schematic representation of prepro-α-factor. (B) The transport and processing of prepro-α-factor are described in text. The different α-factor forms separated by SDS–PAGE are shown on the left.

mating response by binding to a trimeric G protein-coupled receptor on *MATa* cells.

There is substantial evidence that the activity of Kex2 is tightly restricted to a late Golgi compartment. Pro-α-factor that is trapped in the ER or early Golgi compartments of a secretory mutant,[2] or purposely localized to these compartments by appending an ER localization signal,[5,6] is not noticeably cleaved by Kex2. In addition, Kex2 also exhibits a high degree of substrate specificity. This enzyme is the prototype for prohormone-processing enzymes that cleave after pairs of basic residues in an appropriate sequence context. Kex2 protease shows a strong preference for lysine–arginine and arginine–arginine sequences,[7,8] but may also cleave after a proline–arginine sequence in M1 killer toxin.[3,8] These dibasic residues can be found in many proteins that appear to pass through the Kex2 compartment but are not cleaved. Thus, the sequences immediately surrounding the dibasic site and the availability of the dibasic site within the polypeptide are also important determinants of Kex2 cleavage specificity.

Critical to the success of the procedures described here is the ability to fuse pro-α-factor sequences to other proteins and still retain the Kex2 substrate specificity. For example, fusion of the entire pro-α-factor polypeptide onto the C terminus of the ER integral membrane protein Sec12 produced a chimeric protein that was cleaved by Kex2 only when it was mislocalized to the Kex2 compartment.[6] The pro-α-factor portion of the fusion protein was processed normally, resulting in the secretion of biologically active α-factor. This method was used to identify the ER localization signals of Sec12p[9] and has been successfully applied to the isolation and analysis of the *rer* mutants (*rer* stands for retrieval to the ER, or retention in the ER). These mutants have a defect in the proper localization of Sec12p and other ER membrane proteins.[6,10,11]

Expression of Pro-α-Factor Fusion Proteins

It is important to take into consideration the topology of the protein of interest when designing a pro-α-factor fusion protein. The pro-α-factor sequences can be fused to either the N or C terminus of a protein, but the pro-α-factor segment must be translocated into the ER lumen to ultimately

[5] N. Dean and H. R. B. Pelham, *J. Cell. Biol.* **111,** 369 (1990).
[6] S. Nishikawa and A. Nakano, *Proc. Natl. Acad. Sci. U.S.A.* **90,** 8179 (1993).
[7] D. Julius, R. Schekman, and J. Thorner, *Cell* **36,** 309 (1984).
[8] A. Bevan, C. Brenner, and R. S. Fuller, *Proc. Natl. Acad. Sci. U.S.A.* **95,** 10384 (1998).
[9] M. Sato, K. Sato, and A. Nakano, *J. Cell Biol.* **134,** 279 (1996).
[10] K. Sato, S. Nishikawa, and A. Nakano, *Mol. Biol. Cell* **6,** 1459 (1995).
[11] K. Sato, M. Sato, and A. Nakano, *Proc. Natl. Acad. Sci. U.S.A.* **94,** 9693 (1997).

gain access to Kex2 in the lumen of the Golgi complex. To achieve the appropriate topology, advantage may be taken of the signal sequence encoded by *MFα1* (amino acids 1–19) for N-terminal fusions or signal/anchor domains encoded by the protein of interest. The *MFα1* gene contains a unique *Pst*I site in the frame CT GCA G, where GCA encodes amino acid 9 of prepro-α-factor, and a unique *Sal*I site 35 bp 3′ of the stop codon. If these sites are not present in the gene of interest, they can be introduced at the C-terminal codon by site-directed mutagenesis such that the *Pst*I sites are in the same reading frame. The *Pst*I–*Sal*I *MFα1* fragment can then be subcloned into the 3′ end of the gene of interest to produce a C-terminal fusion protein.

Alternatively, it is possible to amplify the pro-α-factor sequences by polymerase chain reaction (PCR) and introduce restriction enzyme sites at the 5′ end of the primers that are compatible with sites within the gene of interest. This has been done to create N-terminal α-factor fusions with two ER membrane proteins, Sec63p and Sec71p. In this case, a *Bst*BI site was introduced just after the start codon of the *SEC63* and *SEC71* genes by site-directed mutagenesis. The *MFα1* sequences were then amplified with *Bst*BI-compatible *Cla*I sites at the 5′ ends of the primers that were engineered to maintain the open reading frame (e.g., 5′-CCATCGAT GAAAC GGCACAAATT-3′ and 5′-GGATCGATGGGTTTTAACTG CAACCA-3′). Even though the α-factor repeat region was no longer at the C terminus, Kex2 still cleaved these fusion proteins when ER retention was lost in an *rer1Δ* mutant.

To detect secretion of α-factor derived from a pro-α-factor fusion protein, it is essential to express the fusion gene in an *MATα* strain in which both the *MFα1* and *MFα2* genes have been deleted. The strain SNY9 listed in Table I meets these criteria and also carries a deletion of the barrier protease gene (*bar1::HIS3*) that helps to stabilize secreted α-factor. Alternatively, it is possible to delete these genes in another strain by homologous recombination, using the knockout plasmids also listed in Table I.

Steady State Halo Assay to Measure α-Factor Secretion

Secretion of α-factor from yeast cells can be detected on agar plates by a halo assay, using a lawn of the *sst2* mutant tester strain that is supersensitive to α-factor (strain BC180 in Table I). Mature α-factor arrests the growth of *sst2* cells in the G_1 phase of the cell cycle, and thus causes formation of a growth inhibition zone (called a "halo") of the *sst2* strain around the α-factor-secreting cells.

TABLE I
STRAINS AND DISRUPTION PLASMIDS

Strain	Genotype
SNY9[a]	MATα mfα1::ADE2 mfα2::TRP1 bar1::HIS3 ura3 leu2 trp1 his3 lys2 ade2
BC180[a]	MATa sst2-Δ2 ura3 leu2 his3 ade2
SEY6210[b]	MATα ura3-52 leu2-3,112 his3-Δ200 trp1-Δ901 lys2-801 suc2-Δ9
TBY130[c]	MATα ura3-52 leu2-3,112 his3-Δ200 trp1-Δ901 lys2-801 suc2-Δ9 Δkex2::URA3

Plasmid	Disruption	Restriction enzymes for cleaving plasmid before transformation
pBAR1ΔH[d]	Δbar1::HIS3	XbaI–SalI
pα1ΔA1[d]	Δmfα1::ADE2	EcoRI
pα2ΔT[d]	Δmfα2::TRP1	BglII–SalI

[a] S. Nishikawa and A. Nakano, *Proc. Natl. Acad. Sci. U.S.A.* **90,** 8179 (1993).
[b] J. S. Robinson, D. J. Klionsky, L. M. Banta, and S. D. Emr, *Mol. Cell. Biol.* **8,** 4936 (1988).
[c] W. T. Brigance, C. Barlowe, and T. R. Graham, *Mol. Biol. Cell* **11,** 171 (2000).
[d] S. Nishikawa, Ph.D. thesis. University of Tokyo, Tokyo, Japan, 1992.

Reagents for Halo Assay: Media

MCA (1.4×): Dissolve 3.35 g of yeast nitrogen base without amino acids (Difco, Detroit, MI), 2.5 g of Casamino Acids (Difco), and 10 g of agar (Sho-ei, Tokyo, Japan) in 350 ml of distilled water. Autoclave for 20 min

MCD (4×): Dissolve 13.4 g of yeast nitrogen base without amino acids, 10 g of Casamino Acids, and 40 g of glucose in 500 ml of distilled water. Sterilize by filtration

Buffer (5×): Dissolve 25 g of succinate in 500 ml of distilled water and adjust to pH 3.5 with NaOH. Autoclave for 20 min

Glucose (2%, w/v): Dissolve 200 g of glucose in 1 liter of distilled water. Autoclave for 20 min

Bacto-agar (2%, w/v): Dissolve 2 g of Bacto-agar (Difco) in 100 ml of distilled water. Autoclave for 20 min

To prepare solid medium for halo assays, autoclave 350 ml of 1.4× MCA, and then add 100 ml of 5× buffer and 50 ml of 20% (w/v) glucose and mix well. Supplement the resulting medium appropriately[12] and dispense into petri plates (20 ml/plate). The plates should be allowed to dry at room temperature for 2 days after pouring. To make lawns of the α-

[12] F. Sherman, *Methods Enzymol.* **194,** 3 (1991).

factor supersensitive tester strain (*sst2*), prepare 2× MCD (pH 3.5) by mixing 250 ml of 4× MCD, 200 ml of 5× buffer, and 50 ml of sterile water with appropriate supplements. Add 4 ml of freshly autoclaved 2% (w/v) Bacto-agar to an equivalent volume of 2× MCD (pH 3.5). Add approximately 0.5–1 × 10⁶ *sst2* cells (from an overnight culture grown to stationary phase) to the mixture, mix several times by inversion, and immediately pour onto the MCD plates buffered at pH 3.5. Gently swirl the plate to ensure an even distribution of the tester strain and top agar, and let stand for 5 min at room temperature to allow the top agar to harden. Colonies to be tested for α-factor production are spotted with sterile toothpicks onto these plates and then incubated at 23 or 30° for 48 hr.

For quantification of secreted α-factor, spot various amounts (1, 5, 10, and 50 ng) of synthetic α-factor (Peptides International, Louisville, KY) on filter paper disks placed on the *sst2* tester lawn. By measuring the radii of the halos, a standard profile of α-factor secretion can be established. As shown in Fig. 2, the presence of as little as 50 pg of α-factor (approximately 30 fmol) is detectable. Using this profile as the standard, the amount of α-factor secreted is calculated by measuring the radii of halos around the colonies being tested. Several independent spots of cells expressing the

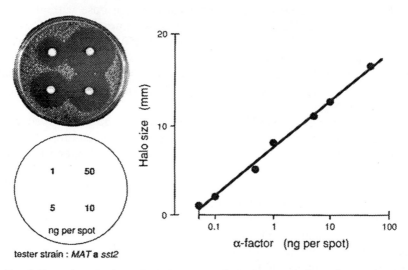

FIG. 2. Detection of α-factor by halo assay. *Left:* An *sst2* strain (BC180) was spread onto an MCD plate buffered at pH 3.5. The indicated amounts of synthetic α-factor were spotted onto filter paper disks on the plate. *Right:* Halo size was measured with a ruler from an edge of a disk to the edge of the halo and plotted against the amount of α-factor. [S. Nishikawa, Ph.D. thesis. University of Tokyo, Tokyo, Japan, 1992.]

fusion protein should be examined to average the amount of α-factor secreted.

Comments. (1) The low pH of the plates increases the sensitivity of the assay; (2) larger halos are produced on plates incubated at 23° rather than at 30°; and (3) it is important to include Casamino Acids in the medium for the halo assay. On conventional minimal medium (SD or MVD), the sensitivity of the halo assay is low because of the poor growth of the *sst2* strain.

Isolation of Mutants that Mislocalize Pro-α-Factor Fusion Protein

A pro-α-factor fusion protein that is efficiently retained in the ER or early compartments of the yeast Golgi will not be cleaved by Kex2. Thus, a strain expressing this fusion protein as the sole source of α-factor will not produce a halo on the *sst2* lawn (Halo⁻ phenotype). In this case, it is possible to screen for mutants that lose the ability to retain the fusion protein in early compartments of the secretory pathway, allowing the fusion protein to come into contact with Kex2. These mutant colonies will then form a halo on an *sst2* lawn (Halo⁺ phenotype). For isolation of the *rer* mutants, cells expressing Sec12-Mfα1p on a multicopy *URA3* plasmid (pSHF9-4) were mutagenized with ethyl methanesulfonate or nitrous acid,[13] spread on selective plates and incubated for 4 days. The surviving colonies were replica plated onto MCD plates (−Ura) buffered at pH 3.5, which had been seeded with 1×10^6 *sst2* cells. For this purpose, the *sst2* strain BC180 was transformed with a *URA3* plasmid so the screen could be performed on −Ura selective medium. Halo⁺ colonies were picked and streaked on plates containing 5-fluoroorotic acid (1 mg/ml) to cure the fusion plasmid. Mutants that became Halo⁻ after plasmid loss were retransformed with pSHF9-4 and examined by the halo assay again. Mutants that were Halo⁺ only when carrying pSHF9-4 were back-crossed three times with a parental strain, and then intercrossed to define complementation groups.

Expression of α-Invertase Fusion Proteins

It is also possible to include a smaller segment of pro-α-factor in a fusion protein and still achieve efficient cleavage by Kex2.[2,14] For example, a 63-base pair (bp) segment of *MFα1* encoding one mature α-factor repeat followed by a Kex2 cleavage site was cloned into the junction of a *PRC1–SUC2* fusion gene such that the reading frame was maintained for all three

[13] C. W. Lawrence, *Methods Enzymol.* **194,** 273 (1991).
[14] T. R. Graham and V. A. Krasnov, *Mol. Biol. Cell* **6,** 809 (1995).

open reading frames (ORFs). *PRC1* and *SUC2* encode the vacuolar protein carboxypeptidase Y (CPY) and the secreted reporter enzyme invertase, respectively. The original fusion protein was directed to the vacuole by the N-terminal CPY fragment; however, the trimeric fusion protein (CPY–α-invertase) containing the Kex2 cleavage site was efficiently cleaved by Kex2 in the late Golgi, allowing for invertase secretion.[2]

To express fusion proteins bearing α-invertase at the C terminus, pAlphaS-308 and pAhplaS-308 (Fig. 3) were constructed by subcloning a 63-bp *MFα1 Hin*dIII fragment into pSEYC308.[14] Alpha indicates the correct orientation of the *Hin*dIII fragment whereas Ahpla indicates the opposite orientation. The latter construct still maintains an open reading frame with *SUC2* (the invertase gene) but does not encode a Kex2 cleavage site and thus serves as a useful negative control. The *Eco*RI, *Sma*I, *Bam*HI, and *Sal*I sites are unique in these vectors and can be used to insert DNA fragments of interest. The pSEYC308 vector is also shown in Fig. 3 and can be used to express invertase fusion proteins that lack the pro-α-factor sequences. GenBank accession numbers for downloading the complete nucleotide sequence of all three vectors are provided in the legend to Fig. 3.

Construction of fusion genes, using the plasmids shown in Fig. 3, requires

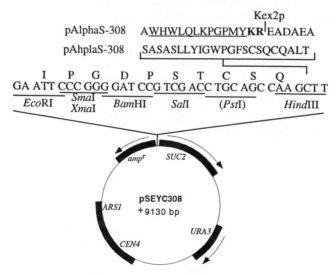

Fig. 3. Plasmids for expression of α-invertase fusion proteins. The construction of pAlphaS-308 (accession number AF138273) and pAphlaS-308 (accession number AF138274) from pSEYC308 (accession number AF138275) is described in text. The open reading frame within the polylinker region is shown. The restriction enzyme sites in the polylinker are all unique except for *Pst*I. The mature α-factor peptide encoded by pAlphaS-308 is underlined, and the Kex2 cleavage site, after the lysine and arginine residues, is indicated.

a restriction enzyme site within the ORF of interest that will maintain the open reading frame with the pro-α-factor sequence. If an endogenous site is not present in the correct reading frame, it is often a simple matter to design a set of PCR primers to amplify a fragment with the appropriate restriction sites at each end. Alternatively, one can introduce a restriction enzyme site by site-directed mutagenesis. A DNA fragment subcloned into these vectors must include a yeast promoter, initiation codon, and signal sequence (or signal/anchor domain) in order to express the fusion protein in the secretory pathway. Use of these vectors also requires a *ura3 suc2* strain in order to introduce the plasmid by a standard lithium acetate yeast transformation procedure[15] and to assay the plasmid-encoded invertase without interference from the endogenous enzyme. Strain SEY6210, listed in Table I, satisfies these requirements and also harbors several other auxotrophic mutations that are useful for transforming this strain with other plasmids. Another advantage of this strain is that a number of isogenic mutant strains in this background are available including a *kex2Δ* strain.

Methods for Determining if α-Invertase Fusion Protein Is Cleaved by Kex2

The reporter enzyme invertase is normally secreted from yeast cells to the periplasmic space, which lies between the cell wall and plasma membrane. Strains secreting invertase can easily be distinguished from nonsecretors by the plate or quantitative liquid invertase assays described in [8] in this volume.[16] Strains expressing the same N-terminal region of a test protein fused with either invertase or α-invertase can be examined for invertase secretion by these assays. If the protein fused with invertase carries sorting information that directs this enzyme to an organelle of the secretory or vacuolar system, most of the invertase activity will be retained inside the cell. In this case, secretion of invertase activity from strains expressing the analogous α-invertase trimeric fusion protein indicates that it had been transported into the Kex2 compartment. In addition, the percentage of invertase secreted should approximate the percentage of the fusion protein that is cleaved by Kex2. To demonstrate Kex2 dependence, invertase should not be secreted from a *kex2Δ* strain expressing the α-invertase trimeric fusion protein, nor from wild-type cells expressing the ahpla–invertase fusion protein.

Note that cleavage of an α-invertase fusion protein by Kex2 leaves a copy of the α-factor peptide fused to the C terminus of the test protein.

[15] H. Ito, Y. Fukuda, K. Murata, and A. Kimura, *J. Bacteriol.* **153,** 163 (1983).
[16] T. Darsow, G. Odorizzi, and S. D. Emr, *Methods Enzymol.* **327,** Chap. 8, 2000 (this volume).

Because antibodies are available that recognize the mature peptide, this should produce an epitope-tagged protein. However, the utility of the α-factor peptide as an epitope tag has not been tested.

If the test protein cannot localize the fusion protein to an intracellular compartment, the invertase activity will be expressed at the cell surface of strains expressing either the invertase or α-invertase fusion proteins. In this case, it is necessary to examine the mobility of the fusion protein by sodium dodecyl sulfate–polyacrylamide gel electrophoresis (SDS–PAGE) to determine if Kex2 cleaves it. Pulse–chase labeling experiments can also be performed to examine the kinetics of Kex2 cleavage by immunoprecipitating any of the fusion proteins described in this chapter. Invertase and pro-α-factor are hyperglycosylated on arrival in the Golgi complex, which causes these proteins to migrate as a smear in SDS gels (Fig. 1). The large decrease in mobility caused by modification of N-linked oligosaccharides by Golgi mannosyltransferases serves as a useful marker to score the arrival of a fusion protein in the Golgi complex. However, this decrease in mobility can also obscure a small increase in mobility caused by Kex2 cleavage of the fusion protein. To examine both the glycosylation status of a fusion protein and the relative mass of the polypeptide by SDS–PAGE, immunoprecipitates are split in half and one portion is treated with endoglycosidase H (endo H) to remove the N-linked oligosaccharides.

Pulse–Chase Immunoprecipitation Procedure

Reagents

SDS–urea buffer: 50 mM Tris-HCl (pH 7.5), 1 mM EDTA, 1% (w/v) SDS, 6 M urea

Tween-20 IP buffer: 50 mM Tris-HCl (pH 7.5), 0.1 mM EDTA, 150 mM NaCl, 0.5% (v/v) Tween 20

Tween-20 urea buffer: 100 mM Tris-HCl (pH 7.5), 200 mM NaCl, 0.5% (v/v) Tween 20, 2 M urea

Chase solution (50×): 50 mM methionine, 5 mM cysteine, and 10% (w/v) yeast extract

Protein A-Sepharose: Swell 0.4 g of protein A-Sepharose (Pharmacia, Piscataway, NJ) in 11.2 ml of 10 mM Tris (pH 7.5), 1 mM NaN$_3$, and bovine serum albumin (BSA, 1 mg/ml) for 2 hr to overnight at 4°. Aspirate the supernatant from the settled beads and add fresh buffer to the same volume.

Procedure

1. Grow yeast in 10–25 ml of synthetic minimal (SD) medium overnight at 30° or at a temperature permissive for the strain. Use an SD

medium that lacks methionine and that will maintain selection for any plasmid(s) carried in the strain.[12]

2. Dilute the overnight culture to 0.25 OD_{600} in the same medium (25 to 50 ml) and culture for 4–5 hr. The cells will usually double only once in this time period when cultured in minimal medium.

3. Measure the OD_{600} of the culture again and pellet the yeast by centrifuging at 5000g for 5 min. Discard the spent medium and resuspend the cells in SD medium to a concentration of 10 OD_{600}. In a typical pulse–chase experiment, we will harvest 2 to 4 OD of cells for each time point. Determine the number of cells required for the experiment and transfer an appropriate culture volume to a 50-ml disposable centrifuge tube, or to a 14-ml disposable tube (Falcon 2059; Becton Dickinson Labware, Lincoln Park, NJ) for volumes less than 1 ml.

4. Preincubate the cells at the desired temperature (typically 30°) for 15 min with vigorous shaking. To initiate the labeling period, add [^{35}S]methionine and cysteine [EXPRE^{35}S^{35}S labeling mix (New England Nuclear, Boston, MA) or Trans^{35}S label (Amersham, Arlington Heights, IL)] to a final concentration of 250 μCi/ml. Typical labeling times are 5 to 10 min.

5. To initiate the chase, the 50× chase solution is added to a 1× final concentration. BSA can also be added to the chase at 1 mg/ml as a carrier when trichloroacetic acid (TCA) precipitating proteins in the medium. Remove equal aliquots (100–400 μl) from the culture at the desired chase times (e.g., 0, 5, 15, and 45 min), and then add TCA to 10% (w/v) final concentration, and place the tubes on ice to terminate the chase.

6. After at least 15 min (and up to overnight) on ice, pellet the cells in a microcentrifuge for 10 min at full speed. Aspirate the supernatant into a radioactive waste trap and wash the pellet with ~1.0 ml of ice-cold acetone. Individually immersing the bottom of the tubes in a bath sonicator will help disperse the pellets. Centrifuge for 4 min at full speed and aspirate off the acetone. Repeat the acetone wash.

7. Dry the washed TCA pellet under vacuum and then add 100 μl of SDS–urea buffer and let sit for at least 15 min at room temperature. Add glass beads (0.1–0.25 mm; Glen Mills, Clifton, NJ) to 80–90% of the sample volume and vortex for 1 min. Heat the samples for 4 min at 95°.

8. Add 900 μl of Tween-20 IP, vortex, and put on ice for 10 min. Centrifuge the samples for 10 min and transfer ~0.85 ml of the supernatant to a fresh tube, being careful not to disturb the glass beads.

9. Add antiserum (usually 1–2 μl per OD of labeled cells) and protein A-Sepharose (75–100 μl) to the supernatants, then lay the tubes on their sides and rock the samples for 4 hr to overnight in a cold room. Agitation should be sufficient to keep the protein A-Sepharose in suspension.

10. Pellet the immune complex by centrifuging for 30 sec at 3000g and wash twice with Tween 20–urea buffer, then once with Tween-20 IP buffer. Dry the pellet for 5–10 min under vacuum.

11. A second immunoprecipitation step is usually required to achieve a "clean" immunoprecipitation. Resuspend the pellets in 100 μl of 50 mM Tris-HCl (pH 7.5), 1% (w/v) SDS and heat at 95° for 4 min. Repeat steps 8, 9, and 10.

12. For endo H treatment of samples, add 64 μl of freshly prepared SDS–2-ME buffer [0.2% (v/v) SDS, 1% (w/v) 2-mercaptoethanol] to dried immunoprecipitates and heat at 95° for 4 min. Add 16 μl of 250 mM sodium citrate buffer, pH 5.5, to each sample. Mix and centrifuge briefly. Transfer 40 μl of each sample to a fresh tube and add 0.5 mU of Endo H (0.5 μl; New England BioLabs, Beverly, MA) and incubate the samples overnight in a 37° air incubator.

13. Add 13.3 μl of 4× sample buffer to each sample and heat at 95° for 4 min. Electrophorese the samples in an SDS–polyacrylamide gel.

Comments. Linkage-specific antibodies that specifically recognize α-1,6-linked mannose and α-1,3-linked mannose residues on yeast glycoproteins are available from several laboratories. These antibodies can be used to score the progression of pro-α-factor or invertase fusion proteins through the Golgi complex because α-1,6-mannose is added in a *cis*-Golgi compartment, and α-1,3-mannose is added in a *medial-* or *trans*-Golgi compartment. To use these antibodies, simply substitute the linkage-specific antibody for the pro-α-factor antibody in the second immunoprecipitation step (step 11).

[10] Lineage Analysis with Retroviral Vectors

By Constance L. Cepko, Elizabeth Ryder, Christopher Austin, Jeffrey Golden, Shawn Fields-Berry, and John Lin

Knowledge of the geneological relationships of cells during development can lead to insight concerning when and where developmental decisions are being made. Hypotheses can be ruled in or out concerning the commitment of cells to particular fates. For example, when analyzing the cell types that result from the marking of a single progenitor cell, insight can be gained as to whether the progenitor was committed to the production of one or multiple cell types. If multiple cell types are found in a clone, it may be concluded that the progenitor that gave rise to these cells was not restricted to the production of only one cell type. Alternatively, if all of the cells that descend from a progenitor are the same type, the hypothesis is supported, but not proved, that the progenitor was committed to making only that cell type. In the latter case, a firm conclusion concerning the commitment of the progenitor can be reached only if the progenitor and/ or progeny are exposed to a variety of environments. If only one cell type is produced despite variations in the environment, commitment of the progenitor to production of one cell type is supported. Analyses of clones generated after marking the progenitors of a tissue at various times in development can greatly aid in charting the stages of production of different cell types, allowing studies concerning cell fate decisions to be focused on particular times and places. In addition, analysis of the proliferation and migration patterns exhibited by clones can increase our understanding of the development of a particular area.

The complexity and inaccessibility of many types of embryos have made lineage analysis through direct approaches, such as time-lapse microscopy and injection of tracers, almost impossible. A genetic and clonal solution to lineage mapping is through the use of retrovirus vectors. The basis for this technique is summarized, and the strategies and current methods in use in our laboratory, are detailed.

Transduction of Genes via Retrovirus Vectors

A retrovirus vector is an infectious virus that transduces a nonviral gene into mitotic cells *in vivo* or *in vitro*.[1] These vectors utilize the same efficient and precise integration machinery of naturally occurring retroviruses to produce a single copy of the viral genome stably integrated into the host chromosome. Those that are useful for lineage analysis have been modified so that they are replication incompetent and thus cannot spread from one infected cell to another. They are, however, faithfully passed on to all daughter cells of the originally infected progenitor cell, making them ideal for lineage analysis.

Retroviruses use RNA as their genome, which is packaged into a membrane-bound protein capsid. They produce a DNA copy of their genome immediately after infection via reverse transcriptase, a product of the viral *pol* gene, which is included in the viral particle. The DNA copy is integrated into the host cell genome and is thereafter referred to as a "provirus." Integration of the genome of most retroviruses requires that the cell go through an M phase,[2] and thus only mitotic cells will serve successfully as

---

[1] J. M. Coffin, S. H. Hughes, and H. E. Varmus, "Retroviruses." Cold Spring Harbor Laboratory Press, Plainview, New York, 1997.
[2] T. Y. Roe, T. C. Reynolds, G. Yu, and P. O. Brown, *EMBO J.* **12,** 2099 (1993).

hosts for integration of most retroviruses. [However, there is a generation of retrovirus vectors based on human immunodeficiency virus (HIV),[3] which can integrate into postmitotic cells. As lineage analysis is designed to ask about the fate of daughter cells, infection of postmitotic cells is not desirable.] Most vectors began as proviruses that were cloned from cells infected with a naturally occurring retrovirus. Although extensive deletions of proviruses were made, vectors retain the *cis*-acting viral sequences necessary for the viral life cycle. These include the packaging sequence (necessary for recognition of the viral RNA for encapsidation into the viral particle), reverse transcription signals, integration signals, viral promoter, enhancer, and polyadenylation sequences. A cDNA can thus be expressed in a vector using the transcription regulatory sequences provided by the virus (although see below for further discussion of this point). Because replication-incompetent retrovirus vectors usually do not encode the structural genes whose products comprise the viral particle, these proteins must be supplied through complementation. The products of the genes *gag, pro, pol,* and *env* are typically supplied by "packaging" cell lines or cotransfection with packaging constructs into highly transfectable cell lines (for review see Cepko and Pear in Ausubel *et al.*[4] Packaging cell lines are stable lines that contain the *gag, pro, pol,* and *env* genes as a result of the introduction of these genes by transfection. However, these lines do not contain the packaging sequence on the viral RNA that encodes the structural proteins. Thus, the packaging lines, or cells transfected with packaging constructs, make viral particles that do not contain the genes *gag, pro, pol,* or *env.*

Retrovirus vector particles are essentially identical to naturally occurring retrovirus particles. They enter the host cell via interaction of a viral envelope glycoprotein (a product of the viral *env* gene) with a host cell receptor. The murine viruses have several classes of Env glycoprotein that interact with different host cell receptors. The most useful class for lineage analysis of rodents is the ecotropic class. The ecotropic Env glycoprotein allows entry only into rat and mouse cells via the ecotropic receptor on these species. It does not allow infection of humans, and thus is considered relatively safe for gene transfer experiments. The first packaging line commonly in use was the ψ2 line.[5] It encodes the ecotropic *env* gene and makes high titers of vectors. However, it can also lead to the production of helper virus (discussed below). A second generation of ecotopic packag-

[3] L. Naldini, U. Blomer, P. Gallay, D. Ory, R. Mulligan, F. H. Gage, I. M. Verma, and D. Trono, *Science* **272,** 263 (1996).

[4] F. M. Ausubel, R. Brent, R. E. Kingston, D. D. Moore, J. G. Seidman, J. A. Smith, and K. Struhl, "Current Protocols in Molecular Biology." Greene Publishing Associates, New York, 1997.

[5] R. Mann, R. C. Mulligan, and D. Baltimore, *Cell* **33,** 153 (1983).

ing lines, ψCRE,[6] GP+E-86,[7] and ψE,[8] has not been reported to lead to production of helper virus to date. A third generation of "helper-free" packaging lines, exemplified by the ecotropic lines Bosc23[9] and Phoenix (G. Nolan, Stanford University, Stanford, CA), were made in 293T cells, and have the advantage over the earlier lines of giving high-titer stocks transiently after transfection. Similarly, contransfection of 293T cells with packaging constructs and vectors can lead to the transient production of high-titer stocks.[10] The first two generations of packaging lines, which are based on mouse fibroblasts, require production of stably transduced lines for production of high-titer stocks.

For infection of nonrodent species, an envelope glycoprotein other than the ecotropic glycoprotein must be used to allow entry into the host cells. The one that endows the greatest host range is the vesicular stomatitis virus (VSV) G glycoprotein, which allows infection of most species, including fish.[11] The G protein apparently also makes for a more stable particle, which allows for greater concentration of the virus preparations. For lineage analysis of avian species, packaging lines and vectors based on avian retroviruses are available.[12-14] In addition, we have found that avian retroviruses with the VSV G protein on their surface were more efficient at infecting chick embryos[15] (Fig. 1; see color insert). Such virions gave the same titer as those with the avian A type Env protein when they were titered on avian cells *in vitro*. However, when injected *in vivo,* the VSV G-carrying particles give an approximately 350-fold more efficient infection, as judged by the number of clones in the retina, than the particles carrying the avian A Env protein. Similar increases in efficiencies were noted throughout the embryo. We interpret these data to mean that cells within the avian embryo do not express high enough levels of the receptor for the A type Env to be readily infected, but are not limited with respect to the ubiquitous

[6] O. Danos and R. C. Mulligan, *Proc. Natl. Acad. Sci. U.S.A.* **85,** 6460 (1988).

[7] D. Markowitz, S. Goff, and A. Bank, *J. Virol.* **62,** 1120 (1988).

[8] J. P. Morgenstern and H. Land, *Nucleic Acids Res.* **18,** 3587 (1990).

[9] W. S. Pear, G. P. Nolan, M. L. Scott, and D. Baltimore, *Proc. Natl. Acad. Sci. U.S.A.* **90,** 8392 (1993).

[10] Y. Soneoka, P. M. Cannon, E. E. Ramsdale, J. C. Griffiths, G. Romano, S. M. Kingsman, and A. J. Kingsman, *Nucleic Acids Res.* **23,** 628 (1995).

[11] J. K. Yee, T. Friedmann, and J. C. Burns, *Methods Cell Biol.* **43,** 99 (1994).

[12] C. F. Boerkoel, M. J. Federspiel, D. W. Salter, W. Payne, L. B. Crittenden, H. J. Kung, and S. H. Hughes, *Virology* **195,** 669 (1993).

[13] F. L. Cosset, C. Legras, Y. Chebloune, P. Savatier, P. Thoraval, J. L. Thomas, J. Samarut, V. M. Nigon, and G. Verdier, *J. Virol.* **64,** 1070 (1990).

[14] A. W. Stoker and M. J. Bissell, *J. Virol.* **62,** 1008 (1988).

[15] C.-M. A. Chen, D. M. Smith, M. A. Peters, M. E. S. Samson, J. Zitz, C. J. Tabin, and C. L. Cepko, *Dev. Biol.* **214,** 370 (1999).

phospholipid receptor for the VSV G protein.[16] We did not see this effect of increased efficiency of infection using murine vectors and murine embryos, but it is possible that different mouse strains vary in this regard. Gaiano *et al.* found that murine virions carrying VSV G gave a slightly different spectrum of cell types following infection of the early mouse brain than virions carrying the murine ecotropic Env.[17]

Two other parameters to be considered when choosing a vector for lineage analysis are the reporter gene and the promoter that drives its expression. The reporters that have been used include cytoplasmic LacZ,[18] nuclear LacZ,[19] human placental alkaline phosphatase (PLAP),[20] and avian Gag.[21] More recently, there are vectors encoding green fluorescent protein (GFP).[22] We have found advantages and disadvantages in the use of each of these reporters. When deciding which reporter to use, we first consider the background activities in the tissue of interest. Although the *lacZ* gene gives a stable, reliable, and specific signal in most cells using 5-bromo-4-chloro-3-indoly1-β-D-galactopyranoside (X-Gal) detection,[23] there is problematic β-galactosidase background in a few tissues. Control staining with X-Gal thus should be done to determine if it is a problem in an area of interest. Changes in the fixation and staining conditions can reduce β-galactosidase background.[24] Similarly, PLAP staining is reliable and stable, with heat treatment of the infected tissue rendering most endogenous alkaline phosphatases inactive. However, in some tissues residual alkaline phosphatases cause a backgroud problem. In such cases, inhibitors of endogenous alkaline phosphatases may solve the problem.[25] When using GFP, the signal from the introduced GFP can be weak, and thus background fluorescence in the tissue can be a problem. We have found that tissue sections prepared on a Vibratome yield bright GFP$^+$ cells, but the same

[16] R. Schlegel, T. S. Tralka, M. C. Willingham, and I. Pastan, *Cell* **32**, 639 (1983).

[17] N. Gaiano, J. D. Kohtz, D. H. Turnbull, and G. Fishell, *Nature Neurosci.* **2**, 812 (1999).

[18] J. Price, D. Turner, and C. Cepko, *Proc. Natl. Acad. Sci. U.S.A.* **84**, 156 (1987).

[19] D. S. Galileo, G. E. Gray, G. C. Owens, J. Majors, and J. R. Sanes, *Proc. Natl. Acad. Sci. U.S.A.* **87**, 458 (1990).

[20] S. C. Fields-Berry, A. L. Halliday, and C. L. Cepko, *Proc. Natl. Acad. Sci. U.S.A.* **89**, 693 (1992).

[21] D. M. Fekete, J. Perez-Miguelsanz, E. F. Ryder, and C. L. Cepko, *Dev. Biol.* **166**, 666 (1994).

[22] M. Chalfie, Y. Tu, G. Euskirchen, W. W. Ward, and D. C. Prasher, *Science* **263**, 802 (1994).

[23] C. L. Cepko, S. Bruhn, and D. Fekete, *Methods Cell Biol.*

[24] W. S. Rosenberg, X. O. Breakefield, C. DeAntonio, and O. Isacson, *Brain Res. Mol. Brain Res.* **16**, 311 (1992).

[25] H. F. Zoellner and N. Hunter, *J. Histochem. Cytochem.* **37**, 1893 (1989).

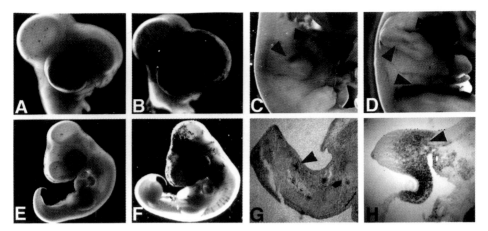

Fig. 1. The VSV G protein endows avian retroviral vectors with a high efficiency of infection *in vivo*. Comparable amounts of an avian replication-incompetent vector, RIA-AP[15], encoding PLAP were injected into developing chick embryos. RIA-AP (G) virion particles contained VSV G protein on their surface and RIA-AP (A) virions contained the avian retroviral A env protein on their surface. Embryos were injected at stage 18 into the eye (A and B), stage 10 into the neural tube (E and F), or stage 18 into the limb bud or heart regions (C, D, G, H). Embryos injected with RIA-AP (A) are shown in panels A, C, and E, and those injected with RIA-AP (G) in panels B, F, D, and H. Sections of infected limbs are shown in G and H. Approximately 3 days postinfection, the embryos were stained to reveal PLAP activity; red arrowheads indicate the limb regions and blue arrowheads the heart.

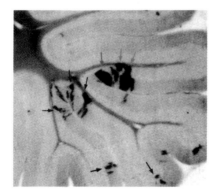

Fig. 4. A 60 μm parasagittal section of a chick cerebellum from a brain infected with CHAPOL and analyzed at P14. The AP+ cells were removed, subjected to PCR, and the PCR products were sequenced. Each pick that yielded a PCR product is labeled by an arrow. The sequence identity of each PCR product is color-coded. The clone indicated by the magenta arrows were not closely clustered, and would have resulted in a splitting error if certain geometric criteria were used in clonal definition. The clones indicated by the blue and the green arrows were located very close to each other, and would have resulted in a lumping error if certain geometric criteria of clonal assignment were used.

tissue sectioned on a cryostat gives much dimmer and more ill-defined GFP⁺ cells.

The second issue to consider concerns how to define the cell types expressing the reporter gene. In some cases, the morphology of the infected cells indicates their identity. In such cases, we recommend the use of LacZ or PLAP with histochemical detection, which is the simplest and most rapid way to find the infected cells. Moreover, the X-Gal precipitate formed by β-galactosidase and the XP/nitroblue tetrazolium (NBT) product of PLAP are stable for months of storage, allowing time to analyze many sections. However, there are differences between LacZ and PLAP that will direct the choice of which to use for morphological identification. LacZ typically does not fill the cell bodies of large cells, such as neurons, as completely as PLAP. Thus, when it is desirable to characterize cells via their morphology, PLAP, which associates with the plasma membrane, is superior to LacZ. PLAP is also the most sensitive of the reporters that we have used, most likely because it is a stable enzyme. However, PLAP can produce such a dense stain that, when clonally related cells are close together, we have been unable to count the number of cells in a clone or distinguish the morphologies of individual cells (see Ref. 26 for an example). In such cases, nuclear LacZ is useful.

If morphological criteria cannot be used to identify the types of cells carrying the reporter gene, one option is to use immunohistochemistry to detect defining cellular antigens. The X-Gal product of LacZ and the reaction product of PLAP make it difficult to detect a fluorescent immunohistochemical product, as they absorb fluorescence. Moreover, the X-Gal product and the XP/NBT precipitate produced by PLAP often are too dark to allow simultaneous detection of another colored precipitate produced by immunohistochemical detection of a cellular antigen. However, occasionally, this will work (e.g., see Refs. 15 and 27). Double immunohistochemical procedures can be used by employing antisera to detect PLAP or LacZ. However, double immunohistochemical procedures are much more time consuming then histochemistal procedures and immunohistochemical detection of LacZ and PLAP is not always as sensitive as the histochemical procedures. For all of these reasons, GFP is a better choice. GFP allows simultaneous detection of the GFP reporter and an immunohistochemical signal (e.g., see Ref. 28). However, as mentioned above,

²⁶ Z. Z. Bao and C. L. Cepko, *J. Neurosci.* **17,** 1425 (1997).
²⁷ E. Y. Snyder, D. L. Deitcher, C. Walsh, S. Arnold-Aldea, E. A. Hartwieg, and C. L. Cepko, *Cell* **68,** 33 (1992).
²⁸ K. Moriyoshi, L. J. Richards, C. Akazawa, D. D. O'Leary, and S. Nakanishi, *Neuron* **16,** 255 (1996).

GFP expression is sometimes weak and, in addition, storage of sections over a long period of time (i.e., months) does not allow for preservation of the GFP signal. The newest reporter to be described, β-lactamase,[29] may offer advantages over GFP in terms of sensitivity, but it has not yet been tested *in vivo,* where it may suffer from leakage of the product from infected cells. If this is a problem, future substrates might overcome this limitation.

The choice of the promoter to drive expression of a reporter gene also requires consideration. To see all the progeny of infected cells, a constitutive promoter should be used. We have had success using the long terminal repeat (LTR) of Moloney murine leukemia virus (Mo-MuLV) for work in rats and mice and the LTR of avian leukosis virus for work in chicks. We compared several alternative promoters located internal to the Mo-MLV LTR in the context of a wild-type LTR promoter and in the context of an LTR promoter with an enhancer deletion, using infections of murine tissue *in vivo.*[30,31] The LTR promoter performed the best of several promoters tested, including the human histone 4, chicken β-actin, cytomegalovirus (CMV) immediate-early, and the simian virus 40 (SV40) early promoters. However, Gaiano *et al.*[17] reported that the Mo-MuLV promoter was relatively inactive in early [embryonic day 8.5 (E8.5) to E9.5] progenitor cells of the CNS. This problem is reminiscent of the failure of the LTR to express in embryonic stem cells and preimplantation embryos, which appears to be due to inhibition of the LTR in stem cells. Gaiano *et al.* found that an internal promoter of Ef1α or CMV/β-actin resulted in more expression in early progenitor cells as well as in later neurons. These findings suggest that stable expression should be tested for in the area of interest, using the infection time and site that will be used for future experiments, before choosing the promoter. However, even after performing such preliminary experiments, and even with the choice of an apparently constitutive promoter, it is important to restrict conclusions about lineal relationships to cells that are marked and not to make assumptions about their relationships to cells that are unmarked. We have found, in some clones, that not all cells express a reporter gene. This is true even in control situations, such as in clones of NIH 3T3 fibroblasts infected with either a LacZ or PLAP vector *in vitro.* This observation has been made with several different promoters and vector designs.

[29] G. Zlokarnik, P. A. Negulescu, T. E. Knapp, L. Mere, N. Burres, L. Feng, M. Whitney, K. Roemer, and R. Y. Tsien, *Science* **279,** 84 (1998).
[30] C. L. Cepko, C. P. Austin, C. Walsh, E. F. Ryder, A. Halliday, and S. Fields-Berry, *Cold Spring Harbor Symp. Quant. Biol.* **55,** 265 (1990).
[31] D. L. Turner, E. Y. Snyder, and C. L. Cepko, *Neuron* **4,** 833 (1990).

Production of Virus Stocks for Lineage Analysis

Any of the aforementioned packaging systems and vectors can be used to produce vector stocks for lineage analysis. All stocks should be assayed for the presence of helper virus. A detailed description of protocols for making stocks, for titering and concentrating them, and for checking for helper virus contamination has been published and is not given here (see Cepko and Pear in Ausubel et al.[4] and Ref. 5). The components for producing such stocks are generally available from the laboratories that have constructed them. For example, we have deposited stable lines that produce the murine replication-incompetent vectors that encode the histochemical reporter genes lacZ and PLAP at the American Type Culture Collection (ATCC, Rockville, MD). The ψ2 and ψCRE producers of BAG,[18] a lacZ virus we have used for lineage analysis, can be obtained and are listed as ATCC CRL 1858 (ψCRE BAG) and 9560 (ψ2 BAG). Similarly, ψ2 producers of DAP,[20] a vector encoding PLAP (described further below), is available as CRL 1949. For reasons that are unclear, the DAP line is more reliable for production of high-titer stocks. Both of these vectors transcribe the reporter gene from the Mo-MuLV LTR promoter and are generally useful for expression of the reporter gene in most tissues (although see discussion of promoters above).

For lineage applications it is usually necessary to concentrate virus in order to achieve sufficient titer. This is typically due to a limitation in the volume that can be injected at any one site. Viruses can be concentrated fairly easily by a relatively short centrifugation step. Virions also can be precipitated with polyethylene glycol or ammonium sulfate, and the resulting precipitate collected by centrifugation. Finally, the viral supernatant can be concentrated by centrifugation through a filter that allows only small molecules to pass [e.g., Centricon (Amicon, Danvers, MA) filters]. Regardless of the protocols used, keep in mind that the murine ecotropic and amphotropic retroviral particles, as well as the avian retroviral particles, are fragile, with short half-lives even under optimum conditions. In contrast, if the VSV G protein is used as the viral envelope protein, the virions are more stable. To prepare the highest titered stock for multiple experiments, we usually concentrate several hundred milliliters of producer cell supernatant. These concentrated stocks are titered and tested for helper virus contamination, and can be stored indefinitely at $-80°$ or in liquid N_2 in small (10- to 50-μl) aliquots. If we anticipate storage for more than several months, we place a drop of mineral oil [e.g., from a fresh tube of polymerase chain reaction (PCR) oil] on top of the stock to prevent dehydration.

Replication-Competent Helper Virus

Replication-competent virus is sometimes referred to as helper virus, as it can complement ("help") a replication-incompetent virus and thus allow it to spread from cell to cell. It can be present in an animal through exogenous infection (e.g., from a viremic animal in the mouse colony), expression of an endogenous retroviral genome (e.g., the *akv* loci in AKR mice), or through recombination events in an infected cell that occurs between two viral RNAs encapsidated in retroviral virions produced during packaging. The presence of helper virus is an issue of concern when using replication-incompetent viruses for lineage analysis as it can lead to horizontal spread of the marker virus, creating false lineage relationships. The most likely source of helper virus is the viral stock used for lineage analysis. The genome(s) that supplies the *gag, pol,* and *env* genes in packaging lines does not encode the ψ sequence, but can still become packaged, although at a low frequency. If it is coencapsidated with a vector genome, recombination in the next cycle of reverse transcription can occur. If the recombination allows the ψ^- genome to acquire the ψ sequence from the vector genome, a recombinant that is capable of autonomous replication is the result. This recombinant can spread through the entire culture (although slowly because of envelope interference). Once this occurs, it is best to discard the producer clone or stock as there is no convenient way to eliminate the helper virus. As would be expected, recombination giving rise to helper virus occurs with greater frequency in stocks with high titer, with vectors that have retained more of the wild-type sequences (i.e., the more homology between the vector and packaging genomes, the more opportunity there is for recombination), and within stocks generated using *gag-pol* and *env* on a complementing genome during transfection (as opposed to two separate genomes, one for *gag-pol* and one for *env*). Note that a murine helper genome itself will not encode a histochemical marker gene as apparently there is not room, or flexibility, within murine viruses that allows them to be both replication competent and capable of expressing another gene such as *lacZ*. The way that spread would occur is by a cell being infected with both the *lacZ* virus and a helper virus. Such a doubly infected cell would then produce both viruses.

When performing lineage analysis, there are several signs that can indicate the presence of helper virus within an individual animal. If an animal is allowed to survive for long periods of time after inoculation, particularly if embryos or neonates are infected, the animal is likely to acquire a tumor when helper virus is present. Most naturally occurring replication-competent viruses are leukemogenic, with the disease spectrum being at least in part a property of the viral LTR. Second, if analysis is performed either short or long term after inoculation, the clone size, clone number, and spectrum of labeled cells may be indicative of helper virus. For example, the eye of a newborn rat or mouse has mitotic progenitors for retinal neurons, as well as mitotic progenitors for astrocytes and endothelial cells.

By targeting the infection to the area of progenitors for retinal neurons, we only rarely see infection of a few blood vessels or astrocytes as their progenitors are outside of the immediate area that is inoculated and they become infected only by leakage of the viral inoculum from the targeted area. However, if helper virus were present, we would see infection of a high percentage of astrocytes, blood vessels, and, eventually, other eye tissues because virus spread would eventually lead to infection of cells outside the targeted area. A correlation between the percentage of such nontargeted cells that are infected and the degree to which their progenitors are mitotically active after inoculation would be expected, because infection requires a mitotic target cell. If tissues other than ocular tissues were examined, we would similarly see evidence of virus spread to cells whose progenitors would be mitotically active during the period of virus spread. In addition, the size and number of "clones" may also appear to be too large for true "clonal" events if helper virus were present. This interpretation of course relies on some knowledge of the area under study.

Determination of Sibling Relationships

When performing lineage analysis, it is critical to define cells unambiguously as descendants of the same progenitor. This can be relatively straightforward when sibling cells remain rather tightly, and reproducibly, grouped. An example of such a straightforward case is the rodent retina, where the descendants of a single progenitor migrate to form a coherent radial array.[31,32] The two analyses described below were applied to the rodent retina, and are applicable in any system where clones are arranged simply and reproducibly. The first assay is to perform a standard virological titration in which a particular viral inoculum is serially diluted and applied to tissue. In the retina, the number of radial arrays, their average size, and their cellular composition were analyzed in a series of animals infected with dilutions that covered a 3-log range. The number of arrays was found to be linearly related to the inoculum size, while the size and composition were unchanged. Such results indicate that the working definition of a clone, in this case a radial array, fulfilled the statistical criteria expected of a single-hit event.

The second assay is to perform a mixed infection with two different retroviruses in which the histochemical reporter genes are distinctive. Two such viruses might encode cytoplasmically localized versus nuclear-localized β-galactosidase (β-Gal). This can work when the cytoplasmically localized

[32] D. L. Turner and C. L. Cepko, *Nature* (*London*) **328**, 131 (1987).

β-Gal is easily distinguished from the nuclear-localized β-Gal.[19,33] We have found that this is not the case in rodent nervous system cells, as the cytoplasmically localized β-Gal quite often is restricted to neuronal cell bodies and is therefore difficult to distinguish from nuclear-localized β-Gal. To overcome this problem, we created the aforementioned DAP virus,[20] which is distinctive from the LacZ-encoding BAG virus. A stock containing BAG and DAP was produced by $\psi2$ producer cells grown on the same dish. The resulting supernatant was concentrated and used to infect rodent retina. The tissue was then analyzed histochemically for the presence of blue (due to BAG infection) and purple (due to DAP infection) radial arrays. If radial arrays were truly clonal, then each should be only one color. Analysis of approximately 1100 arrays indicated that most were clonal. However, five comprised both blue and purple cells. This value will be an underestimate of the true frequency of incorrect assignment of clonal boundaries as sometimes two BAG or two DAP virions will infect adjacent cells and thus not lead to formation of bicolored arrays. A closer approximation of the true frequency can be obtained by using the following formula (for derivation, see Fields-Berry et al.[20]):

$$\% \text{ errors} = \frac{(\text{No. of bicolored arrays}) \left[\frac{(a + b)^2}{2ab} \right]}{\text{No. of total arrays}}$$

where a and b are the relative titers comprising the virus stock. The relative titer of BAG and DAP used in the coinfection was $3:1$ and thus the value for percent errors in clonal assignments was 1.2%.

The value of 1.2% for errors in assignment of clonal boundaries includes errors due to both aggregation and independent virions (e.g., perhaps due to helper virus-mediated spread) infecting adjacent progenitors. The percent errors in other areas of an animal will depend on the particular circumstances of the injection site, and on the multiplicity of infection (MOI, the ratio of infectious virions to target cells). Most of the time the MOI will be quite low (e.g., in the retina it was approximately 0.01 at the highest concentration of virus injected). Concerning the injection site, injection into a lumen, such as the lateral ventricles, should not promote aggregation nor high local MOI, but injection into solid tissue, in which the majority of the inoculum has access to a limited number of cells at the inoculation site, could present problems. By coinjecting BAG and DAP, it is possible to monitor the frequency of these events and thus determine if clonal analysis is feasible.

[33] S. M. Hughes and H. M. Blau, Nature (London) 345, 350 (1990).

An error rate as small as 1.2% does not affect the interpretation of "clones" that are frequently found in a large data set. However, as with any experimental procedure that relies in some way on statistical analysis, rare associations of cell types must be interpreted with some caution and conclusions cannot be drawn independently of other data.

The preceding analysis was performed with viruses that were produced on the same dish and concentrated together. This was done as we felt that the most likely way that two adjacent progenitors might become infected would be through small aggregates of virions. Aggregation most likely occurs during the concentration step, as macroscopic aggregates are often seen after resuspending pellets of virions. Thus, when the two-marker approach is used to analyze clonal relationships, it is best to coconcentrate the two vectors in order for the assay to be sensitive to aggregation, due to this aspect of the procedure. [Although aggregation of virions may frequently occur during concentration, it apparently does not frequently lead to problems in lineage analysis, presumably due to the high ratio of noninfectious particles to infectious particles found in most retrovirus stocks. It is estimated that only 0.1–1.0% of the particles will generate a successful infection. Moreover, most aggregates are probably not efficient as infectious units; it must be difficult for the rare infectious particle(s) within such a clump to gain access to the viral receptors on a target cell.]

Methods for Determining Lineage Analysis

To determine the ratio of two genomes present in a mixed virus stock (e.g., BAG plus DAP), several methods can be used. The first two methods are performed *in vitro,* and are simply an extension of a titration assay. Any virus stock is normally titered on NIH 3T3 cells to determine the amount of virus to inject. The infected NIH 3T3 cells are then either selected for the expression of a selectable marker when the virus encodes such a gene (e.g., *neo* in BAG and DAP), or are stained directly, histochemically, for β-Gal or PLAP activity without prior selection in drugs. If no selection is used, the relative ratio of the two markers can be scored directly by evaluating the number of clones of each color on a dish. Alternatively, selected G418-resistant colonies can be stained histochemically for both enzyme activities and the relative ratio of blue versus purple G418-resistant colonies computed. A third method of evaluating the ratio of the two genomes is to use the values observed from *in vivo* infections. After animals are infected and processed for both histochemical stains, the ratio of the two genomes can be compared by counting the number of clones, or infected cells, of each color. When all the preceding methods were applied to lineage analysis in mouse retina[20] and rat striatum,[34] the value obtained for the ratio of G418-resistant colonies scored histochemically was almost identical to the ratio observed *in vivo*. Directly scoring histochemically stained, non-G418-selected NIH 3T3 cells led to an underestimate of the number of BAG-infected colonies, presumably as such cells often are only faint blue, while DAP-infected cells are usually an intense purple. *In vivo,* this is not generally the case as BAG-infected cells are usually deep blue.

The method of injecting two distinctive viruses is a straightforward and feasible method of assessing clonal boundaries when they are fairly easy to define. This method does not require circumstances in which there is a wide range of dilutions that can be injected to give countable numbers of events, as is required for a reliable dilution analysis. Moreover, it does not rely as critically on controlling the exact volume of injection, as is required in the dilution analysis. The use of a small number of vectors that encode distinctive histochemical products for definition of sibling relationships is appropriate only when there is little migration of sibling cells. In these cases, the arrangement of cells that will be used to identify clonal relationships can be defined, and then this definition can be tested as described above. The fit of the definition with true clonal relationships will be revealed by the percentage of defined "clones" that are of more than one color. When the error is too great, reevaluate the criteria, make a new definition, and again test it by looking for "clones" that are more than one color. Through trial and error, an accurate definition of sibling relations should be possible when migration is not too great.

When clonal relationships with a few distinctive viruses cannot be accurately defined, a much greater number of vectors must be used. A library of retroviral vectors can be employed, each member of which is tagged with a unique small insert of an irrelevant DNA. Each vector is scored by the polymerase chain reaction (PCR). The library/PCR method is tedious but extremely worthwhile when dealing with problematic areas.

Regardless of which method is used to score sibling relationships, one further recommendation to aid in the assignments is to choose an injection site that will allow the inoculum to spread. If injection is done into a packed tissue, the viral inoculum will most likely infect cells within the injection tract and it will be difficult to sort our sibling relationships. For example, a lumen, such as the neural tube, provides an ideal site for injection. Regardless of site, perform the injection such that the virus has clear access to the target population; the virus will bind to cells at the injection site and will not gain access to cells that are not directly adjoining that site.

The procedures described below are those that we have used for infec-

---

[34] A. L. Halliday and C. L. Cepko, *Neuron* **9,** 15 (1992).

tion of rodents and chick embryos, histochemical processing of tissue for β-Gal and PLAP, and preparation and use of a library for the PCR method of defining clonal relationships.

Infection of Rodents

Injection of Virus in Utero in Rodents

The following protocols may be used with rats or mice. Note that clean, but not aseptic, technique is used throughout. We routinely soak instruments in 70% (v/v) ethanol before operations, use the sterile materials noted, and include penicillin–streptomycin (final concentration of 100 units/ml each) in the lavage solution. We have not had difficulty with infection using these techniques.

Materials

Ketamine-HCl injection (ketamine, 100 mg/ml)
Xylazine injection (20 mg/ml)
Animal support platform
Depilatory scalpel and disposable sterile blades
Cotton swabs and balls
Sterile tissue retractors
Tissue scissors
Lactated Ringer solution (LR) containing penicillin–streptomycin (Pen/Strep)
Fiberoptic light source
Virus stock
Automated microinjector
Micropipettes (1–5 μl)
Dexon suture (3-0)
Tissue stapler

1. Mix ketamine and xylazine 1:1 in a 1-ml syringe with a 27-gauge needle; lift the animal's tail and hindquarters with one hand and with the other inject 0.05 ml (mice) or 0.18 ml (rats) of anesthetic mixture intraperitoneally.

One or more additional doses of ketamine alone (0.05 ml for mice and 0.10 ml for rats) is usually required to induce or maintain anesthesia, particularly if the procedure takes more than 1 hr. Respiratory arrest and spontaneous abortion appear to occur more often if a larger dose of the mixture is given initially, or any additional doses of xylazine are given.

2. Remove hair over the entire abdomen, using a depilatory agent (any commercially available formulation, such as Nair, works well); shaving of remaining hair with a razor may be necessary. Wash the skin several times with water, then with 70% (v/v) EtOH, and allow to dry.

3. Place the animal on its back in the support apparatus.

For this purpose, we find that a slab of styrofoam with two additional slabs glued on top to create a trough works well for this purpose. With the trough appropriately narrow, no additional restraint is then needed to hold the anesthetized animal.

4. Make a midline incision in the skin from the xyphoid process to the pubis, using the scalpel, and retract; attaching retractors firmly to the styrofoam support will create a stable working field. Stop any bleeding with cotton swabs before carefully retracting fascia and peritoneum, and incising them in the midline with scissors (care is required here not to incise underlying bowel). Continue incision cephalad along the midline of the fascia (where there are few blood vessels) to expose the entire abdominal contents. If necessary to expose the uterus, gently pack the abdomen with cotton balls or swabs to remove the intestines from the opertive field, being careful not to lacerate or obstruct the bowel. Fill the peritoneal cavity with LR, and lavage until clear if the solution turns at all turbid.

Wide exposure is important to allow the later manipulations. During the remainder of the operation, keep the peritoneal cavity moist and free of blood; dehydration or blood around the uterus increases the rate of postoperative abortion.

5. Elevate the embryos one at a time out of the peritoneal cavity, and transilluminate with a fiberoptic light source to visualize the structure to be injected. For lateral cerebral ventricular injections, for example, the cerebral venous sinuses serve as landmarks.

When deciding on a structure to inject, keep in mind that free diffusion of virus solution throughout a fluid-filled structure lined with mitotic cells is best for ensuring even distribution of viral infection events throughout the tissue being labeled. The neural tube is an example of such a structure; when virus is injected into one lateral ventricle, it is observed to diffuse quickly throughout the entire ventricular system.

6. Using a heat-drawn glass micropipette attached to an automatic microinjector, penetrate the uterine wall, extraembryonic membranes, and the structure to be infected in one rapid thrust; this minimizes trauma and improves survival. Once the pipette is in place, inject the desired volume of virus solution, usually 0.1–1.0 μl. Coinjection of a dye such as 0.005% (w/v) trypan blue or 0.025% (w/v) fast green helps determination of the accuracy of injection and does not appear to impair viral infectivity; coinjection of the polycation Polybrene (80 μg/ml) aids in viral attachment to the

cells to be infected. The type of instrument used to deliver the virus depends on the age of the animal and the tissue to be injected. At early embryonic stages, the small size and easy penetrability of the tissue make a pneumatic microinjector [such as the Eppendorf (Westbury, NY) 5242] best for delivering a constant amount of virus at a controlled rate with a minimum of trauma. Glass micropipettes should be made empirically to produce a bore size that will allow penetration of the uterine wall and the tissue to be infected. At later ages (late embryonic and postnatal), a Hamilton syringe with a 33-gauge needle works best.

When injecting through the uterine wall, all embryos may potentially be injected except those most proximal to the cervix on each side. (injection of these greatly increases the rate of postoperative abortion). In practice, it is often not advisable to inject all possible embryos, if excessive uterine manipulation would be required. At the earliest stages at which this technique is feasable (E12 in the mouse or E13 in the rat), virtually any uterine manipulation may cause abortion, so any embryo that cannot be reached easily should not be injected.

7. Once all animals have been injected, lavage the peritoneal cavity until it is clear of all blood and clots, ensure that all cotton balls and swabs have been removed, and move retractors from the abdominal wall/fascia to the skin.

Filling the peritoneal cavity with LR with Pen/Strep before closing increases survival significantly, probably by preventing maternal dehydration during recovery from anesthesia as well as preventing infection.

8. Using 3-0 Dexon or silk suture material on a curved needle, suture the peritoneum, abdominal musculature, and fascia from each side together, using a continuous locking stitch. After closing the fascia, again lavage with LR–Pen/Strep.

9. Close the skin with surgical staples [such as the Clay-Adams (Parsippany, NJ) Autoclip] placed 0.5 cm apart. Sutures may also be used, but these require much more time (often necessitating further anesthesia, which increases abortion risk) and are frequently chewed off by the animal, resulting in evisceration.

10. Place the animal on its back in the cage and allow anesthesia to wear off. Ideally, the animal will wake up within 1 hr of the end of the operation. Increasing time to awakening results in increasing abortion frequency. Food and water on the floor of the cage should be provided for the immediate postoperative period.

11. Mothers may be allowed to deliver progeny vaginally, or they may be harvested by caesarean section. Maternal and fetal survival are approximately 60% at early embryonic ages of injection, and increase with gestational age to virtually 100% after postnatal injections.

Injection of Virus into Mice by exo Utero Surgery

Injections into small or delicate structures (such as the eye) require micropipettes, which are too fine to penetrate the uterine wall. In addition, it is impossible to precisely target many structures through the rather opaque uterine wall. These problems can be circumvented, however, with a considerable increase in technical difficulty and decrease in survival, by use of the *exo utero* technique.[35] The procedure is similar to that detailed above, with the following modifications to free the embryos from the uterine cavity.

1. The technique works well in our hands only with outbred, virus-antigen free CD-1 and Swiss-Webster mice, but even these strains may have different embryo survival rates when obtained from different suppliers or different colonies of the same supplier.

This variability presumably results from subclinical infections that may render some animals unable to survive the stress of the operation. We have had no success with this technique in rats.

2. After the uterus is exposed and before filling the peritoneum with LR, incise the uterus longitudinally along its ventral aspect with sharp microscissors. The uterine muscle will contract away from the embryos, causing them to be fully exposed, surrounded by their extraembryonic membranes.

3. Only two embryos in each uterine horn can be safely injected, apparently because of trauma induced by neighboring embryos touching each other. Using a dry sterile cotton swab, scoop out all but two embryos, each with its placenta and extraembryonic membranes, and press firmly against the uterus where the placenta was removed for 30–40 sec to achieve hemostasis.

It is important to stop all bleeding before proceeding. From this point on, the embryos must be handled extremely gently, as only the placenta is tethering an embryo to the uterus, and it tears easily.

4. Fill the peritoneal cavity with LR, and cushion each embryo to be injected with sterile cotton swabs soaked in LR. Keeping the embryos submerged throughout the remainder of the procedure is essential to their survival.

5. The injection should then be done with a pneumatic microinjector and heat-pulled glass micropipette. This may usually be done by puncturing the extraembryonic membranes first and then the structure to be injected; for some delicate injections it may be necessary to make an incision in the

---

[35] K. Muneoka, N. Wanek, and S. V. Bryant, *J. Exp. Zool.* **239**, 289 (1986).

extraembryonic membranes, which is then closed with 10-0 nylon suture after the injection.

Infection of Chick Embryos

The following description is an example of an infection protocol used for the chick neural tube. More details of infection protocols for chick embryos can be found in Morgan and Fekete.[36]

1. Fertilized virus-free White Leghorn chicken eggs are obtained from SPAFAS (Norwich, CT) and kept at 4° for 1 week or less, until they are transferred to a high-humidity, rocking incubator (Petersime, Gettysburg, OH) at 38°, which is designated time 0. Line O eggs are the most desirable hosts as they do not encode any endogenous retroviruses that could lead to helper virus generation.[37]

2. To prevent the embryo from sticking to the shell, it is useful to lower the embryo by removing albumin at an early stage. To accomplish this, set the eggs on their sides and rinse with 70% (v/v) ethanol. Poke a hole in both ends of the egg as it lies on its side, using a sharp pair of forceps, scissors, or needle. Use a 5-ml syringe and 21-gauge needle and withdraw 1.5 ml of albumin from the pointed end of the egg. Angle the needle so as not to disrupt the yolk by pointing it down and by not putting it too deep into the egg. Cover both holes with clear tape and return it to the incubator. Alternatively, locate and stage the embryo by cutting a hole on the top (side facing up as it lies on its side) with curved scissors. Remove the shell to make a hole 1/2 to 1 inch in diameter. Locate the embryo and then enlarge the hole to allow easy access to the embryo. If the embryo is to be used later, cover the hole with clear tape and return it to the incubator.

3. After approximately 18 to 42 hr of incubation (Hamburger and Hamilton stage 10–17[38] embryos are injected with 0.1 to 1.0 µl of viral inoculum including 0.25% (w/v) fast green dye.[39] The inoculum is delivered by injection directly into the ventricular system, which is easily accessed at these early times. For delivery, the inoculum is loaded into a glass micropipette, made as described above for rodent injections. A micromanipulator and Stoelting (Wood Dale, IL) microsyringe pump is convenient and delivers the maximum volume in about 1 min. After injection, cover the hole with clear tape and return the egg to the incubator.

[36] B. A. Morgan and D. M. Fekete, *Methods Cell Biol.* **51,** 185 (1996).
[37] S. M. Astrin, E. G. Buss, and W. S. Haywards, *Nature (London)* **282,** 339 (1979).
[38] V. Hamburger and H. L. Hamilton, *J. Exp. Morphol.* **88,** 49 (1951).
[39] D. M. Fekete and C. L. Cepko, *Mol. Cell. Biol.* **13,** 2604 (1993).

Infected chick or rodent embryos can be incubated to any desired point. Chicks can be allowed to hatch and rodents can be delivered by caesarean section and reared to maturity. Embryonic brains are dissected in phosphate-buffered saline (PBS) followed by overnight fixation in 4% (w/v) paraformaldehyde in PBS (pH 7.4) at 4°. Posthatch or postnatal animals are perfused with the same fixative and are incubated overnight in the fixative. They are then washed overnight in three changes of PBS and cryoprotected in 30% (w/v) sucrose. After cryoprotection, the brains are embedded in O.C.T. medium and cut on a Reichart-Jung 3000 cryostat at 60–90 μm. Sections are histochemically stained for the appropriate marker and are mounted with Gelvatol. For details of the histochemical reaction for LacZ or PLAP, see Cepko et al.[40]).

Solutions

PBS (10×)

NaCl	80 g
KCl	2 g
Na_2HPO_4	11.5 g
KH_2PO_4	2 g
H_2O	To 1 liter

Glutaraldehyde fixative (0.5%, v/v): 25% (v/v) stock (Sigma, St. Louis, MO) can be stored at $-20°$ and frozen–thawed many times. Make dilution immediately before use by diluting 1:50 into 1× PBS

Paraformaldehyde fixative (4%, w/v): Combine 4 g of solid paraformaldehyde, 2 mM $MgCl_2$, and 1.25 mM EGTA (0.25 ml of a 0.5 M EGTA stock, pH 8.0) in 100 ml of PBS, pH 7.2–7.4. Heat H_2O to 60°. Add the paraformaldehyde. Add NaOH to dissolve the paraformaldehyde. Cool to room temperature, add a 1:10 volume of 10× PBS, and adjust the pH with HCl. Can be stored at 4° for several weeks

Preparation and Use of Retroviral Library for Lineage Analysis Using Polymerase Chain Reaction and Sequencing

We have developed a direct approach to address lumping and splitting errors[41] by constructing a library of viruses that is analyzed by PCR. In our first libraries, each virus of the library carried 1 member from a pool of approximately 100 DNA fragments from *Arabidopsis thalliana* DNA,

[40] D. L. Spector, R. D. Goldman, and L. A. Leinwand, "Cells: A Laboratory Manual." Cold Spring Harbor Laboratory Press, Cold Spring Harbor, New York, 1977.
[41] C. Walsh and C. L. Cepko, *Science* **255,** 434 (1992).

in addition to the *lacZ* or PLAP gene. Infected cells, recognized by their enzyme activity, were mapped and the positive cells cut from cryosections. The *A. thalliana* DNA was amplified by PCR and characterized by size and restriction enzyme digestion patterns. If the size and restriction digestion pattern of the PCR product from two or more cells were the same, they were considered siblings with a probability calculated on the basis of the number of infections in that brain and the complexity of the library.[41,42] Lineage analysis using such libraries revealed novel lineal relationships in the rat cerebral cortex[41,43,44] and chick diencephalon.[45] However, the limited number of unique members in the library made from *A. thalliana* DNA restrained the analysis to tissues with low infection rates.

More data could be acquired with each experiment and additional questions could be addressed in the central nervous system and other tissues with a more complex library containing a greater number of DNA tags. We therefore have constructed several retroviral vectors, of which CHAPOL [chick alkaline phosphatase with oligonucleotide (oligo) library] is the prototype,[46] that include degenerate oligonucleotides with a theoretical complexity of 1.7×10^7. Studies in the developing nervous system of the chick have been successfully completed with CHAPOL.[47,48]

A summary of the production of CHAPOL and BOLAP (an oligo library in a murine vector) is given here; a detailed description of the construction of CHAPOL can be found elsewhere.[46] For either avian or murine retroviruses, the overall strategy is the same. A population of double-stranded DNA molecules that includes a short degenerate region, $[(G \text{ or } C)(A \text{ or } T)]_{12}$, is generated by PCR amplification of a chemically constructed singel-stranded oligonucleotide population of the same sequence. The oligo preparation is ligated into a retrovirus vector and a preparation of highly competent *Escherichia coli* is transformed. The library is then grown as a pool and a preparation of plasmids from the pool is made. The DNA of the pool is transfected into an avian or mammalian packaging cell line to produce a library of virus particles. The library is injected into an area to be mapped. Infected cells are detected histochemically and each infected cell is recovered for PCR amplification. Each PCR product is then sequenced. Two cells with the same sequence are considered siblings, again

[42] C. Walsh, C. L. Cepko, E. F. Ryder, G. M. Church, and C. Tabin, *Science* **258,** 317 (1992).
[43] C. Walsh and C. L. Cepko, *Nature (London)* **362,** 632 (1993).
[44] C. B. Reid, I. Liang, and C. Walsh, *Neuron* **15,** 299 (1995).
[45] S. A. Arnold-Aldea and C. L. Cepko, *Dev. Biol.* **173,** 148 (1996).
[46] J. A. Golden, S. C. Fields-Berry, and C. L. Cepko, *Proc. Natl. Acad. Sci. U.S.A.* **92,** 5704 (1995).
[47] J. A. Golden and C. L. Cepko, *Development* **122,** 65 (1996).
[48] F. G. Szele and C. L. Cepko, *Curr. Biol.* **6,** 1685 (1996).

with a probability derived from an analysis of the frequency of recovery of each genome (see Walsh et al.[42])

Preparation of CHAPOL

The avian replication-incompetent virus CHAP,[49,50] encoding the human placental alkaline phosphatase (PLAP) gene, is modified to accept the oligo inserts. CHAP is linearized, purified, and mixed with the degenerate oligonucleotides in the presence of ligase and aliquots of the resulting ligation products are used to transform E. coli DH5α. After transformation, all aliquots are pooled. One hundred microliters of the pool is plated at varying dilutions on plates containing ampicillin. The remainder of the pool is divided and added to eight 2-liter flasks containing 1 liter of LB medium with ampicillin (50 μ/ml). The cultures are shaken overnight at 37°. Plasmid DNA is extracted from these cultures by the Triton lysis procedure and purified on CsCl gradients.[51]

CHAPOL DNA is transfected into the avian virus packaging line Q2bn,[14] and the transiently produced virus is collected and concentrated. Aliquots of CaPO$_4$ precipitates of 100 μg of CHAPOL DNA are made in 10 ml of HBS. The precipitate in each aliquot is then distributed equally on ten 10-cm plates of Q2bn and glycerol shock is carried out for 90 sec at room temperature 4 hr later. At 24 hr after glycerol shock, the supernatants are collected and pooled. This is repeated at 48 hr. The supernatants from the 24- and 48-hr harvests are pooled and the titer calculated by infection of QT6 cells and assay of the PLAP activity as described (Cepko and Pear in Ausubel et al.[4]). The stock is filtered through a 0.45-μm pore size filter and concentrated by centrifugation in an SW27 rotor at 4°, 20,000 rpm, for 2 hr. The concentrated stock is titered on QT6 and tested for helper virus, which proves negative. The titer of CHAPOL is determined to be 1.1×10^7 CFU/ml. The same stock has been used for all experiments conducted over a 5-year period and many aliquots remain. We recommend making large stocks and storing them as small aliquots.

Use of CHAPOL

CHAPOL is used to infect the developing brain of chick embryos, using the procedures outlined above. At various times later, the tissue is harvested

[49] S. C. Fields-Berry, in "Genetics," p. 141. Harvard University Press, Cambridge, Massachusetts, 1990.
[50] E. F. Ryder and C. L. Cepko, Neuron 12, 1011 (1994).
[51] J. Sambrook, E. F. Fritsch, and T. Maniatis, "Molecular Cloning: A Laboratory Manual." Cold Spring Harbor Laboratory Press, Cold Spring Harbor, New York, 1989.

and stained for AP activity. The outline of each section is drawn by camera lucida and the location and type of cells labeled on each section are recorded. A single cell or cluster of cells with a small group of surrounding cells are removed with a heat-pulled glass micropipette (Fig. 2) and transferred to a 96-well PCR plate for proteinase K digestion (as described below). After digestion, nested PCR is performed (as described below). The product of each PCR is run on a 1.5% (w/v) agarose gel to determine if a product of the appropriate size has been amplified. The recovery of a PCR product of the proper size occurs from PCR of 30–85% of the picks (the frequency varies depending on the batch of PCRs and the tissue being studied), using CHAPOL. Sequencing of the oligonucleotide insert (as described below) is performed on all reactions that give the expected product on agarose gel analysis (e.g., Fig. 3) and is successful approximately 75% of the time. All sequences are stored in the software program GCG (1991). All common sequences are pulled from the database created in GCG and the corresponding cells labeled. Sections are then aligned to determine the three-dimensional boundaries of clonal expansion. Each type of cell (e.g., neuron, glia) is also recorded to determine the variety of cells that can arise from a single progenitor.

The value of using a complex library of vectors is illustrated by a view of more heavily infected brains (Fig. 4; see color insert), where closely aggregated AP⁺ cells would have been lumped into a single clone on the

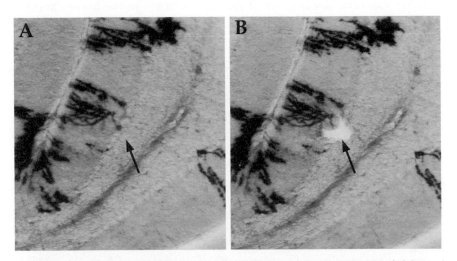

FIG. 2. Picking of AP⁺ cells from tissue sections after infection with CHAPOL. (A) Several AP⁺ cells are present in a chick cerebellar section, including a Purkinje cell (arrow) and many glial cells. (B) After removal of the Purkinje cell with a glass micropipette.

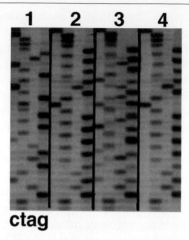

ctag

FIG. 3. The sequence reactions from the PCR products of four representative samples (samples 1–4) are shown. Sequencing of PCR products 1 and 2 each revealed the presence of a unique sequence. Sequencing of PCR products 2 and 4 yielded the same sequence. Sequencing of sample 3 showed the presence of more than one species.

basis of proximity. In addition, even lightly infected brains can give rise to lumping errors (also see Walsh and Cepko[41]).

Two issues are important for determining the value of this type of library of DNA markers: the number of unique members in the library and the distribution of the library members.[42] If only two members exist in the library, for example, then there is a one-in-two chance that the same tag will be selected in two consecutive picks (if they are present in equal concentrations). If 100 members exist in the library at equal concentrations, the chance that 2 picks come up with the same member is reduced to 10^{-2}. The second important variable determining the quality of the library is the distribution of the members within the library. This can be illustrated as follows. Consider a library composed of 10^6 members, with 50% of the library composed of 1 member. If two neighboring or distant cells are found to carry the overrepresented insert, the probability that the two cells arose from separate clones is still 0.5 CHAPOL was found to have an equal distribution in that each of the inserts picked to date ($n > 500$) has occurred independently only once. One further issue to consider is the level of difficulty in using the library. We have found in practice that this method of tag identification is in fact easier than our previous method, based on the analysis of the size and restriction digestion pattern of *A. thalliana* DNA.

These libraries should be useful for application in a wide range of tissues and species. The host range has been expanded to previously uninfectable hosts,[11] and infection of nonneural tissue with CHAPOL has been observed, as one would expect on the basis of experience with avian and murine retroviruses.

Polymerase Chain Reaction and Sequencing from CHAPOL

Proteinase K Digestion. The coverslips are removed from the slides by immersion in sterile H_2O. Single cells or small clusters of cells containing purple NBT precipitate with surrounding unlabeled tissue (approximately 0.5- to 2-mm tissue fragments) are scraped from the slide with a heat-pulled glass micropipette (Fig. 1). The cells are transferred to a 96-well PCR (Hybaid, Teddington, UK) plate with 10 μl of a proteinase K solution [50 mM KCl, 10 mM Tris-HCl (pH 7.5), 2.5 mM MgCl, 0.02% (v/v) Tween 20, proteinase K (200 μg/ml)]. Each well is overlaid with 1 drop of light mineral oil (Sigma) and the plates are heated to 60° for 2 hr, to 85° for 20 min, and to 95° for 10 min in a Hybaid OmniGene thermocycler.

Nested PCR. The first PCR is accomplished by adding 0.15 μl of *Taq* polymerase (Boehringer Mannheim, Indianapolis, IN), 0.15 μl of dNTP mix (Boehringer Mannheim), 0.75 μl each of 10 μM oligonucleotide 0 (5'-TGTGGCTGCCTGCACCCCAGGAAAG-3'), and 10 μM oligonucleotide 5 (5'-GTGTGCTGTCGAGCCGCCTTCAATG-3'), 2 μl of PCR buffer with $MgCl_2$ (Boehringer Mannheim), and 16.2 μl of H_2O to each well of the 10-μl proteinase K solution (final volume, 30 μl). This is cycled at 93° for 2.5 min; [(94° for 45 sec) (72° for 2 min)] for 40 cycles; 72° for 5 min.

The second PCR is performed with 1 μl of reaction product from the first PCR added to 0.25 μl of *Taq* polymerase (Boehringer Mannheim), 0.25 μl of dNTP mix (Boehringer Mannheim), 1 μl each of 10 μM oligonucleotide 2 (5'-GCCACCACCTACAGCCCAGTGG-3') and 10 μM oligonucleotide 3 (5'-GAGAGAGTGCCGCGGTAATGGG-3'), 2 μl of PCR buffer with $MgCl_2$ (Boehringer Mannheim), and 14.5 μl of H_2O (final volume, 30 μl). The reaction is thermocycled at 93° for 2.5 min; [(94° for 45 sec) (70° for 2 min)] for 30 cycles; 72° for 5 min. An aliquot of DNA is run on a 1.5% (w/v) agarose gel [0.75% (w/v) Seakem, 0.75% (w/v) NuSeive; FMC Bioproducts, Rockland, ME] to ensure that the appropriate insert was present (Fig. 2).

Sequencing. Sequencing is performed using the CyclistTM Exo-*Pfu* DNA sequencing kit from Stratagene (La Jolla, CA). Briefly, 5 μl of each d/ddNTP mix is added to 4 wells on a 96-well Hybaid plate. To each of these wells 25% of the following mixture is added: 1 μl of the nested PCR product, 1 μl of 10 μM oligo 3, 3 μl of 10× sequencing buffer, 1 μl of Exo-*Pfu*, 0.75 μl of ^{35}S (10 μCi), 4 μl of DMSO, and 11.25 μl of H_2O. This is cycled at 95° for 5 min; [(95° for 30 sec) (60° for 30 sec) (72° for 1 min)] for 30 cycles. Sequencing reactions are analyzed on a 6% (w/v) acrylamide denaturing gel (Fig. 3).

Note. Reagents, instruments, and glass microscope slides should be handled with scrupulous technique, and UV irradiated when needed to destroy contaminating DNA.

Creation of BOLAP

The BOLAP library was created in a murine retrovirus vector, pBABE,[8] into which the P-ALP1 gene was inserted to create pBABE-AP (S. Fields-Berry and C. L. Cepko, unpublished, 1999; see Fig. 5). BABE-AP-X was generated by inserting PCR-amplified DPL2 into the *Asc*I and *Bgl*II site of BABE-AP. After the ligation, BABE-AP-X was digested with *Asc*I and *Xho*I, phenol–chloroform extracted, and ethanol precipitated. The details are supplied here as an example of how to make such a library. BOLAP is available on request.

DPL1 is prepared by PCR amplification of the following reaction mix: 1 μl of 1:100 diluted DPL (0.6 mg/ml), 1 μl of DPLP (0.44 mg/ml), 1 μl of DPLP5 (0.2 mg/ml), 5 μl of 2.5 mM dNTP mix, 10 μl of 10× PCR buffer (Boehringer Mannheim), 79 μl of water, and 2 μl of *Taq* DNA polymerase (Boehringer Mannheim).

The amplification program is 93° for 2.5 min; (94° for 45 sec, 70° for 2 min) for 30 cycles; and 72° for 5 min.

The product is phenol-chloroform extracted, ethanol precipitated, digested with *Asc*I and *Xho*I, gel purified, and ethanol precipitated. Approximately 3 μg of *Asc*I/*Xho*I-digested BABE-AP-X is ligated to 25 ng of

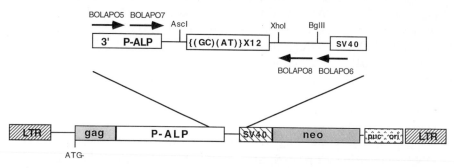

Fig. 5. BOLAP, a murine retroviral preparation encoding an oligonucleotide library. A degenerate oligonucleotide pool (DPL1; see text) was inserted into the BABE-Neo plasmid encoding P-ALP1. The positions and orientations of the oligonucleotides BOLAPO5-8 used to amplify each insert and for sequencing are indicated by the arrows. LTR, long terminal repeat; gag, amino terminus of Mo-MuLV *gag* gene; SV40, simian virus 40 early promoter; *neo,* TnS neomycin resistance gene; puc ori, bacterial origin of replication.

*Asc*I–*Xho*I-digested DPL1 in a 100-μl ligation reaction at 16° overnight. The product is phenol–chloroform extracted, ethanol precipitated, and resuspended in 50 μl of Tris–EDTA (TE). Twenty microliters of ligation product is used to transform ElectroMax DH1OB (GIBCO/BRL, Gaithersburg, MD) competent cells according to the manufacturer instructions. The transformed cells are pooled. One microliter of the pooled cells is serially diluted and plated on Luria–Bertani/ampicillin (LB/amp) plates. The remainder of the pool is split into five 1-liter cultures of TB containing ampicillin and kanamycin sulfate (each at 50 μg/ml) and shaken overnight at 37°. BOLAP plasmid (3.7 mg) is extracted from the overnight cultures by Triton lysis, followed by purification via CsCl gradient. In the case of our preparation of BOLAP, the number of colonies on the LB/amp plates projected a total number of transformants to be 1.28×10^7. A control ligation containing no DPL1 projected the background of vector without DPL1 insert to be 0.9%.

BOLAP Virus Production and Evaluation

Eleven confluent 10-cm dishes of Bosc23 cells[9] are split into fifty 10-cm dishes. The next morning, 350 μg of BOLAP DNA is combined with 1.5 ml of LipofectAMINE (GIBCO-BRL) in 25 ml of Opti-MEM (GIBCO-BRL). The mixture is incubated for 20 min at room temperature and then added to 200 ml of Dulbecco's modified Eagle's medium (DMEM). The 50 plates are washed with DMEM and 5 ml of the DNA–LipofectAMINE–DMEM mixture is added to each plate. The plates are incubated for 5 hr at 37°. Five milliliters of 20% fetal bovine serum in DME is then added to each plate and the plates returned to the incubator overnight. The next morning the supernatant is harvested and replaced with 5 ml of 10% (v/v) fetal calf serum (FCS) in DMEM. The next morning, this supernatant is harvested. The supernatants are pooled, filtered through a 0.45-μm pore size filter unit, and concentrated by centrifugation at 20,000 rpm for 2 hr at 4°. The final concentration of the viral supernatant is 1×10^8 CFU/ml.

To test the complexity of the library, the protocols of Golden *et al.*[46] are followed. One microliter of viral supernatant is diluted into 30 ml of DMEM with 10% (v/v) calf serum. One to 2 μl of this diluted viral stock is used to infect a 30–50% confluent 6-cm dish of NIH 3T3 cells. Five to 6 hr later, the infected cells are trypsinized, diluted 10-fold, and plated on 96-well dishes. Three days later, the plates are washed with phosphate-buffered saline (PBS), fixed with 4% (w/v) paraformaldehyde for 10 min, washed three times with PBS for 5 min, heated to 65° for 25 min, and then stained overnight for AP activity with XP and NBT. The following morning, each well is examined for the presence of a single discrete grouping of AP⁺

cells. The cells in chosen wells are washed with PBS. Ten microliters of proteinase K solution [50 mM KCl, 10 mM Tris-HCl (pH 7.5), 2.5 mM MgCl$_2$, 0.02% (v/v) Tween 20] at 400 μg/ml is added and the cells and solution are scraped/suctioned off and placed in individual wells in a 96-well dish. A drop of mineral oil is placed over each well and the plate is heated to 65° for 2 hr, 85° for 20 min, and 95° for 10 min.

A nested PCR is performed as follows. To each well is added 20 μl of the following reaction mix:

PCR buffer (10×) with Mg^{2+} (Boehringer Mannheim)	2 μl
BOLAPO 5 and 6 (0.6 mg/ml)	0.15 μl
dNTP mixture (25 mM) (Boehringer Mannheim)	0.15 μl
Water	17.4 μl
Taq DNA polymerase (Boehringer Mannheim)	0.4 μl

The reactions are cycled as follows: 93° for 2.5 min, (94° for 45 sec, 67° for 2 min, 72° for 2 min) for 33 cycles, and 72° for 5 min.

One microliter of the above reaction is added to 20 μl of the same reaction mix, substituting BOLAPO 7 and 8 for BOLAPO 5 and 6. The amplification program is 93° for 2.5 min, (94° for 45 sec, 72° for 2 min) for 30 cycles, and 72° for 5 min.

Eight microliters of the second reaction mix is added to 2.5 μl of gel loading buffer and then fractionated on a 3% (w/v) NuSieve GTG/1% (w/v) SeaKem ME agarose gel.

Amplifications that yield a DNA product of the appropriate molecular weight (bp) are sequenced with the Exo-Pfu Cyclist kit (Stratagene). One microliter of nested PCR product is added to the following reaction mix:

BOLAPO 7 (0.6 mg/ml)	0.15 μl
Sequencing buffer (10×)	3 μl
[α-^{35}S] dATP	0.75 μl
DMSO	4 μl
Water	12.1 μl
Exo-Pfu DNA polymerase	1 μl

This reaction is mixed and 5-μl portions are added separately to 5 μl of the four dNTP/ddNTP mixtures. The reactions are overlaid with oil and cycled as follows: 95° for 5 min, (95° for 30 sec, 60° for 30 sec, 72° for 1 min) for 30 cycles.

The reactions are terminated by the addition of 5 ml of stop buffer, and then fractionated by 6% (w/v) acrylamide gel electrophoresis.

At this point, 98 individual clones have been sequenced in our laboratory. All carry a unique DPL1 insert. Of those, the degenerate oligo region is shorter by 24 bases in seven clones and longer in one clone. Occasional variations in the (GC)(AT) sequence were also seen.

Sequences of Oligonucleotides

DPL2:   5'-TAGGAGGCGCGCCTTT-[(GC)(AT)]$_{12}$
            GTTCTCGAGGACACCTGACTGGCTGAGGG
            CTTCCGCGACCCGAGATCTCAGCTTCC-3'

DPL1:   5'-TAGGAGGCGCGCCTTT-[(GC)(AT)]$_{12}$
            GTTACGCGTTAATTAACTCGAGATCT
            CAGCTTC-3'

DPLP: 5'-GAAGCTGAGATCTCGAGTTA-3'

DPLP5:   5'-TAGGAGGCGCGCCTTT-3'

BOLAPO5:   5'-CCAGGGACTGCAGGTTGTGCCCTGT-3'

BOLAPO6:   5'-AGACACACATTCCACAGGGTCGAAG-3'

BOLAPO7:   5'-GGCTGCCTGCACCCCAGGAAAGGAG-3'

BOLAPO8:   5'-GGTCTCGGAAGCCCTCAGCCCAGTC-3'

[11] Use of Pseudotyped Retroviruses in Zebrafish as Genetic Tags

By Shawn Burgess and Nancy Hopkins

Introduction

The zebrafish, *Danio rerio,* has become an important model system for studying vertebrate development.[1–3] Its particular strength lies in the relative ease with which large-scale forward genetics can be performed when compared with other vertebrate models commonly used to study development, such as frogs (*Xenopus laevis*), chickens (*Gallus gallus*), and mice (*Mus musculus*). Using chemicals or X-rays as mutagens, several large-scale genetic screens have been carried out on zebrafish.[4–6] Chemicals or X-rays can

[1] J. S. Eisen, *Cell* **87,** 969 (1996).

[2] M. Granato and C. Nusslein-Volhard, *Curr. Opin. Genet. Dev.* **6,** 461 (1996).

[3] C. Nusslein-Volhard, *Science* **266,** 572 (1994).

[4] W. Driever, L. Solnica-Krezel, A. F. Schier, S. C. Neuhauss, J. Malicki, D. L. Stemple, D. Y. Stainier, F. Zwartkruis, S. Abdelilah, Z. Rangini, J. Belak, and C. Boggs, *Development* **123,** 37 (1996).

[5] P. Haffter, M. Granato, M. Brand, M. C. Mullins, M. Hammerschmidt, D. A. Kane, J. Odenthal, F. J. van Eeden, Y. J. Jiang, C. P. Heisenberg, R. N. Kelsh, M. Furutani-Seiki, E. Vogelsang, D. Beuchle, U. Schach, C. Fabian, and C. Nusslein-Volhard, *Development* **123,** 1 (1996).

[6] P. D. Henion, D. W. Raible, C. E. Beattie, K. L. Stoesser, J. A. Weston, and J. S. Eisen, *Dev. Genet.* **18,** 11 (1996).

be effective in generating a high rate of mutation.[7-9] The major drawback, however, of generating mutations in this fashion is that subsequent identification of the affected gene can be difficult. The only methodology currently available for this task is positional cloning, typically a lengthy and expensive process that can often take an individual several years to accomplish for a single mutation.

In other genetic model systems, such as the fruit fly, *Drosophila melanogaster,* using insertions of foreign DNA as a mutagen has complemented the chemical methods of mutagenesis.[10] This method of mutagenesis has the tremendous advantage of providing a DNA "tag" at the mutated locus that allows rapid cloning of the adjacent flanking genomic region. This can reduce the time necessary to identify the mutated gene from years to weeks. An additional benefit of using foreign DNA as a mutagen is that the content of the DNA can be altered in various ways by the researchers, which can improve the flexibility of the technique dramatically. For example, it is possible to incorporate a "gene trap" construct that contains a splice acceptor at the 5' end of a reporter gene. When the DNA inserts into an intron of a gene, the gene's splicing donor can splice into the artificial intron and activate transcription/translation of the reporter gene. These constructs not only can improve the mutagenicity of the inserted DNA, but they can allow for a way to rapidly screen genes expressed in a specific spatiotemporal fashion. We have developed insertional mutagenesis in zebrafish by using pseudotyped retroviral technology, and we have proven the ease of cloning genes mutated in this fashion.[11,12]

Vesicular stomatitis virus (VSV) is a rhabdovirus with a broad range of species that it is capable of infecting. The envelope glycoprotein of VSV (G protein) also has the ability to form "pseudotypes" with core particles of Moloney murine leukemia virus (Mo-MuLV). The resulting virus contains an active Mo-MuLV retroviral core, with the host specificity of VSV. Such pseudotyped viruses were shown to efficiently infect zebrafish cells in culture.[13] A second advantage of using VSV G protein as the viral

[7] G. Streisinger, F. Coale, C. Taggart, C. Walker, and D. J. Grunwald, *Dev. Biol.* **131,** 60 (1989).

[8] D. J. Grunwald and G. Streisinger, *Genet. Res.* **59,** 103 (1992).

[9] L. Solnica-Krezel, A. F. Schier, and W. Driever, *Genetics* **136,** 1401 (1994).

[10] L. Cooley, C. Berg, R. Kelley, D. McKearin, and A. Spradling, *Prog. Nucleic Acid Res. Mol. Biol.* **36,** 99 (1989).

[11] N. Gaiano, A. Amsterdam, K. Kawakami, M. Allende, T. Becker, and N. Hopkins, *Nature (London)* **383,** 829 (1996).

[12] M. L. Allende, A. Amsterdam, T. Becker, K. Kawakami, N. Gaiano, and N. Hopkins, *Genes Dev.* **10,** 3141 (1996).

[13] J. C. Burns, T. Friedmann, W. Driever, M. Burrascano, and J. K. Yee, *Proc. Natl. Acad. Sci. U.S.A.* **90,** 8033 (1993).

envelope is that this protein allows the virus to be easily concentrated 500 to 1000-fold by high-speed centrifugation. This allows the concentration of virus to titers of 3×10^9 CFU or higher from certain packaging cell lines.

Using these superconcentrated viruses, we have shown that it is possible to inject virus into developing zebrafish embryos to generate mosaic "founder" fish, which then transmit on average 10–20 proviral integrations randomly dispersed throughout the subsequent F_1 generation[11,14,15] (K. Townsend, unpublished results, 1999). Hundreds of thousands of inserts are generated and ultimately fish carrying identical insertions are inbred and their progeny are examined for developmental defects that appear in 25% of the embryos (see Fig. 1). Initial results from experiments done in our laboratory show that 1 in 85 insertions leads to a mutation that affects early zebrafish development. Once the mutation has been shown to be caused by the proviral insertion, it typically takes about 1 month to identify the affected gene product although some genes have been identified in as few as 5 days. We are currently developing other retroviral vectors that will contain gene traps or other useful functions.

This chapter focuses on the techniques and technologies necessary to use pseudotyped retroviruses as a mutagen in zebrafish. Broader aspects, such as screening for mutations or cloning the mutated gene, are beyond the scope of this chapter and are also not unique to using pseudotyped retroviruses with zebrafish; therefore other sources can be consulted for these issues.

Materials Needed

Cell Culture

> Dulbecco's modified Eagle's medium (DMEM) with 10% (v/v) fetal bovine serum (FBS) and penicillin–streptomycin (Pen/Strep)
> Tissue culture plates, 15 cm
> Poly-L-lysine, 0.01% (w/v)
> HEPES-buffered saline (2×): 16.4 g of NaCl, 11.9 g of HEPES acid, 0.21 g of Na_2HPO_4, distilled H_2O to 1 liter; titrate to pH 7.05 with 5 M NaOH and filter sterilize
> $CaCl_2$, 2.5 M
> Plasmid pHCMV-G

---

[14] N. Gaiano, M. Allende, A. Amsterdam, K. Kawakami, and N. Hopkins, *Proc. Natl. Acad. Sci. U.S.A.* **93,** 7777 (1996).

[15] S. Lin, N. Gaiano, P. Culp, J. C. Burns, T. Friedmann, J. K. Yee, and N. Hopkins, *Science* **265,** 666 (1994).

Founders

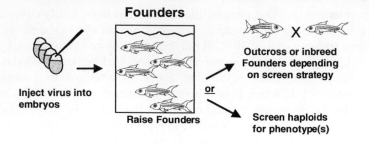

Inject virus into
embryos

Raise Founders

or

Outcross or inbreed
Founders depending
on screen strategy

Screen haploids
for phenotype(s)

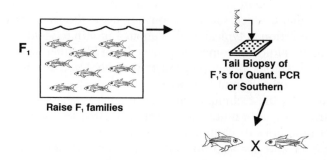

F_1

Raise F_1 families

Tail Biopsy of
F_1's for Quant. PCR
or Southern

Outcross or inbreed F_1's depending
on screen strategy

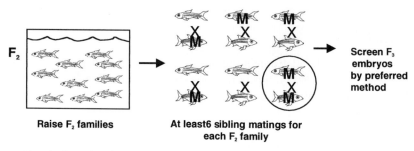

F_2

Raise F_2 families

At least 6 sibling matings for
each F_2 family

Screen F_3
embryos
by preferred
method

FIG. 1. Steps involved in using pseudotyped retroviruses to generate embryonic lethal mutations in zebrafish.

Plasmid containing desired viral genome
Phosphate-buffered saline
Sterile distilled H_2O

Virus Concentration

Bottle top filters, 0.45- and 0.22-μm pore size
SW28 (or equivalent ultracentrifuge rotor) and ultracentrifuge
Ultracentrifuge tubes, 40 ml
Phosphate-buffered saline (PBS)
Fish cell line (we use Pac2 cells)
Six-well tissue culture dishes
Glutaraldehyde, 0.05% (v/v) in PBS
LacZ buffer: 50 mM KCl, 32 mM NaH_2PO_4, 48 mM Na_2HPO_4, 10 mM
    $MgCl_2$, 5 mM potassium ferrocyanide, 5 mM potassium ferricyanide,
    2 mM 5-bromo-4-chloro-3-indolyl-β-D-galactopyranoside (X-Gal)
PBS plus 0.5% (v/v) Triton X-100
Cell culture media (above)
Polybrene (8 mg/ml in distilled H_2O), 1000×

Embryo Preparation

Modified Holtfreter's buffer: 1 liter of distilled H_2O, 0.1 g of $CaCl_2$,
    3.5 g of NaCl, 0.05 g of KCl, and 1 ml of 1 M HEPES (pH 7.5)
Pronase (type XXV protease), 30 mg/ml in distilled H_2O

Virus Injection

Agarose injection ramps (see Methods)
Injection apparatus (see Methods)
Glass microcapillary tubes
Microcapillary needle puller
Scalpel blade
Injection hood w/dissecting scope (see Methods)
Tissue culture dishes, 24 well
Water bath, 28°
Water bath, 31°
Polybrene (10×), 80 μg/ml
Phenol red, 5% (w/v) in PBS
Mineral oil

Virus Evaluation

Scalpel with blade
Lysis buffer: 100 mM NaCl, 10 mM Tris (pH 8.0), 10 mM EDTA, 0.4%
    (w/v) sodium dodecyl sulfate(SDS)

Pronase (0.1 mg/ml) or proteinase K (0.1 mg/ml)

3-Aminobenzoic acid ethyl ester methanesulfate salt (MESAB), 0.04% (w/v) in fish water (Instant Ocean in distilled H_2O, 250 mg/liter)

Flat-bottomed 96-well plates

Ethanol, 95 and 70% (v/v)

Typical Southern blot materials

PhosphorImager and screens (Molecular Dynamics, Sunnyvale, CA) or X-ray film

Polymerase chain reaction (PCR) primers to proviral sequences and to endogenous sequences

PCR reagents

Beakers, 400 ml

Methods

Cell Culture

Pseudotyped retroviruses must be generated in a packaging cell line. For reasons to be discussed shortly, the human liver cell line 293 is excellent for packaging Mo-MuLV core particles with the VSV G protein. The cell line we use is a 293 derivative that has the Mo-MuLV *gag-pol* genes driven to high levels of expression by the human cytolomegalovirus (CMV) promoter. There is also a stable chromosomal integration of the proviral genome whose RNA transcript is packaged into the Mo-MuLV active viral cores. The proviral genome contains full Mo-MuLV long terminal repeats (LTRs), the packaging signal (ψ), and a reporter used for titering purposes, in our case the *lacZ* gene, hence β-galactosidase expression.[14] The Gag-Pol proteins associate with the RNA viral genome to form active core particles. To this cell line, plasmid pHCMV-G (the VSV G protein driven by the CMV promoter) is transfected in transiently. 293 cells have the property of being easily transfected with calcium phosphate precipitates. This is an important property for the cells because VSV G protein is toxic when expressed at the levels necessary to generate high-titer pseudotyped viruses, making it necessary for the expression of the VSV G protein to be transient. The cells are allowed to grow to high density and then transfected with the HCMV-G plasmid. The G protein associates with the Mo-MuLV core particles at the plasma membrane and then the viruses are released into the tissue culture medium and collected. A detailed protocol for generating a large batch of virus follows.

1. Grow the cells in DMEM with 10% (v/v) fetal calf serum and antibiotics (Pen/Strep) at 37° and 5% CO_2. Tissue culture dishes (15 cm) should

be prepared in advance by treating them with 0.01% (w/v) poly-L-lysine (PLL), at 5 ml/plate, for 10 min and then rinsing with PBS. A typical virus preparation uses 15–30 PLL-treated 15-cm plates.

2. Twenty-four to 48 hr before transfection, split the cells into the PLL-treated plates and allow them to grow to ~80% confluence.

3. Two hours before transfection, change the culture medium to fresh, prewarmed DMEM with 10% (v/v) fetal calf serum (FCS) at 20 ml/plate and return the plates to 37° and 5% CO_2.

4. Form six plates' worth of calcium phosphate precipitate at one time by mixing the following ingredients in a 50-ml conical tube:

pHCMV-G plasmid	300 μg
Viral genome (unnecessary if a stable clone is made; see Notes)	120 μg
$CaCl_2$, 2.5 M	1.5 ml
Distilled H_2O	to 15 ml

To this room temperature mixture add 15 ml of ice-cold 2× HEPES-buffered saline (HBS) and quickly but briefly vortex. Let the mixture stand at room temperature for approximately 3 min (the incubation time will vary for every stock of HBS and needs to be determined empirically in a mock transfection), and then add 5 ml of the precipitate to each of six 15-cm plates. Rock the plates several times to evenly distribute the mixture. In 15–20 min the precipitate will begin to settle on the bottom of the plate and there should be fine, uniform grains over the entire surface. Clumping and very large grains suggest an unsuccessful transformation and plates with those characteristics should be discarded.

5. Return the plates to the incubator for 6–8 hr, and then remove the medium and replace it with fresh prewarmed medium at 20 ml/plate.

Virus Concentration

Twenty-four hours after the beginning of the transfection the tissue culture medium should be removed and replaced with fresh medium (15 ml/plate). The old medium does contain virus and should be handled with appropriate precautions, but typically it does not have a virus titer high enough for our purposes. At 48 hr posttransfection the medium is collected, pooled, and filtered through 0.45-μm pore size bottle top filters. After the initial filtering the supernatant can be filtered through a 0.22-μm pore size filter, which improves the cleanliness of the concentrated virus. Culture supernatant (34 ml) is added to each SW28 (or equivalent) ultracentrifuge tube and the medium is centrifuged for 1.5 hr at 21,000 rpm ($r^{min} \approx 40,000g$). The supernatant is then gently poured into a beaker containing bleach and the tube is inverted briefly to allow liquid to collect near the top of the tube. A Pasteur pipette is then used to aspirate away the excess liquid. It is important to be sure that *all* the excess liquid is removed from the sides of the tube (small amounts of medium running back into the pellet can significantly dilute the virus), but also be careful not to overdry the sample. Add 25 μl of phosphate-buffered saline (PBS) to the bottom of the tube and gently pipette up and down to resuspend the sample. Place a small piece of Parafilm over the tube and leave the sample at 4° for 5 hr to overnight. Pool all the samples, aliquot 20 μl per tube, and store at −80°. Collections can be made at 48 and 72 hr after transfection. Collecting supernatants before and after those time points yields virus that is not at high enough titer for our purposes.

To determine the final concentration of virus, some method of detecting infectious events is necessary. The virus we use to infect zebrafish embryos is a derivative of SFG,[14] which contains a *lacZ* gene expressed from the promoter in the Mo-MuLV LTR region. The promoter that drives *lacZ* is weak in fish cells, but it is detectable at a level that can be used as a reference to evaluate the titer of the virus. Pac2 cells (or any zebrafish tissue culture line) are split into six-well dishes, so that 24 hr later the cells are between 10 and 15% confluent. The concentrated virus is serially diluted into tissue culture medium containing Polybrene (8 μg/ml). The medium is removed from the six-well dishes and 300 μl from each dilution is added to the top of the cells. The six-well dishes are returned to the incubator. Every 20–30 min the dishes are gently rocked to prevent the center of each well from drying out. The infection is allowed to continue for 2–3 hr, and then the virus is removed and fresh medium is added to each well. The cells are allowed to grow for 72 hr. The medium on the cells is aspirated off, and then the cells are fixed with 0.05% (v/v) glutaraldehyde in PBS for 10 min. The cells are then washed three times with PBS containing 0.5% (v/v) Triton X-100. A mixture containing 2 mM X-Gal in *lacZ* buffer (see Materials Needed, above) is then added to each well and the dishes are placed at 37° overnight. Each well is then examined under a dissecting microscope and the number of blue-staining colonies is counted. A colony typically consists of a cluster of 2–16 blue-staining cells. The final concentration is then calculated by multiplying the number of colonies by the dilution and the amount of medium (300 μl) placed on the cells.

For our SFG-derived virus, a Pac2 titer of 1×10^7/ml is usually indicative of a good virus stock. This number does not represent the actual titer of the virus, but only what can be detected by β-galactosidase staining (~100-fold lower than the actual titer). It is important to note, however, that for every different viral genome, the titer that indicates a good virus stock must be empirically determined. There is currently no universal detection method or universal standard for the different possible viruses. Further information on how to determine the infectious ability of a virus is discussed below (Virus Evaluation).

Embryo Preparation

The embryos are injected approximately 3 hr after fertilization, and because many need to be injected, the collection of the embryos must be done in a timely manner to allow enough time for all the other preparations to be completed before the embryos reach the correct developmental stage. The desired male and female fish are isolated into separate tanks the evening before injection is to take place. The next morning two males are mixed with four females in the top portion of a two-tiered mouse cage ($14 \times 20 \times 25$ cm, $h \times w \times l$, respectively), one of which has the bottom cut out and replaced with wire screen and is placed into the other box, which is filled with fish water. The fish are allowed to mate for 15 min after the first embryos are released (stop sooner if enough embryos are released in a shorter time). After the fish are removed the embryos are moved into an 80-ml beaker containing 20 ml of modified Holtfreter's solution. From a Pasteur pipette add 6–8 drops of pronase (30 mg/ml) to the embryos and swirl gently. The protease will begin digesting away the chorions of the embryos. Periodically swirl the beaker gently. When the first chorions tear away from the embryos (usually about 2–3 min) wash away the protease by adding fresh Holtfreter's solution to the beaker and gently pouring off. Do not pour off all the buffer as surface tension will damage the embryos. Repeat the wash step four to six more times, swirling the embryos so that the chorions will all fall off, and then place the embryos in a 28° water bath for approximately 2.5–3 hr. To inject as many embryos as possible on a single day, space matings 45 min apart until the fish lose interest in mating. Using young, healthy fish, this strategy usually allows the injector 3–4 hr of injection time in the afternoon with appropriately timed embryos available throughout that period of time.

Virus Injection

At least 1 hr before the injections begin, fill several 24-well dishes with modified Holtfreter's solution and place them in a 31° water bath. In the lid of a 6-cm petri dish pour a small amount of 1.5% (w/v) agarose (made with Holtfreter's buffer) and place a double microscope slide in the lid at an angle, creating a ramp in the agarose (see Fig. 2). After the agarose hardens, carefully remove the slide and fill the lid with modified Holtfreter's buffer that contains Polybrene (8 mg/ml).

The injection apparatus consists of a 20-ml syringe with a 20-gauge needle, connected to polypropylene tubing. The tubing is inserted into the

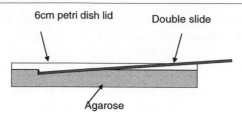

Side View

FIG. 2. Side view of an injection ramp. Agarose is poured into the lid of a 6-cm petri dish. While the agarose is still liquid, a double-width slide is inserted at an angle so that the edges of the slide are touching the lip about halfway up the side of the lid. When the agarose solidifies the slide is removed and the lid is filled with Holtfreter's solution. Embryos are then carefully arrayed along the bottom edge of the ramp for injection.

back of the needle handle and a small silicon gasket is used as an airtight seal at the end of the polypropylene tubing. The needles we use for injection are pulled on a needle puller (Sutter Instrument, Novato, CA). Settings are specific for model P-87; if no model number then delete settings. The end needs to be removed with a scalpel blade. This cut is made under a dissecting scope by lowering the blade straight down onto the needle, close to the point. The needle is then put into the silicon gasket and held in place by a cap that fits over the end of the needle handle.

Efficiently filling the needle with the concentrated virus can sometimes be the most difficult aspect of the injection procedure. The virus aliquot is thawed, and then a 10× stock of Polybrene is added to the virus for a final concentration of 8 μg/ml. As a visual aid to the injector, 1 μl of 0.5% (w/v) phenol red in PBS is added for every 20 μl of concentrated virus. A drop of the virus mixture is placed on a coverslip (5–8 μl) and covered with mineral oil. The virus is then drawn up into the needle by applying a vacuum with the 20-cm^3 syringe. The concentrated virus often contains a significant amount of debris, which can make loading difficult. The urge to cut a larger needle hole, so the virus flows into the needle more readily, should be resisted. The larger hole makes the injections more difficult and also makes it difficult to control the flow of virus out of the needle, causing needless waste.

Once the needle is filled with virus (this can often take 5–10 min), the embryos are placed on the injection ramp in a single row at the bottom edge of the ramp. A 5.25-inch Pasteur pipette is used, but the end should be briefly waved through a Bunsen burner flame to dull the glass edges. The embryos are fragile and the sharp edges of a normal pipette can tear or otherwise damage the embryo. The embryos placed on the ramp should be between the 512-cell stage and the 2000-cell stage. This appears to be the optimum time frame for injection. The majority of the early germ cell divisions happen within a few hours of this point in embryo development,[16] so injecting later offers fewer chances for the provirus to integrate into germ cells. We injected embryos from the 128-cell stage to the 4000-cell stage, and then tested the F_1 offspring for germ line transmission (M. Allende, T. Becker, S. Burgess, and A. Amsterdam, unpublished results, 1997). For our purposes the optimum window for injection probably falls between 512 and 2000 cells.

The injection ramp is then placed on the dissecting microscope which is mounted in the injection hood (Steriguard cabinet, custom built; J. T. Bakes, Phillipsburg, NJ). Virus pseudotyped with VSV G protein can infect humans, and proper precautions should be used in handling these viruses. Injection of virus is accomplished by pressing gently on the 20-cm^3 syringe, which starts the flow of virus from the tip of the needle. The tip is then inserted into the cell mass; avoid letting the tip penetrate all the way into the yolk. Allow a slight pause to let virus flow into the intercellular space, and then remove the needle. Each embryo is injected three or four times. The injector should try to distribute the phenol red color evenly over the entire cell mass. Once every embryo on the ramp has been injected the injector returns to the beginning and repeats the process. Extra attention should be paid to the embryos that have not retained detectable levels of the phenol red dye. Tests in our laboratory have shown there is generally an increase in overall virus infection when the multiple injections are done twice to each embryo.

Once the injections are completed, the embryos are carefully transferred from the injection ramp to the prewarmed 24-well plates, again using a flame-polished Pasteur pipette. We typically put five embryos in each well and leave them in the 31° water bath overnight. The next day embryos that appear to have developed normally (typically 20–50% of the total injected) are transferred into a 1 : 1 mix of modified Holtfreter's buffer and 5× Instant Ocean (250 mg/liter). At 48 hr after injection the embryos are changed to 100% 5× Instant Ocean, and are then raised in the normal fashion.

Virus Evaluation

Not all viruses are created equal. Because the production of pseudotyped virus by this method is dependent on a transient transfection, the harvested virus can vary. The titer of the virus on fish tissue culture cells is the first evaluation available, but it is not adequate by itself. It is also necessary to test injected embryos for level of infectivity of the F_1 offspring

[16] C. Yoon, K. Kawakami, and N. Hopkins, *Development* **124**, 3157 (1997).

from an injected fish to determine the number of fish carrying insertions and the number of different insertions present in each fish.

Initially we had to track the efficacy of the virus by determining the number and frequency of proviral insertions transmitted through the germ line. The founder (injected) fish were raised to sexual maturity and then mated to a nontransgenic wild-type fish. The offspring from these matings were raised for approximately 2 months, until their tails were large enough to provide a sample. The fish were then anesthetized for a few minutes in fish water containing 0.04% (w/v) MESAB. Once the fish stop swimming, each fish was placed on a small square of Parafilm and a scalpel was used to remove a small piece of the tail. The sample was then placed into 1 well of a 96-well plate that contained 100 μl of lysis buffer. The fish was then placed in a labeled 400-ml beaker containing fish water for storage until all the necessary information was obtained and the desired fish could be retrieved. Once all the samples had been acquired, the 96-well plate was sealed around the edges with Parafilm and placed at 55° for at least 3 hr. Using a multitip Pipetman, the sample was pipetted up and down repeatedly; alternatively, the plate was placed on a large vortexer for 5 min. Five microliters of this mixture was placed in 45 μl of distilled H_2O and boiled for 10 min. One microliter of the boiled sample was then placed in a 30-μl PCR containing two primers to the viral genome and two primers to a genomic sequence. The endogenous genomic sequence PCR serves as the positive control for the reaction and the proviral sequences determine if the fish is transgenic with a proviral insertion.

Once all the transgenic fish are identified, it is necessary to use Southern blot analysis to identify all the different insertions. Typically half of the tail cut DNA obtained was precipitated, cut, and run for Southern analysis. Enzymes were chosen that would quickly resolve the number and identity of each independent insertion. Remember that the germ line cells of the founder fish are chimeric, so the F_1 family can represent many different proviral insertions randomly distributed throughout the offspring. On average we observed 11–20 different insertions, each of which was represented in approximately 3% of the total population. Fish carrying each different insertion were then isolated for further genetic analysis.

This is the only way to determine absolutely the efficacy of virus and the identity of the different insertions, but to run a large-scale operation in this fashion would be nearly impossible. So it becomes necessary to monitor the ability of the virus to infect embryos well before the founders reach sexual maturity. We have developed such an assay and correlated it to the analysis described above. The principle is simple: compare the average number of proviral inserts per cell in an embryo with a genomic reference of known copy number. This ratio is determined by quantitative Southern analysis (or more recently by quantitative PCR). The protocol is as follows.

1. Inject embryos with virus and raise normally for 3 days.
2. Pool two embryos in 100 μl of lysis buffer and incubate at 50° for 4 hr to overnight.
3. Vortex each sample and precipitate by adding an equal volume of iso-propanol.
4. Spin down the sample for 5 min and wash with 70% (v/v) ethanol.
5. Resuspend sample in 30 μl of distilled H_2O and digest with restriction enzyme for 4 hr at 37°.
6. Run the samples on a Southern blot; include a genomic "standard." This is a sample of zebrafish genomic DNA that has a single proviral insert (on an agarose gel of 80 lanes we usually run 4 samples in different areas of the gel).
7. Probe the Southern blot with two different probes; the first probe is to proviral sequences and the second probe is to a genomic locus (we use the zebrafish *RAG1* locus).
8. Expose the blot to a phosphorimaging plate and quantitate band strength according to the instructions of the imaging equipment (we use the Molecular Dynamics STORM PhosphorImager and Molecular Dynamics software to analyze the data).
9. The average proviral copy number per cell can then be calculated by the following formula: (sample proviral counts/sample genomic counts)/ ([Σ(standard proviral counts/standard genomic counts)]/number of standards) = average proviral copy number per cell.

This analysis can also be performed as a batch analysis of F_1 embryos. Pools of 50 day 1 embryos from founders can be treated as described above to establish the average copy number of the provirus in the F_1 generation.

Notes

Cell Culture

Every downstream aspect of generating insertional mutations is contingent on producing virus that is efficient in infecting the zebrafish embryos. There are a few key areas that must be carefully tested and optimized for every different virus that is to be produced.

1. The packaging cell line must be tested for its ability to produce high titers of the desired pseudotyped virus. This is more complicated than it

sounds. First, the titer tests must be done on zebrafish-derived cell lines; our experience has shown us that virus titers determined on other cell lines, such as mouse-derived 3T3 cells, are not accurate predictors of the ability of the virus to infect zebrafish embryos. The second problem is that some cell lines will produce virus of high titers on fish lines, but when injected, the viruses are extremely toxic to the embryos, making them effectively useless. So the virus must be titered on fish cell lines and then injected into embryos to test efficacy fully.

2. Because the VSV G protein expressed at high levels is toxic to cells, the G protein is usually expressed in the cells by transient transfection. Under these circumstances there are two factors that must be maximized. First, the transfection needs to transfer the G protein-encoding plasmid to as many cells as possible. The 293 cell line was chosen for a packaging cell line because of its extremely high transfectability rate, but there is still more that must be done to maximize the efficiency. The key to obtaining the highest titers is to transfect the cells at precisely the right time. The cells need to be at a high density on the plate to have as many virus-producing cells as possible, but they cannot be so overgrown that they will no longer take up the plasmid DNA. We usually try to transfect the cells that are between 80 and 90% confluent. The medium should be changed 24 hr before a transfection and again 2 hr before to keep the cells in an active state. Second, it is critical to determine the exact dose of G protein-encoding plasmid for each new virus or transfection condition. A balance must be struck between the cells expressing enough of the G protein to maximize the packaging of the active viral cores, and keeping the level of G protein low enough so that the cells are not killed before they reach maximum output. We typically use 50 μg of plasmid DNA with a calcium phosphate transfection. We have succeeded with some of our packaging lines by using other transfection techniques, particularly lipofection. It is necessary to establish the most effective transfection technique for each virus-packaging cell line; it has become evident that no one technique is optimal under all circumstances with all packaging lines.

3. There are two basic ways of generating virus with a particular viral genome. The first is to express the viral genomic RNA from a plasmid construct that replaces the 5′ U3 region of the virus (containing the viral promoter) with the promoter from cytomegalovirus (CMV), which is much more potent in the human-derived 293 cells than the Moloney promoter.[17] The normal life cycle of the virus will replace the CMV promoter with the normal Moloney LTR before integration. The plasmid construct is cotransfected with the pHCMV-G plasmid at half the concentration (usually

[17] R. K. Naviaux, E. Costanzi, M. Haas, and I. M. Verma, *J. Virol.* **70**, 5701 (1996).

25 μg). Our experience is that this works, but not as consistently or as well as the second option of making a stable clone. To make a stable clone, we generate virus with a transient construct, and then use that virus to infect 293 cells. The cells are then diluted out into single cells in 96-well plates. Each clone is transfected and the virus produced is tested on Pac2 cells (or any zebrafish cell line). Typically 1 or 2 lines in 50 will produce virus at a high enough titer to be useful (at least 10^6 CFU unconcentrated). Although making a stable clone is laborious in the long run for large-scale projects it is essential. More of the virus preparations will be high enough in titer, and infection of embryos will be much better.

Virus Concentration

Virus concentration is a fairly straightforward step. Some anecdotal advice can be given here.

1. Once supernatant has been collected from the producing cells, the remaining steps should be performed as quickly as possible, and the supernatants should be kept on ice at all times. It is tempting to freeze the supernatants and return to the virus concentration at a convenient time, but each freeze–thaw seems to affect the ability of the viruses to infect fish cells by dropping titer.

2. The earlier collection time points after concentration are cleaner and easier to load into the injection needle than later time points. Typically after about day 3 posttransfection, the virus stops infecting the embryos efficiently and our collections after that time point are less useful.

3. It is important to clear as much of the supernatant out of the tubes after concentration as possible. Medium remaining on the side of the tube will run down into the bottom, diluting the virus more than desired.

4. Freezing the virus reduces its effective titer on embryos by nearly 50%, so injecting fresh virus is the best option when at all possible. Typically the virus titer is stable for several (5–7) days at 4°.

Embryo Preparation

It is important to set aside 10–20 embryos that will not be injected, but are otherwise handled in the same way. This will show the quality of the embryos, and the expected survival before the actual injections, which will help in determining the toxicity of the virus being injected.

Digesting the chorions with pronase for too long is extremely detrimental to the overall survival of the embryos. Once the chorions start to fall off, the protease should be washed away immediately. The embryos need to be handled with care from then on because they become fragile once their chorions are removed. We then keep the beaker in a 28° water bath until it is time to inject them (approximately 3 hr).

Virus Injection

The injection of virus into embryos is a learned skill and is perhaps the most difficult technique to convey through writing. But there are several pointers that can help speed the learning process.

1. The pulled needle should be long and thin; in this technique the injection is made into the intercellular space instead of into the cells directly, and so a thin needle slips in between the cells with a minimal amount of disruption. Keep the point of the needle in the cell mass; try not to penetrate the yolk with the point, for that is frequently fatal to the embryo.

2. It is possible to use a micromanipulator to steady the injection needle, but this is not recommended. Our laboratory has found that with practice injections done by hand are much easier, more efficient, and have better survival than when the micromanipulator is used. This may be difficult to believe at first, but with a little practice it will quickly become obvious that it is not only possible, but much better to do the injections without assistance. It is possible and helpful to do two rounds of injection into embryos using only 0.25% (w/v) phenol red in PBS. This will give a much better feel for the injections before a valuable and slightly hazardous reagent such as the pseudotyped viruses is used.

3. Because virus is being injected between the cells, it is best to do several small injections over the entire cell surface of the embryo instead of one large injection. We typically inject three to five times, trying to achieve an even distribution of the dye tracer over the cell mass. After injecting all the embryos on the ramp in such a manner, we return to the beginning and repeat the process, paying extra attention to the embryos that look like they have less dye remaining in the cells.

4. Loading the needle with virus is somewhat trying on a person's patience. The virus after concentration contains a significant amount of particulate matter. These particles tend to clog the end of the needle. We have found the best way to load the needle is to move the syringe plunger almost, but not all the way down, by separating the syringe from the tubing at the louvre lock. This allows a strong vacuum to be applied to the needle, but leaves a small space to apply pressure in the other direction if the needle becomes clogged. Keep moving the needle around in the virus drop, trying to outrun the large particles on a collision course with the needle tip. Once the needle is full, it is then better to disconnect the hose again and move the plunger to about halfway up the syringe. This gives more control over applying pressure during the injection of the virus.

5. Probably the most useful tip concerns how to load the embryos onto the ramp. As previously mentioned, it is important to use a Pasteur pipette whose edges have been dulled by waving the end briefly through a flame. The embryos should then be carefully drawn into the pipette to prevent them from crushing each other as they pass through the opening. Then, to place them on the ramp, it is best not to squeeze the bulb, but to just let gravity pull the embryos down and out of the pipette onto the ramp. This simple technique will allow the embryos to be placed in a neat line.

Virus Evaluation

Virus evaluation is the area that could benefit the most from further improvements in technology. The following points we have observed might be helpful.

1. Using the embryo assay approach is difficult but has proved to be a fairly good predictor of transgenicity. Founder-to-founder variation is considerable, but as a rule of thumb, injections that generate an average of four or five proviral inserts per cell will generate mostly excellent founders. When the copy number is lower than that, more of the founder fish will prove not to carry a sufficient number of insertions in their germ lines.

2. We are currently using a real-time fluorescence PCR strategy to determine viral copy number. This technique has proved to be excellent for this purpose, but involves the use of an expensive piece of equipment, which will not be practical to many laboratories. The major advantage is the speed of the assay. It reduces a 4-day process of Southern blots to a 4-hr process by PCR. The use of real-time PCR has proved to be an extremely accurate and robust technique in our hands.

Section III

Tools for Analysis of Membrane Proteins

[12] A Gene Fusion Method for Assaying Interactions of Protein Transmembrane Segments *in Vivo*

By JENNIFER A. LEEDS and JON BECKWITH

Introduction

Integral membrane proteins are present in all cell membranes and constitute about 20% of the total number of cellular proteins.[1] To function properly, many integral membrane proteins must oligomerize via interactions between their respective transmembrane (TM) segments (reviewed in Ref. 2). These higher order structures may be composed of homooligomers, for example, the homodimeric *Escherichia coli* chemotactic aspartate receptor Tar,[3] the homodimeric human erythrocyte glycoprotein glycophorin A (GpA),[4] and the homopentameric protein phospholamban.[2] TM domain interactions also contribute to heterooligomerization such as in the heterodimeric T cell receptor. However, little is known about TM domain interactions and the rules for integral membrane protein assembly, because few three-dimensional structures of membrane proteins have been solved. Compared with soluble proteins, efforts to solve membrane protein structures by conventional high-resolution biophysical techniques have been complicated by the need for detergent or lipid environments. Such *in vitro* techniques also require the use of equipment not often found in biologically oriented laboratories and thus not accessible to many researchers.

In response to the difficulty of using direct biophysical methods for studying membrane proteins, we have developed a technically simple *in vivo* genetic technique for detecting interactions between TM segments of integral membrane proteins.[5] This system is based on a gene fusion technique that was orginally developed by Hu *et al.*[6] for studying interactions among soluble proteins in *E. coli*.

The soluble domain interaction assay system of Hu *et al.*[6] grew out of the observation that the bacteriophage λ *c*I repressor N-terminal DNA-binding domain (headpiece) can act as a repressor when fused to a heterolo-

---

[1] D. Boyd, C. Schierle, and J. Beckwith, *Protein Sci.* **7**, 201 (1998).

[2] M. A. Lemmon and D. M. Engelman, *Curr. Opin. Struct. Biol.* **2**, 511 (1992).

[3] S. A. Chervitz, C. M. Lin, and J. J. Falke, *Biochemistry* **34**, 9722 (1995).

[4] M. A. Lemmon, J. M. Flanagan, J. F. Hunt, B. D. Adair, B. J. Bormann, C. E. Dempsey, and D. M. Engelman, *J. Biol. Chem.* **267**, 7683 (1992).

[5] J. A. Leeds and J. Beckwith, *J. Mol. Biol.* **280**, 799 (1998).

[6] J. C. Hu, E. K. O'Shea, P. S. Kim, and R. T. Sauer, *Science* **250**, 1400 (1990).

gous C-terminal dimerization domain. The dimerization of λ repressor is required for maintaining the lysogenic state of the λ prophage in *E. coli*, as well as preventing superinfection by homoimmune phage.[7] Because repression is absolutely dependent on dimerization of λ repressor prior to DNA binding,[8] fusion of the λ headpiece to proteins of interest has been shown to be an ideal system for studying dimerization of soluble domains *in vivo*.[6,9–15]

Our laboratory has successfully modified the λ repressor headpiece fusion assay for the study of interactions specifically among integral membrane proteins. Our genetic approach allows one to screen and distinguish among known dimerizing domains, weakly dimerizing mutants, and nondimerizing TM segments. This chapter gives a detailed description of this assay system and discusses its potential uses and limitations.

Materials and Reagents

Bacterial Strains, Bacteriophages, and Plasmids

Escherichia coli 61F: This strain (MC1061/F' $lacI^q$ kan^R) is suggested for construction and expression of λ headpiece fusion plasmids. To maintain the F', 61F should be grown in the presence of kanamycin (20 μg/ml). Other strains may be used provided they are sensitive to λ infection and contain a source of $lacI^q$ that minimizes expression of the fusion genes in the absence of the inducer isopropyl-β-D-thiogalactopyranoside (IPTG)

pJAH01: This plasmid vector contains the gene fragment encoding the λ *c*I repressor headpiece and linker (amino acids 1–132)[16] under the control of the *lac*UV5 promoter. These sequences are cloned into the unique *Eco*RI site in the low copy number plasmid pACYC184,[17] which confers tetracyline resistance. All constructs referred to in

[7] M. Ptashne, "A Genetic Switch." Blackwell Scientific, Cambridge, Massachusetts, 1986.
[8] M. A. Weiss, C. O. Pabo, M. Karplus, and R. T. Sauer, *Biochemistry* **26**, 897 (1987).
[9] C. A. Bunker and R. E. Kingston, *Nucleic Acids Res.* **23**, 269 (1995).
[10] P. A. Battaglia, F. Longo, C. Ciotta, M. F. Del Grosso, E. Ambrosini, and F. Gigliani, *Biochem. Biophys. Res. Commun.* **201**, 701 (1994).
[11] L. Castagnoli, C. Vetriani, and G. Cesareni, *J. Mol. Biol.* **237**, 378 (1994).
[12] O. Amster-Choder and A. Wright, *Science* **257**, 1395 (1992).
[13] J. C. Hu, N. E. Newell, B. Tidor, and R. T. Sauer, *Protein Sci.* **2**, 1072 (1993).
[14] D. Beier, B. Schwartz, T. M. Fuchs, and R. Gross, *J. Mol. Biol.* **248**, 596 (1995).
[15] L. R. Turner, J. W. Olsen, and S. Lory, *Mol. Microbiol.* **26**, 877 (1997).
[16] C. O. Pabo, R. T. Sauer, J. M. Sturtevant, and M. Ptashne, *Proc. Natl. Acad. Sci. U.S.A.* **76**, 1608 (1979).
[17] A. C. Y. Chang and S. N. Cohen, *J. Bacteriol.* **134**, 1141 (1978).

this chapter are derived from pJAH01. The complete sequence of pJAH01 has been deposited to GenBank (accession number AF129432)

λ cI: This phage, unable to produce a functional cI repressor, is used to screen for dimerization of λ cI headpiece–TM fusion proteins. Reduction in the efficiency of plating λ cI correlates with increased TM domain dimerization

λ vir: This phage is used as a positive control for λ infection. λ repressor constructs missing the native C-terminal dimerization/cooperativity domain will not repress λ vir phage[7]

Growth and Assay Media

NZ plate medium (per liter): 15 g of agar, 8 g of NaCl, 10 g of NZ amine (Difco, Detroit, MI), 5 g of yeast extract

NZ top agar (per liter): 10 g of NZ amine, 8 g of NaCl, 8 g of agar

NZ liquid medium (per liter): 10 g of NZ amine, 5 g of yeast extract, 10 g of NaCl

λ suspension medium (SM) (per liter): 5.8 g of NaCl, 2 g of $MgSO_4 \cdot 7H_2O$, 50 ml of 1 M Tris-HCl (pH 7.5)

Stock Solutions

IPTG: 100 mM isopropyl-β-D-thiogalactopyranoside stock solution. Sterilize by filtration and store in aliquots at −20°. Do not refreeze

Tetracyline: 12.5 mg/ml in ethanol–water (1 : 1, v/v). Working concentration, 12.5 μg/ml

Kanamycin: 50 mg/ml in water. Sterilize by filtration and store in aliquots at −20°. Working concentration, 20 μg/ml

Maltose: 20% (w/v) solution. Sterilize by filtration. Do not autoclave. Working concentration, 0.2% (w/v)

$MgSO_4$: 1 M solution. Sterilize by autoclaving. Working concentration, 10 mM

Enzymes and reagents: DNA synthesis, restriction, and modification enzymes are from New England BioLabs (Beverly, MA). Other chemicals are obtained from standard suppliers of molecular biological reagents

General Cloning Scheme

The general cloning scheme is shown in Fig. 1. Candidate TM domains are amplified by polymerase chain reaction (PCR), using primers that insert the appropriate restriction sites and encode amino acids necessary for

1. Amplify TM by PCR

Upstream primer (n =15-20 nt)   **Downstream primer (n =15-20 nt)**

5'-CC<u>ACATGT</u>AAACGAAAGn-3'   5'-CATG<u>CCATGGCCCGGG</u>n-3'
   AflIII   *NcoI SmaI*

2. Cut with *Afl*III and *Nco*I

```
      C   K   R   K            P   G   P
      CATGTAAACGAAAGn-//-nCCCGGGC
          ATTTGCTTTCn-//-nGGGCCCGGTAC
      AflIII                    SmaI    NcoI
```

3. Ligate into *Afl*III site on pJAH01

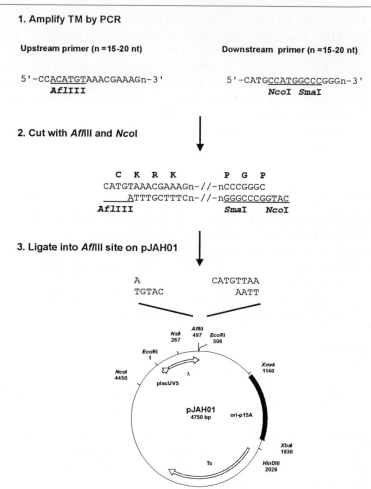

Fig. 1. Cloning scheme for fusion of candidate TM domains to λ *c*I repressor headpiece in the pJAH01 dimerization vector. Suggested upstream and downstream primers for TM domain amplification are shown with *Afl*III, *Nco*I, and *Sma*I restriction sites underlined. The *Afl*III/*Nco*I-digested PCR product encoding the positively charged amino acids KRK, and containing the *Sma*I restriction site (CCCGGG), is ligated into the pJAH01 *Afl*III site.

proper orientation of the fusion protein in the cell membrane. PCR products are ligated into a vector containing the λ *c*I repressor headpiece and native linker. Transformed reporter cells are selected for by growth in the presence of the antibiotic tetracycline.

Cloning Transmembrane Domains into pJAH01

Employing standard PCR methods, the DNA encoding the candidate TM domain from the gene of interest is amplified. Design the primers as suggested in Fig. 1. The upstream primer should contain an *Afl*III recognition sequence (underlined in Fig. 1) as well as two additional 5′ bases for optimal cutting; sequences encoding the positively charged amino acids lysine–arginine–lysine (KRK), which are necessary for directing the N terminus of the TM domain to the cytoplasm and the C terminus of the TM domain toward the periplasm; and 15 to 20 additional bases (shown as "*n*" in Fig. 1) that are identical to the 5′ end of the TM to be amplified. The downstream primer should contain an *Nco*I site (underlined in Fig. 1) as well as four additional 5′ bases for optimal cutting, a *Sma*I site (underlined in Fig. 1) for cloning C-terminal periplasmic domains, and 15–20 additional bases that are the reverse complement to the 3′ end of the TM to be amplified. In the absence of any periplasmic domain, the *Sma*I site encodes a terminal cysteine in all λ headpiece : transmembrane domain fusions. This periplasmically localized cysteine does not contribute to dimerization, but does appear to stabilize the C termini of the dimerized fusion proteins.[18] After digestion of the PCR product with *Afl*III and *Nco*I, the fragment is ligated to the *Afl*III-digested vector pJAH01 and competent *E.coli* 61F is transformed, selecting for tetracycline and kanamycin resistance. The entire fusion system is contained within a unique *Eco*RI fragment and, if necessary, can be moved to alternative vectors.

General Dimerization Assay Scheme

The general dimerization assay scheme is shown in Fig. 2. In principle, induction of a fusion of λ headpiece to a dimerizing TM domain will repress infecting λ cI phage. In practice, repression is detected by subjecting lawns of *E. coli* expressing fusion proteins to infection by λ cI phage. After incubation, the lawns are screened for the presence of plaques and efficiencies of plating λ cI are compared among the induced and uninduced cultures.

Standard Dimerization Assay

Once a headpiece–TM clone is obtained, the assay for the presence of a functional TM dimerization domain is performed as depicted in Fig. 3. Phage λ used in efficiency of plating assays should be grown and titered on *E. coli* 61F by standard methods.[19] From an overnight culture of cells

[18] J. A. Leeds and J. Beckwith, unpublished observations (1999).
[19] J. Sambrook, E. F. Fritsch, and T. Maniatis, "Molecular Cloning: A Laboratory Manual." Cold Spring Harbor Laboratory Press, Cold Spring Harbor, New York, 1992.

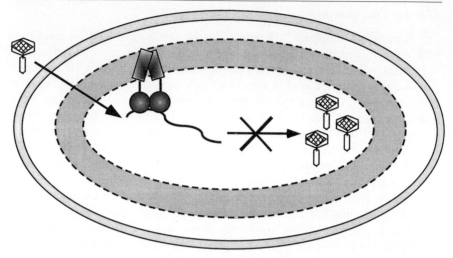

FIG. 2. Diagram of λ cI phage repression by dimerized λ headpiece–TM fusion proteins.

grown in NZ broth plus tetracycline and kanamycin, cultures are back diluted 1:100 into two tubes containing NZ broth plus maltose, MgSO$_4$, kanamycin, and tetracycline. To one tube, IPTG is added to a final concentration of 10 μM. Cultures are grown at 37°, with aeration, to an $A_{600} =$ 0.65–0.70. One hundred microliters of culture is removed and added to 3 ml of NZ top agar already at 50°. The top agar containing uninduced or induced cells is spread immediately onto NZ agar containing tetracycline, kanamycin, and 0 or 10 μM IPTG, respectively. The top agar is allowed to cool and dry for 20–30 min at room temperature. Ten-microliter drops of SM diluent containing 10^8, 10^6, 10^5, 10^4, or 10^3 plaque-forming units (PFU)/ml of λ cI or λ vir are pipetted as individual spots onto the dried top agar. Plates are incubated upright at 37° overnight.

After overnight incubation at 37°, the *E. coli* lawns are observed for plaques. The efficiency of plating is expressed as the number of plaques per number of PFU plated. Count plaques in spots where they are well isolated (i.e., no more than 25 plaques/spot). Semiquantitative comparisons can also be made among the different dilutions without having to count individual plaques. The efficiency of plating λ vir should always be 1. Table I describes the efficiencies of plating λ cI phage on lawns of *E. coli* expressing the vector control (pACYC184), the λ cI headpiece alone (pJAH01), and the wild-type λ cI repressor cloned into pJAH01. Also shown, for comparison, are fusions of λ headpiece to strongly dimerizing TM (GpA TM), a weakly dimerizing TM (GpA TM/G83I), and a nondimerizing TM domain (MalF TM1).

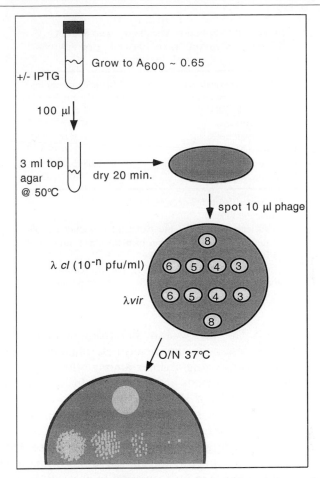

FIG. 3. TM domain dimerization assay scheme. Details of the efficiency of plating assay are described in text.

Importance of Assaying Fusion Protein Expression Levels

The efficiency of plating assay correlates the dimerizing abilities of different TM segments with repressor activities. However, we cannot over-emphasize how important it is to choose the appropriate expression level for assaying dimerization activity. Too high a level of expression could give a repression signal simply by increasing the local concentration of a headpiece construct to a point at which it is able to dimerize and repress. Too low a level of expression could result in no dimer formation. We have found that induction with 10 μM IPTG is sufficient to assay a strongly

TABLE I
EFFECT OF MEMBRANE-ASSOCIATED AND SOLUBLE λ
REPRESSOR CONSTRUCTS ON EFFICIENCY OF PLATING
λ cI PHAGE[a]

Repressor construct	Efficiency of plating[b]
None (pACYC184)	0.72 ± 0.27
Intact λ repressor	$<10^{-6}$
Headpiece alone	1 ± 0^{c}
Headpiece-MalF TM1	0.80 ± 0.28
Headpiece-GpA TM (G83I)	0.08 ± 0.04
Headpiece-GpA TM	$\leq 10^{-6}$

[a] Adapted from J. Leeds and J. Beckwith, *J. Mol. Biol.*
280, 799 (1998).
[b] Efficiency of plating is mean of 5–12 independent repli-
cates, expressed as number of plaques per number of
PFU ± standard deviation.
[c] Four independent assays of *E. coli* expressing headpiece from pJAH01 reported exactly 100% sensitivity
for this strain.

dimerizing TM (GpA TM) and weakly dimerizing mutants of GpA TM.[5]
However, it may be necessary to empirically determine the optimal expression level to compare dimerizing abilities among other fusion partners.

When we make comparisons among various fusion partners induced
with the same concentration of IPTG, we have observed a correlation of
TM dimerizing ability with the steady state levels of fusion protein detected
by Western analyses.[5] That is to say, protein that are less effective repressors
are often present in considerably lower amounts than those that are effective. Our evidence suggests that this correlation results from the susceptibility of undimerized headpiece in our construct to proteolysis.[5,8,20–22] Therefore, in addition to measuring the steady state levels of fusion proteins,
one should also compare protein synthesis and secretion levels among the
fusion partners.

Levels of fusion protein synthesis and secretion can be measured by
fusing the mature *E. coli* alkaline phosphatase protein (PhoA), without its
native signal sequence, to the C termini of the fusion constructs (using the
*Sma*I site described in Fig. 1). Because the mature PhoA is a stable entity,
the fate of the N-terminal portion of the PhoA fusions does not affect the

[20] J. W. Roberts and C. W. Roberts, *Proc. Natl. Acad. Sci U.S.A.* **72,** 147 (1975).
[21] R. T. Sauer, K. Hehir, R. S. Stearman, M. A. Weiss, A. Jeitler-Nilsson, E. G. Suchanek,
and C. O. Pabo, *Biochemistry* **25,** 5992 (1986).
[22] D. A. Parsell and R. T. Sauer, *J. Biol. Chem.* **264,** 7590 (1989).

TABLE II
ALKALINE PHOSPHATASE ACTIVITIES OF
HEADPIECE–TM–PhoA FUSIONS[a]

TM domain	PhoA activity[b]	
	Uninduced	10 μM IPTG
None (pACYC184)	2.14 ± 0.11	1.76 ± 0.31
GpA TM	13.4 ± 0.57	135 ± 10
GpA TM (G83I)	18.2 ± 0.50	158 ± 5.0
MalF TM1	15.7 ± 0.74	147 ± 12

[a] Adapted from J. Leeds and J. Beckwith, *J. Mol. Biol.* **280,** 799 (1998).
[b] In Miller units, mean ± standard deviation.

level of alkaline phosphatase activity produced by the chimeras (Table II). The C-terminal fusion is used to monitor both synthesis and, because PhoA is active only in the cell periplasm,[23] membrane localization of the various headpiece–TM fusion constructs.

Fusion constructs containing the PhoA sequences should not be used for the efficiency of plating assays. Alkaline phosphatase forms a stable homodimer in the periplasm,[24] and is capable of promoting dimerization of the cytoplasmic λ headpiece across the cytoplasmic membrane.[5] The result is that λ headpiece–TM–PhoA fusion constructs show the same level of repression whether or not they contain a dimerizing TM segment.

Suggested Western Blot Procedure

Part of the same culture used for the efficiency of plating assay is retained on ice. Cellular material is resuspended in sodium dodecyl sulfate (SDS) sample buffer containing 2-mercaptoethanol and incubated at 37° for 30 min. Samples are subjected to SDS–polyacrylamide gel electrophoresis followed by Western analysis using an anti-λ repressor primary antibody. To visualize the small, hydrophobic fusion proteins, a 14% (w/v) polyacrylamide gel system is recommended.

Suggested Procedure for Cloning and Assaying C-Terminal Alkaline Phosphatase Fusions

The mature alkaline phophatase domain can be cloned into the *Sma*I site in the λ headpiece–TM construct (Fig. 1). The PCR product

---

[23] S. Michaelis, H. Inouye, D. Oliver, and J. Beckwith, *J. Bacteriol.* **154,** 366 (1983).
[24] C. N. Chang, H. Inouye, P. Model, and J. Beckwith, *J. Bacteriol.* **142,** 726 (1980).

coding for the signal-sequenceless *E. coli* alkaline phosphatase (amino acids 22–471[23]) can be amplified with the upstream primer 5′-TCC<u>CCCGGG</u>CGGACACCAGAAATGCCTG-3′ and the downstream primer 5′-TCC<u>CCCGGG</u>TTATTTCAGCCCCAGAGCGGC-3′ encoding *Sma*I sites (underlined). Alkaline phosphatase activity can be determined by the measuring the rate of *p*-nitrophenyl phosphate hydrolysis in permeabilized cells.[23]

Cloning and Assay Problems

A few problems may arise in the cloning or assaying of candidate TM dimerization domains. Below are some of the anticipated concerns surrounding this assay system. This assay may not be suitable in all instances, in which case we suggest investigating alternative systems that have been described.[25]

Cloning Problems

The first caveat of this assay system is that it requires that the N terminus of the TM face the cell cytoplasm and the C terminus face the periplasm. However, in our experience, reversing the native orientation of candidate TMs has not compromised their ability to dimerize λ headpiece.[5,26] Second, there may be *Afl*III or *Nco*I sites within the TM being analyzed, in which case the TM could be inserted by blunt-ended ligation into an *Afl*III-digested, end-filled vector. What is important is that the positively charged amino acids (KRK) remain on the N-terminal side of the TM to promote proper protein topology.[27]

Assay Problems

In some cases, overexpression of the candidate TM domain could be detrimental or lethal to *E. coli.* If this happens, the expression level of the fusion construct may have to be reduced by lowering the concentration of IPTG. Changing promoters or vectors will also require empirical determination of optimal induction conditions.

Problems with Data Interpretation

For the assay system described here, the relationship between dimerization and plaque-forming ability is not linear. The best approximation is

[25] W. P. Russ and D. M. Engelman, *Proc. Natl. Acad. Sci. U.S.A.* **96,** 863 (1999).
[26] A. Jacq and J. Beckwith, unpublished observations (1999).
[27] G. von Heijne, *Nature (London)* **341,** 456 (1989).

that the relationship is semilogarithmic, where an order of magnitude change in the efficiency of plating λ *c*I reflects a less than 10-fold change in dimerization ability.[5] Therefore, the data gathered by this assay have not been extrapolated to calculate, for example, dissociation constants, although the relationship may be able to be worked out.

Modifications and Future Developments

The system described here could be adapted for more advanced screening of protein dimerization properties. First, the property of conferring resistance to λ *c*I phage makes genetic selections and screenings possible. For example, one can screen for mutants that reduce dimerization by screening for phage-sensitive colonies by the nibbled colony technique.[28] Second, with the construction of appropriate plasmid vectors, it should be possible to screen an appropriately generated library of chromosomal fragments for membrane segments or even periplasmic domains that promote dimerization. Third, with further modification, or by using a two-plasmid system, one should also be able to look at heterodimerization between TM segments within a single protein and between proteins known to form a complex. Our laboratory is in the process of developing vectors and assays for these applications.

Acknowledgment

This work was supported by the National Institutes of Health (grant GM18569-01) to J.A.L.

[28] N. Sternberg, *J. Mol. Biol.* **76,** 1 (1973).

[13] Using *SUC2–HIS4C* Reporter Domain to Study Topology of Membrane Proteins in *Saccharomyces cerevisiae*

By CHRISTIAN SENGSTAG

Introduction

Approximately 2330 proteins encoded by the *Saccharomyces cerevisiae* genome are hydrophobic in nature[1] and thus are presumed to be localized to membranes. These may constitute either peripheral membrane proteins, which associate with a membrane without crossing it, or transmembrane proteins that comprise one to several hydrophobic segments able to span

[1] B. Nelissen, R. De Wachter, and A. Goffeau, *FEMS Microbiol. Rev.* **21,** 113 (1997).

the membrane from one side to the other. Often such transmembrane segments form α helices, which are flanked by charged residues. The hydrophobic nature of membrane proteins often renders attempts toward solubilization difficult and this fact in particular hampers efficient crystallization required for precise structural analysis. For this reason, other strategies have been developed to gain alternative structural information on membrane proteins (reviewed in Ref. 2).

The fusion of a reporter domain to specific sites within a protein followed by biochemical or genetic analysis of the chimeric fusion product has proved a valuable approach to gather such structural information. Other methods that have been applied comprise chemical modification, binding of peptide-specific antibodies, or limited digestion by proteases (reviewed in Ref. 2), techniques that are amenable to intact cells or subcellular fractions thereof and that exploit the fact that membranes constitute a physical barrier to the molecular probe used. Thus, those parts of the protein of interest that are localized within or to one side of the membrane are protected from the molecular probe, whereas parts of the protein that are exposed to the other side are free to become modified. The fusion protein approach, on the other hand, takes advantage of knowledge about the biochemical property of the domain that is coupled to the target protein. Various researchers have used invertase Suc2p as a reporter domain to study the topology of yeast proteins present in the endoplasmic reticulum (ER) membrane. The decision to choose this enzyme as reporter presumably was stimulated by the following facts: (1) A cytoplasmic as well as a secreted form of invertase exists. Thus, invertase represents a protein that is readily translocated through membranes; (2) invertase harbors 14 potential N-glycosylation sites, many of which are glycosylated *in vivo*. In consequence, core glycosylation is recognizable by a clear shift in mobility detected in sodium dodecyl sulfate (SDS) gels; and (3) by introducing a *SUC2* segment into the gene of interest a fortuitous epitope is incorporated that can be used to immunoprecipitate the protein or identify it by Western blotting. Systematic application of this technique has provided structural information about the *S. cerevisiae* ER proteins Sec62p,[3] Sec63p,[4] and Sec61p,[5] as well as about the cytochrome P450Cm1 protein from *Candida maltosa*.[6]

Other fusion protein approaches took advantage of yeast acid phosphatase (Pho5p) as the reporter domain to study the topology of arginine

---

[2] B. Traxler, D. Boyd, and J. Beckwith, *J. Membr. Biol.* **132,** 1 (1993).

[3] R. J. Deshaies and R. Schekman, *Mol. Cell. Biol.* **10,** 6024 (1990).

[4] D. Feldheim, J. Rothblatt, and R. Schekman, *Mol. Cell. Biol.* **12,** 3288 (1992).

[5] B. M. Wilkinson, A. J. Critchley, and C. J. Stirling, *J. Biol. Chem.* **271,** 25590 (1996).

[6] R. Menzel, E. Kargel, F. Vogel, C. Bottcher, and W. H. Schunck, *Arch. Biochem. Biophys.* **330,** 97 (1996).

permease Can1p[7] *in vitro* and that of uracil permease Fur4p *in vivo*.[8] Similarly, the structure of Can1p has been investigated by fusing galactokinase to different sites and analyzing the glycosylation status on *in vitro* transcription/translation in the presence of microsomal membranes.[9] In addition to these approaches, in which the fused domain exclusively served as antigenic determinant to monitor the glycosylation status of the fusion protein, the enzymatic activity of the reporter domain has also been exploited. The topology of a plasma membrane protein, the *S. cerevisiae* α-factor receptor Ste2p, has been investigated by fusing bacterial β-lactamase to various sites. Inclusion of a Kex2 cleavage site at the fusion point resulted in secretion of enzymatically active β-lactamase by hybrids exposing the reporter domain to the extracellular compartment. In contrast, no secreted activity was detected in hybrids exposing β-lactamase toward the intracellular compartment.[10]

None of the described approaches exploited the ability of the fused reporter domain to complement a conditional lethal phenotype, allowing membrane protein topology to be studied *in vivo*. Yet, such a system based on histidinol dehydrogenase activity has been developed and is described below.

SUC2–HIS4C as Genetic Determinant

The *S. cerevisiae HIS4* gene[11] encodes a trifunctional enzyme comprising phosphorybosyl-AMP cyclohydrolase (His4A; EC 3.5.4.19), phosphoribosyl-ATP pyrophosphohydrolase (His4B; EC 3.6.1.31), and histidinol dehydrogenase (His4C; EC 1.1.1.23) activity. This enables His4p to catalyze three individual steps of the histidine biosynthetic pathway. The C-terminal domain (His4C) harbors the histidinol dehydrogenase activity, which converts histidinol to histidine, and *his4Δ* mutants expressing the *HIS4C* moiety alone are able to use histidinol to grow on histidine-deficient media. To do so, the His4C domain must be present in the cytoplasm of the yeast cell, whereas His4C targeted to the ER lumen is inactive. In addition, the host cells expressing *HIS4C* must harbor the dominant *HOL1-1* mutation[12] rendering the cell permeable to exogenously added histidinol. If these

[7] M. Ahmad and H. Bussey, *Mol. Microbiol.* **2,** 627 (1988).
[8] S. Silve, C. Volland, C. Garnier, R. Jund, M. R. Chevallier, and R. Haguenauer Tsapis, *Mol. Cell. Biol.* **11,** 1114 (1991).
[9] G. N. Green, W. Hansen, and P. Walter, *J. Cell Sci. Suppl.* **11,** 109 (1989).
[10] C. P. Cartwright and D. J. Tipper, *Mol. Cell. Biol.* **11,** 2620 (1991).
[11] T. F. Donahue, P. J. Farabaugh, and G. R. Fink, *Gene* **18,** 47 (1982).
[12] R. F. Gaber, M. C. Kielland Brandt, and G. R. Fink, *Mol. Cell. Biol.* **10,** 643 (1990).

requirements are met, *HIS4C*-expressing *his4Δ* mutants are able to grow, albeit slowly, on histidinol medium.

Deshaies and Schekman[13] were the first to develop an elegant direct selection system based on *HIS4C* as reporter to identify mutants defective in secretion. The approach was based on a tripartite fusion protein consisting of the α-factor signal peptide, a fragment of invertase (Suc2p), and the His4C moiety. Expression of the corresponding fusion gene in an appropriate strain (*his4Δ, HOL1-1*) from a high-copy vector gave rise to a fusion protein that was immunologically detectable by use of anti-Suc2p antibodies. However, the fusion protein apparently exhibited no histidinol dehydrogenase activity because no growth of the strain was observed on histidinol medium. Immunoprecipitating the fusion protein from cells previously treated with tunicamycin, an agent that interferes with core glycosylation, produced a shift of the immunoreactive band toward higher mobility on the SDS gel, suggesting that the fusion protein was glycosylated in the absence of tunicamycin. Thus, translocation of the His4C domain to the ER lumen appeared to be incompatible with histidinol dehydrogenase activity, a fact that may either be due to the possibility that the ER membrane is impermeable to the substrate histidinol or the product histidine, or that extensive glycosylation of His4C or aberrant folding of the translocated domain abolished enzyme activity. By selection for temperature-sensitive mutants that were able to grow on histidinol medium, Deshaies and Schekman recovered several secretion mutants. In these mutants, the fusion protein was no longer translocated to the ER. Rather, the His4C domain remained in the cytoplasm, where it converted histidinol to histidine.

Employing *SUC2–HIS4C* Cassette as Topogenic Probe

The above-described results provided sufficient impetus to exploit the *SUC2–HIS4C* moiety for investigating the topology of the *S. cerevisiae* ER membrane protein 3-hydroxy-3-methylglutaryl-coenzyme A (HMG-CoA) reductase (EC 1.1.1.34) encoded by two genes, *HMG1* and *HMG2*. Previous structural analyses have predicted the presence of seven transmembrane-spanning segments (TMs) in the N-terminal half of either protein.[14,15] To substantiate these predictions, plasmids pA and pAΔ7[16] have been constructed. These *URA3*-based 2μ vectors contain fusion genes in which the

[13] R. J. Deshaies and R. Schekman, *J. Cell Biol.* **105**, 633 (1987).
[14] L. Liscum, J. Finer Moore, R. M. Stroud, K. L. Luskey, M. S. Brown, and J. L. Goldstein, *J. Biol. Chem.* **260**, 522 (1985).
[15] M. E. Basson, M. Thorsness, J. Finer Moore, R. M. Stroud, and J. Rine, *Mol. Cell. Biol.* **8**, 3797 (1988).
[16] C. Sengstag, C. Stirling, R. Schekman, and J. Rine, *Mol. Cell. Biol.* **10**, 672 (1990).

SUC2–HIS4C module is fused to TM7 or to the putative lumenal loop between TM6 and TM7, respectively, in *HMG1*. Both plasmids were transformed into strain FC2a, harboring the *his4-401* deletion allele in combination with the *HOL1-1* mutation in a *ura3-52* background. The resulting Ura[+] transformants that harbored pA, but not those harboring pAΔ7, were able to grow on histidinol medium. Moreover, the pAΔ7-encoded fusion protein was core glycosylated, whereas no detectable glycosylation of the pA-encoded fusion protein was identified (Fig. 1).

Because the obtained genetic and biochemical results perfectly matched the structural predictions for at least part of the Hmg1 protein, a more thorough analysis using the Suc2–His4C reporter domain has been performed. To this end, a series of *Xho*I restriction sites have been introduced into the *HMG1–SUC2–HIS4C* construct present on plasmid pA. The sites

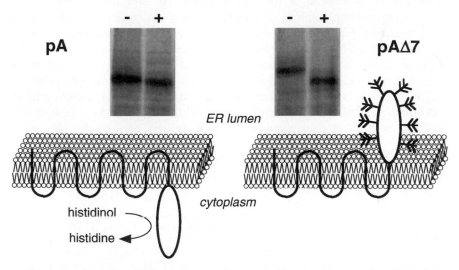

FIG. 1. Topological models for fusion proteins encoded by plasmids pA and pAΔ7. *Left:* In strain FC2a pA, the Suc2–His4C domain (open ellipse) is linked to the seventh transmembrane domain of Hmg1p (boldface wavy line). In this conformation, the reporter domain is exposed to the cytoplasmic side of the ER membrane, where it catalyzes the conversion of histidinol to histidine, conferring the Hol[+] phenotype. The fusion protein is produced in an unglycosylated form because no difference in gel mobility was observed (see *inset*) when immunoprecipitation was performed from tunicamycin-treated (+) or untreated (−) cells. *Right:* In strain FC2a pAΔ7, the fusion protein lacks the seventh transmembrane domain. In consequence, the reporter domain is exposed to the ER lumen, where it becomes extensively core glycosylated (branched arrows). No conversion of histidinol to histidine occurs; thus, pAΔ7 transformants are unable to grow on histidinol (Hol[−] phenotype). Immunoprecipitation of the protein from tunicamycin-treated or untreated cells gave rise to marked difference in size of the detected band (see *inset*). Experimental details are given in Ref. 16.

were created by site-directed mutagenesis at every location where the polypeptide chain was predicted to exit from the membrane. All sites were introduced in the same reading frame, such that the *Xho*I recognition sequence CTCGAG represented two codons of the gene. This later allowed for "domain shuffling."

Serial Analysis of Hmg1p Topology Using the Suc2–His4C Reporter

As extensively described previously,[16] the *SUC2–HIS4C* reporter cassette has been fused to the 3′ end of all seven putative TMs in *HMG1*. The fusion genes were subsequently expressed in yeast and the ability of the transformants to grow on histidinol (Hol⁺ or Hol⁻) was tested. In parallel, the fusion proteins were analyzed by either immunoprecipitation or Western blot analysis to exclude any artifactual Hol⁻ phenotype due to a failure in expression of the fusion genes. As summarized in Table I, deleting increasing numbers of TMs from the 3′ end gave rise to alternating Hol⁻ and Hol⁺ phenotypes. Deletion of TM7 produced a membrane protein comprising six TMs similar to the protein encoded by plasmid pAΔ7. Deletion of an additional TM reconstituted the Hol⁺ phenotype, indicating that the Suc2–His4C domain was presented toward the cytoplasm in the respective fusion protein that comprised five TMs (Fig. 2). Deletion of an additional TM produced a fusion protein with four TMs. As expected, this protein, which was core glycosylated (not shown), conferred a Hol⁻ phenotype.

Expression of fusion genes comprising fewer than four TMs in no case complemented the Hol⁻ phenotype of the host. Western blot analysis sug-

TABLE I

RESULTS OF DELETING PUTATIVE TMs FROM THE 3′ END IN HMG1

Plasmid	Individual TMs present in fusion construct	Total number of TMs	Growth behavior on histidinol
pA	TM1–TM7	7	+
pAD1-6	TM1–TM6	6	−
pAD1-5	TM1–TM5	5	+
pAD1-4	TM1–TM4	4	−
pAD1-3	TM1–TM3	3	−
pAD1-2	TM1–TM2	2	−
pAD1	TM1	1	−
pAD1-3, 6-7	TM1–TM3 and TM6–TM7	5	+
pAD1-2, 5-7	TM1–TM2 and TM5–TM7	5	+
pAD1-2, 4-7	TM1–TM2 and TM4–TM7	6	+
pAD1, 4-7	TM1 and TM4–TM7	5	−

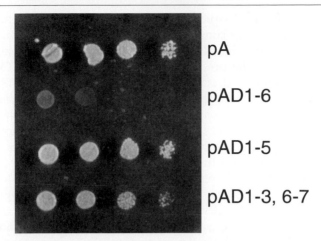

FIG. 2. Growth behavior of FC2a transformants on histidinol medium expressing differently truncated *HMG1–SUC2–HIS4C* fusion genes. Increasing dilutions (from left to right) of cultures from the indicated Ura⁺ transformants were spotted on minimal medium supplemented with histidinol and the plates were incubated at 30° for 5 days.

gested that the Hol⁻ phenotype was based on a decreased stability of certain fusion proteins. To overcome such stability problems, a set of further deletions was constructed, in which internal neighboring pairs of TMs were eliminated. Supposing that every TM would function independently of other TMs in the protein, the elimination of two TMs at a time was expected to produce exclusively fusion proteins exposing the Suc2–His4C reporter to the cytoplasm. Thus, the observation of a Hol⁺ phenotype would provide not only information on the topology, but also on the stability of the fusion protein. Indeed, plasmids pAD1-3, 6-7 (Fig. 2) and pAD1-2, 5-7, where two TMs were deleted simultaneously, conferred a Hol⁺ phenotype to the host strain (Table I). It was thus concluded that the reporter domain was directed to the cytoplasm in both fusion proteins comprising five TMs.

In another case, however, an unexpected behavior was observed: In plasmid pAD1, 4-7 the sequences encoding TM2 and TM3 have been deleted. Although the fusion gene encodes a protein comprising an uneven number of putative TMs, a Hol⁻ phenotype was observed (Table I). Western blot analysis revealed the presence of a stable fusion protein in FC2a transformants. Yet the protein was core glycosylated, arguing for exposure of the reporter domain to the ER lumen. Thus, considering individual TMs as individual entities that act independently of each other turned out to be an illegitimate oversimplification. It appeared therefore more plausible that certain hydrophobic segments acted as TMs only in conjunction with others.

The previously reported structural predictions for TM4 of Hmg1p were indeed on a weaker basis compared with those for the other six TMs.[15] It was therefore possible that the ability of TM4 to act as transmembrane domain depended on the presence of its neighboring domain TM3, and a hypothesis was formulated, stating that in case of a missing TM3 segment, TM4 would not be able to span the membrane. The fusion protein encoded by plasmid pAD1-2, 4-7 in consequence would comprise only five segments functioning as real TMs and thus would produce a Hol⁺ phenotype. This hypothesis was supported by the phenotype conferred by another construct, pAD1-2, 4-7. The encoded fusion protein lacked one putative transmembrane domain (TM3), and thus contained an even number of TMs. Nevertheless, the respective FC2a transformants exhibited a Hol⁺ phenotype. Assuming again that TM4 depended on the presence of TM3, deletion of TM3 alone would interfere with the propensity of TM4 to span the membrane. With this assumption, all experimental discrepancies could be resolved and the structural predictions for Hmg1p could be confirmed.

Experimental Procedures

Topological information for any ER protein can be gained by subcloning defined segments of the gene of interest into plasmid pCS4-14, followed by transformation into strain FC2a and investigation of the Hol phenotype. The individual steps of the analysis are described explicitly below. Plasmid pCS4-14 (Fig. 3) is a derivative of pA, which harbors the *SUC2–HIS4C* cassette on a 2μ plasmid in combination with *URA3*. A unique *Xho*I restriction site was engineered immediately 5′ to the *SUC2* sequence such that the six bases of the recognition sequence CTCGAG encode two amino acids (leucine and glutamate) of the fusion gene. The full sequence of pCS4-14 is known and can be retrieved from GenBank under accession number AF114752. Specific DNA segments from the gene of interest, including the promoter sequence up to the fusion point, may be subcloned into the *Sac*I/*Xho*I sites of pCS4-14, thereby replacing the stuffer fragment derived from *HMG1* sequences.

Subcloning Gene of Interest into pCS4-14

Insert the gene fragment of interest, including its promoter sequence into *Sac*I/*Xho*I-digested pCS4-14, making certain that the correct open reading frame with the *SUC2–HIS4C* sequence is reconstituted such that the *Xho*I recognition site CTCGAG encodes two amino acids, leucine and glutamate. If no restriction site is available to perform the intended gene fusion, the gene fragment can also be amplified by polymerase chain reaction (PCR).

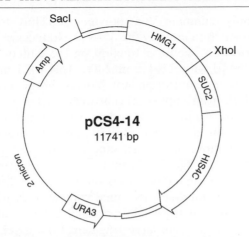

FIG. 3. Structure of plasmid pCS4-14. The plasmid is a derivative of YEp352, a 2μ-based *URA3* plasmid replicating at high copy number in yeast. Gene fusions are created by replacing the sequences between *Sac*I and *Xho*I restriction sites by the DNA fragment of interest. The complete sequence of the plasmid can be retrieved from GenBank under accession number AF114752.

Polymerase Chain Reaction Amplification of Gene Fragment of Interest

Design primer pairs that amplify the gene part of interest and that contain 5′ extensions incorporating *Sac*I and *Xho*I or *Sal*I sites, respectively. Perform the amplification with a mixture of *Taq* DNA polymerase and the proofreading Vent DNA polymerase to prevent incorporation of errors. Calculate the annealing temperature (T_m) for the designed primer pair either by running an appropriate software program (e.g., OLIGO; MedProbe, Oslo, Norway) or by adding up 2° for each hybridizing AT base pair and 4° for each GC base pair. Set up a PCR on ice in a final volume of 50 μl containing 100 ng of plasmid template DNA, 5 μl of 10× Vent buffer [100 mM KCl, 100 mM $(NH_4)_2SO_4$, 200 mM Tris-HCl (pH 8.8), 20 mM $MgSO_4$, 1% (v/v) Triton X-100] (New England BioLabs, Beverly, MA), 5 μl of each primer (10 μM), 5 μl of 2 mM dNTPs, 2.5 units of *Taq* DNA polymerase (Boehringer Mannheim, Indianapolis, IN), and 0.5 unit of Vent DNA polymerase (New England BioLabs). Denature at 94° for 2 min, and then run 25–30 cycles comprising 1 min at 94°, 30 sec at T_m, and 3 min at 72°. Add a final 5-min extension step at 72°. Subject one-tenth of the reaction to agarose gel electrophoresis for analysis of the product. In case of an unsatisfactory result, optimize the annealing temperature by increasing or decreasing in 1° steps. Alternatively, vary the Mg^{2+} concentration. Remove excess primers by running the rest of the PCR over a Micro Spin column (Pharmacia Biotech, Piscataway, NJ). Note that this procedure does not efficiently eliminate Vent polymerase. Thus, to exclude future degradation of restriction ends by the $3' \rightarrow 5'$ exonuclease activity of Vent polymerase, perform a phenol extraction on the eluted fragment: adjust the volume to 200 μl with TE [10 mM Tris (pH 8), 1 mM EDTA], add 100 μl of TE-saturated neutral phenol. Vortex thoroughly for 2 min, and centrifuge at 14,000g in an Eppendorf centrifuge for 2 min. Carefully recover the supernatant. Add 10 μl of 3 M sodium acetate, pH 5.5, and 500 μl of ethanol. Mix, incubate at 15 min at $-80°$, and centrifuge for 15 min at 14,000g. Carefully remove the supernatant, wash the pellet gently with 70% (v/v) ethanol, centrifuge again for 2 min, remove the supernatant, dry the pellet for a few minutes in a desiccator, and resuspend it in about 20 μl of TE. Perform a restriction digest in a final volume of 50 μl and isolate the DNA fragment from an agarose gel. In parallel, digest about 2 μg of pCS4-14 with SacI/XhoI, isolate the large fragment from a gel, and ligate with the digested PCR product in roughly equimolar amounts. Transform *Escherichia coli* and confirm the authenticity of the constructed plasmid by restriction analysis.

Transformation into Yeast Strain FC2a

Although there are several well-suited transformation protocols, the lithium acetate procedure described below appears to produce the most reproducible results. Grow strain FC2a (*MAT*a *his4-101 trp1-1 leu2-3,112, ura3-52 HOL1-1*)[16] in 50 ml of YPD medium [1% (w/v) yeast extract, 1% (w/v) Bacto Peptone, 2% (w/v) glucose] to an OD_{600} of 0.6–0.8. To do so, dilute a freshly grown overnight culture to an OD_{600} of 0.1 and shake at 30° for a few hours. Harvest the exponentially growing cells by centrifugation for 5 min at 3000g. Resuspend the cell pellet in about 20 ml of sterile H_2O and centrifuge again. Resuspend the pellet in 1 ml of sterile H_2O, transfer to an Eppendorf tube, and centrifuge for 2 min at 1500g. Resuspend the pellet in 0.5 ml of 100 mM lithium acetate. Preincubate the cells for 20 min at 30°. Distribute each 50 μl of cells to individual Eppendorf tubes. Add 5 μl of carrier DNA (sheared and heat-denatured salmon sperm DNA, 10 mg/ml) plus 1–2 μl of quick-prep DNA from the respective plasmid. Vortex gently, and then add 300 μl of a solution containing 100 mM lithium acetate and 40% (w/v) polyethylene glycol (PEG) 3350 (Sigma, Buchs, Switzerland) prepared from a 1 M lithium acetate solution and a 50% (w/v) PEG solution; vortex again. Incubate for 30 min at 30°, and then heat shock the cells for 5 min at 42°. Centrifuge the cells for 5 min at 2000g. Discard the supernatant, wash the cells, and resuspend them in 500 μl of YM His Leu Trp medium [0.67% (w/v) yeast nitrogen base, 2% (w/v) glucose, histidine (20 μg/ml), leucine (30 μg/ml), and tryptophan (20 μg/ml)]. Streak about 100 μl on YM His Leu Trp plates containing 2% (w/v) agar and select Ura$^+$ transformants by incubating the plates at 30° for 3 days. Make sure to include in the transformation a positive control (pCS4-14), as well as negative controls (no DNA and pAD1-6, respectively). Pick transformants and restreak for single colonies for purification.

Testing Hol Phenotype of Transformants

Prepare YM-Hol plates resembling YM His Leu Trp plates but containing 6 mM histidinol instead of histidine. Streak individual transformants on YM-Hol plates in a gradient for single colonies and incubate them for about 5 days at different temperatures. Some fusions have been found to confer a thermosensitive phenotype, with growth observed at 23° but not at 30°. To better discriminate between Hol$^+$ and Hol$^-$ clones, spot serial dilutions on the YM-Hol plates (see Fig. 2). Grow individual 2-ml cultures in YM His Leu Trp medium overnight. Harvest the cells by centrifugation and resuspend them in sterile H$_2$O. Adjust the density of the culture to 0.3 OD. Spot 5 μl as well as 5 μl each of 1:10, 1:100, and 1:1000 dilutions on the plates. Include as negative control strain FC2a pAD1-6 and make sure that no growth occurs. If background growth is seen with this strain, the specific batch of histidinol used may be contaminated with trace amounts of histidine. In this case, the batch of histidinol needs further purification.

Purification of Histidinol from Contaminating Histidine

Contaminating traces of histidine can be removed from the histidinol solution by absorbing to anion-exchange material the histidine anions that are formed at high pH. Prepare 10 ml of a 0.5 M histidinol solution in H$_2$O. Adjust to pH 9–10 by adding small amounts (100–200 μl) of 10 M NaOH. Add about 2 ml of AG 1-X8 anion-exchange resin (format form, 140-1454; Bio-Rad, Hercules, CA) and shake for 15 min at room temperature. Remove the resin by centrifugation for 10 min at 3000g, recover the supernatant, and filter sterilize through a 0.2-μm pore size Filtropur S plus syringe filter (Sarstedt, Sevelen, Switzerland). Use as a 0.5 M solution.

Immunoprecipitation of Labeled Fusion Proteins and Analysis of Glycosylation Status

Labeling of Cells. Grow the FC2a transformants to be analyzed in YM His Leu Trp medium (see above) containing 5% (w/v) glucose to exponential phase (OD$_{600}$ of 0.5–1). The high glucose content in the medium represses the secreted form of invertase. Harvest 8 OD units of cells by centrifugation in a clinical centrifuge (5 min at 2000g). Discard the supernatant and wash the cells in 10 ml of medium. Centrifuge again and resuspend in 10 ml of medium. Split the culture and distribute 5 ml each to small Erlenmeyer flasks or to 50-ml Corning (Acton, MA) tubes. Add to one flask or tube 50 μl of 1-mg/ml tunicamycin stock solution (Sigma) to give a final concentration of 10 μg/ml. Incubate at room temperature on a shaker for 15 min. Harvest the cells by centrifugation and resuspend them in 1 ml of YM His Leu Trp medium containing or not containing tunicamycin (10 μg/ml). Transfer the cells to a glass reagent tube, add 100 μCi of Tran^{35}S-Label (specific activity, >1000 Ci/mmol) (ICN Pharmaceuticals, Costa Mesa, CA), and label the cells by shaking the tube at room temperature for 30 min. Kill the cells by adding 1 ml of cold 20 mM NaN$_3$; put the tube on ice for 5 min. Harvest the labeled cells by centrifugation in a clinical centrifuge and resuspend them in the glass tube in 2 ml of cold 10 mM NaN$_3$. Dispose of the supernatant into a radioactive waste container. Centrifuge again and remove all the supernatant. Either freeze the cell pellet or continue with the preparation of the crude extract.

Preparation of Crude Extracts. Prepare disruption buffer containing 8 M urea and 1% (w/v) SDS (stable at room temperature for a few weeks). Just before use, add a mixture of 1000× stock solutions of proteinase inhibitors (PIs; Sigma) to the appropriate volume of 8 M urea–1% (w/v) SDS. To 10 ml of disruption buffer add 10 μl of aprotinin stock (1 mg/ml in H$_2$O; store at 4°) and 10 μl of a stock containing 0.2 M phenylmethylsulfonyl fluoride (PMSF) as well as leupeptin, pepstain, N-tosyl-L-phenylalaninechloromethyl ketone (TPCK), each at 1 mg/ml in DMSO, and store in aliquots at −20°. Add 100 μl of prewarmed (37°) disruption buffer containing PIs to the cell pellet, add 0.25 g of acid-washed glass beads (0.5-mm diameter; Merck, Dietikon, Switzerland), and vortex heavily at room temperature for 90 sec. Place the tubes for 10 min at 50° in a water bath to dissolve membrane proteins in the chaotropic buffer.

Acid-washed glass beads are prepared as follows: Wash the beads in a large Erlenmeyer flask overnight in 1% (v/v) Triton X-100 at room temperature on a shaker. Let the beads settle, and then decant the liquid. Wash the beads several times with distilled H$_2$O, three times with 95% (v/v) ethanol, and again three times with distilled H$_2$O. The beads are then washed twice for 2 hr with 6 M HCl, twice for 2 hr with 6 M HNO$_3$, followed by distilled H$_2$O washes until a neutral pH is reached. Dry and sterilize the beads by baking them in an oven at 180° overnight. The number of beads that fill a 0.5-ml Eppendorf tube up to the conical part approximates the necessary 0.25 g.

After a 10-min incubation at 50°, centrifuge the glass tubes and recover the crude extract dispersed between the glass beads by aspirating it from the bottom of the tube, using a drawn-out Pasteur pipette. Wash the glass beads with another 100 μl of prewarmed disruption buffer containing PIs, vortex again, centrifuge, and recover the crude extract with the drawn-out Pasteur pipette. Pool the two fractions, and then distribute them as four aliquots of 50 μl each. Freeze three of them at $-20°$ for later use and proceed with immunoprecipitation with one aliquot.

Immunoprecipitation. Add to the 50 μl of crude extract 0.5 ml of IP buffer [15 mM NaPO$_4$ (pH 7.5), 150 mM NaCl, 2% (v/v) Triton X-100, 0.1% (w/v) SDS, 0.5% (w/v) sodium deoxycholate] containing proteinase inhibitors. Mix and put on ice. Preadsorb unspecific material by adding 50 μl of *Staphylococcus aureus* cells (10% crude cell suspension, formalin fixed; Sigma) and put for 10 min on a rotator at 4°. Centrifuge at 14,000g for 5 min and recover the supernatant. Add 5 μl of a rabbit invHIS4C antiserum and incubate for 30 min at 4° on the rotator. A specific antiserum[17] has been prepared by D. Sanglard on injection of an *E. coli*-expressed GST-invHIS4C fusion construct into rabbits and a sample may be obtained on request.* Add 30 μl of a protein A-Sepharose CL-4B bead suspension (50%, v/v; Sigma) and rotate at 4° overnight.

Centrifuge for 1 min at 1000g and visualize the tiny white pellet by shining a light from the top through the tube. Remove the supernatant by aspirating it through a hypodermic needle connected to a radioactive waste container, which in turn is connected to a vacuum pump. Wash the pellet three times, each time with 1 ml of IP buffer. (It is not necessary to remove all the supernatant at each step). Wash the pellet three times in buffer II [2 M urea, 0.2 M NaCl, 1% (v/v) Triton X-100, 100 mM Tris (pH 7.5)], followed by a high-salt wash in 500 mM NaCl, 20 mM Tris (pH 7.5), 1% (v/v) Triton X-100. Wash the beads twice in 10 mM Tris (pH 7.5), 50 mM NaCl. Try to remove all liquid after the final wash and resuspend the beads in 50 μl of 2× Lammli buffer [125 mM Tris (pH 6.8), 4% (w/v) SDS, 20% (v/v) glycerol, 4% (v/v) 2-mercaptoethanol, 0.02% (w/v) bromphenol blue]. Dissociate the antiserum–antigen complex by heating to 50° for 10 min. Do not boil membrane proteins; they may aggregate and become insoluble. Centrifuge at 14,000g for 5 min and load a quarter to half on an 8% (w/v) SDS–polyacrylamide gel. Run the gel at the appropriate voltage (for a 17 cm long, 15 cm wide, and 1.5 mm thick gel it takes about 75 min at 15 mA, until the bromphenol blue dye has passed the stacking gel, and then about 2.5 hr at 30 mA until the dye has reached the end of the gel). Fix the gel for 20 min in a solution containing 10% (v/v) acetic acid and 25% (v/v) isopropanol, and then put the gel for 20 min into an amplifier

[17] D. Sanglard, C. Sengstag, and W. Seghezzi, *Eur. J. Biochem.* **216**, 477 (1993).
* Dr. D. Sanglard, University Hospital Lausanne (CHUV), Institute of Microbiology, CH-1011 Lausanne, Switzerland. E-mail: Dominique.Sanglard@chuv.hospvd.ch.

(NAMP100; Amersham, Zurich, Switzerland). Dry the gel on filter paper with a heated vacuum gel dryer. Expose the dry gel to X-ray film.

Other Membrane Proteins Probed with SUC2–HIS4C

We and others have applied *SUC2–HIS4C* reporter domain fusions to investigate the topology of several other yeast membrane proteins. By fusing the *SUC2–HIS4C* cassette to different sites within the *Candida tropicalis* cytochrome P450 genes *alk1* and *alk2,* evidence of each single transmembrane domain responsible for targeting the proteins to the ER membrane was obtained.[17] Moreover, a second hydrophobic segment present in alk1p but absent in alk2p was found unlikely to act as transmembrane domain.

Fusing *SUC2–HIS4C* to three different sites within the *WBP1* gene revealed the presence of a C-terminal transmembrane domain and an orientation where the C terminus was exposed to the cytoplasm.[18] Similarly, the structure of the proteins encoded by *SWP1* and *ALG5,* two genes involved in glycosylation of yeast proteins, has been studied by use of *SUC2–HIS4C* fusions and a cytoplasmic location of either C terminus has been suggested.[19,20] Moreover, C-terminal fusions of *SUC2–HIS4C* to the *STT3* gene and a deletion variant thereof provided evidence of the presence of an internal transmembrane domain and suggested a lumenal location of the C terminus of Stt3 protein.[21] In the experiments with *ALG5* and *STT3,* the genetic analysis, i.e., determination of the Hol[+] and Hol[−] phenotypes, respectively, conferred by the fusion genes was confirmed by a physical analysis in which the glycosylation status of the immunoprecipitated fusion proteins has been investigated. However, in the case of *WBP1* and *SWP1,* the conclusions concerning the topology of the fusion proteins were based solely on the analysis of the Hol phenotype, and it later turned out that the predictions for Swp1 were wrong (S. te Heesen, personal communication, 1999). On the basis of a weak Hol[+] phenotype, it was concluded that the C terminus was located on the cytoplasmic side of the membrane; however, a later analysis by immunoprecipitation revealed that the C terminus in fact was glycosylated. Thus, it is highly recommended that the genetic analysis be complemented by an immunoprecipitation experiment not only

---

[18] S. te Heesen, R. Rauhut, R. Aebersold, J. Abelson, M. Aebi, and M. W. Clark, *Eur. J. Cell. Biol.* **56,** 8 (1991).

[19] S. te Heesen, R. Knauer, L. Lehle, and M. Aebi, *EMBO J.* **12,** 279 (1993).

[20] S. te Heesen, L. Lehle, A. Weissmann, and M. Aebi, *Eur. J. Biochem.* **224,** 71 (1994).

[21] R. Zufferey, R. Knauer, P. Burda, I. Stagljar, S. te Heesen, L. Lehle, and M. Aebi, *EMBO J.* **14,** 4949 (1995).

for fusion constructs conferring a Hol⁻ phenotype but also for those in which a Hol⁺ phenotype has been observed.

Use of *SUC2–HIS4C* to Select for Mutants Defective in Insertion of Integral Membrane Proteins into Endoplasmic Reticulum

In analogy to the selection scheme established by Deshaies and Schekman,[13] which has been described above, Stirling and colleagues[22] have used an *HMG1–SUC2–HIS4C* fusion to select for mutants that did not correctly insert the fusion protein into the ER membrane. In their system, they used plasmid pCS5 encoding a truncated Hmg1p, in which the Suc2–His4C domain was connected to the lumenal loop between TM6 and TM7 of Hmg1p. This construct was similar to the one encoded by plasmid pAΔ7 (see Fig. 1). FC2a transformants harboring pCS5 were unable to grow on histidinol, yet temperature-sensitive mutants that had regained a Hol⁺ phenotype have been identified in a screen. By this procedure, a new allele of *sec61* and a hitherto unidentified gene, *sec65,* have been discovered.[22]

Additional Role of *SUC2* Segment in *SUC2–HIS4C* Cassette

The major reason for incorporating the *SUC2* fragment into the *SUC2–HIS4C* reporter cassette was given by the fact that invertase is a heavily glycosylated protein. Thus, a clear difference in molecular weight could be expected in large fusion proteins depending on their glycosylation status (Fig. 1). However, Green and Walter[23] have identified another important function of this segment. During their studies of arginine permease Can1p, a *HIS4C* fragment was fused to different sites within the *CAN1* gene. Thus, fusion proteins similar to those described for Hmg1–Suc2–His4C (see above) have been produced in a *his4, HOL1-1* mutant strain. However, no Suc2 segment was present in these fusion proteins. Interestingly, a somewhat unpredictable behavior of the His4C domain was found. In one fusion protein, the His4C reporter was coupled to a particular site, which was known from other experiments to be present in the ER lumen. Unexpectedly, however, yeast transformants expressing this specific construct were Hol⁺ and the fusion protein indeed turned out to be unglycosylated. This discrepancy, combined with data from other experiments, then prompted Green and Walter to investigate the anomaly in more detail. It was subsequently found that translocation of the histidinol dehydrogenase

---

[22] C. J. Stirling, J. Rothblatt, M. Hosobuchi, R. Deshaies, and R. Schekman, *Mol. Biol. Cell.* **3,** 129 (1992).
[23] N. Green and P. Walter, *Mol. Cell. Biol.* **12,** 276 (1992).

segment was inhibited by specific N-terminally localized protein sequences. Thus, the specific context within the fusion protein appeared to strongly modulate the behavior of the His4C segment. To overcome such sequence context problems, the authors subsequently incorporated a spacer segment, *SUC2,* between the *CAN1* and the *HIS4C* segments, with the result that in the presence of the Suc2 segment, His4C was translocated through the ER membrane, as had been expected. The data of Green and Walter thus provide strong evidence of the importance of the invertase segment amino terminal to the His4C part of the fusion proteins to provide translocation competence to the His4C domain.

Acknowledgments

I am grateful to S. te Heesen for critical comments on the manuscript. Moreover, I thank my colleagues and friends at U.C. Berkeley, in particular J. Rine, L. Pillus, R. Wright, and C. Stirling, for generously sharing their knowledge with me.

[14] Detecting Interactions between Membrane Proteins *in Vivo* Using Chimeras

By IGOR STAGLJAR and STEPHAN TE HEESEN

Introduction

Knowledge of the precise interactions of a given protein is crucial for the understanding of its functions. While identification of protein interactions initially occurred almost solely through the use of biochemical methods, more recently yeast two-hybrid systems[1-6] have been employed as powerful genetic tools to find proteins that interact specifically with a protein of interest.[3,7-9] The protein of interest is produced in yeast as a fusion to a

[1] S. Fields and O. Song, *Nature (London)* **340,** 245 (1989).
[2] J. Gyuris, E. A. Golemis, H. Chertkov, and R. Brent, *Cell* **75,** 791 (1993).
[3] O. Nayler, W. Strätling, J. P. Bourquin, I. Stagljar, L. Lindemann, H. Jasper, A. M. Hartmann, F. O. Fackelmayer, A. Ullrich, and S. Stamm, *Nucleic Acids Res.* **26,** 3542 (1998).
[4] A. Aronheim, E. Zandi, H. Henneman, S. Elledge, and M. Karin, *Mol. Cell. Biol.* **17,** 3094 (1997).
[5] E. A. Golemis and R. Brent, *Mol. Cell. Biol.* **2,** 3006 (1992).
[6] E. Zandi, T. N. Tran, W. Chamberlain, and C. S. Parker, *Genes Dev.* **15,** 1299 (1997).
[7] T. Durfee and K. Becherer, *Genes Dev.* **7,** 555 (1993).
[8] J.-P. Bourquin, I. Stagljar, P. Meier, P. Moosmann, J. Silke, T. Baechi, O. Georgiev, and W. Schaffner, *Nucleic Acid Res.* **25,** 2055 (1996).
[9] I. Stagljar, J. P. Bourquin, and W. Schaffner, *BioTechniques* **21,** 430 (1996).

DNA-binding domain (DBD), typically the amino-terminal end of the yeast transcription factor encoded by *GAL4* or the bacterial repressor protein encoded by *LexA*. Interaction of this DBD–protein fusion (a "bait") with a transcriptional activation domain-fused partner protein (either a defined partner, or a novel protein screened from a library) allows the activation of reporter genes (*lacZ*, *HIS3*, and *LEU2*) responsive to the cognate DBD. However, such *GAL4*- and *LexA*-based yeast two-hybrid systems do not operate at the endoplasmic reticulum (ER) or cell membrane and cannot detect interactions dependent on posttranslational modifications occurring within the ER, and interactions between integral membrane proteins.

In this chapter we describe a method that can detect interactions of membrane proteins with other soluble or membrane proteins *in vivo*.[10] The method employs the previously developed split-ubiquitin technique.[11] This technique is based on the ability of the N- and C-terminal fragments of ubiquitin (Nub and Cub)* to assemble into quasi native ubiquitin (split-ubiquitin). Ubiquitin-specific proteases (UBPs) do not recognize Nub or Cub alone, but recognize the reconstituted split-ubiquitin. The UBPs cleave a reporter protein attached to the C terminus of Cub from the assembled split-ubiquitin. Analogous to the two-hybrid system, the libration of the reporter serves as a readout indicating the reconstruction of ubiquitin. The system is designed in such a way that productive association of Nub and Cub is prevented unless the two ubiquitin halves are linked to proteins that interact *in vivo* (Fig. 1) (for more detail see Refs. 10–12).

Principle of Method

Figure 2 outlines the principle of the split-ubiquitin system applicable to membrane proteins. As a model, we use three defined ER membrane proteins involved in N-glycosylation. Wbp1p is an essential component of the yeast oligosaccharyltransferase complex[13] and is used as a bait protein. Wbp1p is known to interact with Ost1p,[14,15] another protein of the oligosac-

[10] I. Stagljar, C. Korostenky, N. Johnsson, and S. te Heesen, *Proc. Natl. Acad. Sci. U.S.A.* **95,** 5187 (1998).

[11] N. Johnsson and A. Varshavsky, *Proc. Natl. Acad. Sci. U.S.A.* **91,** 10340 (1994).

* Cub, C-terminal part of ubiquitin (amino acids 35–76); Nub, amino acids 1–37 of ubiquitin (either Nub*I* or Nub*A* or Nub*G*); Nub*I*, wild-type Nub, isoleucine at position 13; Nub*A*, Ile-13 replaced by alanine; Nub*G*, Nub Ile-13 replaced by glycine.

[12] M. Dünnwald, A. Varshavsky, and N. Johnsson, *Mol. Biol. Cell* **10,** 329 (1999).

[13] S. te Heesen, B. Janetzky, L. Lehle, and M. Aebi, *EMBO J.* **11,** 2071 (1992).

[14] S. Silberstein and R. Gilmore, *FASEB J.* **10,** 849 (1996).

[15] S. Silberstein, P. G. Collins, D. J. Kelleher, P. J. Rapiejko, and R. Gilmore, *J. Biol. Chem.* **128,** 525 (1995).

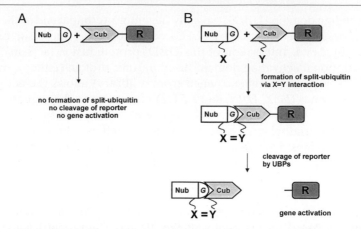

Fig. 1. The split-ubiquitin technique. (A) Nub*G* and Cub–reporter do not spontaneously associate to form split-ubiquitin. Hence, no cleavage of the reporter and no activation of reporter genes occur. (B) Nub*G* and Cub are linked to the interacting proteins X and Y. The X–Y complex brings Nub*G* and Cub into close proximity. Nub*G* and Cub reconstitute into split-ubiquitin, which is cleaved by the UBPs, yielding the free reporter R. The reporter can then activate reporter genes in the nucleus.

charyltransferase complex. Alg5p encodes dolicholphosphoglucose synthetase, an enzyme activity separate from that of the oligosaccharyltransferase.[16] Alg5p is not known to interact physically with Wbp1p or Ost1p.

The C terminus of Wbp1p is fused to the Cub interaction module, followed by a modified transcription factor, protein A-LexA-VP16 (PLV) (Fig. 2A). The protein A sequence allows sensitive and easy detection of the fusion protein as well as the cleaved reporter. Expression of the bait fusion Wbp1-Cub-PLV results in an enzymatically active Wbp1 fusion protein that does not activate reporter genes in the nucleus (Fig. 2B). Coexpression of Ost1-Nub*G* results in the interaction of Ost1p with Wbp1p and in the formation of the split-ubiquitin heterodimer. The formation of the heterodimer leads to cleavage of PLV from Wbp1-Cub-PLV. The liberated PLV can enter the nucleus and bind to LexA sites, leading to active transcription of the *lacZ* and *HIS3* reporters (Fig. 2D). Coexpression of Nub*G*-Alg5p does not lead to assembly of the split-ubiquitin heterodimer because Nub*G* does not spontaneously associate with Cub, and Wbp1p does not interact with Alg5p (Fig. 2C). Therefore, the specific interaction of Wbp1p with Ost1p as part of the oligosaccharyltransferase complex in the membrane of the endoplasmic reticulum can be detected via the activation of reporter genes in the nucleus.[10]

[16] S. te Heesen, L. Lehle, A. Weissmann, and M. Aebi, *Eur. J. Biochem.* **224,** 71 (1994).

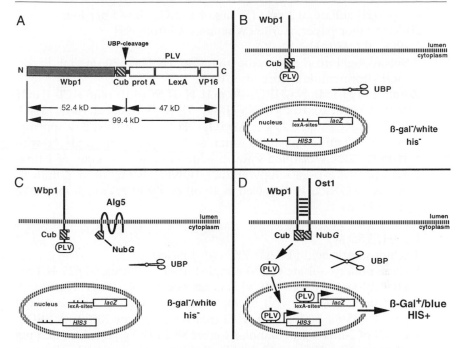

FIG. 2. Membrane protein-based two-hybrid system using Wbp1p, Ost1p, and Alg5p. (A) Scheme of the Wbp1-Cub-PLV fusion protein. Ubiquitin-specific protease(s) (UBP) cleave once at the C terminus of Cub, resulting in a Wbp1-Cub peptide of about 52 kDa and a soluble transcription factor PLV (protein A-LexA-VP16) of 47 kDa. (B) Because no split-ubiquitin can be formed, the Wbp1-Cub-PLV protein alone is stable. The PLV transcription factor is unable to activate genes, so that the cells lack β-galactosidase activity. (C) Coexpression of Wbp1-Cub-PLV with the noninteracting NubG-Alg5p does not lead to formation of the split-ubiquitin and therefore not to cleavage by UBP (closed scissors) and gene activation. (D) Interaction between Wbp1 and Ost1 results in a local increase in Cub and NubG concentration, leading to the formation of the split-ubiquitin heterodimer. The heterodimer is recognized and cleaved by the UBP (open scissors), liberating PLV. PLV enters the nucleus and binds to the LexA-binding sites of the *lacZ* and *HIS3* reporter genes, resulting in β-galactosidase activity. Cells are blue in the presence of X-Gal and grow on agar plates lacking histidine.

Materials and Reagents

Dropout medium (10×) without leucine and/or tryptophan for selection of plasmids: Dissolve 300 mg of L-leucine, 300 mg of L-lysine-HCl, 300 mg of L-tyrosine, 1500 mg of L-valine, 200 mg of adenine sulfate, 200 mg of L-arginine-HCl, 200 mg of L-histidine-HCl, 200 mg of L-methionine, 200 mg of L-tryptophan, 1000 mg of L-leucine, 500 mg of L-phenylalanine, and 2000 mg of L-threonine in 1 liter of deionized water

Copper(II) sulfate, 0.2 mM (50 mg of $CuSO_4 \cdot 5H_2O$ per liter)

III MM Filter paper, sterile (Whatman, Clifton, NJ)

$NaPO_4$, 0.1 M (pH 7.0)

5-Bromo-4-chloro-β-D-galactopyranoside (X-Gal, 20 mg/ml in dimethylformamide); store at $-20°$

Z-buffer: 113 mM Na_2HPO_4–40 mM NaH_2PO_4–10 mM KCl–1 mM $MgSO_4$ (pH 7.0): 16.1 g of $Na_2HPO_4 \cdot 7H_2O$, 5.5 g of $NaH_2PO_4 \cdot H_2O$, 0.75 g of KCl, and 0.25 g of $MgSO_4 \cdot 7H_2O$ are dissolved in approximately 900 ml of bidistilled water, and then adjusted to pH 7.0 with H_3PO_4; double-distilled water is added to a final volume of 1 liter

2-Nitrophenyl-β-D-galactopyranoside (ONPG, 4 mg/ml in Z-buffer): 40 mg of ONPG is dissolved in 10 ml of Z-buffer, which may take some hours

Na_2CO_3, 0.1 M

NaOH, 1.85 M

Trichloroacetic acid (TCA), 50% (w/v) in water

Sodium dodecyl sulfate (SDS)-sample buffer: 8 M urea, 0.0625 M Tris-HCl, 2% (w/v) SDS, 5% (v/v) 2-mercaptoethanol, 10% (v/v) glycerol, 8 M urea, 0.025% (w/v) bromphenol blue: 4.8 g of urea is mixed with 0.625 ml of 1 M Tris-HCl (pH 6.8), 1 ml of 20% (w/v) SDS, 0.5 ml of 2-mercaptoethanol, 1.1 ml of 88% (v/v) glycerol, and 2.5 mg of bromphenol blue; add double-distilled water to a final volume of 10 ml (dissolve at 37°)

Peroxidase–IgG

Methods

To test interactions between two proteins (here depicted as X and Y), construction of genes encoding hybrid proteins containing in-frame fusions to the Cub-PLV or Nub$G$ are made.

Step 1: Construction of Ubiquitin-Hybrid Proteins

The bait protein (Y) is fused to Cub-PLV via an XhoI site (codons CTC and GAG encoding leucine and aspartate, respectively). The sequence of the plasmid pRS305(Δwbp1-Cub-PLV) is available on request (Table I lists vectors and strains used). The Wbp1 sequence can be removed by cleavage with XhoI. The low copy number vector pRS305[6] is used for the bait construct. For stable expression of the construct, the vector must be linearized and integrated into the genome, e.g., into the *leu2* locus.

The prey protein (X) is fused to Nub either as an N- or C-terminal fusion on a high copy number plasmid. The native split-ubiquitin subunit

TABLE I
VECTORS AND STRAINS USED IN SPLIT-UBIQUITIN SYSTEM APPLICABLE TO
MEMBRANE PROTEINS

Plasmid	Characteristics	Ref.
pRS304	TRP1, no yeast origin of replication	6
pRS304(Δost1-Nub)	Ost1 (aa 102–476)-Nub, TRP1	5
pRS305	LEU2, no yeast origin of replication	6
pRS305(Δwbp1-Cub-PLV)	Wbp1p (aa 251–430)-Cub-PLVLEU2	5
pRS314	CEN6/ARSH4, TRP1	6
pRS314(NubI-Alg5)	NubI-Alg5, CEN6/ARSH4, TRP1	5
pAS2	TRP, 2-μm origin, GAL4 DB (reference plasmid)	8
pOST1-Nub	OST1-Nub, 2-μm origin, TRP1	5

Yeast strain	Genotype	Ref.
L40	*MAT*a *trp1 leu2 his3 LYS2::lexA-HIS3 URA3::lexA-lacZ*	8
YG0673	*MAT*a *trp1 leu2 his3 LYS2::lexA-HIS3 URA3::lexA-lacZ* wbp1:pRS305(Δwbp1-Cub-PLV)	5

NubI fused to X can be used as a positive control for expression. The prey construct can also be integrated into the genome.

Step 2: Yeast Transformation

The ubiquitin fusion proteins are transformed into strain L40 by a standard transformation protocol using lithium acetate.[17] Two plasmids can be simultaneously introduced into the L40 strain. However, we had more success in transforming sequentially first with the linearized bait pRS305(Δwbp1-Cub-PLV) followed by the prey plasmid (pNub-OST1). Transformants are plated on dropout plates lacking uracil, leucine, and tryptophan and incubated at 30° until colonies appear.

Step 3: Assays for β-Galactosidase

Transformants are allowed to grow at 30°, until they are large enough to assay for β-galactosidase activity, usually for 2–4 days. Two different assays can be used to detect or quantitate β-galactosidase activity.

Protocol 1: Filter Assay for Detection of β-Galactosidase Activity. Yeast cells expressing the bait together with the prey are grown for 2 days at 30°

[17] D. Gietz, J. A. St, R. A. Woods, and R. H. Schiestl, *Nucleic Acids Res.* **20,** 1425 (1992).

on sterile Whatman filters placed onto dropout agar plates lacking leucine and tryptophan. If the *CUP1* promotor is used, the plates are supplemented with 0.2 mM CuSO$_4$. Dropout medium is used because cells tend to grow poorly in standard minimal medium. The cells are lifted from the agar with the filter and lysed by submersion in liquid nitrogen for 3 min. After allowing the filter to warm to room temperature, the filters are overlaid with 1.5% (w/v) agarose in 0.1 M NaPO$_4$ buffer (pH 7.0) containing X-Gal (0.4 mg/ml). The filter is then incubated at 30° until a blue color appears. This may require from 20 min up to 24 hr. Published strains and plasmids are available for comparison and as controls.

Protocol 2: Quantitative β-Galactosidase Assay. Yeast transformants expressing Wbp1-Cub-PLV together with Nub fusion proteins are grown in liquid dropout medium lacking uracil, leucine, and tryptophan to an OD$_{546}$ of ~1. After pelleting cells from 1 ml of the culture, the samples are washed once in Z-buffer and resuspended in 300 μl of Z-buffer. Cells (100 μl) are taken and lysed by three freeze–thaw cycles. Z-buffer (700 μl) containing 0.27% (v/v) 2-mercaptoethanol and 160 μl of ONPG (4 mg/ml in Z-buffer) is added. The samples are incubated for 1–20 hr at 30°. To terminate the reaction, 400 μl of 0.1 M Na$_2$CO$_3$ is added, the samples are centrifuged, and the OD$_{420}$ is measured. The specific β-galactosidase activity is calculated by the following formula:

$$\text{β-Galactosidase units} = 1000 \times [\text{OD}_{420}/(\text{OD}_{546} \times \text{min})]$$

where OD$_{546}$ is the cell density and min is minutes of incubation.

Step 4: Detection of Protein–Protein Interaction by Cleavage of Y-Cub-PLV in Vivo

The interaction between X and Y leads to the cleavage of the PLV reporter, which can be determined by Western blot analysis and probing with peroxidase–IgG.[10]

The yeast cells expressing Wbp1-Cub-PLV together with Nub fusion proteins are grown in liquid dropout medium lacking uracil, leucine, and tryptophan to an OD$_{546}$ of ~1. The cells are pelleted and resuspended in 50 μl of 1.85 M NaOH per 3 OD units of cells, and incubated for 10 min on ice. An equal volume of 50% (w/v) trichloroacetic acid is added and the proteins are precipitated by centrifugation for 5 min. The supernatant is removed and the pellet resuspended in 50 μl of SDS-sample buffer containing 8 M urea. Twenty microliters of 1 M Tris base is added per 3 OD units of cells and the proteins are dissolved at 37° (heating to 95° sometimes results in clumping of membrane proteins). The samples are centrifuged for 2 min, and 10 μl of extract is used for SDS–polyacrylamide gel electrophoresis (PAGE)/Western blotting analysis. The amount of protein loaded is verified by Coomassie staining of the SDS gels after the transfer. The membranes are then probed with peroxidase–IgG at 1 : 2000–1 : 5000 dilution. The protein A fusion proteins are then detected by enhanced chemiluminescence [Pierce (Rockford, IL) or Amersham (Arlington Heights, IL)].

Concluding Remarks

The cellular localization of a protein is crucial for its specific interaction with other proteins. Therefore, it is important to determine the interaction at the natural place of the protein inside of the cell. The modified split-ubiquitin method allows membrane proteins to be tested in their native localization.

Interactions between two membrane proteins can be tested with one protein of interest fused to Cub-PLV and the other to Nub. The Nub fusion can also be a soluble protein, thus allowing detection of interactions between a membrane-bound receptor and soluble proteins. We have successfully tested Nub fusions both at the N terminus and the C terminus of proteins. The expression of PLV fusions can be monitored conveniently with the protein A sequence of the PLV cassette, using immunoprecipitation or Western blotting and probing with peroxidase–IgG.

The "bait" protein carrying Cub-PLV should not be overexpressed because overexpression results in false positives. Also, a soluble Cub-PLV without any Nub results in gene activation (I. Stagljar and S. te Heesen, unpublished, 1999). Therefore, the "bait" fusion protein must be anchored to a lipid bilayer. Soluble proteins of interest may be tested by fusing them to a membrane protein anchor. The Nub fusion protein should be overexpressed, because the fusion protein might have to compete with a wild-type protein for interactions within the protein complex. Any Ost1-Nub expressed from the low copy number vector pRS314 does not result in β-galactosidase activity unless wild-type Ost1p has been removed (data not shown).

The split-ubiquitin system can also be used to determine the topology of a membrane protein. Cleavage occurs only if both Nub and Cub localize to the cytosolic side of the membrane harboring the ubiquitin-specific proteases. For this purpose, a membrane protein of interest would be tagged with Cub-PLV at various locations and coexpressed with a soluble Nub*I* fusion protein. An extension of the system might be the identification of peptides or molecules that disrupt or alter the interactions between proteins. We are currently testing the screening for interactions between membrane proteins fused to Cub-PLV as the "bait" and cDNA or genomic library with N-terminal and C-terminal fusions to Nub$G$ as well as the PLV-Cub cassette fused to the N terminus of proteins. In summary, a new system has been generated, which will be useful in the detection and study of membrane protein interactions.

Acknowledgment

The authors thank Prof. Ulrich Hübscher and Prof. Markus Aebi for their support of this project. We are thankful to Jonne Helenius for critical reading of the manuscript.

[15] Alkaline Phosphatase Fusion Proteins for Molecular Characterization and Cloning of Receptors and Their Ligands

By JOHN G. FLANAGAN and HWAI-JONG CHENG

Introduction

A large number of so-called orphan receptors have been identified, which appear by their structure to be cell surface receptors for polypeptide ligands, yet have no known ligand. This has implied an opportunity to identify corresponding ligands that would be novel cell–cell signaling molecules. One strategy to identify such molecules has been to use the extracellular domain of the orphan receptor, fused to an alkaline phosphatase (AP) or immunoglobulin Fc tag, which can serve as an affinity probe to identify the corresponding ligand(s).[1,2] This approach appears to be widely applicable, and has been used to identify a large number of novel ligands (e.g., Refs. 1–8). An analogous strategy, using ligands fused to AP or Fc tags, has similarly been used to identify novel receptors.[9–11]

[1] J. G. Flanagan and P. Leder, Cell 63, 185 (1990).
[2] A. Aruffo, I. Stamenkovic, M. Melnick, C. B. Underhill, and B. Seed, Cell 61, 1303 (1990).
[3] R. J. Armitage, W. C. Fanslow, L. Strockbine, T. A. Sato, K. N. Clifford, B. M. Macduff, D. M. Anderson, S. D. Gimpel, T. Davis-Smith, C. R. Maliszewski, E. A. Clark, C. A. Smith, K. H. Grabstein, D. Cosman, and M. K. Spriggs, Nature (London) 357, 80 (1992).
[4] S. D. Lyman, L. James, T. V. Bos, P. de Vries, K. Brasel, B. Gliniak, L. T. Hollingsworth, K. S. Picha, H. J. McKenna, R. R. Splett, F. A. Fletcher, E. Maraskovsky, T. Farrah, D. Foxworthe, D. E. Williams, and M. P. Beckmann, Cell 75, 1157 (1993).
[5] T. D. Bartley, R. W. Hunt, A. A. Welcher, W. J. Boyle, V. P. Parker, R. A. Lindberg, H. S. Lu, A. M. Colombero, R. L. Elliott, B. A. Guthrie, P. L. Holst, J. D. Skrine, R. J. Toso, M. Zhang, E. Fernandez, G. Trail, B. Varnum, Y. Yarden, T. Hunter, and G. M. Fox, Nature (London) 368, 558 (1994).

In [2] in this volume[12] we describe the production of probes consisting of a soluble receptor or ligand fused to an AP tag. In this chapter we describe the use of such fusion proteins to identify and clone novel ligands or receptors. We also describe methods for the molecular characterization of ligands, receptors, and their interactions.

To clone a new ligand or receptor, the first step is usually to identify a good source. Traditionally, it had been thought that control molecules are likely to be present at low concentrations in tissues, and that cell lines may be a preferable source.[13] However, it has emerged that many receptors and ligands are expressed in tissues or embryos at high levels, and moreover this expression is often highly localized. An efficient procedure to identify a good source for cloning is therefore to test embryonic or adult tissues by *in situ* staining, using a receptor or ligand–AP fusion protein as a probe to identify locations of high expression.[6] An alternative approach is to screen cell lines. This can be done by quantitative binding of the AP fusion probe to the cell surface,[1] or alternatively by using a receptor–AP fusion protein to coimmunoprecipitate its ligand, which can then be visualized by gel electrophoresis.[14]

Once a good source has been identified, the next step is typically to use an expression cloning procedure. A cDNA library is prepared, expressed in a cell line, and screened for binding of the receptor– or ligand–AP fusion protein.[6] This type of procedure can be used to identify not only cell surface receptors or ligands, but also soluble ligands in the secretory pathway.[8] To clone soluble ligands, an alternative approach is to coimmunoprecipitate enough ligand protein to directly obtain amino acid sequence, and then use this information to make corresponding oligonucleotides, and isolate

---

[6] H.-J. Cheng and J. G. Flanagan, *Cell* **79,** 157 (1994).

[7] S. Davis, N. W. Gale, T. H. Aldrich, P. C. Maisonpierre, V. Lhotak, T. Pawson, M. Goldfarb, and G. D. Yancopoulos, *Science* **266,** 816 (1994).

[8] S. Davis, T. H. Aldrich, P. F. Jones, A. Acheson, D. L. Compton, V. Jain, T. E. Ryan, J. Bruno, C. Radziejewski, P. C. Maisonpierre, and G. D. Yancopoulos, *Cell* **87,** 1161 (1996).

[9] L. A. Tartaglia, M. Dembski, X. Weng, N. H. Deng, J. Culpepper, R. Devos, G. J. Richards, L. A. Campfield, F. T. Clark, J. Deeds, C. Muir, S. Sanker, A. Moriarty, K. J. Moore, J. S. Smutko, G. G. Mays, E. A. Woolf, C. A. Monroe, and R. I. Tepper, *Cell* **83,** 1263 (1995).

[10] Z. G. He and M. Tessier-Lavigne, *Cell* **90,** 739 (1997).

[11] A. L. Kolodkin, D. V. Levengood, E. G. Rowe, Y. T. Tai, R. J. Giger, and D. D. Ginty, *Cell* **90,** 753 (1997).

[12] J. G. Flanagan, H.-J. Cheng, D. A. Feldheim, Q. Lu, M. Hattori, and P. Vanderhaeghen, *Methods Enzymol.* **327,** Chap. 2, 2000 (this volume).

[13] R. Levi-Montalcini, *EMBO J.* **6,** 1145 (1987).

[14] M.-K. Chiang and J. G. Flanagan, *Growth Factors* **12,** 1 (1995).

cDNA clones by polymerase chain reaction or nucleic acid hybridization.[14]

We also describe in this chapter methods that can be used for the quantitative analysis of ligand–receptor interactions (e.g., see Refs. 1, 2, 15, and 16). These methods involve AP tagging one component of the ligand–receptor pair. The AP-tagged probe can then be tested for binding to its partner(s) on a cell surface. Alternatively, the binding partner can be fused to a second tag, such as an Fc tag, and binding can be tested in a cell-free system. These methods can allow confirmation that members of a candidate ligand–receptor pair actually do interact directly. They can also be used to investigate binding parameters such as affinities, kinetics, and the effect of cofactors or mutations. Because binding is quantitated by measuring the intrinsic enzyme activity of the AP tag, procedures are generally easier, safer, and less expensive than alternatives such as radiolabeling, BIAcore (Biacore, Uppsala, Sweden) measurements, or monitoring Fc fusions with secondary antibodies. The sensitivity is also as good as, or better than, these alternative methods, with detection into the subfemtomole range.

There is relatively little information on how the binding characteristics measured by these various methods compare with one another. Steric hindrance by the AP tag, which is attached through a flexible linker, does not generally seem to be a problem, although it is possible that it might interfere in individual cases. An important point to bear in mind is that the AP tag, like the Fc tag, is dimeric, and is expected to produce a fusion protein with a pair of ligand or receptor moieties, both facing away from the tag in the same direction. This is expected to increase the avidity for binding partners that are also oligomeric, or are bound to a cell surface or matrix (because the unfavorable entropy change of binding should be less than for monomers). The significance of this effect may vary case by case. For molecules that normally bind as monomers, it might introduce an artifact. On the other hand, for molecules that in their native state bind as dimers, or are attached to a cell surface or matrix, testing them as an artificial soluble monomer might be less informative than testing them as a dimer with a tag.

The first two sections of this chapter describe techniques for quantitative ligand–receptor binding measurements on cell surfaces or in solution. This is followed by a section describing characterization of ligands or receptors by coimmunoprecipitation. These approaches can be useful in several contexts: determining ligand–receptor binding characteristics; assessing the molecular properties of a ligand or receptor, especially when a good antibody is

[15] Z. E. Wang, G. M. Myles, C. S. Brandt, M. N. Lioubin, and L. Rohrschneider, *Mol. Cell. Biol.* **13,** 5348 (1993).
[16] A. M. Koppel, L. Feiner, H. Kobayashi, and J. A. Raper, *Neuron* **19,** 531 (1997).

not available; and identifying novel ligands or receptors that have not previously been characterized. In a final section of the chapter we discuss expression cloning of ligands or receptors.

Quantitative Measurement of Ligand–Alkaline Phosphatase or Receptor–Alkaline Phosphatase Binding to Cell Surfaces

In this method, the soluble receptor– or ligand–AP fusion probe is added to cultured cells bearing the cognate ligand or receptor on their surface. The cells are then washed, lysed, and assayed for bound AP activity (Fig. 1A and B).

As a preliminary to cloning, this is the method of choice for screening cell lines for potential expression of a receptor or ligand that is cell surface associated (Fig. 1A). When applied to cultured cells, this method is much more sensitive than *in situ* staining,[12] and also has the advantage that the results are quantitative. On the other hand, *in situ* staining is generally more suitable to identify a novel ligand or receptor in an embryo or tissue, where the ability to localize the signal can improve sensitivity and provide more information (see [2] in this volume[12]).

This procedure can also be used to study quantitative aspects of ligand–receptor interactions. In the most common application, the affinity of the interaction can be determined by varying the concentration of AP fusion protein in solution, and then quantitating the amount bound to the cells (Fig. 1B). A full discussion of binding analyses is beyond the scope of this chapter. Briefly, the most common way to calculate affinity is by a Scatchard plot (Fig. 1B). To obtain reliable results, it is important to use a wide range of concentrations extending from well below the K_D value to near saturation levels. A straight line in the Scatchard plot indicates a single affinity class of sites, whereas a curved plot may indicate more than one class of sites. In addition to measuring affinities, quantitative binding to cell surfaces can also be used to determine interaction kinetics, the effects of mutations in a structure–function analysis of receptor or ligand, or the effects of binding conditions or cofactors (see, e.g., Refs. 14 and 15).

The protocol below is for adherent cells. However, it can be performed equally well with suspension cells, by placing them in a microcentrifuge tube at the start, and performing washes with a low-speed spin for 1 min at 5000 rpm in a microcentrifuge.

Procedure for Quantitation of Alkaline Phosphatase Fusion Protein Binding to Cell Surfaces

1. Grow the cells to be tested until they are almost confluent, or have just reached confluence. This can be done in a six-well tissue culture plate.

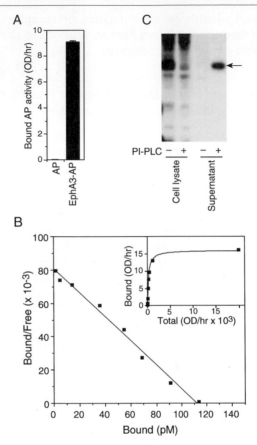

FIG. 1. Ligand characterization using a soluble receptor–AP probe. Note that all experiments here show characterization of endogenous (not artificially overexpressed) ligand produced by BRL-3A cells. The probe is a fusion protein consisting of the ectodomain of the EphA3 receptor fused to AP.[6] (A) Quantitation of binding of receptor–AP probe to the surface of cultured cells. Cells were treated with supernatant containing EphA3–AP fusion protein, washed, and assayed for bound AP activity. Unfused AP is used as a control. This type of simple binding experiment can be used to screen cell lines for expression of a novel ligand or receptor that has not previously been identified. (B) Measurement of receptor–ligand binding affinity. The inset graph shows the increase in binding as the concentration of EphA3–AP is increased. A specific receptor–ligand interaction is expected to be saturable. The same data are converted into a Scatchard plot (bound versus bound/free). This plot appears to be linear over the range tested, indicating a single class of binding site, with 50,000 sites per cell and $K_D = 1.3 \times 10^{-9}\,M$. (C) Analysis of ligand by coimmunoprecipitation. Cultured BRL-3A cells were labeled with [^{35}S]methionine, and then the cell lysate and supernatant were separately incubated with EphA3–AP fusion protein. After immunoprecipitation with an anti-AP antibody, and electrophoresis on an SDS–polyacrylamide gel, the coimmunoprecipitated ligand polypeptide (arrow) can be visualized by autoradiography. This type of procedure can be used to identify a novel ligand or receptor that has not previously been identified. It can

2. Wash the cells once with 5 ml of cold HBAH [Hanks' balanced salt solution, bovine serum albumin (0.5 mg/ml), 0.1% (w/v) NaN$_3$, 20 mM HEPES (pH 7.0)]. During the wash steps, the cells can dry and fall off quickly if all the medium is aspirated out of the plate. The problem is mainly seen around the edges, and thus is more severe in smaller plates. To minimize this effect, pipette the medium out, but leave just enough to provide a thin covering. With experience this can be done quickly with a vacuum aspirator by withdrawing the tip of the pipette as soon as the liquid level reaches the bottom of the well at its center.

3. Add 1 ml of AP fusion protein solution and incubate at room temperature for 90 min. Swirl briefly to mix at approximately the 30- and 60-min time points.

4. Remove the AP fusion protein solution with a pipette. Wash the cells six times with 5 ml of cold HBAH. For each wash, incubate HBAH with the cells for 5 min and gently swirl the medium by hand or on a platform shaker.

5. Aspirate out all the remaining HBAH completely, and lyse the cells with 300 μl of Triton-Tris buffer [1% (v/v) Triton X-100, 10 mM Tris-HCl (pH 8.0)] at room temperature. It usually takes a few minutes at most for the cells to dissolve.

6. Collect all the lysate in an Eppendorf tube, rinse the plate with an additional 200 μl of Triton-Tris, and pool this with the first lysate. Vortex for 30 sec, allow to sit at room temperature for 5 min, and vortex again.

7. Spin down the lysate at maximum speed in a microcentrifuge, and transfer the supernatant to another Eppendorf tube.

8. Heat inactivate the supernatant in a 65° water bath for 10 min.

9. Put the supernatant on ice to cool it.

10. Take 100 μl of supernatant and add an equal amount of 2× AP substrate buffer to check the AP activity as described in [2] in this volume.[12]

If there are traces of background alkaline phosphatase activity even after heat inactivation, this can be removed by immunoprecipitating the fusion protein, as described in [2] in this volume,[12] omitting steps 2, 3, and 6, and finally adding a suspension of the antibody-Sepharose beads to an equal volume of 2× AP substrate solution. However, it is usually unneces-

---

also be valuable in molecularly characterizing ligands or receptors, especially when a high-quality antibody is not available. For example, in this case, even though no antibody was yet available, it was possible to show that the ligand polypeptide is linked to the cell surface by a glycosyl phosphatidylinositol linkage, which can be cleaved by the enzyme PI-PLC.

sary to perform this immunoprecipitation step. The results should always be compared with negative controls run in parallel, to control for any traces of endogenous alkaline phosphatase activity, and because the AP enzyme substrate p-nitrophenyl phosphate has a low rate of spontaneous hydrolysis. It is also desirable to compare the fusion protein results with unfused AP as a control.

Quantitative Analysis of Ligand–Receptor Binding in Solution

Ligand–receptor binding can be analyzed, as described above, by testing the binding of a soluble AP fusion protein to a cell surface. However, multiple cellular components will be present that in principle could contribute to binding, or cause misleading results. For example, if a novel cDNA is isolated by expression cloning on the basis that it confers binding activity for a receptor–AP fusion protein, in principle this cDNA could encode either a ligand for that receptor, or a molecule that upregulates the true ligand. To demonstrate a direct interaction between ligand and receptor, binding assays can be performed in a cell-free system. More generally, this procedure provides a method to characterize ligand–receptor interactions in solution, in the absence of other cellular components.

A convenient way to perform this procedure is to fuse the receptor to an AP tag, and fuse the ligand to an immunoglobulin Fc tag, or vice versa. This provides for a simple assay, where the ligand can be immobilized on anti-Fc beads, and receptor binding can then be monitored by measuring the associated AP activity. Instead of an Fc tag, other tags can also be used such as Myc or hemagglutinin (HA) epitopes, or a hexahistidine (His_6) tag.

Procedure to Analyze Ligand–Receptor Binding in Cell-Free System

1. Prepare ligand–immunoglobulin fusion protein. The ligand (or receptor) cDNA can be fused to the sequence of the human IgG_1 Fc region in a vector such as pcDNAI.[2,17] Prepare the Fc fusion protein in COS cells or 293T cells as described for the preparation of AP fusion protein.[12]

2. Incubate ligand–Fc fusion proteins with an equal volume of protein A-conjugated Sepharose beads (Pharmacia, Piscataway, NJ) on a rotator at room temperature for 1 hr. Protein A binds the Fc region.

3. Wash the beads twice with HBAH.

4. Add a 15-μl aliquot of beads to 500 μl of receptor–AP fusion protein and incubate on a rotator at room temperature for 2 hr.

---

[17] A. D. Bergemann, H.-J. Cheng, R. Brambilla, R. Klein, and J. G. Flanagan, *Mol. Cell. Biol.* **15,** 4921 (1995).

5. Wash the beads five times with HBAH and once with HBS (150 mM NaCl, 20 mM HEPES, pH 7.0).

6. Incubate the beads with 100 μl of HBS in a 65° water bath for 10 min, and then transfer to ice.

7. Add an equal amount of 2× AP substrate buffer to measure the AP activity.[12]

Coimmunoprecipitation of Binding Partners with Receptor– or Ligand–Alkaline Phosphatase Fusion Proteins

This procedure involves incubating the receptor– or ligand–AP fusion probe in solution with a cell supernatant or lysate containing the cognate binding partner, usually radiolabeled. The fusion protein is then immunoprecipitated with an antibody against AP, and any coprecipitated binding partner can then be identified on an electrophoretic gel (Fig. 1C).

As a preliminary to cloning, this method can be used to screen cell lines, and this can be particularly useful for soluble ligands. If a soluble ligand is identified in this way, it can be cloned by expression methods, as described below. Alternatively, the coimmunoprecipitation procedure can be scaled up to obtain peptide sequence,[14] as described at the end of the protocol.

In addition to cloning projects, this coimmunoprecipitation method can also be extremely useful to characterize the molecular weight and other properties of a ligand or receptor (Fig. 1C). This can be especially useful when no antibody is available for direct immunoprecipitation to characterize the molecule of interest.[18] In this regard, it is worth noting that raising a high-quality antibody against an extracellular receptor or ligand can take months or even years, whereas AP fusion proteins can be prepared in as little as 1 or 2 weeks.

Procedure for Coimmunoprecipitation with Alkaline Phosphatase Fusion Proteins

1. For a six-well tissue culture plate, label cells with 2 ml of labeling solution (Dulbecco's modified Eagle's medium [DMEM] without methionine, containing 10% (v/v) dialyzed serum and 400 μCi of [^{35}S]methionine) at 37° for 3–7 hr.

2. Collect the supernatant and concentrate to about 200 μl on a Centricon-10 (Amicon, Danvers, MA).

[18] J. G. Flanagan, D. C. Chan, and P. Leder, *Cell* **64,** 1025 (1991).

3. Wash the cells five times with 5 ml of HBAH buffer, and lyse in 200 μl of Triton-Tris buffer containing 1 mM phenylmethylsulfonyl fluoride (PMSF). The nuclei are spun out by centrifugation for 10 min in a microcentrifuge at maximum speed.

4. Incubate the concentrated supernatants and the cell lysates (separately) with an equal volume of AP fusion protein solution on a rotator at room temperature for 90 min.

5. Add to each tube 20 μl of Sepharose beads coupled with excess anti-AP antibodies and incubate for 30 min on a rotator at room temperature. The preparation of anti-AP beads is described in [2] in this volume.[12]

6. Wash the beads twice in TBS [150 mM NaCl, 25 mM Tris (pH 8.0)]–0.1% (v/v) Nonidet P-40 (NP-40), three times in modified radioimmunoprecipitation assay (RIPA) buffer [0.5% (v/v) NP-40, 0.5% (w/v) sodium deoxycholate, 0.1% (w/v) sodium dodecyl sulfate (SDS), 0.1% (w/v) NaN$_3$ 144 mM NaCl, 50 mM Tris-HCl (pH 8.0)], and once in TBS–0.1% (v/v) NP-40. Use ice-cold buffers and do this step quickly. Spins are at 5000 rpm for 1 min.

7. Add an equal volume of loading buffer and heat the sample for 2 min at 100°. Any molecules that bind to the AP fusion protein are analyzed on an SDS–polyacrylamide gel.

If a putative ligand is found by radioactive detection, the procedure can be scaled up for microsequencing. Microsequencing usually requires amounts of protein in the microgram range, although advances in methodologies such as sequencing by mass spectrometry may reduce this requirement. In any case, the amounts required are not large, and with a typical ligand production in the nanograms per milliliter range, it should be possible to isolate enough ligand from 1 liter or so of conditioned medium. The ligand concentration in conditioned medium can be estimated in gels of immunoprecipitated protein by silver staining or any other sensitive staining method. The following provides, in brief, an example of a scale-up procedure for sequencing a ligand.[14] A volume of 0.2 ml of Sepharose beads coupled to 0.4 mg of anti-AP antibody is incubated with a saturating amount of AP fusion protein and is then washed and cross-linked with dimethyl pimelimidate[19] to prevent the AP fusion protein from leaching off the beads. One liter of conditioned medium containing 1% (v/v) calf serum is collected and concentrated to 30 ml with an Amicon pressure cell, and is then incubated on a rotator with the AP fusion protein cross-linked to beads. The beads are then loaded on a column, washed with modified RIPA buffer and then with 10 mM sodium phosphate, pH 6.8, and eluted with

[19] E. Harlow and D. Lane, "Antibodies: A Laboratory Manual." Cold Spring Harbor Laboratory Press, Cold Spring Harbor, New York, 1988.

100 mM glycine, pH 2.5. The eluted sample is precipitated with trichloroace-tic acid (TCA), run on an SDS–polyacrylamide gel, and transferred to a polyvinylidene difluoride (PVDF) membrane for microsequencing.[14] Pep-tide sequences obtained by this type of procedure can be compared with the sequence databases. If the peptide does not correspond to any known sequence, it can be used to design oligonucleotides for polymerase chain reaction (PCR) or library screening.

Expression Cloning

As a first step, before making or obtaining a cDNA expression library, it is generally well worth while to devote some effort to finding a good source material of RNA. Enriching the representation of the target molecule in this way at the outset can save work later at the screening stage, and can help to achieve a successful outcome. Sometimes a good guess about a ligand source can be made on the basis of the biology of the receptor. A more direct approach is to use a receptor– or ligand–AP fusion protein to detect expression of the putative binding partner in cells or tissues. This can be done by testing tissues or embryos with receptor– or ligand–AP fusions as *in situ* probes, as described in [2] in this volume.[12] Alternatively, it can be done by screening cell lines, using the quantitative cell surface binding or coimmunoprecipitation methods described earlier in this chapter.

If a good source of ligand expression can be identified with an AP fusion protein probe, then one can have a reasonable confidence that an expression cloning approach should work. A potential problem with all expression cloning approaches is that in principle they might fail if the target molecule consists of more than one polypeptide chain. However, in practice there are many examples of ligands or receptors that have been successfully expression cloned even though they are heterooligomers in their native state, because each chain of the oligomer may be itself have a sufficiently high binding affinity.

Once a good source of RNA has been identified, a cDNA library can be prepared. The library is then expressed in a cell line, and is screened for binding of the receptor– or ligand–AP fusion protein. This expression cloning approach can be used to clone cell surface molecules (Fig. 2). It can also be used to clone soluble molecules, by detecting them while they are in the secretory pathway.[8]

*Procedure to Clone Ligand Using Receptor–Alkaline Phosphatase
    Fusion Probe*

1. Prepare total RNA and then poly(A)$^+$ RNA from a tissue or cell line with high expression of the ligand. Prepare double-stranded cDNA

A. Primary screen   B. Tertiary screen

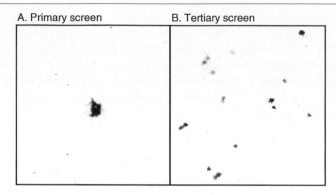

FIG. 2. Screening a COS cell expression library with receptor–AP probe. COS cells transfected with library pools were treated with mixed supernatants containing EphA3–AP and EphA4–AP fusion proteins, and then the cells were washed, fixed, and stained for bound AP activity. (A) A single positively stained cell in the primary round of screening. (B) A field of cells in the third round of screening. Magnification is lower in (B) than in (A). Unstained cells are not visible here, but the cells were nearly confluent at the time of staining.

and insert into an appropriate expression vector to make a cDNA library for transient expression. Size-selecting the cDNA to remove molecules smaller than the predicted size of the molecule to be cloned is likely to greatly improve the quality of the library. Synthesis of cDNA usually results in a large amount of short fragments; moreover, smaller cDNAs are present at higher molar concentrations for the same weight, and also are likely to replicate preferentially after insertion into the vector. It is therefore worthwhile to eliminate cDNAs that are smaller than the molecule to be cloned. In our experience gel purification methods work well and gives an efficient separation of large and small molecules. Several manufacturers provide kits for RNA purification, cDNA synthesis, and library construction. A number of expression vectors are available; we have had good results with the CDM8 vector[20] (InVitrogen, San Diego, CA), which contains the simian virus 40 (SV40) origin and can replicate efficiently in cells containing the SV40 T antigen, such as COS cells or 293T cells.

2. Transfect the library into competent *Escherichia coli* to make pools. Use a pool size of approximately 1000 to 2000 clones per pool. It is important not to make the pool size too large. If the pool size is in the range of 1000 to 2000 clones per pool, a positive pool should contain several positive cells and should be obvious at the screening stage, whereas larger pool sizes can cause difficulties in telling the difference between true-positive and false-positive pools. For each pool, plate the transfected *E. coli* on a nitrocellulose

---

[20] A. Aruffo and B. Seed, *Proc. Natl. Acad. Sci. U.S.A.* **84**, 8573 (1987).

filter on an agar plate, so that a replica filter can be made. Keep the original filter on agar at 4°. Grow the bacteria on the replica and use this for the next step.

3. Collect the colonies from the replica and prepare plasmid DNA. For a 15-cm filter, we scrape off the bacteria in 15 ml of LB medium, and make a conventional alkaline-SDS DNA minipreparation.

4. To screen the library, 4 μg of DNA of each pool is transiently transfected into a 10-cm plate of COS cells. This can be done using LipofectAMINE (GIBCO-BRL, Gaithersburg, MD) or any other high-efficiency transient transfection procedure.

5. Forty-eight hours after the start of transfection, affinity probe the cells *in situ* with the receptor– or ligand–AP fusion probe. Details of *in situ* staining on cultured cells are given in [2] in this volume.[12] Different sizes of plates can be used, such as 6-well plates. However, smaller plates can give problems with edge effects such as loss of cells or differences in washing efficiency, whereas a 10-cm dish gives a large, uniform central area. For library screening it is important to ensure a uniform density of cells over all parts of the plate, and to stain cells that are just under confluence, or just recently confluent. Overconfluent cells can pile up, trapping the fusion protein probe and sometimes causing unpredictable background staining. It is also important to ensure uniform heat inactivation by placing the plates exactly horizontal so there is an even depth of liquid over the cells.

6. Staining should be monitored periodically against a white background under a dissecting microscope (Fig. 2). We prefer initially screening the plate under a dissecting microscope rather than a high-powered microscope, because it is faster, and it is usually easier to distinguish false positives such as clumps of cells, dead cells, and so on. Staining of positive cells may become visible in about 30 min, or may take a few hours. False positives will tend to appear after 15 hr. A positive pool should usually show 10 or more positive cells. The distribution of staining within the cell may also be a diagnostic feature: Cell surface staining should be distributed over the whole cell, whereas detection in the secretory pathway is likely to be concentrated in the perinuclear region. True positive cells should be distributed evenly over the plate. Sometimes a cluster of two or a few cells may result from division of a genuine positive cell, but a large number of stained cells in only one region of the plate is more likely to result from locally inadequate heat inactivation, or too high a cell density.

7. When a positive pool is obtained, take the original bacterial colony filter, make another replica, and cut the replica into 10 segments that will be 10 subpools (make marks on the original to confirm which colonies correspond to each subpool). Always keep the original filter and make DNA from the replica.

8. Collect DNA from each subpool and screen as described above (steps 3 to 6) to identify a positive subpool.

9. Pick individual colonies from the region of the original filter corresponding to the positive subpool (about 100 to 200 colonies). Transfer to a 96-well plate so there are one or two colonies in each well, with each well containing 200 μl of growth medium. Grow bacterial colonies overnight. Take an aliquot of bacterial culture from each well, and pool aliquots of bacterial culture from each column of wells and each row of wells. Screen again (steps 3 to 6). By matching up the positive row and the positive column it should be possible to identify the positive well and subsequently isolate a positive clone in a final round of screening. In the case of a true-positive clone, each successive screening should have an increased number of positive cells per plate. We have not encountered problems with false positives due to plasmids encoding or upregulating a heat-stable AP activity, but any such plasmids can be ruled out by a control in which unfused AP, or no AP reagent, is added instead of the receptor– or ligand–AP fusion probe.

[16] Surface Chimeric Receptors as Tools in Study of Lymphocyte Activation

By BRYAN A. IRVING and ARTHUR WEISS

Introduction

Cell surface receptors function to translate cues from the local extracellular environment to the cell interior, where they initiate appropriate cellular responses. Many receptors have evolved to perform this function within the confines of a single polypeptide, utilizing intrinsic enzymatic activity or association with signal transducing molecules to communicate to the cell. In contrast, other receptors rely on the cooperation of multiple subunits to facilitate optimal ligand binding and signaling function. Many such oligomeric receptors are present on the surface of lymphocytes and include cytokine receptors, Fc receptors, and the antigen receptors on B and T cells (Fig. 1). The most structurally complex of these, the T cell antigen receptor (TCR), is composed of as many as seven distinct integral membrane proteins, each of which is required for assembly and expression of the receptor at the cell surface.[1] The stringent constraints on assembly

---

[1] J. D. Ashwell and R. D. Klausner, *Annu. Rev. Immunol.* **8,** 139 (1990).

T Cell Antigen Receptor   B Cell Antigen Receptor   Fc$_\varepsilon$R1

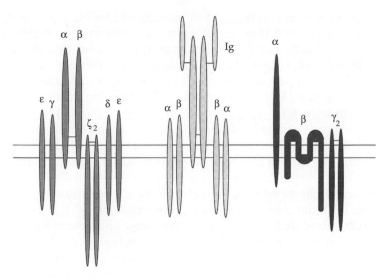

FIG. 1. Schematic representation of three oligomeric receptors expressed in the immune system.

and expression within this group of receptors—frequently imposed by the subunit extracellular and/or transmembrane domains—had hindered progress in determining the function of individual receptor components.

The lack of an approach to address the functional significance of this structural complexity inspired the advent of a technique that is the subject of this chapter, that is, the use of chimeric surface proteins as tools in dissecting the function of oligomeric receptors and signaling molecules. By replacing the extracellular and transmembrane domains of the TCR invariant chains with heterologous sequences, we and others successfully expressed the cytoplasmic sequences from these chains at the cell surface with the surprising finding that they each contained signal-transducing potential. This technique has since been expanded to allow the specific manipulation of virtually any cytoplasmic domain or protein of choice for analysis of its functional properties. In this chapter, we showcase the variety of chimeras that can be made, using examples from the literature to illustrate the wide utility of this technique. We also discuss theoretical considerations and technical approaches for engineering chimeric receptors in addition to experimental methods for analyzing their surface expression and function in lymphocytes.

Chimeric Receptors to Isolate Individual Subunits from
  Oligomeric Receptors

For the sake of illustration and to provide a historical perspective, the
T cell antigen receptor (Fig. 1) will serve as a prototype to highlight the
utility of the chimera technology. For years it was known that the clonally
variant TCR α and β chains defined the antigen specificity of the receptor,
recognizing foreign peptides presented on the surface of major histocom-
patability complex (MHC) molecules.[2,3] The function of the remaining
invariant subunits, CD3 γ, δ, ε, and ζ, remained unclear, although it was
assumed that one or more likely served to couple ligand binding to intracel-
lular signal transduction events. The structural basis for the interaction
between the TCR chains had been shown to reside in their transmembrane
domains, where oppositely charged residues in the normally hydrophobic
environment of the membrane—positive in the α/β heterodimer, and negative in the invariant chains—facilitate proper assembly.[4-6] It is this feature,
in part, that prevented the individual expression of the receptor subunits.
To circumvent the problem, it was reasoned that replacement of the transmembrane and extracellular sequences with those of another transmembrane protein would relieve the structural constraints, allowing isolated
expression of the cytoplasmic portion of the individual chains; the heterologous extracellular domain would also serve to distinguish the chimeric
receptor from its endogenous counterpart. This notion proved to be true.
A chimeric protein consisting of the cytoplasmic domain of the ζ chain
fused to the transmembrane and extracellular domains of CD8 could be
expressed independently of the TCR.[7] Importantly, ligation of the CD8–ζ
chimera with monoclonal antibodies revealed the capacity of ζ to directly
activate signal transduction pathways that were identical to those activated
by an intact TCR. A similar strategy linking ζ to the extracellular domain
of CD4 was successful, suggesting a broader application of the approach.[8]

Soon after the initial success, a number of chimeras were generated by
utilizing a variety of extracellular and transmembrane domains linked to

---

[2] Z. Dembic, W. Haas, S. Weiss, J. McCubrey, H. Kiefer, H. von Boehmer, and M. Steinmetz, *Nature* (*London*) **320,** 232 (1986).
[3] T. Saito, A. Weiss, J. Miller, M. A. Norcross, and R. N. Germain, *Nature* (*London*) **325,** 125 (1987).
[4] H. Clevers, B. Alarcon, T. Wileman, and C. Terhorst, *Annu. Rev. Immunol.* **6,** 629 (1988).
[5] L. Tan, J. Turner, and A. Weiss, *J. Exp. Med.* **173,** 1247 (1991).
[6] N. Manolios, J. S. Bonifacino, and R. D. Klausner, *Science* **249,** 274 (1990).
[7] B. A. Irving and A. Weiss, *Cell* **64,** 891 (1991).
[8] C. Romeo and B. Seed, *Cell* **64,** 1037 (1991).

ζ (Fig. 2A) or the remaining TCR invariant chains (Fig. 2B).[9–12] These chimeras revealed an unexpected redundancy of signaling capacity of the CD3 and ζ chains. The ability to analyze functionally the cytoplasmic domain of an isolated TCR chain facilitated structure–function studies that revealed an activation motif present singly in each of the CD3 chains and triplicated in the ζ chain.[11,13,14] Moreover, this motif was also contained within subunits of several other oligomeric receptors of the immune system and was thus designated ITAM, or immunoreceptor tyrosine-based, activation motif. With the use of chimeric receptors (some of which are depicted in Fig. 2B) the functional integrity of the motifs present in subunits of the B cell antigen receptor and Fc receptors was confirmed.[8,9,15,16] Furthermore, this technology allowed the confirmation of functional ITAMs in several plasma membrane molecules encoded by viruses that cause pathology within the hematopoietic lineage.[17–19]

In addition to providing an assessment of the signaling capacity of various cytoplasmic domains, the chimeric proteins have facilitated discovery of the mechanisms underlying this capacity. For example, studies utilizing chimeric antigen receptor chains have identified the molecular basis for the direct association of the ZAP-70 and Syk tyrosine kinases with the ITAM motifs[14,20,21]; defining these association requirements without this approach would have been arduous, at best, given the multiplicity of ITAMs in antigen receptors. Moreover, the CD8–ζ chimeric protein greatly simpli-

[9] F. Letourneur and R. D. Klausner, *Proc. Natl. Acad. Sci. U.S.A.* **88,** 8905 (1991).

[10] F. Letourneur and R. D. Klausner, *Science* **255,** 79 (1992).

[11] C. Romeo, M. Amiot, and B. Seed, *Cell* **68,** 889 (1992).

[12] L. K. Timson Gauen, A. N. Kong, L. E. Samelson, and A. S. Shaw, *Mol. Cell. Biol.* **12,** 5438 (1992).

[13] A. M. Wegener, F. Letourneur, A. Hoeveler, T. Brocker, F. Luton, and B. Malissen, *Cell* **68,** 83 (1992).

[14] B. A. Irving, A. C. Chan, and A. Weiss, *J. Exp. Med.* **177,** 1093 (1993).

[15] K. M. Kim, G. Alber, P. Weiser, and M. Reth, *Eur. J. Immunol.* **23,** 911 (1993).

[16] U. Wirthmueller, T. Kurosaki, M. S. Murakami, and J. V. Ravetch, *J. Exp. Med.* **175,** 1381 (1992).

[17] P. Beaufils, D. Choquet, R. Z. Mamoun, and B. Malissen, *EMBO J.* **12,** 5105 (1993).

[18] H. Lee, J. Guo, M. Li, J. K. Choi, M. DeMaria, M. Rosenzweig, and J. U. Jung, *Mol. Cell. Biol.* **18,** 5219 (1998).

[19] M. Lagunoff, R. Majeti, A. Weiss, and D. Ganem, *Proc. Natl. Acad. Sci., U.S.A.* **96,** 5704 (1999).

[20] M. Iwashima, B. A. Irving, N. S. van Oers, A. C. Chan, and A. Weiss, *Science* **263,** 1136 (1994).

[21] A. C. Chan, N. S. van Oers, A. Tran, L. Turka, C. L. Law, J. C. Ryan, E. A. Clark, and A. Weiss, *J. Immunol.* **152,** 4758 (1994).

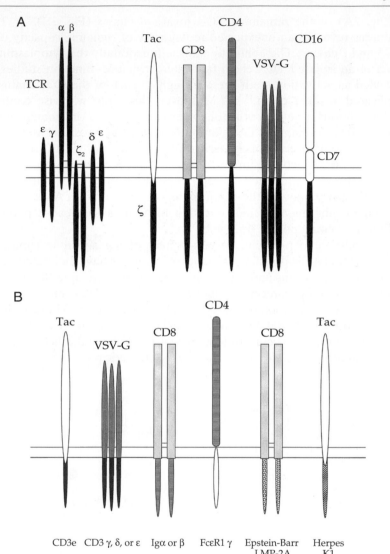

Fig. 2. Schematic representation of chimeric proteins used to analyze the function of individual subunits of oligomeric receptors. (A) The ζ chain in its native context of the T cell antigen receptor and in the context of several heterologous extracellular and transmembrane domains. (B) Chimeric receptors designed to test the function of the activation motif (ITAM) contained within the cytoplasmic domains of the designated receptor subunits or virally encoded proteins.

CD8/TCRαβ   Ig/ζ   Kinase Chimeras   Phosphatase Chimeras

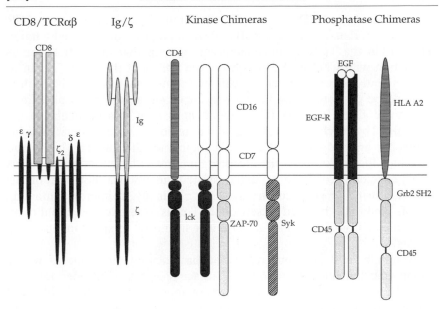

FIG. 3. Expanded utility of surface chimeric receptors. Schematic representation of a variety of chimeric receptors used in the study of lymphocyte activation.

fied the purification of ZAP-70 for its subsequent cloning.[22] In addition, by relieving association and assembly constraints, chimeric receptors have allowed expression of individual subunits in nonlymphoid cells to address tissue-specific requirements for their function.[20,23] In numerous instances, the chimera technology has proved to be extremely valuable for studying the function of individual components of multisubunit receptors.

Expanded Utility of Surface Chimeric Receptors in Analysis of Lymphocyte Function

Transmembrane chimeras have the potential for utility in a myriad of contexts in addition to their role in simplifying the complexity of oligomeric receptors (Fig. 3). Like their cytoplasmic counterparts, surface chimeras can be used to map critical domains in protein interactions. A chimeric heterodimer consisting of the extracellular domains of CD8 linked to the transmembrane and brief cytoplasmic domains of the TCR α/β heterodimer helped define the structural requirements for associations between the

[22] A. C. Chan, M. Iwashima, C. W. Turck, and A. Weiss, *Cell* **71,** 649 (1992).
[23] C. G. Hall, J. Sancho, and C. Terhorst, *Science* **261,** 915 (1993).

antigen-binding and CD3 chains (Fig. 3).[5] Similarly, a fusion between TCR α and the interleukin 2 receptor (IL-2R) α chain (Tac) defined eight amino acids within the TCR α transmembrane domain that are critical for its association with CD3 δ.[6] Furthermore, the ability of the CD3 ζ chain to independently activate a T cell in a chimeric context suggests another utility. It should be possible, using ITAM-bearing chimeras, to specifically direct a cytotoxic T cell in an MHC-independent fashion to a particular target determined by the choice of extracellular domain utilized. Chimeric proteins linking ζ to antibody heavy chains (Fig. 3), immunoglobulin (Ig) heavy and light chain fusions, or CD4 [the receptor for human immunodeficiency virus (HIV) envelope protein] have demonstrated the practical potential of this approach using *in vitro* culture systems.[8,24] Finally, chimeras containing ITAMs could also be used as sensors in T cells to test the potential of an extracellular domain to bind a particular ligand or to spontaneously aggregate.

Although until now our discussion has focused on chimeras whose cytoplasmic domains derive from cell surface receptors, a broader potential exists. In an innovative use of the chimera approach, the extracellular and transmembrane domains of CD16 and CD7, respectively, were fused to the sequences encoding the cytoplasmic tyrosine kinases implicated in TCR signaling, i.e., Lck, Fyn, ZAP-70, and Syk.[25] By expressing the kinases in this fashion, they could be manipulated independently of TCR engagement and when aggregated in appropriate combinations, were sufficient to activate proximal signaling. Importantly, these experiments provided evidence of differential activation requirements for Syk and ZAP-70, with only the latter requiring phosphorylation by a Src family kinase for its activation. Another study, utilizing a fusion linking Lck to the extracellular and transmembrane domains of the coreceptor CD4 (Fig. 3), provided evidence of a kinase-independent role for Lck in coreceptor function.[26] A similar approach fusing the epidermal growth factor receptor (EGF-R) extracellular and transmembranse domains to the cytoplasmic domain of the CD45 tyrosine phosphatase has led to the speculation that dimerization of transmembrane phosphatases may inhibit rather than activate their catalytic functions.[27,28] These experiments highlight an important feature of chimeric receptors, that is, that despite the presence of endogenous receptors (in

[24] M. R. Roberts, L. Qin, D. Zhang, D. II. Smith, A. C. Tran, T. J. Dull, J. E. Groopman, D. J. Capon, R. A. Byrn, and M. H. Finer, *Blood* **84,** 2878 (1994).
[25] W. Kolanus, C. Romeo, and B. Seed, *Cell* **74,** 171 (1993).
[26] H. Xu and D. R. Littman, *Cell* **74,** 633 (1993).
[27] D. M. Desai, J. Sap, J. Schlessinger, and A. Weiss, *Cell* **73,** 541 (1993).
[28] R. Majeti, A. M. Bilwes, J. P. Noel, T. Hunter, and A. Weiss, *Science* **279,** 88 (1998).

this case, kinases), they allow selective manipulation of a subset of receptors, made distinct, at least in part, through their modified extracellular domains.

Exemplifying yet another novel use of cell surface chimeras is a tripartite fusion designed to direct the dephosphorylation of a particular substrate whose phosphorylation is believed to be important for signaling by the TCR.[29] The chimera depicted in Fig. 3 brings the CD45 phosphatase domain into proximity with the SH2 domain of Grb2, a domain known to recruit the phosphorylated form of a critical linker molecule; the chimera is tethered to the plasma membrane via the "ligand neutral" MHC HLA-A2 transmembrane and extracellular domains. In fact, expression of this chimera in a T cell line appeared to selectively dephosphorylate its intended substrate and inhibit T cell activation by the TCR. Similarly, this approach could be used to test the functional consequences of bringing a kinase into close proximity with a specific substrate. These examples serve to illustrate the value of surface chimeric receptors as tools in studying a wide range of biological questions.

Genetic Engineering of Surface Chimeric Receptors: Theoretical Considerations

The initial decision in the development of any chimeric protein regards the selection of the receptors that will serve as sequence donors. The selection of the cytoplasmic component is driven by the nature of the scientific question being posed, and therefore, does not merit further discussion here. Suffice to say, one can use the cytoplasmic domain, or portion thereof, from any transmembrane protein, or any cytoplasmic domain whose proximity with the membrane may impact the biologic process being studied. The choice of an extracellular and transmembrane domain is influenced by a number of factors that include the desired valency of the chimera, for example, monomer or dimer, the availability of reagents against the candidate domain, and the absence of its expression in the cell that will serve as host for the chimera. This final constraint can be avoided if species-specific reagents are available. Judging from the wide variety of extracellular domains that have been utilized successfully—those of CD4, CD8, CD16, CD28, IL-2Rα (Tac), and the vesicular stomatitis viral G protein (VSV-G)—the choice seems almost limitless. This may reflect the ability of the lipid bilayer to separate or buffer the divergent structures of the cytoplasmic and extracellular components; thus, one is afforded the choice of nearly any ectodomain existing in nature.

---

[29] D. G. Motto, M. A. Musci, S. E. Ross, and G. A. Koretzky, *Mol. Cell. Biol.* **16**, 2823 (1996).

Although surface chimeric receptors appear to tolerate a variety of extracellular and transmembrane domains, there are important things to consider when making the selections. Perhaps most important to consider is whether interactions may occur between the extracellular/transmembrane domains being considered and the receptor whose function is under investigation. This becomes especially important if the goal is to isolate a particular chain from that receptor. Some early chimeras constructed to study CD3 ζ contained sequences that included its transmembrane domain linked to a heterologous extracellular domain.[8,9] Because antigen receptor subunits associate in part through their transmembrane domains, interpretation of the results was complicated by the possible incorporation of the chimera into endogenous TCR complexes. Similarly, when truncation mutations of the ζ cytoplasmic sequences were made in its chimeric context, heterodimers were found to exist between the chimeric proteins and endogenous ζ.[11] Although these heterodimers represented a minor fraction of the total dimers, this result serves to illustrate the potential problems that could arise through interactions conferred by the transmembrane and extracellular sequences. To circumvent some of these problems, functional analysis of chimeras composed of subunits from oligomeric receptors are often analyzed in cells deficient in expression of the receptor from which they are derived.

One interesting phenomenon we have uncovered in comparing chimeric receptor activity in stable and transient transfectants relates to the potential of the extracellular domain to interact with proteins expressed on the host cell. Much of our initial analysis of CD8–ζ and its mutants was performed in stable transfectants of Jurkat; under these conditions no signaling was detected by the chimera unless stimulated with anti-CD8 monoclonal antibodies. In contrast, if CD8–ζ is expressed transiently, no ligation is necessary to elicit a robust response. This signaling most likely arises through intra- or intercellular interactions between CD8 and class I MHC, which is expressed at high levels on Jurkat cells. Consistent with this hypothesis, a Tac–ζ chimeric receptor expressed transiently at the same level fails to exhibit this elevated signaling. Because stable clones expressing CD8–ζ were not difficult to obtain, a likely explanation is that the cell is able to reset its signaling equilibrium in the face of constitutive signaling by a receptor. Therefore, in selecting domains for chimeras, one needs to consider potential interactions between the extracellular domain and proteins the chimera will encounter on the host cell or other cells used during its functional analysis.

Another consideration when designing a chimeric receptor is its valency. Many of the ITAM-bearing chimeras were constructed in monomeric forms despite the fact that most subunits that contain ITAMs exist in pairs in their native context. Apparently in this case altering the valency had no deleterious functional consequences, perhaps because these chimeras were aggregated with monoclonal antibodies. However, there may be circumstances when dimerization is essential or more relevant if a chimera is designed to function independently of ligation or "extrinsic dimerization." Such a chimera might incorporate a dimeric cytoplasmic domain that is known to constitutively associate with a negative-regulatory enzyme; without manipulation, this chimera could exhibit dominant-negative properties, but may require its dimeric state to do so. Alternatively, evidence suggests that expressing the normally monomeric cytoplasmic domain of CD45 in dimeric form would likely inhibit its function, consequently impairing TCR signaling as well.[27,28] Thus, in some circumstances, maintaining the native valency of the cytoplasmic component may be advantageous when designing its new context.

Finally, there are features of the transmembrane domain that need to be considered when designing a surface chimeric receptor. Except in rare cases, transmembrane domains consist of 20–30 exclusively hydrophobic residues that are believed to assume an α-helical conformation in the lipid bilayer. The boundaries of these domains are usually defined by algorithm-derived hydropathy plots and are by no means exact. It is not uncommon to find posttranslationally modified residues within the presumed boundaries of the lipid bilayer, demonstrating that they are, in fact, accessible to enzymes. This should be considered when engineering the juxtamembrane junctions of the chimeric protein. The cytoplasmic junction is often marked by a stop transfer sequence of positively charged residues that facilitate anchoring of the protein in the membrane. It is important either to retain these charged residues from the transmembrane donor or to include them as the N-terminal boundary of the sequences derived from the cytoplasmic donor. Because the boundary between a transmembrane and extracellular domain is less well characterized, it is suggested to extend beyond the transmembrane domain into the extracellular domain if such a junction is desired. This was the strategy utilized in the construction of the CD16–CD7–kinase tripartite fusions (Fig. 3), which included the transmembrane and stalk region of CD7 linked to CD16.[25]

One last consideration worthy of discussion relates to the fact that the plasma membrane is not a homogeneous mix of lipids, but rather, it contains microdomains that harbor distinct lipid constituents. Evidence is accumulating that membrane subdomains enriched in sphingolipids and cholesterol, called rafts, are important in the activation of some immuno-receptors.[30–33] Several proteins critical for TCR signal transduction are

[30] K. A. Field, D. Holowka, and B. Baird, *J. Biol. Chem.* **272,** 4276 (1997).

concentrated in these rafts by means of acylation on membrane-juxtaposed cysteine residues. Among these are the Lck and Fyn kinases, Ras, and the transmembrane linker molecule LAT, each of whose localization in the rafts is essential for its function.[34,35] Therefore, if chimeras are to be constructed that utilize such raft-targeted proteins, care should be taken to include the membrane-proximal cysteine residues required for acylation (the CD16–CD7–Lck and Fyn chimeric proteins referred to earlier contained the entire coding sequences of the Src kinases and therefore were likely to be properly acylated). Last, it may be possible to assess the effect of ectopically expressing any cytoplasmic domain in the rafts by linking it to the transmembrane and membrane-proximal cysteines of the LAT protein. It should be noted that while considerations regarding rafts may be important in some chimera strategies, they are likely to be rare because only a minor subset of proteins permanently resides in this membrane compartment.

Genetic Engineering of Surface Chimeric Receptors:
  Technical Approaches

Two basic genetic engineering approaches can be applied in the construction of surface receptor chimeras. The first involves DNA amplification by polymerase chain reaction (PCR), using internal primers that overlap to create the junction of the chimera (overlap PCR). The second approach allows the direct ligation of receptor components generated by PCR. Because it is unlikely that a restriction site exists at the desired position of the junction in either receptor sequence, they are usually introduced via the PCR primers. However, if a site does exist at an appropriate position within either of the receptor segments, PCR amplification of only a single chain is required; this can overcome complications arising from mutations, which are not infrequently introduced during PCR amplification (discussed below).

We have successfully applied both strategies in constructing receptor chimeras. The overlap PCR approach is schematically depicted in Fig. 4, with the construction of CD8–ζ serving to illustrate the technique. In this approach, the components of each receptor are amplified in separate reac-

[31] P. S. Kabouridis, A. I. Magee, and S. C. Ley, *EMBO J.* **16,** 4983 (1997).

[32] R. Xavier, T. Brennan, Q. Li, C. McCormack, and B. Seed, *Immunity* **8,** 723 (1998).

[33] C. Montixi, C. Langlet, A. M. Bernard, J. Thimonier, C. Dubois, M. A. Wurbel, J. P. Chauvin, M. Pierres, and H. T. He, *EMBO J.* **17,** 5334 (1998).

[34] M. D. Resh, *Cell. Signal.* **8,** 403 (1996).

[35] W. Zhang, R. P. Trible, and L. E. Samelson, *Immunity* **9,** 239 (1998).

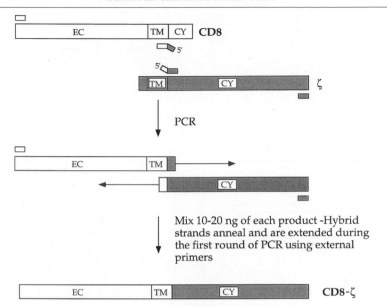

Fig. 4. Schematic representation of the overlap PCR strategy for the production of chimeric receptors. Construction of the CD8-ζ chimera serves to illustrate the technique, with the CD8-derived sequences represented in white, and those derived from ζ in gray. EC, Extracellular domain; TM, transmembrane domain; CY, cytoplasmic domain.

tions, with each internal primer containing 5′ sequences that correspond to the partner chain and therefore do not anneal to the template. The internal primers encoding the C-terminal junctional sequences from CD8 and the N-terminal junctional sequences of the ζ fragment were designed to overlap, such that annealing the two PCR products yielded a hybrid template. Once the two products from the first PCR are mixed, the hybrid template is formed in the first PCR cycle after denaturation and is extended during the first synthesis phase. From this newly synthesized hybrid template, the complete chimera is amplified with only the external primers. These 5′ and 3′ primers anneal N terminal of the CD8 initiation codon and C terminal of the ζ stop codon, respectively. Restriction sites can be included within the external primers to facilitate cloning of the final product.

This technique offers the advantage of complete freedom in designing the junction of the chimera without the need for amino acid modifications. However, one disadvantage is the potential for reduced PCR efficiency during the primary amplification rounds brought about by the disparately sized primers or the nonannealing sequences within the internal primers. To circumvent these problems we have reduced the number of overlapping

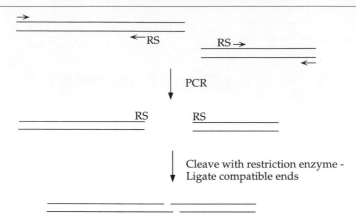

FIG. 5. Introducing restriction sites for a chimeric receptor junction. Restriction sites (RS) are introduced via the internal PCR primers as depicted.

nucleotides to 10–15, finding this degree of overlap to be sufficient for effective production of final product.

The second approach, widely used for the engineering of chimeric proteins, involves the introduction of a restriction site into one or both components at any desired position (depicted in Fig. 5). This is accomplished by incorporating the restriction site into the 5′ end of the internal PCR primers, adding at most two additional amino acids at the chimeric junction. As previously mentioned, if a site preexists in one component at a position appropriate for the junction, PCR of only a single segment need be performed. Once cleaved with the designated restriction enzyme, the two receptor fragments are ligated directly and cloned into a vector for sequencing. This method is especially useful if a given extracellular and transmembrane domain will serve as a recipient for a number of different cytoplasmic domains. For this purpose, one introduces a restriction site C terminal of the stop transfer sequence in the coding region of the protein chosen for the extracellular/TM domain. This then allows the subcloning of a PCR-derived cytoplasmic fragment into the appropriate site.

We have found a technique based on the method of mutagenesis described by Kunkel[36] to be rapid and efficient for the introduction of restriction sites (outlined in Fig. 6). The materials are available commercially (Quikchange from Stratagene, La Jolla, CA). It involves engineering two complementary primers, each of which contains the mutation (restriction site) at its center flanked by approximately 15 nucleotides. These primers

[36] T. A. Kunkel, *Proc. Natl. Acad. Sci. U.S.A.* **82,** 488 (1985).

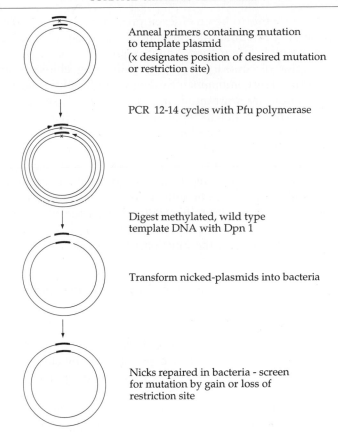

Anneal primers containing mutation
to template plasmid
(x designates position of desired mutation
or restriction site)

PCR 12-14 cycles with Pfu polymerase

Digest methylated, wild type
template DNA with Dpn 1

Transform nicked-plasmids into bacteria

Nicks repaired in bacteria - screen
for mutation by gain or loss of
restriction site

FIG. 6. Schematic representation of a rapid and efficient protocol for the introduction of restriction sites or mutations. The procedure relies on methylation of the wild-type plasmid for its subsequent digestion with *Dpn*I.

are annealed to a plasmid derived from a bacterial strain that expresses Dam (DNA adenosine methylation) methylases, ensuring that the wild-type DNA template is methylated. Twelve to 14 rounds of PCR are then performed, resulting in the synthesis of nicked, complementary strands that both contain the mutation. This procedure relies on a property that distinguishes the wild-type template DNA from the mutated strands synthesized *in vitro:* in this case, it is methylation. The PCRs are then subjected to restriction digestion by *Dpn*I, an enzyme that cleaves methylated DNA, theoretically leaving only a highly enriched population of mutated, nicked plasmids intact. These plasmids are transformed into bacteria, where they are repaired and can now be screened for the presence of the restriction site. We have found this to be an efficient procedure, with 80–100% of the screened plasmids containing the introduced mutation. (The efficiency is reduced with plasmids larger than 7 kilobases.)

Both strategies discussed above for engineering chimeras are PCR based. Therefore, a few comments regarding sporadic mutations that result during amplification are warranted. In our experience, the error rate of Taq polymerase is higher than its theoretical 1 mutation per 1000 base pairs. Polymerases are now available (*Pfu* and Vent) that exhibit proofreading activity, thereby minimizing the number of mutations introduced by PCR; we currently use these whenever sequence fidelity is desired. Mutations can also be reduced by increasing the amount of original template used in each reaction and by reducing the number of amplification cycles. Finally, whenever possible, one should minimize the size of the PCR-derived chimeric fragment that requires sequence conformation by utilizing preexisting restriction sites that flank the junction.

Analysis of Expression and Function of Chimeric Receptors

Expression of Chimera

Because the most minor alterations in a protein can affect its conformation, stability, trafficking, etc., surface expression of any chimeric receptor must be confirmed in a reasonably quantitative fashion. This is most frequently done by flow cytometry with monoclonal antibodies against the specific extracellular domain employed. One major advantage of the chimera approach involves the selection of the extracellular domain, which, in part, can be based on the availability and cost of specific reagents. Therefore analysis of surface expression rarely presents a problem. If a chimeric receptor is designed to utilize an extracellular domain against which no reagents exist, the addition of an N-terminal epitope tag will allow its detection at the cell surface. Commonly used epitopes are the influenza hemagglutinin (HA),[37] FLAG,[38] and Myc tags.[39]

Either transient or stable expression approaches may be applied to the analysis of chimeric receptors. Transient transfection or infection of T cell lines has been widely used for rapid functional analysis of chimeras. We routinely utilize transient transfection, by electroporation, of the Jurkat leukemic line, which has a relatively high transfection efficiency of 40–60%. As an expression vector, we have found the EF bos plasmid, which employs

[37] V. A. Canfield, L. Norbeck, and R. Levenson, *Biochemistry* **35,** 14165 (1996).
[38] A. Knappik and A. Pluckthun, *BioTechniques* **17,** 754 (1994).
[39] G. I. Evan, G. K. Lewis, G. Ramsay, and J. M. Bishop, *Mol. Cell. Biol.* **5,** 3610 (1985).

the EF 1-α promoter, to be highly effective.[40] Others have used recombinant vaccinia virus to deliver chimeric receptors to a variety of cell lines.[8,12] With viral packaging lines becoming more efficient, it is also possible to infect primary lymphocytes with viruses encoding any gene of choice.[41] Transient expression represents a rapid and valuable approach, provided the assays to be employed are sensitive, and transfection efficiency and surface expression are reliably efficient and reproducible. This is especially true if it is possible to enrich for cells that express the chimera by cell sorting or direct analysis in a flow cytometer. Disadvantages of transient transfection include the low transfection efficiency of many lines, and the substantial cell death that can occur, resulting in low cell yield for analysis.

For many purposes such as biochemically defining protein interactions of low stoichiometry, or identifying substrates of receptor-activated enzymes, stable transfection presents the advantage. Although more time consuming to obtain, stably transfected clones are valuable for their homogeneity in expression and viability, in addition to their limitless expansion potential. Moreover, with a uniform, healthy population of cells one can utilize procedures that rely on cell integrity such as cell surface iodination or biotinylation and metabolic pulse–chase labeling.

Analysis of Chimera Function

Early Events. Once the integrity of the chimera in the plasma membrane has been confirmed, its function can be analyzed. In this section, we highlight in brief a few assays that have been used to assess the signaling capacity of chimeric receptors in lymphocytes; we focus on those with which we are most familiar, that is, those that evaluate signals that emanate from the T cell antigen receptor.

Antigen receptors in both B and T cells initiate signal transduction through the activation of Src and Syk family kinases. This results in the phosphorylation of a number of substrates on tyrosine residues.[42–44] A direct consequence of these phosphorylations is the formation and recruitment of signaling complexes at the membrane, which initiate mobilization of intracellular calcium and activation of the Ras/mitogen-associated protein (MAP)-kinase cascade. Therefore, if a chimeric receptor is expected to mimic, or impinge early on, TCR or B cell receptor (BCR) signaling, it is often tested for its ability to elicit an increase in intracellular calcium. Cells

[40] M. Woodrow, N. A. Clipstone, and D. Cantrell, *J. Exp. Med.* **178**, 1517 (1993).
[41] G. P. Nolan and A. R. Shatzman, *Curr. Opin. Biotechnol.* **9**, 447 (1998).
[42] A. Weiss and D. R. Littman, *Cell* **76**, 263 (1994).
[43] C. M. Pleiman, D. D'Ambrosio, and J. C. Cambier, *Immunol. Today* **15**, 393 (1994).
[44] M. Reth, *Annu. Rev. Immunol.* **10**, 97 (1992).

are loaded with Indo-1 or Fluo-3, fluorescent dyes whose fluorescence properties are altered on binding of Ca^{2+}.[45] After receptor stimulation, the blue-to-violet ratio is monitored in a fluorimeter or flow cytometer (the latter is best for transient transfections because data can be acquired exclusively from cells that have been efficiently transfected). If the chimera is coupled efficiently to the activation of phospholipase C, a peak calcium response is observed within 60–90 sec of stimulation.

The integrity of early tyrosine phosphoryation events can also be examined; this entails examining the extent and kinetics of receptor-induced phosphorylation of ZAP-70/Syk, LAT, phospholipase C, and possibly the chimera itself. More generally, one can analyze the induced pattern of tyrosine phosphoproteins as a first quantitative and qualitative measure of signaling by a chimera and/or its parent receptor. This is done by analyzing lysates from stimulated or unstimulated cells for tyrosine phosphoproteins by Western blot with anti-phosphotyrosine antibodies. Any alterations observed in the substrate pattern of phosphorylation may be indicative of a unique signaling property of the chimera. Furthermore, the identification of a candidate substrate, whose phosphorylation is uniquely modified by stimulation of the chimera, could be facilitated by reagents that are available against many identified substrates.

Another technique useful for analysis of very proximal chimera function is the *in vitro* kinase assay. Because many receptors initiate signaling through their direct association with kinases, an *in vitro* kinase assay is often used for defining and characterizing such an association. Unstimulated or stimulated chimeric receptors are immunoprecipitated, followed by their incubation with $[\gamma\text{-}^{32}P]ATP$, allowing the transfer of labeled phosphate by an associated kinase to proteins contained in the immunoprecipitate; in some cases an exogenous protein is added as a substrate. The reactions are analyzed by SDS–PAGE followed by autoradiography. Once the relevant kinases have been identified, this assay can also be a valuable tool in studying the functional consequence of kinase interactions provided that specific exogenous substrates exist for the kinases.

Analysis of Distal Events of T Cell Activation. T cell activation results from the sustained delivery of receptor-mediated signals to the nucleus, where they act to initiate a differentiation program. Among the genes induced early in this program is the interleukin 2 gene, whose product is important in promoting T cell expansion and effector function.[46] Because expression of IL-2 requires the integration of a number of TCR-derived

[45] A. Minta, J. P. Kao, and R. Y. Tsien, *J. Biol. Chem.* **264,** 8171 (1989).
[46] K. A. Smith, *Science* **243,** 1169 (1988).

signals[47] it has commonly been used as a stringent test for the fidelity of signaling by ITAM-containing chimeric receptors. Secretion of IL-2 can be measured directly by ELISA, or indirectly by analyzing the ability of cell supernatants to drive the proliferation of an IL-2-dependent cell line.[48,49] As an alternative, we and others often use a luciferase assay that reports on the relative transcriptional activity of the IL-2 promoter or a multimer of one of its transcriptional elements, the NFAT site.[50] Like the IL-2 gene, the NFAT element requires the convergence of signaling pathways for its activation.[40] Chimeras and reporter plasmids are introduced into cells by transient transfection, the receptors are stimulated for 5–8 hr, and the cell extracts are analyzed for luciferase activity in a luminometer. The assay provides a rapid, sensitive, and quantitative means of assessing distal signaling by chimeric receptors.

Finally, another assay that has been widely used tests the ability of chimeric receptors to mediate cytolysis of a target cell.[8] The chimera is introduced into a cytotoxic T cell line that has the capacity to kill any cell that appropriately activates its antigen receptor. A target cell bearing on its surface a ligand for, or an antibody against, the extracellular domain of the chimera is loaded with ^{51}Cr, and then incubated with the cytotoxic T cell. Specific lysis is measured by counting the ^{51}Cr released into the medium. Given the success of chimeras containing ITAMs to mediate cytolysis, this assay now offers potential therapeutic utility by providing a means to specifically target cytotoxic T cells to tumors or virally infected cells in an MHC-independent manner.

Conclusion and Summary

In this chapter we have described a powerful technology that has allowed the functional dissection of individual subunits from oligomeric receptors. We have focused primarily on chimeras derived from antigen receptors or their downstream signaling components to illustrate the wide utility of the approach; however, the technology has been applied to numerous multimeric receptors of the immune system including cytokine receptors,[51] Fc receptors,[52,53] and natural killer (NK) cell inhibitory receptors.[54] Although the significance of the structural complexity of oligomeric receptors

[47] G. R. Crabtree, *Science* **243,** 355 (1989).

[48] S. Gillis, M. M. Ferm, W. Ou, and K. A. Smith, *J. Immunol.* **120,** 2027 (1978).

[49] T. Mosmann, *I. Immunol. Methods* **65,** 55 (1983).

[50] J. P. Shaw, P. J. Utz, D. B. Durand, J. J. Toole, E. A. Emmel, and G. R. Crabtree, *Science* **241,** 202 (1988).

[51] M. A. Goldsmith, S. Y. Lai, W. Xu, M. C. Amaral, E. S. Kuczek, L. J. Parent, G. B. Mills, K. L. Tarr, G. D. Longmore, and W. C. Greene, *J. Biol. Chem.* **270,** 21729 (1995).

is by no means understood, it is certain that valuable benefits must be derived from the integrated function of their subunits. In the case of antigen receptors, the multiplicity of ITAMs likely allows the cell to distinguish subtle variations in ligand affinities with exquisite sensitivity. Clearly, an isolated subunit that is ligated with antibodies cannot confer such complex function. For instance, it cannot reveal the subtle changes in signal transduction that likely occur on stimulation with altered antigenic peptide ligands or during a complex cell–cell interaction. However, before the intricacies of integrated receptor function can be appreciated, the potential or unique functional properties contributed by each individual receptor component must first be understood. Providing a tool to acquire this kind of understanding has been the greatest asset of this technology. Acknowledging its limitations, the use of surface chimeric receptors remains an invaluable approach toward our understanding the complex function of oligomeric receptors.

[52] M. H. Jouvin, M. Adamczewski, R. Numerof, O. Letourneur, A. Valle, and J. P. Kinet, *J. Biol. Chem.* **269**, 5918 (1994).
[53] T. Muta, T. Kurosaki, Z. Misulovin, M. Sanchez, M. C. Nussenzweig, and J. V. Ravetch, *Nature (London)* **368**, 70 (1994).
[54] N. Gupta, A. M. Scharenberg, D. N. Burshtyn, N. Wagtmann, M. N. Lioubin, L. R. Rohrschneider, J. P. Kinet, and E. O. Long, *J. Exp. Med.* **186**, 473 (1997).

[17] Use of Chimeric Receptor Molecules to Dissect Signal Transduction Mechanisms

By WARREN J. LEONARD

Introduction

The use of chimeric receptors has become a valuable technique in receptor biology as a means to elucidate the function and signaling properties of receptors or domains of receptors. This approach is particularly valuable in (1) determining the signaling potential of a particular cytoplasmic domain, (2) learning whether homodimerization of a single chain is sufficient for signaling or whether two or more different chains must be ligated together, or (3) investigating the functional equivalence of regions of two different receptor molecules. Chimeric receptors are often used in cells that can respond to the stimulus of primary interest. For example, for studying signaling via the interleukin 2 receptor (IL-2R), chimeric receptors have been used in an IL-2-responsive cell line, CTLL-2.[1] In this setting it is possible to compare stimulation of the endogenous receptor components with IL-2 with the triggering of, for example, c-Kit–IL-2R chimeric constructs with stem cell factor (the ligand for c-Kit).[1] Alternatively, it is possible to use cells such as 32D cells, which do not respond to IL-2 but have all the necessary cellular machinery to respond to IL-2 once the IL-2 receptor β chain (IL-2Rβ) is expressed.[2,3] Experiments using the 32D cell background provide valuable information, as this system eliminates any possible contribution of endogenous IL-2Rβ protein to signaling. Although much can be learned from chimeric receptor approaches, it is evident that the results obtained must be interpreted with caution because receptors and their domains are being studied in an artificial "nonnative" context. Accordingly, any conclusions derived from the behavior of chimeric receptor constructs should be confirmed in a more physiological context. This issue is discussed further below.

Use of Chimeric Receptors in Study of Many Types of Receptors, Including Cytokine Receptors, Antigen Receptors, and Receptor Tyrosine Kinases

This chapter details how chimeric receptor constructs can be used to analyze the significance of extracellular, cytoplasmic, and transmembrane domains in signaling via IL-2 receptors. The IL-2 system is a particularly good illustrative system for this methodology, as it allows for the analysis of the roles of homodimers, heterodimers, and various receptor domains. In addition, a number of different methodological approaches have been used for studying this receptor system. It should be noted, however, that chimeric receptor constructs have been used for a large number of systems, including, for example, not only cytokine receptors but also antigen receptors such as the T cell antigen receptor[4] and IgE receptor,[5] and in studies of receptor tyrosine kinases.[6] In addition to chimeras in which domains of

[1] B. H. Nelson, J. D. Lord, and P. D. Greenberg, *Nature (London)* **369,** 333 (1994).
[2] H. Otani, J. P. Siegel, M. Erdos, J. R. Gnarra, M. B. Toledano, M. Sharon, H. Mostowski, M. B. Feinberg, J. H. Pierce, and W. J. Leonard, *Proc. Natl. Acad. Sci. U.S.A.* **89,** 2789 (1992).
[3] Y. Nakamura, S. M. Russell, S. A. Mess, M. Friedmann, M. Erdos, C. Francois, Y. Jacques, S. Adelstein, and W. J. Leonard, *Nature (London)* **369,** 330 (1994).
[4] B. A. Irving and A. Weiss, *Cell* **64,** 891 (1991).
[5] G. Alber, L. Miller, C. L. Jelsema, N. Varin-Blank, and H. Metzger, *J. Biol. Chem.* **266,** 22613 (1991).
[6] S. Raffioni, D. Thomas, E. D. Foehr, L. M. Thompson, and R. A. Bradshaw, *Proc. Natl. Acad. Sci. U.S.A.* **96,** 7178 (1999).

receptors are exchanged, some investigators have fused kinase domains to receptors, including the IL-2 receptor.[7] These types of studies can potentially clarify the functional equivalence of a nonreceptor kinase to a receptor tyrosine kinase or evaluate if a receptor-associated kinase will function equivalently if the catalytic domain is directly "fused" into the cytoplasmic domain of a receptor chain that does not normally contain an intrinsic kinase domain.

Interleukin 2 Receptors

There are three different classes of IL-2 receptors, binding IL-2 with high, intermediate, or low affinity. These are formed by different combinations of three different receptor components: the IL-2 receptor α chain (IL-2Rα), IL-2Rβ, and IL-2Rγ.[8] IL-2Rγ has now been renamed as the common cytokine receptor γ chain, γ_c, based on its being shared by the receptors for IL-2, IL-4, IL-7, IL-9, and IL-15.[9] High-affinity receptors contain IL-2Rα, IL-2Rβ, and γ_c; intermediate-affinity receptors contain IL-2Rβ and γ_c; and low-affinity receptors contain only IL-2Rα.[8,9] The functional forms of IL-2 receptors are the intermediate- and high-affinity forms. This suggested that heterodimerization of the cytoplasmic domains of IL-2Rβ and γ_c might be vital for signaling. Chimeric receptor molecules have been used to address this question.[1,3] In one approach,[3] a variety of constructs were prepared (Fig. 1), including some constructs in which the extracellular domain of IL-2Rα was linked to the cytoplasmic domains of either IL-2Rβ or γ_c, and these constructs were transfected separately or together into target cells. Monoclonal antibodies to IL-2Rα, which can dimerize the extracellular domains of these constructs, could then be used to evaluate the functional consequences of homodimerization or heterodimerization of the IL-2Rβ and γ_c cytoplasmic domains. In addition, the potential importance of the IL-2Rβ transmembrane domain for signaling could be investigated.

Construction of Chimeric Interleukin 2 Receptor Constructs

For the set of studies summarized here, the principal vector was a eukaryotic expression vector, denoted pME18S, in which expression of a cDNA is directed by the SRα promoter.[10] Construction of plasmids is as follows.

---

[7] B. H. Nelson, B. C. McIntosh, L. L. Rosencrans, and P. D. Greenberg, *Proc. Natl. Acad. Sci. U.S.A.* **94,** 1878 (1997).

[8] J.-X. Lin and W. J. Leonard, *Cytokine Growth Factor Rev.* **8,** 313 (1997).

[9] W. J. Leonard, "Fundamental Immunology," 4th Ed. (W. E. Paul, ed.), p. 741. Lippincott Raven, Philadelphia, Pennsylvania, 1999.

[10] G.-X. Xie, K. Maruyama, and A. Miyajima, *Neurosci. Protocols* 95-020-01-01-17 (1995).

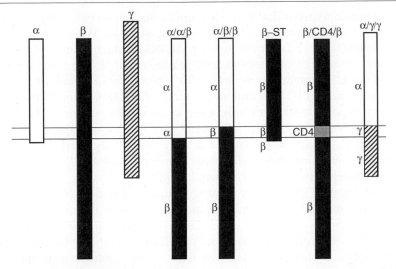

FIG. 1. Schematic of the α, β, and γ chains of the IL-2 receptor and various chimeric receptor or truncation mutant constructs ($\alpha/\alpha/\beta$, $\alpha/\beta/\beta$, β–ST, β/CD4/β, and $\alpha/\gamma/\gamma$). [Modeled after a schematic published in Y. Nakamura, S. M. Russell, S. A. Mess, M. Friedmann, M. Erdos, C. Francois, Y. Jacques, S. Adelstein, and W. J. Leonard, *Nature* (*London*) **369,** 330 (1994).]

pME18Sα was constructed by ligating the *Nci*I-to-*Xba*I fragment from the IL-2Rα cDNA pIL-2R3[11] (filled in at only the *Nci*I site) to pME18S digested with *Eco*RI and *Xba*I, filled in at the *Eco*RI site.

pME18Sβ was prepared by polymerase chain reaction (PCR), using the IL-2Rβ cDNA pBSβ1[12] as a template and appropriate primers to generate an *Eco*RI restriction site, followed by a perfect Kozak consensus sequence,[13] and then by the 5' end of the IL-2Rβ coding region, extending to an *Xho*I site in the cDNA. The 3' section of IL-2Rβ was prepared as an *Xho*I-to-*Hin*dIII fragment (filled in at the *Hin*dIII site) from another IL-2Rβ cDNA, pBSβ2.[12] These two fragments were ligated by a three-way ligation into pME18S digested with *Xho*I and *Eco*RI (fill in only at the *Xho*I site).

pME18Sγ was cloned from a cDNA library prepared in pME18S in which inserts are cloned between the *Bst*XI sites of the vector.

To generate a chimeric construct containing the extracellular and trans-

[11] W. J. Leonard, J. M. Depper, G. R. Crabtree, S. Rudikoff, J. Pumphrey, R. J. Robb, M. Kronke, P. B. Svetlik, N. J. Peffer, T. A. Waldmann, and W. C. Greene, *Nature* (*London*) **311,** 626 (1984).

[12] J. R. Gnarra, H. Otani, M. G. Wang, O. W. McBride, M. Sharon, and W. J. Leonard, *Proc. Natl. Acad. Sci. U.S.A.* **87,** 3440 (1990).

[13] M. Kozak, *Nucleic Acids Res.* **15,** 8125 (1987).

membrane domains of IL-2Rα and cytoplasmic domain of IL-2Rβ (denoted $\alpha/\alpha/\beta$), we used a three-way ligation cloning strategy. We first isolated the small *Hind*III–*Bcl*I fragment from pME18Sα, which contains the SRα promoter plus most of the extracellular domain of IL-2Rα, and the large *Nco*I–*Hind*III fragment from pME18Sβ, which contains most of the cytoplasmic domain of IL-2Rβ and much of the pME18S vector. Two complementary oligonucleotides encoding the extracellular portion of IL-2Rα 3′ of the *Bcl*I site, the IL-2Rα transmembrane region, and the cytoplasmic domain of IL-2Rβ 5′ to the *Nco*I site were then synthesized. These oligonucleotides were annealed, digested with *Bcl*I and *Nco*I, and pME18S $\alpha/\alpha/\beta$ was generated by a three-way ligation.

To generate a chimeric construct containing the extracellular domain of IL-2Rα and the transmembrane and cytoplasmic domains of IL-2Rβ (denoted $\alpha/\beta/\beta$), we cleaved pME18S $\alpha/\alpha/\beta$ with *Aat*II, which digests in the IL-2Rα extracellular domain and the IL-2Rβ cytoplasmic domain. The region between these two cleavage sites, which contains principally the IL-2Rα transmembrane domain, is thereby deleted. We then generated by PCR the IL-2Rα extracellular domain 3′ of the *Aat*II site and ending with an overhang complementary to the IL-2Rβ transmembrane domain. A second PCR product contains a 5′ overhang complementary to the IL-2Rα extracellular domain, and the region from IL-2Rβ corresponding to the entire IL-2Rβ transmembrane domain extending 3′ of the *Aat*II site in the IL-2Rβ cytoplasmic domain. These two products were combined and PCR was performed with oligonucleotides of the IL-2Rα extracellular and IL-2Rβ cytoplasmic regions 5′ and 3′ of their respective *Aat*II sites. The final PCR product was digested with *Aat*II and ligated into *Aat*II-digested pME18S $\alpha/\alpha/\beta$.

The IL-2R $\alpha/\gamma/\gamma$ chimera was generated by coupled PCRs. The extracellular portion of the IL-2Rα chain was produced by PCR, using one primer 5′ of the *Nde*I site and a second 30-base-long primer comprising the 15 nucleotides immediately 5′ to the IL-2Rα transmembrane domain and the first 15 nucleotides inside the γ_c transmembrane domain. A second PCR was performed with a 30-base-long primer complementary to that of the first reaction and a primer 3′ of the *Xba*I site in the pME18Sγ expression vector. The two reactions were mixed and annealed slowly, followed by PCR between the upstream primer in the IL-2Rα chain and the downstream primer in the pME18Sγ expression vector. The resulting product was digested with *Nde*I and *Xba*I and then subcloned into the pME18Sα expression vector between these sites.

To prepare an IL-2Rβ construct in which its transmembrane domain is replaced with that from CD4 (β/CD4/β), the Bio-Rad (Hercules, CA) Mutagene kit and oligonucleotide 5′-CCTGCAGCCCTTGGGAAGGA-

CACCGCCT GATTGTGCTGGGGGGCGTCGCCGGCCTCCTGCTT-
TTTCATTGGGCT AGGCATCTTCTTCAACTGCAGGAACACCGG-
GCCATGGCTG-3′ were used with full-length IL-2Rβ as the template.
Other mutagenesis kits are available as well. The IL-2Rβ–ST truncation
mutant, which lacks all residues of the IL-2Rβ cytoplasmic domain distal
to amino acid 323, was generated by a similar site-directed mutagenesis in
which a premature stop codon is inserted at amino acid 323.

Confirming the correctness of constructs by DNA sequencing is particularly important when PCR is used. Although errors are rare, it is vital to
ascertain that plasmids will direct the expression of exactly the construct
that one expects. Given the amount of time required to establish stable
transfectants and perform functional assays, the DNA sequencing should
be performed before making transfectants. Overall, the constructs described
above were prepared by a combination restriction digestion and cut-and-
paste approach, by coupled PCR methodology (probably the most common
approach currently used for generating chimeric receptors), and by site-
directed mutagenesis. Thus, the IL-2 receptor chimeric constructs provide
an illustration of many of the standard approaches that can be used to
generate these types of molecules.

Evaluation of Functional Properties of Chimeric Receptor Molecules

To evaluate the functional properties of the chimeric receptors, either
transient or stable transfectants can be used. For stable transfectants, ideally
at least three independent transfectants should be established and analyzed
for similar functional properties. For the IL-2 chimeric receptor constructs,
stable transfectants were established in 32D cells, a murine myeloid progenitor cell that can proliferate in response to IL-3 but that also responds
to IL-2 when IL-2Rβ is stably expressed (see Fig. 2A).[2] The responsiveness of 32D-IL-2Rβ cells to IL-2 proves that except for IL-2Rβ, these
cells constitutively express all the cellular proteins that are required for
proliferation in response to IL-2.

32D cells were maintained in RPMI 1640 medium supplemented with
10% (v/v) fetal bovine serum (FBS), 10 mM 2-mercaptoethanol (2-ME, a
supplement that is often required for murine lymphocytes), 2 mM glutamine, penicillin (100 U/ml), and streptomycin (100 μg/ml) (herein denoted
as RPMI complete medium). In addition, 5% WEHI-CB conditioned medium (WEHI-CM) was added as a source of IL-3. WEHI-CM is prepared
by growing WEHI-3B cells to near confluence in RPMI complete medium,
and harvesting the supernatant of these cells, which constitutively produce
IL-3.

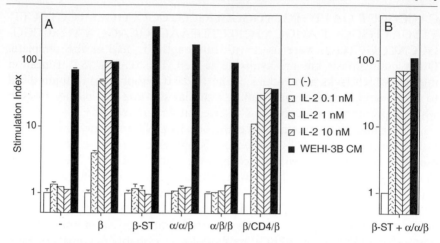

FIG. 2. IL-2 induces proliferation of 32D cells transfected with IL-2Rβ, β/CD4/β, or a combination of β–ST and α/α/β. [Slightly modified from a figure published in Y. Nakamura, S. M. Russell, S. A. Mess, M. Friedmann, M. Erdos, C. Francois, Y. Jacques, S. Adelstein, and W. J. Leonard, *Nature (London)* **369,** 330 (1994).]

    32D cells were stably transfected with various constructs, as shown in Fig. 1. The transfectants were created by cotransfecting plasmids with either pCDNA3Neo (which confers G418 resistance) or TG76 (which confers resistance to hygromycin B), using 10 μg of receptor or chimeric receptor plasmid and 1 μg of the selectable marker plasmid. In general, it is preferable to use an expression vector that also incorporates expression of a selectable marker as part of a single plasmid, but the cotransfection approach also works well. Electroporation was performed using a Bio-Rad Gene Pulser according to the manufacturer's instructions at 250–300 V, 960 μF, average time constant approximately 30 msec. After 24 hr, the cells were aliquoted into 24-well plates and selected with G418 or hygromycin. It is essential to titrate the G418 and hygromycin prior to transfection for any particular cell line in order to establish a reasonably low dose that will kill untransfected cells but allow survival of transfected cells. Too high a drug concentration may kill even transfected cells. For G418, 1 mg/ml is a standard effective concentration for 32D cells, but there are different clones of "32D" that may differ in their resistance. Although we used electroporation to generate these transfectants, other methods, such as retroviral transduction, can also be used for the same purpose.

    Once resistant clones are identified, an essential step is to confirm expression of the transfected plasmids. Although the 1 : 10 ratio of selectable marker plasmid to the expression plasmid strongly favors the likelihood that resistant clones will also contain the chimeric receptor plasmid, this must be confirmed. Even when a single expression vector also directs expression of a selectable marker, it is essential to confirm expression of the chimeric receptor protein. Because of the excellent antibodies to human IL-2Rα and IL-2Rβ, successful expression of IL-2Rβ, $\alpha/\alpha/\beta$, $\alpha/\beta/\beta$, $\alpha/\gamma/\gamma$, β–ST, and β/CD4/β can readily be analyzed by flow cytometry. Staining for IL-2Rα was performed with anti-Tac and 7G7/B6 monoclonal antibodies, while staining for IL-2Rβ was performed with Mikβ1 monoclonal antibody. Moreover, to confirm the proper molecular weight of the expressed proteins, Western blotting was performed with rabbit R3134 antiserum to the extracellular domain of IL-2Rα or ErdA antiserum to IL-2Rβ, which recognizes the cytoplasmic domain of IL-2Rβ.[3] Western blotting can be performed with either ^{125}I-labeled protein A or an enhanced chemiluminescence (ECL; Amersham, Arlington Heights, IL) approach with a goat anti-rabbit IgG. For the β–ST construct, which lacks most of the IL-2Rβ cytoplasmic domain including the ErdA epitope(s), instead of Western blotting, IL-2Rβ can be detected by immunoprecipitation. Cells can be surface iodinated with Na[125]I and lactoperoxidase, followed by immunoprecipitation with Mikβ1 monoclonal antibody. For cell surface iodination, 5–10×10^6 cells were washed and suspended in 100 μl of Hanks' balanced salt solution and 1 mCi of carrier-free Na[125]I is added. Lactoperoxidase (4 μl of a solution at 1 U/10 μl) and H_2O_2 [7.5 μl of a 0.03% (v/v) solution] was added and cells are incubated for 4 min at room temperature. An additional 2 μl of lactoperoxidase and 7.5 μl of H_2O_2 were added and the cells were washed for an additional 4 min. Cells were then washed in phosphate-buffered saline (PBS) and immunoprecipitations performed.

The expected sizes are as follows: IL-2Rβ and β/CD4/β, 70–75 kDa; β–ST, 45–50 kDa; $\alpha/\alpha/\beta$ and $\alpha/\beta/\beta$, 85–90 kDa; $\alpha/\gamma/\gamma$, 65 kDa. In this fashion, independent clones expressing the transfected components or combinations thereof were identified and expression levels could be compared.[3]

Cells were then stimulated with either IL-2 (which is believed to bind monomerically to IL-2Rα) or anti-Tac or 7G7/B6 MAb to IL-2Rα (which should form dimers of $\alpha/\alpha/\beta$, dimers of $\alpha/\gamma/\gamma$, or heterodimers of $\alpha/\alpha/\beta$ and $\alpha/\gamma/\gamma$, depending on which proteins are expressed within a cell). In addition, as a positive control for maximal proliferation, cells were stimulated with WEHI-3B conditioned medium as a source of IL-3. These proliferation assays were performed as follows. Cells were maintained in RPMI complete medium supplemented with 5% (v/v) WEHI-CM as a source of IL-3. Cells (2×10^4 cells in 200 μl) were washed and then cultured for 16 hr in triplicate wells in 96-well microculture plates (Corning, Acton, MA) in RPMI complete medium not supplemented or supplemented with 5% (v/v) WEHI-CM, 0.1 nM IL-2, 1 nM IL-2, or 10 nM IL-2. [3H]Thymidine

(0.5 μCi) was then added to each well for 4 hr; cells are then harvested with a cell harvester and incorporated thymidine determined.

As shown in Fig. 2A, as expected, parental 32D cells proliferate vigorously in response to IL-3, but do not proliferate in response to IL-2; however, when 32D cells are transfected with wild-type IL-2Rβ, IL-2 responsiveness is achieved. No proliferation is seen with the β–ST truncation, proving the importance of the IL-2Rβ cytoplasmic domain. Interestingly, no proliferation is detected with either the $\alpha/\alpha/\beta$ or $\alpha/\beta/\beta$ chimeric receptor (Fig. 2A), even though they can effectively bind IL-2. This demonstrates that cell surface binding of IL-2 cannot trigger proliferation through all molecules containing the IL-2Rβ cytoplasmic domain, and thus suggested that the extracellular domain of IL-2Rβ plays an essential role in IL-2 signaling. The chimeric β/CD4/β molecule can mediate proliferation (Fig. 2A). This demonstrates that the transmembrane domain of IL-2Rβ can be replaced by a similar region from another protein, minimizing the likelihood that the IL-2Rβ transmembrane domain is mediating specific essential interactions. Thus, from the preceding results, it is clear that the cytoplasmic domain of IL-2Rβ is essential, whereas the transmembrane domain can be replaced. However, it is also demonstrated that the context in which the cytoplasmic domain of IL-2Rβ is expressed is important, given that neither the $\alpha/\alpha/\beta$ nor the $\alpha/\beta/\beta$ construct, by itself, could mediate a proliferative response to IL-2. We therefore hypothesized that a combination of β–ST and $\alpha/\alpha/\beta$ might be able to mediate proliferation. This combination of chimeric receptors would provide all regions of IL-2Rβ, but the cytoplasmic domain of IL-2Rβ would be in *trans* on IL-2Rα rather than in *cis* on IL-2Rβ. Indeed, 32D cells expressing these two proteins together mediate vigorous proliferation in response to IL-2 (Fig. 2B). Thus, both the extracellular and cytoplasmic domains of IL-2Rβ are essential for IL-2 responsiveness.

Because the functional forms of the IL-2 receptor are the intermediate- and high-affinity forms of the receptor, both of which contain IL-2Rβ and γ_c, it has been hypothesized that the extracellular domain of IL-2Rβ is vital for allowing the proper recruitment of γ_c in response to IL-2. To investigate this possibility, 32D transfectants expressing either the $\alpha/\alpha/\beta$ or $\alpha/\gamma/\gamma$ construct are tested for their responsiveness to IL-2, WEHI-CM, and antibodies that bind to the extracellular domain of IL-2Rα (7G7/B6 and anti-Tac). No proliferation is seen to any of these stimuli except WEHI-CM (the positive control) (Fig. 3A). Thus, even antibody-mediated homodimerization of the IL-2Rβ or γ_c cytoplasmic domains does not induce a signal. In contrast, however, when $\alpha/\alpha/\beta$ and $\alpha/\gamma/\gamma$ are coexpressed in the same cells, either 7G7/B6 or anti-Tac monoclonal antibodies can trigger proliferation (Fig. 3B). Thus, heterodimerization of the IL-2Rβ and γ_c

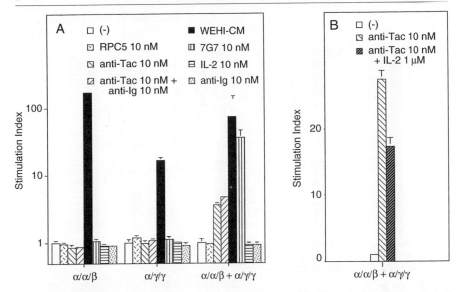

FIG. 3. (A) and (B) Anti-IL-2Rα antibodies stimulate proliferation of 32D cells transfected with α/α/β and α/γ/γ, [Slightly modified from a figure published in Y. Nakamura, S. M. Russell, S. A. Mess, M. Friedmann, M. Erdos, C. Francois, Y. Jacques, S. Adelstein, and W. J. Leonard, *Nature (London)* **369,** 330 (1994).]

cytoplasmic domains is sufficient to induce signaling. As anticipated, IL-2, which binds monovalently rather than bivalently to IL-2Rα, cannot trigger proliferation, presumably because of its inability to mediate heterodimerization. One additional type of experiment was performed, using cells cotransfected with IL-2Rβ and α/γ/γ. In these cells, stimulation with a bispecific antibody, in which one Fab recognizes IL-2Rα and the other recognizes IL-2Rβ, is able to stimulate proliferation (summarized in Fig. 4).

Conclusions and Caveats

The studies we have summarized allow a number of conclusions. First, they indicate the versatility of studies with chimeric receptors. The preceding studies allowed us to demonstrate the importance of the extracellular and intracellular domains of IL-2Rβ and that the transmembrane domain could be replaced with the transmembrane domain of another protein. Second, they demonstrate the importance of heterodimerization of the cytoplasmic domains of IL-2Rβ and γ_c. Third, they indicate that, while physiologically important, the extracellular domain of IL-2Rβ is dispensable if an alternative means of heterodimerizing the IL-2Rβ and γ_c cytoplasmic

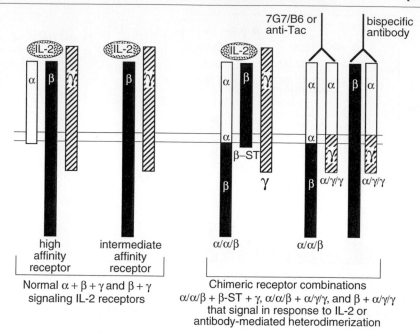

Fig. 4. *Left:* Schematic of the normal signaling high- and intermediate-affinity IL-2 receptors. *Right:* Combinations of chimeric receptor constructs that respond to IL-2, to anti-IL-2Rα monoclonal antibodies, or to a bispecific antibody in which one Fab binds to IL-2Rα and the other binds to IL-2Rβ. [Modeled after a schematic published in Y. Nakamura, S. M. Russell, S. A. Mess, M. Friedmann, M. Erdos, C. Francois, Y. Jacques, S. Adelstein, and W. J. Leonard, *Nature (London)* **369,** 330 (1994).]

domains can be achieved. Fourth, they provide functional data indicating that IL-2 binds monomerically to the extracellular domain of IL-2Rα. Fifth, they indicate a strategy for mapping critical residues within the cytoplasmic domains, namely by using similar chimeric molecules with mutations in residues of interest. In this regard, Nelson *et al.* have used an alternative type of chimeric system for IL-2 signaling [in which the IL-2Rβ and γ_c cytoplasmic domains are fused to the extracellular domains of granulocyte-macrophage colony-stimulating factor (GM-CSF) receptor chains], and they have evaluated the importance of tyrosine residues in γ_c in this system.[14] This type of approach complements evaluating the role of tyrosine by mutating them in the natural rather than chimeric receptor context, as has been done for IL-2Rβ.[15]

[14] B. H. Nelson, J. D. Lord, and P. D. Greenberg, *Mol. Cell. Biol.* **16,** 309 (1996).
[15] M. C. Friedmann, T.-S. Migone, S. M. Russell, and W. J. Leonard, *Proc. Natl. Acad. Sci. U.S.A.* **93,** 2077 (1996).

A related methodology has been used to evaluate IL-4 signaling. IL-4 is known to signal through either of two types of receptors, IL-4Rα plus γ_c (the type I IL-4 receptor) or via IL-4Rα plus IL-13Rα1 (type II IL-4 receptor).[9] These groups created chimeric constructs of IL-4Rα and triggered homodimerization.[16,17] Interestingly, such homodimerization allowed signaling to occur, a finding with potentially important implications for mapping functional domains. However, other compelling data indicate that IL-4 cannot induce homodimerization of IL-4Rα[18,19] and that it induces only formation of heterodimers with γ_c or IL-13Rα1. Thus, the fact that homodimerization of IL-4Rα triggers a signal does not necessarily mean such dimerization physiologically occurs. In conclusion, the use of chimeric receptors offers an extremely powerful methodology for mapping functional domains of receptors and clarifying more about the signaling potential of these molecules. However, it is vital to confirm in physiological settings the conclusions that are drawn from these artificial systems.

Acknowledgment

I thank Dr. Jian-Xin Lin for critical comments.

[16] W. Kammer, A. Lischke, R. Moriggl, B. Groner, A. Ziemiecki, C. B. Gurniak, L. J. Berg, and K. Friedrich, *J. Biol. Chem.* **271**, 23634 (1996).
[17] S. Y. Lai, J. Molden, K. D. Liu, J. M. Puck, M. D. White, and M. A. Goldsmith, *EMBO J.* **15**, 4506 (1996).
[18] R. C. Hoffman, B. J. Castner, M. Gerhart, M. G. Gibson, B. D. Rasmussen, C. J. March, J. Weatherbee, M. Tsang, A. Gustchina, C. Schalk-Hihi, L. Reshetnikova, and A. Wlodawer, *Protein Sci.* **4**, 382 (1995).
[19] T. Hage, W. Sebald, and P. Reinemer, *Cell* **97**, 271 (1999).

[18] Fusion Protein Toxins Based on Diphtheria Toxin: Selective Targeting of Growth Factor Receptors of Eukaryotic Cells

By JOHANNA C. VANDERSPEK and JOHN R. MURPHY

Introduction

The construction of diphtheria toxin-based fusion protein toxins provides a means for the selective targeting and elimination of eukaryotic cells that express on their surface the targeted receptor.[1] Biochemical and genetic

[1] J. R. Murphy and J. C. vanderSpek, *Semin. Cancer Biol.* **6**, 259 (1995).

evidence strongly suggests that native diphtheria toxin can be divided into three functional domains: the N-terminal, 21, 164-Da, enzymatically active fragment that, once delivered to the cytosol of a eukaryotic cell, catalyzes the NAD$^+$-dependent ADP-ribosylation of elongation factor 2 (EF2); and a 37,198-Da C-terminal fragment that carries both the native receptor-binding domain and a hydrophobic domain that facilitates the delivery of ADP-ribosyltransferase across the eukaryotic membrane.[2,3] The subsequent solution and refinement of the diphtheria toxin X-ray structure confirmed these earlier observations and provided precise structural definition to each of the functional domains.[4,5] Since Yamizumi *et al.* demonstrated that the delivery of a single molecule of the ADP-ribosyltransferase to the cytosol of a eukaryotic cell was sufficient to cause the death of that cell,[6] it became attractive to design and construct both conjugate toxins (i.e., chimeric proteins in which the catalytic domain was chemically cross-linked to a targeting ligand)[7] and fusion protein toxins (i.e., fusion proteins in which the native receptor-binding domain was deleted and replaced with a polypeptide hormone or growth factor).[8] In both instances, the intent was to develop highly toxic targeted toxins that could potentially be used as experimental therapeutics for the treatment of refractory human disease. Indeed, the first of the targeted protein fusion toxins, DAB$_{389}$IL-2 (ONTAK; Ligand Pharmaceuticals, Inc., San Diego, CA)[9] has been approved by the Food and Drug Administration for the treatment of refractory cutaneous T cell lymphoma and is now in clinical use.

Receptor-Binding Domain Substitution

Once bound to their respective cell surface receptors the diphtheria toxin-based fusion protein toxins are internalized into the cell by receptor-mediated endocytosis. Productive delivery of the catalytic domain to the target cell cytosol requires passage through an acidic early endosomal compartment.[10] As the lumen of the early endosome becomes acidified through

---

[2] A. M. Pappenheimer, Jr., *Annu. Rev. Biochem.* **46,** 69 (1977).

[3] P. Bacha, J. R. Murphy, and S. Reichlin, *J. Biol. Chem.* **258,** 1565 (1983).

[4] S. Choe, M. J. Bennett, G. Fujii, P. M. Curmi, K. A. Kantardjieff, R. J. Collier, and D. Eisenberg, *Nature (London)* **357,** 216 (1992).

[5] M. J. Bennett, S. Choe, and D. Esienberg, *Proc. Natl. Acad. Sci. U.S.A.* **91,** 3127 (1994).

[6] M. Yamizumi, E. Mekada, T. Uchida, and Y. Okada, *Cell* **15,** 245 (1978).

[7] D. B. Cawley, H. R. Herschman, D. G. Gilliland, and R. J. Collier, *Cell* **22,** 563 (1980).

[8] J. R. Murphy, W. Bishai, M. Borowski, A. Miyanohara, J. Boyd, and S. Nagle, *Proc. Natl. Acad. Sci. U.S.A.* **83,** 8258 (1986).

[9] D. P. Williams, C. E. Snider, T. B. Strom, and J. R. Murphy, *J. Biol. Chem.* **265,** 11885 (1990).

[10] C. A. Waters, P. A. Schimke, C. E. Snider, K. Itoh, K. A. Smith, J. C. Nichols, T. B. Strom, and J. R. Murphy, *Eur. J. Immunol.* **20,** 785 (1990).

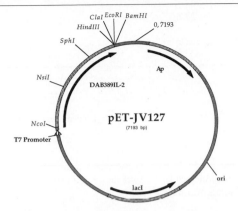

Fig. 1. Plasmid pET-JV127 encodes the diphtheria toxin-based fusion protein toxin DAB$_{389}$IL-2. Digestion with *Sph*I and *Hind*III removes the IL-2 receptor-binding domain encoding a portion of the gene fusion and allows the construction of a unique fusion protein toxin by cassette exchange.

the action of a vATPase, the transmembrane domain of the fusion protein spontaneously denatures and forms a pore, or channel, in the endosomal membrane and facilitates the translocation of the catalytic domain to the cytosol of the target cell. Because passage through an acid compartment is an essential step in the intoxication process, prior knowledge that a given surrogate ligand is internalized in an early endosomal compartment that becomes acidified is important. Accordingly, the selection of a surrogate receptor-binding domain for the construction of a novel fusion protein toxin requires knowledge of the distribution of the receptor for that ligand, its route of internalization, and its intracellular trafficking pattern.

Materials and Methods

Genetic Construction of Diphtheria Toxin-Based Fusion Toxins

The construction of a diphtheria toxin-based fusion protein toxin is generally performed by cassette exchange. The parental plasmid for construction is pET-JV127, which carries a fusion gene encoding the catalytic and transmembrane domains of diphtheria toxin linked to a synthetic gene encoding interleukin 2 (IL-2) (Fig. 1).[11] The junction between the diphtheria toxin transmembrane and receptor-binding domain portions of the gene

[11] J. C. vanderSpek, J. A. Mindell, A. Finkelstein, and J. R. Murphy, *J. Biol. Chem.* **268,** 12077 (1993).

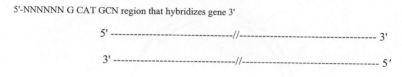

27 nt

5'-NNNNNN G CAT GCN region that hybridizes gene 3'

5' -----------------------------//----------------------------------- 3'

3' -----------------------------//----------------------------------- 5'

3' region that hybridizes gene TTCGAA NNNNNN-5'

27 nt

FIG. 2. Strategy used for PCR amplification of a gene encoding a surrogate receptor-binding domain. The introduction of 5' *Sph*I and 3' *Hin*dIII restriction endonuclease sites allows for vectorial cloning into pET-JV127.

fusion contains a unique *Sph*I site. Following a translation stop codon at the 3' end of the receptor-binding domain a unique *Hin*dIII site is introduced. As a result, an *Sph*I-to-*Hin*dIII cassette exchange allows the rapid substitution of receptor-binding domains in the construction of novel diphtheria toxin-based fusion protein toxins. During the construction of pET-JV127, the pET11d vector (Novagen, Madison, WI) was modified such that the *Sph*I and *Hin*dIII restriction sites are unique. Accordingly, the gene encoding a new surrogate receptor-binding domain should be designed such that there is a unique *Sph*I site at the 5' end and a unique *Hin*dIII site at the 3' end. In general, this may be accomplished by using the polymerase chain reaction (PCR) to introduce these sites. Because the surrogate receptor-binding domain is positioned on the C-terminal end of the fusion protein, it is also important to position the *Sph*I site such that it will be in the correct translational reading frame with respect to the C-terminal end of the diphtheria transmembrane domain. In the pET-JV127 vector the junction at the 5' end consists of the following sequence: ACG CAT GCA (*Sph*I site is underlined and translational reading frame is designated). To ensure *Sph*I digestion after PCR amplification an additional six nucleotides are added to the 5' end. The 3' primer should encode a translational stop codon followed by a *Hin*dIII restriction endonuclease site (six nucleotides should be added 3' to the *Hin*dIII site to ensure digestion after amplification). A scheme for primer design is shown in Fig. 2.

The reaction mixture for PCR amplification is as follows:

Double-distilled H$_2$O	78 μl
Reaction buffer (10×)	10 μl
Vector carrying gene of interest (1 ng)	1 μl
*Sph*I primer (250 ng)	2 μl
*Hin*dIII primer (250 ng)	2 μl

dNTP mix (200 μM, each nucleotide)   8 μl

Pfu DNA polymerase   1 μl

The reaction mixture is placed in a 0.5-ml Eppendorf tube and overlaid with 50 μl of mineral oil (United States Biochemical, Cleveland, OH). The *Pfu* DNA polymerase is from Stratagene (La Jolla, CA). Nucleotides are purchased as 10 mM stock solutions from Perkin-Elmer (Norwalk, CT) and mixed before use. A second reaction mixture is also set up in which 10 ng of vector is used. The amplification conditions used for PCR are as follows: 95° for 30 sec (1 cycle); and then 95° for 30 sec, 37° for 1 min, and 68° for 2 min (25 cycles).

After amplification, 5 μl from the reaction is removed from under the mineral oil and analyzed by electrophoresis on an agarose minigel. If both the 1- and 10-ng reactions contain the desired amplified product, the DNA is cleaned and precipitated by standard methods. If only one reaction yields product, the amplification is repeated and the two products are then pooled, cleaned, and precipitated. After precipitation, the DNA is digested with *Sph*I and *Hin*dIII, cleaned again, and then ligated into pET-JV127 that has also been digested with *Sph*I and *Hin*dIII and treated with calf intestinal phosphatase. If the DNA encoding the desired targeting ligand contains internal *Sph*I and/or *Hin*dIII sites, these sites must be mutated prior to the amplification reaction to introduce unique *Sph*I and *Hin*dIII sites at the 5′ and 3′ ends of the insert. After ligation, plasmids are transformed into competent *Escherichia coli* TOP10 and plated on Luria Bertani–ampicillin (LB–Amp) plates. Colonies are selected, the plasmid DNA is prepared with Qiagen (Valencia, CA) minipreparation kits, and the newly inserted cassette portion of the gene fusion is sequenced to ensure there are no PCR misincorporations.

Expression and Purification of Diphtheria Toxin-Based Fusion Protein Toxins

Once assured that the new plasmid encodes the catalytic and transmembrane domains of diphtheria toxin, fused in-frame to the gene encoding the new surrogate receptor-binding domain, the recombinant plasmid is transformed into *E. coli* HMS174 and HMS174(DE3) for protein expression (Novagen, Madison, WI). The production of recombinant protein in *E. coli* HMS173(DE3) is induced on addition of isopropyl-β-D-thiogalactosidase (IPTG) to a final concentration of 1 mM. When basal levels of fusion protein toxin production are deleterious to the host bacteria, *E. coli* HMS174 is the host strain of choice and recombinant protein expression is induced on infection of the culture with coliphage CE6, a λ phage derivative that encodes the T7 RNA polymerase. In these host–vector systems the final yield of fusion protein toxin may approach 20% of total cellular protein.

After induction of gene expression, the diphtheria toxin-based fusion protein toxins generally accumulate in the cytosol of recombinant *E. coli* as insoluble inclusion bodies. Inclusion body preparations are then harvested, solubilized in denaturing buffer, and the recombinant protein is either diluted in or dialyzed against refolding buffer. After refolding, monomeric forms of the fusion protein toxins may be separated from aggregates by ion-exchange chromatography on DEAE-Sephadex.

Induction of Protein Expression with Isopropyl-β-D-thiogalactopyranoside

A 7.0-ml culture of HMS174(DE3) transformed with a derivative of pET-JV127 containing the gene of interest is grown in LB–Amp [1% (w/v) Bacto Tryptone, 0.5% (w/v) yeast extract, 1% (w/v) NaCl, pH 7.5, supplemented with ampicillin (100 μg/ml)] with shaking at 37°. The following day, the culture is added to 500 ml of LB–Amp in a 2.0-liter flask and grown at 37°, with vigorous shaking, to an OD_{600} (optical density at 600 nm) between 0.6 and 0.8 A 1.0-ml aliquot is removed and placed in a microcentrifuge tube and the bacterial cells are pelleted by centrifugation and resuspended in 200 μl of sodium dodecyl sulfate–polyacrylamide gel electrophoresis (SDS–PAGE) loading buffer for gel electrophoresis and immunoblot analysis. IPTG is added to a final concentration of 1 mM to the remaining culture and the culture is incubated, with shaking, for an additional 2 hr. Another 1.0-ml aliquot should be removed for SDS–PAGE and immunoblot analysis. The bacteria are then pelleted by centrifugation at 5500g for 10 min. The supernatant fraction is discarded and the bacterial pellet may be stored at $-20°$. Gel electrophoresis and immunoblot analysis using anti-diphtheria toxoid antibodies [Biogenesis (Sandown, NH), Biochemed (Huntington Beach, CA), or Biodesign (Kennebunk, ME)] or antibody against the surrogate receptor-binding domain should be performed to compare the before and after induction samples to determine if the desired diphtheria toxin-based fusion protein toxin was expressed.

Induction of Protein Expression with λ CE6

When the fusion protein toxin to be expressed is inhibitory to the bacterial host, a host–vector induction system with a lower basal level of expression should be used. A 7.0-ml culture of HMS174 transformed with the derivative of pET-JV127 containing the gene of interest is grown in LB–Amp–maltose [maltose is added to LB–Amp to a final concentration of 0.2% (w/v) just before use]. The following day the culture is used to inoculate 500 ml of LB–Amp–maltose and grown at 37°, with vigorous shaking, to an OD_{600} of 0.3, at which time glucose is added to a final concentration of 4 mg/ml. The culture is then incubated until an OD_{600} of 0.8 is reached. At this point, 1.0 ml of the culture is removed for analysis, and $MgSO_4$ is then added to a final concentration of 10 mM along with 2×10^9 plaque-forming units (PFV) of λ CE6 per milliliter of culture medium. The culture is incubated with shaking at 37° for an additional 2–3 hr. One milliliter of the culture is removed for analysis, and the bacteria are harvested by centrifugation at 5500g. Bacterial pellets may be stored at $-20°$. Polyacrylamide gel and immunoblot analysis should be performed as described above to determine if the desired protein was expressed.

Preparation of λ CE6

Induction of recombinant protein expression by bacteriophage λ CE6, which encodes the RNA polymerase that binds the T7 promoter in the pET system, requires relatively large amounts of CE6. To prepare the coliphage CE6, inoculate one colony of E. coli LE392 into 5.0 ml of LB–maltose and incubate overnight, with shaking, at 37°. The following day 2.5 ml of the overnight culture is used to inoculate 22.5 ml of LB–maltose broth, and the culture is incubated at 37°, with shaking, until an OD_{600} between 0.3 and 0.6 is reached. Three hundred microliters of log-phase LE392 culture is next infected with λ CE6 (Novagen). [We generally use four different multiplicities of infection (MOI) of λ CE6 to establish the correct ratio of phage to bacteria to ensure that complete bacterial lysis will occur in at least one of the flasks. The concentration of λ CE6 are $10,^6$ 10^7, 10^8, and 10^9 PFU per 300 μl of log-phase LE392.] The cultures are incubated at room temperature for 20 min to allow the phage to adsorb to the bacterial host. Each tube of LE392–λ CE6 is then used to inoculate one of four flasks containing 500 ml of LB–maltose–$MgSO_4$ [0.2% (w/v) maltose, 10 mM $MgSO_4$]. The cultures are then incubated overnight at 37° with shaking, and bacterial lysis should occur after 12–16 hr. The following day, 2 ml of chloroform is added to each flask and the cultures are shaken for 15 min at 37°. The cultures from the four flasks are combined and whole cells and debris are removed by centrifugation at 4000g for 30 min. The supernatant fluid is decanted into a clean flask and DNase I and RNase A are each added to a final concentration of 1 μg/ml (Boehringer Mannheim, Indianapolis, IN). The mixture is agitated gently at room temperature for 30 min, and polyethylene glycol (PEG) 8000 and NaCl are added to a final concentration of 10% (w/v) and 1 M, respectively. The mixture is incubated overnight at 4° with gentle shaking to precipitate the λ CE6.

The following day, phage are harvested by centrifugation in 500-ml bottles at 4000g, 4°, for 30 min. The supernatant fluid is decanted, and the pellet is resuspended in 5 ml of SM buffer [0.1 mM NaCl, 8 mM

$MgSO_4 \cdot 7H_2O$, 50 mM Tris-HCl (pH 7.5), 0.1% (w/v) gelatin] (the SM buffer is sterilized by autoclaving and is stored at room temperature until use). Transfer the λ CE6 phage preparation to a chloroform-resistant centrifuge tube and add an equal amount of chloroform. Vortex well. Centrifuge at 6000g for 5 min and remove the upper, aqueous phase, which contains the purified CE6 phage particles (we generally extract the λ CE6 preparations two or three times until no PEG 8000 remains in the interface). Titer the phage on lawns of $E.$ $coli$ LE392, according to standard protocols. Determine the total number of plaque-forming units and adjust the concentration with SM–glycerol buffer [SM containing 8% (v/v) glycerol] to 10^{12} PFU/ml. Purified CE6 are stored at $-70°$ in 1-ml aliquots until use.

Purification of Diphtheria Toxin-Based Fusion Toxins from Inclusion Body Preparation

Expression of diphtheria toxin-based fusion proteins from the T7 promoter usually results in the formation of insoluble inclusion bodies in recombinant $E.$ $coli.$ After the bacteria are harvested by centrifugation at 5500g for 10 min the pellet is resuspended in 20 ml of ice-cold STET [50 mM Tris-HCl (pH 8.0), 10 mM EDTA, 8% (w/v) sucrose, 5% (v/v) Triton X-100; store at 4°] and transferred to a 30-ml screw-cap centrifuge tube. The suspension is then homogenized with a tissue homogenizer. Avoid foaming and keep the tube on ice to prevent digestion by bacterial proteases. Add 500 μl of lysozyme stock solution (10 mg/ml; store in 1.0-ml aliquots at $-20°$) and incubate on ice for 1 hr. Homogenize the solution once again, and then lyse the bacteria by sonication [e.g., Fisher (Pittsburgh, PA) 550 sonic dismembrator at 9% total output, 2.5 min total time, 1.5 sec on, 1.0 sec off]. After sonication, the suspension is centrifuged at 13,000g for 20 min at 4°. The supernatent fraction is decanted and saved for future analysis (if the pellet is viscous and it is not possible to decant the supernatant fluid, the sample should be homogenized and centrifuged again). The crude inclusion body preparation is next resuspended in 10 ml of ice-cold STET and then homogenized and sonicated as described above. Once again the supernatant fraction is decanted and saved for SDS–PAGE and immunoblot analysis.

In general, we have found that the diphtheria toxin-based fusion protein toxins can be purified to greater than 90% purity simply by washing the inclusion body preparations three or four times with STET buffer.

Purification of Soluble Diphtheria Toxin-Based Fusion Protein Toxins

When the fusion protein toxin is found by SDS–PAGE and immunoblot analysis to be in the first supernatant fraction, the sample may be purified or partially purified over an anti-diphtheria toxoid immunoaffinity column. The sample should be centrifuged in 30- to 50-ml tubes at 17,000g for 30 min and the supernatant fraction filtered through a 0.45-μm pore size membrane filter. The sample is then loaded onto an immunoaffinity column that has been equilibrated with loading buffer [50 mM KH$_2$PO$_4$, 50 mM NaOH, 10 mM EDTA, 750 mM NaCl, and 0.1% (v/v) Tween 20; final pH 8.0]. The void volume fraction should be collected for SDS–PAGE and immunoblot analysis. The sample should be washed in loading buffer until a baseline is reached and then eluted in elution buffer [100 mM KH$_2$PO$_4$, 4 M guanidine-HCl, 0.1% (v/v) Tween 20; final pH 7.2]. Further purification is generally required and high-performance liquid chromatography (HPLC) sizing may prove useful.

When an impure inclusion body requires further purification it should be homogenized and sonicated in 20.0 ml of loading buffer, using more stringent sonication conditions (Fisher 550 sonic dismembrator, 20% total output, 2 sec on, 1 sec off, 15 min total time). The sample is then centrifuged at 17,000g for 30 min, and then filtered through a 0.45-μm pore size membrane filter. The sample is then loaded onto and eluted from the immunoaffinity column as described above.

Refolding Denatured Fusion Protein Toxin Preparations into Biologically Active Conformation

The purified inclusion body preparation is dissolved in 5.0 ml of denaturing buffer (7 M guanidine, 0.1 M Tris-HCl, 10 mM EDTA, pH 8.0). Dithiothreitol (DTT) is added to 6 mM immediately before use. Inclusion bodies are solubilized by sonication (Fisher 550 sonic dismembrator at 9% total output, 5 min total time, 1.5 sec on, 1.0 sec off), and diluted to about 10 μg/ml in refolding buffer (50 mM Tris-HCl, 50 mM NaCl, pH 8.0) at 4°. Immediately prior to use, reduced glutathione and oxidized glutathione are added to 5 and 1 mM, respectively. The mixture is stirred gently overnight at 4°. The following day, the fusion protein toxin is concentrated to approximately 50 ml, using a 10,000 molecular weight cutoff membrane filter (Omega series, Ultrasette tangential flow device; Filtron Technology, Northborough, MA). The concentrate is centrifuged at 10,000g for 30 min at 4° in order to remove any insoluble material, and the total protein concentration is determined. The partially purified and refolded fusion protein is then analyzed by PAGE, both under denaturing and nondenaturing conditions and by immunoblot analyses. Native PAGE analysis is performed to determine if aggregate forms of the fusion toxin are present. If aggregates are present they may be removed by DEAE-Sepharose ion-exchange chromatography. In general, the final yield of fusion protein toxin obtained from a 500-ml culture is in the range of 2–5 mg. After purification, aliquots are stored at −70° until use. Many of the diphtheria toxin-based fusion protein toxins are sensitive to repetitive freeze–thaw cycles and aliquots should be prepared such that they are thawed only once.

Purification of Monomeric Fusion Protein Toxin by DEAE-Sepharose Ion-Exchange Chromatography

When native gel electrophoresis indicates the presence of aggregate forms of the fusion protein toxin, the sample is centrifuged at 17,000g for 30 min and then filtered through a 0.45-μm pore size membrane filter. The sample is loaded onto a DEAE-Sepharose column (DEAE-Sepharose Fast Flow; Pharmacia Biotech, Piscataway, NJ) that has been equilibrated in 10 mM phosphate buffer, pH 7.2 [we use approximately 30 ml of resin in a Bio-Rad (Hercules, CA) chromatography column]. The column is washed with the 10 mM phosphate buffer until a baseline absorbance ($A_{280\ nm}$) is maintained, and the protein is eluted with a linear KCl gradient (0–0.8 M) in phosphate buffer. Each fraction is analyzed by PAGE under native and denaturing conditions and by immunoblot. The fractions containing the purified fusion protein toxin are then aliquoted and stored at −70°.

Cytotoxicity Assays

The biologic activity of a given diphtheria-based fusion protein toxin is determined *in vitro* with continuous cell lines that express the targeted cell surface receptor. In general, cells are seeded at a concentration of 5×10^4 cells per well in 96-well plates. The fusion protein toxin is diluted in complete medium and added to each well such that the final volume is 200 μl, and the fusion protein toxin concentration varies from 10^{-7} to 10^{-12} M. The cells are then incubated for 18 hr at 37° in a 5% CO_2 atmosphere. If the target cells grow in suspension, microtiter plates are centrifuged at 170g for 5 min to pellet the cells, and the medium is carefully removed by aspiration. Cells are resuspended in 200 μl of leucine-free minimal essential medium (BRL Life Technologies, Gaithersburg, MD) supplemented with [^{14}C]leucine (1.0 μCi/ml, 280 mCi/mmol; Du Pont-New England Nuclear, Boston, MA), 2 mM L-glutamine, penicillin (50 IU/ml), and streptomycin (50 μg/ml). The cells are incubated at 37° for 90 min and the plates are centrifuged at 170g for 5 min. The medium is carefully removed by aspiration and the cells are lysed by the addition of 60 μl of 0.4 M KOH per well. Total protein is then precipitated by the addition of 140 ml of 10% (w/v) trichloroacetic acid (TCA) per well. Total precipitated protein is then collected on glass fiber filters (GF/A; Whatman, Clifton, NJ) and radioactivity is measured by liquid scintillation. All assays are performed in quadruplicate and medium alone serves as the control. Results are calculated as percent control incorporation of [^{14}C]leucine.

Discussion

The diphtheria toxin-based fusion toxin proteins that we have worked with in this laboratory include those targeted by the following ligands: α-MSH, IL-2, mIL-4, IL-6, CD4, sIL-15, SP, GRP, and IL-7. Of these, only the CD4 and sIL-15 receptor-targeted fusion protein constructs could not be purified from inclusion bodies and required purification over anti-diphtheria toxoid immunoaffinity columns. In both instances, these fusion protein toxins did not form inclusion bodies and were found in the first supernatant fraction after lysis of recombinant *E. coli*. All of the above-mentioned diphtheria toxin-based fusion proteins were constructed as the DAB_{389} form and possess wild-type catalytic and transmembrane domains of diphtheria toxin. For constructs in which deletions, exchanges, or even single amino acid substitutions were made in these domains, modification of the purification schemes described above were usually required.

[19] Green Fluorescent Protein-Based Sensors for Detecting Signal Transduction and Monitoring Ion Channel Function

By Micah S. Siegel and Ehud Y. Isacoff

Introduction

Measuring signal transduction is a fundamental problem in studying information processing in the nervous system. To address this problem, we have designed a family of detectors that are chimeras between signal transduction proteins and fluorescent proteins. The prototype sensor is a novel, genetically encoded probe that can be used to measure transmembrane voltage in single cells. In this chapter, we describe a modified green fluorescent protein (GFP) fused to a voltage-sensitive K$^+$ channel so that voltage-dependent rearrangements in the K$^+$ channel induce changes in the fluorescence of GFP. The probe has a maximal fractional fluorescence change that is comparable to some of the best organic voltage-sensitive dyes. Moreover, the fluorescent signal is expanded in time in a way that makes the signal 30-fold easier to detect than that of a traditional linear dye. Sensors encoded into DNA have the advantage that they may be introduced into an organism noninvasively and targeted to specific brain regions, cell types, and subcellular compartments.

Fluorescent Dyes as Sensors

Fluorescent indicator dyes have revolutionized our understanding of cellular signaling by providing continuous measurements of physiological events in single cells and cell populations. There are two major impediments to progress with indicator dyes: (1) the lack of direct, noninvasive assays for most cellular communication events and (2) the difficulty in observing selected cell populations. In practice, these dyes must be synthesized chemically and introduced as hydrolyzable esters or by microinjection.[1–3] Delivering indicator dyes to specific cell populations has proved to be a difficult problem. In the absence of such localization, optical measurements in neural tissue usually cannot distinguish whether a signal originates from activity in neurons or glia; neither can it distinguish which types of neurons are involved.

Chimeric Protein Sensors

To increase our understanding of the signaling events that govern development, sensory transduction, and learning and memory, it is necessary to expand the range of signals that can be detected and to develop means of targeting sensors to specific cellular locations. One general approach to this problem is to construct sensors out of proteins, the biological molecules that transduce and transmit cellular signals. This approach would harness the high sensitivity and specificity of biological systems, which can detect an enormous range of signals with exquisite sensitivity. Moreover, it could permit the detection of signaling events throughout a signal transduction cascade: from the earliest stage of membrane transduction, through the network of signaling relays and amplification steps, and finally to downstream events that occur anywhere from the nucleus back to the plasma membrane.

Because protein-based sensors are encoded in DNA, they can be placed under the control of cell-specific promoters introduced *in vivo* or *in vitro*

[1] L. Cohen and S. Lesher, *Soc. Gen. Physiol. Ser.* **40**, 71 (1986).
[2] R. Y. Tsien, *Annu. Rev. Neurosci.* **12**, 227 (1989).
[3] D. Gross and L. M. Loew, *Methods Cell Biol.* **30**, 193 (1989).

by gene transfer techniques, and even targeted to specific subcellular compartments by protein signals recognized by the protein-sorting machinery in the cell. In addition to solving the problem of sensor production and targeting, DNA-encoded sensors have the advantage that, unlike organic dyes, they can be rationally tuned by modification of their functional domains with mutations that are known to adjust their dynamic range of operation. Finally, by creating "biological spies" out of native proteins, it is possible to avoid the introduction of foreign substances that could interfere with cellular physiology.

We have designed a family of protein-based detectors for measuring signal transduction events in cells (Fig. 1A). These sensors are chimeras between a signal transduction protein fragment (detector) and a fluorescent protein (reporter). This chapter describes a prototypical example of this class of chimeric proteins: a GFP sensor that we have engineered to measure fast membrane potential changes in single cells embedded within a population of cells.

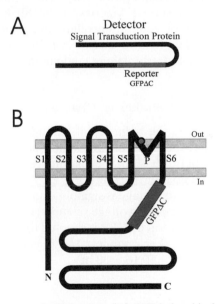

FIG. 1. GFP-based sensors of cell signaling. (A) GFP is fused in-frame into the middle of a signal transduction protein (detector) so that conformational rearrangements in the detector perturb the fluorescence of GFP. The GFP has been sensitized to the rearrangements of the detector protein by deletion of the last eight residues in the C terminal, which are disordered in the crystal structure.[20] (B) FlaSh is a chimeric protein in which GFP has been fused into the Shaker potassium channel.

Fluorescent Shaker K[+] Channel: Chimeric Protein to
  Measure Voltage

In outline, we have engineered a chimeric protein that is a modified
GFP[4] fused in-frame at a site just after the sixth transmembrane segment
(S6; Fig. 1B) of the voltage-activated Shaker K[+] channel.[5–7] The detailed
description and characterization of this sensor protein have been described
previously.[8] (Henceforth this chimeric protein is called FlaSh, for Fluorescent Shaker.) To prevent FlaSh from loading down target cells with an
additional potassium current, we engineered a point mutation into the pore
of the channel. This mutation prevents ion conduction but preserves the
gating rearrangements of the channel in response to voltage changes.

Our idea was that the voltage-dependent rearrangements in the channel
could be transmitted to GFP, resulting in a measurable change in its spectral
properties. The Shaker–GFP chimeric protein reports changes in membrane potential by a change in its fluorescence emission. In addition, the
fluorescence response is amplified in time over the electrical event, drastically increasing the optical signal power per event. Temporal amplification
in FlaSh is due to the response kinetics of the Shaker channel. Taken
together, the properties of genetic encoding and temporal amplification
allow the sensor to be delivered to specific cells in which action potentials
may be detected with standard imaging equipment.

Blueprint for Fluorescent Shaker K[+] Channel

We have characterized the behavior of FlaSh in single *Xenopus laevis*
oocytes by cRNA injection and voltage-clamp fluorimetry. Voltage steps
from a holding potential of -80 mV evoke fluorescent emission changes
and gating currents, but no ionic currents (Fig. 2). The relation of the steady
state fluorescence change to voltage is sigmoidal and correlates closely with
the steady state gating charge-to-voltage relation (Fig. 3), indicating that in
FlaSh, the fluorescent emission of GFP is coupled to the voltage-dependent
rearrangements of the Shaker channel. The dynamic range of FlaSh is
steep, from approximately -50 to -30 mV.

[4] M. Chalfie, Y. Tu, G. Euskirchen, W. W. Ward, and D. C. Prasher, *Science* **263**, 802
  (1994).
[5] B. L. Tempel, D. M. Papazian, T. L. Schwarz, Y. N. Jan, and L. Y. Jan, *Science* **237**, 770
  (1987).
[6] A. Baumann, A. Grupe, A. Ackermann, and O. Pongs, *EMBO J.* **7**, 2457 (1988).
[7] A. Kamb, J. Tseng-Crank, and M. A. Tanouye, *Neuron* **1**, 421 (1988).
[8] M. S. Siegel and E. Y. Isacoff, *Neuron* **19**, 735 (1997).

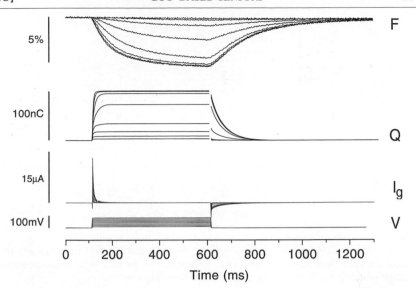

FIG. 2. Cell membrane potential modulates fluorescence output in FlaSh. Simultaneous two-electrode voltage-clamp recording and photometry show current and fluorescence changes in response to voltage steps (V) between -60 and 10 mV, in 10-mV increments. The holding potential was -80 mV. FlaSh exhibits on-and-off gating currents (I_g) but no ionic current because of the W434F mutation of Shaker. Integrating the gating current gives the total gating charge (Q) moved during the pulse. FlaSh fluorescence (F) decreases reversibly in response to membrane depolarizations. Traces are the average of 20 sweeps. Fluorescence scale, 5% $\Delta F/F$.

Physiological Impact on Target Cell

To prevent FlaSh from altering the physiology of cells in which it is expressed, we made the W434F point mutation in the Shaker pore to prevent ion conduction. This mutation blocks conduction by locking a gate in the pore into a closed conformation. Normally this gate closes slowly during sustained depolarization, producing slow inactivation. Other gating processes and rearrangements remain normal in the mutant channel, including activation in response to depolarization, opening of the activation gate, ball-and-chain (N-type) inactivation, and the rearrangement that consolidates slow inactivation and changes the fluorescence of GFP.[8–12]

The use of the W434F mutation works to prevent ion conduction in

[9] E. Perozo, R. MacKinnon, F. Bezanilla, and E. Stefani, *Neuron* **11,** 353 (1993).
[10] F. Bezanilla, E. Peroza, D. M. Papazian, and E. Stefani, *Science* **254,** 679 (1991).
[11] Y. Yang, Y. Yan, and F. J. Sigworth, *J. Gen. Physiol.* **109,** 779 (1997).
[12] E. Loots and E. Y. Isacoff, *J. Gen. Physiol.* **112,** 377 (1998).

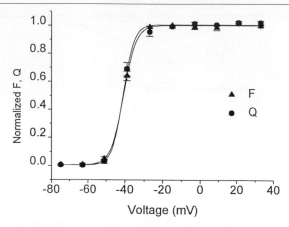

FIG. 3. FlaSh fluorescence is correlated with Shaker activation gating. The voltage dependence of the normalized steady state gating charge displacement (Q) overlaps with the normalized steady state fluorescence change (F). Both relations were fit by a single Boltzmann
equation (solid lines). Values are means $\pm$ SEM from five oocytes.

nonexcitable cells, such as *Xenopus* oocytes, where FlaSh subunits are the
only subunits from the Shaker K$^+$ channel subfamily that are expressed,
so that FlaSh channels form as nonconducting (permanently inactivated)
homotetramers. However, in excitable cells such as neurons and muscle,
where native subunits from the Shaker subfamily (which carry the wild-
type W at position 434) are expressed, FlaSh subunits may co-assemble with
those native subunits to form heterotetramers. In such heterotetramers, the
slow inactivation gate will shut more quickly than in wild-type (W434)
homotetramers, meaning that the properties of the K$^+$ conductances in the
cell will be altered. Although the effect of altering one class of voltage-
gated K$^+$ channels pharmacologically is often subtle, such heterotetrameric
channels may nevertheless affect the functional properties of the cells, a
side effect of sensor expression that would be better avoided.

Our approach to circumventing coassembly between FlaSh and native
channels in excitable cells is to link four FlaSh cDNAs in tandem in such
a way that the four subunits of the channel are covalently attached.[13] This
approach has been used earlier to force subunits to assemble in a known

[13] E. Y. Isacoff, Y. N. Jan, and L. Y. Jan, *Nature (London)* **345,** 530 (1990).

stoichiometry.[14–16] The expectation is that linked FlaSh constructs should assemble into FlaSh homotetramers even in excitable cells because of the higher likelihood of intramolecular assembly between linked FlaSh subunits than intermolecular assembly with native channel subunits.

Modifying Dynamic Range of Fluorescent Shaker K^+ Channel
through Mutagenesis

Given the narrow range of voltage over which Shaker channels gate, and over which FlaSh modulates its brightness (Fig. 3), it is clear that some voltage signals will be reported more efficiently, while others may be missed altogether. Because mammalian neurons tend to rest at about −70 mV, small excitatory and inhibitory postsynaptic potentials will not fall within the dynamic range of FlaSh (−50 to −30 mV). However, suprathreshold excitatory postsynaptic potentials and action potentials should be reported.

We examined the response of FlaSh to physiologically realistic voltage traces. Voltage transients measured in response to light in a variety of salamander retinal cell types[17] were applied via a voltage clamp to oocytes expressing FlaSh. FlaSh fluorescence reflected the dynamics of on-bipolar cell transients quite well, because the response of the on-bipolar cell is within the dynamic range of FlaSh (−50 to −30 mV). However, FlaSh did not capture the response of wide-field amacrine cells, because the main response of these cell types occurs from −70 to −50 mV, outside of its dynamic range.

One advantage of using the Shaker channel is that many mutations have been described that produce unique alterations in its voltage dependence and kinetics. This provides flexibility in tuning FlaSh to an operating range that best suits the signals of interest. For example, we have made versions of FlaSh with a more negative operating range based on mutations identified by Lopez et al.[18] These provide a good optical sensor for measuring the voltage waves from wide-field amacrine cells.[19] Similarly, mutations

[14] R. S. Hurst, M. P. Kavanaugh, J. Yakel, J. P. Adelman, and R. A. North, J. Biol. Chem. 267, 23742 (1992).
[15] E. R. Liman, J. Tytgat, and P. Hess, Neuron, 9, 861 (1992).
[16] D. T. Liu, G. R. Tibbs, and S. A. Sigelbaum, Neuron 16, 983 (1996).
[17] B. Roska, E. Nemeth, and F. S. Werblin, J. Neurosci. 18, 3451 (1998).
[18] G. A. Lopez, Y. N. Jan, and L. Y. Jan, Neuron 7, 327 (1991).
[19] G. Guerrero, M. S. Siegel, B. Roska, C. Dean, and E. Y. Isacoff, in preparation (2000).

that alter the kinetics of the rearrangement that consolidates channel inactivation predictably alter the kinetics of the fluorescence output of FlaSh.[19]

Designing Green Fluorescent Protein-Based Sensors

In general, we have found that the polymerase chain reaction (PCR) works well for inserting GFP into signal transduction proteins and that channel proteins, in particular, are surprisingly tolerant to GFP insertions. In the case of Shaker, we find that GFP can be inserted at the N terminal, at the C terminal, and also at a variety of internal sites. In our experience, these chimeric proteins are usually fluorescent, and the signal transduction protein usually functions. The tolerance of many proteins to GFP insertion is probably due to the structure of GFP, in which the N and C termini emerge in close proximity to one another on the same side of the barrel.[20] Sometimes we find that the insertion of GFP can alter the kinetics of the signal transduction protein (Siegel and Isacoff, unpublished data, 1999).

We have found that it can be helpful to examine sequences from a wide variety of homologous proteins (e.g., Shaker, Shab, Shaw). Often homology can give insight into those regions in the protein sequence that are highly conserved through evolution, and that therefore might be intolerant to GFP insertions. For example, we never were able to make functional proteins in which GFP was fused into regions near the fourth transmembrane segment (S4) in Shaker. We had hoped that this region would be interesting because of the high homology between S4 transmembrane segments in a variety of channel proteins, and because the Shaker S4 segment is known to undergo conformational rearrangements in response to transmembrane voltage. Unfortunately, the Shaker channel was intolerant to GFP insertions in three regions before (external side of the membrane) and after (internal side of the membrane) the S4 region. These fusion proteins were not fluorescent, nor did the channel function, implying that they interfered with protein folding, assembly, or stability.

Designing Green Fluorescent Protein-Based Sensors

We have found it useful to consider endogenous targeting sequences that are present in the signal transduction protein of interest. For example, the Shaker channel contains a PDZ interaction domain at its C terminus that targets Shaker preferentially to postsynaptic specializations.[21–23] This PDZ

[20] M. Ormo, A. B. Cubitt, K. Kallio, L. A. Gross, R. Y. Tsien, and S. J. Remington, *Science* **273,** 1392 (1996).
[21] F. J. Tejedor, A. Bokhari, O. Rogero, M. Gorczyca, J. Zhang, E. Kim, M. Sheng, and V. Budnik, *J. Neurosci.* **17,** 152 (1997).
[22] K. Zito, R. D. Fetter, C. S. Goodman, and E. Y. Isacoff, *Neuron* **19,** 1007 (1997).
[23] K. Zito, D. Parnas, R. D. Fetter, E. Y. Isacoff, and C. S. Goodman, *Neuron* **22,** 719 (1999).

interaction domain can target heterologous proteins to the synapse[22, 23] and therefore can be used to target other GFP-based sensors. In contrast, some mutations in the PDZ interaction domain prevent targeting[22] and can be used to ensure that FlaSh is distributed more uniformly throughout the cell.

Practical Issues in Screening for Green Fluorescent Protein-Based Sensors

One strategy for producing GFP-based sensors is first to find a fusion protein that exhibits a small fluorescence change in response to a signal transduction event. The next step is to modify the fusion protein in order to increase the size of its fluorescence change. In our experience, when looking for the presence of a small fluorescence change, it is extremely important to have a well-controlled, reproducible method of activating the signal transduction cascade. For example, in the case of FlaSh, it was important to test the system with rapid voltage jumps. In our experience with ligand-gated receptors, it has been useful to be able to wash ligands into the solution quickly and reproducibly, and then to remove them quickly and reproducibly.

Reproducibility in the stimuli is important because it is often desirable to average a number of stimulus presentations to find a weak fluorescence change in a noisy background. When the onset of the stimuli is ambiguous (e.g., with a ligand wash), it is possible to confuse bleaching, arc-lamp jitter, and other optical artifacts for a genuine fluorescence change.

It is worthwhile to test a positive control in parallel with the fusion protein of interest. In our experience, most GFP fusion proteins do not exhibit a fluorescence change. It is obviously important to know whether the fusion protein is a failure or the test apparatus is broken. We have found that organic calcium-sensitive dyes can serve as a useful control in this context.

Optical Issues in Screening Green Fluorescent Protein-Based Sensors

We have sometimes found that different GFP mutants can behave differently when fused into the same signal transduction protein. For example, wild-type GFP inserted into the Shaker potassium channel (e.g., FlaSh, as described above) exhibits a fluorescence change on membrane depolarization. However, the mutant S65TGFP[24] inserted into Shaker at the same

[24] R. Y. Tsien, *Annu. Rev. Biochem.* **67,** 509 (1998).

location produces a fusion protein that is brighter than FlaSh but gives a fluorescence change that is approximately 10-fold smaller.

It is important to be aware of different excitation and emission spectra of different GFP mutants, and to choose appropriate filters for looking at the fluorescent sensor. For FlaSh, we use an HQ-GFP filter (Chroma Technology, Brattleboro, VT) with the following filters: excitation, 425–475 nm, dichroic, 480-nm long-pass; emission, 485–535 nm. We have found that FlaSh is also visible with a standard fluorescein filter set (e.g., Chroma Technology HQ-FITC filter set). The relative fluorescence change with an HQ-FITC set is similar to that with an HQ-GFP set; however, the total fluorescence emission is smaller because HQ-GFP is better matched to the excitation spectra of wild-type GFP.

Finally, we have found that some GFP fusion proteins contain multiple optical excitation peaks, as would be expected from the excitation spectra of wild-type GFP. It would be desirable to detect changes in the relative excitation levels of these peaks. It is useful in these cases to test the resulting fusion proteins with a variety of filters, each tuned to a specific peak.

Future Improvements in Fluorescent Shaker K+ Channel

One advantage to using GFP is that many mutations have been discovered that alter the fluorescence absorption or emission properties of GFP.[24] Several groups have used variants of GFP to monitor the spatial location of different proteins within the same cell. The same approach could be used to monitor different signal transduction pathways within the same cell, by creating chimeric sensor proteins with GFP variants. This would require that the GFP variants be distinguishable in the same way, e.g., fluorescence absorption, emission, or lifetime (see filter discussion above). For example, sensors with cyan-fluorescent protein could potentially be distinguished from sensors with yellow-fluorescent protein.

Discussion and Conclusion

FlaSh is not a typical fluorescent voltage probe. Traditional "fast" voltage-sensitive dyes have been designed to respond quickly and linearly to membrane potential.[1,3,25] In contrast, FlaSh provides a different solution to the underlying problem of detecting fast voltage transients: FlaSh gives long, stereotypical fluorescent pulses in response to brief voltage spikes. The temporally expanded response from FlaSh provides advantages for detecting individual electrical events, as the area under the response to

[25] J. E. Gonzalez and R. Y. Tsien, *Chem. Biol.* **4,** 269 (1997).

single spikes is approximately 30-fold larger than the area under the input spike (converting units appropriately).

In general, it is best to consider ways to visualize fast, discrete cellular events (e.g., action potentials) by exploiting the slower kinetics of a chimeric protein. The sensor protein can often be constructed so that fast physiological events give rise to a slowed fluorescence signature that is matched to standard imaging hardware (e.g., video-rate cameras).

The success of FlaSh suggests that a modular approach could be used to produce optical sensors to detect other signaling events. We are using this approach to engineer signaling proteins so that changes in their biological activity are converted into changes in fluorescence emission. These chimeric sensor proteins have two domains: a "detector" that undergoes a conformational rearrangement during a cell signaling event (e.g., a channel, receptor, enzyme, G protein) and a "reporter" fluorophore (e.g., GFP). Ideally, GFP is fused to the detector protein near a domain that undergoes a conformational rearrangement when the detector protein is activated. Movement in the detector domain alters the environment of GFP or places stress on the structure of GFP, thus altering its spectral properties. These constructs typically include a variant of GFP as a reporter, a signal transduction protein as a detector, and a subcellular targeting peptide.

A thorough characterization of the sensor should make it possible to show definitively that the fluorescence reports on detector protein activity, rather than on other physiological signals, such as changes in ion concentration, pH, or oxidative–reductive state. We showed that the fluorescence output depends on the site of GFP insertion, perfectly follows the voltage dependence of channel gating, and is modified in parallel with channel gating by mutations that alter the dynamic range or kinetics of the channel.

Protein-based sensor proteins analogous to FlaSh may enable the noninvasive detection of activity in a variety of proteins, including receptors, G proteins, enzymes, and motor proteins. The developmental timing and cellular specificity of expression can be directed by placing the construct under the transcriptional control of a specific promoter. The combined ability to tune the sensor module via mutagenesis and to target the sensor to specific locations affords powerful advantages for the study of signal transduction events in intact tissues.

Acknowledgments

We thank Botond Roska for preparing voltage traces from the salamander retina; and Scott Fraser, Henry Lester, Carver Mead, Gilles Laurent, Norman Davidson, Sanjoy Mahajan, John Ngai, and members of the Isacoff laboratory for helpful discussions. Research was supported by the McKnight and Klingenstein, and Whitehall (#J9529) Foundations. M. S. S. is a Howard Hughes predoctoral fellow in the biological sciences.

[20] Metabolic Labeling of Glycoproteins with Chemical Tags through Unnatural Sialic Acid Biosynthesis

By Christina L. Jacobs, Kevin J. Yarema, Lara K. Mahal, David A. Nauman, Neil W. Charters, and Carolyn R. Bertozzi

Introduction

The importance of protein glycosylation is becoming increasingly apparent, and much effort has been directed to the investigation of glycoproteins with the aim of identifying the biological functions of the pendant glycans. Chemical synthesis now provides access to oligosaccharides and their analogs for structure–function analyses *in vitro*. However, it has been shown that the function of cell surface glycoproteins can depend heavily on the spatial context of their presentation. Thus, biological investigations of isolated glycoprotein molecules and those displayed on a cell surface can show drastically different results.[1,2] For this reason, the extension of a chemical modification route to cell surface oligosaccharides would be a valuable addition to the tools of cell surface biochemistry. Traditional methods for manipulating the structures of carbohydrates on the cell surface include exogenous modification by glycosidases and glycosyltransferases,[3] and chemical or enzymatic oxidation of sugars followed by labeling with nucleophiles or reducing agents.[4] Although all these methods are useful in certain systems, their potential for modification of oligosaccharides on cells *in vivo* is limited, because the materials used are either toxic or not bioavailable.

An alternative strategy for alteration of cell surface carbohydrate structure involves introduction of a modified monosaccharide precursor into a cell, possibly resulting in its metabolic conversion and incorporation into unnatural cell surface oligosaccharides. Opportunities for unnatural oligosaccharide biosynthesis are limited by the substrate specificities of the biosynthetic enzymes, which in many cases permit only conservatively modified precursors such as deoxy or fluoro sugars to complete the biosynthetic transformations.[5] One exception to this rule is the biosynthesis of sialosides.

[1] K. J. Yarema and C. R. Bertozzi, *Curr. Opin. Chem. Biol.* **2,** 49 (1998).
[2] S. Tsuboi, Y. Isogai, N. Hada, J. K. King, O. Hindsgaul, and M. Fukuda, *J. Biol. Chem.* **271,** 27213 (1996).
[3] L. K. Mahal and C. R. Bertozzi, *Chem. Biol.* **4,** 415 (1997).
[4] G. A. Lemieux and C. R. Bertozzi, *Trends Biotechnol.* **16,** 506 (1998).
[5] R. J. Bernacki and W. Korytnyk, *in* "The Glycoconjugates," Vol. IV (M. I. Horowitz, ed.), p. 245. Academic Press, New York, 1982.

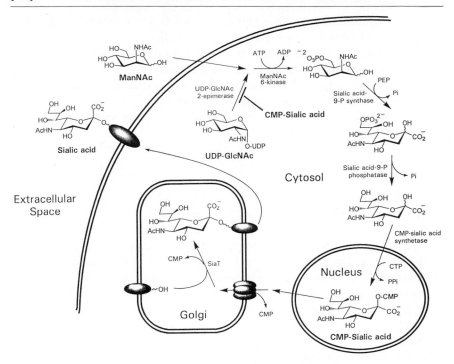

FIG. 1. The sialic acid biosynthetic pathway.

Sialic acid glycoconjugates are synthesized *in vivo* through a series of enzymatic steps from the first committed precursor *N*-acetylmannosamine (ManNAc), as depicted in Fig. 1.[6] The biosynthesis of sialoglycoconjugates is initiated by the phosphorylation of ManNAc, and proceeds through the condensation with phosphoenolpyruvate (PEP) to yield sialic acid 9-phosphate. After hydrolysis of the phosphate monoester by a specific phosphatase, sialic acid is converted into its activated donor form by condensation with CTP to give CMP-sialic acid. The activated sugar donor makes its way into the Golgi compartment by way of a specific transport protein. At this point, a membrane-bound sialyltransferase (SiaT) catalyzes the transfer of the sugar residue onto the terminus of an oligosaccharide chain, and the resulting glycoprotein or glycolipid is delivered to the cell surface. This pathway is regulated by CMP-sialic acid, which acts as a feedback inhibitor of UDP-GlcNAc-2-epimerase, the enzyme that produces ManNAc

[6] L. Warren, *in* "Bound Carbohydrates in Nature," p. 29. Cambridge University Press, New York, 1994.

from endogenous UDP-N-acetylglucosamine (UDP-GlcNAc).[7] ManNAc can also be obtained from the extracellular milieu.

Several individual enzymes of the sialic acid pathway have been shown to have relaxed substrate specificities, allowing substitutions at various positions on the sugar.[8–10] Eckhardt and co-workers demonstrated that the CMP-sialic acid Golgi transporter recognizes primarily the CMP portion of the molecule, possibly allowing modification of the sugar portion.[11] In accord with these observations, Kayser and co-workers demonstrated that unnatural mannosamine analogs with lengthened N-acyl chains can be converted into sialic acids and incorporated into cell surface and secreted glycoproteins in cell culture and in laboratory rats.[12] Furthermore, N-acyl modifications were shown to affect binding affinity between cell surface sialic acids and viral protein receptors, and therefore modulate viral infectivity.[13–15]

On the basis of these observations, we hypothesized that the permissivity of the sialic acid pathway could be used as a vehicle for the introduction of novel reactive functional groups onto cell surface glycoproteins, providing a chemical handle for further modification of cell surface glycoconjugates. Our first efforts in this area focused on the introduction of a ketone group into cell surface sialic acid residues.[16] Ketones are chemically orthogonal to normal cell surface components. They are not found in proteins, lipids, or other cell surface biomolecules and do not react appreciably with groups present on those molecules. Furthermore, the ketone group undergoes aqueous chemoselective ligation reactions with aminooxy and hydrazide groups, forming stable oxime and hydrazone adducts, respectively. By installing ketone groups on the cell surface via sialoside biosynthesis and

[7] B. Potvin, T. S. Raju, and P. Stanley, *J. Biol. Chem.* **270,** 30415 (1995).

[8] M. A. Sparks, K. W. Williams, C. Lukacs, A. Schrell, G. Priebe, A. Spaltenstein, and G. M. Whitesides, *Tetrahedron* **49,** 1 (1993).

[9] R. E. Kosa, R. Brossmer, and H.-J. Gross, *Biochem. Biophys. Res. Commun.* **190,** 914 (1993).

[10] S. L. Shames, E. S. Simon, C. W. Christopher, W. Schmid, G. M. Whitesides, and L.-L. Yang, *Glycobiology.* **1,** 187 (1991).

[11] M. Eckhardt, B. Gotza, and R. Gerardy-Schahn, *J. Biol. Chem.* **274,** 8779 (1999).

[12] H. Kayser, R. Zeitler, C. Kannicht, D. Grunow, R. Nuck, and W. Reutter, *J. Biol. Chem.* **267,** 16934 (1992).

[13] M. Herrmann, C. W. von der Lieth, P. Stehling, W. Reutter, and M. Pawlita, *J. Virol.* **71,** 5922 (1997).

[14] O. T. Keppler, P. Stehling, M. Herrmann, H. Kayser, D. Grunow, W. Reutter, and M. Pawlita, *J. Biol. Chem.* **270,** 1308 (1995).

[15] O. T. Keppler, M. Herrmann, C. W. von der Lieth, P. Stehling, W. Reutter, and M. Pawlita, *Biochem. Biophys. Res. Commun.* **253,** 437 (1998).

[16] L. K. Mahal, K. J. Yarema, and C. R. Bertozzi, *Science* **276,** 1125 (1997).

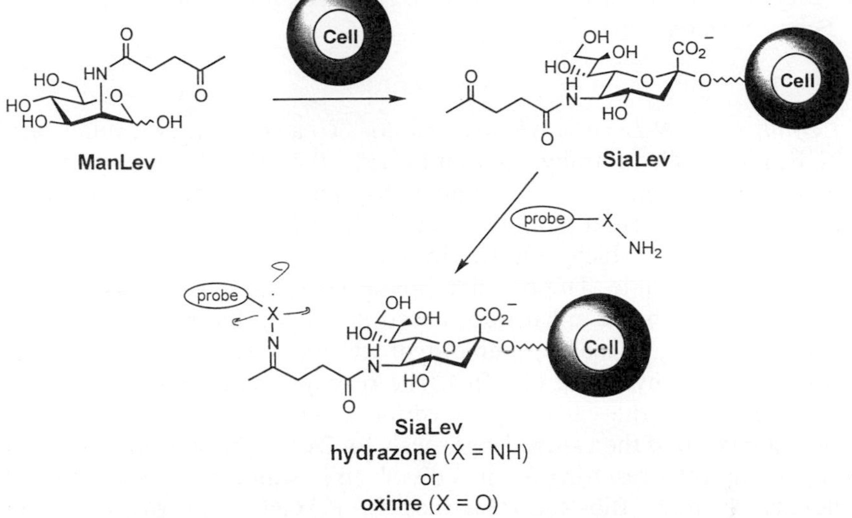

FIG. 2. Metabolic labeling strategy. Ketone groups were incorporated into cell surface sialosides (SiaLev) by metabolic labeling with ManLev. The cells were then labeled with a hydrazide- or aminooxy-derivatized probe to form hydrazone or oxime linkages to cell surface sialoglycoproteins.

subsequently attaching chemically defined structures by chemoselective ligation, novel structures can be attached to cell surface glycoproteins. This method provides a more biologically relevant context of display for synthetic oligosaccharides than traditional immobilization techniques. Furthermore, other novel synthetic compounds including molecular probes can be attached to the cell surface in this fashion.

In this chapter, we describe our strategy for the metabolic introduction of ketone groups into cell surface sialoglycoconjugates, using the unnatural sialic acid precursor N-levulinoylmannosamine (ManLev). As depicted in Fig. 2, ManLev is converted to N-levulinoyl sialic acid (SiaLev) in cultured cells, permitting the chemical attachment of hydrazide- or aminooxy-derivatized probes to cell surface glycoproteins. The basic procedures for cell surface labeling and quantitation are detailed, so as to make this strategy available for use in other laboratories. In addition, we provide additional qualitative and quantitative characterization of this metabolic engineering system. Finally, two applications of the technique are discussed: the display of new carbohydrate epitopes on cell surfaces and the delivery of magnetic resonance probes to cell surfaces.

Metabolic Delivery of Ketones to Cell Surfaces and Cell
Surface Detection

Methods

Synthesis of N-Levulinoylmannosamine. To a solution of levulinic acid
(1.1 Eq) in distilled tetrahydrofuran (THF, ~0.2 M final concentration) is
added triethylamine (1.1 Eq) under N_2, and the reaction mixture is
stirred.[16,17] After 10 min, isobutyl chloroformate (1.1 Eq) is added dropwise
by syringe, during which time triethylammonium salt ($Et_3NH^+Cl^-$) forms
as a white precipitate. The reaction mixture is stirred vigorously for 5 hr.
The resulting carbonic anhydride is used directly in the next step.

To a solution of mannosamine hydrochloride (1.0 Eq) in H_2O (~0.4
M) is added triethylamine (1.3 Eq). The reaction mixture is stirred for 30
min, after which the carbonic anhydride solution is added slowly. The
reaction mixture is then stirred vigorously for 24 hr. The solvent is removed
in vacuo, and the resulting foam is dissolved in water and passed through
columns of cation [Bio-Rad (Hercules, CA) AG50W-X8, pyridine form]
and anion (Bio-Rad AGI-X2, acetate form) exchange resin, eluting with
water. (Removal of the triethylammonium salts from the reaction mixture
in this manner is found to improve the efficiency of subsequent column
chromatography.) The resulting yellow amorphous foam is purified by silica
gel chromatography, eluting with a gradient of $CHCl_3$-methanol (20:1 to
5:1, v/v) to yield the desired product as a white amorphous foam. The
product is characterized as a mixture of anomers. The best results are
obtained when the reactions are performed on no greater than a 5 g scale,
largely because of difficulties in purification. Yields typically range from
50 to 75%. R_f = 0.63 [butanol-acetic acid-H_2O, 5:3:2 (v/v)]; ^{1}H NMR
(300 MHz, D_2O): δ 2.19 (s, 3), 2.20 (s, 3), 2.47-2.61 (m, 4), 2.75-2.90 (m,
4), 3.38 (ddd, 1, J = 2.3, 3.4, 9.8), 3.49 (app t, 1, J = 9.8), 3.56-3.63 (m, 1),
3.74-3.88 (m, 6), 4.01 (dd, 1, J = 4.7, 9.8), 4.26 (dd, 1, J = 1.5, 4.6), 4.41
(dd, 1, J = 1.5, 4.4), 4.98 (d, 1, J = 1.6), 5.07 (d, 1, J = 1.5); ^{13}C NMR (100
MHz, D_2O): δ 29.2, 29.3, 29.4, 38.2, 38.3, 53.2, 54.1, 60.5, 66.6, 66.9, 68.9,
72.1, 72.1, 76.4, 93.1, 93.2, 175.9, 176.7, 214.1, 214.3; high-resolution mass
spectrum (FAB[+]), calculated for $C_{11}H_{20}NO_7$ (MH[+]) 278.1240, found
278.1238.

Synthesis of 1,3,4,6-Tetra-O-acetyl-N-levulinoylmannosamine. Prior to
purification, ManLev prepared according to the preceding procedure is
concentrated *in vacuo* and dissolved in pyridine-acetic anhydride (1:1,
v/v) to a final concentration of 0.1 M.[18] After acetylation is complete, the

---

[17] ManLev has been licensed and will be commercially available through Molecular Probes
(Eugene, OR).
[18] C. L. Jacobs, unpublished results (1998).

reaction is slowly quenched with methanol on ice and concentrated. The residue is dissolved in ethyl acetate and washed with 1 M citric acid or NaHSO$_4$ solution to remove any remaining pyridine and/or triethylamine. The reaction mixture is purified by silica gel chromatography, eluting with a gradient of hexane–ethyl acetate (1:1, v/v) to 100% ethyl acetate. The desired product is characterized as a mixture of anomers ($\sim$1:2 α/β), and the average yield for this reaction is 60–70%. R_f = 0.63 (CHCl$_3$–methanol, 9:1, v/v); ^{1}H NMR (300 MHz, CDCl$_3$): δ 1.87–2.08 (m, 30), 2.31–2.53 (m, 4), 2.67–2.72 (m, 4), 3.69–3.75 (ddd, 1, J = 2.4, 5.0, 8.8), 3.91–4.03 (m, 4), 4.11–4.20 (m, 2), 4.50 (ddd, 1, J = 1.6, 4.2, 9.3), 4.64 (ddd, 1, J = 1.8, 3.3, 9.3), 4.92–5.10 (m, 2), 5.17 (dd, 1, J = 4.4, 10.2), 5.76 (d, 1, J = 1.7), 5.89 (d, 1, J = 1.9), 6.48 (d, 1, J = 9.4), 6.72 (d, 1, J = 9.3); ^{13}C NMR (100 MHz, CDCl$_3$): δ 20.6, 20.7, 29.6, 29.8, 38.5, 38.6, 49.0, 49.1, 62.1, 62.3, 65.4, 65.6, 67.0, 70.1, 71.1, 73.1, 90.6, 91.72, 168.2, 168.5, 169.6, 170.0, 170.7, 172.5, 172.8, 207.7, 208.0; high-resolution mass spectrum (FAB$^+$), calculated for C$_{19}$H$_{28}$NO$_{11}$ (MH$^+$) 446.1662, found 446.1664.

Tissue Culture and Cell Surface Labeling. Mammalian cell culture is performed according to standard procedures.[19] We recommend maintaining cell stocks in the log phase of growth, especially immediately preceding a cell surface labeling experiment. More dense populations that have entered the plateau phase experience a decrease in their metabolic rate, and therefore will display decreased and possibly erratic incorporation of an unnatural sugar into cell surface glycoconjugates. Furthermore, certain cell types, such as Jurkat cells, undergo "shedding" of their cell surface proteins and/ or oligosaccharides when subjected to high cell densities. Accordingly, the most reproducible levels of metabolic incorporation are observed for Jurkat cells when cell densities are maintained between 100,000 and 800,000 cells/ ml during each 3-day growth period.

Solutions of ManLev and ManNAc in phosphate-buffered saline (PBS) at pH 7.4 are filter sterilized. Normally, a stock solution of 500 mM is diluted into medium to a final concentration between 5 and 40 mM immediately prior to incubation with cells. Cells incubated with ManNAc or only PBS are used as controls. Cells are typically incubated with the unnatural sugar for 2 or 3 days, but longer incubations (on the order of weeks) of ManLev with mammalian cells cause no observable deleterious effects.[19]

Solutions of peracetylated ManLev (Ac$_4$ManLev) in ethanol or dimethylsulfoxide (DMSO) at a stock concentration of 50 mM are also filter sterilized and added to medium prior to use in cell culture. These solutions are diluted into medium to give a final concentration between 5 and 50

[19] K. J. Yarema, L. K. Mahal, R. E. Bruehl, E. C. Rodriguez, and C. R. Bertozzi, *J. Biol. Chem.* **273**, 31168 (1998).

μM. Control cells are grown in the presence of ethanol or DMSO only, to account for changes in cell growth due to these solvents. Because the more lipophilic acetylated monosaccharide can passively diffuse through the plasma membrane, much lower concentrations of the sugar can be used to obtain levels of cell surface ketone expression comparable to those found with unprotected ManLev. However, cellular toxicity was observed at concentrations of $Ac_4ManLev$ above 50 μM, as evidenced by inclusion of propidium iodide and/or trypan blue.[18] Therefore, although acetylated ManLev is a far more efficient and economical reagent for cell surface labeling, it is inappropriate for use in some applications where toxicity at high dosages is a concern.

Flow Cytometry Assay. The most straightforward quantitative assay for the presence of cell surface ketones proceeds by labeling with biotin hydrazide, to form hydrazone linkages with the ketones, followed by flow cytometry analysis with FITC–avidin.[16] Cells to be assayed for cell surface ketone expression in this manner are incubated with ManLev or $Ac_4ManLev$ for 2 to 3 days as described above. Generally, concentrations of 5 mM ManLev or 10 μM $Ac_4ManLev$ are sufficient to generate a significant signal above background levels. Cells (typically 200,000 to 1 million per sample) are washed with 1.0 to 10 ml of biotin-staining buffer [PBS (pH 6.5), 0.1% (v/v) fetal bovine serum (FBS)] three times, taking care to resuspend cells thoroughly after pelleting, and then resuspended in 0.4 to 1.6 ml of biotin-staining buffer. For most cell types, centrifugation at 4000g for 1 min is sufficient to pellet cells without disrupting cellular structure. The washed cells are added to new sample tubes that contain 100 to 400 μl of biotin hydrazide [BH (Sigma, St. Louis, MO), 5 mM in biotin-staining buffer], resulting in a final concentration of 1 mM BH. The cell samples are then incubated at room temperature for 1.5 to 2 hr. This incubation is carried out at pH 6.5 as a compromise between the acidic conditions that are optimal for hydrazone formation and a neutral environment that is most conducive to cell viability. After biotin labeling is complete, the cell samples and buffers used are kept at 4° to prevent membrane recycling. Biotinylated cells are then washed three times with avidin-staining buffer [PBS (pH 7.4), 0.1% (v/v) FBS, 0.1% (w/v) NaN_3] to remove unreacted BH. The cells are resuspended in 100 μl of avidin-staining buffer and added to 100 μl of FITC-labeled avidin (11.2 μg/ml; Sigma) to give a final 5.6-μg/ml concentration of FITC–avidin. (Avidin concentrations between 0.5 and 50 μg/ml give similar flow cytometry signals.) The cells are incubated with FITC–avidin in the dark at 4° for 10 to 15 min, and then washed three times with avidin-staining buffer and resuspended in 300 μl of avidin-staining buffer in preparation for analysis by flow cytometry. Because the amount of BH added to each sample is in gross excess of the eventual quantity of FITC–avidin used, it is imperative that all excess BH be removed from the sample before avidin labeling. To this end, a second round of FITC–avidin staining and washes can be carried out. It is assumed that this procedure allows more consistent results because any remaining free BH in a cell pellet will be scavenged by the first addition of FITC–avidin, allowing for a quantitative second round of cell staining. Flow cytometry analysis is performed on a Coulter EPICS XL-MCL cytometer (Coulter, Miami, FL), using a 488-nm argon laser.

For adherent cells such as HeLa cells, all washing and staining steps (BH and FITC–avidin) can be carried out on cells still attached to the tissue culture plate. This procedure is faster than labeling in suspension, requires fewer manipulations of the cells, and provides similar or improved signal-to-noise ratio and error in the flow cytometry assay.[20] After staining with FITC–avidin and washing the cells twice with avidin-staining buffer, the cells are treated with a solution of trypsin–EDTA (0.25%, w/v) until they first appear round under the microscope (5–10 min at room temperature). They are then gently removed from the plate with a pipette and transferred to Eppendorf tubes for a final wash and resuspension. The labeling efficiency for cells in an adherent state is comparable to the labeling efficiency for cells that are suspended by trypsin treatment prior to labeling, indicating that the entire surface of the adhered cell is available for binding to BH and FITC–avidin.

Typical fluorescence intensities (arbitrary units) for the most commonly used cell lines in our laboratory are given in Table I, including background fluorescence levels measured with control cells that were incubated with ManNAc. It should be noted that these numbers are useful only as a guideline, because actual flow cytometry data will depend on instrumental parameters. In addition, genetic variability within a cell line could affect the efficiency of SiaLev expression. We have found that the ketone-associated fluorescence signal is linearly dependent on BH concentration up to 5 mM.[21] Above this level, BH is no longer sufficiently soluble in PBS to allow for its use in cell labeling. Provided that the concentration of the ManLev stock solution has been standardized and the cells used are healthy and grown at an appropriate density, we find the flow cytometry assay to be highly precise, with less than 4% deviation between duplicate samples split from the same cell population and less than 10% deviation from day to day over a period of weeks, using the same cell stock.

Expression of cell surface ketones via ManLev metabolism is cell type specific to some degree. Several types of cell lines have been utilized in initial

[20] D. A. Nauman, unpublished results (1999).
[21] K. J. Yarema, and S. Goon, unpublished results (1999).

TABLE I
APPROXIMATE FLOW CYTOMETRY SIGNALS FOR JURKAT, HeLa,
AND HL-60 CELLS

Cell line	Background fluorescence[a] (arbitrary units)	Fluorescence[b] (arbitrary units)
Jurkat	1–2	~30 at 5 mM
		80 at 25 mM
HeLa[c]	1–2	100–200 at 10 mM
		~230 at 20 mM
HL-60	0.5–1	7 at 5 mM
		26 at 25 mM

[a] Background fluorescence was determined for cells treated with ManNAc and then labeled with BH and stained with FITC–avidin.

[b] All values were determined with cells grown with the indicated concentration of ManLev for 2 days, followed by labeling with 1 mM BH and staining with FITC–avidin.

[c] Values shown for HeLa cells were obtained by the double-FITC labeling procedure described in Methods. Substantially lower values were observed with a single labeling step. Background fluorescence was determined with cells that had been labeled while still adherent. A background fluorescence of 4–6 units was commonly seen for cells labeled in suspension after treatment with trypsin.

experiments, differing in both species of origin and tissue type.[16,19,21–24] A summary of the qualitative behavior of each cell line is given in Table II. Interestingly, rodent-derived cell lines (murine and CHO) showed the lowest numbers of ketones per cell after treatment with ManLev. The murine cell line S49 was, however, comparable to human Jurkat cells with respect to both size and sialic acid content. A possible explanation for the low signal observed is that the murine enzymes are less permissive for unnatural substrates.

Fluorescence Microscopy. Cells that have been treated with ManLev and stained with BH and FITC–avidin can be evaluated qualitatively for fluorescence by fluorescence microscopy. This procedure is simplest with an adherent cell line such as HeLa. For ease of visual analysis, cells should be grown to about 50% coverage of the plate (rather than to confluence) after incubation with ManLev. After several days of cell growth in the presence of ManLev or ManNAc, the medium is removed by careful aspira-

[22] L. K. Mahal, unpublished results (1999).

[23] J. H. Lee, T. J. Baker, L. K. Mahal, J. Zabner, C. R. Bertozzi, D. F. Wiemer, and M. J. Welsh, *J. Biol. Chem.* **274**, 21878 (1999).

[24] N. W. Charters, unpublished results (1998).

TABLE II

QUALITATIVE LEVELS OF CELL SURFACE KETONES AS DETERMINED BY FLOW CYTOMETRY
OR FLUORESCENCE MICROSCOPY

Cell line	Species	Tissue type	Ketone
Jurkat	Human	T cell lymphoma	+++
HeLa	Human	Cervical epithelial carcinoma	++++
HL-60	Human	Neutrophil lymphoma	++
SK-N-SH	Human	Neuroblastoma	++
NT-2	Human	Differentiated neuronal	+[a]
293-T	Human	Embryonic kidney	++
HUVEC	Human	Primary umbilical vein endothelial	++++
RPMI-7666	Human	Non-transformed leukocyte	+++
HEK293	Human	Embryonic kidney	++++
MDCK	Canine	Kidney epithelial	++
NIH 3T3	Murine	Fibroblasts	++
P19	Murine	Differentiated neuronal	+[a]
D1B	Murine	Erythroleukemia	+
S49	Murine	T cell lymphoma	+
LEC.29	Chinese hamster	Ovarian	−
COS-7	Green monkey	Kidney derived	++

[a] Analysis via flow cytometry was not possible for these cell lines due to physical frailty of the cells. Examination of cells by fluorescence microscopy was performed instead.

tion and cells are washed on the plate twice with biotin-staining buffer (see above). Cells are then labeled with 2 mM BH (the higher concentration allows for increased signal, provided that all excess BH is washed away after labeling is complete) for 1.5 to 2 hr at room temperature. The labeled cells are next washed twice on the plate with cold avidin-staining buffer and labeled with FITC–avidin (5.6 μg/ml, final concentration) for 10 min on ice and in the dark. The cells are then washed and resubjected to FITC–avidin labeling to increase the fluorescence staining. After the final round of washes is completed, the cells are examined by both bright-field and fluorescence microscopy. Cells are viewed at 100× total magnification, using an inverted stage microscope (Nikon, Garden City, NY) equipped with a digital camera (Diagnostic Instruments, Sterling Heights, MI). We have found that using a low gain setting and increasing the exposure time leads to the best signal-to-noise ratio when comparing ManLev-treated cells with controls.

Characterization of Cellular Metabolism of *N*-Levulinoylmannosamine

The following section summarizes the findings of our laboratory regarding the details of ManLev metabolism in human and other mammalian cell lines. This information should prove useful for adaptation of the technique to other applications.

Inhibition of N-Levulinoylmannosamine Metabolism by N-Acetylmannosamine. Cells incubated with both ManLev and ManNAc concurrently exhibited reduced levels of ketone expression as indicated by flow cytometry analysis.[19] Depending on the cell line, a median inhibitory concentration (IC_{50}) value of 1 to 4 mM (or 4 to 20 μM for Ac₄ManNAc) was observed, independent of the concentration of ManLev added to the medium (1 to 30 mM). Although ManNAc is the preferred substrate for the sialic acid biosynthetic pathway, it is possible to "flood" the cell with exogenously added unnatural precursor, because the intracellular ManNAc concentration is maintained at micromolar levels.

Time and Dose Dependence of Cell Surface Ketone Expression. The dose dependence of ketone display for Jurkat cells, as determined by flow cytometry, saturated at ManLev concentrations of 25 to 30 mM.[19] Similar results were observed with HeLa cells, while HL-60 cells did not show saturation of ketone-dependent fluorescence even at a concentration of 40 mM ManLev.

In Jurkat cells, ketone expression was characterized by a lag phase of 15 to 20 hr after addition of ManLev to cell culture, during which time the level of expression was below the lower limit of detection for the flow cytometry assay (approximately 10,000 ketones per cell). A steady increase in the number of cell surface ketones was seen for approximately 5 days, after which time a steady state was reached and sustained indefinitely, provided that the appropriate concentration of ManLev was maintained in the growth medium. The concentration of ManLev used did not affect the kinetics of expression; however, higher concentrations of ManLev led to higher equilibrium expression levels. On removal of ManLev from the growth medium, the half-life for depletion of cell surface ketones was just under 48 hr for Jurkat cells. HeLa and HL-60 cells showed a similar time-dependent decrease in ketone expression. For each of these cell lines, fluorescence intensity returned to background levels in approximately 5 days. These results correlate well with predictions based solely on the effects of cell division. Accordingly, addition of NaN₃ to the medium at concentrations sufficient to inhibit cell growth also substantially retarded the depletion rate for cell surface ketones. It is also possible that SiaLev residues are actively removed from glycoconjugates during the membrane recycling process, although we have observed that SiaLev cannot be enzymatically removed from the cell surface with any commercially available sialidase.

Unlike the behavior seen when using the unprotected sugar, ketone expression of Ac₄ManLev-fed cells did not saturate. Instead, the dose–

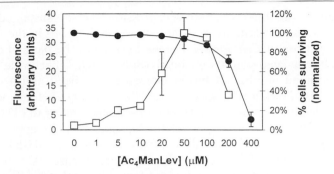

FIG. 3. Dose response and toxicity of Ac₄ManLev. Ketone incorporation (□) was measured by the flow cytometry assay. The fraction of surviving cells (●) was determined by staining with propidium iodide (10 μl/sample) and quantitation by flow cytometry. Cells were assayed after a 2-day incubation with Ac₄ManLev.

response curve peaked at approximately 50 to 100 μM. This behavior is attributed to the toxicity of Ac₄ManLev above these levels; above concentrations of 50 μM, both growth inhibition and toxicity were observed as quantified by cell count, trypan blue inclusion observed under light microscopy, and viability measured by flow cytometry with propidium iodide (see Fig. 3).[18] In comparison, Ac₄ManNAc showed no growth inhibitory activity up to 200 μM, and the analog 1,3,4,6-tetra-O-acetyl-N-pentanoylmannosamine (Ac₄ManPent), which does not contain a ketone group on the N-acyl side chain (Fig. 4), exhibited modest growth inhibition with Jurkat cells.[21]

At concentrations of Ac₄ManLev up to 50 μM, the number of ketones on the cell surface was comparable to that obtained by treatment with unprotected ManLev at a concentration approximately 250-fold higher.[18] The time dependence of Ac₄ManLev metabolism was similar to that reported for unprotected ManLev.[18] No signal above background was observed by flow cytometry until 18 to 24 hr after addition of Ac₄ManLev. A linear increase in signal was seen until about 4 days, at which point a

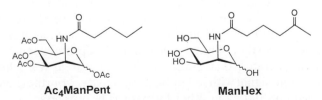

FIG. 4. Structures of other unnatural mannosamine derivatives: Ac₄ManPent and ManHex.

steady state was attained. The ketone expression level was sustained without significant decline as long as Ac$_4$ManLev was maintained in the medium.

Quantification of Cell Surface Ketone Levels. On the basis of comparisons with biotinylated polystyrene beads (Spherotech, Libertyville, IL) with a predetermined number of biotin molecules per bead and a diameter similar to that of Jurkat cells, we estimated the number of ketones accessible to chemoselective ligation with BH and subsequent binding to FITC–avidin to be approximately 1.8×10^6 per cell for Jurkat cells.[16] This approximation was confirmed by Scatchard analysis of ManLev-treated, BH-labeled cells with [^{35}S]streptavidin. We calculate that there are at least 2.0×10^6 ketones per cell expressed on ManLev-saturated Jurkat and HL-60 cells and more than 10^7 ketones per cell on saturated HeLa cells.[19] The actual number of ketones per cell is likely much higher, because not all SiaLev residues are labeled with BH at the concentration used for cell surface staining, and not all BH molecules may be accessible for binding to FITC–avidin.

Substrate Specificity. In an effort to determine the extent of permissivity of the sialic acid pathway, other unnatural mannosamine derivatives were synthesized and analyzed for conversion to cell surface sialic acids. One such compound, *N*-(5-oxohexanoyl)mannosamine or ManHex (shown in Fig. 4), was found to be incorporated as the corresponding sialic acid at drastically lower levels than ManLev. The treatment of Jurkat cells with ManHex resulted in 2 to 6% of the number of cell surface ketones observed using equal concentrations of ManLev.[18] This result indicates that the steric requirements of at least one enzyme or transport protein in the sialic acid pathway are quite strict, because the *N*-acyl chain of ManHex is only one methylene unit longer than that of ManLev. Further work in this area is currently in progress in our laboratory.

Linkage Analysis. To delineate the possible uses of ManLev as an unnatural sialic acid precursor, it is important to know which types of glycoproteins incorporate SiaLev residues; several different techniques have been used to elucidate this information. Initially, we investigated the incorporation of SiaLev into both N- and O-linked glycoproteins. We found cell surface ketone expression to be inhibited in a dose-dependent fashion by both tunicamycin (a specific inhibitor of N-linked glycosylation) and α-benzyl GalNAc (a competitive inhibitor of O-linked protein glycosylation).[16] In addition, high levels of ketone expression were observed when using cell lines with mostly N-linked glycoproteins such as Jurkat, as well as cell lines with mostly O-linked mucin-like glycoproteins, such as HL-60. We have demonstrated that SiaLev residues occur in α-2,3 linkages by correlating ketone signal with lectin-binding activity.[21] Further investigation of the inclusion of SiaLev in other linkages is in progress.

Applications of Metabolic Cell Surface Ketone Expression

Cell Surface Glycoprotein Remodeling. The landscape of oligosaccharide structures on cell surfaces is highly heterogeneous and difficult to control, which makes study of the biological roles of specific carbohydrate epitopes particularly difficult.[19] Therefore, we exploited cell surface ketones as sites for attachment of chemically defined carbohydrate structures and showed that the method can be used to modulate the ability of cell surface glycoproteins to interact with a carbohydrate-binding receptor.

The two synthetic carbohydrate derivatives shown in Fig. 5, aminooxygalactose (Gal-ONH$_2$) and lactose hydrazide (Lac-NHNH$_2$), were prepared.[19] These sugars each possess a nucleophilic functional group that is capable of condensing with a ketone group under aqueous conditions. These analogs were chosen for their potential binding activity with ricin, a plant-derived lectin. The reactions of both Gal-ONH$_2$ and Lac-NHNH$_2$ with cell surface ketone groups were monitored in two ways. First, cells treated with ManLev were incubated simultaneously with the desired sugar (between 0 and 10 mM) and the BH probe (1 mM). We evaluated the ability of the synthetic carbohydrates to compete with BH for cell surface ketones by detecting a dose-dependent decrease in fluorescence on staining with FITC–avidin as described above. We observed that Gal-ONH$_2$ is a more potent competitor than Lac-NHNH$_2$, by about twofold, perhaps because aminooxy groups are more nucleophilic than hydrazides at pH 6.5.

The second method for detecting the presence of the synthetic carbohy-

Gal-ONH$_2$ Lac-NHNH$_2$

Eu-DTPA-ONH$_2$

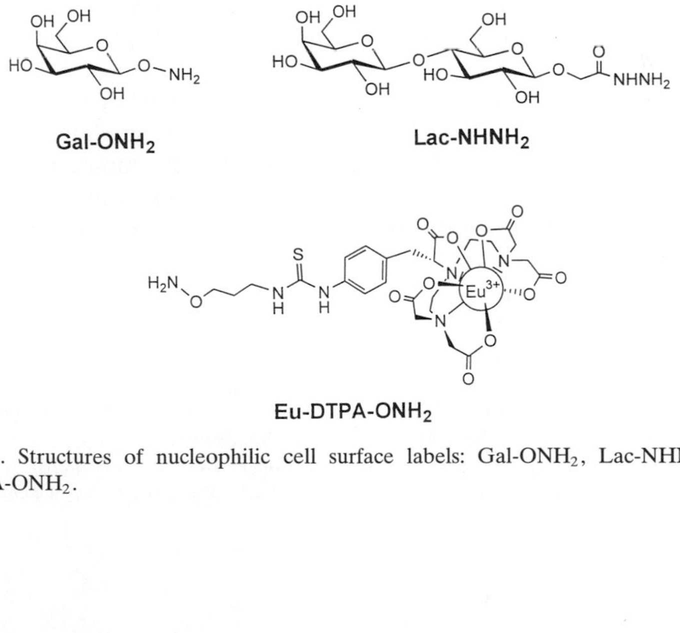

FIG. 5. Structures of nucleophilic cell surface labels: Gal-ONH$_2$, Lac-NHNH$_2$, and Eu-DTPA-ONH$_2$.

drates involved the use of FITC–ricin as a flow cytometry probe. Jurkat cells were treated with 20 mM ManLev for 6 days and then labeled with either Gal-ONH$_2$ or Lac-NHNH$_2$ (25 mM) for 2 hr at pH 6.5. These cells exhibited elevated ricin-binding activity compared with control cells that were treated with ManNAc or not labeled with the synthetic carbohydrates. These results demonstrate that SiaLev residues can be used as handles for chemical remodeling of the cell surface carbohydrate landscape and that newly installed carbohydrate epitopes are sterically accessible for recognition and binding to protein receptors. Methodology for generating aminooxy- or hydrazide-derivatized sugars from both free and protected oligosaccharides has been developed in our laboratory,[25] allowing the introduction of almost any naturally or synthetically available carbohydrate to the cell surface via condensation with metabolically delivered ketone groups.

Tumor Targeting. Overexpression of sialylated glycoforms is a common phenotype for several types of cancer, including cancers of the liver, lung, breast, prostate, and blood.[26–30] Therefore, diagnostic or chemotherapeutic techniques based on differences in cell surface sialic acid levels may be broadly applicable to a variety of cancers. We have applied unnatural sialic acid biosynthesis to a strategy for selective delivery of an aminooxy-derivatized lanthanide ion chelate (Eu-DTPA-ONH$_2$; Fig. 5) to highly sialylated cells for the purpose of magentic resonance imaging (MRI) enhancement.[31] In an *in vitro* model, we incubated ManLev-treated cells with Eu-DTPA-ONH$_2$ and then assayed for fluorescence of the Eu^{3+} ion. Highly sialylated wild-type Jurkat cells were compared with Jurkat cells that had been previously selected for low levels of α-2,3-linked sialic acid, thus gaining resemblance to noncancerous cells. Both cell populations were treated with 50 μM Ac$_4$ManLev for 3 days and stained with the Eu^{3+} chelate. This experiment showed a positive correlation between the amount of α-2,3-linked sialic acids and the extent of chelate labeling; the increase in fluorescence for the normal Jurkat cells was determined to arise from targeting of approximately 3×10^6 chelates per cell, which is within the range of ions needed for detection of a relaxation enhancement by MRI.

[25] E. C. Rodriguez, L. A. Marcaurelle, and C. R. Bertozzi, *J. Org. Chem.* **63,** 7134 (1998).

[26] R. Takano, E. Muchmore, and J. W. Dennis, *Glycobiology* **4,** 665 (1994).

[27] S. Sell, *Hum. Pathol.* **21,** 1003 (1990).

[28] E. P. Scheidegger, P. M. Lackie, J. Papay, and J. Roth, *Lab. Invest.* **70,** 95 (1994).

[29] J. Roth, C. Zuber, P. Wagner, D. J. Taatjes, C. Weisgerber, P. U. Heitz, C. Goridis, and D. Bitter-Suermann, *Proc. Natl. Acad. Sci. U.S.A.* **85,** 2999 (1988).

[30] G. Yogeeswaran, B. S. Stein, and H. Sebastian, *Cancer Res.* **38,** 1336 (1978).

[31] G. A. Lemieux, K. J. Yarema, C. L. Jacobs, and C. R. Bertozzi, *J. Am. Chem. Soc.* **121,** 4278 (1999).

This application of metabolic cell surface labeling may have promise in developing more general and sensitive *in vivo* MRI diagnostics for certain types of cancer.

Acknowledgments

This work was supported in part by grants from the National Institutes of Health (GM58867-01), the Burrows Wellcome Fund, the University of California Cancer Research Coordinating Committee, and the W. M. Keck Foundation. C.L.J. was supported by an NSF fellowship, L.K.M. was supported by an ACS Medicinal Chemistry fellowship, and K.J.Y. was supported by a grant from the Laboratory Directed Research and Development Program of Lawrence Berkeley National Laboratory under Department of Energy Contract number DE-AC03-76SF00098.

Section IV

Signals for Addressing Proteins to Specific Subcellular Compartments

[21] Using Sorting Signals to Retain Proteins in Endoplasmic Reticulum

By HUGH R. B. PELHAM

It is sometimes desirable to express an altered form of a secreted or membrane protein such that it remains in the endoplasmic reticulum (ER). For example, this approach can be used to accumulate high levels of a particular protein,[1] to test whether the location of a transmembrane protein is important for its activity,[2] to modify the properties of the ER,[3] or to prevent export of a protein by expressing a retained version of a polypeptide with which it interacts.[4] The expression system used will depend on the application, but in most cases the desired localization is readily achieved by appending an ER sorting signal[5] to the protein of interest. The strategy differs somewhat depending on whether the protein is soluble or an integral membrane component. "Soluble" refers to secretory proteins or extracytoplasmic domains of membrane proteins. Although it is possible to target cytosolic proteins to the lumen of the ER by adding an N-terminal signal sequence from a secretory protein, such proteins frequently fail to fold into a functional form because of cryptic glycosylation sites, oxidizable cysteine residues, or incompatibility with the folding environment (chaperones, ionic conditions) of the ER.

Soluble Proteins

Natural occupants of the ER lumen are retained by a sorting system that recognizes a C-terminal signal, typically KDEL.[6] The KDEL receptor binds proteins in the ER–Golgi intermediate compartment, or in the early Golgi, and returns with them to the ER. Thus, proteins bearing the signal can reach post-ER compartments, and may receive some Golgi-specific modifications, although usually only early ones.[7,8]

[1] J. Haseloff, K. R. Siemering, D. C. Prasher, and S. Hodge, *Proc. Natl. Acad. Sci. U.S.A.* **94,** 2122 (1997).
[2] J. Sparkowski, J. Anders, and R. Schlegel, *EMBO J.* **14,** 3055 (1995).
[3] B. Chaudhuri, S. E. Latham, and C. Stephan, *Eur. J. Biochem.* **210,** 811 (1992).
[4] L. Buonocore and J. K. Rose, *Nature (London)* **345,** 625 (1990).
[5] R. D. Teasdale and M. R. Jackson, *Annu. Rev. Cell Dev. Biol.* **12,** 27 (1996).
[6] S. Munro and H. R. B. Pelham, *Cell* **48,** 899 (1987).
[7] H. R. B. Pelham, *EMBO J.* **7,** 913 (1988).
[8] N. Dean and H. R. B. Pelham, *J. Cell Biol.* **111,** 369 (1990).

Any standard mutagenesis procedure can be used to add an ER retention signal. Because the sequence required is short, it is easy to use synthetic oligonucleotides to encode it. Polymerase chain reaction (PCR)-based procedures, as detailed elsewhere in this volume, are particularly convenient.

Although the common signal in mammalian proteins is KDEL, the receptor actually binds the tetrapeptide HDEL more tightly,[9,10] and this serves as an excellent signal. It can also be used in yeast and plants, and probably most other eukaryotic species. There is, however, some variation in the first two positions of the signal,[11] and for unusual organisms it may be wise to check known sequences of ER proteins in the species in question. The hsp70 homolog BiP is an abundant ER protein that has been sequenced from many species.

The context of the signal may be important in some cases. Octapeptides bind the receptor more tightly than tetrapeptides, but there is little specificity outside the terminal four residues. A spacer can help to ensure accessibility of the signal, which is crucial, but this depends entirely on the structure of the protein and usually a short sequence is sufficient. ER proteins typically have extended acidic stretches near the terminus, and the acidic Myc epitope recognized by the monoclonal antibody (MAb) 9E10 provides an excellent spacer.[6] Addition of a restriction site facilitates removal of the signal, and we have often used the sequence EQKLISEEDLNSEHDEL (Myc epitope underlined) with an *Eco*RI site at the NS sequence (G AAT TTC). Other spacers will certainly also suffice, but problems can occur. For example, replacement of the second serine residue in the preceding sequence with phenylalanine, when appended to chick lysozyme, abolished retention even though FEHDEL is the natural terminus of yeast BiP; this was presumably due to an interaction between the phenylalanine residue and the lysozyme protein.[10] The Flu tag (YPYDVPDYA) should be used with caution for secretory proteins. It is a substrate for tyrosine sulfation in the *trans*-Golgi network (TGN) of animal cells, a modification that interferes with the binding of the 12CA5 MAb (S. Munro, personal communication, 1999). Similarly, glycosylation sites (NxS/T) and cysteines (which may be oxidized to disulfides) are not recommended.

A control construct is desirable to check that retention in the ER is due to recognition of the signal rather than to misfolding of the modified protein. This can be provided by an Myc tag alone (e.g., by filling in and religating the *Eco*RI site in the preceding sequence), or by varying the terminal tetrapeptide. The terminal leucine is crucial, although a valine in

[9] A. A. Scheel and H. R. B. Pelham, *J. Biol. Chem.* **273,** 2467 (1998).
[10] D. W. Wilson, M. J. Lewis, and H. R. B. Pelham, *J. Biol. Chem.* **268,** 7465 (1993).
[11] H. R. B. Pelham, *Trends Biochem. Sci.* **15,** 483 (1990).

this position may sometimes substitute.[10] Addition of residues is sufficient to block binding to the receptor, and addition of GL (originally chosen for arbitrary reasons) to make the sequence KDELGL or HDELGL provides an effective null signal.[6,10] Because specific mechanisms exist to recognize and retain poorly folded proteins,[12] secretion of such a control construct provides indirect evidence that modification of the C terminus of the protein per se has not prevented folding (although, of course, this does not guarantee retention of activity).

As with any receptor-mediated process, it is possible to saturate the KDEL retention system. In mammalian cells such as COS cells this is surprisingly difficult, and even with strong promoters a well-recognized substrate will be efficiently retained. High levels of accumulation have also been reported for HDEL-tagged proteins in plant cells. However, complete retention should not be assumed. In yeast, the situation is different—even wild-type yeast strains can secrete measurable amounts of endogenous BiP (Kar2p), and expression of additional HDEL protein easily saturates the system. Retention can be improved by overexpression of the receptor (Erd2p),[13] although this might have other effects on the secretory pathway.

Membrane Proteins

Retention of membrane proteins in the ER can be achieved in several ways.[5] For proteins whose C terminus is on the lumenal side of the membrane, the KDEL system can be used exactly as described above[14] (and naturally occurring membrane proteins that use this system are known). Some steric problems with access to the receptor might be expected, but in practice this does not seem critical, possibly because the anchoring of ligand and receptor in the same membrane encourages binding. Whether large, multispanning proteins would present greater problems has not been explored.

The alternative approach is to place a retrieval signal on the cytoplasmic side of the membrane. The classic example of this for type I membrane proteins is the C-terminal signal KKXX or KXKXX,[15] which works in animal cells and yeast,[16,17] and probably many other eukaryotic species. The terminal location is important. The length of the cytoplasmic tail can

[12] C. Hammond and A. Helenius, *Curr. Opin. Cell Biol.* **7,** 523 (1995).

[13] J. C. Semenza, K. G. Hardwick, N. Dean, and H. R. B. Pelham, *Cell* **61,** 1349 (1990).

[14] B. L. Tang, S. H. Wong, S. H. Low, and W. Hong, *J. Biol. Chem.* **267,** 7072 (1992).

[15] M. R. Jackson, T. Nilsson, and P. A. Peterson, *EMBO J.* **9,** 3153 (1990).

[16] F. M. Townsley, and H. R. B. Pelham, *Eur. J. Cell Biol.* **64,** 211 (1994).

[17] E. C. Gaynor, S. te Heesen, T. R. Graham, M. Aebi, and S. D. Emr, *J. Cell Biol.* **127,** 653 (1994).

also have effects, but these are difficult to predict. In general, there seems considerable flexibility provided the signal is at least five residues from the membrane, and distances of hundreds of residues (i.e., whole domains) are possible. Polytopic membrane proteins with cytoplasmic C termini can be retained in the same way. A detailed discussion of the sequence requirements is available elsewhere.[5]

The KK motif is known to bind to COPI coat proteins and to mediate retrieval of proteins from post-ER compartments.[5] However, the extent of progress down the secretory pathway varies depending on the particular sequence. The prediction of the behavior of a given protein is complicated by the frequent presence of multiple signals within the cytoplasmic tail and elsewhere. Location and itinerary may thus depend on a balance of forward, retrograde, and possible retention signals. For example, the terminal sequence KKFF found on the intermediate compartment marker ERGIC53 seems to comprise both a KK retrieval signal and an FF ER exit signal.[18,19] Changing the signal to KKAA results in a steady state ER location for ERGIC53.

For efficient ER localization many possible sequences will suffice, and KKSS seems an adequate signal in both yeast and animal cells.[15,16] One strategy is to replace the entire C-terminal cytoplasmic tail of the protein of interest with that of a known ER protein. The sequence of the adenovirus E19 protein is a classic (and short) example[15]: . . .KYKSRRSFIDEKKMP.

For type II proteins, or polytopic proteins whose N terminus is cytoplasmic, a slightly different consensus has been defined, summarized as XXRR.[20] Again, the precise requirements in terms of context and distance from the membrane are flexible and not completely defined. The classic animal cell example is the N terminus of the "invariant chain" of the MHC class II molecule (normally positioned some 36 residues from the membrane)[20]: MHRRRSRSCR. . . .

Other cytoplasmic sequences have also been defined as retention/retrieval sequences. Those that have been studied seem also to mediate binding to COPI coats,[21] and may have the advantage that they do not have to have a terminal location. An example is the sequence . . .GQRDLYSGL. . . .[22] This is found on the ε subunit of the T cell receptor, and its activity is masked when assembly of this multisubunit protein is complete. This illustrates a general caveat: many membrane

[18] C. Itin, R. Schindler, and H. P. Hauri, *J. Cell Biol.* **131,** 57 (1995).

[19] F. Kappeler, D. R. Klopfenstein, M. Foguet, J. P. Paccaud, and H. P. Hauri, *J. Biol. Chem.* **272,** 31801 (1997).

[20] M. P. Schutze, P. A. Peterson, and M. R. Jackson, *EMBO J.* **13,** 1696 (1994).

[21] P. Cosson, Y. Lefkir, C. Demollière, and F. Letourneur, *EMBO J.* **17,** 6863 (1998).

[22] A. Mallabiabarrena, M. A. Jimenez, R. Rico, and B. Alarcon, *EMBO J.* **14,** 2257 (1995).

proteins are oligomeric, and accessibility of an engineered signal may be compromised by association with other subunits. This may be less of a problem when tails are extended or changed radically in sequence, which is likely to discourage normal interactions. Masking of ER retention signals is a natural device to ensure retention of individual subunits in the ER until assembly occurs.

The efficiency of retention varies between individual proteins but for well-retained proteins the capacity of the system is large. However, massive expression of any membrane protein can have effects on Golgi and ER morphology, and these should be checked if unexpected phenomena occur. Some proteins readily associate in the ER or nuclear envelope, leading to aberrant multilayered structures.[23] We have observed similar behavior for some chimeric membrane proteins that have green fluorescent protein appended to their cytoplasmic domains.

Finally, as with soluble proteins, malfolding can be a reason for retention of membrane proteins in the ER, either through direct aggregation or through chaperone association. Usually, this is not the desired targeting mechanism, and as with the KDEL system it is a good idea to compare the retained protein with a control construct that differs in a minimal way, for example by replacement of one or both lysines in the KKXX signal with another hydrophilic residue. Exit of this control construct from the ER provides evidence that other introduced changes (such as epitope tags or other extensions or truncations) have not resulted in a folding problem.

[23] M. L. Parrish, C. Sengstag, J. D. Rine, and R. L. Wright, *Mol. Biol. Cell* **6**, 1535 (1995).

[22] Directing Proteins to Nucleus by Fusion to Nuclear Localization Signal Tags

By HEIKE KREBBER and PAMELA A. SILVER

Introduction

In all eukaryotic cells the nucleus is separated from the cytoplasm by the nuclear envelope. Molecules that travel from one compartment to the other use the channels created by large proteinaceous structures termed nuclear pore complexes (NPCs). Proteins enter the nucleus from the cytoplasm through a 25-nm active channel by a facilitated process that often requires the presence of a nuclear localization signal (NLS) on the transported protein as well as soluble transport factors.[1–5] The necessary transport factors include "receptors" that recognize the NLS and a small GTPase termed Ran that mediates the association of the receptor with its cargo. There exist several classes of NLSs that require the action of different NLS receptors. The receptors thus far characterized are members of a growing family of proteins related to the founding member termed importin (or karyopherin) β. (See Refs. 2 and 6–8 for reviews.)

Often during the course of characterization of a novel protein, various mutated forms are expressed and examined for interaction with cellular partners. A potential problem might result when the truncated protein of interest does not localize to the correct cellular compartment, such as the nucleus, because necessary targeting information has been lost. However, this localization problem can be overcome by fusion to an NLS tag. In addition, it has become increasingly clear that movement in and out of the nucleus offers an important mechanism for gene regulation. Thus, it might be desirable to constitutively target a particular gene regulator to the nucleus by fusing it to a hyperactive NLS and prevent its exit by removing potential nuclear export signals.

The ability of an NLS to target a normally nonnuclear protein to the nucleus can be tested in a number of ways. These include (1) creation of gene fusions encoding chimeric proteins and subsequent localization by immunofluorescence of the expressed protein, (2) microinjection of chimeric proteins and/or peptide–protein conjugates into the cytoplasm of *Xenopus* oocytes followed by extraction of nuclei and analysis of their content,[9,10] and (3) use of fluorescently tagged NLS-containing proteins as substrates in a semiintact *in vitro* nuclear transport system assayed by fluorescence microscopy for nuclear uptake of the test protein.[11]

In this chapter, we present a comprehensive description of various NLSs and how they are used by the cell to mediate nuclear protein import. In addition, we present details on how to use some of these NLSs to specifically

[1] D. Gorlich and I. W. Mattaj, *Science* **271,** 1513 (1996).
[2] E. A. Nigg, *Nature (London)* **386,** 779 (1997).
[3] A. H. Corbett and P. A. Silver, *Microbiol. Mol. Biol. Rev.* **61,** 193 (1997).
[4] K. Weis, *Trends Biochem. Sci.* **23,** 185 (1998).
[5] M. Ohno, M. Fornerod, and I. W. Mattaj, *Cell* **92,** 327 (1998).
[6] P. A. Silver, *Cell* **64,** 489 (1991).
[7] C. Dingwall and R. A. Laskey, *Trends Biochem. Sci.* **16,** 478 (1991).
[8] D. Gorlich, *Curr. Opin. Cell Biol.* **9,** 412 (1997).
[9] P. A. Silver, L. P. Keegan, and M. Ptashne, *Proc. Natl. Acad. Sci. U.S.A.* **81,** 5951 (1984).
[10] D. Kalderon, B. L. Roberts, W. D. Richardson, and A. E. Smith, *Cell* **39,** 499 (1984).
[11] F. Melchior, D. J. Sweet, and L. Gerace, *Methods Enzymol.* **257,** 279 (1995).

target proteins to the nucleus in both mammalian and yeast cells. Here we focus on the use of the well-studied "classic" NLS. However, there are now a number of distinct NLSs that employ alternative transport receptors and their use is described as well.

Classic Nuclear Localization Signal

Proteins containing the classic NLS are imported into the nucleus in a two-step process. First, the NLS is recognized in the cytoplasm by the NLS receptor importin α (Srp1p in *Saccharomyces cerevisiae*). This complex then binds to importin β in the presence of RanGDP and docks to the nuclear pore for its import. The classic NLS consists of basic amino acid residues that are either continuous (monopartite) or interrupted by several nonbasic residues (bipartite). The canonical short NLS from the simian virus 40 (sv40) large T antigen is the 7-amino acid stretch **PKKKRKV**. It is an example of a monopartite NLS.[10] A mutation in the underlined lysine renders the sequence nonfunctional.[10,12] This short lysine- and arginine-rich sequence is sufficient to confer nuclear localization even when it has been conjugated as a synthetic peptide to serum albumin.[10,13-16] The bipartite classic NLS consists of two stretches of basic amino acid residues separated by a spacer region.[1,3,17] The classic example of this type of NLS is found in nucleoplasmin (**KRPAATKKAGQAKKKK**).[18]

Nuclear Localization Signal–Peptide Coupling

For direct chemical coupling of the NLS to a nonnuclear protein such as albumin, the most commonly used NLS is that of the SV40 T antigen (PPKKKRKV) (Fig. 1). These peptides can be used, when coupled to biotinylated human serum albumin (HSA) or a different protein of interest, as import substrates. As a negative control a mutant peptide conjugate (PPKTKRKV) (Fig. 1), which has been shown to remain cytoplasmic due to its nonfunctional NLS, should be used.[10,13,15,16] A cysteine should be included at the N terminus of each peptide for the cross-linking procedure. For preparation of the conjugates, HSA (Calbiochem, La Jolla, CA) or

[12] R. E. Lanford and J. S. Butel, *Cell* **37,** 801 (1984).
[13] D. S. Goldfarb, J. Gariepy, G. Schoolnik, and R. D. Kornberg, *Nature (London)* **322,** 641 (1986).
[14] R. E. Lanford, P. Kanda, and R. C. Kennedy, *Cell* **46,** 575 (1986).
[15] P. A. Silver, I. Sadler, and M. A. Osborne, *J. Cell Biol.* **109,** 983 (1989).
[16] M. S. Moore and G. Blobel, *Cell* **69,** 939 (1992).
[17] L. Gerace, *Cell* **82,** 341 (1995).
[18] J. Robbins, S. M. Dilworth, R. A. Laskey, and C. Dingwall, *Cell* **64,** 615 (1991).

Wildtype peptide conjugate:

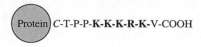

Mutant peptide conjugate:

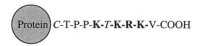

Fig. 1. Peptides containing the wild-type and mutant NLS of the SV40 T antigen, coupled to the protein of interest. The amino-terminal cysteine and tyrosine as well as the threonine in the mutant NLS sequence (in italics) are not part of the T antigen sequence.

the protein of interest is dissolved in 0.1 M sodium bicarbonate, pH 8.5 (conjugation buffer 1), and mixed with the bifunctional cross-linker, m-maleimidobenzoyl-N-hydroxysuccinimide ester (MBS; Pierce, Rockford, IL) (10 mg/ml in dimethylformamide), to obtain final concentrations of HSA (or protein of interest) and MBS of 10 and 2.5 mg/ml, respectively. After a 1-hr incubation at room temperature, free cross-linker is removed by using a Bio-Gel P6-DG column (Bio-Rad, Hercules, CA) equilibrated in 0.1 M sodium phosphate, pH 6.0 (conjugation buffer 2). HSA–MBS is then pooled and 5 mg is mixed with 2.5 mg of the NLS-peptide (10 mg/ml in conjugation buffer 2) and incubated for 1 hr at room temperature. Uncoupled peptides are then removed by using a Bio-Gel P6-DG column equilibrated in phosphate-buffered saline. Conjugate-containing fractions are pooled and adjusted to a concentration of 4 μM by using a Microcon 30 concentrator (Amicon, Danvers, MA), aliquoted, and stored at $-80°$.[19]

For the production of biotinylated conjugates, HSA (10 mg/ml in conjugation buffer 1), or the protein of interest, is incubated with 0.4 mg of N-hydroxysuccinimide-LC-biotin II (Pierce) for 1 hr at 4°. Biotinylated HSA is separated from free biotin by several rounds of concentration with a Centriprep 10 (Amicon).[19] Peptide conjugation is then performed as described above. The number of peptides per conjugate is estimated by comparing its mobility with that of HSA–MBS by sodium dodecyl sulfate–polyacrylamide gel electrophoresis (SDS–PAGE). Optimally, not more than 5–15 peptides per protein molecule should be added in order to avoid highly positively charged transport ligands that accumulate in the nucleoli.[11]

[19] A. E. Pasquinelli, M. A. Powers, E. Lund, D. Forbes, and J. E. Dahlberg, *Proc. Natl. Acad. Sci. U.S.A.* **94,** 14394 (1997).

Generation of Nuclear Localization Signal Tag

The second standard approach for generating NLS-tagged proteins is to create gene fusions encoding NLS-containing chimeric proteins. Often, the most facile way to generate such a fusion is to anneal two complementary oligonucleotides encoding a classic NLS with some spacer amino acid residues and include a restriction side of choice to assay the presence of the NLS-encoding DNA. General rules for cloning and protein expression must be considered, such as using a suitable promoter, using the Kozak sequence including a start codon, and also providing a stop codon, if necessary. An example of oligonucleotides encoding an N-terminal SV40 NLS is given in Fig. 2. Restriction sites (*Cla*I and *Xma*I) are provided that are ready for ligation after the complementary DNA strands are annealed (shaded box). Oligonucleotides of this size should be purified by PAGE to avoid annealing of incomplete primers that reduce the cloning efficiency. Each oligonucleotide should be resuspended in H_2O to a concentration of 500 μM. The vector in which the annealed primers are to be cloned must be cut with the appropriate restriction enzymes, purified by gel separation, and finally resuspended to 1 ng/μl ($\sim 10^{-4}\,\mu M$). (Do not dephosphorylate the linearized vector, because the oligonucleotides are not phosphorylated.) To anneal the oligonucleotides 4 μl of 5× annealing buffer [200 mM Tris-HCl (pH 7.5), 100 mM MgCl$_2$, 250 mM NaCl] is mixed with 1.4 μl of primer 1, 1.4 μl of primer 2, and 13.2 μl of H_2O to a total volume of 20 μl. For denaturation of any secondary structure within the primers, incubate the annealing mix at 65° for 5 min. For the annealing, place the tube for 2 min at 37°. Dilute the annealing mix 1:500 and use this for the ligation. To ligate the annealed oligonucleotides to the cut vector, 1 μl of the 1:500-diluted annealing mix is mixed with 1 μl of the linearized vector (1 mg/μl), 6 μl of H_2O, 1 μl of a 10× ligation buffer, and 1 μl of ligase to a final volume of 10 μl. Incubate the ligation mix overnight at 16°. Use 5 μl of the ligation mix to transform competent *Escherichia coli* cells. Alternatively, the SV40 NLS-encoding sequence can be obtained from already existing vectors if

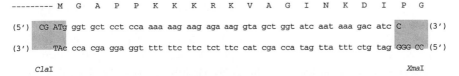

```
--------- M   G   A   P   P   K   K   K   R   K   V   A   G   I   N   K   D   I   P   G

(5')  CG ATg  ggt gct cct cca aaa aag aag aga aag gta gct ggt atc aat aaa gac atc C        (3')

(3')      TAc cca cga gga ggt ttt ttc ttc tct ttc cat cga cca tag tta ttt ctg tag GGG CC  (5')

     ClaI                                                                      XmaI
```

FIG. 2. Example of oligonucleotides encoding an N-terminal SV40 NLS. *Top line:* Amino acid sequence. *Bottom two lines:* DNA sequence. Restriction sites (*Cla*I and *Xma*I) are provided that are ready for ligation after the complementary DNA strands are annealed (shaded box).

the restriction sites are convenient for that purpose. By the use of the following restriction sites in certain vectors, the resulting fragment encodes the SV40 NLS: pPS1372 using ClaI and XmaI[20] and pKW430 using HindIII and EcoRI.[21]

Expression of Classic Nuclear Localization Signal Fusion Proteins

To identify potential NLSs in a certain protein, the sequence of interest can be tested in a so-called "sufficiency experiment." In this assay a potential NLS is expressed as a fusion to a cytoplasmic protein and analyzed for its ability to localize to the nucleus. This kind of experiment has been carried out in mammalian cells as well as in yeast cells. In the next four sections of this chapter we have listed some of the vectors that have been used successfully for this purpose.

Expression of Classic Nuclear Localization Signal Fusion Proteins in Mammalian Cells. A widely used mammalian expression vector that allows constitutive expression of a protein of interest is pcDNA3 (InVitrogen, Carlsbad, CA). It utilizes a cytomegalovirus (CMV) promoter in conjunction with the bovine growth hormone polyadenylation transcription termination signal sequence to allow high-level mRNA expression of inserted genes. It also contains a neomycin resistance gene (*neo*) to allow the selection of stable mammalian cell lines. A useful tool has been generated by Toby and colleagues with their pGTN (go to nucleus) vector, which was derived from pcDNA3 by cloning DNA encoding an NLS sequence upstream of a green fluorescent protein (GFP) open reading frame in the polylinker region. In this way, the following cassette was created: HindIII-ATG-SV40 NLS (PPKKKRKVA)-EcoRI-XbaI-GFP-ApaI.[22] This vector can now easily be used to exchange the GFP open reading frame for that of another protein, using EcoRI or XbaI and ApaI, or to insert a protein of interest between the NLS and the GFP open reading frame, using the EcoRI and XbaI sites. It should be noted that because the CMV promoter used in pcDNA3 is potent, proteins are expressed at high levels. This potential overexpression is often linked to substantial cellular toxicity when cells are incubated for 24–48 hr after transfection of pGTN. To avoid this toxicity, the incubation time can be reduced to 12–15 hr, when protein levels are lower.[22]

A similar vector derived from pRc/CMV2 (InVitrogen) is described by Liu and colleagues.[23] In their pCMV-lacZ vector, KpnI, SmaI, and BamHI sites can be used for in-frame fusions of the protein of interest

[20] T. Taura, H. Krebber, and P. A. Silver, *Proc. Natl. Acad. Sci. U.S.A.* **95,** 7427 (1998).
[21] K. Stade, C. S. Ford, C. Guthrie, and K. Weis, *Cell* **90,** 1041 (1997).
[22] G. Toby, S. F. Law, and E. A. Golemis, *BioTechniques* **24,** 637 (1998).
[23] M. T. Liu, T. Y. Hsu, J. Y. Chen, and C. S. Yang, *Virology* **247,** 62 (1998).

to an N-terminal NLS and a C-terminal β-galactosidase (β-Gal). In their studies they cloned an NLS of the Epstein–Barr virus DNase ([291]AWNLKDV**RKRK**LGPGH[306]) downstream of the CMV promoter and upstream of a *Kpn*I site, resulting in the following vector: placZ-291–306-*Sac*II-pCMV-NLS (291–306)-*Kpn*I-*Sma*I-*Bam*HI-*lacZ*. The NLS is sufficient to localize β-Gal to the nucleus of mammalian cells.[23] If the coding sequence of the protein of interest is cloned in frame between the NLS and β-Gal-sequences, using the *Kpn*I, *Sma*I, or *Bam*HI site, the resulting tagged protein can then be tested for nuclear localization.

The bipartite NLSs can also be used to direct a cytoplasmic protein into the nucleus. Because monopartite NLSs are shorter than their bipartite counterparts, they may have less effect on the activity of protein to which they are fused. However, in some cases the bipartite NLS may be more active for driving a particular protein to the nucleus. The bipartite NLS (MAV**KR**PAATKKAGQA**KKK**) of *Xenopus laevis* nucleoplasmin (NP) has been used to direct GFP into the nucleus of HeLa cells (NP-GFP).[24,25] A functional bipartite NLS ([21]**KR**PAEDMEEEQAF**KRSR**[37]) from "heterogeneous nuclear ribonucleoprotein K" (hnRNP K) has been shown to promote the nuclear import of PK in HeLa cells.[26] A useful plasmid was constructed by using a double-stranded oligonuceotide encoding this NLS and inserting it, using *Eco*RI and *Xho*I, into Myc-PK-pcDNAI.[50]

Expression of Classic Nuclear Localization Signal Fusion Proteins in Saccharomyces cerevisiae. We have constructed the following vectors to generate N-terminal NLS fusions and C-terminal GFP fusions with the protein of interest in the yeast *S. cerevisiae*. pPS1372 is a *CEN URA3* plasmid and pPS1525 is a *CEN TRP1* version with the same insert. Both plasmids express an SV40 NLS (MGA**PPKKKRKVA**GINKDIPG) and the protein kinase A inhibitor (PKI) nuclear export signal (NES) (**LALKLAGLDINKT**GIDKNIGEF) with spacer regions (not in boldface letters), which have been shown to be necessary for proper function of the NLS[28] and two GFP proteins in tandem, resulting in a 63-kDa fusion protein. By exchanging the NES with the individual protein of interest, using the *Xma*I and *Eco*RI sites, an NLS- and GFP-tagged fusion protein is generated that is directed into the nucleus and is visible in living cells when viewed by fluorescence microscopy. The cassette is the same in both constructs: *Xba*I-pADH-*Cla*I-NLS-*Xma*I-NES-*Eco*RI-GFP-*Xho*I-

[24] S. Chatterjee and U. Stochaj, *BioTechniques* **24,** 668 (1998).
[25] W. Barth and U. Stochaj, *Biochem. Cell. Biol.* **74,** 363 (1996).
[26] W. M. Michael, H. Siomi, M. Choi, S. Pinol-Roma, C. S. Nakielny, Q. Liu, and G. Dreyfuss, *CSHSQB* **60,** 633 (1995).
[27] W. M. Michael, M. Choi, and G. Dreyfuss, *Cell* **83,** 415 (1986).
[28] S. Chatterjee, M. Javier, and U. Stochaj, *Exp. Cell Res.* **236,** 346 (1997).

GFP-*Hind*III. Moore *et al.*[29] generated a plasmid encoding the integrase NLS: *Eco*RI-pGAL-*Bam*HI-GFP-*lacZ-Not*I-IN NLS ([591]TTINSKK-RSLEDNETEI[607])-*Hin*dIII, which can be used for similar purposes. Further, a mutant version of this NLS, which is not able to direct the fusion protein GFP-LacZ into the nucleus, has been cloned into the same vector: *Eco*RI-pGAL-*Bam*HI-GFP-*lacZ-Not*I-IN mutant NLS ([591]TTINS**GGR**S-LEDNETEI[607])-*Hin*dIII.

Three sequences of *S. cerevisiae* Upf3p, a protein that is required for nonsense-mediated mRNA decay, have been shown to function to direct β-Gal into the nucleus.[30] The first and the third type of NLS are bipartite sequences, [15]**KK**GRGNRYHN**KNRGKSK**[31] (NLS1) and [284]**KK**FKEEEA-SAKIP**KKKR**[300] (NLS3), whereas the second NLS is the monopartite type [58]RRNN**KKRNREYYNYKRK**[74] (NLS2). Interestingly, by fusion of β-Gal to NLS3 of Upf3p, a mostly nucleolar signal was obtained, suggesting that NLS3 can function to direct a reporter protein to the nucleolus.[30] All NLSs were polymerase chain reaction (PCR) amplified and cloned in frame, using *Bam*HI and *Sal*I restriction sites, into a 2μ plasmid (pSJ101), resulting in pLS95 (*GAL10*-NLS1-*lacZ*), pLS96 (*GAL10*-NLS2-*lacZ*), and pLS97 (*GAL10*-NLS3-*lacZ*). In this way an NLS-β-Gal fusion protein is expressed from a galactose-inducible promoter in a *GAL2*+ strain (YM4126) in the presence of 2% (w/v) galactose.[30]

Expression of Classic Nuclear Localization Signal Fusion Proteins in Schizosaccharomyces pombe. The following plasmids encoding NLSs were used in the wild-type strain JY266 (*h*+ *leu1-32*) by Kudo and colleagues. These pREP1 plasmids contain the wild-type *nmt1* promoter, which promotes transcription when thiamine is absent in the medium. For suppression of the *nmt1* promoter of pREP plasmids, thiamine hydrochloride was added to minimal medium at a final concentration of 10 μg/ml.[31,32] Cells were then grown in the absence of thiamine for 12 hr at 30°, for induction of the proteins. Two plasmids derived from pREP1 have been used in nuclear transport studies in *Schizosaccharomyces pombe:* pR1GF1 encoding *Nde*I-GST-*Nhe*I-GFP-*Sma*I and pR1GsvNLSF1 encoding GST-NLS-GFP.[32]

Expression of Classic Nuclear Localization Signal Fusion Proteins in Escherichia coli. In microinjection experiments or the *in vitro* nuclear import assay, either coupled NLS-peptides or recombinant proteins can be used.

[29] S. P. Moore, L. A. Rinckel, and D. J. Garfinkel, *Mol. Cell. Biol.* **18,** 1105 (1998).
[30] R. L. Shirley, M. J. Lelivelt, L. R. Schenkman, J. N. Dahlseid, and M. R. Culbertson, *J. Cell Sci.* **111,** 3129 (1998).
[31] N. Kudo, S. Khochbin, K. Nishi, K. Kitano, M. Yanagida, M. Yoshida, and S. Horinouchi, *J. Biol. Chem.* **272,** 29742 (1997).
[32] N. Kudo, B. Wolff, T. Sekimoto, E. P. Schreiner, Y. Yoneda, M. Yanagida, S. Horinouchi, and M. Yoshida, *Exp. Cell Res.* **242,** 540 (1998).

We have therefore listed some useful constructs for this purpose. Recombinant GST-NLS can be obtained with the GST-T NLS plasmid.[33] It was constructed by ligation of a *Bam*HI–*Eco*RV-cleaved double-stranded oligonucleotide encoding two copies of the SV40 T NLS into *Bam*HI- and *Sma*I-digested pGEX2 TK (Pharmacia, Piscataway NJ).[33] A similar vector, including a C-terminal GFP fusion, was constructed in pGEX2 TK (Pharmacia) and resulted in the GST-NLSc-GFP vector[34] GST-*Bam*HI-*Sma*I-NLS (PKKKRKV)-*Sma*I-GFP-*Eco*RI.

Nonclassic Nuclear Localization Signals

M9 Signal

To take advantage of a nuclear transport pathway different from that used by the classic NLS, a different type of NLS, such as the so-called M9 sequence, can be used. M9 was originally defined as a stretch of approximately 40 amino acids in the C terminus of the RNA-binding protein hnRNP A1. Subsequently, several groups demonstrated that a nuclear transporter termed transportin (also called karyopherin β2), a protein with significant ($\sim$22%) homology to importin β, could mediate the nuclear import of M9-containing proteins *in vitro*.[33,35,36] Importantly, the transportin-mediated nuclear import of M9-signal containing substrates is different from the importin β-mediated import of basic NLS proteins, because no importin α subunit is required. However, similar to the importin α/importin β pathway, transportin-dependent import is known to require Ran as a cofactor.[35,37,38] Another difference between the classic NLS and the M9 signal is that the M9 sequence contains a signal for import and for export, whereas the classic NLS exclusively promotes import. Attempts to separate both signals in M9 have failed. Mutations that slow import also reduce the rate of export. Thus, one should keep in mind that the M9 signal functions as a shuttling sequence.

[33] V. W. Pollard, W. M. Michael, S. Nakielny, M. C. Siomi, F. Wang, and G. Dreyfuss, *Cell* **86**, 985 (1996).

[34] H. Eguchi, T. Ikuta, T. Tachibana, Y. Yoneda, and K. Kawajiri, *J. Biol. Chem.* **272**, 17640 (1997).

[35] N. Bonifaci, J. Moroianu, A. Radu, and G. Blobel, *Proc. Natl. Acad. Sci. U.S.A.* **94**, 5055 (1997).

[36] R. A. Fridell, R. Truant, L. Thorne, R. E. Benson, and B. R. Cullen, *J. Cell Sci.* **110**, 1325 (1997).

[37] S. Nakielny, M. C. Siomi, H. Siomi, W. M. Michael, V. Pollard, and G. Dreyfuss, *Exp. Cell Res.* **229**, 261 (1996).

[38] R. Truant, R. A. Fridell, R. E. Benson, H. Bogerd, and B. R. Cullen, *Mol. Cell. Biol.* **18**, 1449 (1998).

```
             261          271          281          291          301
(260-289)  Y NDFGNYNNQS SNFGPMKGGN FGGRSSGPY

(264-305)    GNYNNQS SNFGPMKGGN FGGRSSGPYG GGGQYFAKPR NQGGY

(268-305)        NQS SNFGPMKGGN FGGRSSGPYG GGGQYFAKPR NQGGY  (38aa M9-signal)
                     A
```

Fig. 3. M9 shuttling signal as it has been characterized by Weighardt et al.[40] (positions 260–289), Siomi and Dreyfuss[39] (positions 264–305), and Michael et al.[27] (positions 268–305). One amino acid exchange (G274 → A) results in a loss of transportin binding and shuttling activity.[41]

Expression of Fusion Proteins that Contain an M9 Nuclear Localization Signal in Mammalian Cells. The glycine-rich M9 sequence has been shown to be sufficient to localize the cytoplasmic reporter proteins β-Gal and pyruvate kinase (PK) to the nucleus.[39] Further characterization of this sequence by Michael and colleagues[27] has localized the minimum M9 sequence by deletion mutagenesis to between residues 268 and 305 of hnRNP A1, whereas Weighardt and colleagues,[40] using a similar approach, localized the hnRNP A1 NLS to between residues 260 and 289 (Fig. 3). Further analysis revealed that the most critical residues for transportin binding are located between positions 269 and 276. In particular, a mutation of Glyc-274 to alanine in the M9 signal completely prevents transportin binding as well as the ability to function as an NLS.[41] To use the M9 sequence to direct proteins into the nucleus, we suggest the use of the 38-amino acid sequence characterized by Michael and colleagues[27] (Fig. 3, in boldface), because a single amino acid change (G274 → A) results in a loss of transport activity and can therefore be used as a negative control (Fig. 3).

The following M9-containing plasmids are available and have been used successfully to localize proteins to the nucleus in mammalian cells. One based on pcDNA3 (InVitrogen) contains *Bst*XI-*myc*-*Eco*RI-PK-*Kpn*I-M9(264–305)-*Not*I.[39] The Myc tag comprises the peptide sequence MEQ-KLISEEDL, which is recognized by the 9E10 anti-Myc monoclonal antibody. PK in this construct is derived from codon 12 to codon 443 of the chicken muscle pyruvate kinase. Recombinant GST-M9 for use in *in vitro* import reactions can be obtained with a plasmid constructed by ligation of

[39] H. Siomi and G. Dreyfuss, *J. Cell Biol.* **129,** 551 (1995).
[40] F. Weighardt, G. Biamonti, and S. Riva, *J. Cell Sci.* **108,** 545 (1995).
[41] H. P. Bogerd, R. E. Benson, R. Truant, A. Herold, M. Phingbodhipakkiya, and B. R. Cullen, *J. Biol. Chem.* **274,** 9771 (1999).

161	171	181	191	201	211	221	231	241	251	261	271

LQPQLGTQNA MQTDAPATPS PISAFSGVVN AAAPPQFAPV DNSQRFTQRG GGAVGKNRRG GRGGNRGGRN NNSTRFNPLA KALGMAGESN MNFTPTTKKE GRCRLFPHCP L

AAEEL

FIG. 4. Import signal of *S. cerevisiae* Nab2p. The exchange of the boxed amino acid residues (RGGRN → AAEEL) results in loss of Kap104p binding activity and thus a loss of nuclear import. This mutant has been termed M3.[38]

an *Eco*RI–*Xho*I fragment encoding amino acid residues 268–305 of the human hnRNP A1 into pGEX2TK.[33]

Nab2p-Nuclear Localization Signal in Saccharomyces cerevisiae. For the yeast homolog of transportin, Kap104p, two hnRNPs have been identified as import substrates: Nab2p and Hrp1p.[42,43] Interestingly, neither protein contains any sequence with evident homology to the M9 NLS of hnRNP A1. For Nab2p, a 110-amino acid segment, extending from residues 161 to 271 (Fig. 4), has been shown to promote the import of GFP.[38] Truant and colleagues further demonstrated that the yeast Nab2p NLS substrate is imported by yeast Kap104p, but not mammalian transportin. Further, it is imported by mammalian importin β directly, without importin α functioning as an adapter. Thus, the yeast Nab2p NLS substrate functions in both the yeast and the mammalian systems; however, the function of the M9 NLS substrate is restricted to mammalian cells.[38] To use Nab2p NLS to import a protein of interest, it can be either PCR amplified or directly cloned, using existing constructs. By cutting with *Eco*RI and *Xho*I, the Nab2p NLS or the mutant form (M3) can be released from a yeast 2μ *LEU2* plasmid that expresses a GFP fusion from the phosphoglycerate kinase promoter: *Bgl*II-GFP-*Eco*RI-Nab2p NLS-*Xho*I.[38] These constructs can also be used as templates in PCRs, to obtain not only Nab2p NLS but also the import-defective mutant M3 with restriction enzymes of individual choice.

Import of Ribosomal Proteins

Although ribosomal proteins are often small enough theoretically to diffuse through NPCs into the nucleus, the import of ribosomal proteins is active and receptor mediated.[44] The import of several ribosomal proteins has been studied in some detail. Among them are S7, L5, L13, L23a, and

[42] J. T. Anderson, S. M. Wilson, K. V. Datar, and M. S. Swanson, *Mol. Cell. Biol.* **13,** 2730 (1993).

[43] M. Henry, C. Z. Borland, M. Bossie, and P. A. Silver, *Genetics* **142,** 103 (1996).

[44] S. Jakel and D. Gorlich, *EMBO J.* **17,** 4491 (1998).

```
        41              51              61              71

VHSHKKKKI  RTSPTFRRPK  TLRLRRQPKY  PRKSAPRRNK  LDHY  (32-74)

                       SPTFRRPK  TLRLRRQPKY  PRKSAPRRNK  LDHY  (43-74)

                                             RKSAPRRNK  LDHY  (62-74)
```

Fig. 5. Binding domain of the ribosomal protein L23a. The upper sequence (positions 32–74) is the minimal binding sequence that binds as efficiently to importin β, transportin, RanBP5, or RanBP7 as the full-length protein. The middle sequence (positions 43–74) is still capable of binding to all four transporters; however, the binding efficiency is slightly reduced, whereas the lower sequence does not bind any importer[44] and thus can function as a negative control.

L25.[44–47] All are transported into the nucleus by members of the importin β family. Interestingly, ribosomal proteins seem not to rely on just one member of the importin family. In mammalian cells, S7, L5, and L23a have been shown to be imported alternatively by at least four different importins: importin β, transportin, RanBP5, and RanBP7. Their interaction with importin β, however, is direct and independent of importin α.[44] A core domain in L23a has been shown to bind to all four transport receptors (Fig. 5). This domain might therefore be used as a general import signal that can take advantage of different parallel-acting importers.

A short stretch of only four amino acid residues of the human ribosomal protein S7 ([115]KRPR[118]) has been shown to be sufficient for targeting β-Gal into the nucleus.[47] Annilo and colleagues used the mammalian expression vector pRc/CMV2 from InVitrogen to generate a useful tool for NLS sequence analysis. In their resultant pKHlacZ expression vector any NLS can be expressed as a fusion protein to β-Gal, when cloned downstream of an ATG start codon into a NotI and a HindIII site.[47] Interestingly, 17 additional amino acid residues ([89]RRILPKPTRKSRTKNKQKRPR[118]) resulted in a nuclear and nucleolar accumulation of the reporter construct.[47]

L25 transport has been investigated in S. cerevisiae, where it is transported by Yrb4p (also termed Kap123p) and possibly also by Pse1p.[45,48] A fusion of the first 49 amino acids of L25 with the lacZ gene is sufficient to

[45] G. Schlenstedt, E. Smirnova, R. Deane, J. Solsbacher, U. Kutay, D. Gorlich, H. Ponstingl, and F. R. Bischoff, EMBO J. 16, 6237 (1997).
[46] N. R. Yaseen and G. Blobel, Proc. Natl. Acad. Sci. U.S.A. 94, 4451 (1997).
[47] T. Annilo, A. Karis, S. Hoth, T. Rikk, J. Kruppa, and A. Metspalu, Biochem. Biophys. Res. Commun. 249, 759 (1998).
[48] M. P. Rout, G. Blobel, and J. D. Aitchison, Cell 89, 715 (1997).

```
      331              341              351              361
```

YDRRGRPG DRYDGMVGFS ADETWDSAID TWSPSEWQMA Y (323–361)

GFS ADETWDSAID TWSPSEWQMA Y (338–361)

FIG. 6. KNS shuttling signal contains 39 amino acid residues (residues 323–361). This domain is sufficient to import a cytoplasmic reporter protein. The 24-amino acid partial sequence of the KNS (positions 338–361) is not sufficient to import a reporter construct into the nucleus and thus can be used as a negative control.[50]

drive β-Gal into the nucleus. Within this sequence lie at least two additive NLSs.[49]

Nuclear Protein Import Independent of Cytosolic Factors

Heterogeneous Nuclear Ribonucleoprotein K Nuclear Shuttling Domain

Work has established that the human hnRNP K protein contains a stretch of amino acids termed the KNS (hnRNP K nuclear shuttling domain), which can confer bidirectional transport across the nuclear envelope. However, unlike the other NLSs described thus far, the KNS does not appear to utilize soluble transport factors, including Ran, and may in fact directly contact the NPC.[50] Instead, KNS-mediated nuclear protein import appears to require only ATP hydrolysis.[50] The amino acid sequence is shown in Fig. 6. The vector that has been used to test KNS function in transient transfection experiments in HeLa cells is pCDNAI (InVitrogen) and contains the following cassette: (5')HindIII-myc-BamHI-PK-EcoRI-KNS (residues 323–361)-XhoI(3'). The Myc tag is recognized by the MAb 9E10 and was cloned in frame with codon 12 to 443 of the pyruvate kinase (PK)-coding sequence.[50] As a negative control, a partial sequence of the KNS can be used in the same vector (KNS, residues 338–361).[50] For in vitro import assays, recombinant protein can be produced with the pGEX-5X-1 prokaryotic expression vector containing (5')EcoRI-GST-[322]KNS[463]-XhoI(3').[50]

Human Immunodeficiency Virus Tat

Another example of Ran and import receptor-independent import is the HIV-1 Tat protein.[51] The Tat-NLS is capable of targeting a large heterol-

[49] P. J. Schaap, J. van't Riet, C. L. Woldringh, and H. A. Raue, J. Mol. Biol. 221, 225 (1991).
[50] W. M. Michael, P. S. Eder, and G. Dreyfuss, EMBO J. 16, 3587 (1997).
[51] A. Efthymiadis, L. J. Briggs, and D. A. Jans, J. Biol. Chem. 273, 1623 (1998).

ogous protein to the nucleus and nucleolus[51–53] through a pathway that is dependent on ATP hydrolysis but independent of exogenous cytosol. Thus Tat-NLS-directed import does not require transport components that mediate conventional NLS-dependent nuclear accumulation, including importin β, transportin, and Ran.[51] The plasmid expressing the Tat-NLS-β-Gal fusion protein was derived by oligonucleotide insertion, encoding the Tat-NLS ([48]GRKKRRQRRRAP[59]) into the SmaI site of pPR2. In this way the Tat-NLS is fused to the N terminus of β-Gal.

Final Remarks

In summary, there are a number of ways to direct proteins from the cytoplasm to the nucleus via fusion to different types of NLSs. The best understood and by far the most commonly used way is by the classic NLS. In addition, the classic NLS consists of a short sequence, which can easily be attached to the protein of interest and may cause fewer structural problems than longer tags. However, in some cases, for example, working in importin α or importin β mutant backgrounds or when the fusion between the classic NLS and the protein of interest leads to a misfolded and thus nonfunctional protein, it might be desirable to use a different NLS and a different pathway to direct a protein into the nucleus. For that purpose, the M9 signal or in yeast the Nab2p signal should be considered. The mammalian ribosomal L23a NLS or the yeast L25 NLS might also be an alternative way to direct a cytoplasmic protein into the nucleus, especially because both NLSs seem not to be restricted to just one import receptor. However, ribosomal NLSs have not been studied in great detail, especially in vivo. The same is true for the mammalian KNS and viral Tat-NLS, which seem to function independently of import receptors and Ran. Generally, it is best to keep in mind that, in some cases, the M9 signal, the Nab2p signal, and the KNS can also act as shuttling sequences and therefore also promote nuclear export.

Acknowledgments

We thank Holger Bastians, Anne McBride, Elissa Lei, and Tetsuya Taura for helpful discussion and for critical reading of the manuscript.

[52] H. Siomi, H. Shida, M. Maki, and M. Hatanaka, J. Virol. **64,** 1803 (1990).
[53] S. Kubota, H. Siomi, T. Satoh, S. Endo, M. Maki, and M. Hatanaka, Biochem. Biophys. Res. Commun. **162,** 963 (1989).

[23] Identification, Analysis, and Use of Nuclear Export Signals in *Saccharomyces cerevisiae*

By KARSTEN WEIS

Introduction

In eukaryotic cells, many proteins contain signal sequences that are recognized by cellular transport factors in order to ensure delivery to the appropriate target compartment. One of the major intracellular trafficking paths is the transport of proteins into and out of the nucleus. Transport events between the cytoplasm and the nucleus occur through the nuclear pore complex, a multiprotein complex that is embedded in the nuclear envelope.[1] The nuclear pore provides the only known gate through which macromolecules can enter or leave the nucleus. It contains an aqueous channel, and small molecules and proteins up to 40–60 kDa in size are able to pass nonselectively. However, larger proteins and ribonucleoparticles are actively and selectively transported through the pore with an astonishing degree of accuracy. To ensure fast delivery into the target compartment and uptake against a concentration gradient, transport requires both the presence of targeting signals in the transported macromolecules and the consumption of energy. Several different nuclear localization signals (NLSs) and nuclear export signals (NESs) have been identified.[2] These signals are by definition both necessary and sufficient to grant entry into or exit out of the nucleus. For many of these signals, it was shown that they are recognized by soluble transport receptors that have been classified as importins and exportins (alternative names for these factors include karyopherins or transportins in metazoans and KAPs in yeast). Importins and exportins are members of a large protein family that were shown to shuttle continuously between the cytoplasm and the nucleus. They interact with components of the nuclear pore complex and deliver their cargo into the correct compartment with the help of the small regulatory GTPase: Ran (reviewed in Weis[3]).

In addition to the enormous number of constitutive transport events, many transport processes between the cytoplasm and the nucleus were shown to be highly regulated. A controlled localization allows eukaryotic cells to adjust biological processes in a fast and efficient manner without

[1] M. Ohno, M. Fornerod, and I. W. Mattaj, *Cell* **92,** 327 (1998).
[2] E. A. Nigg, *Nature* (*London*) **386,** 779 (1997).
[3] K. Weis, *Trends Biochem. Sci.* **23,** 185 (1998).

the need for *de novo* protein synthesis. For example, the controlled import of a transcription factor into the nucleus can be used to activate a transcriptional program in response to external stimuli. Conversely, the regulated export of a transcription factor can be used to repress transcription of given target genes.

To date nine different import and four different export pathways have been identified and characterized in yeast.[3] Unfortunately, for only a few of these transport pathways have the signals that are recognized by the individual import or export receptors been unambiguously mapped and identified. The best characterized import signal is the "classical" NLS, which can be either a monopartite or a bipartite stretch of positively charged amino acids, originally identified in the simian virus 40 (SV40) large T antigen and in nucleoplasmin, respectively.[4] The first identified export signal was a leucine-rich NES characterized in the human immunodeficiency virus type 1 (HIV-1) Rev protein and in the small protein kinase A inhibitor (PKI).[5,6] Although consensus sequences for both these motifs have been defined it still remains impossible to predict functional import or export signals on the basis of their primary amino acid sequence. One of the challenges therefore remains to understand how individual transport receptors specifically recognize their cargoes and how the interaction between cargoes and their receptors can be regulated in response to environmental changes.

Nuclear Export

Presently, four individual exportin-mediated nuclear export pathways have been identified in yeast (Table I). Exportins bind to their cargo in the nucleus in the presence of Ran-GTP; however, they release their substrate in the cytoplasm after GTP hydrolysis by Ran (reviewed in Weis[3]). For Crm1/ Xpo1, Cse1 and Msn5 protein sequences have been mapped that are required for a specific interaction with these receptors. In addition, it was shown that Los1 is able to bind directly to tRNA in the presence of Ran-GTP.[7] Crm1/Xpo1 recognizes the leucine-rich sequence motif and as described above a consensus sequence has been defined.[8] *In vivo* substrates for Crm1/Xpo1 in yeast include the Yap1 transcription factor.[9] The Msn5

---

[4] D. Görlich, *Curr. Opin. Cell Biol.* **9,** 412 (1997).

[5] W. Wen, J. L. Meinkoth, R. Y. Tsien, and S. S. Taylor, *Cell* **82,** 463 (1995).

[6] U. Fischer, J. Huber, W. C. Boelens, I. W. Mattaj, and R. Lührmann, *Cell* **82,** 475 (1995).

[7] K. Hellmuth, D. M. Lau, F. R. Bischoff, M. Kunzler, E. Hurt, and G. Simos, *Mol. Cell. Biol.* **18,** 6374 (1998).

[8] H. P. Bogerd, R. A. Fridell, R. E. Benson, J. Hua, and B. R. Cullen, *Mol. Cell. Biol.* **16,** 4207 (1996).

[9] C. Yan, L. H. Lee, and L. I. Davis, *EMBO J.* **17,** 7416 (1998).

TABLE I
Exportins in *Saccharomyces cerevisiae*

Exportin	Cargo	Ref.
Xpo1/Crm1[a]	L-rich NES	Stade *et al.* (1997)[b]
Cse1[a]	Importin α	Solsbacher *et al.* (1998)[c]; Künzler and Hurt (1998)[d]; Hood and Silver (1998)[e]
Msn5/Ste21	Pho4, Far1, Msn2, Msn4, Mig1	Kaffman *et al.* (1998)[f]; reviewed in Hopper (1999)[g]
Los1	tRNA	Hellmuth *et al.* (1998)[h]

[a] Encode essential genes.
[b] K. Stade, C. F. Ford, C. Guthrie, and K. Weis, *Cell* **90,** 1041 (1997).
[c] J. Solsbacher, P. Maurer, F. R. Bischoff, and G. Schlenstedt, *Mol. Cell. Biol.* **18,** 6805 (1998).
[d] M. Künzler and E. C. Hurt, *FEBS Lett.* **433,** 185 (1998).
[e] J. K. Hood and P. A. Silver, *J. Biol. Chem.* **273,** 35142 (1998).
[f] A. Kaffman, N. M. Rank, E. M. O'Neill, L. S. Huang, and E. K. O'Shea, *Nature (London)* **396,** 482 (1998).
[g] A. K. Hopper, *Curr. Biol.* **9,** R803 (1999).
[h] K. Hellmuth, D. M. Lau, F. R. Bischoff, M. Kunzler, E. Hurt, and G. Simos, *Mol. Cell. Biol.* **18,** 6374 (1998).

interaction motif is less clearly delimited; however, it seems that Msn5 specifically interacts with phosphorylated proteins and it was shown that MSN5 is required for several regulated export events[10] (reviewed in Hopper[11]).

Export Assays in Yeast

To characterize a functional NES motif it is essential to demonstrate that a putative signal is necessary in the context of the full-length protein, i.e., point mutations within the motif are expected to interfere with its nuclear export. In addition, it should be demonstrated that the signal itself is sufficient to target a nuclear reporter protein to the cytoplasm.

Nuclear Export Signal–Nuclear Localization Signal Shuttling Assay

The analysis of export signals in yeast is hampered by the small size of budding yeast, preventing, for example, export studies based on microinjection experiments. To observe protein export *in vivo,* a reporter protein must first be imported into the nucleus. Fortunately, it was shown that the rate of export can be higher than the rate of import and, at steady state, a protein containing both an SV40-type NLS and a leucine-rich NES is located in the cytoplasm.[12] An NLS–NES-containing shuttling reporter

[10] A. Kaffman, N. M. Rank, and E. K. O'Shea, *Genes Dev.* **12,** 2673 (1998).
[11] A. K. Hopper, *Curr. Biol.* **9,** R803 (1999).
[12] K. Stade, C. F. Ford, C. Guthrie, and K. Weis, *Cell* **90,** 1041 (1997).

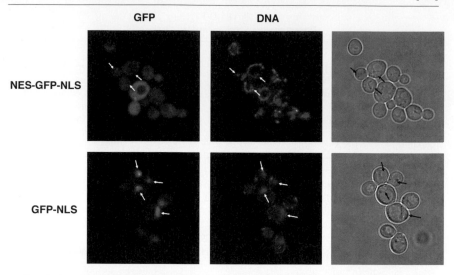

Fig. 1. An export assay in budding yeast. Yeast cells expressing either an NES-GFP-NLS reporter or a GFP-NLS fusion protein were examined by both fluorescence and bright-field microscopy. The GFP and DAPI signals were visualized in the FITC or UV channel, respectively. Under these conditions, DAPI preferentially stains mitochondrial DNA. Nuclei are indicated by arrows. Nuclear exclusion of the GFP signal can be seen in the cells expressing the NES-GFP-NLS reporter. No GFP signal can be detected in vacuoles.

protein can thus be used to identify and characterize putative nuclear export sequences. Figure 1 shows wild-type yeast cells expressing either a green fluorescent protein (GFP) reporter that contains both a functional NES and NLS, or a GFP-NLS reporter in which the NES sequence was mutated. Whereas in cells expressing the NES-GFP-NLS protein the nuclei are devoid of GFP signal, the GFP-NLS reporter rapidly accumulates inside nuclei in the absence of a functional NES. This suggests that in an *in vivo* competition in yeast, the NES motif is stronger than the NLS motif. The NES-GFP-NLS fusion protein shuttles continuously between the cytoplasm and the nucleus but is enriched in the cytoplasmic compartment, because its export rate is higher than its import rate. A series of expression constructs has been generated, allowing for the expression of various GFP-NLS constructs. Putative NES motifs can be introduced into these constructs. It is then possible to analyze the subcellular localization of the fusion proteins and to functionally characterize and delineate export motifs. Figure 2 shows two examples of such expression constructs. pKW417 contains a URA3 selectable marker and expression of the GFP reporter is driven by the constitutive alcohol dehydrogenase (ADH) promoter. Identical constructs with different selectable markers (HIS3, TRP1) or galactose-inducible pro-

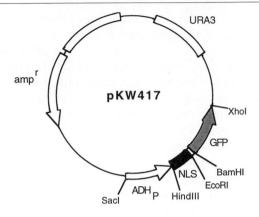

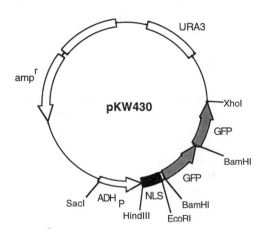

Fɪɢ. 2. Plasmid maps of export reporter constructs. Plasmids pKW417 and pKW430 allow for the expression of NLS-NES-GFP export reporter constructs. Putative NES motifs can be cloned into the *Eco*RI–*Bam*HI cloning site (reading frame: . . .CCC-**GAA-TTC**-. . . . . . -**GGG-ATC-C**AC. . . ; restriction sites in boldface). The subcellular localization of the fusion proteins can be analyzed and used to characterize nuclear export signals.

moters are also available. Putative nuclear export signals can be introduced into the *Eco*RI/*Bam*HI restriction sites, which are located upstream of the GFP gene. It is important to note that the NLS-GFP construct encodes a protein of approximately 40 kDa. Because proteins that are smaller than 40 to 60 kDa can freely pass through the nuclear pore, this reporter protein is still able to diffuse between the nuclear and cytoplasmic compartments.

Depending on the size of the introduced export motif it may therefore be necessary to use larger fusion proteins. Vectors containing two GFP moieties have been constructed for that purpose (e.g., pKW430; Fig. 2).

Plasmids containing the putative NES sequences fused to the NLS-GFP moiety are transformed into wild-type yeast cells and transformants are selected by growth on complete minimal medium dropout plates lacking uracil (CM-URA). Single colonies are obtained after reselection on CM-URA plates and used to inoculate liquid CM-URA medium. Cells are grown to an OD_{600} of approximately 0.5 to 1. One milliliter of the cell suspension is centrifuged for 5 min at 6000g at room temperature. Cells are carefully resuspended in 50 μl of sterile H_2O and can either be viewed directly under a fluorescence microscope or fixed by the addition of 50 μl of 7.4% (w/v) paraformaldehyde solution [7.4% (w/v) paraformaldehyde in 2× phosphate-buffered saline (PBS)]. The use of fresh paraformaldehyde reduces the amount of autofluorescence significantly. To label nuclei, the DNA dye 4′,6-diamidino-2-phenylindole (DAPI) can be added 10 min prior to the centrifugation step. Because DAPI enters yeast cells poorly, final DAPI concentrations of up to 100 μg/ml are recommended. Nevertheless, DAPI will preferentially stain the mitochondrial DNA under these conditions (see Fig. 1). To circumvent this problem, alternative DAPI labeling protocols can be considered. For example, it has been observed that DAPI penetrates cells quite easily after a quick freeze in dry ice (M. Goodin, personal communication). For this protocol, cells are pelleted as described above and then frozen for 10 min in powdered dry ice. Cells are thawed and the freezing step is repeated. Cells are then carefully resuspended in 1× PBS containing 0.4 M sorbitol plus DAPI (1 μg/ml) and incubated for 10 min at room temperature. To remove the excess DAPI, cells are washed three times in 0.4 M sorbitol–1× PBS and then resuspended in a final volume of 50 μl of 0.4 M sorbitol–1× PBS.

In Vivo Export Assay with Mutant Strains

If a protein is localized in the cytoplasm or the cytoplasmic localization of the protein can be induced by external stimuli, the shuttling behavior of a given protein can be analyzed in mutant strains defective for known export factors. Conditional alleles for the export factors XPO1/CRM1 and CSE1 are available (xpo1-1[12]; cse1-1[13]). Because the MSN5 gene is not essential for viability, export studies can be performed in a strain in which

[13] Z. Xiao, J. T. McGrew, A. J. Schroeder, and M. Fitzgerald-Hayes, Mol. Cell. Biol. 13, 4691 (1993).

XPO1   xpo1-1

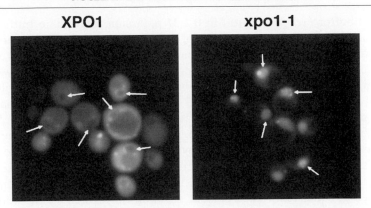

FIG. 3. Export of leucine-rich NES is inhibited in xpo1-1 cells. The export of an NES-GFP-NLS reporter was analyzed in either wild-type (XPO1) or mutant cells (xpo1-1) and the intracellular localization was determined after shifting the cells for 5 min to 37° (i.e., the nonpermissive temperature for xpo1-1). Under these conditions leucine-rich NES export is inhibited and the shuttling reporter accumulates inside nuclei. Nuclei are indicated by arrows.

the gene for MSN5 has been deleted ($msn5\Delta^{14}$). The *xpo1-1* allele shows a rapid onset of its export phenotype and export defects can be seen in less than 5 min after the shift to the nonpermissive temperature (Fig. 3). If the protein of interest is available as a GFP fusion protein, its export behavior can be observed live with a fluorescence microscope. Ideally, the microscope stage should be heated to 37° (i.e., the nonpermissive temperature for the *xpo1-1* allele) for these experiments. If a heated microscope stage is not available it should be noted that the export phenotype in the *xpo1-1* strain is rapidly reversible and rapid manipulation of the cells is essential. This can be achieved by having a heating block or water bath close to the microscope. Yeast cells that were incubated at the nonpermissive temperature can then be spotted onto prewarmed glass slides and observed immediately. Alternatively, cells can be fixed with paraformaldehyde as described above.

If the protein of interest is not tagged with GFP and its location needs to be determined by indirect immunofluorescence, logarithmically growing cells can be fixed directly with formaldehyde. Formaldehyde is added to the growth medium (5% final concentration) and cells are incubated for another 1 hr at the nonpermissive temperature. Cells are then transferred to Eppendorf tubes, pelleted at 6000g, washed two times in 500 μl of

[14] A. Kaffman, N. M. Rank, E. M. O'Neill, L. S. Huang, and E. K. O'Shea, *Nature (London)* **396,** 482 (1998).

sorbitol buffer (1.2 M sorbitol in 0.2 M potassium phosphate buffer, pH 7.2), carefully resuspended in 100 μl of sorbitol buffer containing zymolyase (40 g/ml) and 25 mM 2-mercaptoethanol, and subsequently incubated for 40 min at 37°. One hundred microliters of paraformaldehyde [8% (w/v) in PBS] with 10 mM MgCl$_2$ is added and mixed carefully. One hundred microliters of the cell suspension is applied to a 10-mm glass coverslip that has been pretreated with poly-L-lysine (1 mg/ml; Sigma, St. Louis, MO). Cells are fixed and settled for 2 min, the liquid is aspirated, and the cells are air dried for about 2 min. The coverslips are transferred onto ice (e.g., in a tissue culture dish) and 100 μl of ice-cold methanol is added. They are incubated for 6 min on ice. The liquid is removed and the sample is air dried at room temperature. Cells are then rehydrated by the addition of PBS containing 0.05% (v/v) Tween 20. Cells can then be blocked and primary and secondary antibodies can be added according to standard immunofluorescence protocols.

Conclusion

The study of nuclear–cytoplasmic transport has made tremendous progress and we have learned a great deal about the factors that are involved in these trafficking events. It has become increasingly clear that many proteins shuttle between the cytoplasm and the nucleus. In addition, it was shown that many of these import and export events are highly regulated. To understand the biological significance of these regulatory steps, it is essential that we characterize the signals in the transported proteins and analyze the pathways that mediate the corresponding trafficking events. Herein, protocols are presented to identify and characterize nuclear export signals in the budding yeast *Saccharomyces cerevisiae*. These protocols should be useful to analyze NES motifs in proteins that can be localized to the cytoplasm, at least under certain growth conditions. If a protein is inside the nucleus at steady state other techniques are available to study the potential shuttling behavior, for example, with the help of a well-established heterokaryon assay.[15] Once export signals and pathways are identified the location of proteins or protein complexes can be controlled and used to conditionally regulate the activity of nuclear factors.

Acknowledgments

I thank M. Goodin for providing the DAPI staining protocol. This work was supported by NIH Grant GM58065-01.

[15] J. Flach, M. Bossie, J. Vogel, A. Corbett, T. Jinks, D. A. Willins, and P. A. Silver, *Mol. Cell. Biol.* **14**, 8399 (1994).

[24] Directing Proteins to Mitochondria by Fusion to Mitochondrial Targeting Signals

By KOSTAS TOKATLIDIS

Introduction

Specific targeting mechanisms have been identified for the various subcellular compartments in eukaryotic cells and it is now established that these processes are mediated by dedicated translocation machineries.[1] One of the most critical sites of extensive protein translocation across membranes occurs in mitochondria, where virtually all the mitochondrial proteins (about 1000 different polypeptides) are imported from the cytosol and are then sorted within the organelle. This event is orchestrated by distinct translocation complexes in the outer and inner mitochondrial membranes.[2,3]

Most mitochondrial proteins are synthesized in the cytosol as larger precursors carrying a positively charged, N-terminal presequence that is cleaved by a specific protease on import into the innermost compartment, the mitochondrial matrix. In this pathway, the precursor is bound by cytosolic chaperones and then delivered to a set of receptors on the outer surface of the organelle. Then, the polypeptide chain is passed through the TOM complex in the outer membrane and the TIM23 complex in the inner membrane. Insertion into the inner membrane is driven electrophoretically by the electrochemical potential across the membrane, finally, the precursor is pulled completely across the membrane into the matrix by an ATP-powered translocation motor attached to the inner side of the TIM23 complex. It appears that the specificity of this process is ensured by a cascade of binding steps of the positively charged presequence to acidic receptor domains of increasing avidity, strategically localized along the import pathway ("acid chain hypothesis").[4,5] Variants of this "general import pathway" apply for proteins targeted to the outer membrane (OM) or the intermembrane space.[6–9] It seems that the presence of a "prese-

[1] G. Schatz and B. Dobberstein, *Science* **27,** 1519 (1996).
[2] G. Schatz, *J. Biol. Chem.* **271,** 31763 (1996).
[3] W. Neupert, *Annu. Rev. Biochem.* **66,** 863 (1997).
[4] G. Schatz, *Nature (London)* **388,** 121 (1998).
[5] K. Dietmeier, *et al., Nature (London)* **388,** 195 (1998).
[6] B. S. Glick, *et al., Cell* **69,** 809 (1992).
[7] U. Bömer, *et al., J. Biol. Chem.* **272,** 30439 (1997).

quence-like" motif, i.e., positively charged segment with the propensity to form an amphipathic helix, either internally or at the C terminus of the polypeptide chain can direct correct targeting to the matrix.[10,11] Until recently it was generally assumed that the general import pathway also applies to proteins that lack a presequence and are inserted at the inner membrane (like the mitochondrial metabolite transporters). However, it is now established that import of many integral membrane proteins requires the function of the novel chaperone-like complex of TIM9/10 in the intermembrane space that facilitates their passage across this aqueous compartment and delivers them to still another new complex, the inner membrane-embedded TIM22 complex.

Many of the essential features of protein import into mitochondria, both in terms of identification of the components involved and the molecular mechanisms, have been dissected by using hybrid proteins. It is now well established that almost every protein when fused to a mitochondrial matrix-targeting signal can be directed to the mitochondrial matrix. Such hybrid proteins consist of a nonmitochondrial, "passenger" protein fused to the matrix-targeting signal of a mitochondrial protein. Hybrid precursors have some considerable advantages over authentic mitochondrial precursors: (1) As the effects of the "mature," presequence-free part of the mitochondrial proteins on the protein import pathways are still unclear, using nonmitochondrial passenger proteins simplifies significantly the experimental manipulation; (2) they usually have increased stability and can be purified in large amounts as recombinant proteins that are import competent; (3) they allow, by appropriate choice of the "passenger" protein, the introduction of an enzymatic activity, an affinity tag, or any other biologically functional domain that can be studied in the context of the mitochondrion; and (4) they may facilitate protein purification and the study of protein folding and protein–protein interactions in the context of the organelle. Such hybrid proteins have been used extensively for both *in vivo* and *in vitro* import studies. *In vivo* studies of import into mitochondria have mainly focused on the use of green fluorescent protein (GFP) chimeras with a mitochondrial presequence at the N terminus in mammalian[12] and yeast cells.[13,14] Although this methodology provides a useful means for monitoring targeting events

---

[8] V. Haucke, *et al., Mol. Cell. Biol.* **17,** 4024 (1997).

[9] A. Hönlinger, *et al., EMBO J.* **15,** 2125 (1996).

[10] H. Fölsch, B. Guiard, W. Neupert, and R. A. Stuart, *EMBO J.* **15,** 479 (1996).

[11] H. Fölsch, B. Gaume, M. Brunner, W. Neupert, and R. A. Stuart, *EMBO J.* **17,** 6508 (1996).

[12] M. Yano, *et al., J. Biol. Chem.* **272,** 8459 (1997).

[13] R. George, T. Beddoe, K. Landl, and T. Lithgow, *Proc. Natl. Acad. Sci. U.S.A.* **95,** 2296 (1998).

[14] H. Sesaki and R. E. Jensen, *J. Cell Biol.* **147,** 699 (1999).

in living cells, a detailed dissection of the mechanism, the molecular interactions, and energetics of the process is more feasible using an *in vitro* import system with isolated mitochondria. This chapter describes the use of hybrid proteins for *in vitro* targeting to isolated mitochondria from the yeast *Saccharomyces cerevisiae*.

Mitochondria from different mammalian cell tissues such as liver or brain have been used for *in vitro* import experiments in addition to those isolated from simple yeast cells. In general, intact organelles that maintain a functional electrochemical potential across their inner membrane can be obtained. However, unless the import process must be studied in the context of a particular tissue, yeast mitochondria is usually the system of choice because of a number of distinct advantages: (1) Yeast cells can be grown in large amounts and allow preparation of mitochondria easily and inexpensively; (2) yeast mitochondria can be stored in aliquots at −80° in liquid nitrogen[15] without losing their membrane potential and import capacity; (iii) yeast cells are amenable to well-established genetic manipulation, thus allowing isolation of mitochondria from mutant cells lacking specific subunits of the protein import complexes; and (4) with the completion of the yeast genome, identification of novel proteins and their corresponding genes potentially involved in protein import is enormously facilitated. In general terms, the key features of the translocation process in mitochondria appear to be the same in yeast and higher eukaryotic cells.

Import Assay

Hybrid Precursor

The choice of the passenger protein is important and is guided by the specific aims of the mitochondrial targeting experiments. In general, it should have an activity that is easy to monitor experimentally. This could be an enzymatic activity, specific binding to monospecific polyclonal or monoclonal antibodies, the existence of an affinity tag [hexahistidine tag, glutathione *S*-transferase (GST) tag, etc.], or a specific binding to a ligand. If an additional tag must be fused for purification or other purposes, it is advisable to do so at the COOH end in order to avoid any steric interactions with the presequence that is usually fused at the NH_2 end. Two additional considerations must be borne in mind: (1) The final intramitochondrial location of the protein (one of its four subcompartments) may not constitute an optimal environment for the expression of the activity of the hybrid protein; and (2) if the hybrid precursor is used in a radioactive form produced in a cell-free translation system, the yield is low (femtomoles); this

---

[15] M. Kozlowski and W. Zagorski, *Anal. Biochem.* **172,** 382 (1988).

does allow monitoring of the hybrid protein because of its radioactivity, but makes it difficult to measure an enzymatic activity or perform some biochemical analyses. A way around this problem is the production of the precursor in a purified, recombinant form and use of large amounts in scaled-up import experiments.[16] One passenger protein that has been used extensively, both as a radioactive precursor and in scaled-up import experiments, is mouse dihydrofolate reductase (DHFR). This is a soluble protein of about 21 kDa that specifically binds the ligand methotrexate. This allows easy purification from *Escherichia coli* via affinity chromatography and the generation of translocation-arrested intermediates (see below). We have been using routinely commercially available vectors with a hexahistidine (His$_6$) tag at the COOH end of DHFR and an isopropyl-β-D-thiogalactopyranoside (IPTG)-inducible promoter that allow cloning of a presequence in front of DHFR and one-step purification of the hybrid onto Ni-NTA beads (pQE series of vectors; Qiagen, Chatsworth, CA). Such DHFR hybrids are import competent and provide an efficient means of purification of protein complexes after import into mitochondria.

The presequence is usually chosen as a strong matrix targeting signal such as the presequence of subunit IV of the cytochrome oxidase complex[17] (COXIV), or subunit 9 of ATPase.[18]

For *in vitro* translation, vectors that allow *in vitro* synthesis of a single mRNA have mostly been used with a variety of promoters such as T5,[19] T7,[20] SP6.[21,22] Coupled transcription–translation is also possible and indeed faster and simpler (see below). Routinely, we use the pGEM series of vectors (Promega, Madison, WI), which allow for coupled transcription–translation either from the T7 or the SP6 promoter.

In Vitro Translation and Import

The choice of the cell-free translation system is important, as the efficiency of translation and the presence of cytosolic targeting factors can vary widely. Reticulocyte lysate[23] has been used almost exclusively for import into mitochondria. Wheat germ lysate has been used for import studies into other biological membranes (in particular the endoplasmic reticulum), but as it apparently lacks heat shock protein 70 (hsp 70) chaper-

[16] K. Tokatlidis, *et al., Nature (London)* **384,** 585 (1996).
[17] E. C. Hurt, B. Pesold-Hurt, and G. Schatz, *EMBO J.* **3,** 3149 (1984).
[18] N. Pfanner, H. K. Müller, M. A. Harmey, and W. Neupert, *EMBO J.* **6,** 3449 (1987).
[19] D. Stueber, *et al., EMBO J.* **3,** 3143 (1984).
[20] W.-J. Chen and M. G. Douglas, *Cell* **49,** 651 (1987).
[21] P. A. Krieg and D. A. Melton, *Nucleic Acids Res.* **12,** 7057 (1984).
[22] D. A. Melton, *et al., Nucleic Acids Res.* **18,** 7035 (1984).
[23] H. R. B. Pelham and R. J. Jackson, *Eur. J. Biochem.* **67,** 247 (1976).

one molecules that interact with some mitochondrial precursors[24,25] this is not an optimal system. An attractive alternative is the use of a yeast translation system,[26-28] which has the obvious advantages of working in a homologous system with yeast mitochondria. However, such a system is not commercially available, it is less reproducible than the reticulocyte system, and often translation is less efficient.

Preparation of mitochondria from lactate-grown cells is the method of choice for isolation of respiring, functional mitochondria. Daum and co-workers[29] were the first to describe such a procedure, which was later improved by Glick[30,31] by the addition of a Nycodenz (Nycomed, Finland) purification step, which separates mitochondria from peroxisomes.[32] This procedure yields highly purified, nearly homogeneous mitochondria with a typical yield of about 150 mg of Nycodenz-purified mitochondria from 10 liters of D273-10B yeast cells. Mitochondria prepared in this way are stored in 1- to 10-mg aliquots at 25 mg/ml, in liquid nitrogen ($-80°$). We are using routinely frozen mitochondria for *in vitro* import reactions, but they can also be used for immunoblot and immunoprecipitations as well as electron microscopy. The structure and function of the organelle are apparently preserved in this way and mitoplasts (mitochondria whose outer membrane has been selectively ruptured) and inner membrane vesicles can be prepared without any significant rupture or loss of the electrochemical potential of the inner membrane.

A typical protocol for *in vitro* translation of the precursor using coupled reticulocyte lysate from Promega is as follows:

Reticulocyte lysate	25 µl
H_2O	17 µl
TNT reaction buffer (Pranega, USA)	2 µl
[^{35}S]Methionine (1000 Ci/mmol, 10 mCi/ml)	2 µl
RNA polymerase (1 U/µl)	1 µl
Amino acid mixture minus methionine	1 µl
Ribonuclease inhibitor (40 U/µl)	1 µl
Purified plasmid (1 µg/µl)	1 µl
Total:	50 µl

[24] H. Murakami, D. Pain, and G. Blobel, *J. Cell Biol.* **107,** 2051 (1988).
[25] R. J. Deshaies, B. D. Koch, M. Werner-Washburne, E. A. Craig, and R. Schekman, *Nature (London)* **332,** 228 (1988).
[26] M. Fujiki nd K. Verner, *J. Biol. Chem.* **266,** 6841 (1993).
[27] K. Verner and M. Weber, *J. Biol. Chem.* **264,** 3877 (1989).
[28] U. Fünfschilling and S. Rospert, *Mol. Biol. Cell* **10,** 3289 (1999).
[29] G. Daum, P. C. Boehni, and G. Schatz, *J. Biol. Chem.* **257,** 13028 (1982).
[30] B. S. Glick *Methods Cell Biol.* **34,** 389 (1991).
[31] B. S. Glick and L. A. Pon, *Methods Enzymol.* **260,** 213 (1995).
[32] A. S. Lewin, V. Hines, and G. M. Small, *Mol. Cell. Biol.* **10,** 1399 (1990).

Components are added in the order presented, on ice, and the mixture is then incubated for 60 min at 30°. The amount of template DNA used may vary and it should be optimized. The mixture is shielded from light to prevent photooxidation and disulfide bond formation or, alternatively, 20 mM dithiothreitol (DTT) may be added. Ribosomes are removed by centrifugation at 100,000g for 15 min at 4°.

The precursor synthesized in the reticulocyte lysate is either bound to cytosolic chaperones or folded.[33] In both cases, when presented to mitochondria, it will unfold on the mitochondria surface on import. To alleviate any influence on the import process of the chaperones contained in the lysate, the precursor must be unfolded by urea and presented as such to isolated mitochondria. To this end, 100 μl of saturated $(NH_4)_2SO_4$ solution in 50 mM HEPES-KOH, pH 7.4, is added to 50 μl of the translation reaction supernatant, mixed well, and incubated on ice for 30 min. This allows precipitation of the protein, which is then recovered as a pellet by centrifugation at 20,000g for 15 min at 4°. Unfolding is induced by resuspending the pellet in 50 μl of 8 M urea in 50 mM HEPES-KOH, pH 7.4. The mixture is then kept on ice for 30 min. This unfolded precursor can then be used for import in the same way as the native precursor.

The precursor is added to a suspension of mitochondria in "import buffer" (see below) supplemented with ATP as an energy source and NADH, a substrate of mitochondrial respiration to generate the membrane potential. The volume of the translation mixture used is 5–20% of the total final volume when native precursor is used. When a urea-denatured precursor is used, it should be diluted at least 20 times in the import mixture so that the final concentration of urea is less than 0.4 M; higher urea concentrations are inhibitory for import. Typical import reactions are performed in a final volume of up to 400 μl, containing 100 to 200 μg of mitochondria. For multiple import reactions, a stock of mitochondria mixture up to 10 ml can be prepared (this can be kept on ice up to 30 min before import) and then divided in 400-μl aliquots in Eppendorf tubes for each individual import reaction. Mitochondria are prepared as follows (for a 1-ml total volume):

Import buffer (2×)	500 μl
ATP, 0.1 M	20 μl
NADH, 0.5 M	5 μl
Purified mitochondria (25-mg/ml stock)	20 μl
H_2O	455 μl

The import buffer is prepared as a 2× stock, kept frozen at −20°, and is made up as follows (2× concentrate):

[33] C. Wachter, G. Schatz, and B. S. Glick, *Mol. Biol. Cell* **5,** 465 (1994).

Sorbitol (deionized), 1.2 M
HEPES, 100 mM
KCl, 100 mM
MgCl$_2$, 20 mM
Na$_2$-EDTA (pH 7.0), 5 mM
KH$_2$PO$_4$, 4 mM
Fatty acid-free bovine serum albumin (BSA), 2 mg/ml
L-Methionine, 10 mM
Adjust to pH 7.1 with KOH

It is important that the BSA be essentially fatty acid free because otherwise residual fatty acids in the BSA stock intercalate in the mitochondrial outer membrane lipid bilayer and block import.

The rate of import varies with the temperature, the concentration of mitochondria, and the concentration of precursor. Usually import is performed for up to 20 min, at 15 to 30°. It is important to establish the linear range of import as this varies widely for different precursors and may be different for the same precursor when presented to mitochondria in a native or urea-denatured state.

For a typical import reaction we proceed as follows.

1. Keep the precursor mixture and the mitochondrial mixture on ice until import.

2. Preincubate the mitochondria mixture in the import reaction for 2–3 min.

3. Add the precursor mixture in a dropwise fashion, mixing gently on a Vortex mixer. Good mixing is essential, particularly in the case of denatured precursor because of its tendency to aggregate irreversibly once it is diluted into the mitochondrial mixture.

4. Incubate the mixture for up to 20 min, and stop the import reaction by transferring the tubes in ice. This is sufficient to stop completely the import of most precursors, but for some that may still import (albeit slowly) at 4°, it is advisable to disrupt the membrane potential by adding an uncoupler such as valinomycin at 1 μg/ml.

5. Mitochondria are then reisolated by centrifugation at 20,000g for 5 min at 4°.

Import is defined here as translocation across the outer membrane. The most efficient method to test this after an import experiment is by checking the protection of the radioactive precursor against externally added protease. This treatment degrades any precursor molecules that are nonspecifically attached to the outer surface of mitochondria or precursor molecules that simply aggregate and cosediment with the mitochondria on centrifugation. We usually proceed as follows.

1. Reisolated mitochondria are resuspended at 0.5 mg/ml (in the same volume as the original import reaction) in 0.6 M sorbitol–20 mM HEPES-KOH, pH 7.4, containing trypsin at 0.1 mg/ml.

2. The mixture is incubated for 30 min on ice.

3. Soybean trypsin inhibitor is added at 1 mg/ml to inactivate trypsin and the mixture is incubated for an additional 15 min on ice.

4. Mitochondria are reisolated by centrifugation at 20,000g for 5 min at 4° for further analysis.

Protease K may also be used, but it must be inactivated by adding phenylmethylsulfonyl fluoride (PMSF) at 1 mM for 15 min on ice. Protease resistance against externally added trypsin indicates precursor molecules have completely crossed the outer membrane. As a control that the imported precursor can be degraded by trypsin, one aliquot is treated with 1% (v/v) Triton X-100 during trypsin digestion: this treatment solubilizes mitochondria and allows trypsin to access molecules inside the organelle. For most matrix-targeted precursors, an important operational test for import is the cleavage to the mature form by the matrix processing peptidase, which can be easily monitored by the presence of a shorter radioactive fragment on sodium dodecyl sulfate–polyacrylamide gel electrophoresis (SDS–PAGE). However, one must be cautious that this fragment represents the mature, matrix-targeted form, which is indicated only if it is protected against externally added protease.

Another important control for the specificity of the import reaction is to determine dependence on the electrochemical potential. All matrix-targeted precursors tested so far require the membrane potential; as a result, import into mitochondria with their potential disrupted by an uncoupler is completely blocked. To test this, import is performed in the same way but using uncoupled mitochondria that have been treated with valinomycin (a strong membrane potential uncoupler) at 1 μg/ml.

Although for most applications it is sufficient to monitor the import of a hybrid precursor into mitochondria according to the above described general guidelines, the yeast import system described here offers unparalleled possibilities for a much more detailed analysis of the energetic requirements of the import process. This is important for two reasons: (1) Different precursors have different requirements for ATP hydrolysis either outside mitochondria or inside the mitochondrial matrix; and (2) differences in the ATP requirement for import can be used as an operational criterion for judging the intramitochondrial sorting and final localization of hybrid precursors. The outer mitochondrial membrane is semipermeable, allowing free diffusion of small molecules of up to about 1 kDa and therefore also ATP. This means there is an equilibrium of ATP in the intermembrane space and the exterior of mitochondria. On the other hand, the inner membrane is tightly sealed and ATP can only be shuttled by the function of the specific ATP/ADP carrier (AAC). We can therefore operationally make the distinction between "external" ATP (outside the inner membrane) and "internal" ATP (in the matrix). It is generally accepted that a requirement for external ATP reflects an interaction with ATP-dependent cytosolic chaperones, whereas a requirement for internal ATP reflects an interaction with the mitochondrial, matrix-localized ATP-dependent hsp70 and passage across the inner membrane. A detailed analysis of ATP requirements of different precursors has been reported[33] together with detailed experimental treatments on how to manipulate levels of ATP inside or outside mitochondria.[34] Briefly, external ATP can be depleted by treatment with hexokinase or a combination of glycerol and glycerokinase. Synthesis of ATP in the matrix is inhibited by addition of oligomycin, which blocks the proton channel of the F_0 particle of the F_1F_0ATP-synthase, or efrapeptin, which has a similar effect. A combination of the two was most effective in our hands. The ADP/ATP translocator is blocked by the addition of either carboxyatractyloside, which binds to the intermembrane space side of AAC, or bongkrekic acid, which binds to the matrix side of AAC. Both reagents have the same effect of completely blocking any ATP shuttling across the inner membrane. Atractyloside can also be used instead of carboxyatractyloside, but is less efficient and must be used at 5- to 10-fold higher concentrations.

The translocation system of the inner membrane can import precursors even in the absence of the outer membrane.[35] It is therefore feasible to study the import of hybrid precursors into mitoplasts (mitochondria whose outer membrane has been selectively disrupted but whose inner membrane is intact), or even inner membrane vesicles that are essentially free of outer membrane material. For import into mitoplasts the same general protocol is used, whereas for import into inner membrane vesicles a potential must be maintained by supplementing the import buffer with ascorbate and cytochrome c.[36] A detailed protocol for the preparation of inner membrane vesicles has been given elsewhere[37] and mitoplasts can generally be prepared by a simple osmotic shock according to the following procedure.

1. Nycodenz-purified mitochondria are reisolated by centrifugation at 20,000g for 5 min at 4° and resuspended at 10 mg/ml in import buffer.
2. Nine volumes of 20 mM HEPES-KOH, pH 7.4, is added, mixed well

[34] B. S. Glick, *Methods Enzymol.* **260**, 224 (1995).
[35] M. Ohba and G. Schatz, *EMBO J.* **6**, 2117 (1987).
[36] S. Hwang, T. Jascur, D. Vestweber, L. Pon, and G. Schatz, *J. Cell Biol.* **109**, 487 (1989).
[37] T. Jascur, *Methods Cell Biol.* **34**, 359 (1991).

but gently over a Vortex mixer, and put on ice for 30 min with occasional gentle agitation to ensure constant and efficient mixing.

3. The mitoplasts are reisolated by centrifugation at 20,000g for 5 min at 4°. The supernatant contains soluble intermembrane space proteins, while the pellet contains intact the inner membrane, the matrix, and pieces of the ruptured membrane that stick to the inner membrane.

Import into the mitoplast pellet follows all the rules of import into intact mitochondria as far as matrix-targeted precursors are concerned. However, import of precursors such as the AAC, which requires the chaperone-like complex of TIM9/10 in the intermembrane space,[38–41] is dramatically affected, presumably because of the nearly quantitative release of these proteins in the supernatant.

Analysis after Import

After import into intact mitochondria or submitochondrial particles (mitoplasts or inner membrane vesicles), standard methods for SDS–PAGE followed by fluorography or autoradiography are performed to monitor the radioactive precursor and analyze its import. Additional useful information on the exact intramitochondrial localization of the precursor is obtained by subfractionation after import.

Subfractionation of Mitochondria

One useful way of distinguishing between soluble and integral membrane proteins is by sodium carbonate extraction. This treatment can be performed either in intact mitochondria, or in mitoplasts (care must be taken, however, in the preparation of mitoplasts as this may render the inner membrane more fragile). This can be achieved by resuspending well the mitochondria (or mitoplasts) in ice-cold, freshly made 0.1 M Na_2CO_3, pH 11.5,[42] for 30 min on ice, followed by centrifugation at 100,000g for 30 min at 4°. The supernatant contains the soluble proteins and the pellet contains the membrane proteins.

A complementary method to monitor the release of soluble proteins of the matrix and the intermembrane space in sonication of mitochondria: three or four rounds of sonication (using an S-tip sonicator or water bath sonicator for small volumes, but a probe sonicator for volumes larger than

[38] C. M. Koehler, *et al.*, *Science* **279**, 369 (1998).
[39] C. M. Koehler, *et al.*, *EMBO J.* **17**, 6477 (1998).
[40] C. Sirrenberg, *et al.*, *Nature* (*London*) **391**, 912 (1998).
[41] A. Adam, *et al.*, *EMBO J.* **18**, 313 (1999).
[42] Y. Fujiki, A. L. Hubbard, S. Fowler, and P. B. Lazarow, *J. Cell Biol.* **93**, 97 (1982).

1 ml), followed by snap-freezing in liquid nitrogen and quick thawing are used. This is followed by centrifugation as for carbonate extraction to separate soluble from membrane proteins. Although this method can be used efficiently for intact mitochondria it does not work well for mitoplasts.

The previous two methods do not distinguish between a protein that is inserted in the inner membrane, but facing either the intermembrane space or the matrix. In both cases, the protein will partition into the membrane-associated material. To distinguish between the two topologies, mitoplasts should be made in the presence of a protease (usually protease K): if the protein is facing the intermembrane space, it will be degraded by the protease; if it is facing the matrix it should be protected, as the protease cannot penetrate the inner membrane, which remains tightly sealed during mitoplasting. In general terms, proteins that are more prone to sticking to the inner membrane may sometimes be released more easily by addition of salt or low concentrations (up to 2 M) of urea.

One potential problem in the generation of mitoplasts by osmotic shock is the high (10-fold) dilutions used. An alternative method is the titration of the detergents digitonin[43] and octylpolyoxyethylene.[44] By this method, components of the four mitochondrial compartments are gradually solubilized with increasing concentrations of the detergent used.

To determine accurately the submitochondrial location of a protein, usually a combination of some of the preceding methods is necessary. In all cases, careful immunoblot controls using marker proteins for each compartment must be performed in the same experiment. Such convenient markers are porin for the outer membrane, cytochrome b_2 for the intermembrane space, AAC for the inner membrane, and α-ketoglutarate dehydrogenase (KDH) for the matrix.

Further Assays

For some hybrid precursors it is of interest to demonstrate specific interactions with mitochondrial proteins and, in particular, components of the import machinery. This is best accomplished by cross-linking and immunoprecipitation with monospecific antibodies. Such studies are even more informative when using a translocation intermediate of the hybrid precursor at a certain stage of import. In this way, interactions with specific import components can be assigned to physical proximity at specific steps during the import process. To address these questions, a number of methodologies have been developed to generate translocation intermediates: (1) The presursor is imported under low ATP concentrations outside mitochon-

[43] C. Schnaitman and J. W. Greenawalt, *J. Cell Biol.* **38,** 158 (1968).
[44] D. Vestweber and G. Schatz, *J. Cell Biol.* **107,** 2037 (1988).

dria. This has allowed, for example, the accumulation of an intermediate of AAC in interaction with the receptor complex on the outer surface of mitochondria.[45] This is a simple approach but the translocation intermediate thus created is not stable; and (2) tight folding of the "passenger" protein (or mature part of the precursor) outside mitochondria can provide a more stable translocation intermediate. In fact, this has been achieved for DHFR hybrids by incubation with methotrexate, a specific ligand.[46] The resulting precursor was immobilized across the translocation channel, which allowed the identification of the first component of the mitochondrial translocation machinery. A similar approach was applied by using bovine pancreatic trypsin inhibitor as a fusion protein,[44] or metallothionein (with copper as a specific ligand).[47] In the same vein, an irreversible intermediate can be created between the hybrid protein and a specific antibody.[48]

In all cases, the hybrid precursor can then be cross-linked to the neighboring molecules by a variety of cross-linking reagents, either homobifunctional (that carry the same reactive group on both sides), or heterobifunctional (that carry two reactive groups with different specificities). Most commercially available cross-linkers react with either lysine or cysteine residues. In any cross-linking experiment, it is advisable to perform a great number of initial screenng experiments using various cross-linkers (both membrane permeable and membrane impermeable) with different spacer lengths. Almost always, it is essential to try a wide concentration range of the cross-linker to establish optimal conditions. A useful feature for a cross-linker is the possibility to cleave it, so as to establish specificity of the cross-linking reaction. Usually this is achieved by cleavage with dithiothreitol or 2-mercaptoethanol of a disulfide bond in the middle of the cross-linker. One particularly interesting possibility is the creation of a disulfide bond between the radioactive hybrid protein and its nearest neighbors. This is a risky approach because it requires the presence of two closely apposed cysteine residues, one on the hybrid molecule and the other on the target molecule, but if it works the essentially zero-length spacing ensures extremely high yields of cross-linking. We undertook this approach and were able to cross-link a cytochrome b_2–DHFR hybrid to TIM11, a novel protein of the mitochondrial inner membrane with a yield of greater that 80%.[16] This allowed us to scale up the import reaction 1 million times and obtain chemical amounts of the cross-linked product.[16] However, this is a difficult task, as the efficiencies of cross-linking usually obtained are on the order of 10% or less.

[45] N. Pfanner and W. Neupert, *J. Biol. Chem.* **262,** 7528 (1987).
[46] M. Eilers and G. Schtz, *Nature (London)* **322,** 228 (1986).
[47] W.-J. Chen and M. G. Douglas, *J. Biol. Chem.* **262,** 15605 (1987).
[48] M. Schwaiger, V. Herzog, and W. Neupert, *J. Cell Biol.* **105,** 235 (1987).

Final Remarks

Most of what we know about protein import into mitochondria is based on the use of matrix-targeting signals to make hybrid proteins for import into isolated mitochondria. This approach has been extremely successful, but it is now imperative to study in more detail the import signals of presequence-free precursors that contain internal and probably multiple targeting signals. This is a far more complicated task, but the existing methodology developed using hybrid precursors should be extremely valuable.

The experimental system described here provides a simple, chemically defined system with clear advantages to study mechanistic aspects of the import process of hybrid precursors; the use of hybrid proteins to monitor targeting events in living cells certainly complements these experiments and will provide insightful information for this fundamental process.

Acknowledgments

I thank Jeff Schatz for discussions, help, and encouragement. Work in my laboratory is supported by the Wellcome Trust, the Royal Society, and the UK Biotechnology and Biological Sciences Research Council (BBSRC).

[25] Targeting Proteins to Plasma Membrane and Membrane Microdomains by N-Terminal Myristoylation and Palmitoylation

By WOUTER VAN'T HOF and MARILYN D. RESH

Introduction

The fatty acids myristate and palmitate are covalently linked to a wide variety of viral and cellular proteins.[1] Sequence motifs located at the N terminus are responsible for conferring N-myristoylation of all known N-myristoylated proteins. In addition, signals for protein palmitoylation are often located at or near the N terminus. Many myristoylated proteins and nearly all dually fatty acylated proteins are attached to the inner leaflet of

---

[1] M. D. Resh, *Cell. Signal.* **8,** 403 (1996).

the cell plasma membrane. The N-terminal sequence motifs that direct fatty acylation have been shown to be both necessary and sufficient for conferring plasma membrane association. In this chapter, we discuss the use of N-terminal fatty acylation sequences for targeting chimeric proteins to the plasma membrane.

There are three general types of N-terminal fatty acylation motifs that are known to confer membrane binding. Type 1 motifs contain a "myristate plus basic" signal and are found in Src, human immunodeficiency virus type 1 (HIV-1) Gag, and MARCKS.[1] These proteins are modified by covalent attachment of the 14-carbon fatty acid myristate to the N-terminal glycine residue. Myristate inserts hydrophobically into the lipid bilayer, but is not sufficient by itself to confer stable membrane binding. Additional membrane-binding energy is provided by electrostatic interactions between positively charged amino acid residues in a basic patch of the protein and negatively charged head groups of acidic phospholipids in the membrane. Type 2 motifs contain the consensus sequence "Met-Gly-Cys": the N-terminal methionine is removed, Gly-2 is N-myristoylated, and Cys-3 is palmitoylated. Type 3 motifs are not myristoylated, but instead contain two cysteines near the N terminus that confer dual palmitoylation. The presence of two fatty acids in the type 2 and 3 motifs provides strong membrane binding through hydrophobic interactions with the lipid bilayer. Thus, two signals are required for membrane association: myristate plus basic, myristate plus palmitate, or two palmitates. When these two signals are present within the first 10–25 amino acids of the protein, it is possible to use the N-terminal sequence of the proteins as a cassette to confer membrane binding to other proteins.

In addition to providing plasma membrane binding, palmitoylated sequences have also been shown to confer targeting of proteins to specific plasma membrane microdomains. These domains, known as rafts, detergent-resistant membranes (DRMs), or caveolae, consist of subspecialized regions of the plasma membrane that are enriched in glycosphingolipids and cholesterol. Many signaling molecules are also localized to rafts, especially myristoylated and palmitoylated proteins and dually palmitoylated proteins (Table I[2–18]). Type 2 and 3 N-terminal motifs are therefore useful for conferring plasma membrane microdomain localization to proteins.

[2] D. Murray, L. Hermida-Matusumoto, C. A. Buser, J. Tsang, C. T. Sigal, N. Ben-Tal, B. Honig, M. D. Resh, and S. McLaughlin, *Biochemistry* **37,** 2145 (1998).
[3] A. Aronheim, D. Engelberg, N. Li, N. Al-Alawi, J. Schlessinger, and M. Karin, *Cell* **78,** 949 (1994).
[4] A. Klippel, C. Reinhard, W. M. Kavanaugh, G. Apell, M.-A. Escobedo, and L. T. Williams, *Mol. Cell. Biol.* **16,** 4117 (1996).

TABLE I
EXAMPLES OF N-TERMINAL FUSIONS OF FATTY ACYLATED SEQUENCES[a]

Fatty acyl protein sequence	Modification	Acceptor protein	Subcellular localization	Ref.
Src	Myr + basic	β-Gal	PM	2
	Myr + basic	Sos	PM	3
	Myr + basic	PI3K	PM	4
	Myr + basic	CD45	PM	5
	Myr + basic	RhoB	PM	6
	Myr + basic	RSV Gag	PM	7
HIV-1 Gag	Myr + basic	ΔSrc	PM	8
	Myr + basic	GFP	PM	9
Fyn	Myr + Palm	β-Gal	PM/DRM	10
	Myr + Palm	GFP	PM/DRM	11
Lck	Myr + Palm	CAT	PM/DRM	12
$G_i\alpha1$	Myr + Palm	GFP	PM/caveolae	13
GAP43	2 Palm	MARCKS	PM	14, 15
	2 Palm	β-Gal	PM/DRM	16
	2 Palm	ΔFyn	PM/DRM	17
RGS4	2 Palm	GFP	PM	18

Abbreviations: Myr, Myristate; β-Gal, β-galactosidase; PM, plasma membrane; PI3K, phosphatidylinositol-3-kinase; GFP, green fluorescent protein; Palm, palmitate; DRM, detergent-resistant membranes; CAT, chloramphenicol acetyltransferase.

[a] N-terminal sequences from the indicated proteins were fused in frame to the indicated acceptor proteins. A "Δ" denotes that the N-terminal sequence of the acceptor protein was removed prior to fusion with the indicated targeting sequence.

Applications of Use of Membrane-Targeting Motifs

Fusion of membrane-targeting sequences can be employed in several types of applications. First, fatty acylation motifs can be used to target

[5] B. B. Niklinska, D. Hou, C. June, A. M. Weissman, and J. D. Ashwell, *Mol. Cell. Biol.* **14,** 8078 (1994).
[6] P. F. Lebowitz, J. P. Davide, and G. C. Prendergast, *Mol. Cell. Biol.* **15,** 6613 (1995).
[7] J. W. Wills, R. C. Craven, and J. A. Achacoso, *J. Virol.* **63,** 4331 (1989).
[8] W. Zhou, L. J. Parent, J. W. Wills, and M. D. Resh, *J. Virol.* **68,** 2556 (1994).
[9] W. Zhou and M. D. Resh, *J. Virol.* **70,** 8540 (1996).
[10] A. Wolven, H. Okamura, Y. Rosenblatt, and M. D. Resh, *Mol. Biol. Cell* **8,** 1159 (1997).
[11] W. van't Hof and M. D. Resh, *J. Cell Biol.* **145,** 377 (1999).
[12] P. Zlatkine, B. Mehul, and A. I. Magee, *J. Cell Sci.* **110,** 673 (1997).
[13] F. Galbiati, D. Volonte, D. Meani, G. Milligan, D. M. Lublin, M. P. Lisanti, and M. Parenti, *J. Biol. Chem.* **274,** 5843 (1999).

cytosolic proteins to the inner face of the plasma membrane. This approach has been exploited for studies of signaling pathways that involve recruitment of cytosolic proteins to activated, membrane-bound receptors. Attachment of fatty acylation signals to normally soluble signaling molecules causes these molecules to bind constitutively to the plasma membrane and results in activation of signal transduction independent of receptor activation (Table I). Fatty acylation motifs can also be used to replace or bypass existing membrane-targeting signals within a given protein. For example, replacement of the N-terminal sequence of Src with that of HIV-1 Gag showed that the myristate plus basic motifs of the two proteins are functionally interchangeable.[8] A different approach was taken for MARCKS, a protein whose membrane binding is mediated by N-terminal myristoylation and a central basic domain. Phosphorylation of MARCKS by protein kinase C results in a myristoyl-electrostatic switch that releases the protein from the membrane. When the N terminus of MARCKS is replaced by a dually palmitoylated sequence from GTPase-activating protein 43 (GAP43), the resultant chimera can still be phosphorylated by protein kinase C but is not released from the membrane, presumably because the presence of two fatty acids stably anchors the protein to the lipid bilayer.[14] Finally, N-terminal fusions can be used to test the potential membrane-targeting ability of a given sequence. For example, attachment of the N-terminal sequences of Fyn or inhibitory G protein α chain ($G_i\alpha$) to soluble carrier proteins was shown to result in dual fatty acylation and targeting of the chimeric proteins to detergent-resistant membrane subdomains.[10,13,17]

Design of Fatty Acid-Modified Chimeric Proteins

In this section we discuss the design of chimeras containing N-terminal fatty acylation motifs fused in frame to heterologous proteins. The properties and applications of several soluble marker proteins, i.e., β-galactosidase (β-Gal), glutathione S-transferase (GST), and green fluorescent protein (GFP), are described in detail in Chapters 6 and 7 of this volume. We present examples for the application of three fatty acylated membrane targeting motifs: (1) myristate plus polybasic amino acid clusters, (2) myris-

---

[14] J. T. Seykora, M. M. Myat, L. A. H. Allen, J. V. Ravetch, and A. Aderem, *J. Biol. Chem.* **271,** 18797 (1996).

[15] M. M. Myat, S. Anderson, L.-A. II. Allen, and A. Aderem, *Curr. Biol.* **7,** 611 (1997).

[16] S. Arni, S. A. Keilbaugh, A. G. Ostermeyer, and D. A. Brown, *J. Biol. Chem.* **273,** 28478 (1999).

[17] W. van't Hof and M. D. Resh, *J. Cell Biol.* **136,** 1023 (1997).

[18] S. P. Srinivasa, L. S. Bernstein, K. J. Blumer, and M. E. Linder, *Proc. Natl. Acad. Sci. U.S.A.* **95,** 5584 (1998).

tate plus palmitate, and (3) dual palmitoylation. We address important parameters we have recognized in selecting protein sequences of appropriate length, containing the minimal necessary information for functional expression of the different motifs. Because the enzymology dictates that protein myristoylation be confined to the amino terminus, the design of all myristoylated fusion proteins is restricted to creating N-terminal fusions. The methodology for the construction of C-terminal fusion proteins carrying prenylated membrane-targeting motifs is described elsewhere.[19]

Selection of Protein Sequences of Appropriate Length

Myristate Plus Polybasic Motif of Src. The amino terminus of Src contains a myristate and a stretch of six positively charged amino acid residues, composed of Lys-5, -7, -9 and Arg-14, -15, and -16 (MGSSKSKPKDPSQRRR).[20,21] Although the entire motif itself is present within the first 16 residues, it is advisable to add a spacer segment between the last arginine and the beginning of the carrier protein. For instance, Src(20)–β-Gal is more efficiently targeted to the plasma membrane than Src(16)–or Src(18)–β-Gal constructs. The addition of a spacer likely enhances the function of the polybasic motif by facilitating exposure of the positive charges. It is relevant to note that Ser-17 is a phosphorylation site in Src. Src(20)–β-Gal, when expressed in cells, is phosphorylated, but this seems to have little effect on the overall level of membrane binding.[2]

Myristate Plus Palmitate Motifs. The N-terminal Met-Gly-Cys motif defines a minimal sequence for dual fatty acylation with myristate and palmitate.[1] However, addition of this motif alone will not guarantee dual acylation of fusion proteins because the presence of certain residues within the first 10–15 amino acids can drastically affect the efficiency of fatty acid modification. The consensus sequence for N-myristoylation is GXXXS/T. Gly-2 is critical for myristate attachment and a serine or threonine at position 6 is preferred by N-myristoyl transferase (NMT).[22] Additional amino acids are important as well. For instance, the lysines at positions 7 and 9 are essential for efficient myristoylation of the Src family kinases Src and Fyn.[11] The presence of N-terminal cysteine residues, in proximity to a myristate group embedded in the lipid bilayer, favors palmitoylation, but

[19] P. A. Solski, L. A. Quilliam, S. G. Coats, D. J. Der, and J. E. Buss, *Methods Enzymol.* **250,** 435 (1995).

[20] C. Buser, C. T. Sigal, M. D. Resh, and S. McLaughlin, *Biochemistry* **33,** 13093 (1994).

[21] C. T. Sigal, W. Zhou, C. Buser, S. McLaughlin, and M. D. Resh, *Proc. Natl. Acad. Sci. U.S.A.* **91,** 12253 (1994).

[22] D. R. Johnson, R. S. Bhatnagar, L. J. Knoll, and J. I. Gordon, *Annu. Rev. Biochem.* **63,** 869 (1994).

surrounding amino acids may reduce or inhibit this process. For example, introduction of a cysteine at position 3 in the myristoylated N terminus of the HIV Gag protein (i.e., changing the N-terminal sequence from MGAR to MGCR) does not result in Gag palmitoylation (W. van't Hof and M. D. Resh, unpublished results, 1999). This is likely caused by the presence of the arginine residue at position 4 in Gag. In contrast, when cysteine is introduced into position 3 of Src, the mutant protein is palmitoylated.

To achieve efficient dual fatty acylation, it is advisable to select at least the first 10 amino acids of a known dually acylated protein for fusion onto a heterologous carrier. However, more than 10 amino acids may be used, if use is being made of specific downstream restriction enzyme sites for subcloning purposes. For example, we have constructed Fyn(16)–β-Gal and Fyn(25)–GST, containing the first 16 or 25 amino acids of Fyn, fused in frame to β-Gal and GST, respectively; these chimeras are targeted to the plasma membrane.[10]

Dual Palmitoylation. The N terminus of the GAP43 (neuromodulin) protein is dually acylated on cysteines at positions 3 and 4. The first 10 and 20 amino acids of GAP43 is sufficient to confer dual palmitoylation of Fyn[17] and β-Gal,[16] respectively.

Single Palmitoylation. The $G_S\alpha$ heterotrimeric G protein subunit is not myristoylated, but it contains a single cysteine at position 3 that is palmitoylated. Replacement of the first 10 amino acids of Fyn with those of $G_S\alpha$ is not sufficient to promote palmitoylation of the $G_S\alpha$ (10)–Fyn chimera.[17] Others have shown that coexpression with $\beta\gamma$ complexes is essential for palmitoylation of certain α subunits.[23] Thus, the amino terminus of $G_S\alpha$ does not provide an independent fatty acylation signal when taken out of the context of its interactions with $\beta\gamma$ subunits.

Design of Negative Controls. Negative controls are essential in testing the specific effect of fatty acylation motifs on the behavior of the fusion protein. This is generally accomplished by replacing essential amino acids with other structurally similar, but nonfunctional, amino acids. For myristoylation, the most commonly used negative control is a G2A mutant, in which the essential glycine at position 2 is replaced by alanine. For myristate plus basic motifs, the arginines and lysines in the polybasic motif are replaced by asparagines. Negative controls for palmitoylation are generated by replacing cysteines with serine or alanine residues.

Construction of Chimeras by Polymerase Chain Reaction

As already mentioned, the chimeric proteins discussed here are restricted to N-terminal fusions. Sense oligonucleotide primers are designed

[23] J. Morales, C. S. Fishburn, P. T. Wilson, and H. R. Bourne, *Mol. Biol. Cell* **9,** 1 (1998).

to encode a unique restriction enzyme site with a 5′ overhang, the first 10 amino acids of the protein carrying the fatty acylation signal of interest, followed in frame by amino acids 2–6 of the carrier protein of choice. This generally results in primers of 50 to 60 nucleotides in length, containing a minimum overlap with the template sequence of 15 base pairs. Antisense oligonucleotide primers are designed to be complementary to downstream regions of the carrier protein so that the polymerase chain reaction (PCR) products contain restriction enzyme digestion sites that are useful for subcloning into the desired vectors. The annealing temperature in the thermocycling cycles is usually selected to be between 45 and 50°. To test for generation of nonspecific DNA fragments at these relatively low annealing temperatures, controls are performed with primers but without template, or with template DNA but without primers. The selection of template DNA and plasmid vectors depends on the desired application of the constructs. For example, we routinely use pGEM3Z as a template for PCRs, followed by subcloning of the cDNA product into pSP65 or pGEM vectors for *in vitro* translation studies, and pCMV5 for transfection experiments in COS-1 cells.

Examples of Construction of Fatty Acylated Chimeric Proteins

Myristate Plus Polybasic Motif: Construction of Src(20)–β-galactosidase. For this fusion, the first 20 amino acids of v-Src are fused in frame to β-Gal.[2] The sense primer contains an *Xba*I site, and 24 nucleotides starting 243 bases upstream from the v-Src ATG start site. Inclusion of the v-Src upstream region is not necessary for *in vivo* expression, but does seem to enhance the level of *in vitro* translation. The antisense primer contains codons for amino acids 13–20 of v-Src and a *Bam*HI site. Amplification by PCR is achieved with a v-Src pGEM3Z template. The PCR product is digested with *Xba*I and *Bam*HI and ligated into pVALO, a eukaryotic expression vector containing β-galactosidase under the control of the β-globin promotor.

Myristate Plus Palmitate Motif: Construction of Fyn(16)–β-Galactosidase or Green Fluorescent Protein. The first 16 amino acids of Fyn, a dually fatty acylated Src family kinase, can be used to target heterologous proteins to the plasma membrane and to rafts.[10,11] A Fyn(16)–β-Gal chimera can be constructed by a strategy similar to that described for Src–β-Gal. Although the construct contains 16 amino acids from Fyn, the first 10 amino acids will probably suffice. Alternatively, the Fyn sequence can be fused to another carrier protein such as GFP. Fyn(16)–GFP is generated by PCR, using a sense oligonucleotide encoding an *Eco*RI site followed by the first 16 amino acids of human Fyn fused in frame to amino acids 2–6 of enhanced GFP (eGFP). An antisense oligonucleotide encoding the seven C-terminal amino acids of eGFP, a stop codon, and an *Xba*I site is used. These primers are used with pEGFP-N1 as a template to generate a 1.6-kb DNA fragment that is digested with *Eco*RI and *Xba*I, followed by ligation into pCMV5.

Dual Palmitoylation Motif: Construction of GAP43(10)–Fyn. In this example, the N-terminal myristoylation–palmitoylation motif of Fyn is replaced with a dual palmitoylation motif from GAP43. A sense oligonucleotide is synthesized to encode a *Bam*HI site, the upstream region of pGEM3Z, the first 10 amino acids of human GAP43, and amino acids 11–15 of human Fyn.[4] An antisense oligonucleotide corresponding to a region of the SH3 domain of Fyn is used. These primers are used with pGEM3Z-Fyn as a template to generate chimeric cDNAs by PCR. The PCR product is digested with *Bam*HI and *Bst*XI to generate a 100-base pair fragment that is used to replace the corresponding fragment of *Bam*HI/*Bst*XI-digested wild-type Fyn in pSP65. The construct is then digested with *Eco*RI and *Sal*I, followed by ligation of the GAP43(10)–Fyn chimeric cDNA into *Eco*RI- and *Sal*I-cut pCMV5.

Analysis of Fatty Acylated Fusion Proteins

Analysis of the fatty acylation profiles of chimeric proteins expressed in cells can be performed by radiolabeling with commercially available ^{3}H labeled myristate and palmitate. Alternatively, ^{125}I-labeled synthetic analogs, containing ^{125}I at the ω-carbon of the fatty acid, can be exploited. The chemical synthesis and radioiodination of the iodo fatty acid analogs has been described in detail.[24,25] The use of ^{125}I-labeled fatty acids has several advantages. Exposure times to visualize fatty acylated proteins on film are significantly reduced, the ^{125}I signal can be quantitated by phosphorimaging, and the ^{125}I label from the iodo fatty acids is not incorporated into the polypeptide backbone after addition to cells for long incubation times.

Cell Labeling with Radiolabeled Fatty Acids

Routinely, adherent cells are grown to subconfluency, starved for 1 hr at 37° in Dulbecco's modified Eagle's medium (DMEM) containing 2% (v/v) dialyzed fetal bovine serum (FBS) and radiolabeled for 2–6 hr with

---

[24] L. Berthiaume, S. M. Peseckis, and M. D. Resh, *Methods Enzymol.* **250,** 454 (1995).
[25] S. M. Peseckis, I. Deichaite, and M. D. Resh, *J. Biol. Chem.* **268,** 5107 (1993).

radiolabeled fatty acid at 25–50 μCi/ml.[26] [9,10-^{3}H(N)]Myristic acid and [9,10-^{3}H]palmitic acid (Du Pont New England Nuclear, Boston, MA) are stored in ethanol. For use in cell labeling, the desired amount of fatty acid is transferred to a glass vial, dried under a gentle stream of nitrogen, resuspended in a small volume of ethanol (less than 5% of the final volume), and diluted in DMEM containing 2% (v/v) dialyzed FBS. Alternatively, dried ^{3}H-labeled fatty acid can be resuspended in medium by brief sonication. The iodo fatty acids 13-iodotridecanoic acid (IC13) and 16-iodohexadecanoic acid (IC16) are stored lyophilized in glass vials at $-20°$. Before use they are dissolved in ethanol, diluted in DMEM–2% (v/v) dialyzed FBS, and added to cells. Iodo fatty acids stick to plastic and when possible glass pipettes should be used for their transfer. As a control for total expression levels, the protein backbone is labeled with Tran^{35}S label (ICN, Irvine, CA). For this, cells are starved for 1 hr at 37° in DMEM minus methionine and cysteine, supplemented with 2% (v/v) dialyzed FBS, and labeled for 2–6 hr at 37° with Tran^{35}S label at 25 μCi/ml. Alternatively, total protein levels can be analyzed by Western blotting.

When signals for fatty acylation are weak using ^{3}H-labeled fatty acids, it is preferable to use more cells and more radiolabel (100–250 mCi/ml), rather than to increase the incubation time. Prolonged incubation will result in increased fatty acid degradation and the ^{3}H label from [9,10-^{3}H]myristate and palmitate will become incorporated into amino acids and the polypeptide backbone of newly synthesized proteins.

Immunoprecipitation and Sodium Dodecyl Sulfate–Polyacrylamide Gel Electrophoresis. After labeling, cells are washed several times with cold STE [50 mM Tris-HCl (pH 7.4), 150 mM NaCl, 1 mM EDTA], lysed in lysis buffer [50 mM Tris-HCl (pH 8), 150 mM NaCl, 1% (v/v) Nonidet P-40, 0.5% (w/v) sodium deoxycholate, 2 mM EDTA] supplemented with protease inhibitors [1.5 μg/ml each of leupeptin, aprotinin, and pepstatin A (Boehringer Mannheim, Indianapolis, IN), 20 μg/ml each of 4-(2-aminoethyl)-benzene sulfonyl fluoride (AEBSF), N$^\alpha$-tosyl-Lys-chloromethyl ketone (TLCK), tolylsulfonyl phenylalanyl chloromethyl ketone (TPCK), and benzamidine (Calbiochem, La Jolla, CA)] for 10–15 min at 4°, and scraped off the dish. Lysates are clarified by centrifugation for 15 min at 100,000g at 4°, followed by immunoprecipitation with antibody directed against the carrier protein domain or with glutathione beads in the case of a GST fusion construct. Immunoprecipitates are washed and resuspended in 2× sodium dodecyl sulfate (SDS) sample buffer containing 0.1 M dithiothreitol (DTT) and analyzed by SDS–polyacrylamide gel electrophoresis (PAGE). Because of the labile nature of

[26] A. Wolven, W. van't Hof, and M. D. Resh, *Methods Mol. Biol.* **84,** 261 (1998).

the thioester bond, [^{3}H]palmitate- or IC16-containing samples should not be boiled before loading on the gel because this will result in significant loss of the label from the protein.

Analysis of Fatty Acid Incorporation into Protein

Acid Hydrolysis. This treatment is used to assay fatty acids attached to proteins through amide bonds, a characteristic of myristoylation. SDS–polyacrylamide gels containing radiolabeled proteins are wrapped in Saran wrap and exposed to film overnight at 4°. ^{3}H-Labeled samples can be flanked on either side with ^{35}S-labeled protein samples to facilitate subsequent band identification. The autoradiograph is used as a guide to excise the desired radiolabeled protein from the gel and the gel slice is placed in a glass resealable reaction vial (Reacti-Vial reaction vials; Pierce, Rockford, IL). The gel piece is first rehydrated for 30 min at room temperature in 0.5 ml of water, the solution and backing paper are removed, and the gel is crushed with a glass rod and hydrolyzed for 18 hr at 100° in 0.5 ml of 6 N HCl. Next, the solution is cooled to room temperature and neutralized with 0.5 ml of 6 N NaOH, and the hydrolysate is extracted two or three times with 1.0-ml portions of chloroform. The combined chloroform layers are dried under a nitrogen stream, and released fatty acids are analyzed by thin-layer chromatography (TLC).

Alkaline Hydrolysis. Alkaline hydrolysis is used to detect fatty acid incorporated by thioester bonds that characterize palmitoylation.[27] Radiolabeled protein in an SDS–polyacrylamide gel is localized and excised as described above. The crushed gel slice is treated for 24 hr at room temperature with 0.5 ml of 1.5 M NaOH, after thorough mixing. The hydrolysate is neutralized with 0.75 ml of 1 N HCl, extracted with chloroform, and dried under nitrogen. Released fatty acids are analyzed by TLC. Alternatively, polyacrylamide gels containing palmitic acid-labeled and Tran^{35}S-labeled samples are soaked overnight at room temperature in neutral hydroxylamine or in Tris buffer. The gels are dried and exposed to film; radiolabeled palmitate should be removed by the alkaline treatment.

Reversed-Phase Thin-Layer Chromatography Analysis. The dried radiolabeled fatty acids from the preceding two sections are dissolved in a few microliters of chloroform and applied to 250-μm reversed phase silica F (RPS-F) plates (Analtech, Newark, DE). For positive identification of the fatty acids extracted from protein, radiolabeled fatty acid standards must be included in the reversed-phase TLC (RP-TLC) analysis. Each plate is developed with a mobile phase of 1 : 1.75 : 1.75 (v/v/v) water–acetic acid–acetonitrile. One development is sufficient to separate saturated fatty acids

---

[27] L. Berthiaume and M. D. Resh, *J. Biol. Chem.* **270,** 22399 (1995).

containing 11 to 17 carbons, with shorter length fatty acids moving faster toward the top of the plate. Plates are dried with a hair dryer and subject to autoradiography when ^{125}I label is used, or to fluorography using EN^{3}HANCE spray (Du Pont New England Nuclear, Boston, MA) for ^{3}H-labeled samples.

Analysis of Intracellular Membrane Distribution

Subcellular Fractionation and Marker Enzyme Analysis

Analysis of Cytosol (S100) and Total Membranes (P100). Radiolabeled cells are washed and scraped from the dishes in ice-cold STE and centrifuged for 5 min at 1500 rpm in a refrigerated tabletop centrifuge. Cells are resuspended in hypotonic buffer [10 mM Tris-HCl (pH 7.2), 0.2 mM MgCl$_2$, 100 mM Na$_3$VO$_4$] supplemented with protease inhibitors, incubated on ice for 10 min, and homogenized by 25 strokes in a 1.5-ml Dounce homogenizer with a tight pestle. The homogenate is adjusted to 250 mM sucrose and 1 mM EDTA and centrifuged for 45 min at 100,000g (50,000 rpm in a TLA 100.2 or 100.3 rotor) in an Optima TL ultracentrifuge (Beckman, Fullerton, CA). The total cellular membrane pellet (P100) is resuspended in lysis buffer supplemented with protease inhibitors [e.g., radioimmunoprecipitation assay (RIPA) buffer: 10 mM Tris (pH 8.0), 150 mM NaCl, 1 mM EDTA, 1% (v/v) Triton X-100, 0.1% (w/v) SDS, 1% (w/v) deoxycholate] and the soluble cytosolic fraction (S100) is supplemented with 0.2 volume of 5× concentrated lysis buffer. Samples are lysed for 15 min on ice, clarified by centrifugation for 15 min at 100,000g, and analyzed by immunoprecipitation, SDS–PAGE, and autoradiography.

Analysis of Linear Density Gradients. The intracellular distribution of chimeric proteins can be analyzed in greater detail by density gradient centrifugation. In our experience, the best fractionation results are obtained with the FuGene transfection reagent (Boehringer Mannheim). To optimize recovery of intact cellular organelles, isotonic homogenization conditions (250 mM sucrose) may be selected. Roughly 5–10 mg of cellular protein is used for each gradient, equivalent to six subconfluent 100-mm plastic dishes. Cells are scraped from the dishes, spun down for 5 min at 1500 rpm, resuspended in 1 ml of isotonic homogenization buffer with protease inhibitors, and homogenized with 15 strokes in a 1.5-ml Dounce homogenizer with a tight pestle. Depending on the cell type and the transfection method used, the number of strokes needed to break all cells open while maintaining organelle integrity may vary. We advise testing this in a pilot experiment in which cells are homogenized by 10, 15, 20, or 25 strokes. Other homogenization methods may be preferable, depending on the type of cells used. Nuclei and unbroken cells are removed from the homogenate by centrifugation for 10 min at 1000g. A ~30% loss of organelle marker activity in the nuclear pellet (P1 fraction) is standard. If there is a more significant loss (>50%), the homogenate should be pretreated with DNAse I (5 μg/ml) for 10 min at room temperature to reduce cross-linking of organelles by DNA from sheared nuclei. In addition, a small volume of homogenization buffer can be used to wash the homogenizer and to resuspend the nuclear pellet, followed by a respin. The postnuclear supernatants (PNSs) are centrifuged for 45 min at 100,000g. The resulting P100 fraction is diluted in 0.5 ml of isotonic buffer, resuspended by 5–10 strokes in a Dounce homogenizer with a loose pestle, and layered on top of a continuous iodixanol density gradient [25.0 to 2.5% (w/v) OptiPrep; Nycomed, Oslo, Norway]. Iodixanol is stored as a 50% (w/v) stock solution and diluted in 0.25 M sucrose–TE. Gradients are centrifuged for 3 hr at 180,000g (38,000 rpm in an SW 40 Ti rotor) and 0.5 to 0.7-ml samples are collected, starting from the bottom of the gradient, for a total of 15–18 fractions.

Analysis of Marker Enzyme Activities. Fluorescent, 4-methylumbelliferone (4MU) enzyme substrates (Sigma, St. Louis, MO) provide sensitive and inexpensive tools for measurement of marker enzyme activities of organelles in the pathway of membrane protein synthesis. Assays are routinely performed in duplicate in 96-well microtiter plates (Costar, Boston, MA). From each fraction, 20–50 μl is taken and assays are started by adding 200 ml of the following assay mixes to each sample: Endoplasmic reticulum (α-glucosidase II): 150 mM citrate-phosphate (pH 6.5), 1 mM 4 MU-α-D-glucoside; Golgi complex (α-mannosidase II): 150 mM phosphate buffer (pH 6.0), 0.02% (v/v) Triton X-100, 1 mM 4 MU-α-D-mannopyranoside; plasma membrane (alkaline phosphatase): 100 mM Tris-NaOH (pH 9.5), 100 mM NaCl, 5 mM MgCl$_2$, 1 mM 4 MU-phosphate; lysosomes (acid phosphatase): 180 mM sodium acetate-acetic acid (pH 5), 1 mM 4 MU-phosphate. Reactions are allowed to proceed for 2–4 hr at 37° and stopped by addition of 100 μl of ice-cold 1 M glycine-NaOH (pH 10), 1 M Na$_2$CO$_3$. Fluorescence is measured at λ_{ex} 360 nm, λ_{em} 460 nm using a CytoFluor 2350 fluorescence measurement system (Millipore, Bedford, MA). If signals are low, reactions may be allowed to proceed overnight at room temperature. Assays can also be performed with analogous *p*-nitrophenyl substrates and measured by spectrophotometry.

Fluorescence Microscopy

The intracellular distribution of the chimeric constructs can be determined by standard immunofluorescence with antibodies against known markers as references. Adherent cells are grown on glass coverslips, fixed with 3.7% (v/v) formaldehyde in phosphate-buffered saline (PBS), and permeabilized with 0.2% (v/v) Triton X-100 in PBS. Coverslips are then incubated in primary antibody solution in PBS–10% (v/v) calf serum, washed, incubated with secondary antibody, washed, mounted on slides, and observed under a microscope with epifluorescence optics.[10] An example of an immunofluorescence analysis of Fyn(16)–β-Gal expressed in COS-1 cells is depicted in Fig. 1. The carrier protein, β-Gal, is entirely soluble (Fig. 1A), whereas Fyn(16)–β-Gal exhibits a striking pattern of fluorescence at the plasma membrane (Fig. 1B). Mutation of one or both of the palmitoylation sites (Fig. 1C–F) results in loss of plasma membrane fluorescence and redistribution of much of the chimeric protein to intracellular membranes. Similar results are obtained with a Fyn(16)–GFP construct. As described in Chapters 6 and 7 of this volume, GFP fusions can be studied in live cells, without the need for treating cells with fixative.

Plasma Membrane Microdomain Analysis

Extraction of Transfected Cells with Nonionic Detergent. Cells on dishes are rinsed with ice-cold STE and incubated on a rocker for 10 min at 4° with 1.0 ml of 1% (v/v) Triton X-100 in MNE buffer [10 mM morpholine-ethanesulfonic acid (MES, pH 6.5), 150 mM NaCl, 5 mM EDTA]. The detergent-soluble phase is removed from the dish and supplemented with 0.25 ml of 5× lysis buffer. The remaining detergent-resistant fraction on the dishes is subsequently scraped from the plate in 1 ml of 1× lysis buffer. Both fractions are subsequently clarified for 15 min at 100,000g and analyzed by immunoprecipitation and SDS–PAGE.

Purification of Rafts or Detergent-Resistant Membranes. Cells are treated with 1% (v/v) Triton X-100 as described above and the detergent-insoluble fraction is scraped from the dish into 1 ml of MNE buffer, and mixed with 1 ml of 80% (w/v) sucrose in MNE.[28] The lysate is transferred to the bottom of an SW55 Ti centrifugation tube and overlaid with 2 ml of 30% (w/v) sucrose–MNE and 1 ml of 5% (w/v) sucrose–MNE, respectively, followed by centrifugation for >16 hr at 4° at 200,000g and collection of 0.4-ml fractions. DRMs or rafts float to the 30–5% sucrose interface as marked by the distribution of glycophosphatidylinositol (GPI)-linked proteins.

Application of Fumagillin and 2-Hydroxy-myristate as Negative Controls

Pretreatment of cells with fumagillin, an inhibitor of methionine aminopeptidase II,[29] or with 2-OH-myristate,[30] results in the inhibition of myristoylation and can be used for negative controls. 2-OH-myristate (Sigma) is stored as a 100 mM stock solution in dimethyl sulfoxide (DMSO) and

[28] W. Zhang, R. P. Trible, and L. E. Samelson, *Immunity* **9,** 239 (1998).
[29] N. Sin, L. Meng, M. Q. W. Wang, J. J. Wen, W. G. Bornmann, and C. M. Crews, *Proc. Natl. Acad. Sci. U.S.A.* **94,** 6099 (1997).
[30] M. J. S. Nadler, M. L. Harrison, C. L. Ashendel, J. M. Cassady, and R. L. Geahlen, *Biochemistry* **32,** 9250 (1993).

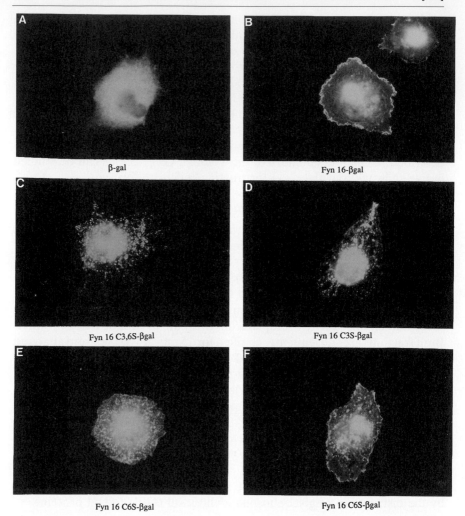

FIG. 1. Immunofluorescence of Fyn(16)–β-Gal chimeras in transiently transfected COS-1 cells. COS cells transfected with β-Gal (A), Fyn(16)–β-Gal (B), Fyn(16)–C3,6S–β-Gal (C), Fyn(16)–C3S–β-Gal (D), or Fyn(16)–C6S–β-Gal (E and F) were fixed, permeabilized, and stained with anti-β-Gal antibodies followed by fluorescein-conjugated anti-mouse IgG. [Reprinted from *Molecular Biology of the Cell* (1997, Volume 8, pp. 1159–1173), with permission of the American Society for Cell Biology.]

added to cells for 2–12 hr at 37° at 0.1–1.0 mM. 2-OH-myristate is dissolved in DMEM containing 1% (w/v) defatted BSA, sonicated briefly, and filtered. For overnight incubations 5% (v/v) FBS should be added.

[26] Analysis of Function and Regulation of Proteins That Mediate Signal Transduction by Use of Lipid-Modified Plasma Membrane-Targeting Sequences

By Gary W. Reuther, Janice E. Buss, Lawrence A. Quilliam, Geoffrey J. Clark, and Channing J. Der

Introduction

Protein function can be modulated by regulating the level of protein expression and by modulating intrinsic catalytic function. In addition, the dynamic translocation between distinct subcellular compartments is another widely appreciated mechanism of regulation of many proteins involved in signal transduction. For example, cytoplasmic protein kinases become activated by association with plasma membrane-associated proteins such as activated receptor tyrosine kinases or heterotrimeric large and monomeric small GTPases.[1,2] Transcription factors can be regulated by reversible translocation to the nucleus.[3,4] This mode of regulation of protein function has introduced a powerful approach to study the function of many signaling molecules. Because constitutively activated mutants of signaling proteins provide important reagents to study protein function, investigators have exploited the use of heterologous sequences to create mislocated and chronically activated proteins. In particular, peptide sequences that signal for covalent modification by lipids have been used to relocate cytoplasmic proteins to membranes.[5] This has been done for a variety of cell signaling proteins including Raf,[6,7] phosphatidylinositol 3-kinase (PI 3-kinase),[8–10]

---

[1] D. K. Morrison and R. E. Cutler, Jr., *Curr. Opin. Cell Biol.* **9,** 174 (1997).

[2] S. L. Campbell, R. Khosravi-Far, K. L. Rossman, G. J. Clark, and C. J. Der, *Oncogene* **17,** 1395 (1998).

[3] T. Hunter and M. Karin, *Cell* **70,** 375 (1992).

[4] C. S. Hill and R. Treisman, *Cell* **80,** 199 (1995).

[5] P. A. Solski, L. A. Quilliam, S. G. Coats, C. J. Der, and J. E. Buss, *Methods Enzymol.* **250,** 435 (1995).

[6] D. Stokoe, S. G. Macdonald, K. Cadwallader, M. Symons, and J. F. Hancock, *Science* **264,** 1463 (1994).

[7] S. J. Leevers, H. F. Paterson, and C. J. Marshall, *Nature (London)* **369,** 411 (1994).

[8] A. Klippel, C. Reinhard, M. Kavanaugh, G. Apell, M.-A. Escobedo, and L. T. Williams, *Mol. Cell. Biol.* **16,** 4117 (1996).

[9] S. Wennstrom and J. Downward, *Mol. Cell. Biol.* **19,** 4279 (1999).

[10] A. D. Ma, A. Metjian, S. Bagrodia, S. Taylor, and C. S. Abrams, *Mol. Cell. Biol.* **18,** 4744 (1998).

and AKT.[11–13] A second application of these membrane-targeting sequences has been to determine whether the function of a particular protein sequence is involved in promoting membrane association. For example, the loss of function caused by deletion of the pleckstrin homology (PH) domain from Dbl family proteins can be restored by addition of a plasma membrane-targeting sequence.[14–16] Thus, at least one function of PH domains involves promoting membrane association. This chapter describes approaches for using myristoylation and prenylation signal sequences to create membrane-targeted chimeric proteins.

Lipid-Modifying Signal Sequences and Protein Lipid Modifications

Locations of Lipid-Modified Proteins

Protein signal sequences have been identified for four types of lipid modifications of proteins. One of these, found only on proteins located on the extracellular face of the plasma membrane, is the attachment of glycophosphatidylinositol (GPI). The signals for attachment of a GPI group and the use of this signal to direct heterologous proteins to the cell surface have been previously described and are not a topic of this chapter.[17]

The other lipid modifications for which signal sequences have been identified are found on cytoplasmic proteins that are often attached to the cytoplasmic face of membranes or vesicles. These lipid modifications include attachment of myristate, palmitate, or isoprenoids (Fig. 1). While the modification of a protein with a lipid appears often to be the first and crucial step to initiate protein binding to any type of membrane, the targeting of proteins to specific membranes is less understood.

Lipid Modification Stability

The attachment of the fatty acid myristate and the addition of an isoprenoid are the two most well-characterized types of lipid modifications. These

[11] G. Kulik, A. Klippel, and M. J. Weber, *Mol. Cell. Biol.* **17,** 1595 (1997).
[12] S. G. Kennedy, A. J. Wagner, S. D. Conzen, J. Jordan, A. Bellacosa, P. N. Tsichlis, and N. Hay, *Genes Dev.* **11,** 701 (1997).
[13] M. Andjelkovic, D. R. Alessi, R. Meier, A. Fernandez, N. J. Lamb, M. Frech, P. Cron, P. Cohen, J. M. Lucocq, and B. A. Hemmings, *J. Biol. Chem.* **272,** 31515 (1997).
[14] I. P. Whitehead, S. Campbell, K. L. Rossman, and C. J. Der, *Biochim. Biophys. Acta* **1332,** F1 (1997).
[15] C. E. Tognon, H. E. Kirk, L. A. Passmore, I. P. Whitehead, C. J. Der, and R. J. Kay, *Mol. Cell. Biol.* **18,** 6995 (1998).
[16] I. P. Whitehead, H. Kirk, C. Tognon, G. Trigo-Gonzalez, and R. Kay, *J. Biol. Chem.* **271,** 18388 (1995).
[17] N. M. Hooper, *Curr. Opin. Cell Biol.* **2,** 617 (1992).

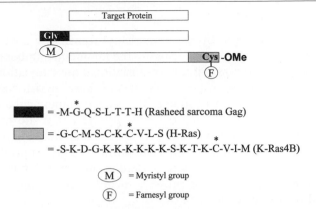

FIG. 1. Membrane targeting of heterologous proteins can be done by myristoylation or farnesylation. Myristoylation signaling peptide sequences are added to the amino terminus of proteins, whereas farnesylation signaling peptide sequences are added to the carboxy terminus of proteins. A myristoyl group is attached to the glycine residue of the myristoylation sequence and a farnesyl group is attached to the cysteine residue of the CAAX motif of the farnesylation sequence. These residues are marked with asterisks. Note that addition of a farnesyl isoprenoid to the cysteine residue of the CAAX sequence is followed by two additional posttranslational modifications: proteolytic removal of the terminal −AAX residues and carboxyl methylation (OMe) of the now carboxy-terminal farnesylated cysteine residue. A sequence reminiscent of the amino-terminal myristoylation signal sequence of the Rasheed rat sarcoma virus Gag protein and the H-Ras and K-Ras4B prenylation signal sequences are shown. These sequences are commonly used to generate membrane-targeted proteins.

provide an apparently permanent and stoichiometric modification of the acceptor protein.[18,19] These lipids are attached to proteins through chemically stable amide or thioether bonds, respectively. Unlike these modifications, substantial turnover of palmitate modifications has been reported in several proteins.[20,21] In this modification, the palmitate is attached through the chemically fragile thioester linkage. Therefore, investigators must exercise care not to expose the sample to common reducing agents such as mercaptoethanol or dithiothreitol.[22]

[18] D. A. Towler, J. I. Gordon, S. P. Adams, and L. Glaser, *Annu. Rev. Biochem.* **57,** 69 (1988).
[19] A. D. Cox and C. J. Der, *Curr. Opin. Cell Biol.* **4,** 1008 (1992).
[20] A. I. Magee, L. Gutierrez, I. A. McKay, C. J. Marshall, and A. Hall, *EMBO J.* **6,** 3353 (1987).
[21] M. E. Linder, P. Middleton, J. R. Hepler, R. Taussig, A. G. Gilman, and S. M. Mumby, *Proc. Natl. Acad. Sci. U.S.A.* **90,** 3675 (1993).
[22] A. I. Magee, J. Wootton, and J. de Bony, *Methods Enzymol.* **250,** 330 (1995).

Myristoylation Signals

The signals recognized by the *N*-myristoyl transferase have been previously described.[18] Eight amino acids are the minimum number of residues that provide an effective signal for cotranslational myristoylation of a protein within the cell.[23,24] Conserved residues in these myristoylation signal sequences include a glycine residue as the second amino acid and a serine or threonine residue at position 6.[18] The glycine is the site of myristate attachment (Fig. 1); however, studies have described additional important residues. These include basic amino acids within the first approximately 30 residues of a protein, which are important in membrane binding. These sequences are found in various proteins including Src family tyrosine kinases[25] and retrovirus Gag proteins.[26]

Amino-terminal acylation with both myristate and palmitate has been described for Src family protein tyrosine kinases[27] and heterotrimeric G protein α subunits.[21] A glycine in the second position and a nearby cysteine residue (e.g., Met-Gly-Cys-X-X-Ser/Thr) appear to be sufficient to cause this double lipid modification. Myristoylation signal sequences derived from the Rasheed sarcoma virus Gag protein[28] and the Src tyrosine kinase[29] are capable of targeting proteins to the plasma membrane.

Successful examples of myristoyl targeting include the targeting of ARF1,[30] protein kinase C,[31,32] Csk,[33] Fyn,[34] and Ras.[35,36] In addition, some of the first myristoylated proteins to be recognized were the fusion proteins between retroviral Gag proteins and the products of cellular protooncogenes.[37] These chimeras are naturally occurring examples of the attachment of an amino-terminal myristoylation signal onto a heterologous protein.

[23] J. M. Kaplan, H. E. Varmus, and J. M. Bishop, *Mol. Cell. Biol.* **10,** 1000 (1990).

[24] J. E. Buss, C. J. Der, and P. A. Solski, *Mol. Cell. Biol.* **8,** 3960 (1988).

[25] L. Silverman and M. D. Resh, *J. Cell Biol.* **119,** 415 (1992).

[26] W. Zhou, L. J. Parent, J. W. Wills, and M. D. Resh, *J. Virol.* **68,** 2556 (1994).

[27] M. D. Resh, *Cell* **76,** 411 (1994).

[28] J. E. Buss, P. A. Solski, J. P. Schaeffer, M. J. MacDonald, and C. J. Der, *Science* **243,** 1600 (1989).

[29] B. M. Willumsen, A. D. Cox, P. A. Solski, C. J. Der, and J. E. Buss, *Oncogene* **13,** 1901 (1996).

[30] J.-X. Hong, R. S. Haun, S.-C. Tsai, J. Moss, and M. Vaughan, *J. Biol. Chem.* **269,** 9743 (1994).

[31] J. Sambrook, E. F. Fritsch, and T. Maniatis, "Molecular Cloning: A Laboratory Manual," 2nd Ed. Cold Spring Harbor Laboratory Press, Cold Spring Harbor, New York, 1989.

[32] G. James and E. Olson, *J. Cell Biol.* **116,** 863 (1992).

[33] L. M. Chow, M. Fournel, D. Davidson, and A. Veillette, *Nature (London)* **365,** 156 (1993).

[34] G. L. Timson, A. N. Kong, L. E. Samelson, and A. S. Shaw, *Mol. Cell. Biol.* **12,** 5438 (1992).

[35] P. M. Lacal, C. Y. Pennington, and J. C. Lacal, *Oncogene* **2,** 533 (1988).

[36] J. E. DeClue, W. C. Vass, A. G. Papageorge, D. R. Lowy, and B. M. Willumsen, *Cancer Res.* **51,** 712 (1991).

[37] A. Schultz and S. Oroszlan, *Virology* **133,** 431 (1984).

Examples of these include Gag-ErbB,[38] Gag-Ras,[39] Gag-Fos,[40] and Gag-Abl.[37]

Palmitoylation Signals

Palmitoylation is a posttranslational covalent modification that often occurs with other lipid modifications. For example, the Src family kinases are modified by both palmitate and myristate[27] while Ras proteins are palmitoylated and isoprenylated.[41] While a putative signal sequence for amino-terminal myristoylation and palmitoylation is known, no well-defined signal for carboxy-terminal palmitoylation has been determined.[29] Palmitoylation can also occur in the cytoplasmic region of transmembrane proteins, such as the β_2-adrenergic receptor.[42] However, the amino acids that define the signal for palmitoylation at these sites and whether such a signal is functional in other proteins are not yet known.

Prenylation Signals

Two different isoprenoids have been shown to be posttranslational modifications of eukaryotic proteins. These are the C_{15} farnesyl group and the C_{20} geranylgeranyl group. The carboxy-terminal signal sequence for isoprenylation is the CAAX motif (where C is cysteine, A is often an aliphatic amino acid, and X is any amino acid) found at the carboxy terminus of the target protein.[43,44] The rules for an effective CAAX prenylation signal have been determined by both *in vitro* and *in vivo* methods.[19,45] The cysteine as the fourth residue from the carboxy terminus is required, because it is the site of isoprenoid attachment (Fig. 1). The amino acid in the X position of the CAAX motif defines whether a farnesyl transferase (FTase) or a geranylgeranyl transferase I (GGTaseI) will recognize and subsequently modify the protein with the appropriate isoprenyl group. FTase preferentially recognizes CAAX sequences where the terminal X residue is serine, methionine, cysteine, alanine, or glutamine. GGTaseI

---

[38] A. Bruskin, J. Jackson, J. M. Bishop, D. J. McCarley, and R. C. Schatzman, *Oncogene* **5,** 15 (1990).

[39] S. Rasheed, G. L. Norman, and G. Heidecker, *Science* **221,** 155 (1983).

[40] N. Kamata, R. M. Jotte, and J. T. Holt, *Mol. Cell. Biol.* **11,** 765 (1991).

[41] J. F. Hancock, H. Paterson, and C. J. Marshall, *Cell* **63,** 133 (1990).

[42] B. Mouillac, M. Caron, H. Bonin, M. Dennis, and M. Bouvier, *J. Biol. Chem.* **267,** 21733 (1992).

[43] Y. Reiss, J. L. Goldstein, M. C. Seabra, P. J. Casey, and M. S. Brown, *Cell* **62,** 81 (1990).

[44] Y. Reiss, S. J. Stradley, L. M. Gierasch, M. S. Brown, and J. L. Goldstein, *Proc. Natl. Acad. Sci. U.S.A.* **88,** 732 (1991).

[45] K. Kato, A. D. Cox, M. M. Hisaka, S. M. Graham, J. E. Buss, and C. J. Der, *Proc. Natl. Acad. Sci. U.S.A.* **89,** 6403 (1992).

recognizes CAAX sequences terminating in leucine. However, it has been appreciated that some Ras proteins (e.g., N-Ras, K-Ras4A, and K-Ras4B) can undergo alternative prenylation *in vivo* by GGTaseI when FTase activity is inhibited.[46]

After prenylation the AAX residues are proteolytically cleaved and the carboxy terminus is methylated (Fig. 1).[46] The CAAX tetrapeptide sequence alone is sufficient to signal prenylation and has been used as a foundation for the design of peptidomimetic compounds as cell-permeable inhibitors of protein prenylation.[46] The CAAX tetrapeptide sequence alone is usually not sufficient to target proteins to membranes.[41] Choy and colleagues showed that CAAX-signaled modifications alone cause the targeting and stable association of Ras proteins to endosome membranes.[47] Lysine/arginine-rich sequences (e.g., K-Ras4B) or palmitoylated cysteine residues (e.g., H-Ras) serve as essential second signals positioned immediately upstream of the CAAX sequence to facilitate efficient plasma membrane association.[41,48]

Selection of Target Protein Lipid Modification

The most commonly used lipid modification membrane-targeting sequences are an amino-terminal myristoylation sequence and a carboxy-terminal farnesylation sequence. Presently, no well-defined sequence has been identified for introduction of an internal lipid modification. Therefore, it must be determined if the acceptor protein can best tolerate addition of a new domain at the amino or carboxy terminus without disruption of an essential region or activity of the acceptor protein. The signal sequences described here have been designed to introduce the minimal number of amino acids required for effective lipid modification and membrane binding. This is done in order to limit effects on the structure of the target protein and to limit possible interactions of the introduced amino acid sequences with other proteins. To obtain the most clearly interpretable results if targeting by the lipid has been successful, it is best to use a cytosolic version of the acceptor protein that lacks competing targeting signals of its own.

Verification of Lipid Attachment

Lipid modification of a protein can be analyzed *in vitro* as well as *in vivo*. *In vitro* transcription and translation in reticulocyte lysates or assays

[46] A. D. Cox, L. G. Toussaint, J. J. Fiordalisi, K. Rogers-Graham, and C. J. Der, "Farnesyltransferase and Geranylgeranyltransferase Inhibitors: The Saga Continues" *in* "Farnesyltransferase Inhibitors in Cancer Therapy"(S. M. Sebti and A. D. Hamilton, eds.). Humana Press, Totowa, New Jersey, 2000.
[47] E. Choy, V. K. Chui, J. Silletti, M. Feoktistov, T. Morimoto, D. Michaelson, I. E. Ivanov, and M. R. Philips, *Cell* **98,** 69 (1999).
[48] J. F. Hancock, A. I. Magee, J. E. Childs, and C. J. Marshall, *Cell* **57,** 1167 (1989).

using a more purified acyltransferase can be utilized.[49–51] These experiments are carried out in the presence of a radioactive lipid substrate. The protein is then analyzed for incorporation of radioactivity. Successful *in vitro* modification merely verifies that the selected signal can be recognized by an acyltransferase in the context of the acceptor protein. It does not, however, address whether the chimeric protein is lipid modified or membrane associated within an intact cell. Therefore, it must be verified *in vivo* that the added signaling sequence is functional. Metabolic labeling with [^{3}H]myristate can be used to validate the success of a myristoylation signal sequence. [^{3}H]Mevalonate, which becomes incorporated into farnesyl pyrophosphate within the cell, can be used to verify lipid modification of a CAAX-containing sequence.[22,52,53]

Unexpected failures of certain CAAX sequences to be prenylated *in vivo* and the unanticipated successful prenylation of other motifs derived from nonprenylated proteins demonstrate that we do not yet fully comprehend the mechanisms that control protein prenylation within the complex setting of the living cell.[54]

Detection of Membrane Binding

The construction of a chimeric lipid-modified protein is undertaken because the chimera provides a unique method for studying the effect of membrane targeting on the biological function of the protein within a living cell. To determine the effect of the lipid modification, the subcellular localization of the protein must be determined. A variety of techniques can be used to determine the subcellular location of the lipid-modified protein. These include subcellular fractionation[23,24,41] as well as immunofluorescence.[23,41] It is worth noting that organic solvents used for fixation for immunofluorescence studies may disrupt some protein interactions with membranes, especially if the lipid moiety is the only interaction the protein has with the membrane.[55]

Use of Lipid Modifications on Proteins Involved in Cellular
Signal Transduction

Those who have studied cellular signal transduction have frequently used the technique of targeting proteins to membranes by lipid modification.

[49] L. J. Knoll, D. R. Johnson, M. L. Bryant, and J. I. Gordon, *Methods Enzymol.* **250,** 405 (1995).

[50] A. L. Wilson and W. A. Maltese, *Methods Enzymol.* **250,** 79 (1995).

[51] R. Khosravi-Far and C. J. Der, *Methods Enzymol.* **255,** 46 (1995).

[52] B. M. Willumsen, *Methods Enzymol.* **250,** 269 (1995).

[53] M. E. Linder, C. Kleuss, and S. M. Mumby, *Methods Enzymol.* **250,** 314 (1995).

[54] A. D. Cox, S. M. Graham, P. A. Solski, J. E. Buss, and C. J. Der, *J. Biol. Chem.* **268,** 11548 (1993).

[55] K. H. Muntz, P. C. Sternweis, A. G. Gilman, and S. M. Mumby, *Mol. Biol. Cell* **3,** 49 (1992).

In particular, membrane-targeted versions of regulators and effectors of Ras have been created.[56] Ras itself is targeted to the inner face of the plasma membrane by carboxy-terminal prenylation signal sequences.[46] Proteins that control Ras GDP/GTP cycling or are downstream effectors of active Ras (Ras-GTP) function can be rendered chronically activated by membrane targeting. Thus, membrane-targeted versions of Ras guanine nucleotide exchange factors (GEFs: RasGRF,[57] SOSI,[57,58] and RasGRP[15]) can lead to chronic activation of Ras, while membrane-targeted versions of Ras GTPase-activating proteins (p120 Ras GAP)[59,60] lead to downregulation of Ras function. Because active Ras-GTP associates with downstream effectors promoting their association with the plasma membrane,[2,61] membrane-targeted effector proteins (e.g., Raf-1[6,7] and PI 3-kinase[8–10]) or effector targets (e.g., AKT/PKB[11–13]) have been shown to be constitutively activated variants of these normally cytoplasmic proteins (Fig. 2). In addition, lipid modification sequences have been used to replace the normal membrane-binding domain of a protein in order to assess the importance of membrane localization to the activity of the protein. This has been done to a variety of Dbl family proteins.[14] In the following section, we summarize the results seen with Raf-1, PI 3-kinase, and AKT as well as describe the effect of lipid-modifying sequences on Dbl family proteins.

Membrane Targeting of Raf Serine/Threonine Kinase

It was discovered that the Raf serine/threonine kinase is activated by plasma membrane-bound protein Ras following mitogenic stimulation of cells.[1,2] Because it was known that Ras and Raf interacted and that Raf translocated to the plasma membrane during activation, it was possible that the relocalization of Raf to the plasma membrane was an important step in its activation. Therefore, researchers designed a Raf protein that would be constitutively associated with the plasma membrane rather than being present only in the cytosol.[6,7] To do this, the polylysine and CAAX motif of K-Ras4B were placed at the carboxy terminus of Raf (Raf-CAAX) to direct the prenylation and subsequent plasma membrane localization of Raf. This experiment resulted in the constitutive activation of Raf. This

[56] M. S. Boguski and F. McCormick, *Nature* (*London*) **366,** 643 (1993).
[57] L. A. Quilliam, S. Y. Huff, K. M. Rabun, W. Wei, D. Broek, and C. J. Der, *Proc. Natl. Acad. Sci. U.S.A.* **91,** 8512 (1994).
[58] A. Aronheim, D. Engelberg, N. Li, N. Al-Alawi, J. Schlessinger, and M. Karin, *Cell* **78,** 949 (1994).
[59] G. J. Clark, L. A. Quilliam, M. M. Hisaka, and C. J. Der, *Proc. Natl. Acad. Sci. U.S.A.* **90,** 4887 (1993).
[60] D. C. S. Huang, C. J. Marshall, and J. F. Hancock, *Mol. Cell. Biol.* **13,** 2420 (1993).
[61] A. B. Vojtek and C. J. Der, *J. Biol. Chem.* **273,** 19925 (1998).

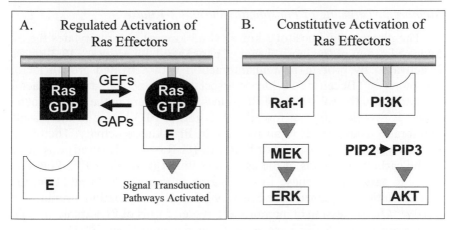

FIG. 2. Effectors of Ras can be activated by membrane targeting. (A) Ras proteins are targeted to the inner surface of the plasma membrane by their carboxy-terminal CAAX farnesylation signal sequence. Ras proteins function as GDP/GTP-regulated switches that cycle between inactive GDP-bound and active GTP-bound states. Guanine nucleotide exchange factors (GEFs) stimulate GDP/GTP exchange and are activators of Ras. GTPase-activating proteins (GAPs) act as negative regulators of Ras by stimulating the hydrolysis of bound GTP, promoting the formation of Ras-GDP. Activated Ras-GTP preferentially binds a spectrum of downstream effector targets (E), including Raf-1 and PI 3-kinase. The regulated association of effectors with Ras promotes translocation of the effectors to the plasma membrane, where additional membrane-associated events occur that lead to their activation and the subsequent activation of downstream signaling pathways. (B) Membrane-targeted forms of Raf-1 and PI 3-kinase function as constitutively activated proteins. Constitutively activated Raf-1 activates the MEK/ERK mitogen-activated protein kinase cascade. Constitutively activated PI 3-kinase promotes the conversion of phosphatidylinositol 4,5-bisphosphate (PIP_2) to phosphatidylinositol 3,4,5-trisphosphate (PIP_3) and PIP_3 promotes the activation of AKT, a downstream effector of PI 3-kinase. Activated growth factor receptors also bring signaling molecules to the plasma membrane, resulting in the activation of downstream signaling pathways. Therefore, selective targeting of signaling proteins to the plasma membrane results in the constitutive activation of specific cell signaling pathways. This activation occurs in the absence of other signals elicited by activated Ras or growth factor stimulation of the cell.

result suggested that the relocalization of Raf from the cytosol to the plasma membrane was sufficient to lead to its activation. While the full mechanism of Raf activation is still not understood, it is clear that membrane localization plays an important step. It is possible that a membrane-bound protein or lipid is involved in the activation of Raf. The Raf-CAAX protein has been used extensively to specifically activate Raf-dependent signal transduction pathways. The experimental design of targeting Raf to the plasma membrane played an important role in enhancing our knowledge of the mechanism of Raf activation.

Membrane Targeting of Phosphatidylinositol 3-Kinase

The phosphatidylinositol 3-kinase (PI 3-kinase) phosphorylates the 3'-hydroxyl group of the inositol ring of phosphatidylinositides.[62] PI 3-kinase is activated by protein–protein interaction with activated growth factor receptors.[63–65] Because PI 3-kinase associates with activated growth factor receptors and the substrates of PI 3-kinase are present in the plasma membrane, it was thought that forced expression of PI 3-kinase in the plasma membrane would result in an increase in PI 3-kinase activity. Therefore, the prenylation CAAX motif of K-Ras4B (H-Ras was also used) was fused to the carboxy terminus of the catalytic subunit (p110α) of PI 3-kinase in order to target PI 3-kinase to the plasma membrane.[8,9] In addition, the myristoylation sequence of c-Src was placed at the amino terminus of p110γ.[66] On expression of membrane-targeted forms of PI 3-kinase in cells, a large increase in PI 3-kinase activity was detected *in vivo*.[8,9] These membrane-targeted forms of PI 3-kinase led to activation of a variety of signal transduction pathways that are known to be activated in a PI 3-kinase-dependent manner.[8,9] These membrane-targeted PI 3-kinase proteins have been widely used by researchers in order to mimic the effects of PI 3-kinase activation in the absence of growth factor stimulation.

Membrane Targeting of AKT Serine/Threonine Kinase

The serine/threonine kinase AKT was originally identified as the product of a viral oncogene.[67] This protein, vAKT, was shown to be myristoylated and associated with the plasma membrane while normal AKT was not.[68] An important reagent used in investigating the function of AKT has been a membrane-targeted version of the kinase.[11–13] AKT was shown to be activated when targeted to the plasma membrane by fusing the myristoylation sequence of c-Src to its amino terminus.[11–13] AKT has been shown to

[62] S. J. Leevers, B. Vanhaesebroeck, and M. D. Waterfield, *Curr. Opin. Cell. Biol.* **11,** 219 (1999).

[63] A. Kazlauskas and J. A. Cooper, *EMBO J.* **9,** 3279 (1990).

[64] J. A. Escobedo, D. R. Kaplan, W. M. Kavanaugh, C. W. Turck, and L. T. Williams, *Mol. Cell Biol.* **11,** 1125 (1991).

[65] M. Otsu, I. Hiles, I. Gout, M. J. Fry, F. Ruiz-Larrea, G. Panayotou, A. Thompson, R. Dhand, J. Hsuan, and N. Totty, *Cell* **65,** 91 (1991).

[66] M. Lopez-Ilasaca, P. Crespo, P. G. Pellici, J. S. Gutkind, and R. Wetzker, *Science* **275,** 394 (1997).

[67] A. Bellacosa, J. R. Testa, S. P. Staal, and P. N. Tsichlis, *Science* **254,** 274 (1991).

[68] N. N. Ahmed, T. F. Franke, A. Bellacosa, K. Datta, M. E. Gonzalez-Portal, T. Taguchi, J. R. Testa, and P. N. Tsichlis, *Oncogene* **8,** 1957 (1993).

elicit an antiapoptotic or anticell death signal.[12,69–73] This membrane-bound form of AKT was shown to have elevated kinase activity and elicited a strong antiapoptotic signal.[11,12,74] It has also been shown that AKT is activated by interacting with lipids in the plasma membrane.[75–77] This may be one reason why membrane-targeted AKT, as well as vAKT, are more active proteins than the normal untargeted AKT. The myristoylated AKT protein has been used to specifically mimic the effects of activated AKT, particularly in the study of antiapoptotic effects of the kinase.

Replacement of Plasma Membrane Localization Domains of Dbl Family Proteins with Heterologous Membrane-Binding Sequences

Lipid modifications have also been used to replace the regular membrane- or lipid-binding domain [e.g., pleckstrin homology (PH) or cysteine-rich domain] of a protein in order to further characterize the role of membrane or lipid binding in the function of that protein. For example, mutation of the normal plasma membrane-binding sequence of a protein may alter a specific function of that protein. The attachment of a heterologous membrane-targeting sequence to relocate the protein back to the plasma membrane can be used to verify the importance of membrane localization to that function. Dbl family proteins are GEFs for members of the Rho family of Ras-related proteins.[14] All Dbl family proteins possess tandem Dbl homology (DH) and PH domains (Fig. 3). While the GEF function of Dbl family proteins is catalyzed by the DH domain, the PH domain is believed to regulate DH domain function. Many Dbl family proteins are transforming and deletion of the PH domain causes a loss of transforming activity (Fig. 3).[14] The addition of the plasma membrane-targeting sequence of Ras onto PH domain-deficient mutants restores the

[69] J. Downward, *Curr. Opin. Cell Biol.* **10,** 262 (1998).
[70] E. M. Eves, W. Xiong, A. Bellacosa, S. G. Kennedy, P. N. Tsichlis, M. R. Rosner, and N. Hay, *Mol. Cell Biol.* **18,** 2143 (1998).
[71] H. Dudek, S. R. Datta, T. F. Franke, M. J. Birnbaum, R. Yao, G. M. Cooper, R. A. Segal, D. R. Kaplan, and M. E. Greenberg, *Science* **275,** 661 (1997).
[72] A. Kauffmann-Zeh, P. Rodriguez-Viciana, E. Ulrich, C. Gilbert, P. Coffer, J. Downward, and G. Evan, *Nature (London)* **385,** 544 (1997).
[73] S. R. Datta, H. Dudek, X. Tao, S. Masters, H. Fu, Y. Gotoh, and M. E. Greenberg, *Cell* **91,** 231 (1997).
[74] A. Bellacosa, T. O. Chan, N. N. Ahmed, K. Datta, S. Malstrom, D. Stokoe, F. McCormick, J. Feng, and P. Tsichlis, *Oncogene* **17,** 313 (1998).
[75] A. Klippel, W. M. Kavanaugh, D. Pot, and L. T. Williams, *Mol. Cell. Biol.* **17,** 338 (1997).
[76] T. F. Franke, D. R. Kaplan, L. C. Cantley, and A. Toker, *Science* **275,** 665 (1997).
[77] D. Stokoe, L. R. Stephens, T. Copeland, P. R. J. Gaffney, C. B. Reese, G. F. Painter, A. B. Holmes, F. McCormick, and P. T. Hawkins, *Science* **277,** 567 (1997).

Dbl Protein   **Transformation**

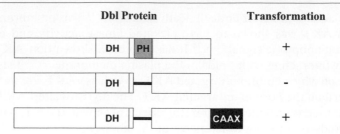

FIG. 3. The PH domain of Dbl family proteins functions as a membrane-targeting domain and can be substituted by other membrane-targeting sequences. Dbl family proteins function as GEFs for members of the Rho family of Ras-related proteins. The GEF activity of Dbl family proteins is catalyzed by the DH domain. The PH domain is believed to regulate DH domain function. While many Dbl family proteins are transforming, deletion of the PH domain results in a loss of transforming activity. The plasma membrane-targeting sequence of Ras (CAAX) restores the transforming activity of some PH domain-deficient mutant Dbl family proteins. Thus, experimentally targeting these PH domain-deficient Dbl proteins back to the plasma membrane confirms a role in plasma membrane localization for the PH domain.

transforming activity of at least some Dbl family proteins (Fig. 3). This supports a membrane-targeting role for the PH domain. This membrane targeting is thought to facilitate the interaction of the DH domain with its membrane-associated GTPase substrates. RasGRP,[15] Lfc,[16] and Vav[78] are examples of such studies.

Methods

Attachment of Amino-Terminal Myristoylation Signal

The addition of an amino-terminal myristoylation signal can be done in a number of ways. The polymerase chain reaction (PCR) can be used to fuse coding sequences of a myristoylation sequence to that of the protein to be modified. However, because it is often desirable to modify more than one protein, a reusable vector containing the coding sequence for an amino-terminal myristoylation site can be created. In this approach, any gene can be cloned into this vector in frame with the myristoylation signal sequence. The gene, now fused to the coding sequence for the myristoylation signal, can then be excised and placed in an appropriate vector for expression of the chimeric protein.

[78] K. Abe, I. P. Whitehead, J. P. O'Bryan, and C. J. Der, *J. Biol. Chem.* **274**(43), 30410 (1999).

In preparation for making the myristoylation cassette (Fig. 4), two complementary oligonucleotides, which encode the signal for myristoylation (in this illustration, the first eight amino acids of the Rasheed sarcoma virus Gag protein[28]), are designed and synthesized. Other myristoylation signal sequences can also be used.[35] The oligonucleotides are annealed and cloned into the pBluescript vector (Stratagene, La Jolla, CA). This ligation is facilitated by the synthesis of *Eco*RI overhangs at the ends of the oligonucleotides. In addition, an *Nde*I restriction site is included just downstream of the myristoylation signal-encoding nucleotides in order to provide a cloning site for acceptor target genes. Site-directed mutagenesis of the ATG start codon of the cDNA of the target protein into an *Nde*I restriction site allows the cDNA to be properly cloned in frame downstream of the myristoylation signal. Site-directed mutagenesis kits are commercially avail-

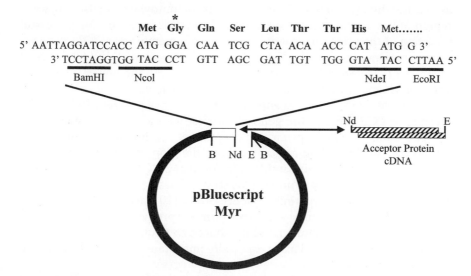

FIG. 4. An example of a vector designed to attach an amino-terminal myristoylation signal to acceptor proteins. The nucleotide sequences of the complementary oligonucleotides are shown annealed and cloned into pBluescript (Stratagene) via the *Eco*RI site of the vector. These oligonucleotides encode a sequence similar to the amino-terminal myristoylation signal from the Rasheed rat sarcoma virus Gag protein. An asterisk marks the glycine that is the site of myristate attachment. The 5′ *Eco*RI site is destroyed on ligation of the annealed oligonucleotides with the vector. *Bam*HI, *Nco*I, *Nde*I, and *Eco*RI sites are underlined. After mutating the start codon of the cDNA of the target protein into an *Nde*I site, the cDNA can be cloned into this pBluescript-Myr vector as an *Nde*I–*Eco*RI fragment. Abbreviations in the vector: B, *Bam*HI; Nd, *Nde*I; E, *Eco*RI. The cDNA, now attached to the myristoylation-encoding sequence, can be subcloned as a *Bam*HI fragment. The specific restriction sites and vector used can be altered to suit the needs of the investigator.

able; however, any method of mutagenesis is suitable. An *Nco*I site can be included in designing the oligonucleotides to assist in confirmation of cDNA ligation. A *Bam*HI site is also included in the oligonucleotide so that the final chimeric gene can be excised and easily subcloned as a *Bam*HI fragment. The exact myristoylation signal used and the restriction endonuclease sites that are incorporated into the oligonucleotides can be altered to suit the needs of the investigator.

Attachment of Carboxy-Terminal CAAX Signal Sequences for Prenylation by Two-Step Polymerase Chain Reaction

Carboxy-terminal sequences from both H-Ras and K-Ras4B have been successfully used to target proteins to the plasma membrane.[6–9,15,16,48,57,59,60,79] Although certain tetrapeptide Cys-XXX motifs may be sufficient to act as a signal for prenylation, an additional signal (such as the nearby cysteines of H-Ras that are palmitoylated or the adjacent region of basic amino acid residues of K-Ras4B) is necessary for successful targeting of the acceptor protein to the plasma membrane.[29,80] Because we have experience in using the K-Ras4B targeting sequence[15,57,59] (amino acid residues 171 to 188 of K-Ras4B, which includes both the polybasic domain and CAAX motif), it is used in the examples below.

The most direct and rapid approach to generate a protein with a CAAX motif is to PCR amplify the gene of interest with a 5' primer specific for the gene and a long 3' primer that contains the last approximately 18 nucleotides of the coding region (without the stop codon) of the gene of interest, the 3'-most 54 nucleotides (18 codons) of K-Ras4B, a stop codon, and a restriction site for cloning. This strategy has been used to attach the K-Ras4B membrane-targeting sequence onto p120 RasGAP.[59] While this is a rapid method for attaching a carboxy-terminal prenylation signal, multiple PCR products must be cloned and sequenced to ensure that no errors were created during the PCR. This is especially important and tedious when the gene of interest is large.

An alternative strategy that has been used to membrane-target several proteins is four-primer or two-step PCR. This procedure involves two rounds of PCR amplification. The products of the first amplification are used as templates for the second amplification, as outlined in Fig. 5. The example given is for the addition of the targeting signals from K-Ras4B to the GRB-2 adaptor protein.[81] This same strategy has also been successfully used to generate CDC25-CAAX and SOS1-CAAX chimeras.[57]

[79] L. M. L. Chow, M. Fournel, D. Davidson, and A. Velliette, *Nature (London)* **365,** 156 (1993).
[80] J. F. Hancock, K. Cadwallader, H. Paterson, and C. J. Marshall, *EMBO J.* **10,** 4033 (1991).
[81] L. A. Quilliam, unpublished data (1994).

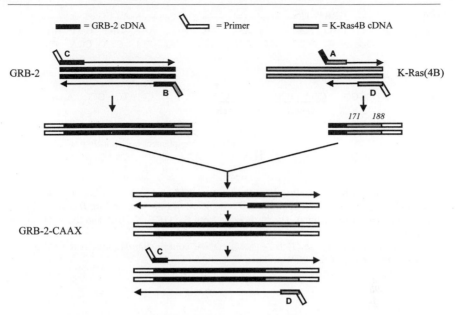

Fig. 5. A schematic representation of the attachment of a prenylation signal by two-step PCR. The PCR was performed with primers C and B (see Fig. 6) to generate a full-length GRB-2 cDNA fused to nucleotides encoding amino acids 171 to 176 of the K-Ras4B cDNA. A second PCR was performed with primers A and D (see Fig. 6) to generate a DNA fragment that contains 3' nucleotides of GRB-2 fused upstream of the prenylation-encoding nucleotides of K-Ras4B (encoding amino acids 171 to 188). These PCR products were then used as templates in a third PCR with primers C and D to effectively join the two products. In the first round of PCR the templates anneal and are extended by the polymerase. This DNA is then amplified by the remaining PCR cycles. This results in the synthesis of a GRB-2 cDNA that contains the prenylation signal (CAAX)-encoding nucleotides of K-Ras4B. This cDNA encodes the entire GRB-2 protein with the addition of amino acids 171 to 188 of K-Ras4B at the carboxy terminus. These amino acids of K-Ras4B contain the polybasic region as well as the CAAX motif.

Two complementary oligonucleotides (primers A and B) are generated (Fig. 6) containing 17 bases encoding the last 6 codons of GRB-2 in frame with the nucleotides encoding codons 171–176 of K-Ras4B. Primer C contains a *Bam*HI restriction site preceded by four bases, enabling cleavage of the PCR product with *Bam*HI (Fig. 6). This restriction site is placed just upstream of a Kozak sequence, providing for efficient translation in mammalian cells, and the first six codons of GRB-2. There is an *Nco*I restriction site within this Kozak sequence. Primer D contains a *Bam*HI

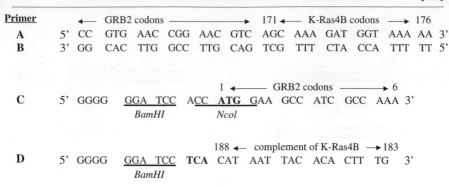

Fig. 6. The PCR primers used for the construction of GRB-2-CAAX are shown. The nucleotide sequences of the oligonucleotides depicted in Fig. 5 are shown. *Bam*HI and *Nco*I restriction sites are underlined. The start codon from the GRB-2 cDNA and the STOP codon after the K-Ras4B nucleotides are shown in bold face. Primers C and D, which are used in the final PCR to amplify the GRB-2-CAAX cDNA, contain *Bam*HI sites at their 5' ends to facilitate subcloning of the PCR product.

restriction site, a termination codon, and sequence complementary to K-Ras4B codons 183–188 (Fig. 6).

A PCR containing the complementary strand chimeric primer B and primer C with GRB-2 cDNA as template generates GRB-2 with a K-Ras4B (codon 171–177) extension (Fig. 5). Similarly, a fusion of K-Ras4B carboxy-terminal sequences containing codons 171–188 with upstream GRB-2 sequences is generated by PCR, using primers A and D with K-Ras4B cDNA as template (Fig. 5).

These PCR products are then used as templates for a second round of PCR. The first extension reaction does not use the oligonucleotide primers, but relies on the annealing of the two complementary templates, which function as primers themselves. This generates template for further amplification by primers C and D, as depicted in Fig. 5. The final PCR product contains the full GRB-2-coding sequence fused in frame at its carboxy terminus with the last 18 codons of K-Ras4B. These codons encode the polybasic region and CAAX motif (see Fig. 7) required for membrane targeting via lipid modification. PCR products are digested with *Bam*HI and subcloned as a *Bam*HI fragment into a mammalian expression vector. Again, multiple PCRs should be performed and sequenced to ensure that no errors were generated during the PCR amplification. While these PCR methods are rapid, they may not necessarily be the best for large genes or if multiple proteins are to be targeted by the same lipid modification sequences.

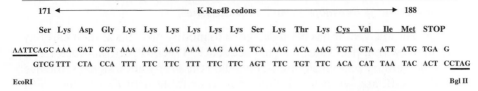

171 ◄———————————————— K-Ras4B codons ————————————► 188

Ser Lys Asp Gly Lys Lys Lys Lys Lys Lys Ser Lys Thr Lys <u>Cys Val Ile Met</u> STOP

AATTCAGC AAA GAT GGT AAA AAG AAG AAA AAG AAG TCA AAG ACA AAG TGT GTA ATT ATG TGA G
    GTCG TTT CTA CCA TTT TTC TTC TTT TTC TTC AGT TTC TGT TTC ACA CAT TAA TAC ACT CCTAG

EcoRI   Bgl II

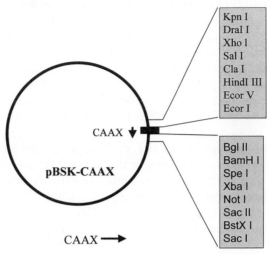

FIG. 7. The construction of a reusable prenylation vector. *Top:* Sequences of the complementary oligonucleotides synthesized to encode the K-Ras4B prenylation site. The amino acids of K-Ras4B are indicated with the CAAX motif underlined. The oligonucleotides are designed such that when they are annealed the 5′ end has an overhang that is compatible with *Eco*RI and the 3′ end has an overhang that is compatible with *Bgl*II (underlined). *Bottom:* The annealed oligonucleotides were ligated into pBSK (Stratagene) that has been digested with *Eco*RI and *Bgl*II, creating pBSK-CAAX. The full prenylation sequence (the polybasic region and the CAAX motif) of K-Ras4B is indicated by CAAX on the map of the vector. After mutating the STOP codon of the cDNA for the target protein to be prenylated into an *Eco*RI site, the cDNA can be cloned into the *Eco*RI site of pBSK-CAAX. The *Eco*RI site that is introduced at the STOP codon must result in the codons of the cDNA being in frame with the codons of the prenylation sequence. In this example, the codons of the cDNA must be in frame with GAA of the *Eco*RI site. Once cloned into pBSK-CAAX, the new cDNA fused to the prenylation-encoding sequence can be subcloned into an appropriate expression vector. The exact restriction sites and vectors that are used can be altered to suit the needs of the investigator.

Construction of Reusable Prenylation Cassette

A reusable vector has been designed to modify proteins at their carboxy terminus with a prenylation signal sequence. To construct this vector two complementary oligonucleotides, encoding the prenylation signal sequence of K-Ras4B, are synthesized with ends compatible with *Eco*RI and *Bgl*II restriction sites (Fig. 7). These oligonucleotides are annealed and cloned into pBSK (Stratagene) (Fig. 7). The oligonucleotides are designed such that the *Eco*RI site just 5' of the CAAX-encoding prenylation sequence can be used as a cloning site for the cDNA of the target protein to be prenylated. Because the CAAX prenylation signal sequence is at the carboxy terminus of proteins, the STOP codon of the target protein cDNA must be destroyed in order to allow translation through the attached prenylation site. Utilizing site-directed mutagenesis, the cDNA sequence of the target protein must be altered to change the STOP codon to an *Eco*RI site or to introduce an *Eco*RI site just 5' of the STOP codon. Special attention must be paid to ensure that the *Eco*RI site introduced will allow the target gene to be cloned in frame with the *Eco*RI site at the 5' end of the prenylation signal-encoding sequence within the vector.

As an example, we have placed a CAAX prenylation sequence of K-Ras4B onto the carboxy terminus of the breakpoint cluster region gene (BCR).[82,83] The STOP codon of BCR was mutated to an *Eco*RI site such that the last codon of BCR was placed in frame with the CAAX-encoding sequence of the pBSK-CAAX vector. Because there is an *Eco*RI site at the 5' end of the BCR cDNA this altered BCR cDNA was cloned into the pBSK-CAAX vector as an *Eco*RI fragment and oriented by restriction analysis. The BCR-CAAX cDNA was then cloned into the pCMV5 mammalian expression vector.

The idea of attaching the BCR protein with a prenyl group was to relocalize BCR to the plasma membrane to study the effect of subcellular localization on the function and activity of BCR. To determine if the CAAX sequence on BCR was functional, the pCMV5-BCR[84] and pCMV5-BCR-CAAX expression plasmids were transiently transfected into 293T cells. The cells were subjected to a crude fractionation procedure in which the proteins of the cytosol are separated from proteins that are associated with membranes. Cells are resuspended in two cell pellet volumes of 20 mM Tris (pH 8.0)–140 mM NaCl for 1 hr on ice. After homogenization, sucrose

[82] N. Heisterkamp, K. Stam, J. Groffen, A. de Klein, and G. Grosveld, *Nature* (*London*) **315,** 758 (1985).
[83] Y. Maru and O. N. Witte, *Cell* **67,** 459 (1991).
[84] G. W. Reuther, H. Fu, L. D. Cripe, R. J. Collier, and A. M. Pendergast, *Science* **266,** 129 (1994).

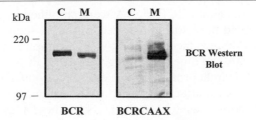

FIG. 8. The BCR protein is targeted to membranes by addition of a K-Ras4B farnesylation signal sequence. A carboxy-terminal farnesylation signal sequence was added to the BCR protein by using the pBSK-CAAX vector as described in text and in Fig. 7. Using the pCMV5 expression vector, the cDNAs for BCR and BCR-CAAX were transiently transfected into 293T cells. Lysates of these cells were subjected to a crude subcellular fractionation that separates cytosolic and membrane components. Fractions were analyzed by Western blotting with a monoclonal antibody that recognizes BCR, followed by detection with enhanced chemiluminescence (Amersham). The cytosolic (C) and membrane (M) fractions are indicated. Molecular weight markers are indicated in kilodaltons.

is added to 250 mM, EDTA is added to 1 mM phenylmethylsulfonyl fluoride (PMSF) is added to 1 mM, and leupeptin is added to 25 μg/ml. Samples are centrifuged at 1000g for 10 min at 4°. The supernatant is removed and the pellet is resuspended in the same buffer and centrifuged again. This supernatant is combined with the previous supernatant and a small sample is saved as a total protein sample. The supernatant is then spun at 100,000g for 2 hr at 4°. The resulting supernatant is removed and saved as the cytosolic protein sample. The pellet from the high-speed centrifugation is resuspended in the same buffer supplemented with 0.5% (v/v) Triton X-100 and 0.5% (w/v) sodium deoxycholate and retained as the membrane fraction. After determining protein concentrations, equal amounts of protein from the cytosolic and membrane extracts were analyzed by Western blotting with a monoclonal antibody that recognizes BCR. Figure 8 shows that relative to the amount of BCR in the cytosol, there was much more BCR-CAAX protein associated with membranes than there was normal untargeted BCR associated with membranes. In fact, in this experiment nearly all of the expressed BCR-CAAX was associated with membranes. This indicates that the targeting of BCR to cellular membranes by attaching a carboxy-terminal prenylation sequence was successful.

As described earlier, it is also possible to attach a carboxy-terminal prenylation site to a target sequence by PCR. However, in the example described here for BCR, the use of PCR would be more time consuming. BCR is a large protein (160 kDa) and its cDNA is about 4.5 kilobases.[83]

If at some point the entire BCR cDNA were amplified by PCR, the entire PCR product would have to be analyzed by DNA sequencing to ensure that no errors were introduced during the PCR amplification. This would likely be more costly and time consuming. However, PCR may be amenable for attaching small target proteins with a carboxy-terminal prenylation site. PCR may also work for attaching prenylation sequences to long cDNAs if convenient restriction sites exist in the cDNA. These sites may be used to replace the 3' end of the cDNA with a PCR product containing a fusion with prenylation-encoding sequences. In addition, the pBSK-CAAX vector described here, or a similar vector designed to suit more specifically the needs of the investigator, can be used to attach prenylation signal sequences to multiple target proteins.

Summary

It is now established that the function of many signaling molecules is controlled, in part, by regulation of subcellular localization. For example, the dynamic recruitment of normally cytosolic proteins to the plasma membrane, by activated Ras or activated receptor tyrosine kinases, facilitates their interaction with other membrane-associated components that participate in their full activation (e.g., Raf-1). Therefore, the creation of chimeric proteins that contain lipid-modified signaling sequences that direct membrane localization allows the generation of constitutively activated variants of such proteins. The amino-terminal myristoylation signal sequence of Src family proteins and the carboxy-terminal prenylation signal sequence of Ras proteins have been widely used to achieve this goal. Such membrane-targeted variants have proved to be valuable reagents in the study of the biochemical and biological properties of many signaling molecules.

Acknowledgments

We thank Jennifer Parrish for preparation of figures. G.W.R. was supported by a Cancer Research Institute/Merrill Lynch Fellowship. Our research studies were supported by NIH Grants to L.A.Q. (CA63139), G.J.C. (CA72644), and C.J.D. (CA42978, CA55008, and CA63071) and by the Roy J. Carver Charitable Trust (J.E.B.) and the American Cancer Society (RPG97-008-01-BE) (L.A.Q.).

[27] Glycerolphosphoinositide Anchors for Membrane-Tethering Proteins

By John D. Fayen, Mark L. Tykocinski, and M. Edward Medof

Introduction

The surface of eukaryotic cells is composed of a lipid bilayer that is decorated with a wide variety of proteins. Most of these proteins are anchored in the plasma membrane by means of 18–22 sequential hydrophobic amino acid residues. Conventional type I (C-terminal anchored) and type II (N-terminal anchored) transmembrane proteins are attached by a single hydrophobic domain. For certain proteins (e.g., tetraspans, G proteins, ATP-binding complex proteins), multiple internal noncontiguous stretches of hydrophobic amino acids are present that cause the proteins to span the cell membrane several times.

In contrast, some proteins are anchored by lipid moieties added as a posttranslational modification. On the inner leaflet of the plasma membrane there are lipoprotein families that are anchored via fatty acid(s). A prominent example of this class of proteins is the Src family tyrosine kinases ($p56^{lck}$, Fyn), which are myristoylated[1–4] (and when activated, palmitoylated[5,6]) at their N termini, allowing for their expression in a cytoplasmic direction. Members of another family of lipid-containing proteins that, conversely, are attached to the outer leaflet of the membrane bilayer bear glycosylinositol phospholipid (GPI) anchors at their C termini. A schematic diagram outlining the essential components of the GPI anchor is presented in Fig. 1 (see Fig. 1 on page 356). This GPI modification provides for the extracellular expression of these proteins in a fashion similar to conventional type I transmembrane proteins.

In spite of the fact that they share the same mechanism of C-terminal membrane attachment, GPI-anchored proteins consist of a diverse group of surface molecules belonging to many different protein families. Among GPI-anchored proteins are immunoglobulin supergene family members [CD90/Thy-1, CD16/FcγRIII, CD58/LFA-3, CD66E, carcinoembryonic

[1] M. D. Resh, *Oncogene* **5,** 1437 (1990).
[2] A. I. Magee, L. Gutierrez, C. J. Marshall, and J. F. Hancock, *J. Cell Sci.* **11S,** 149 (1989).
[3] J. E. Buss, M. P. Kamps, and B. M. Sefton, *Mol. Cell. Biol.* **4,** 2697 (1984).
[4] F. R. Cross, E. A. Garber, D. Pellman, and H. Hanafusa, *Mol. Cell. Biol.* **4,** 1834 (1985).
[5] A. M. Shenoy-Scaria, L. K. T. Gauen, J. Kwong, A. S. Shaw, and D. M. Lublin, *Mol. Cell. Biol.* **13,** 6385 (1993).
[6] J. Kwong and D. M. Lublin, *Biochem. Biophys. Res. Commun.* **207,** 868 (1995).

antigen (CEA), CD56/neural cellular adhesion molecule (NCAM)], complement control supergene family members [CD55/decay accelerating factor (DAF)], Ly-6 supergene family members [CD59/membrane inhibitor of reactive lysis (MIRL), Ly-6C/CD87], tumor necrosis factor receptor family members, and histocompatibility proteins (Qa-2). These proteins serve numerous functions including downregulation of the complement activation on self cell surfaces (CD55/DAF, CD59/MIRL), promotion of intercellular adhesion (CD58/LFA-3, CD56/NCAM, CD66E/CEA), modification of DNA (5'-nucleotidase/CD73), and binding of specific ligands, in the latter case functioning either as an isolated unit (folate receptor) or as part of a multicomponent receptor complex [CD14/lipopolysaccharide (LPS)-binding receptor, and ciliary neurotrophic factor receptor (CNTFR)]. With the exception of greater lateral mobility in the membrane,[7,8] anchoring by a GPI unit rather than a transmembrane polypeptide according to all data available so far does not affect the extracellular activities of these proteins. In contrast, the GPI unit may be important in the intracellular functions of some of these proteins.

The property of GPI-anchored proteins that endows them with the ability to mediate extracellular functions, despite their restricted localization on the outer leaflet of the plasma membrane, is that they are (at least in part) nonrandomly associated in the membrane in microdomains.[9–11] Within these domains, they are organized in cholesterol/sphingolipid-rich areas termed "rafts," which are a site of accumulation of G proteins, Src family tyrosine kinases, and other specialized proteins.[12–16] In some cell types including muscle and fat cells, these rafts are part of caveolae, non-clathrin-coated plasma membrane invaginations.[17,18] In addition to this specialized organization, GPI anchors confer other intracellular properties. In

[7] F. Zhang, B. Crise, B. Su, Y. Hou, J. K. Rose, A. Bothwell, and K. Jacobson, *J. Cell Biol.* **115,** 75 (1991).

[8] P. Y. Chan, M. B. Lawrence, M. L. Dustin, L. M. Ferguson, D. E. Golan, and T. A. Springer, *J. Cell Biol.* **155,** 245 (1991).

[9] R. Varma and S. Mayor, *Nature (London)* **394,** 798 (1998).

[10] D. A. Brown and J. K. Rose, *Cell* **68,** 533 (1991).

[11] T. Friedrichson and T. V. Kurzchalia, *Nature (London)* **394,** 802 (1998).

[12] I. Stefanova, V. Horejsi, I. J. Ansotegui, W. Knapp, and H. Stockinger, *Science* **254,** 1016 (1991).

[13] T. Cinek and V. Horejsi, *J. Immunol.* **149,** 2262 (1992).

[14] A. M. Shenoy-Scaria, J. Kwong, T. Fujita, M. W. Olszowy, A. S. Shaw, and D. M. Lublin, *J. Immunol.* **149,** 3535 (1992).

[15] G. Arreaza, K. A. Melkonian, M. LaFevre-Bernt, and D. A. Brown, *J. Biol. Chem.* **269,** 19123 (1994).

[16] I. Parolini, M. Sargiacomo, M. P. Lisanti, and C. Peschle, *Blood* **87,** 3783 (1996).

[17] A. Schlegel, D. Volonte, J. A. Engelman, F. Galbiati, P. Mehta, X. L. Zhang, P. E. Scherer, and M. P. Lisanti, *Cell. Signal.* **10,** 457 (1998).

[18] R. G. Anderson, *Annu. Rev. Biochem.* **67,** 199 (1998).

polarized epithelial cells they function as signals for apical expression,[10,19–22] and in all cell types they cause recycling at the cell surface via a pathway different from that of conventional membrane proteins.[23,24]

Finally, a unique property of GPI-anchored proteins is that when purified, these proteins are able to spontaneously reincorporate into cell surface membranes and, once incorporated, exert their native physiological functions.[25] Such transfer of GPI-anchored proteins into cells has been demonstrated in a number of experimental situations. It is mediated by the lipid groups in the phospholipid moiety of GPI structures. The exploitation of this latter property of GPI-linked proteins has great practical value in that it can be used for modifying cells via a process termed "protein engineering" or "cell surface painting," an alternative to engineering at the DNA level that can offer a number of advantages.[26]

Given (1) the wide range of functional activities of proteins bearing GPI anchors and (2) the absence of effects on the extracellular molecular activities of these proteins, engineering of proteins bearing C-terminal GPI anchors in place of their normal transmembrane and intracellular sequences can be capitalized on as an important tool for investigating protein function. Not only does reincorporation of GPI-anchored proteins enable modification of cell surface phenotype independent of gene transfection techniques, it has the capacity to provide new insights not readily obtainable by traditional gene transfection methods. In this review, we summarize methodologies that are available for preparing engineered GPI-modified proteins and highlight some of the ways in which these proteins have been used experimentally to date.

Engineering Chimeric Glycosylinositol Phospholipid-Modified Fusion Proteins

Signals Governing Glycosylinositol Phospholipid Modification

The preparation of chimeric cDNAs encoding nascent proteins directing GPI anchor addition requires a knowledge of the signals that trigger the

[19] M. P. Lisanti, M. Sargiacomo, L. Graeve, A. R. Saltiel, and E. Rodriguez-Boulan, *Proc. Natl. Acad. Sci. U.S.A.* **85**, 9557 (1988).

[20] M. P. Lisanti, E. W. Caras, M. A. Davitz, and E. Rodriguez-Boulan, *J. Cell Biol.* **109**, 2145 (1989).

[21] M. P. Lisanti, I. W. Caras, T. Gilbert, D. Hanzel, and E. Rodriguez-Boulan, *Proc. Natl. Acad. Sci. U.S.A.* **87**, 7419 (1990).

[22] D. A. Brown, B. Crise, and J. K. Rose, *Science* **245**, 1499 (1989).

[23] F. R. Maxfield and S. Mayor, *in* "ADP-Ribosylation in Animal Tissue" (F. Haag and F. Koch-Nolte, eds.), p. 355. Plenum Press, New York, 1997.

[24] G. A. Keller, M. W. Siegel, and I. W. Caras, *EMBO J.* **11**, 863 (1992).

[25] M. E. Medof, T. Kinoshita, and V. Nussenzweig, *J. Exp. Med.* **160**, 1558 (1984).

[26] M. E. Medof, S. Nagarajan, and M. L. Tykocinski, *FASEB J.* **10**, 574 (1996).

GPI modification. These signals direct the transfer of a GPI donor unit that is preassembled in the lumen of the endoplasmic reticulum (ER) to an acceptor amino acid near the C terminus of the nascent polypeptide (see below).[27-30] There are two essential components of the signal. The first is a stretch of hydrophobic amino acid residues distal to the site of attachment of the anchor. These residues temporarily tether the N-terminal processed nascent polypeptide to the lumenal side of ER prior to the GPI anchor substitution reaction. They are cleaved concomitant with anchor attachment. Although many different kinds of hydrophobic sequences have been shown to function in this process, some data suggest that this domain possesses intimal structural features that significantly influence the efficiency of GPI addition.[30a] The second component of the GPI modification signal is the cleavage/anchor attachment site itself (designated the ω site), which is located usually 10 to 12 residues upstream of the hydrophobic region. In mammalian cells, the ω site residue must be one of the following six small amino acids: G, A, S, C, D, or N.[27,31-34] Specific rules also dictate the presence of a restricted set of residues at $\omega + 1$ and $\omega + 2$ positions immediately downstream of the ω cleavage/GPI attachment site.[32,34] The rules that must be followed at the $\omega + 2$ site roughly parallel those at the ω site, whereas those at $\omega + 1$ are less stringent. The preceding rules controlling peptide cleavage at the C-terminal site in some ways mirror those governing peptide cleavage of the N-terminal cleavage site.[32] A final directive governing GPI anchor attachment is that (with some exceptions) it does not occur if a hydrophilic region is present distal to the cleavable membrane-anchoring C terminus.[27,35,36]

On the basis of these rules, GPI modification of a selected polypeptide can be conferred by positioning the cDNA sequence encoding that polypeptide upstream from (and in frame with) the 3′ mRNA end sequence of any GPI-anchored protein. As mentioned above, not all GPI modification

[27] G. A. M. Cross, *Annu. Rev. Cell Biol.* **6,** 1 (1990).

[28] M. G. Low, *FASEB J.* **3,** 1600 (1989).

[29] M. A. J. Ferguson and A. F. Williams, *Annu. Rev. Biochem.* **57,** 285 (1988).

[30] M. G. Low and A. R. Saltiel, *Science* **239,** 268 (1988).

[30a] R. Chen, J. J. Knez, and M. E. Medof, in preparation (2000).

[31] P. Moran, H. Raab, W. J. Kohr, and I. W. Caras, *J. Biol. Chem.* **266,** 1250 (1991).

[32] L. D. Gerber, K. Kodukula, and S. Udenfriend, *J. Biol. Chem.* **267,** 12168 (1992).

[33] P. Moran and I. W. Caras, *J. Cell Biol.* **115,** 1595 (1991).

[34] K. E. Coyne, A. Crisci, and D. M. Lublin, *J. Biol. Chem.* **268,** 6689 (1993).

[35] J. Berger, A. D. Howard, L. Brink, L. Gerber, J. Hauber, B. R. Cullen, and S. Udenfriend, *J. Biol. Chem.* **263,** 10016 (1998).

[36] J. Berger, R. Micanovic, R. J. Greenspan, and S. Udenfriend, *Proc. Natl. Acad. Sci. U.S.A.* **86,** 1457 (1989).

signals are equipotent. Some GPI signaling sequences are inefficient inducers of C-terminal GPI anchor attachment.[27,31,32,37] For this reason, the choice of 3' fusion partner for the chimeric cDNA should be restricted to those that have been shown to direct GPI anchoring most effectively. These include the 3' mRNA end sequences of 5' nucleotidase/CD73, DAF/CD55, Campath/CD52, Thy-1/CD90, and placental alkaline phosphatase (PLAP).[30a]

Chimeric cDNAs encoding GPI-modified polypeptides can be prepared by several approaches. If appropriate restriction sites are present, the cDNA encoding the polypeptide of interest can be directly ligated upstream of a selected GPI-signaling sequence derived from a natural GPI-anchored protein. Alternatively, a chimeric polypeptide-GPI signaling sequence can be engineered by polymerase chain reaction (PCR)-based splice-by-overlap extension[38] (see Fig. 2). This latter technique (which allows the juxtaposition of the two fusion partners without intervening sequences encoding restriction endonuclease sites) has been successfully employed for the assembly of several chimeric cDNAs.[39,40] It is critical that the polypeptide fusion partner selected for GPI anchoring have an N-terminal signal peptide that initially directs it into the lumen of the ER because, as alluded to above, the final GPI modification enzymes responsible for transfer are located in this intracellular site.[41]

Minimal 3' DAF cDNA sequences that can direct GPI anchoring have been delineated. These sequences encode 37 or as few as 30[30a] C-terminal DAF residues and include the hydrophobic residues that tether the nascent fusion polypeptide chain to the ER membrane prior to anchor substitution. After C-terminal peptide cleavage/GPI anchor attachment at the ω Ser-319, the processed GPI-anchored fusion protein retains only nine or as few as two C-terminal residues derived from DAF.[42] The former minimal DAF sequence, originally engineered for the expression of a GPI-modified form of the herpes simplex virus glycoprotein D1,[42] has been employed to prepare GPI-modified forms of several other proteins, including human growth hormone, CD8, granulocyte-macrophage colony-stimulating factor (GM-CSF), class I MHC, B7-1, and B7-2 (see below). A similar minimal GPI-

[37] M. E. Lowe, *J. Cell Biol.* **116,** 799 (1992).
[38] R. M. Horton, H. D. Hunt, S. N. Ho, J. K. Pullen, and L. R. Pease, *Gene* **77,** 61 (1989).
[39] R. Knorr and M. L. Dustin, *J. Exp. Med.* **186,** 719 (1997).
[40] H. J. Meyerson, J.-H. Huang, J. D. Fayen, H.-M. Tsao, R. R. Getty, N. S. Greenspan, and M. L. Tykocinski, *J. Immunol.* **156,** 574 (1996).
[41] I. W. Caras, *J. Cell Biol.* **113,** 77 (1991).
[42] I. W. Caras, G. N. Weddell, M. A. Davitz, V. Nussenzweig, and D. W. Martin, *Science* **238,** 1280 (1987).

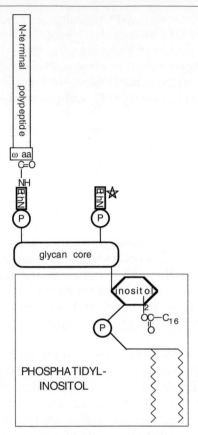

FIG. 1. Schematic for the glycosylphosphatidylinositol anchor. The core structure of the GPI anchor is depicted. Two acyl chains attached to glycerol are intercalated into the outer leaflet of the cytoplasmic membrane. A third acyl chain, indicated at position 2 of the inositol ring, is found in some GPI anchors, and is the structural basis for the "third foot" that is sometimes intercalated into the membrane in selected cell types. The central glycan core is composed of three mannose rings that can be modified by a wide array of side groups. The ethanolamine phosphate capping the GPI structure tethers the polypeptide chain by means of an amide linkage to the C-terminal ω residue. A second phosphoethanolamine group (indicated by the star) is sometimes found modifying the mannose groups of the glycan core.

anchoring 3' cDNA sequence derived from PLAP has been widely employed.[43]

It must be noted that other than the GPI addition event itself, the signals that shuttle polypeptide chains through intracellular compartments

[43] A. Y. Lin, B. Devaux, A. Green, C. Sagerstrom, J. F. Elliott, and M. M. Davis, *Science* **249,** 677 (1990).

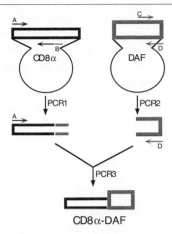

CD8α-DAF

FIG. 2. Engineering a chimeric cDNA encoding a GPI-modified polypeptide by PCR-base splice-by-overlap extension. The engineering of a chimeric cDNA encoding a GPI-modified form of the extracellular domain of CD8 is used for illustration purposes. Sequence from the C terminus of CD55/DAF is used to signal GPI modification. In the first round of PCRs, sequence encoding the extracellular domain of CD8 is amplified with primers A and B (PCR1). Note that the 3′ primer B includes nucleotides complementary to the GPI-modifying sequence of DAF. The resulting product of PCR1 is a chimeric cDNA in which a portion of the GPI-encoding DAF sequence is positioned downstream from CD8. A sequence encoding the minimal GPI-modifying sequence of DAF is amplified in PCR2 with primers C and D. In PCR3, the PCR products of the previous reaction are mixed and amplified with primers A and D. The complementary partial DAF sequence amplified in PCR1 allows annealing to the corresponding sequence of the DAF product in PCR2. The product of PCR3 is a chimeric cDNA in which the two cDNAs are directly spliced in frame without any intervening restriction sites.

and trigger the full set of posttranslational modifications of GPI-anchored proteins are incompletely understood. This is illustrated by considering the posttranslational fate of the β chain of the T cell receptor (TCR), which is normally expressed as a heterodimer with a disulfide-attached α chain. When transfected into cells without a companion α chain, the TCR β chain is degraded in the ER, because of the activity of its five-residue cytoplasmic retention signal.[44] Transfection of a truncated TCR β chain (without the retention residues) results in cell surface transport. The surface-expressed truncated TCR β chain, surprisingly, is GPI anchored. Cotransfection of the α chain and the truncated TCR β chain induces coexpression of a heterodimer in which the β chain bears a conventional transmembrane anchor. Thus, the TCR β chain possesses a cryptic GPI modification signal that is unmasked only when its companion α chain and its ER retention

[44] L. M. Bell, K. R. Solomon, J. P. Gold, and K. N. Tan, *J. Biol. Chem.* **269,** 22758 (1994).

signals are eliminated.[44] A similar pattern of GPI modification can be demonstrated for the B cell surface receptor IgD.[45,46] Thus, the complete posttranslational GPI anchor attachment modification is ultimately accomplished by an accumulated series of positive and negative signals recognized in the ER and Golgi compartments.

Selection of Cell Type to Be Used for Production of Glycosylinositol Phospholipid-Modified Proteins

Natural GPI-anchored proteins have been found in all eukaryotic organisms that have been studied, spanning both the plant and animal kingdoms. This indicates that the cascade of enzymes that direct GPI anchor biosynthesis is phylogenetically ancient. Accordingly, within the animal kingdom, it includes parasites and extends to the phylum protozoa. Among parasites, GPI-modified variant surface glycoproteins (VSGs), which comprise the major surface coat protein of trypanosomes, have been characterized in detail.[47] GPI-modified proteins also have been described in insects.[48] Within chordates, GPI-anchored acetylcholinesterase (AChE) derived from the *Torpedo* electrical organ also has been extensively investigated.[49] The ability to carry out GPI anchoring in the plant kingdom includes both higher[50] and lower plant forms.[51] It also includes yeast[52,53] and slime molds.[54,55]

There is correspondingly a wide range of laboratory cell lines capable of carrying out GPI anchoring, and consequently usable for engineered GPI-anchored protein production. Among mammalian cell lines, CHO, COS, MDCK, and HeLa cell lines have been most widely employed. For high-level expression of engineered GPI-anchored proteins and for specialized purposes (see below), transfectable S2 Schneider insect cells[56,57] and

[45] R. N. Mitchell, A. C. Shaw, Y. K. Weaver, P. Leder, and A. K. Abbas, *J. Biol. Chem.* **266,** 8856 (1991).

[46] J. Wienands and M. Reth, *Nature (London)* **356,** 246 (1992).

[47] M. A. Ferguson, J. S. Brimacombe, S. Cottaz, R. A. Field, L. S. Guther, S. W. Homans, M. J. McConville, A. Mehlert, K. G. Milne, and J. E. Ralton, *Parasitology* **108**(Suppl.), S45 (1994).

[48] M. D. Ganfornina, D. Sanchez, and M. J. Bastiani, *Development* **121,** 123 (1995).

[49] H. Soreq, A. Gnatt, Y. Lowenstein, and L. F. Neville, *Trends Biochem. Sci.* **17,** 353 (1992).

[50] A. M. Takos, I. B. Dry, and K. L. Soole, *FEBS Lett.* **405,** 1 (1997).

[51] N. Morita, H. Nakazato, H. Okuyama, Y. Kim, and J. G. A. Thompson, *Biochim. Biophys. Acta* **1290,** 53 (1996).

[52] G. Sipos, F. Reggiori, C. Vionnet, and A. Conzelmann, *EMBO J.* **16,** 3494 (1997).

[53] A. Conzelman, A. Puoti, R. L. Lester, and C. Desponds, *EMBO J.* **11,** 457 (1992).

[54] H. Sadeghi, A. M. daSilva, and C. Klein, *Proc. Natl. Acad. Sci. U.S.A.* **85,** 5512 (1988).

[55] J. Stadler, T. W. Keenan, G. Bauer, and G. Gerisch, *EMBO J.* **8,** 371 (1989).

[56] J.-H. Huang, R. R. Getty, F. V. Chisari, P. Fowler, N. S. Greenspan, and M. L. Tykocinski, *Immunity* **1,** 607 (1994).

[57] J. P. Incardona and T. L. Rosenberry, *Mol. Cell. Biol.* **7,** 595 (1996).

baculovirus-infectable Sf-9 insect cells,[58–61] as well as *Pichia* yeast[62] have been used. Moreover, work has demonstrated that GPI anchor attachment to engineered proteins can be effected in *in vitro* translation systems, albeit with diminished efficiency.[63] With further improvements in GPI anchor attachment efficiency by *in vitro* methods, e.g., using the isolated transamidase enzyme complex, it is possible that such *in vitro* translation systems may offer an alternative to cell-based systems for engineered GPI-modified protein production.

The structural requirements for GPI modification of proteins are generally similar across the animal and plant kingdoms.[27] The signals that trigger GPI modification, however, differ subtly in different organisms. The GPI-signaling sequences of trypanosomal VSGs and malarial circumsporozoite protein, for example, fail to initiate GPI anchor attachment in the mammalian COS cell line, although this line responds to the DAF signaling sequence.[64] The failure of the VSG C-terminal region to initiate GPI modification is due to differences in $\omega \to \omega + 2$ site signal requirements.[64]

The biochemical structure of the GPI anchor is similar across the animal and plant kingdoms[65] (see Fig. 1). In particular, part of the phospholipid, the linear core glycan, and the terminal phosphoethanolamine (P-EthN) anchor attachment component are fully conserved. However, branch-chain sugars that attach to the central glycan differ in different organisms.[65] In addition, glycerol-associated acyl/alkyl chains in the phospholipid vary. Yeast anchors, for example, start out with phosphatidylinositol (PI) and, in most cases, then are modified to inositolphosphoceramides and/or inositolphosphoglyceramides with longer fatty acid acyl groups.[52,53] The lipid groups can modify the properties of the GPI-anchored protein in the membrane.[66,67] Whether carbohydrate alterations in the core GPI anchor structure affect the biological properties of GPI-anchored proteins is unstudied.

[58] J. Fayen, H. Meyerson, D. Kaplan, and M. Tykocinski, *FASEB J.* **5** (4 Part I), A615 (1991). [Abstract]
[59] H. Chaabihi, D. Fournier, Y. Fedon, J. P. Bossy, M. Ravallec, G. Devauchelle, and M. Cerutti, *Biochem. Biophys. Res. Commun.* **203,** 734 (1994).
[60] S. Estrada-Mondaca, A. Lougarre, and D. Fournier, *Arch. Insect Biochem. Physiol.* **38,** 84 (1998).
[61] A. Davies and B. P. Morgan, *Biochem. J.* **295,** 889 (1993).
[62] N. Morel and J. Massoulie, *Biochem. J.* **328,** 121 (1997).
[63] T. L. Doering and R. Schekman, *Biochem. J.* **328,** 669 (1997).
[64] P. Moran and I. W. Caras, *J. Cell Biol.* **125,** 333 (1994).
[65] J. R. Thomas, R. A. Dwek, and T. W. Rademacher, *Biochemistry* **29,** 5413 (1990).
[66] E. I. Walter, W. D. Ratnoff, K. E. Long, J. W. Kazura, and M. E. Medof, *J. Biol. Chem.* **267,** 1245 (1992).
[67] K. E. Long, R. Yomtovian, M. Kida, J. J. Knez, and M. E. Medof, *Transfusion* **33,** 294 (1993).

Characterization of Engineered Glycosylinositol
Phospholipid-Anchored Proteins

After assembly of a chimeric cDNA encoding a GPI-modified version of a polypeptide and transfection into a cell line, analyses are needed to confirm that the engineered protein is appropriately GPI anchored.

Most commonly, the initial approach for verifying GPI anchor attachment exploits phosphatidylinositol-specific phospholipase C (PI-PLC), a natural enzyme that is expressed by several bacterial strains, including *Bacillus thurigiensus*.[68] This enzyme cleaves cell surface GPI anchors at the junction between the bilayer-associated diacylglycerol and the phosphoinositol ring, resulting in release of the polypeptide chain with its attached anchor glycan from the cell surface. Measurement of the amount of GPI-anchored protein (e.g., by flow cytometry following cell staining) should show a significant decrement after PI-PLC digestion. It should be noted, however, that this digestion may not be complete; significant amounts (although lower than on untreated cell lines) of the GPI-anchored protein sometimes remain after PI-PLC digestion. Moreover, GPI anchors present on certain cell lines are resistant to PI-PLC cleavage. In particular, the GPI anchors present on erythrocytes,[69–71] on the erythroleukemic K562 cell line,[72–74] and on murine L929 fibroblasts[75] contain acylated inositol rings. This acyl group, by virtue of occupying the 2-position on the inositol ring, blocks formation of cyclic inositol monophosphate, which is required for the action of PI-PLC. Natural GPI-modified proteins for which monoclonal antibodies (MAbs) are available for detection (e.g., DAF, PLAP) can serve as positive controls to document PI-PLC enzymatic activity in these analyses.

Metabolic labeling of transfected cells provides a second path for documentation of GPI anchor attachment. Immunoprecipitation with specific antibody of an ^{35}S-labeled transfectant should identify a single band. The

[68] M. G. Low, J. Stiernberg, G. L. Waneck, R. A. Flavell, and P. W. Kincade, *J. Immunol. Methods* **113,** 101 (1988).
[69] E. I. Walter, W. L. Roberts, T. L. Rosenberry, W. D. Ratnoff, and M. E. Medof, *J. Immunol.* **144,** 1030 (1990).
[70] W. L. Roberts, J. J. Myher, A. Kuksis, and T. L. Rosenberry, *Biochem. Biophys. Res. Commun.* **150,** 271 (1988).
[71] W. L. Roberts, J. J. Myher, A. Kuksis, and T. L. Rosenberry, *J. Biol. Chem.* **263,** 18766 (1988).
[72] J. Toutant, M. Richards, J. Krall, and T. Rosenberry, *Eur. J. Biochem.* **187,** 31 (1990).
[73] S. Hirose, G. M. Prince, D. Sevlever, L. Ravi, T. L. Rosenberry, E. Ueda, and M. E. Medof, *J. Biol. Chem.* **267,** 16968 (1992).
[74] R. P. Mohney, J. J. Knez, L. Ravi, D. Sevlever, T. L. Rosenberry, S. Hirose, and M. E. Medof, *J. Biol. Chem.* **269,** 6536 (1994).
[75] N. Singh, D. Singleton, and A. M. Tartakoff, *Mol. Cell. Biol.* **11,** 2362 (1991).

electrophoretic mobility of the GPI-anchored protein may or may not differ from that of its conventionally anchored counterpart, depending on the protein itself, the extent of truncation, and the size of the lipid-modified fusion partner. The biochemical constituents of the anchor core structure provide another avenue for metabolic labeling of engineered GPI-anchored proteins. The most commonly employed constituent is EthN,[27,28,65] which in mammalian anchors modifies the first and sometimes the second mannose residues as well as the third mannose, where it serves as the terminal constituent that attaches to the ω amino acid via an amide linkage. Immunoprecipitation of the GPI-modified fusion protein from cells radiolabeled with [^{3}H]EthN again should identify a single band with electrophoretic mobility identical to that obtained by ^{35}S labeling.

Glycosylinositol Phospholipid-Modified Fusion Protein Applications

Tethering Isolated Protein Domains

Like DNA transfection techniques, GPI-modified proteins provide a way for investigating the function(s) of individual polypeptide domains, but advantages are offered by the appended anchor. The engineering of a chimeric cDNA in which a single protein domain is fused to a GPI-anchoring signal sequence allows the evaluation of the function of that domain after incorporation of the isolated protein into any cell target. Insights into the functional activity of a polypeptide provided by this technique complement those provided by site-directed mutagenesis and epitope-specific antibody-blocking studies. Advantages of the GPI-anchoring technique are that the protein can be used readily for multiple cell types and that more than one GPI-anchored protein or domain can be simultaneously employed.[26] Because the polypeptide of interest must have an N-terminal signal sequence, a substitude signal sequence (e.g., that of oncostatin M) can be appended to the N-terminal sequence of the polypeptide.

This approach has allowed insights into the function of isolated domains in proteins with adhesin function. With this technique, the isolated internal I domain from the LFA-1 α chain has been shown to mediate binding to intercellular adhesion molecule 1 (ICAM-1).[39] Similarly, the v-homology domain of CD8α has been shown to mediate intercellular adhesion to HLA class I-bearing cells.[40] Cell surface tethering by a GPI anchor similarly can be applied to evaluate the properties of soluble proteins.[76,77] Using this approach, a GPI-anchored version of M-CSF has been shown to function as

[76] M. C. Weber, R. K. Groger, and M. L. Tykocinski, *Exp. Cell Res.* **210,** 107 (1994).
[77] L. Pang, P. Sivaram, and I. J. Goldberg, *J. Biol. Chem.* **271,** 19518 (1996).

an artificial adhesin, mediating intercellular adhesion to an M-CSF receptor-bearing cell line.[76]

In a particularly innovative use of this strategy by Rice et al.,[78] a PCR-mutated cDNA library of the soluble tissue plasminogen activator (t-PA), in which amplification was performed under conditions of low Taq polymerase fidelity, was ligated into an expression vector upstream of the GPI-signaling sequence of DAF. The library of GPI-tethered t-PA mutants then was expressed on the surface of transfected cells, thereby providing suspension cells as a flow cytometric substrate to assay for polypeptides that otherwise would be naturally soluble. Transfectants that failed to bind a t-PA MAb were subsequently identified by fluorescence-activated cell sorting (FACS) of MAb-stained cells. The mutagenized t-PA-encoding plasmids from the epitope loss mutants were recovered and sequenced to localize the MAb binding epitope.[78] In this way, the use of chimeric GPI-anchoring technology to direct the tethering of a soluble polypeptide to cell membranes, in conjunction with flow cytometry, allowed the identification of low-frequency mutagenic events.

Glycosylinositol Phospholipid Modification of Major Histocompatibility Complex Proteins

As mentioned above, the murine Qa-2 gene encodes an MHC class I-like protein that has a natural GPI anchor.[79,80] Qa-2 binds small peptides in a fashion similar to other members of the MHC family, but to date has not been shown to possess antigen-presenting capacity. GPI-reanchoring technology has been used to study conventionally anchored proteins encoded by the MHC locus. Data from several laboratories have demonstrated that the GPI-modified MHC class I fusion proteins can be chaperoned through the ER and Golgi network and expressed at the cell surface along with a nine-amino acid-binding groove resident and its noncovalently attached light chain, β_2-microglobulin.[56,81-87]

[78] G. C. Rice, D. V. Goeddel, G. Cachianes, J. Woronicz, E. Y. Chen, S. R. Williams, and D. W. Leung, Proc. Natl. Acad. Sci. U.S.A. 89, 5467 (1992).

[79] I. Stroynowski, M. Soloski, M. G. Low, and L. Hood, Cell 50, 759 (1987).

[80] G. L. Waneck, D. H. Sherman, P. W. Kincade, M. G. Low, and R. A. Flavell, Proc. Natl. Acad. Sci. U.S.A. 85, 577 (1988).

[81] D. W. Mann, I. Stroynowski, L. Hood, and J. Forman, J. Immunol. 142, 318 (1989).

[82] C. J. Aldrich, L. C. Lowen, D. Mann, M. Nishimura, L. Hood, I. Stroynowski, and J. Forman, J. Immunol. 146, 3082 (1991).

MHC proteins are multifunctional surface proteins. Not all functions of GPI-anchored MHC remain intact. A GPI-anchored form of MHC class I has activity equivalent to that of its conventionally anchored counterpart as an inhibitory ligand of natural killer cell attack.[87] However, its antigen-presenting function is significantly altered.[81–86] In general, these MHC class I fusion proteins are able to present alloantigen and processed peptides to pre-existing CD8+ T cell lines, albeit with decreased capacity as compared with their native transmembrane counterparts.[56,82–86] Moreover, GPI-anchored MHC class I fusion proteins present endogenous antigens and unprocessed proteins poorly or not at all.[82–85] In addition, they cannot prime naive T cells[82–85] or induce skin graft rejection reactions.[83] Investigations into the mechanism(s) underlying these differences in reactivity may help to provide insights into the intracellular antigen-loading compartments of class I MHC molecules and into differences in trafficking between GPI-modified and conventionally anchored proteins.[23,24]

Glycosylinositol Phospholipid Modification of Viral Receptor Proteins

In virology, engineering GPI-modified proteins has proved useful in investigating the domains of cell surface receptors that facilitate viral binding and entry into cells. For the human rhinovirus, a GPI-anchored variant of its receptor, ICAM-1, has been found to efficiently mediate both viral binding and productive infection in transfected COS cells. This indicates that neither the cytoplasmic nor transmembrane domains of ICAM-1 are essential for receptor-mediated infection by this virus.[88] In contrast, with the measles virus, a GPI-anchored variant of its receptor, CD46/MCP (membrane cofactor protein), mediates infection with greatly reduced efficiency,[89] although the GPI-anchored receptor binds the virus as well as the

[83] E. Simpson, P. J. Robinson, P. Chandler, M. M. Millrain, H.-P. Pircher, D. Brandle, P. Tomlinson, J. Antoniou, and A. Mellor, *Immunology* **81,** 132 (1994).

[84] U. M. A. Motal, C. L. Sentman, X. Zhou, P. J. Robinson, J. Dahmen, and M. Jondal, *Eur. J. Immunol.* **25,** 1121 (1995).

[85] R. Zamoyska, T. Ong, G. Kwan-Lim, P. Tomlinson, and P. J. Robinson, *Int. Immunol.* **8,** 551 (1996).

[86] A. Cariappa, D. C. Flyer, C. T. Rollins, D. C. Roopenian, R. A. Flavell, D. Brown, and G. L. Waneck, *J. Immunol.* **26,** 2215 (1996).

[87] C. L. Sentman, M. Y. Olsson-Alheim, U. Lendahl, and K. Karre, *Eur. J. Immunol.* **26,** 2127 (1996).

[88] D. E. Staunton, A. Gaur, P.-Y. Chan, and T. A. Springer, *J. Immunol.* **148,** 3271 (1992).

[89] T. Seya, M. Kurita, K. Iwata, Y. Yanagi, K. Tanaka, K. Shida, M. Hatanaka, M. Matsumoto, S. Jun, A. Hirano, S. Ueda, and S. Nagasawa, *Biochem. J.* **322,** 135 (1997).

wild-type receptor.[89–91] The use of the GPI-reanchoring methodology in virology has been most widely employed to study CD4, the receptor for the human immunodeficiency virus (HIV). Although its intracellular signaling properties in T cell activation are significantly impaired,[92] a GPI-anchored CD4 molecule mediates binding in transfected human cells with an efficiency approaching that of its conventionally anchored counterpart, but productive infection is greatly diminished.[93–95] In summary, for viral receptors, C-terminal GPI anchor substitution for their endogenous transmembrane anchors is useful not only for mapping receptor domains but also for investigating the components critical for initiating viral entry into and further interaction with the cellular machinery. It thereby is a valuable tool for characterizing the early events of infection.

Cooperative Activities of Different Surface Molecules in Complex Biological Reactions

Many cell surface and intercellular reactions require the combined actions of multiple cell surface proteins. One example among such cell surface reactions is protection of self cells from activation of autologous complement on their surface membranes. Three cell surface proteins function for this purpose, two of which, the decay-accelerating factor (DAF) and membrane cofactor protein (MCP), function together in the critical C3 amplification step of the cascade. DAF is naturally GPI anchored, while MCP is conventionally anchored. Studies of complement are best done using defined sheep erythrocyte (E^{sh}) intermediates bearing purified complement components deposited in desired amounts. The incorporation of GPI-anchored DAF and of a GPI-reanchored variant of MCP into such intermediates has allowed the conclusion that the former molecule, which dissociates the C3 convertase complexes $\overline{C4b2a}$ and $\overline{C3bBb}$, greatly facilitates the action of the latter molecule. This cooperatively promotes the cleavage of the residual bound C4b and C3b fragments, so that they are unable to take up new C2 and factor B zymogens, respectively.

[90] G. Varior-Krishnan, M.-C. Trescol-Biemont, D. Naniche, C. Rabourdin-Combe, and D. Gerlier, *J. Virol.* **68,** 7891 (1994).

[91] M. Manchester, A. Valsamakis, R. Kaufman, M. K. Liszewski, J. Alvarez, J. P. Atkinson, D. M. Lublin, and M. B. A. Oldstone, *Proc. Natl. Acad. Sci. U.S.A.* **92,** 2303 (1995).

[92] B. P. Sleckman, Y. Rosenstein, V. E. Igras, J. L. Greenstein, and S. J. Burakoff, *J. Immunol.* **147,** 428 (1991).

[93] D. C. Diamond, R. Finberg, S. Chaudhuri, B. P. Sleckman, and S. J. Burakoff, *Proc. Natl. Acad. Sci. U.S.A.* **87,** 5001 (1990).

[94] M. Jasin, K. A. Page, and D. R. Littman, *J. Virol.* **65,** 440 (1991).

[95] R. A. Brodsky, S. M. Jane, E. F. Vanin, H. Mitsuya, T. R. Peters, T. Shimada, M. E. Medof, and A. W. Nienhuis, *Hum. Gene Ther.* **5,** 1231 (1994).

Methods for Incorporating Purified Glycosylinositol Phospholipid-
Modified Proteins: Accomplishing Protein Transfer

The phenomenon of GPI-modified protein incorporation into cells was first documented for CD55/DAF.[25] In these early studies, purified CD55/ DAF or CD55/DAF within crude human erythrocyte (E^{hu}) stromal extracts selectively reincorporated into E^{sh} and, following incorporation, protected the E^{sh} cells from complement attack. Incorporation was found to be temperature dependent (with maximal transfer at 37°) and inhibitable by lipid-binding proteins such as albumin or lipoproteins.[25] Subsequently, the incorporation of other purified natural GPI-anchored proteins, including Thy-1,[96] LFA-3/CD58,[97] CD16,[98] and the complement regulatory proteins heat-stable antigen (CD24)[99] and CD59/MIRL,[100] was described.

Most GPI-anchored proteins contain PI with two glycerol-associated alkyl/acyl chains, i.e., they are "two-footed," GPI-anchored proteins. In contrast, as alluded to above, GPI-anchored proteins isolated from E^{hu} contain a third lipid group, i.e., a palmitate on inositol, and hence, are "three-footed."[69–71] In a systematic study of the function of these two types of anchors and "one-footed" anchors using (1) three-footed E^{hu}-derived DAF, (2) two-footed HeLa cell-derived DAF, and (3) two E^{hu} DAF proteins with experimentally modified one-footed anchors, the effect of the second and third lipid on the incorporation process was analyzed. It was found that while the uptake of the three anchor types was comparable, the stability of reincorporated protein was dependent on having at least a two-footed anchor and greatly enhanced for the three-footed DAF species.[66] The capacity to membrane reincorporate is entirely a property of the C-terminal GPI anchor and, in all studies so far, appears to be independent of its attached N-terminal polypeptide.

The factors that influence spontaneous incorporation into surface membranes apply to GPI-reanchored proteins in a fashion identical to natural GPI-anchored proteins. This was first shown with CD4 · DAF[95] and HLA A2.1 · DAF[56] and subsequently in work done with several of the other GPI-reanchored proteins described above. In the latter study, a modified version

[96] F. Zhang, W. G. Schmidt, Y. Hou, A. F. Williams, and K. Jacobson, *Proc. Natl. Acad. Sci. U.S.A.* **89,** 5231 (1992).
[97] P. Selvaraj, M. L. Dustin, R. Silber, M. G. Low, and T. A. Springer, *J. Exp. Med.* **166,** 1011 (1987).
[98] S. Nagarajan, M. Anderson, S. N. Ahmed, K. W. Sell, and P. Selvaraj, *J. Immunol. Methods* **184,** 241 (1995).
[99] Y. Hitsumoto, H. Ohnishi, A. Nakano, F. Hamada, S. Saheki, and N. Takeuchi, *Int. Immunol.* **5,** 805 (1993).
[100] S. J. Piddlesden and B. P. Morgan, *J. Neuroimmunol.* **48,** 169 (1993).

of the human class I protein HLA A2.1 produced in insect cells and loaded with a nominal antigenic peptide *in vitro* was incorporated into antigen-presenting cells and the incorporated protein shown to stimulate specific CD8[+] T cells lines.[56] Similarly, a GPI-modified "second signal" protein, B7-1, purified from a mammalian overexpression system, was incorporated into tumor cell surfaces and shown to provide a costimulatory signal critical for eliciting T cell effector functions.[101,102]

The incorporation of purified (or unpurified) GPI-anchored proteins requires procedures that optimize membrane protein extraction from GPI-anchored protein-expressing cells. For applications that require purified GPI-anchored proteins, a subsequent purification method is required. GPI-anchored proteins have limited solubility in many cold nonionic detergents.[10] We have found that octylglucoside optimally extracts these proteins from membrane preparations even at 0°. We have found that 3-[(3-cholami-dopropyl)-dimethyl-ammonio]-1-propanesulfonate (CHAPS) and deoxycholate (DOC) are other useful detergents in that, like octylglucoside, they also are dialyzable so that their concentrations in the final products can be lowered to allow for the use of higher input concentrations of the GPI-anchored protein when mixed with cells. Many investigators have found that detergent solubilization with Triton X-100 at 20–37° is adequate[25]; the efficiency of extraction with this detergent, however, is less than for other detergents and this detergent is not dialyzable.

For the purification of GPI-anchored proteins from crude detergent extracts, immunoaffinity chromatography with MAbs or ligands reactive with the N-terminal polypeptide is the procedure of choice. The engineering of an artificial epitope (e.g., polyhistidine amino acid stretch) into the junctional site between the N-terminal polypeptide and the GPI-encoding C-terminus (see Fig. 3) can be used for purification if MAbs or specific ligands are not available.[103]

The incorporation of GPI-modified proteins, like that of native proteins,[25,66,98] into cellular membranes is not only dependent on incorporation time, temperature, and absence of competing lipid-binding proteins[66,102,104,105] but also is proportional to the concentration of GPI-an-

[101] R. S. McHugh, S. N. Ahmed, Y.-C. Wang, K. W. Sell, and P. Selvaraj, *Proc. Natl. Acad. Sci. U.S.A.* **92,** 8059 (1995).

[102] E. B. Brunschwig, J. D. Fayen, M. E. Medof, and M. L. Tykocinksi, *J. Immunother.* **22,** 390 (1999).

[103] J. Fayen and M. Tykocinski, *FASEB J.* **9** (4 Part II), A1055 (1995). [Abstract]

[104] C. W. van den Berg, T. Cinek, M. B. Hallett, V. Horejsi, and B. P. Morgan, *J. Cell Biol.* **131,** 669 (1995).

[105] D. D. Premkumar, Y. Fukuoka, E. Brunschwig, D. Sevlever, and M. E. Medof, submitted (2000).

A.   Assembly of chimeric *His-DAF cDNA*

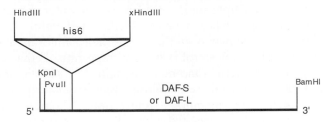

B.   His-tagged GPI-modified Fusion Proteins

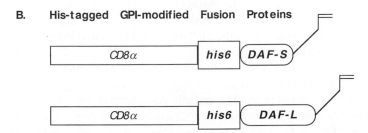

FIG. 3. Engineering a polyhistidine epitope tag into the junctional site of a GPI-modified fusion protein. Sequence encoding a hexahistidine tag was inserted as a linker into the *Hind*III site upstream from GPI modification sequence derived from the C terminus of CD55/DAF (A). Subsequently, the cDNA encoding the extracellular domain of CD8α was inserted upstream from the polyhistidine linker site. The resulting chimeric cDNA encodes a protein in which a polyhistidine tag is positioned at the junctional site between the extracellular domain of CD8α and the DAF GPI-modifying sequence (B). Cells transfected with this chimeric cDNA express GPI-anchored CD8α at their surface. GPI-modified CD8α-His6 fusion protein in transfectant lysates reversibly binds to a nickel-Sepharose matrix, indicating recognition of the internal hexahistidine tag.

chored protein added. Several investigators have found a direct relationship between the input quantity of GPI-modified protein and the quantity of incorporated material, extending to the incorporation of proteins incubated at concentrations in excess of 10 μg/ml.[25,66,95,98] The usual efficiency of uptake is 2 to 5%.[25,66] The incorporation of GPI-anchored proteins occurs relatively rapidly, with detection of the inserted GPI protein within minutes of starting the incubation. Incorporation performed under optimal conditions usually reaches a steady state within 2 hr.[25,66,98] Temperature is a critical variable for GPI protein incorporation,[25,66,98,102] with optimal incorporation probably at 30°, a compromise between optimal uptake and detergent-mediated cell damage at 37°. The temperature dependence reflects the requirement of membrane fluidity for GPI reinsertion.

Although extracellular molecular interactive functions of incorporated proteins are transferred following *in vitro* addition to cells, some properties may be altered. As indicated above (see Introduction), GPI-anchored proteins do not exhibit simple static residence on the outer membrane leaflet. Reincorporated GPI-anchored proteins, like their naturally synthesized counterparts, transit slowly from the site of their membrane insertion into detergent-insoluble, cholesterol-rich membrane rafts[9,10] that comprise, in some cells, caveolar membrane microdomains.[17,18] Kinetic analyses have shown that the transit to these membrane microdomains progressively increases for 24 hr after membrane reinsertion.[105] Some of the complex intracellular properties associated with GPI-anchored proteins, including their ability to participate in intracellular signaling, require transit into these specialized membrane microdomains.[104,105] The full reconstitution of the properties of reincorporated GPI-modified proteins thus is not complete until this transition has occurred.

Interestingly, free GPI-anchored complement regulatory proteins in seminal fluid have been found to reincorporate into the membranes of coincubated cells and protect them from complement attack, suggesting that the transfer of these proteins may occur *in vivo* and represent a physiological event in seminal fluid.[106] *In vivo* incorporation has also been implicated in infections, where it has been reported that (1) *Schistosoma mansoni* schistosomula bound to vascular endothelial cells can acquire DAF on the surface membranes[66,107] and that (2) bovine erythrocytes can acquire the dense antigenic GPI-anchored membrane coat proteins (variant surface glycoproteins) of *Trypanosoma brucei* during infection with this organism.[108]

In summary, the engineering of GPI-reanchored proteins in which a cDNA encoding the N-terminal extracellular domain of the protein of interest is appended to a 3' mRNA end sequence directing GPI modification allows the expression of proteins with desired biochemical and cellular biologic properties on any cell type, enabling insights into many fields. "Protein painting" of such fusion proteins offers the possibility of selective alteration of cell surface phenotype without gene manipulation.[26] Such an approach also offers several potential applications as a therapeutic tool.

Acknowledgment

Supported by NIH Grant HL55773.

[106] I. A. Rooney, J. E. Heuser, and J. P. Atkinson, *J. Clin. Invest.* **97**, 1675 (1996).
[107] M. Fatima, M. Horta, and F. J. Ramalho-Pinto, *J. Exp. Med.* **174**, 1399 (1991).
[108] M. R. Rifkin and F. R. Landsberger, *Proc. Natl. Acad. Sci. U.S.A.* **87**, 801 (1990).

[28] Fusions to Members of Fibroblast Growth Factor Gene Family to Study Nuclear Translocation and Nonclassic Exocytosis

By I. Prudovsky, M. Landriscina, R. Soldi, S. Bellum, D. Small, V. Andreeva, and T. Maciag

Biological Activities of Fibroblast Growth Factors

The fibroblast growth factor (FGF) gene family includes at least 19 genes encoding proteins that regulate proliferation, differentiation, motility, and survival of cells of neuroectodermal and mesodermal origin.[1-3] The two best studied and most abundantly expressed representatives of the family are the prototypes FGF-1 and FGF-2, and these are involved in the regulation of a wide variety of developmental, pathological, and regenerative processes in mammalian organisms, including humans.[3] Among these processes are mesodermal induction, limb formation, angiogenesis, skeletal muscle growth and regeneration, the growth and metastasis of certain types of solid tumors, atherosclerosis, and restenosis.[1] Other members of the family usually show a more limited pattern of expression and their functions are more restricted. The expression and function of FGF-5 in hair follicles and its involvement in the regulation of hair growth[4] and the function of FGF-7 in epidermal morphogenesis and the reepithelization of wounds[5] serve as examples.

The FGF proteins are relatively small, usually about 200 amino acids. Within every FGF a conservative core of 120 amino acids can be identified, where protein sequences from different members of the family are 22 to 66% identical.[1] FGFs are old, in evolutionary terms, being found in organisms as

[1] G. Szebenyi and J. Fallon, *Int. Rev. Cytol.* **45,** 45 (1999).

[2] T. Nishimura, Y. Utsunomiya, M. Hoshikawa, H. Ohuchi, and N. Itoh, *Biochim. Biophys. Acta* **1444,** 148 (1999).

[3] R. E. Friesel and T. Maciag, *Thromb. Haemost.* **82,** 748 (1999).

[4] J. Hebert, T. Rosenquist, J. Gotz, and G. Martin, *Cell* **78,** 1 (1994).

[5] S. Werner, H. Smola, X. Liao, M. T. Longaker, T. Krieg, P. H. Hofschneider, and L. T. Williams, *Science* **266,** 819 (1994).

primitive as round worms.[6,7] Studies of the molecular evolution of FGFs demonstrated that the diversification of FGF types was most probably due to gene duplication and occurred most prominently at least twice: during the transition from invertebrates to vertebrates and then from fish to quadruped vertebrates,[8] thus stressing the role of FGFs in the development of the vertebrate skeleton, limbs, central nervous system, and circulatory system. The effects of the FGFs are mediated by transmembrane protein kinase receptors (FGFRs) and four FGFRs are currently known. Alternative splicing of FGFRs dramatically increases the variety of these molecules, changing their affinity for different members of the FGF family and modifying cell responses to FGFs.[9] FGF binding to FGFRs activates the Ras–MAPK (mitogen-associated protein kinase) pathway, which is also induced by other growth factors, such as platelet-derived growth factor (PDGF) and epidermal growth factor (EGF), and is crucial for the stimulation of cell proliferation.[9] Another FGF-activated pathway is mediated by Src tyrosine kinase and the F actin-binding protein cortactin, and this pathway is involved in the regulation of cell migration.[10] Both exogenous and endogenous FGFs can translocate into cell nuclei.[11-15] Although the precise functions of FGFs in the nuclei are not known, there are numerous examples demonstrating that certain biological effects of FGFs may be dependent on the nuclear translocation of FGF.[16-18] Indeed, FGFR–PDGF receptor

[6] R. B. J. Wilson, J. R. Ainsworth, K. C. M. Anderson, C. Baynes, M. Berks, J. Bonfield, T. Copsey, T. Cooper, A. Coulson, M. Craxton, S. Dear, Z. Du, R. Durbin, A. Favello, A. Fraser, L. Fulton, A. Gardner, P. Green, T. Hawkins, L. Hillier, M. Jier, L. Johnston, M. Jones, J. Kershaw, J. Kirsten, N. Laissier, P. Latreille, J. Lightning, C. Lloyd, B. Mortimore, M. O'Callaghan, J. Parsons, C. Percy, L. Rifken, A. Roopra, D. Saunders, R. Shownkeen, M. Sims, N. Smaldon, A. Smith, M. Smith, E. Sonnhammer, R. Staden, J. Sulston, J. Thierry-Mieg, K. Thomas, M. Vaudin, K. Vaughan, R. Waterston, A. Watson, L. Weinstock, J. Wilkinson-Sprout, and P. Wohldman, *Nature (London)* **368**, 32 (1994).

[7] J. Hodgkin, R. H. A. Plasterk, and R. H. Waterston, *Science* **270**, 410 (1995).

[8] F. Coulier, P. Pontarotti, R. Roubin, H. Hartung, M. Goldfarb, and D. Birnbaum, *J. Mol. Evol.* **44**, 43 (1997).

[9] P. Klint and L. Claesson-Welsh, *Front. Biosci.* **4**, 165 (1999).

[10] T. M. LaVallee, I. A. Prudovsky, G. A. McMahon, X. Hu, and T. Maciag, *J.Cell Biol.* **141**, 1647 (1998).

[11] R. Z. Florkiewicz and A. Sommer, *Proc. Natl. Acad. Sci. U.S.A.* **86**, 3978 (1989).

[12] B. Bugler, F. Amalric, and H. Prats, *Mol. Cell. Biol.* **11**, 573 (1991).

[13] P. Kiefer, P. Acland, D. Pappin, G. Peters, and C. Dickson, *EMBO J.* **13**, 4126 (1994).

[14] X. Zhan, X. Hu, R. Friesel, and T. Maciag, *J. Biol. Chem.* **268**, 9611 (1993).

[15] A. Gualandris, C. Urbinati, M. Rusnati, M. Ziche, and M. Presta, *J. Cell. Physiol.* **161**, 149 (1994).

[16] P. Kiefer and C. Dickson, *Mol. Cell. Biol.* **15**, 4364 (1995).

[17] J. Shi, S. Friedman, and T. Maciag, *J. Biol. Chem.* **272**, 1142 (1997).

[18] N. Quarto, F. P. Finger, and D. B. Rifkin, *J. Cell. Physiol.* **147**, 311 (1991).

(PDGFR) chimera studies have emphasized the contribution of FGF as a ligand and for the regulation of myoblast proliferation.[19]

The FGFs are heparin-binding proteins and bind cell surface heparin sulfate proteoglycans (HPSGs), which not only protect FGFs from proteolysis but also present FGFs to FGFRs.[20,21] The use of FGF–reporter gene constructs has proved useful for the identification of FGF domains responsible for the regulation of import and export trafficking events.

Fibroblast Growth Factor Release

FGFs-3 through -8, FGF-10, FGF-15, FGF-17 (*Caenorhabditis elegans;* EGL-17), and FGF-18 display at their amino termini consensus signal sequences enabling them to be secreted through the endoplasmic reticulum (ER)–Golgi pathway.[1,2,22] The rest of the presently known members of the family, including the FGF prototypes, lack signal sequences and thus require export through alternative pathways. While early studies suggested that cell death and reversible cell membrane damage were major processes for FGF prototype export to target cells, more recent studies have shown that cell stress including hypoxia and heat shock, which induce neither cell death nor cell permeabilization, results in the release of FGF-1[23] (C. Mouta Carriera and T. Maciag, unpublished results, 2000). It is noteworthy that hypoxia and hyperthermia are characteristic of areas of tissue inflammation, where the function of FGF-1 is required for tissue remodeling, and imply the existence of a mechanism for the stress-induced release of FGF-1 that would provide a tightly regulated source of exogenous FGF-1. Both in the case of heat shock and hypoxia, FGF-1 is released as biologically inactive homodimers[23] (C. Mouta Carriera and T. Maciag, unpublished results, 1999). The homodimer can be converted to functional monomers by reducing agents such as dithiothreitol, and the residue Cys-30 is solely responsible for the formation of FGF-1 homodimers.[24] *In vivo,* reduced glutathione, which is usually present in areas of inflammation,[25] may be responsible for

[19] A. J. Kudla, N. C. Jones, R. S. Rosenthal, K. Arthur, K. L. Clase, and B. B. Olwin, *J. Cell Biol.* **142,** 241 (1998).

[20] A. Sommer and D. B. Rifkin, *J. Cell. Physiol.* **138,** 215 (1989).

[21] D. M. Ornitz, A. Yayon, J. G. Flanagan, C. M. Svahn, E. Levi, and P. Leder, *Mol. Cell. Biol.* **12,** 240 (1992).

[22] N. Ohbayashi, M. Hoshikawa, S. Kimura, M. Yamasaki, S. Fukui, and N. Itoh, *J. Biol. Chem.* **273,** 18161 (1998).

[23] A. Jackson, S. Friedman, X. Zhan, K. A. Engleka, R. Forough, and T. Maciag, *Proc. Natl. Acad. Sci. U.S.A.* **89,** 10691 (1992).

[24] F. Tarantini, S. Gamble, A. Jackson, and T. Maciag, *J. Biol. Chem.* **270,** 29039 (1995).

[25] A. Hirai, Y. Minamiyama, T. Hamada, M. Ishii, and M. Inoue, *J. Invest. Dermatol.* **109,** 314 (1997).

the reduction of the FGF-1 dimer. While the stress-induced release of FGF-1 was not attenuated by inhibitors of the classic ER–Golgi secretion pathway and exocytosis,[26] it was repressed by deoxyglucose, cycloheximide, and actinomycin D, indicating that the release mechanism required energy and depended on stress-induced protein synthesis and transcription.[23,26,27]

The presence of a phosphatidylserine (pS)-binding domain in FGF-1 and the transition of FGF-1 to the molten globule conformation at 42°[24,28] suggest the possibility of direct contact of this protein with the cell membrane. More recently it has been demonstrated that FGF-1 is released as a member of a multiprotein complex that includes synaptotagmin 1 and S100A13, and that these two calcium- and pS-binding proteins are critical for the stress-induced release of FGF-1.[27,29,30]

FGF-2, which also lacks a classical signal peptide sequence, may be constitutively released from certain types of cells.[31,32] This release occurs at normal physiological temperature and is not enhanced by heat shock. It is not sensitive to the inhibition of the ER–Golgi pathway but is attenuated by methylamine, an inhibitor of exocytosis. Also, unlike FGF-1, FGF-2 is released in a monomeric form. Thus, it appears that the constant release of FGF-2 and stress-induced release of FGF-1 may proceed through two distinct non-ER–Golgi pathways. It is of interest to identify within FGF-1 and FGF-2 the domains responsible for the regulation of their release because this information would be valuable for the identification of organelles and proteins participating in the export of the FGF prototypes.

Nuclear Association of Fibroblast Growth Factors

Unlike FGF-1, whose open reading frame is flanked by termination codons and whose mRNA translation results in a single 18-kDa protein, the FGF-2 transcript contains several upstream CUG translational start

[26] A. Jackson, F. Tarantini, S. Gamble, S. Friedman, and T. Maciag, *J. Biol. Chem.* **270,** 33 (1995).

[27] T. LaVallee, F. Tarantini, S. Gamble, C. Mouta Carreira, A. Jackson, and T. Maciag, *J. Biol. Chem.* **273,** 22217 (1998).

[28] H. Mach and C. R. Middaugh, *Biochemistry* **34,** 9913 (1995).

[29] F. Tarantini, T. LaVallee, A. Jackson, S. Gamble, C. Mouta Carreira, S. Garfinkel, W. H. Burgess, and T. Maciag, *J. Biol. Chem.* **273,** 22209 (1998).

[30] C. Mouta Carreira, T. LaVallee, F. Tarantini, A. Jackson, J. Tait Lathrop, B. Hampton, W. H. Burgess, and T. Maciag, *J. Biol. Chem.* **273,** 22224 (1998).

[31] P. Mignatti, T. Morimoto, and D. B. Rifkin, *Proc. Natl. Acad. Sci. U.S.A.* **88,** 11007 (1991).

[32] P. Mignatti, T. Morimoto, and D. B. Rifkin, *J. Cell. Physiol.* **151,** 81 (1992).

sites that yield larger forms of these proteins.[11,33,34] A similar alternative upstream CUG start site is also present in the FGF-3 transcript,[13] and a higher molecular weight form of the FGF-3 protein also exists. Interestingly, intracellular high molecular weight forms of FGF-2 and FGF-3 demonstrated a predominantly nuclear localization, with especially high concentrations of FGF-3 in nucleoli.[16,34,35] However, nuclear translocation is not limited only to the intracellular high molecular weight forms of FGF-2 and FGF-3. Indeed, it has been demonstrated that 10–15% of exogenous FGF-1, which is internalized by target cells, traffics to the nucleus during the G_1 period of the cell cycle.[14] Similar behavior was shown for the exogenous low molecular weight forms of FGF-2.[36,37] The traffic of exogenous FGF-1 to the nucleus is accompanied by the migration of FGFR-1 to a perinuclear locale,[38] and the nuclear translocation of exogenous FGF-1 depends on the presence in the extracellular domain of FGFR-1 of a terminal immunoglobulin-like loop that is dispensable for receptor–ligand binding.[39] Interestingly, in three independent situations in which FGF-1, FGF-2, and FGF-3 have either trafficked or been forced to traffic to the nucleus, the primary result has been the repression rather than the stimulation of cell proliferation.[16–18] Thus the identification of the FGF sequences that control their nuclear translocation is important because these mechanisms may yield new insight into the biological activities of these proteins.

Types of Protein Chimeras Used for the Study of Fibroblast Growth Factor Translocations

The major types of chimera used to identify the sequences that regulate release and nuclear translocation of FGFs are the products of FGF fusion with the bacterial β-galactosidase (β-Gal) reporter protein. A number of considerations prompted us to choose β-Gal. Proteins fused to β-Gal can be easily detected by a histochemical color reaction based on the use at neutral pH of the chromogenic substrate 5-bromo-4-chloro-3-indolyl-β-D-

[33] H. Prats, M. Kaghad, A. C. Prats, M. Klagsbrun, J. M. Lelias, P. Liauzun, P. Chalon, J. P. Tauber, F. Amalric, and J. A. Smith, *Proc. Natl. Acad. Sci. U.S.A.* **86,** 1836 (1989).

[34] M. Renko, N. Quarto, T. Morimoto, and D. B. Rifkin, *J. Cell. Physiol.* **144,** 108 (1990).

[35] R. Z. Florkiewicz, A. Baird, and A. M. Gonzalez, *Growth Factors* **4,** 265 (1991).

[36] G. Bouche, N. Gas, H. Prats, V. Baldin, J. P. Tauber, J. Teissie, and F. Amalric, *Proc. Natl. Acad. Sci. U.S.A.* **84,** 6770 (1987).

[37] V. Baldin, A. M. Roman, I. Bosc Bierne, F. Amalric, and G. Bouche, *EMBO J.* **9,** 1511 (1990).

[38] I. Prudovsky, N. Savion, X. Zhan, R. Friesel, J. Xu, J. Hou, W. L. Mckeehan, and T. Maciag, *J. Biol. Chem.* **269,** 31720 (1994).

[39] I. Prudovsky, N. Savion, T. M. LaVallee, and T. Maciag, *J. Biol. Chem.* **271,** 14198 (1996).

galactopyranoside (X-Gal). While mammalian cells and tissues are β-Gal negative under these conditions, it should be noted that senescent cells and certain types of differentiated cells may display endogenous β-Gal activity at neutral pH.[40,41] Thus, the use of control β-Gal-transfected cells is important for the interpretation of experiments using β-Gal fusion chimeras. There are also a number of monoclonal and polyclonal antibodies against β-Gal, and these are useful for immunohistochemistry, immunoblot analysis, and immunoprecipitation studies that enable the investigator to distinguish the FGF:β-Gal chimeras from endogenous FGFs. In addition, β-Gal fusion products can be easily isolated from conditioned media by affinity chromatography using the high-affinity binding of β-Gal to β-galactose. Finally, and most importantly, the expression of the β-Gal protein does not influence the intracellular localization or the release of fusion partners.[17] When transfected into mammalian cells, β-Gal displays cytosolic localization and, because of its relatively large mass (117 kDa), it does not translocate to the nucleus by simple diffusion.

The protein fusion approach to the study of FGF translocations is not limited to the use of reporter genes. Indeed, known nuclear localization signal (NLS) sequences from other proteins have been used for fusion with the FGFs to direct their nuclear translocation and a comparative study of FGF-1 and FGF-2 nuclear localization and release also included the shuffling of homologous regions/domains between these two proteins. Finally, Wiedlocha et al.[41a] produced chimeras of FGF-1 and diphtheria toxin (DT) that allowed the direct delivery of FGF-1 into the cytosol of Vero cells. This approach avoided the participation of the FGFR in the FGF-1 internalization pathway because Vero cells do not contain functional FGFRs, and this enabled an evaluation of the role of internalized FGF-1 in the activation of cell proliferation in the absence of FGFR-mediated signaling events.

Construction of FGF:β-Gal Chimeras

To construct the different forms of FGF:β-Gal chimeras, we use synthetic FGF-1 and FGF-2 cDNAs,[42,43] both encompassing amino acid resi-

[40] G. Dimri, X. Lee, G. Basile, M. Acosta, G. Scott, C. Roskelley, E. Medrano, M. Linskens, I. Rubel, O. Pereira-Smith, M. Peacocke, and J. Campisi, Proc. Natl. Acad. Sci. U.S.A. 92, 9363 (1995).
[41] Y. Yegorov, S. Akimov, R. Hass, A. Zelenin, and I. Prudovsky, Exp. Cell. Res. 243, 207 (1998).
[41a] A. Wiedlocha, P. O. Falnes, I. H. Madhus, K. Sandvig, and S. Olues, Cell 76, 1039 (1994).
[42] R. Forough, K. Engleka, J. A. Thompson, A. Jackson, T. Imamura, and T. Maciag, Biochim. Biophys. Acta 1090, 293 (1991).
[43] T. Imamura, S. A. Friedman, S. Gamble, Y. Tokita, S. R. Opalenik, J. A. Thompson, and T. Maciag, Biochim. Biophys. Acta 1266, 124 (1995).

dues 21 to 154 and containing a Kozak sequence and translation and termination codons, because the nucleotide sequence encoding for the first 20 amino acids of FGF-1 strongly downregulates the translation of the FGF-1 mRNA and FGF-1 devoid of this domain exhibits biological activities identical to those of the full-length FGF-1 translation product.[42] The plasmid pMC1871 (Pharmacia, Piscataway, NJ) is a source of β-Gal cDNA and common restriction sites that divide both the FGF-1 and FGF-2 cDNAs into four cassettes make possible the convenient shuffling of homologous sequences between these two proteins as well as the creation of deletion mutants. This approach in combination with site-specific mutagenesis is used to create a variety of FGF : β-Gal chimeras. For the transfection of mammalian cells, the chimeras are recloned into the pMEXneo expression vector.[44]

Expression of Fibroblast Growth Factor Chimeras in Mammalian Cells

NIH 3T3 cells (American Type Culture Collection, Rockville, MD) grown in Dulbecco's modified Eagle's medium (DMEM; GIBCO, Grand Island, NY) supplemented with 10% (v/v) calf serum are used in all experiments for the expression of FGF : β-Gal chimeras. Transfection is performed with a calcium phosphate transfection kit (Stratagene, La Jolla, CA). Transfectants are selected in complete cell culture medium containing G418 (800 μg/ml). The expression of transfected FGF : β-Gal chimeras is confirmed by immunoblot analysis with affinity-purified polyclonal antibodies developed against recombinant human FGF-1 and/or with commercial rabbit anti-β-Gal antibodies (5 Prime $\rightarrow$ 3 Prime, Boulder, Co). Secondary peroxidase-conjugated goat anti-rabbit antibodies (Bio-Rad, Hercules, CA), the ECL detection kit (Amersham, Arlington Heights, IL), and autoradiography are used to visualize the bound primary antibodies. Individual clones of cells expressing the FGF : β-Gal chimeras are maintained in cell culture medium containing G418 (400 μg/ml). Because, in many cases, the expression of the FGF : β-Gal chimeras strongly decreases cell adhesion, the transfectants are grown on cell culture dishes precoated with human fibronectin (10 μg/cm^2).

Detection of Intracellular Locale of Fibroblast Growth Factor

Immunofluorescence

NIH 3T3 cell transfectants are grown on fibronectin-coated (10 μg/cm^2) chamber slides (Nunc, Roskilde, Denmark) and the cells are fixed for 5 min

---

[44] D. Martin-Zanca, R. Oskam, G. Mitra, T. Copeland, and M. Barbacid, *Mol. Cell. Biol.* **9,** 24 (1989).

with 3% (v/v) formaldehyde in phosphate-buffered saline (PBS), washed in PBS, permeabilized for 5 min in PBS plus 0.5% (v/v) Nonidet P-40 (NP-40), blocked for 30 min in PBS containing 5% (w/v) bovine serum albumin (BSA) and 5% (v/v) goat serum, incubated for 3 hr with rabbit anti-β-Gal antibodies diluted 500-fold in PBS containing 1.5% (w/v) BSA and 1.5% (v/v) goat serum (staining buffer), washed with PBS, incubated for 1 hr with biotinylated goat anti-rabbit IgG antibodies (Life Technologies, Gaithersburg, MD) diluted 1 : 250 in staining buffer, washed with PBS, incubated for 30 min in streptavidin–fluorescein isothiocynate (FITC) (Life Technologies) diluted 1 : 250 in staining buffer, embedded in 50% (v/v) glycerol, and examined under a fluorescence microscope. To verify the nuclear localization of certain FGF : β-Gal chimeras the immunofluorescence preparations are examined with a laser confocal microscope with a 10-μm confocal slit.

X-Gal Staining

In addition to the immunofluorescence studies, the localization of the FGF : β-Gal chimeras is identified at the cytologic level by a colorimetric enzymatic reaction. To monitor the enzymatic activity of β-Gal, the transfectant monolayers are grown on cell culture dishes precoated with human fibronectin (10 μg/cm^2), fixed with 2% (v/v) formaldehyde plus 0.2% (v/v) glutaraldehyde in PBS for 5 min at 4°, washed with PBS, and incubated for at least 20 min at 37° with X-Gal (1 mg/ml) in PBS containing 2 mM MgCl$_2$, 5 mM potassium ferrocyanide, and 5 mM potassium ferricyanide. The intracellular localization of the insoluble blue complex formed by the enzymatic activity of β-Gal is examined under a light microscope.

Cell Fractionation

The nonquantitative immunofluorescence and X-Gal staining experiments are verified by obtaining biochemical information about the intracellular distribution of the FGF : β-Gal chimeras. Thus cell fractionation in combination with immunoblot analysis is implemented, including the use of two different methods for the isolation of nuclei. The procedures for both methods are performed at 4°, if not otherwise indicated.

In the first method, cells are scraped from dishes into cold PBS, quickly centrifuged (5 min, 1000g), resupended in a hypotonic buffer [2 mM NaHCO$_3$, pH 7.0, containing 1 mM MgCl$_2$, 1 mM EDTA, 2 mM phenylmethylsulfonyl fluoride (PMSF) and aprotinin (2 μg/ml)] and lysed with a Dounce homogenizer. Nuclear pellets are precipitated by centrifugation at 800g for 20 min. Cytosolic lysates are further clarified by centrifugation at 14,000g for 10 min. The nuclear pellets are washed twice with 10 mM HEPES, pH 7.4, containing 0.25 M sucrose, 3 mM NaCl, 2 mM PMSF, and aprotinin (2 μg/ml) and then washed three times with 10 mM Tris-HCl, pH 7.4, containing 1 mM EDTA, 2 mM MgCl$_2$, 0.25 M sucrose, 2 mM PMSF, and aprotinin (2 μg/ml). The nuclear proteins are extracted with standard 2× sodium dodecyl sulfate (SDS) sample buffer [6.25 mM Tris-HCl, pH 6.8, containing 10% (v/v) glycerol, 3 mM SDS, 1% (v/v) 2-mercaptoethanol, and 0.75 mM bromphenol blue], resolved in parallel with the cytosol fractions (mixed 1:1 with the sample buffer) by SDS–7.5% (w/v) polyacrylamide gel electrophoresis (7.5% SDS–PAGE), and analyzed by immunoblotting as described above.

In the second method of cell fractionation, scraped cells are resuspended in 1 ml of buffer A [10 mM HEPES, pH 7.5, containing 300 mM sucrose, 60 mM KCl, 15 mM NaCl, 0.5 mM EDTA, 0.5% (v/v) Triton X-100, 1 mM PMSF, and aprotinin (2 μg/ml)] and kept on ice for 10 min. After lysis by rapid pipetting, the lysates are centrifuged for 10 min at 800g (pellet/nuclear fraction) and the supernatant is clarified by centrifugation at 14,000g for 10 min (cytosol fraction). The nuclear pellets are resuspended in buffer A and then centrifuged through buffer A containing 350 mM sucrose at 800g for 10 min to remove the remains of the cytosol fraction. The nuclear pellets are resuspended in buffer A containing 0.4% (w/v) SDS, 8 mM MgCl$_2$, and 10 units of DNase and incubated for 5 min at 37°, followed by a 15-min incubation at room temperature to destroy DNA and release nuclear proteins. The nuclear fractions are clarified by centrifugation at 14,000g for 10 min. The FGF:β-Gal chimeras from both the cytosol and nuclear fractions are concentrated by immunoprecipitation with rabbit anti-β-Gal antibodies (1 μl/ml of lysate with 1 hr of rotation) followed by the addition of protein A–Sepharose beads (CL-4B; Pharmacia-LKB) (20 μl/ml of lysate with 1 hr of rotation). The Sepharose beads are washed three times with RIPA buffer [10 mM Tris-HCl, pH 7.5, containing 150 mM NaCl, 1% (v/v) Triton X-100, 1 mM EDTA, 0.5% (w/v) deoxycholic acid, and 0.1% (w/v) SDS], extracted by boiling for 3 min in 2× SDS sample buffer, and resolved by 7.5% SDS–PAGE for immunoblot analysis.

Detection of Fibroblast Growth Factor Chimeras Released under Heat Shock Conditions

Subconfluent monolayers of NIH 3T3 transfectants are used and 1 day prior to heat shock, the cells are incubated with cell culture medium lacking G418 in order to eliminate the possible leakage, potentially mediated by G418, of FGF-1 from the cells. Immediately prior to the heat shock, the cell culture medium is removed and the transfectants fed serum-free DMEM containing heparin (10 μg/ml).[26] The cells are incubated in this medium for 2 hr at 42° and control cell transfectants are incubated at 37° under identical conditions for 2 hr. The conditioned media are collected and treated with 0.1% (w/v) dithiothreitol (DTT) for 2 hr at 37° in order to reduce the FGF-1 dimers to their monomer form, which binds to heparin–Sepharose with high affinity. The reduced cell culture medium is filtered through a 0.22-μm pore size filter (Corning, Acton, MA) and processed over a 1-ml heparin–Sepharose 4B (Pharmacia-LKB) column previously equilibrated with TE (10 mM Tris-HCl, pH 7.4, containing 10 mM EDTA). The column is washed with 10 to 20 ml of TE, and bound material is eluted with TE containing 1.5 M NaCl, concentrated through a Centricon 10 membrane (Amicon, Danvers, MA), eluted from the Centricon with 100 μl of 2× SDS sample buffer, and resolved by 7.5% SDS–PAGE for immunoblot analysis.

When an FGF:β-Gal chimera has a mutation that reduces its affinity for immobilized heparin,[45] an alternative method for its isolation must be utilized. In this case, the DTT-treated and filtered medium can be processed over a 1-ml column of β-Gal–agarose (Sigma, St. Louis, MO) equilibrated with buffer A [10 mM Tris-HCl, pH 7.6, containing 0.25 M NaCl, 10 mM MgCl$_2$, 1 mM EDTA, 10 mM 2-mercaptoethanol and 0.1% (v/v) Triton X-100], the column washed with 20 ml of buffer A, followed by a wash with 2 ml of buffer A without Triton X-100, and bound material eluted with 2 ml of 0.1 M Na$_2$B$_4$O$_7$, pH 10, into a vial already containing 0.5 ml of 10 mM NaH$_2$PO$_4$, pH 7.2, and 150 mM NaCl. We recommend that the sample be concentrated in a Centricon 30 and eluted from the Centricon with 2× SDS sample buffer prior to resolution by 7.5% SDS–PAGE for immunoblot analysis.

Identification of Nuclear Localization and Cytosolic Retention Sequences in Fibroblast Growth Factors

As described earlier, endogenously expressed FGF-1 displays a cytosolic localization and does not translocate to the nucleus even though it contains a putative NLS (NYKKPKL) near its amino-terminal end. To evaluate the potential of any NLS, its minimum sequence should be fused to a reporter gene, stable transfectants obtained, and the localization of the reporter within the nucleus confirmed by microscopic and biochemical methods. Using polymerase chain reaction (PCR) methods,[46] the NYKKPK sequence is fused in frame at the amino terminus of β-Gal, the resultant construct

[45] P. Wong, B. Hampton, E. Szylobryt, A. M. Gallagher, M. Jaye, and W. H. Burgess, *J. Biol. Chem.* **270,** 25805 (1995).

[46] X. Zhan, X. G. Hu, S. Friedman, and T. Maciag, *Biochem. Biophys. Res. Commun.* **188,** 982 (1992).

is introduced by transfection into NIH 3T3 cells, and stable transfectants are obtained under G418 selection. Indeed, histochemical analysis using anti-β-Gal antibodies reveals that the transfectants exhibit nuclear localization of the chimeric protein, in contrast with the control NIH 3T3 cells transfected with β-Gal, which display cytosolic localization of the reporter protein.[46] However, the product of the β-Gal fusion at the carboxy terminus of full-length FGF-1 reveals a rather distinct cytosolic localization despite the presence of the putative NLS.[46] These data suggest that FGF-1 may contain sequence(s) responsible for its retention in the cytosol. Indeed, histochemical cell fractionation and immunoblot analysis demonstrate that the replacement of the native FGF-1 NLS with the NLS from yeast histone-2B fails to direct the FGF-1:β-Gal chimera to the nucleus, even though the histone-2B NLS provides efficient nuclear translocation of β-Gal.[46] These data suggest that FGF-1 may contain a sequence that represses the function of its NLS and maintains the presence of FGF-1 in the cytosol.

In a further set of experiments, the domains of FGF-1 are analyzed in order to characterize the domain responsible for cytosolic retention. As previously discussed, the synthetic human FGF-1 gene is prepared by ligation of four cassettes encoding the FGF-1 open reading frame (ORF), and this enables different combinations of these cassettes to be fused with the reporter gene, β-Gal (Fig. 1). Analysis of the intracellular localization of

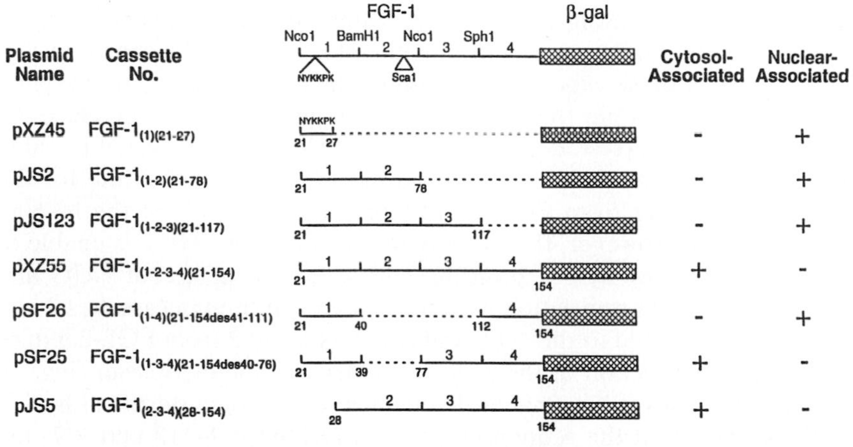

Plasmid Name	Cassette No.		Cytosol-Associated	Nuclear-Associated
pXZ45	FGF-1$_{(1)(21-27)}$		-	+
pJS2	FGF-1$_{(1-2)(21-78)}$		-	+
pJS123	FGF-1$_{(1-2-3)(21-117)}$		-	+
pXZ55	FGF-1$_{(1-2-3-4)(21-154)}$		+	-
pSF26	FGF-1$_{(1-4)(21-154des41-111)}$		-	+
pSF25	FGF-1$_{(1-3-4)(21-154des40-76)}$		+	-
pJS5	FGF-1$_{(2-3-4)(28-154)}$		+	-

FIG. 1. Representation of stable FGF-1:β-Gal expression constructs and cellular distribution of the fusion proteins in NIH 3T3 cells as determined by immunofluorescence and immunoblot analysis of cell fractionation samples. The first set of parentheses in the subscript refers to the FGF-1 cassette number and the second set of parentheses in the subscript refers to the FGF-1 amino acid sequence. The notation "des" refers to sequences within a specific cassette that were deleted.

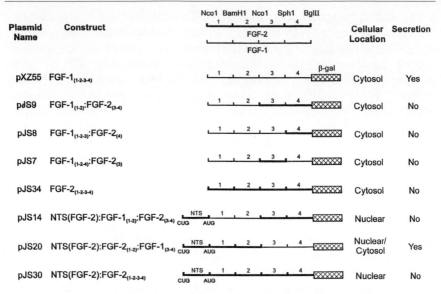

FIG. 2. Schematic representation of FGF-1, FGF-2, and FGF-1 : FGF-2 chimera reporter gene constructs and their intracellular distribution and appearance in the extracellular compartments in response to heat shock. Symbols are the same as in Fig. 1.

these FGF-1 : β-Gal chimeras within the individual NIH 3T3 cell transfectants is performed using histochemistry, immunocytochemistry, and cell fractionation in combination with immunoblot analysis. As shown in Fig. 1, the results of these experiments indicate that a cytosolic retention signal (CRS) is located within the carboxy-terminal half of FGF-1 encompassed by cassettes 3 and 4 (residues 84–154). Using a similar strategy, it is possible to demonstrate that cassettes 3 and 4 from FGF-2 can substitute for the respective cassettes from FGF-1 and repress the function of the NLS in FGF-1 (Fig. 2). However, the carboxy-terminal half of FGF-2 is unable to repress the NLS of FGF-2 positioned between the upstream CUG and conventional AUG translation initiation sites, even though cassettes 3 and 4 from FGF-1 ligated to this NLS and cassettes 1 and 2 from FGF-2 induce a significant redirection of the reporter β-Gal into the cytoplasm (Fig. 2). Additional studies using smaller deletions in the carboxy-terminal half of FGF-1 suggest that the sequences between residues 99–112 and 137–154 are both important for the cytosolic retention of FGF-1 (Fig. 3).

In an attempt to identify individual amino acid residues critical for the cytosolic retention of FGF-1, FGF-1 : β-Gal chimeras are obtained with asparagine at position 109 mutated to valine. This particular amino acid is chosen because Asp-109 is conserved among FGF family mem-

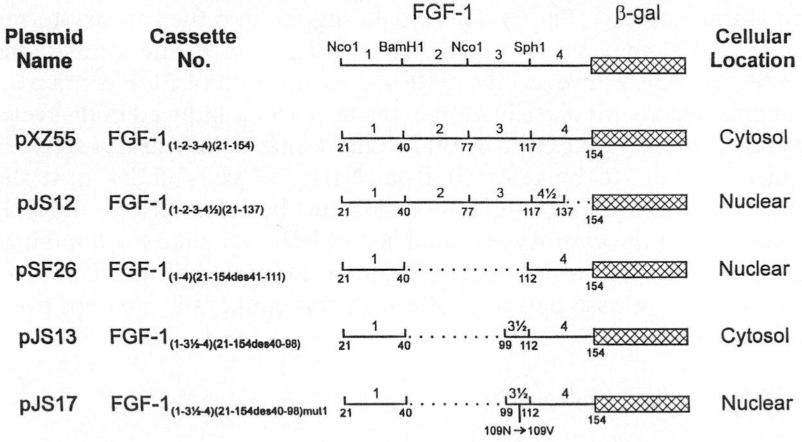

FGF-1   β-gal

Plasmid Name	Cassette No.	Ncol 1 BamH1 2 Ncol 3 Sph1 4	Cellular Location
pXZ55	FGF-1(1-2-3-4)(21-154)	1 / 2 / 3 / 4 ; 21 40 77 117 154	Cytosol
pJS12	FGF-1(1-2-3-4½)(21-137)	1 / 2 / 3 / 4½ ; 21 40 77 117 137 154	Nuclear
pSF26	FGF-1(1-4)(21-154des41-111)	1 . . . . 4 ; 21 40 112 154	Nuclear
pJS13	FGF-1(1-3½-4)(21-154des40-98)	1 . . . . . 3½ 4 ; 21 40 99 112 154	Cytosol
pJS17	FGF-1(1-3½-4)(21-154des40-98)mut1	1 . . . . . 3½ 4 ; 21 40 99 |112 154 ; 109N→109V	Nuclear

FIG. 3. Identification of cytosol retention sequences in the carboxy-terminal half of FGF-1, using specific deletion mutants. Symbols are the same as in Figs. 1 and 2.

bers.[47] Indeed, the FGF-1 N109V mutation in the cytosol-localized chimera lacking residues 41 to 88 can direct this chimera to the nucleus. This result suggests that Asp-109 is crucial for the cytosolic retention of FGF-1.

Studies of Nonclassical Fibroblast Growth Factor Release Pathway Using Chimeric Proteins

Prior to the identification of sequences in FGF-1 that are involved in the regulation of stress-induced nonclassic release, it was important to determine whether intranuclear FGF-1 could be released in response to heat shock. Preliminary experiments have demonstrated that the FGF-1:β-Gal chimera, like FGF-1, is released into the conditioned medium in response to heat shock. However, the FGF-1:β-Gal chimera containing the simian virus 40 (SV40) large T antigen NLS and displaying nuclear localization fails to be released under identical conditions.[17] This result argues that intranuclear FGF-1 : β-Gal is not accessible to the FGF-1 release pathway. Also, the FGF:β-Gal chimeras containing at their amino termini the FGF-2 NLS positioned between the CUG and AUG codons, and also displaying nuclear localization, are not released in response to heat shock[17] (Fig. 2). In addition, a chimera containing the FGF-2 NLS, FGF-2 cassettes 1 and 2, and FGF-1 cassettes 3 and 4, and demonstrating both nuclear and cytosolic localization has been found to be released in response to

[47] W. H. Burgess and T. Maciag, *Annu. Rev. Biochem.* **58,** 575 (1989).

temperature stress[17] (Fig. 2). These data suggest that the carboxy-terminal half of FGF-1 may be critical for its participation in the stress-induced release pathway. However, the cytosolic localization of FGF is not a sufficient prerequisite for participation in the heat shock-induced FGF-1 release pathway. Indeed, the FGF-2:β-Gal chimera that demonstrates a cytosolic localization fails to be released from NIH 3T3 cells under heat shock conditions (Fig. 2). The shuffling of cassettes between FGF-1 and FGF-2 has shown that the carboxy-terminal half of FGF-2, unlike the homologous region of FGF-1, does not direct the amino-terminal half of FGF-1 to the stress-induced release pathway, although this chimera is also not translocated to the nucleus.

Conclusions and Perspectives

The reporter gene chimera strategy proved to be an informative approach for the study of nuclear translocation and nonclassic exocytosis of FGFs. This strategy enabled us to determine that the intracellular localization of both FGF-1 and FGF-2 is regulated by the interplay of the amino-terminal NLS and the CRS located in the carboxy-terminal domain. Indeed, the experiments with chimeras containing the NLS of the FGF-2 CUG-initiated product indicated that the carboxy-terminal half of FGF-2 has a lower cytosolic retention potential than the corresponding domain from FGF-1 and that unlike FGF-2, the carboxy-terminal half of FGF-1 may direct the protein to enter the heat shock-induced release pathway, which does not recognize nuclear forms of FGF-1.

It is anticipated that the reporter gene strategy will provide a more detailed delineation of the FGF-1 and FGF-2 sequences responsible for the regulation of their release and intracellular trafficking, and may aid in the identification of the mechanisms for these nonconventional processes. In addition, the gene chimera approach will undoubtedly be used to study the other members of the growing FGF family, perhaps using the cassette shuffle strategy. Likewise, transgenic mice expressing various types of FGF:β-Gal chimeras should prove to be informative experimental models in verifying the physiologic significance of the *in vitro* data. Indeed, FGF:β-Gal transgenics may enable the study of the stress-induced FGF release pathway *in vivo* as well as nuclear translocation of FGFs in different tissues, at different stages of ontogenesis, and within cells with different proliferative potential, both normal and malignant.

Acknowledgments

We thank Ms. A. M. Blier for expert administrative assistance. M.L. was supported by a fellowship from the Catholic University of Rome and this effort was supported in part by NIH Grants HL35627, HL32348, and AG07450 to T.M.

Section V

Application of Chimeras in Monitoring and Manipulating Cell Physiology

[29] Posttranslational Regulation of Proteins by Fusions to Steroid-Binding Domains*

By Didier Picard

Introduction

It is often helpful or even a necessity in biological studies to be able to regulate not only the expression but also the activity of a protein of interest. Intracellular receptors by virtue of being ligand-regulated transcription factors have naturally been examined as regulatory tools. The application that is the focus of this chapter exploits the normal regulatory activity of the hormone-binding domain (HBD) of vertebrate steroid receptors for the regulation of heterologous proteins *in cis*. On fusion to a heterologous protein, the HBD can function as an autonomous regulatory cassette and subject various protein functions to hormonal control.[1,2] Thus, one or several functional activities of the heterologous moiety may be turned off in the absence of steroid ligand. While the chimeric protein can be expressed constitutively, it becomes active only on addition of hormone. Within at most a few minutes the hormonally-induced conformational change in the HBD results in the relief of its protein inactivation function and activation of the chimeric protein (see Fig. 1).

Other methods to obtain conditional expression and/or activity of proteins of interest have, of course, also been described. These include the elegant use of the insect hormone receptor for ecdysone for the specific transcriptional regulation of gene expression in mammalian cells and in transgenic mice.[3] However, several features and advantages make the application of HBDs unique and have made it increasingly popular since its first description more than 10 years ago.[1] We have previously provided a comprehensive review and comparison of this method with others[4-6]; this chapter therefore focuses on the most recent developments and the practical

---

* This chapter has been adapted from D. Picard, *in* "Nuclear Receptors: A Practical Approach" (D. Picard. ed.), p. 261. Oxford University Press, Oxford, 1999. [By permission of Oxford University Press.]

[1] D. Picard, S. J. Salser, and K. R. Yamamoto, *Cell* **54**, 1073 (1988).

[2] M. Eilers, D. Picard, K. R. Yamamoto, and J. M. Bishop, *Nature (London)* **340**, 66 (1989).

[3] D. No, T. P. Yao, and R. M. Evans, *Proc. Natl. Acad. Sci. U.S.A.* **93**, 3346 (1996).

[4] D. Picard, *Curr. Opin. Biotechnol.* **5**, 511 (1994).

[5] D. Picard, *Trends Cell Biol.* **3**, 278 (1993).

[6] T. Mattioni, J.-F. Louvion, and D. Picard, *in* "Protein Expression in Animal Cells" (M. Roth, ed.), p. 335. Academic Press, San Diego, California, 1994.

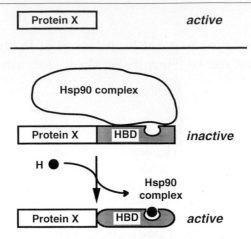

FIG. 1. Schematic representation of the regulation of a heterologous protein (protein X) by fusion to the hormone-binding domain (HBD) of a steroid receptor. The HBD along with the hormone-reversible Hsp90 complex functions as a posttranslational regulatory cassette. Inactivation mediated by the Hsp90 complex ceases on hormone (H) binding. [Adapted from D. Picard, *Curr. Opin. Biotechnol.* **5,** 511 (1994); by permission of Current Biology, Ltd.]

aspects. To facilitate access to information, much of it is presented in protocols and tables.

Key Features

Many of the key features of this approach to regulating protein activity are unique not by themselves, but in combination. They can be summarized as follows.

1. A wide variety of heterologous proteins can be regulated with HBDs (see Table I,[7–59] and for updates http://www.picard.ch).

[7] R. W. Sablowski and E. M. Meyerowitz, *Cell* **92,** 93 (1998).

[8] T. Aoyama, C. H. Dong, Y. Wu, M. Carabelli, G. Sessa, I. Ruberti, G. Morelli, and N.-H. Chua, *Plant Cell* **7,** 1773 (1995).

[9] R. M. Umek, A. D. Friedman, and S. L. McKnight, *Science* **251,** 288 (1991).

[10] C. Müller, E. Kowenz-Leutz, S. Grieser-Ade, T. Graf, and A. Leutz, *EMBO J.* **14,** 6127 (1995).

[11] R. Simon, M. I. Igeno, and G. Coupland, *Nature (London)* **384,** 59 (1996).

[12] P. Jansen-Dürr, A. Meichle, P. Steiner, M. Pagano, K. Finke, J. Botz, J. Wessbecher, G. Draetta, and M. Eilers, *Proc. Natl. Acad. Sci. U.S.A.* **90,** 3685 (1993).

[13] H. Pelczar, O. Albagli, A. Chotteau-Lelièvre, I. Damour, and Y. de Launoit, *Biochem. Biophys. Res. Commun.* **239,** 252 (1997).

[14] M. Schuermann, G. Hennig, and R. Müller, *Oncogene* **8,** 2781 (1993).

[15] G. Superti-Furga, G. Bergers, D. Picard, and M. Busslinger, *Proc. Natl. Acad. Sci. U.S.A.* **88,** 5114 (1991).

TABLE I
FUSION PROTEINS

Protein X[a]	HBD[b]	Regulated as[c]:	Ref.
Transcription factors			
APETALA3	GR	Transcription factor in *Arabidopsis*	7
Athb-1	GR	*Arabidopsis* transcription factor in to-bacco	8
C/EBP	ER, GR	Transcription factor	9
C/EBPβ (NF-M)	ER	Transcription factor, differentiation factor	10
CONSTANS	GR	Putative transcription factor in *Arabidopsis*	11
E1A	GR	Transcription factor	1
	ER	Oncoprotein	12
E7 (of HPV16)	ER	Oncoprotein	d
Erm (Ets family)	ER	Transcription factor	13
c-Fos, v-Fos, FosB-L, FosB-S	ER, GR	Oncoprotein, transcription factor	14, 15
Gal4	ER, GR, MR, PR	Transcription factor in yeast and tissue culture cells	16, e
Gal4-KRAB	PR[f]	Transcriptional repressor	17
Gal4-p65[g]	PR[f]	Transcription factor	18
Gal4-VP16	ER, GR, PR[f]	Transcription factor in yeast, tissue culture cells, transgenic mice, and plants	17, 19–23
GATA-1, -2, -3	ER	Transcription factor, promoter of proliferation	24
Gcn4	ER, MR	Transcription factor	25
c-Jun	ER	Transcription factor	26
JunD	ER	Transcription factor	27
v-Jun (DBD[h])	ER	As DNA-binding factor	28
LexA-VP16	ER	Transcription factor in yeast	e
v-Myb	ER	Transcription factor	29
c-Myc	ER, GR	Oncoprotein	2
MyoD	ER, TR, GR	Transcription factor in tissue culture and frog embryos	30, 31
p53	ER	Regulator of proliferation	32
Pax-5	ER	Transcription factor	33
R (of maize)	GR	Transcription factor in *Arabidopsis*	34
v-Rel, c-Rel	ER	Oncoprotein, transcription factor	35, 36
STAT6	ER[f]	Transcription factor	37
TLS-CHOP	ER	Oncoprotein	38
Xbra	GR	Transcription factor in frog embryos	39
Kinases			
Abl	ER, GR	Oncoprotein, tyrosine kinase	40
Akt (PKB)	ER[f]	Serine/threonine kinase	41
erbB1	ER	Tyrosine kinase	d

(*continued*)

TABLE I (*continued*)

Protein X[a]	HBD[b]	Regulated as[c]:	Ref.
MEK1	ER[f]	Oncoprotein, dual kinase	42
MEKK3	ER	Activation of SAPK pathway	43
Raf-1	ER	Oncoprotein, serine/threonine kinase	44
A-Raf, B-Raf	ER	Oncoproteins	45
Ste11	ER, MR, PR	Serine/threonine kinase in yeast	6, e
Src	ER	Tyrosine kinase	d; see also Ref. 46
Recombinases			
Cre	ER,[f] PR,[f] GR[f]	Recombinase in tissue culture cells and in transgenic mice	47–53
Flp	ER, GR, AR	Recombinase in tissue culture cells and yeast	54, 55
Miscellaneous			
Fas	ER, RAR	Apoptosis	56
β-Galactosidase	ER, PR	α-Complementation in yeast	57, e
Rep (of AAV)	ER	Replication	i
Rev (of HIV)	GR	*trans*-Activation (RNA-binding protein)	58
Rex (of HTLV-I)	ER	Rex functions, localization	59

[a] Proteins are alphabetically grouped within classes.
[b] HBDs are from the following receptors: AR, ER, GR, MR, PR, RAR, and TR (androgen, estrogen, glucocorticoid, mineralocorticoid, progesterone, retinoic acid, and thyroid receptors, respectively).
[c] Unless otherwise indicated assays were done in vertebrate tissue culture cells.
[d] J. M. Bishop, personal communication (1993).
[e] Picard laboratory, unpublished results (1999).
[f] Mutant HBDs that respond only to antihormones were used in some experiments.
[g] GAL4 fusion protein contains activation domain of the NF-κB component p65.
[h] DBD, DNA-binding domain
[i] A. Salvetti, personal communication (1999).

2. Only one DNA construct, the one for expression of the chimeric protein (protein X–HBD fusion protein), needs to be made and expressed.

3. Activation is rapid and reversible.

4. Intermediate levels of activity can be obtained at subsaturating hormone concentrations.

[16] D. Picard, *in* "Nuclear Receptors: A Practical Approach" (D. Picard, ed.), p. 261. Oxford University Press, Oxford, 1999.
[17] Y. Wang, J. Xu, T. Pierson, B. O'Malley, and S. Tsai, *Gene Ther.* **4**, 432 (1997).
[18] M. M. Burcin, G. Schiedner, S. Kochanek, S. Y. Tsai, and B. W. O'Malley, *Proc. Natl. Acad. Sci. U.S.A.* **96**, 355 (1999).

TABLE II
CHOICE OF HORMONE-BINDING DOMAINS AND LIGANDS

HBD	Amino acids from C terminus	Ligand[a]	Selected refs.
GR	255 to 310		
wild type		Dex or RU-486	1
I747T[b,c]		Dex or RU-486	53
MR	300	Aldosterone	25
ER[b]	314[d]		
G400V		E2 or OHT	2
G521R[c]		OHT	51
G400V-L539A-L540A[c]		OHT or ICI	48
G400V-M543A-L544A[c]		OHT	48
AR	295	DHT	55
PR			
Wild type (chicken)	290	Progesterone	e
hPRB891 (human)[c]	252	RU-486	19
delta C19 (human)[c]	~250	RU-486	17

[a] Dex, Dexamethasone; E2, 17β-estradiol; OHT, 4-hydroxytamoxifen; ICI, ICI 182780; DHT, 5α-dihydrotestosterone. Ligands can be obtained from Sigma except ICI (Zeneca Pharmaceuticals). Reasonable final concentrations: 10 μM Dex or RU-486 (10 μM deoxy-corticosterone for yeast) for GR, 10 nM aldosterone (10 μM deoxycorticosterone for yeast), 0.1 μM E2, OHT, ICI, DHT, progesterone, and RU-486 for PR. Stock solutions (1000× in ethanol) are stored at $-20°$.

[b] Amino acid positions of mutants are based on numbering in human receptor.

[c] These mutant HBDs do not respond to the physiological agonist.

[d] Amino acids corresponding to human ER amino acids 304–551 may be sufficient [M. Nichols, J. M. Rientjes, and A. F. Stewart, *EMBO J.* **17,** 765 (1998)].

[e] Picard laboratory, unpublished results (1999).

5. Side effects through the hormonal activation of endogenous steroid receptors can be avoided by choosing a wild-type or mutant HBD with the appropriate specificity (see Table II).

6. Several proteins can be regulated independently and specifically in the same cell or organism by using different HBDs (see Table II).

[19] Y. Wang, B. W. O'Malley, Jr., S. Y. Tsai, and B. W. O'Malley, *Proc. Natl. Acad. Sci. U.S.A.* **91,** 8180 (1994).

[20] J.-F. Louvion, B. Havaux-Copf, and D. Picard, *Gene* **131,** 129 (1993).

[21] T. Aoyama and N.-H. Chua, *Plant J.* **11,** 605 (1997).

[22] Y. Wang, F. J. DeMayo, S. Y. Tsai, and B. W. O'Malley, *Nature Biotechnol.* **15,** 239 (1997).

[23] S. Braselmann, P. Graninger, and M. Busslinger, *Proc. Natl. Acad. Sci. U.S.A.* **90,** 1657 (1993).

[24] K. Briegel, K.-C. Lim, C. Planck, H. Beug, J. D. Engel, and M. Zenke, *Genes Dev.* **7,** 1097 (1993).

7. The approach works in a large variety of tissue culture cell lines and species ranging from yeast to plants to mammals.

8. The approach works in transgenic organisms and has the potential to work in gene therapeutic applications.

Basic Approach to Making and Using Hormone-Binding Domain Fusion Proteins

For the following the assumption is made that readers are familiar with standard molecular biology techniques or have access to corresponding

[25] C. P. Fankhauser, P.-A. Briand, and D. Picard, *Biochem. Biophys. Res. Commun.* **200,** 195 (1994).

[26] I. Fialka, H. Schwarz, E. Reichmann, M. Oft, M. Busslinger, and H. Beug, *J. Cell Biol.* **132,** 1115 (1996).

[27] M. K. Francis, D. G. Phinney, and K. Ryder, *J. Biol. Chem.* **270,** 11502 (1995).

[28] U. Kruse, J. S. Iacovoni, M. E. Goller, and P. K. Vogt, *Proc. Natl. Acad. Sci. U.S.A.* **94,** 12396 (1997).

[29] O. Burk and K.-H. Klempnauer, *EMBO J.* **10,** 3713 (1991).

[30] S. M. Hollenberg, P. F. Cheng, and H. Weintraub, *Proc. Natl. Acad. Sci. U.S.A.* **90,** 8028 (1993).

[31] P. J. Kolm and H. L. Sive, *Dev. Biol.* **171,** 267 (1995).

[32] K. Roemer and T. Friedmann, *Proc. Natl. Acad. Sci. U.S.A.* **90,** 9252 (1993).

[33] S. L. Nutt, A. M. Morrison, P. Dorfler, A. Rolink, and M. Busslinger, *EMBO J.* **17,** 2319 (1998).

[34] A. M. Lloyd, M. Schena, V. Walbot, and R. W. Davis, *Science* **266,** 436 (1994).

[35] G. Boehmelt, A. Walker, N. Kabrun, G. Mellitzer, H. Beug, M. Zenke, and P. J. Enrietto, *EMBO J.* **11,** 4641 (1992).

[36] M. Zurovec, O. Petrenko, R. Roll, and P. J. Enrietto, *Oncogene* **16,** 3133 (1998).

[37] Y. Kamogawa, H. J. Lee, J. A. Johnston, M. McMahon, A. O'Garra, and N. Arai, *J. Immunol.* **161,** 1074 (1998).

[38] H. Zinszner, R. Albalat, and D. Ron, *Genes Dev.* **8,** 2513 (1994).

[39] M. Tada, M. A. O'Reilly, and J. C. Smith, *Development* **124,** 2225 (1997).

[40] P. Jackson, D. Baltimore, and D. Picard, *EMBO J.* **12,** 2809 (1993).

[41] A. D. Kohn, A. Barthel, K. S. Kovacina, A. Boge, B. Wallach, S. A. Summers, M. J. Birnbaum, P. H. Scott, J. C. Lawrence, Jr., and R. A. Roth, *J. Biol. Chem.* **273,** 11937 (1998).

[42] H. Greulich and R. L. Erikson, *J. Biol. Chem.* **273,** 13280 (1998).

[43] H. Ellinger-Ziegelbauer, K. Brown, K. Kelly, and U. Siebenlist, *J. Biol. Chem.* **272,** 2668 (1997).

[44] M. L. Samuels, M. J. Weber, J. M. Bishop, and M. McMahon, *Mol. Cell. Biol.* **13,** 6241 (1993).

[45] C. A. Pritchard, M. L. Samuels, E. Bosch, and M. McMahon, *Mol. Cell. Biol.* **15,** 6430 (1995).

[46] T. Mattioni, B. J. Mayer, and D. Picard, *FEBS Lett.* **390,** 170 (1996).

[47] D. Metzger, J. Clifford, H. Chiba, and P. Chambon, *Proc. Natl. Acad. Sci. U.S.A.* **92,** 6991 (1995).

[48] R. Feil, J. Wagner, D. Metzger, and P. Chambon, *Biochem. Biophys. Res. Commun.* **237,** 752 (1997).

protocols. Only a few basic reagents are required. Most noteworthy is the need for a recombinant plasmid with the open reading frame (ORF) of the HBD of choice (see Table II), preferably with several unique restriction sites 3' of the stop codon. Please refer to the original literature (see also Table II) to obtain information on where to obtain the HBDs of interest. With applications in tissue culture cells steroid contaminants in serum and the steroid-like action of the pH indicator phenol red can be a concern. They can be avoided by treating the serum with charcoal (see below) and by using commercially available medium without phenol red. Table II provides information on the preparation of stock solutions of activating ligands.

Method

1. Choose the HBD on the basis of the complement of endogenous steroids and steroid receptors in the *in vivo* system of interest. Alternatively, choose one of the antihormone-specific HBD mutants such as the tamoxifen- and RU-486-specific estrogen receptor (ER)[47,48] and progesterone receptor (PR)[17,19,49] mutants, respectively (Table II). These mutants are particularly helpful for applications in transgenic mammals because the soaring estrogen and progesterone concentrations during pregnancy would activate fusion proteins with wild-type HBDs.

2. Decide between an N- and a C-terminal fusion. The latter is usually slightly easier to construct because HBD ORFs naturally come with a stop codon. A disadvantage of placing HBD coding sequences at the 5' end of the heterologous protein is that translational initiation at an internal AUG

[49] C. Kellendonk, F. Tronche, A.-P. Monaghan, P.-O. Angrand, F. Stewart, and G. Schütz, *Nucleic Acids Res.* **24,** 1404 (1996).

[50] J. Brocard, X. Warot, O. Wendling, N. Messaddeq, J. L. Vonesch, P. Chambon, and D. Metzger, *Proc. Natl. Acad. Sci. U.S.A.* **94,** 14559 (1997).

[51] R. Feil, J. Brocard, B. Mascrez, M. LeMeur, D. Metzger, and P. Chambon, *Proc. Natl. Acad. Sci. U.S.A.* **93,** 10887 (1996).

[52] Y. Zhang, C. Riesterer, A. M. Ayrall, F. Sablitzky, T. D. Littlewood, and M. Reth, *Nucleic Acids Res.* **24,** 543 (1996).

[53] J. Brocard, R. Feil, P. Chambon, and D. Metzger, *Nucleic Acids Res.* **26,** 4086 (1998).

[54] M. Nichols, J. M. Rientjes, C. Logie, and A. F. Stewart, *Mol. Endocrinol.* **11,** 950 (1997).

[55] C. Logie and A. F. Stewart, *Proc. Natl. Acad. Sci. U.S.A.* **92,** 5940 (1995).

[56] H. Takebayashi, H. Oida, K. Fujisawa, M. Yamaguchi, T. Hikida, M. Fukumoto, S. Narumiya, and A. Kakizuka, *Cancer Res.* **56,** 4164 (1996).

[57] T. Abbas-Terki and D. Picard, *Eur. J. Biochem.* **266,** 517 (1999).

[58] T. J. Hope, X. J. Huang, D. McDonald, and T. G. Parslow, *Proc. Natl. Acad. Sci. U.S.A.* **87,** 7787 (1990).

[59] S. Rehberger, F. Gounari, M. DucDodon, K. Chlichlia, L. Gazzolo, V. Schirrmacher, and K. Khazaie, *Exp. Cell Res.* **233,** 363 (1997).

could lead to the loss of hormonal control. But this choice should be based primarily on the type of fusion that the heterologous protein might tolerate and on achieving maximal proximity of the HBD to an important function of the heterologous protein (on the linear protein map).

3. Introduce missing sequence features by polymerase chain reaction (PCR) or site-directed mutagenesis. For an N-terminal fusion these are restriction site 1–initiator codon–ORF of HBD–restriction site 2 for in-frame fusion, and for a C-terminal fusion restriction site 1 for in-frame fusion–ORF–stop codon–restriction site 2. The consensus sequence CCACCATGG provides an efficient context for the initiator codon for expression in mammalian cells.

4. Construct an expression vector for the fusion protein.

5. Express the fusion protein constitutively in cells that are maintained in the absence of the activating ligand to keep the fusion protein inactive. Medium for tissue culture cells: for fusion proteins with wild-type HBDs except that of the glucocorticoid receptor (GR), it is preferable to use medium without phenol red and supplemented with charcoal-stripped serum. Medium for yeast: both rich and minimal media are appropriate, but certain sources of galactose should be avoided.[60]

6. Induce the activity of the fusion protein by the addition of the appropriate ligand (Table II). Conversely, turn off the fusion protein by removing the activating ligand (by washing the cells several times with medium or saline).

Stripping Serum of Steroids with Charcoal

1. Add 2 g of acid-washed activated charcoal (Sigma, St. Louis, MO) per 100 ml of serum in a glass beaker. Stir the suspension for 90 min in a cold room.

2. Spin out the bulk of the charcoal in 50-ml conical tubes (e.g., spin at 6000 rpm for 10 min at 4°).

3. Filter the supernatant through a regular paper filter, either through a pleated filter in a funnel or with suction through a Büchner flask.

4. Sterilize by filtration through a cellulose acetate filter (0.22-μm pore size).

5. Store in aliquots at $-20°$.

The conditions (time and temperature) of step 1 are a compromise between insufficient stripping of steroids and removal of other components. They should be sufficient, but more stringent methods do exist.[61] Note that

---

[60] J. W. Liu and D. Picard, *FEMS Microbiol. Lett.* **159,** 167 (1998).
[61] S. Masamura, S. J. Santner, D. F. Heitjan, and R. J. Santen, *J. Clin. Endocrinol. Metab.* **80,** 2918 (1995).

steps 2 and 3 are necessary because fine charcoal particles easily clog regular paper as well as cellulose acetate filters.

Mechanism

From a purely practical point of view the regulatory mechanism might seem irrelevant. However, the models not only help explain how the mechanism works, but also help in designing the fusion protein and in predicting whether it will be regulated. One of the most striking findings that needs to be considered is the impressive range of protein activities and structures that HBDs can subject to hormonal control (see Table I). We have previously argued that this could be explained only by a relatively nonspecific mechanism[1,62] and have proposed that protein inactivation could be due to steric interference mediated by the molecular chaperone Hsp90–steroid receptor complex.[4–6] The Hsp90 complex assembles on the unliganded HBD and is released on ligand binding (for review, see Ref. 63).

Practical Implications of Proposed Mechanism

Several predictions of the steric interference model are of practical importance.

1. The regulatory machinery depends on the hormone-reversible formation of the Hsp90 complex, limiting the application of this technique to the cytosolic and nuclear compartments of eukaryotic cells.

2. The effectiveness of steric interference, despite remarkable flexibility, will depend on how the Hsp90 complex is positioned relative to a key function(s) of the heterologous protein.

3. Several functions of the heterologous moiety may be subjected to hormonal control.

4. Protein functions that are difficult to interfere with sterically may not be regulatable by fusion to an HBD.

5. Hormonal regulation may be difficult to observe under cell-free conditions because of the inherent instability of the Hsp90–HBD complex.

By and large these predictions are supported by currently available data, but a comprehensive and thorough effort to provide formal proof has not been made. It is difficult to predict to what degree a particular fusion protein will be regulated. The relative positioning of heterologous function and Hsp90–HBD complex is certainly critical but structural information

---

[62] K. R. Yamamoto, P. J. Godowski, and D. Picard, *Cold Spring Harbor Symp. Quant. Biol.* **53,** 803 (1988).

[63] W. B. Pratt and D. O. Toft, *Endocr. Rev.* **18,** 306 (1997).

about the whole complex is not yet available. The influence of the spacer sequences between the heterologous function and the HBD has been examined only with E1A[1] and by varying the HBD moiety with Flp.[54,64] However, these studies are hampered by the fact that there is hardly a neutral protein sequence that could be inserted for a systematic analysis of this issue.

The number of different functions of the heterologous moiety that become hormone dependent may be a particularly important parameter in determining to what degree the fusion protein is regulated. For example, if nuclear localization of a fusion protein is necessary but hormone dependent, a slight leakiness in the regulation of another function may be more tolerable. Unfortunately, HBDs other than that of the GR are not particularly efficient at regulating nuclear localization functions.[65] The subcellular localization of a fusion protein in the absence of hormone is typically unpredictable as it is probably determined by a competition between targeting signals in the heterologous moiety and the inactivation function of the HBD. Nuclear localization has been reported to be regulated, for example, in the case of an E1A–GR[1] and a Cre–ER[50] fusion protein.

It should be emphasized that the failure of a heterologous protein to become subjected to hormonal control is typically a negative result and could have any one of a number of possible causes. In keeping with the above-mentioned model (point 4), it could be speculated that the active sites of enzymes that bind small molecules could not be regulated by fusion to an HBD. Indeed, β-galactosidase, galactokinase, URA3, and dihydrofolate reductase HBD fusion proteins remain active in the absence of hormone (D. Picard, unpublished results, 1989; see also Ref. 66).

Problems and Troubleshooting

Potential Complications

Be aware of the fact that HBDs are not just neutral regulatory cassettes with a hormone-reversible "protein inactivation function." They carry additional functions that may be undesirable in fusion proteins. Apart from their considerable size (about 250 amino acids or 30 kDa), HBDs can contribute hormone-dependent dimerization,[67–69] nuclear local-

[64] M. Nichols, J. M. Rientjes, and A. F. Stewart, *EMBO J.* **17,** 765 (1998).

[65] D. Picard, V. Kumar, P. Chambon, and K. R. Yamamoto, *Cell Regul.* **1,** 291 (1990).

[66] D. I. Israel and R. J. Kaufman, *Proc. Natl. Acad. Sci. U.S.A.* **90,** 4290 (1993).

[67] V. Kumar and P. Chambon, *Cell* **55,** 145 (1988).

[68] Ö. Wrange, P. Eriksson, and T. Perlmann, *J. Biol. Chem.* **264,** 5253 (1989).

[69] S. E. Fawell, J. A. Lees, R. White, and M. G. Parker, *Cell* **60,** 953 (1990).

ization,[70–72] and transcriptional activation functions (reviewed in Refs. 73 and 74).

The dimerization activity of the HBD is relatively weak. As illustrated by several cases mentioned in Table I, for example, Fos and Myc fusion proteins, the HBD dimerization function does not appear to interfere with the heterodimerization with their respective obligatory partner proteins (Jun and Max, respectively, for the aforementioned examples). It has been speculated that dimerization imposed by the HBD may be responsible for the hormone-induced oncogenic activation of a c-Abl–ER fusion protein.[40] There are as yet no point mutations available that are known to abolish dimerization selectively *in vitro* as well as *in vivo*. Likewise, there are as yet no point mutations that selectively eliminate the hormone-dependent nuclear localization function. However, only the HBD of the GR has been shown to work as an autonomously active nuclear localization signal.[65,70]

When the HBD is used to regulate a transcription factor, the contribution from the activation function 2 (AF2) associated with HBDs may be a concern. However, despite the large number of reports on using HBDs to regulate heterologous transcription factors (see Table I), alterations of their *trans*-activation properties by AF2 have only rarely been encountered.[14,28,75] Unless the *trans*-activation function of the heterologous moiety is particularly weak, this may usually be a negligible potentiality. As a fail-safe prophylactic measure, AF2 activity can be eliminated by point mutations or by using certain antihormones instead of the normal agonists to activate the fusion protein (see Table II).

Unexpected Behavior

In the context of a practically oriented chapter it may be appropriate to point out observations that are difficult to explain within the framework that is presented above or that are unexpected at first sight. It may be argued that the regulation by HBDs depends on the formation of the hormone-reversible Hsp90 complex. Only the five vertebrate steroid receptors—the GR, ER, mineralocorticoid receptor (MR), PR, and androgen receptor (AR)—are known to form these complexes stably.[63] Therefore, only the HBDs of these five intracellular receptors would be expected to

[70] D. Picard and K. R. Yamamoto, *EMBO J.* **6,** 3333 (1987).
[71] A. Guiochon-Mantel, H. Loosfelt, P. Lescop, S. Sar, M. Atger, M. Perrot-Applanat, and E. Milgrom, *Cell* **57,** 1147 (1989).
[72] T. Ylikomi, M. T. Bocquel, M. Berry, H. Gronemeyer, and P. Chambon, *EMBO J.* **11,** 3681 (1992).
[73] H. Gronemeyer, *Annu. Rev. Genet.* **25,** 89 (1991).
[74] C. K. Glass, D. W. Rose, and M. G. Rosenfeld, *Curr. Opin. Cell Biol.* **9,** 222 (1997).
[75] S. Kim, P. H. Brown, and M. J. Birrer, *Oncogene* **12,** 1043 (1996).

work for the regulation of heterologous proteins. Three reports challenge this view to date. A chimeric *trans*-activator could be regulated by the ligand-binding domain of the ecdysone receptor,[76] MyoD by that of the thyroid receptor,[30] and Fas by that of the retinoic acid receptor.[56] It remains to be seen whether this is more generally applicable or whether these are special cases that involve a different mechanism. Indeed, we have been unable to regulate the transcription factor Fos or the tyrosine kinase Abl with the HBD of the ecdysone receptor (data not shown).

It is pointed out above that not all functions of the heterologous moiety may be subject to hormonal control. This can lead to rather paradoxical results. Abl–ER fusion proteins are oncogenic in the presence of estrogens and growth inhibitory on removal of hormone.[6,77] Thus, in the absence of hormone they are not simply inactive but display another function. We have speculated that the growth inhibitory function remains constitutively active and is overridden by the transforming activity in the presence of hormone. Fusion proteins with members of the Jun family provide another example of unusual, yet explainable, behavior. Fusion proteins of the ER HBD with JunD[27] or a dominant-negative truncation mutant of c-Jun[75] inhibit AP-1 (Jun/Fos) activity in the absence of hormone. Both *trans*-activate on addition of hormone, the latter apparently through the HBD-associated *trans*-activation function. The ligand-independent activity of these fusion proteins is probably due to the constitutive formation of heterodimers with Fos while DNA binding, and thus *trans*-activation, remains ligand dependent.[27]

Although it is the specific activity, and not the level of a fusion protein, that is expected to be regulated, and although this has been confirmed for a large number of examples, there are apparent exceptions. An example is that of a Mek1 fusion protein, where ligand primarily induces the accumulation.[42] Just as ligand alters the accumulation of wild-type steroid receptors, it is not surprising that the ligand-induced conformational changes and alterations in protein complex composition can affect the accumulation of some chimeras. Note that in this case the regulation is still at the protein level even though the mechanistic details and notably the kinetics of activation are different.

Troubleshooting

Some of the most common questions and problems are discussed in Table III.

[76] K. S. Christopherson, M. R. Mark, V. Bajaj, and P. J. Godowski, *Proc. Natl. Acad. Sci. U.S.A.* **89,** 6314 (1992).
[77] T. Mattioni, P. J. Jackson, O. Bchini-Hooft van Huijsduijnen, and D. Picard, *Oncogene* **10,** 1325 (1995).

TABLE III
TROUBLESHOOTING: FREQUENTLY ASKED QUESTIONS

Question	Answer/possible solution
Does the cell line express endogenous steroid receptor XYZ?	Databases with this type of information are not currently available, but as a rule of thumb GR is expressed in almost all tissues and cell lines
Is it necessary to use medium without phenol red and with charcoal-stripped serum?	It depends on the HBD that is used (see text)
Why is the fusion protein constitutive *in vitro* (e.g., in a gel shift experiment)?	*In vitro* conditions are usually not compatible with maintaining the HBD–Hsp90 complex in the inactivated form
What can be done to reduce the basal level?	Check step 5 under Method in text. If the basal level is likely to be inherent to the fusion protein itself, try moving the HBD closer to a key function. According to one report,[a] adding a pure antihormone prior to induction might help. Reducing the expression level may also help if the basal activity is due to proteolytic fragments lacking the HBD
Why doesn't the fusion protein work at all?	It may not tolerate being fused to a 30-kDa protein domain. Try fusing the HBD to the other end of the protein
Should the fusion protein be expected to be cytoplasmic in the absence of hormone?	Not necessarily. In most cases, the subcellular localization is primarily determined by the heterologous moiety
Does it work only with a cytoplasmic protein?	No, almost any cytosolic or nuclear localization is probably fine. Membrane anchoring is all right as long as the HBD remains in the aforementioned compartments (see examples in Table I)
Can the HBD be fused to the N terminus of the protein?	Yes, but there are as yet few reported cases
If an antihormone is used to activate the fusion protein, can it be assumed that the endogenous receptor remains inactive?	Not necessarily: it depends on the ligand, and on the cell and promoter context
Does it work in animals?	Applications in transgenic mice,[b] in virally transduced mice,[c] and in frog embryos[d] have been reported

[a] H. Pelczar, O. Albagli, A. Chotteau-Lelièvre, I. Damour, and Y. de Launoit, *Biochem. Biophys. Res. Commun.* **239,** 252 (1997).

[b] R. Feil, J. Brocard, B. Mascrez, M. LeMeur, D. Metzger, and P. Chambon, *Proc. Natl. Acad. Sci. U.S.A.* **93,** 10887 (1996); J. Brocard, X. Warot, O. Wendling, N. Messaddeq, J. L. Vonesch, P. Chambon, and D. Metzger, *Proc. Natl. Acad. Sci. U.S.A.* **94,** 14559 (1997); Y. Wang, F. J. DeMayo, S. Y. Tsai, and B. W. O'Malley, *Nature Biotechnol.* **15,** 239 (1997); F. Schwenk, R. Kuhn, P. O. Angrand, K. Rajewsky, and A. F. Stewart, *Nucleic Acids Res.* **26,** 1427 (1998); P. S. Danielian, D. Muccino, D. H. Rowitch, S. K. Michael, and A. P. McMahon, *Curr. Biol.* **8,** 1323 (1998); C. Kellendonk, F. Tronche, E. Casanova, K. Anlag, C. Opherk, and G. Schütz, *J. Mol. Biol.* **285,** 175 (1999).

[c] M. M. Burcin, G. Schiedner, S. Kochanek, S. Y. Tsai, and B. W. O'Malley, *Proc. Natl. Acad. Sci. U.S.A.* **96,** 355 (1999).

[d] P. J. Kolm and H. L. Sive, *Dev. Biol.* **171,** 267 (1995); M. Tada, M. A. O'Reilly, and J. C. Smith, *Development* **124,** 2225 (1997).

Applications in Whole Organisms

Yeast

Steroid receptors from vertebrates can function as hormone-dependent transcription factors in the budding yeast *Saccharomyces cerevisiae*. We were therefore encouraged to test the application of HBDs in this organism and initially found that the function of the chimeric *trans*-activator Gal4-VP16 fused to an HBD is hormone dependent.[20] Subsequently, HBDs were shown to be able to regulate the kinase Ste11,[6] the recombinase Flp,[54] and the α peptide for β-complementation of β-galactosidase.[57] HBDs from several steroid receptors are applicable but it may not always be possible to activate the GR HBD efficiently because of its relatively poor ligand-binding properties in yeast (Ref. 20; and J.-F. Louvion and D. Picard, unpublished results, 1994).

Despite these successful applications, it is too early to say whether the approach is generally applicable in yeast. Indeed, when the yeast transcription factor Gcn4 is fused to an HBD, it is tightly regulated in mammalian cells,[25] but the same fusion protein is constitutively active in yeast (J.-F. Louvion and D. Picard, unpublished results, 1994). Conversely, it appears that the hormonal regulation of the α peptide of β-galactosidase does not work in mammalian cells (O. Donzé, P.-A. Briand and D. Picard, unpublished results, 1999). It remains to be seen whether the underlying mechanisms are the same in yeast and in mammalian cells.

Given that the activity of transcription factors based on Gal4 can be regulated with HBDs, they provide excellent tools for conditional expression in yeast. Although several systems for inducible gene expression exist (see, e.g., Ref. 78), they typically require switching to a different growth medium, which results in substantial physiological changes. In contrast, most steroid hormones are gratuitous signals in yeast with no or minimal effects on general physiology. One of the most popular inducible systems relies on the galactose induction of constructs with the *GAL1* or *GAL10* promoters. Therefore, Gal4 fusion proteins can be used to regulate the expression of standard galactose-inducible vectors with steroid hormones instead of a switch in carbon source. We have reported[16] a further improvement of our original Gal4–HBD fusion constructs.[20] The new fusion proteins consist of the HBDs of either the ER or MR fused to Gal4, which is almost full length and lacks only the interaction domain for the Gal80 inhibitor.[16] Both fusion proteins have negligible activity in the absence of hormone; the activity is even lower than that of the original Gal4–ER–VP16

---

[78] J. C. Schneider and L. Guarente, *Methods Enzymol.* **194,** 373 (1991).

fusion proteins. The following provides the protocol, adapted from an earlier version,[20] for using hormonally regulated Gal4 fusion proteins for conditional gene expression in budding yeast. The assumption is made that readers are familiar with basic yeast techniques.

Reagents

Regulator plasmid, for example, recombinant pHCA/GAL4(848).ER or pHCA/GAL4(848).MR (Ref. 16)

Coding sequences for the protein of interest in pYES2 (InVitrogen, Carlsbad, CA) or under the control of a galactose-inducible promoter in any other yeast vector

An appropriate *S. cerevisiae* strain that carries a *his3* mutation

Growth medium with 2% (w/v) raffinose instead of glucose or galactose as carbon source; in certain strains glucose may also be used as carbon source

Ligand stock solutions according to Table II

Method

1. Transform the regulator plasmid into a yeast strain carrying the coding sequences for the protein of interest under the control of a galactose-inducible promoter. Select for transformants on minimal medium lacking histidine (at this point use the regular carbon source, glucose).
2. Prior to and during induction experiments grow transformants with raffinose as the carbon source. If the strain does not grow well with raffinose, test one of the following growth media.
   a. In addition to raffinose, add 2% (v/v) glycerol and 1% (v/v) ethanol.
   b. Substitute glucose for raffinose and test whether induction works anyway (see step 3).
3. Activate the GAL4(848).ER or GAL4(848).MR fusion proteins, and thereby induce expression of the protein by adjusting the growth medium to 0.1 μM β-estradiol or 10 μM deoxycorticosterone, respectively. Note that solid or liquid media with hormones can be stored for several weeks at 4°.

Plants

As in yeast, many vertebrate steroids are gratuitous signals in plants, allowing the use of HBDs as a completely heterologous regulatory system. Two applications of the GR HBD have become popular (see also Table I): (1) the direct posttranslational regulation of transcription factors by fusion to the GR HBD,[7,8,11,34] and (2) the regulation of expression of a gene of interest with the hormone-inducible Gal4–VP16–GR fusion protein.[21] Hormone has been applied by adding it to growth medium or by spraying it onto target organs such as the leaves.

So far, only the GR HBD has been used, presumably mostly for historical reasons. There is no reason that other HBDs or their mutant derivatives should not work in plants. Future experiments will show whether the study of plants can also benefit from the available diversity of regulatory domains.

Vertebrates

The same classes of steroids are essentially shared by all vertebrates. Thus, it is their concentrations during development or at the time of analysis that determine whether a particular wild-type HBD can be used. Perhaps, not surprisingly, the approach has first been shown to work in frog embryos, whose endogenous glucocorticoid concentrations are apparently low enough to permit regulation of heterologous proteins by the GR HBD.[31,39]

Applications in transgenic mice could be seriously envisaged only with the advent of mutant HBDs (Table II). Specifically inducible Cre recombinase fusion proteins have elicited considerable interest as powerful tools for spatiotemporally controlled somatic mutagenesis. Cre has been regulated either with the tamoxifen-responsive ER mutant G521R[50,51,79,80] or with an RU-486-responsive PR truncation mutant.[81] It is encouraging that these HBD fusion proteins indeed fail to be activated by endogenously present signals both during development (pregnancy) and in the adult, and that they can be activated in a variety of tissues by administration of the appropriate ligand. However, it has been difficult to achieve Cre-mediated recombination with 100% efficiency,[50,51,79,80] perhaps because of insufficient intracellular availability of the drug, and the drug dose required for efficient activation of the Cre fusion protein during development *in utero* is close to that which is incompatible with the maintenance of pregnancy.[80] Other mutant HBDs, such as those mentioned in Table II, have yet to be tested in animals. To date, the only other HBD fusion protein that has been expressed in transgenic mice is the RU-486-inducible transcriptional regulator Gal4–PR–VP16.[22]

HBD fusion proteins may also find gene therapeutic applications. Pre-

[79] F. Schwenk, R. Kuhn, P. O. Angrand, K. Rajewsky, and A. F. Stewart, *Nucleic Acids Res.* **26**, 1427 (1998).
[80] P. S. Danielian, D. Muccino, D. H. Rowitch, S. K. Michael, and A. P. McMahon, *Curr. Biol.* **8**, 1323 (1998).
[81] C. Kellendonk, F. Tronche, E. Casanova, K. Anlag, C. Opherk, and G. Schütz, *J. Mol. Biol.* **285**, 175 (1999).

liminary experiments have already been done in tissue culture[18,82] and in mice by transgenesis[22] and adenoviral transduction.[18] However, whenever the systemic administration of inducing ligands will be necessary, either to mice or to humans, new HBD mutants with novel ligand specificities may become a necessity. Ultimately, it would be desirable to have an HBD mutant that (1) does not respond to physiological ligands and (2) is activated by a specific ligand that has no effect on any endogenous target.

Acknowledgments

I am grateful to members of my laboratory, notably Jean-François Louvion, Toufik Abbas-Terki, and Olivier Donzé, for their unpublished data included in this chapter. I also thank all those colleagues who have communicated both positive and negative results obtained with HBD fusion proteins, and Karl Matter for critical reading of the manuscript.

[82] M. Kokubun, A. Kume, M. Urabe, H. Mano, M. Okubo, R. Kasukawa, A. Kakizuka, and K. Ozawa, *Gene Ther.* **5,** 923 (1998).

[30] Tet Repressor-Based System for Regulated Gene Expression in Eukaryotic Cells: Principles and Advances

By Udo Baron and Hermann Bujard

Introduction

The tetracycline (Tc) responsive regulatory systems[1,2] (Fig. 1) have been widely applied to control gene activities in eukaryotes. They were shown to function in cultured cells from mammals, plants, amphibians, and insects as well as in whole organisms including yeast, *Drosophila,* plants, mice, and rats (for compilations see Ref. 3). The basic regulatory circuits of the Tet systems and the elements constituting these circuits are based on simple principles. Nevertheless, a number of parameters that may not be too obvious must be reconciled when setting up one of the systems for a particular purpose. In the following, we therefore discuss in some detail

[1] M. Gossen and H. Bujard, *Proc. Natl. Acad. Sci. U.S.A.* **89,** 5547 (1992).
[2] M. Gossen, S. Freundlieb, G. Bender, G. Müller, W. Hillen, and H. Bujard, *Science* **268,** 1766 (1995).
[3] Homepage Bujard laboratory: http://www.zmbh.uni-heidelberg.de/Bujard/homepage.html

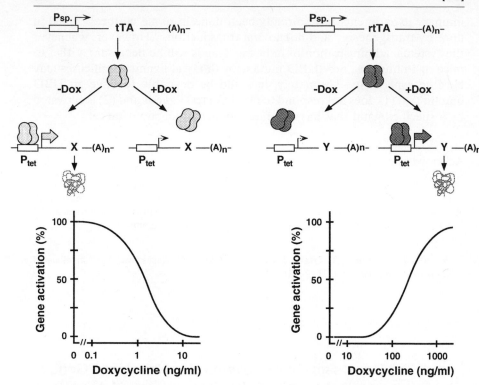

FIG. 1. Schematic outline of the Tet regulatory systems. *Top Left:* The mode of action of the Tc-controlled *trans*-activator (tTA). tTA, a fusion protein between the Tet repressor of the Tn*10* Tc resistance operon from *E. coli* and the C-terminal portion of VP16 from herpes simplex virus, binds in the absence of the effector molecule doxycycline (Dox) to multiple *tet* operator sequences (*tet*O) placed upstream of a minimal promoter and activates transcription of gene *x*. Addition of Dox prevents tTA from binding and thus the initiation of transcription. *Bottom left:* The dose–response effect of Dox on tTA-dependent gene expression. Gene activity is maximal in the absence of the antibiotic whereas increasing effector concentrations gradually decrease expression to background levels at concentrations ≥ 10 ng/ml. *Top right:* The mechanism of action of the reverse Tc-controlled *trans*-activator (rtTA). rtTA is identical to tTA with the exception of 4 amino acid substitutions in the TetR moiety, which convey a reverse phenotype. rtTA requires Dox for binding to *tet*O sequences in order to activate transcription of gene *y*. *Bottom right:* The dose–response effect of Dox on rtTA-dependent transcription activation. There is no gene expression in the absence of the antibiotic. By increasing the effector concentration of Dox beyond 100 ng/ml, rtTA-dependent gene expression is gradually stimulated.

the principles underlying the Tet systems and point out essential technical and methodological implications. On the other hand, we refrain from giving detailed experimental protocols, because we would merely reiterate previously described procedures[4,5] of which most are common practice in cell biology.

Tet Regulatory Systems and Their Elements

The Tet systems are based on fusions between the repressor (TetR) of the Tn*10* tetracycline resistance operon of *Escherichia coli* and domains that interact with the eukaryotic transcription machinery in a positive (activators) or negative (silencers) way. These fusion proteins have largely maintained the properties of their parent domains: They bind to operator sequences (*tet*O) of the *tet* operon in a tetracycline-dependent manner, and if such operator sequences are placed within the context of an RNA polymerase II promoter, transcription from this promoter may be influenced positively or negatively through the binding of the respective TetR fusions. Hence the activity of such promoters becomes susceptible to tetracycline (Tc) or one of its many derivatives.

In one of the systems, the tetracycline-controlled transcriptional activator, tTA, a fusion protein between TetR and a transcriptional activating domain, binds to P_{tet}, a fusion between a minimal promoter and an array of *tet*O sequences, and stimulates the onset of transcription. Tc and particularly doxycycline (Dox) prevent binding and consequently abolish transcription because the minimal promoter by itself is inactive (Figs. 1 and 2). A complementary system based on certain TetR mutants functions in an opposite fashion: rtTA (reverse tetracycline controlled *trans*-activator) requires Dox to bind to *tet*O (and thus to P_{tet}) and therefore will activate transcription of P_{tet} only in the presence of the effector (Dox or anhydrotetracycline, ATc).

The basic elements of the Tet systems are outlined in Fig. 2. The original tTA and rtTA contain a C-terminal portion of the activator protein VP16 of herpes simplex virus (HSV) while later versions of these *trans*-activators are fusions between TetR and multiples of 13-amino acid "minimal activation domains."[6] The advantage of the latter constructs is discussed below. The most widely used tTA/rtTA-responsive promoter is P_{tet}-1 and derivatives thereof. They are fusions between an array of seven *tet*O sequences

[4] S. Freundlieb, U. Baron, A. L. Bonin, M. Gossen, and H. Bujard, *Methods Enzymol.* **283,** 159 (1997).
[5] S. Freundlieb, U. Baron, and H. Bujard, Controlling gene activities via the tet regulatory systems. *In* "Cell Biology—A Laboratory Handbook" (J. E. Celis, ed.), 2nd Ed., Vol. 4, pp. 230–238. Academic Press, San Diego, California, 1997.
[6] U. Baron, M. Gossen, and H. Bujard, *Nucleic Acids Res.* **25,** 2723 (1997).

A

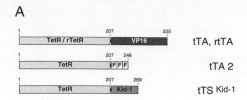

B

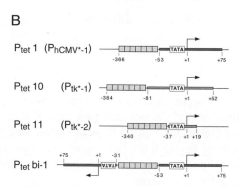

Fig. 2. Tetracycline-controlled fusion proteins and their target promoters. (A) Fusions between TetR/rTetR with domains capable of either activating or silencing transcription, respectively. tTA is a fusion protein between TetR, consisting of 207 amino acids, and the 128-amino acid carboxy-terminal portion of the *trans*-activator protein VP16 from herpes simplex virus. In tTA2, the VP16 moiety of tTA is replaced by three acidic minimal activation domains (F), each consisting of only 13 amino acids. tTS[Kid-1] is a fusion between TetR and a 61-amino acid KRAB domain, a transcriptional repression unit derived from the human kidney protein Kid-1. In tTS[Kid-1], the silencing moiety is connected to TetR via the nuclear localization signal derived from SV40 Tag. (B) Representation of tetracycline-responsive promoters. Those promoters are fusions between heptamerized *tet*O sequences (indicated as gray boxes) and promoters derived from either the hCMV IE promoter (P_{tet}-1) or the HSV tk promoter (P_{tet}-10 and P_{tet}-11), respectively. In the bidirectional promoter P_{tet}bi-1 heptamerized *tet*O sequences are flanked by two divergently orientated hCMV-derived minimal promoters. Positions spanning promoter regions and *tet* operator sequences are indicated with respect to the start site of transcription (+1). These bidirectional promoters permit coregulation of the two genes.

and a minimal promoter sequence derived from the human cytomegalovirus immediate-early (IE) promoter (P_{hCMV}). P_{tet}-3 is an analogous construct in which the promoter moiety stems from the *tk* promoter of HSV. In the P_{tet}-bi series, the *tet*O sequences are flanked by two minimal promoters. The resulting bidirectional promoter allows coregulation of two genes. In many instances, it has been of advantage to regulate a gene of interest together with one that encodes a reporter function. In this context the luciferase gene is particularly suited because of the extraordinary sensitivity of the luciferase assay, which allows one not only to readily monitor the degree of the control but also to accurately assess levels of induction.

More recently, we remodeled both systems such that they can be combined in one cell, thereby creating a dual regulatory system allowing one to switch between the activities of two genes in a mutually exclusive manner (Fig. 3). To this end, tTA and rtTA were equipped with new DNA-binding

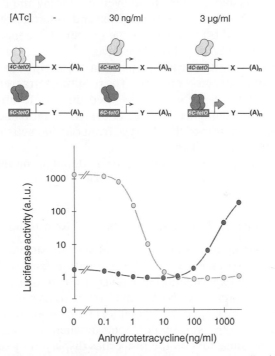

FIG. 3. Switching between the expression of two genes by tetracycline(s). *Top:* Mode of action of the dual regulatory system. The two *trans*-activators tTA2-1 (light gray) and rtTA2-1 (dark gray) discriminate efficiently between their target promoters P_{tet}-4 and P_{tet}-6, containing either heptamerized 4C-*tet*O or 6C-*tet*O sequences, respectively. tTA2-1 activates gene x in the absence of anhydrotetracycline (ATc or Dox) under conditions in which rtTA2-1 is kept inactive. In the presence of the effector substance (3 μg/ml), rtTA2-1 can bind to its target promoter P_{tet}-6 and thus stimulates transcription of gene y. Because of the different susceptibilities of tTA2-1 and rtTA2-1 to the effector substance (ATc or Dox), not only can both genes be activated in a mutually exclusive fashion, depending on the presence/absence of the antibiotic, but both genes can be turned off at intermediate concentrations of the effector substance. *Bottom:* Dose–response analysis of the effect of ATc on the activity of tTA2-1 and rtTA2-1. In the presence of ATc at the concentrations indicated, HeLa cells stably producing tTA2-1 and rtTA2-1 were transiently transfected with a plasmid carrying the luciferase gene controlled either by P_{tet}-4 (light gray curve) or by P_{tet}-6 (dark gray curve), respectively.

specificities enabling them to discriminate better than 1000-fold between their cognate promoters, P_{tet}-4 and P_{tet}-6, respectively, harboring altered tetO sequences.[7] In addition, the dimerization specificity of tTA was modified to prevent formation of rtTA/tTA heterodimers. Accordingly, only two populations of homodimeric $trans$-activators are generated when both $trans$-activator genes are expressed simultaneously in one cell. On the basis of these new specificities of tTA and rtTA, two genes can now be expressed in a mutually exclusive manner. Moreover, exploiting the differential susceptibility of tTA and rtTA toward Dox and other tetracyclines, both genes can be switched off by choosing the proper Dox concentration or by applying appropriate tetracycline derivatives.

TetR fusions were also used to position transcriptional silencing domains near or within promoter regions. The resulting tetracycline-controlled transcriptional silencers (tTSs) may be used directly to affect promoters equipped with tetO sequences[8] or, as discussed in more detail below, to shield minimal promoters of the P_{tet} type from outside activation[9] that may cause elevated basal activities.

All these elements have been described in detail in the references given and most of them are commercially available.[10]

Tightness of Control, Expression Levels, and Regulation Factors

When considering properties of the Tet system, such as degree (tightness) of control, expression levels, and regulation factors, it is crucial to distinguish between experimental situations in which the systems are stably inserted in the genome of cells in culture or of transgenic organisms and situations in which the respective regulatory elements are in a nonintegrated state, as in transient expression experiments or when contained in episomes. We would like to focus first on the properties displayed by the Tet systems on integration into the chromosome of cultivated cells and transgenic mice. Table I shows a collection of cell lines that were derived from HtTA-1 cells (HeLa cells that constitutively synthesize tTA[1]) by introducing stably a P_{tet}-1 (Fig. 2)-controlled luciferase gene. Obviously, the cell lines show differences in luciferase levels in the "off" state (i.e., basal activity in the presence of Tc) as well as in the induced state (i.e., when Tc is removed from the culture). Because all these cell lines are derived from the same

---

[7] U. Baron, D. Schnappinger, V. Helbl, M. Gossen, W. Hillen, and H. Bujard, *Proc. Natl. Acad. Sci. U.S.A.* **96**, 1013 (1999).

[8] U. Deuschle, W. K. Meyer, and H. J. Thiesen, *Mol. Cell. Biol.* **15**, 1907 (1995).

[9] S. Freundlieb, C. Schirra-Müller, and H. Bujard, *J. Gene Med.* **1**, 1 (1999).

[10] Clontech Laboratories, Inc., Palo Alto, California 94303.

TABLE I

DEPENDENCE OF TARGET GENE EXPRESSION IN HeLa CELLS ON SITE OF INTEGRATION[a]

| Clone | Luciferase activity (RLU/μg protein) | | Activation factor |
	With Tc	Without Tc	
T_7	$1{,}074 \pm 75$	$79{,}197 \pm 2{,}119$	7.3×10^1
T_{11}	2.4 ± 0.4	$34{,}695 \pm 1{,}127$	1.3×10^4
X1	≤ 2	$257{,}081 \pm 40{,}137$	$\geq 1 \times 10^5$
X2	≤ 2	$104{,}840 \pm 20{,}833$	$\geq 5 \times 10^4$
X7	75 ± 7	$125{,}745 \pm 18{,}204$	1.6×10^3
	Luciferase activity (RLU/μg protein)		
	With Dox	Without Dox	Activation factor
HR5-CL11	165,671	100	1,660
HR5-CL14	44,493	43	1,030

[a] The HeLa cell clone HtTA-1, which constitutively expresses tTA, was cotransfected with the P_{tet}-1-*luc* expression unit and pHMR272 encoding hygromycin resistance.[1] Resistant clones were examined for luciferase activity and several clones were subcloned. Luciferase activity was quantified in the presence (1 μg/ml) and absence of tetracycline (Tc). Luciferase activities of ≤ 2 RLU/μg protein are instrumental background. The HeLa cell line HR5, which constitutively synthesizes rtTA, was transfected with the luciferase reporter unit in an analogous way, and HR5-CL11 and HR5-CL14 were subcloned and analyzed as described above. Details are as described in Ref. 2. The number of integrated copies of the P_{tet}-1-*luc* constructs was not determined.

parent cell line (HtTA-1), it can be safely assumed that they all contain comparable amounts of *trans*-activator. The differences observed must, therefore, be due to the site of insertion and the number of integrated copies of the P_{tet}-1-controlled luciferase gene (P_{tet}-1-*luc* unit). Interestingly, there are several cell lines (e.g., X1 and X2) in which, in the presence of Tc, no luciferase activity is detectable [2 relative light units (RLU)/μg of protein is instrumental background] although high luciferase levels are induced on removal of Tc. These findings allow several conclusions to be drawn: First, as expected, the minimal promoter moiety of P_{tet}-1 is susceptible to outside influence at the insertion site; accordingly, locus-specific basal activity that is independent of tTA activation may be seen as exemplified in cell lines T_7, T_{15}, and X7; second, there are loci where P_{tet} has no detectable activity but where it can nevertheless be activated by tTA to high levels (e.g., cell lines X1 and X2); third, tetracycline at 0.1 to 1.0 μg/ml is capable of completely abolishing the function of tTA.

TABLE II

INTEGRATION SITE-DEPENDENT BASAL ACTIVITY OF P$_{tet}$-1 IN TRANSGENIC MICE[a]

Mouse line	Luciferase activity (RLU/μg total protein)				
	Liver	Pancreas	Kidney	Muscle	Brain
L7 (F3)	0.03 ± 0.03	0.06 ± 0.03	0.05 ± 0.09	0.63 ± 0.65	0.6 ± 0.4
L0	0.3 ± 0.1	1.0 ± 0.2	3.5 ± 0.4	2.3 ± 0.7	68.1 ± 19.5
L8	1.4 ± 0.8	2.1 ± 0.7	2.5 ± 0.4	2644.3 ± 1781.1	362.0 ± 158.1

[a] Several mouse lines in which the P$_{tet}$-1-*luc* expression unit is stably integrated[13] were analyzed for luciferase activity in various organs. In all mouse lines, luciferase activity was inducible to high levels on breeding with tTA or rtTA animals. The L7 line has proved to be a useful indicator line for the examination of tTA function in a large number of organs and cell types.

The X1 and X2 cell lines of Table I show that the tTA system can indeed be set up in a way that permits a tight control of gene expression. Actually, taking the instrumental background of our luciferase measurements (2 RLU/μg protein) representing the detection limit as a real value, the luciferase concentration in X1 cells in the presence of Tc would amount to seven molecules of enzyme per cell, indicating that only a fraction of the cell population produces a luciferase-encoding mRNA at any time. Obviously, there are "silent but activatable" (s/a) loci where, on integration of a P$_{tet}$-controlled expression unit, an extremely tight regulation can be achieved via tTA and its effectors. Such loci may be targeted quite readily because in our experience about 5–15% of the integration events in HeLa cells occur at such sites.[11] A similar situation is observed when the P$_{tet}$-1-*luc* expression unit is integrated into the mouse genome via pronucleus injection. As shown in Table II, mouse lines are generated that exhibit different basal luciferase activities within the various tissues. However, in the case of the L7 mouse line, the basal activities are again extremely low throughout the various organs corresponding, for example, to less than 1 luciferase molecule per 10 cells in liver, pancreas, and lung.[12,13] When L7 animals are crossed with mice producing tTA in hepatocytes, luciferase activity can be regulated over more than six orders of magnitude,[13] demonstrating that s/a loci can also be identified and utilized in transgenic animals.

[11] U. Baron and H. Bujard, unpublished observations (1999).

[12] A. Kistner, Ph.D. thesis. Universität Heidelberg, Heidelberg, Germany, 1996.

[13] A. Kistner, M. Gossen, F. Zimmermann, J. Jerecic, C. Ullmer, H. Lübbert, and H. Bujard, *Proc. Natl. Acad. Sci. U.S.A.* **93**, 10933 (1996).

This reasoning is impressively supported by the work of Lee *et al.*,[14] who placed the diphtheria toxin A gene under P_{tet}-1 control and succeeded in generating several mouse lines that would breed normally despite the incorporation of a gene encoding a most toxic product. When animals of such lines were crossed with tTA-producing mice in the presence of Dox, offspring were obtained that showed no abnormalities as long as the antibiotic was present in the drinking water, whereas removal of Dox was lethal. These remarkable findings by Fishman and colleagues demonstrate that s/ a loci appear to exist that maintain in all essential tissues a silent state throughout development and in the adult state of the animal. These loci appear however, accessible for activation by tTA.

In summary, control of gene activity can be achieved with the Tet systems whenever the experimental strategy permits screening or selection for events in which the expression unit carrying the target gene is integrated into a proper chromosomal site.

The high regulation factors found for the tTA/P_{tet}-1 system in cells and animals suggest that P_{tet}-1 may be a promoter suitable for the efficient expression of genes in eukaryotic cells. This is indeed the case, as the fully tTA-activated P_{tet}-1 was found to be several times more efficient than the commonly used P_{hCMV} in a number of cell lines.[11,15] A high level of gene expression was also observed when the Tet regulatory system was integrated into adenovirus-based vectors.[16] Thus, when properly set up, precise control as well as efficient expression of a gene may be achieved with the tTA/ P_{tet}/Dox system.

In transient expression, i.e., when the target gene controlled by P_{tet} is not integrated in the chromosome, an intrinsic basal activity is observed that can reach considerable levels depending on the cell type and experimental conditions.[9] This background activity is mainly due to the high copy number of the template in the cell and the lack of chromatin-mediated repression in the nonintegrated state of the vector DNA. Moreover, it appears that some cell lines cause higher background activities than others, probably because of different abundancies of transcription factors. Consequently, regulation factors measured in transient expression experiments are smaller and not comparable with those found in carefully selected stable cell or mouse lines.

When performing transient expression experiments, it is important to

[14] P. Lee, G. Morley, Q. Huang, A. Fischer, S. Seiler, J. W. Horner, S. Factor, D. Vaidya, J. Jalife, and G. I. Fishman, *Proc. Natl. Acad. Sci. U.S.A.* **95**, 11371 (1998).
[15] D. X. Yin, L. Zhu, and R. T. Schimke, *Anal. Biochem.* **235**, 195 (1996).
[16] B. Massie, F. Couture, L. Lamoureux, D. D. Mosser, C. Guilbault, P. Jolicoeur, F. Bélanger, and Y. Langelier, *J. Virol.* **72** 2289 (1998).

consider the synergistic nature of P_{tet} activation.[1,17] Because more than one *trans*-activator appears to be required for full promoter activation, high concentrations of P_{tet} may titrate out tTA/rtTA, leading to suboptimal promoter activation. At the same time, the high copy number of the P_{tet}-controlled transcription unit will raise the unregulated background activity. Therefore, in an experiment in which cells are cotransfected with tTA/ rtTA encoding DNA and with DNA carrying the gene of interest under P_{tet} control, the stoichiometry and the absolute amount of DNA must be reconciled. Hence, the ratio should be 10:1 to 100:1 in favor of the *trans*-activator-encoding sequences and the total amount of vector DNA trans-ferred should be kept low.[5] With these precautions, regulation factors of up to 1000-fold can be obtained in transient expression experiments in HeLa cells.

All the principles discussed here in the context of the tTA system apply also for rtTA-based regulation, particularly for the new generation of rtTAs, which are discussed below.

Effector Substances

Choice of the effector substance is critical to the use of either Tet system and, therefore, several aspects should be emphasized. For tetracyclines, as for other pharmacologically active substances, dosage above a certain concentration is deleterious to cells. However, for tetracyclines the concen-trations required to fully deactivate/activate either system are orders of magnitude below the levels that cause cytotoxicity. Whereas for full activa-tion of rtTA Dox at 1 μg/ml suffices, the tTA is already inactivated at 20 ng/ml.[18] Comparing the effector properties of different tetracyclines revealed that Tc efficiently inactivates tTA while it only marginally activates rtTA. Dox and ATc, however, are potent effector substances for both rtTA and tTA. Because of its suitability for both regulatory systems, its well-characterized pharmacological properties, and its commercial availability at low cost, Dox is currently our effector substance of choice. Nevertheless, in the future there may be tetracycline derivatives available that will prove useful for special applications.

During the analysis of cell lines such as X1 and X2 (Table I), we noticed striking variations in the level of expression of the target gene. Because these variations coincided with changes of the growth medium, we suspected that different batches of sera caused these problems. This was indeed

---

[17] M. Gossen and H. Bujard, *BioTechniques* **19,** 4 (1995).
[18] M. Gossen, Ph.D. thesis. Universität Heidelberg, Heidelberg, Germany, 1993.

the case.[4] Therefore, it is important either to use commercially available tetracycline-free medium[10] or to examine new batches of sera for the absence of tetracyclines. This can be readily achieved by using X1 cells (Table I).

Tetracycline-Controlled *trans*-Activators and Kinetics of Induction

The *trans*-activators we have originally described are fusions between TetR and a 128-amino acid C-terminal portion of VP16 of HSV. VP16 is known to be an exceptionally strong transcriptional activator that functions in many cellular environments and, indeed, there is ample evidence that tTA and rtTA are capable of activating transcription from P_{tet} constructs in a broad spectrum of cultured cells as well as in many cell types within transgenic organisms including *Drosophila melanogaster,* mice, and plants.[3]

The high activation potential of tTA and rtTA and the unusually specific interaction between TetR and *tet*O allow P_{tet} to be fully induced at low intracellular concentrations of the *trans*-activators. This feature is the basis of the remarkable specificity of the system, where the operator sequences within P_{tet} compete efficiently with potential secondary binding sites for tTA/rtTA and whose occupancy could yield unpredictable effects. Moreover, low intracellular concentrations of tTA/rtTA require only low concentrations of effector molecules for their inactivation/activation, a particular advantage considering that any substance will interfere with the metabolism of a cell above a certain threshold concentration.

On the other hand, overexpression of any transcription factor above a specific threshold will unbalance intracellular equilibria of interacting partners, a phenomenon called "squelching,"[19] which may severely upset metabolic pathways and even lead to cell death. Strong *trans*-activators such as tTA and rtTA must, therefore, not be produced above a certain intracellular concentration. Obviously, our HtTA-1[1] and HR5[2] cell lines, which have been kept for years without selection pressure, stably produce their respective *trans*-activators below the critical threshold, which is estimated to be about 10,000 molecules per cell. Interestingly, in these cells the tTA/rtTA-encoding genes are controlled by the strong P_{hCMV}. Apparently, the low levels of tTA/rtTA expression are likely due to the site of integration in the genome. Hence, when generating a tTA- or rtTA-producing cell line, it is important to propagate several positive clones over a period of time and to screen or select for those that stably maintain their phenotype while exhibiting strong activation potential. The many tTA- and

[19] G. Gill and M. Ptashne, *Nature (London)* **334,** 721 (1988).

rtTA-producing cell lines published, of which numerous are commercially available,[10] as well as the increasing number of tTA/rtTA mouse lines (Table III) demonstrate the feasibility of identifying conditions in which *trans*-activators are produced within a concentration window that prevents any detectable perturbation of the metabolism of a cell.

Whenever selection for a suitable expression level of a tTA/rtTA-encoding gene is not feasible, for example, when the *trans*-activator gene should be placed under the control of a specific promoter via homologous recombination, it may be necessary to use *trans*-activators with a graded activation potential. Such *trans*-activators have indeed been generated by fusing TetR with various combinations of minimal activation domains[6] (Fig. 2). Their activation strength spans a greater than 1000-fold range. Accordingly, most of them are tolerated at higher intracellular concentrations as activation and squelching potential apparently correlate.

The availability of the tTA and rtTA system raises questions concerning under which conditions one should be preferred over the other. At the level of cell culture, both systems are nearly equivalent: both permit the regulation of gene activity over several orders of magnitude and the kinetics of induction are fast. Nevertheless, if a gene is to be kept inactive most of the time and turned on only occasionally, the rtTA system seems more

TABLE III
tTA- AND rtTA-PRODUCING MOUSE LINES[a]

Promoter controlling the *trans*-activator gene	tTA	rtTA	Tissue specificity
hCMV IE	+	+	Skeletal muscle and other tissues
Rat insulin	+	+	Pancreas, β cells
α myosin heavy chain	+		Heart muscle cells
LAP	+		Hepatocytes
Albumin	+		Hepatocytes
α-CamkII	+	+	Brain
Prion	+		Brain
Enolase	+		Brain
PMP22	+		Brain, Schwann cells
MMTV	+		Salivary gland
CC10	+	+	Lung airway
Keratin 5	+		Epidermis (basement layer, hair follicles)

[a] The number of mouse lines that express tTA or rtTA in specific tissues/cell types is steadily increasing. This compilation contains only the published mouse lines (cited in Ref. 3); numerous additional lines exist in various laboratories. The general availability of such lines will in the future greatly facilitate the study of gene function *in vivo*.

appropriate. In contrast, whenever an active gene is to be occasionally turned off, the tTA system is preferable. In both situations, Dox is present only during the period in which the activity of a gene is to be altered. As is discussed below, in transgenic organisms saturation with and depletion of different compartments within an organism by the administration of Dox follows complex pathways that will largely determine the kinetics of activation or inactivation of a gene. There, the choice between the two regulatory systems will be of relevance.

When compared quantitatively, tTA and rtTA are not fully complementary. For example, HeLa cell lines with a regulatory range of $>10^5$ and no detectable background activity could readily be obtained with tTA as exemplified with the X1 line (Table I). In contrast, analogously generated rtTA-based cell lines such as HR5-CL11 (Table I) generally exhibited a low but distinct basal activity in the absence of Dox. The regulation factors measured in HR5-derived cell lines, therefore, rarely exceeded a value of 10^3. Interestingly, as is discussed below, such differences between the tTA and rtTA systems were not observed in transgenic mouse lines.

Further analysis of rtTA has revealed that this *trans*-activator has indeed a low residual affinity for *tet*O in the absence of Dox, and that this affinity can cause an elevated basal activity of P_{tet}-1 depending on the intracellular concentration of rtTA.[11] Moreover, this background activity was increased significantly when acidic minimal activation domains were fused to rTetR.[20] These findings have sparked a reinvestigation of the sequence space of TetR, which yielded several new mutant rtTAs. Some of these reverse *trans*-activators, described elsewhere,[21] have remarkable properties that make them fully compatible with tTA. In addition, some exhibit an increased susceptibility toward Dox and thus are active at significantly lower concentrations of Dox when compared with the original rtTA.

The induction of the tTA/rtTA systems in cell culture is fast because it depends largely on the diffusion constant of the antibiotic. As shown previously,[1] cells are not only rapidly saturated with Dox, the antibiotic can also be washed out of a culture efficiently. Thus, with X1 cells that are kept under Dox, luciferase activity can already be monitored after a 10-min wash of the culture with Dox-free medium.[1] More recently, experiments designed to determine the half-life of short-lived intron RNA[22] provided proof that tTA-mediated activation is fully shut off in less than 5 min after supplying Dox to cultured cells. The same kinetics can safely be assumed

[20] U. Baron, Ph.D. Thesis. Universität Heidelberg, Heidelberg, Germany, 1998.
[21] S. Urlinger, U. Baron, M. Thellmann, M. T. Hazan, H. Bujard, and W. Hillen, *Proc. Natl. Acad. Sci. USA,* in press (2000).
[22] J. Q. Clement, L. Qian, N. Kaplinsky, and M. F. Wilkinson, *RNA* **5,** 206 (1999).

for the activation of the rtTA system in cell culture. It is important to keep in mind that despite the quick response of the Tet systems toward their effectors supplied from outside, the appearance of a phenotype based on an alteration of the activity of a gene depends decisively on other parameters such as the half-life of the mRNA and the protein encoded by the gene of interest.

The new family of rtTAs together with improved versions of tTA[21] will considerably expand the applicability of the tetracycline-controlled *trans*-activators, particularly when taking into account further developments of effector molecules that are presently under way.

Doxycycline-Controlled Activation and Repression

As discussed above, precise regulation of P_{tet} can be achieved whenever cross-talk between the minimal promoter contained within P_{tet} and outside elements such as nearby *cis*-acting enhancers can be avoided. This is most readily achieved by integrating the Tet regulatory system in cell lines or transgenic organisms in a two-step procedure whereby the tTA/rtTA expression unit is integrated in a locus that warrants proper intracellular *trans*-activator concentrations while the P_{tet}-controlled expression unit carrying the target gene is placed in an s/a locus. The requirements for this strategy preclude procedures in which both expression units are transferred simultaneously either within a single construct or by cointegration of, for example, two plasmids. In both scenarios enhancers essential for the expression of the *trans*-activator gene will be in close proximity to P_{tet} and thus are prone to be the cause of a basal activity. While in some experiments such background activities are acceptable—and success with this short-cut strategy has been reported in a number of publications—they cannot be tolerated in others. To overcome these limitations via autoregulation of tTA or rtTA is problematic, as discussed elsewhere.[4,23]

Two approaches come to mind for shielding a P_{tet}-controlled expression unit from outside activation: separation of the unit by insulators as characterized in *Drosophila* chromatin[24] or active repression of P_{tet} by tetracycline-controlled transcriptional silencers. Whereas insulators that function efficiently and, like in *Drosophila,* are totally locus independent, have not unequivocally been identified in mammalian systems so far, the development of Tc-controlled transcriptional silencers offers an obvious approach. Thus, fusions between TetR and transcriptional silencing domains (Fig. 2A) that bind to P_{tet} in the absence of Dox will protect this promoter from

[23] G. L. Gallia and K. Khalili, *Oncogene* **16,** 1879 (1998).
[24] R. Kellum and P. Schedl, *Mol. Cell. Biol.* **12,** 2424 (1992).

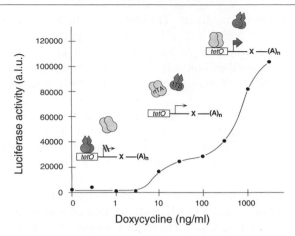

FIG. 4. Combined Tc-dependent repression and activation of P_{tet}-1. At the Dox concentrations indicated, HeLa HR5 cells[2] constitutively producing rtTA were transiently cotransfected with a plasmid carrying the luciferase gene controlled by P_{tet}-1 and a plasmid encoding tTS^{Kid-1}. At effector concentrations <10 ng/ml, tTS^{Kid-1} binds to P_{tet}-1 and represses transcription, whereas at Dox concentrations >100 ng/ml rtTA binds to the target promoter and thus stimulates gene expression. At intermediate effector concentrations, a basal activity is seen because neither tTS^{Kid-1} nor rtTA can bind to their target promoter P_{tet}-1. *Insets:* Three different states of activity of the combined activation/repression system.

outside activation. In the presence of the effector Dox, tTS will be replaced by rtTA and the previously repressed promoter will be activated. This principle, which was first demonstrated in *Saccharomyces cerevisiae*,[25] can efficiently reduce background activities, e.g., in cell lines in which the P_{tet} expression unit is not integrated in an s/a locus. Indeed, we have shown that tTS is capable of reducing a locus-induced activity of P_{tet} by up to 3 orders of magnitude.[9]

The application of tTS will also be of particular advantage in transient experiments in which P_{tet}-controlled genes are not repressed by chromatin as in the integrated state and in which, depending on the cell type, rather high background activities may exist.[9] In such situations, tTS can reduce the basal transcription by orders of magnitude[9] and thus expand the regulatory range of the system (Fig. 4). The application of tTS will facilitate the generation of stable cell lines in which genes encoding toxic products should be subject to Tet control. Obviously, on integration into s/a loci such genes can be strongly controlled. The failure to generate such cell lines is therefore frequently due to the toxic effects of gene products that are synthesized during the time period preceding the integration of the P_{tet}-controlled target

---

[25] E. Garí, L. Piedrafita, M. Aldea, and E. Herrero, *Yeast* **13,** 837 (1997).

gene into an appropriate locus. Transfection with low amounts of DNA encoding the target gene under P_{tet} control together with excess DNA that transiently warrants high concentrations of tTS in the cell should considerably enhance the chance for a successful integration event and thus for the establishment of stable cell lines that safely control a potentially toxic gene.

The combination of rtTA and tTS may permit the development of tightly regulatable systems that can be contained within one vector, e.g., of viral origin. The design of such constructs may, however, not be as trivial as appears at first glance. Both rtTA and, e.g., tTS[Kid] act over sizable distances and cross-talk between the various regulatory elements is expected. Moreover, so long as we know little about the mechanism of transcriptional silencing, e.g., of the KRAB domain contained in tTS[Kid], it will be difficult to predict how generally these regulators, which appear to act at the level of chromatin structure, can be utilized.

Expression Control in Transgenic Organisms

The Tet regulatory systems have been transferred into a variety of organisms including plants, *Drosophila,* mice, and rats (see Ref. 3). Here, we focus on the mouse system, for which considerable data has accumulated. Thus, several well-characterized mouse lines have been described that synthesize tTA or rtTA in specific cell types depending on the promoter that directs the transcription of the *trans*-activator gene (Table III). Similarly, mouse lines that contain various target genes under P_{tet} control were generated. By crossing animals from "tTA or rtTA lines" with individuals of "target gene lines," double-transgenic animals are obtained in which the expression of the target gene is susceptible to Dox. As discussed above, the degree to which regulation is controlled will depend on the site where the P_{tet}-controlled target gene is integrated into the genome. This is exemplified in Table II, which shows three mouse lines in which the P_{tet}-1-*luc* expression unit gives rise to different basal expression levels. The L7 line exhibits remarkably low background activity throughout the organs examined. However, when crossed with various tTA- or rtTA-producing lines, high luciferase levels can be achieved (Fig. 5[12,13,26]) and regulation factors of up to 5 orders of magnitude were measured in several organs.[13] Because of these properties, the L7 line is widely used for the identification of functional tTA/rtTA founder animals. Moreover, using the bidirectional promoter P_{tet}bi-1 (Fig. 2), indicator mouse lines were generated that coregu-

---

[26] M. Mayford, M. E. Bach, Y. Y. Huang, L. Wang, R. D. Hawkins, and E. Kandel, *Science* **274**, 1678 (1996).

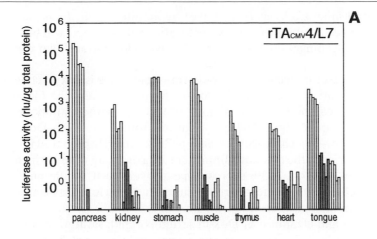

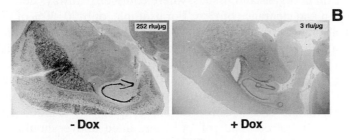

FIG. 5. Tet regulation in transgenic mice. (A) Mice expressing the rtTA gene under the control of P_{hCMV} ($rtTA^{CMV}$-4) were crossed with animals of the L7 line, which contain the P_{tet}-1-*luc* expression unit.[13] Luciferase activity was measured in several organs of mice supplied with Dox-containing (leftmost light gray columns) and plain drinking water (dark gray columns). Luciferase background activity of the L7 line is represented by the white (rightmost) columns. Missing columns indicate nonmeasurable luciferase activity. (B) Coregulation of luciferase and β-galactosidase via P_{tet}-bi-1 is demonstrated in a cross between the nzl-2 animals[12] and individuals that express the tTA gene under the control of the αCamKII promoter restricting tTA synthesis largely to certain regions in the brain.[26] Histology of brain portions of mice kept without (*left*) and with Dox (200 μg/ml) in the drinking water. Staining of β-Gal shows regions of tTA activity (*left*) that are highly reduced in the presence of Dox (*right*). *Insets:* Luciferase activity per microgram of protein.

late luciferase and β-galactosidase[12] or luciferase and eGFP. By crossing such animals with tTA/rtTA founders, it will be possible not only to monitor the degree and range of regulation but also to identify histologically the cell type in which tTA or rtTA is active[12] (Fig. 5).

Coregulation of a target gene with the luciferase gene via P_{tet}bi-1 also enables screening for founder animals that contain the expression unit integrated in an s/a locus (see above) by monitoring luciferase activity in the presence and absence of Dox. Integration into such loci will be essential for genes encoding products that are effective at low intracellular concentrations.

The effector molecule of choice in transgenic mice is doxycycline, which penetrates various tissues well, crosses the placenta and the blood–brain barrier, and can be found in the milk of feeding mothers at about 50% of its serum concentration. There are various routes by which Dox may be administered—injection, slow-release pellets, properly prepared food—but in our experience the amount administered via drinking water (which should contain, in addition, 2–5% sucrose) has proved most simple and reliable and, therefore, is the preferred route unless injection into a special compartment of the animal is required for kinetic reasons.

Uptake and clearance of Dox in serum is rapid: saturation within 3 hr, depletion with a half-life time of 6 hr.[12] These values do not, however, reflect the situation in organs and tissues, where complex pathways determine the distribution and clearance of Dox. For example, only 20–30% of the serum concentration of Dox is found in the brain and unequal distribution prevails throughout the animal. This must be taken into account when partial activation is to be achieved in a specific tissue or organ.[13] However, for simply switching gene activities via the tTA system, Dox concentrations of 20–200 μg/ml in the drinking water are sufficient, whereas for activating rtTA, particularly in the brain, concentrations of Dox between 200 μg/ml and 1 mg/ml are required. The rtTAs[21] mentioned above should, however, function at considerably lower Dox concentrations.

Whenever a target gene is supposed to be activated or inactivated rapidly, a number of parameters must be reconciled. In case of activation, the rtTA system is preferable, because saturation of a system by Dox is an intrinsically faster process than depletion. For the same reason, the tTA system would be preferable when a gene activity must be turned off quickly. However, should a gene be efficiently reactivated by Dox depletion (tTA system), it is important to keep it inactive at the lowest effective concentration of Dox, which for all organs except the brain is 20 μg/ml in drinking water.[13]

Taking these parameters into account, full induction of luciferase activity was monitored in various organs of the mouse within 24 hr, using the rtTA system.[13] In contrast, by activating the luciferase gene under tTA control via depletion of Dox from the system, 3 to 10 days were required depending on the organ.[11,13] This time course remains constant when repeated cycles of activation and deactivation are invoked over a period of several months.[11] Considerably slower kinetics of gene activation via Dox depletion were observed when animals were bred under a high concentration of Dox in order to keep gene function downregulated until the adult state is reached.[14] This slow clearance is most likely due to the time required for the depletion of Dox depots accumulated in various tissues during the development of the animal. By using the rtTA system, this problem can be avoided.

Finally, remarkable cell type-specific regulation was achieved with a variety of promoters. In one example, in which tTA was produced exclusively in hepatocytes, it could be shown that all hepatocytes in the organ were subject to Dox regulation,[11,13] indicating that position effect variegation when observed in the context of Tet regulation is not an intrinsic feature of the system itself.

Concluding Remarks

Regulating, from the outside, individual gene activities within a complex genetic environment like a eukaryotic cell by such means as provided by the Tet systems requires the reconciliation of numerous parameters that govern the onset of transcription at RNA polymerase II promoters. On the other hand, the Tet regulatory systems excel in their simplicity, a feature that has certainly facilitated their widespread application. Thus, to establish efficient Tc-dependent regulation, it suffices to transfer to a target cell an equivalent of only 1200 bp of "foreign" DNA, approximately 750 bp encoding, e.g., tTA2 (Fig. 2A) and 300–400 bp for P_{tet} (Fig. 2B). Moreover, tetracycline and doxycycline as well as other representatives of this class of compounds are nontoxic at the effective concentrations and do passively enter cells. This enables quantitative regulation of the transcription of a gene at the single-cell level in a seamless manner, homogeneously throughout a population of cells, by varying the concentration of the effector substance.[27]

Because of their simplicity and limited space requirements, the Tet systems have also been implemented in viral vectors, particularly of retroviral[28] (see also Ref. 4) and adenoviral[16] origin. Several vectors were developed that harbor both the *trans*-activator as well as the P_{tet}-controlled target gene expression unit with the aim of transferring a controllable gene into a cell in one step. The use of such vectors is, however, limited because of the cross-talk between the promoter driving the *trans*-activator gene and P_{tet}, resulting in basal activities of the target gene. Moreover, sequence elements within the vector backbone that act like enhancers may contribute to this background expression. While this may be acceptable in some cases,

[27] A. M. Kringstein, F. M. V. Rossi, A. Hofmann, and H. M. Blau, *Proc. Natl. Acad. Sci. U.S.A.* **95,** 13670 (1998).

[28] D. Lindemann, E. Patriquin, S. Flug, and R. Mulligan, *Mol. Med.* **3,** 466 (1997).

it cannot be tolerated in others. Hence, for a high degree of regulation double vector systems[16,28] in which the two expression units are kept apart are still preferable.

A solution to the cross-talk dilemma, which is not actually limited to the Tet systems, may eventually come from advances in our understanding of insulator elements[29] and transcriptional silencers (see above). Transcriptional silencers, on the other hand, have been shown to effectively protect P_{tet} from outside activation. However, most of these silencers act over considerable distances and thus will introduce additional loops of cross-talk when all expression units establishing the Tet regulatory system are placed in close proximity to one another. The result will be difficult to predict in homeostatic states of transcriptional activity. Consequently, the most reliable strategy for establishing Tet regulation remains the transfer of the two essential expression units into target cells in separate steps or in the case of viral vectors via two separate entities, followed by selection or screening procedures that allow the identification of clones that function well at the cellular as well as the organismal level. These strategies are feasible for most experimental approaches including the generation of cell lines for *ex vivo* gene therapy. Many studies will be facilitated by the steadily increasing pool of cell lines and transgenic organisms that express tTA or rtTA constitutively at suitable intracellular concentrations. Of particular interest will be mouse lines in which a *trans*-activator is stably produced in subsets of cells. It is hoped that this rapidly expanding "zoo" of tTA and rtTA mouse lines will be made increasingly available to the research community by distribution centers as is the case for the mouse lines from our laboratory, which are available through the Jackson Laboratories (Bar Harbor, ME).

Previous[30-32] and ongoing work[21] has shown that the Tet repressor has an unexpected large sequence space that permits wide variation in the specificity of the TetR protein while maintaining its essential basic features: stringent recognition of DNA sequences and ligands as well as allosteric properties that render the DNA binding sensitive to ligands. The fact that several mutant proteins have been identified that have a reverse phenotype,[2,21] i.e., they bind DNA only in the presence of the ligand, is not only of practical but also theoretical interest, particularly because the multiple mutations required for this phenotype are distributed throughout the molecule, so far without yielding much mechanistic insight. These muta-

[29] A. C. Bell and G. Felsenfeld, *Curr. Opin. Genet. Dev.* **9,** 191 (1999).

[30] D. Schnappinger, P. Schubert, K. Pfleiderer, and W. Hillen, *EMBO J.* **17,** 535 (1998).

[31] V. Helbl, C. Berens, and W. Hillen, *J. Mol. Biol.* **276,** 319 (1998).

[32] V. Helbl and W. Hillen, *J. Mol. Biol.* **276,** 313 (1998).

tions can be superimposed onto sequence variants that modify the specificity of DNA recognition, and such proteins may still be fused to various activation domains although numerous combinations appear nonfunctional and thus are forbidden.[7,20]

Moreover, proper modifications of the dimerization surface of TetR have allowed the production in one cell of two populations of homodimeric *trans*-activators that recognize different promoters and that interact differentially with different tetracyclines. In view of these findings and with the genetic and biochemical approaches at hand, there is reason for optimism that the Tet repressor/operator/inducer system may be adaptable to still further applications, thus expanding its usefulness in basic and applied research.

Acknowledgments

We thank M. Hasan for critically reviewing the manuscript and S. Reinig for patience in preparing it. This work was supported by the Volkswagen-Stiftung, the BioRegio program of the Bundesministerium für Bildung und Forschung, and the Fonds der Chemischen Industrie Deutschlands.

[31] Coumermycin-Induced Dimerization of GyrB-Containing Fusion Proteins

By Michael A. Farrar, Steven H. Olson, and Roger M. Perlmutter

Introduction

Cellular responses to external stimuli in general require multiple signal transduction cascades that act together to produce a coordinated biological outcome. Considerable progress has been made in identifying the proteins that make up these various signal transduction pathways. However, much less is known about the specific roles that individual pathways play in eliciting biological responses, or the interactions among signaling pathways simultaneously recruited by the same stimulus. Addressing such questions requires the ability to regulate the activity of one signaling pathway without affecting other pathways. Under ideal circumstances this would be achieved by using methods that mimic, insofar as possible, the process whereby a particular pathway or enzyme is activated *in vivo*.

One mechanism by which many signal-transducing proteins become activated is through dimerization or higher order clustering. This was initially demonstrated for receptor tyrosine kinases[1] and has subsequently been observed in numerous other biological systems.[2] These observations suggested that forced dimerization of proteins, e.g., protein kinases or transcription factors, would lead to their activation and the subsequent initiation of downstream signaling cascades. The validity of this approach was initially demonstrated with antibodies to cross-link individual components of multisubunit transmembrane receptors, thereby allowing an assessment of their functions in the absence of other receptor components.[3-5] More recently, small synthetic molecules have been used to promote specific dimerization.[6] The initial approach involved using dimeric versions of the immunosuppressant FK506 to juxtapose chimeric proteins engineered to contain an FK506-binding protein, FKBP12. Here we review an analogous method that makes use of the symmetrically dimeric antibiotic coumermycin.

General Properties of Coumermycin

Coumermycin is a *Streptomyces* product that consists of two identically substituted coumarin rings joined by a methyl pyrrole linker (Fig. 1). Novobiocin, a related antibiotic, can be regarded as a monomeric version of coumermycin (Fig. 1). Both coumermycin and novobiocin block bacterial cell growth by binding to, and inhibiting the action of, the B subunit of bacterial DNA gyrase.[7] Structure–function studies have localized the binding site for these drugs to the N-terminal 220-amino acid domain (the GyrB domain) of *Escherichia coli* gyrase.[8] Equilibrium binding and gel-filtration studies have established that coumermycin binds GyrB with a stoichiometry of 1:2 while novobiocin binds in a 1:1 ratio.[9] Novobiocin and coumermycin bind GyrB with high affinity. The K_D for novobiocin or coumermycin bound

[1] C. H. Heldin, *Cell* **80,** 213 (1995).

[2] J. D. Klemm, S. L. Schreiber, and G. R. Crabtree, *Annu. Rev. Immunol.* **16,** 569 (1998).

[3] B. A. Irving, A. C. Chan, and A. Weiss, *J. Exp. Med.* **177,** 1093 (1993).

[4] C. Romeo, M. Amiot, and B. Seed, *Cell* **68,** 889 (1992).

[5] A.-M. K. Wegener, F. Letourneur, A. Hoeveler, T. Brocker, F. Luton, and B. Malissen, *Cell* **68,** 83 (1992).

[6] D. M. Spencer, T. J. Wandless, S. L. Schreiber, and G. R. Crabtree, *Science* **262,** 1019 (1993).

[7] A. Maxwell, *Mol. Microbiol.* **9,** 681 (1993).

[8] E. J. Gilbert and A. Maxwell, *Mol. Microbiol.* **12,** 365 (1994).

[9] J. A. Ali, A. P. Jackson, A. J. Howells, and A. Maxwell, *Biochemistry* **32,** 2717 (1993).

Coumarin Antibiotics

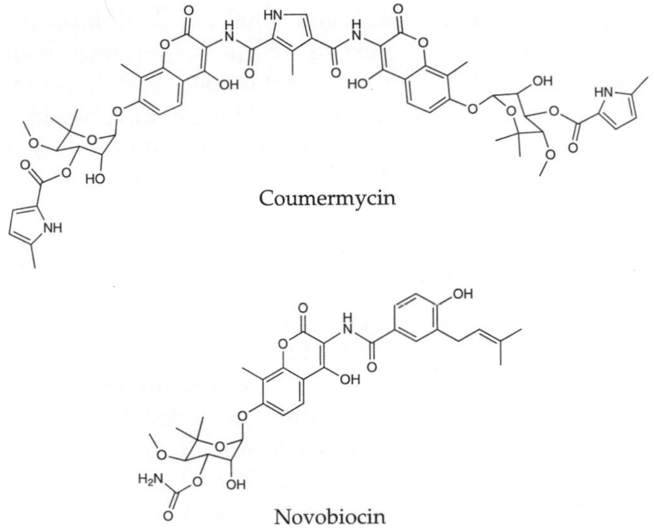

Coumermycin

Novobiocin

FIG. 1. Structure of coumermycin and novobiocin.

to GyrB has been determined to lie in the range of $3–5 \times 10^{-8}\ M$.[10] In addition, the crystal structure of novobiocin bound to GyrB has been solved,[11,12] which allows modeling of the potential structure of a $GyrB_2$: coumermycin complex (Fig. 2; see color plate). Thus we have a substantial understanding of how the coumarin antibiotics interact with GyrB.

Coumermycin exhibits several favorable features that encourage broad use as a chemical dimerizer in eukaryotic systems. First, its target is a prokaryotic enzyme; no high-affinity endogenous binding targets exist in normal eukaryotic cells. Coumermycin does interact with and inhibit eukaryotic type II topoisomerases, but only at concentrations more than 100-fold higher than those required to bind to GyrB.[13] Extensive testing of coumermycin in both rodents and dogs, at concentrations that exhibit sig-

---

[10] N. A. Gormley, G. Orphanides, A. Meyer, P. M. Cullis, and A. Maxwell, *Biochemistry* **35,** 5083 (1996).

[11] R. J. Lewis, O. M. P. Singh, C. V. Smith, T. Skarzynski, A. Maxwell, A. J. Wonacott, and D. B. Wigley, *EMBO J.* **15,** 1412 (1996).

[12] G. A. Holdgate, A. Tunnicliffe, W. H. Ward, S. A. Weston, G. Rosenbrock, P. T. Barth, I. W. Taylor, R. A. Pauptit, and D. Timms, *Biochemistry* **36,** 9663 (1997).

[13] K. G. Miller, L. F. Liu, and P. T. Englund, *J. Biol. Chem.* **256,** 9334 (1981).

nificant antibacterial activity, has not revealed any overt toxicity.[14] Hence coumermycin concentrations that are sufficient to bind and inhibit bacterial DNA gyrases *in vivo* do not alter eukaryotic topoisomerase II function. Furthermore, coumermycin exhibits good pharmacokinetic properties, with a reported serum half-life in mice of 5.5 hr.[14] Therefore, coumermycin should be useful for applications in animal models as well as for studies in cell culture systems. Finally, both coumermycin and novobiocin are readily available from commercial sources [Sigma (St. Louis, MO) and Fluka (Rokonkoma, NY)] at low cost.

Generation of GyrB Fusion Proteins

Coumermycin-induced dimerization has been used to activate a number of signaling molecules including the serine/threonine kinases Raf[15] and ASK1,[16] the tyrosine kinase Jak2,[17] the transcription factors STAT3,[18] STAT5a[18] and STAT5b,[19] and the adhesion molecule L-selectin.[20] The GyrB domain used in these studies consists of amino acids 2–220 of bacterial DNA gyrase. In addition, an 18-amino acid flexible linker (GSEGKSSGSGSESKVTDS), used previously to connect heavy and light chains in single-chain antibodies,[21] was added to the carboxy-terminal end of GyrB. This linker sequence can be used to concatenate additional GyrB domains in tandem, thereby permitting oligomerization of such GyrB fusion proteins. In applications requiring only dimerization, the linker sequence should not be required. However, because no experiments have been done to formally rule out that the linker sequence fortuitously promotes coumermycin-mediated dimerization of the GyrB domain, the most conservative approach would include the linker region in all GyrB constructs.

In most cases the GyrB domain has been added to the carboxy terminus of the protein that is to be dimerized. However, the optimal location for the GyrB domain can only be determined empirically. Proximity to the domain to be activated appears to be desirable, although the possibility of steric interference due to the introduction of the GyrB domain within the

[14] J. C. Godfrey and K. E. Price, *Adv. Appl. Microbiol.* **15,** 653 (1972).
[15] M. A. Farrar, J. Alberola-Ila, and R. M. Perlmutter, *Nature (London)* **383,** 178 (1996).
[16] Y. Gotoh and J. A. Cooper, *J. Biol. Chem.* **273,** 17477 (1998).
[17] M. G. Mohi, K. Arai, and S. Watanabe, *Mol. Biol. Cell* **9,** 3299 (1998).
[18] A. M. O'Farrell, Y. Liu, K. W. Moore, and A. L. Mui, *EMBO J.* **17,** 1006 (1998).
[19] M. A. Farrar and R. M. Perlmutter, unpublished results (1999).
[20] X. Li, D. A. Steeber, M. L. K. Tang, M. A. Farrar, R. M. Perlmutter, and T. F. Tedder, *J. Exp. Med.* **188,** 1385 (1998).
[21] M. W. Pantoliano, R. E. Bird, S. Johnson, E. D. Asel, S. W. Dodd, J. F. Wood, and K. D. Hardman, *Biochemistry* **30,** 10117 (1991).

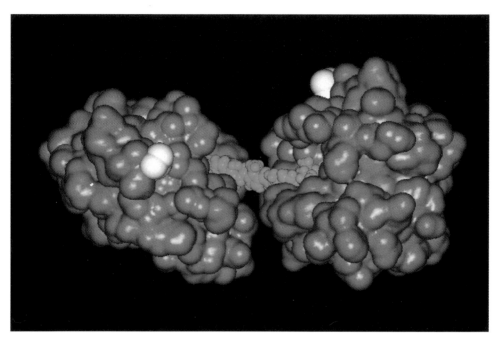

Fig. 2. Model of a coumermycin-induced GyrB dimer. This model was based on the crystal structure of the Novobiocin:GyrB complex by G. A. Holdgate *et al.* [*Biochemistry* **36,** 9663–9673 (1997)] and involved superimposing the structure of coumermycin on that of novobiocin. Coumermycin is shown in light blue while the GyrB domains are depicted in orange. The white balls represent the respective amino termini of the two GyrB domains.

target protein must also be considered. Therefore, evaluation of several different chimeric constructs to achieve optimal activation is recommended. The importance of the location of the GyrB domain within a chimeric protein has been most clearly demonstrated by Mohi *et al.*[17] for the tyrosine kinase Jak2. They found that placing the GyrB domain centrally within the Jak2 protein, immediately adjacent to the Jak2 kinase domain, permitted elicitation of coumermycin-induced increases in kinase activity and subsequent biological responses. In contrast, constructs containing GyrB at the amino-terminal end of Jak2 were inactive. These results indicate that for at least some GyrB fusion proteins appropriate localization of the GyrB domain, within the fusion protein, is critical for successful activation by coumermycin-mediated dimerization.

An additional factor that needs to be considered when trying to activate proteins by dimerization is their cellular location. For example, activation of the serine/threonine kinase Raf takes place at the plasma membrane.[22] Nonetheless, coumermycin-induced dimerization of cytoplasmic Raf–GyrB fusion proteins does lead to the activation of Raf targets, e.g., Mek.[15] Similar results have been obtained by forcing Raf dimerization, using the FK1012 system.[23] However, more distal events in this signaling cascade, notably the Map kinases Erk1 and Erk2, are not efficiently activated after coumermycin-induced dimerization of the cytoplasmic form of Raf.[19] In contrast, if Raf–GyrB fusion proteins are localized to the plasma membrane by adding an N-terminal myristoylation sequence, then potent activation of the entire downstream Raf signaling pathway can be achieved after coumermycin addition.[19] Thus, placing GyrB fusion proteins in the appropriate cellular location may be important for obtaining optimal biological responses after coumermycin-induced dimerization.

Evaluation of GyrB Fusion Proteins

On assembling a GyrB fusion construct and transfecting the appropriate cell type (or generating the appropriate transgenic mouse lines) it is important to confirm that the chimeric protein that is to be activated is actually expressed and that the GyrB domain has folded properly. This can be done with novobiocin coupled to agarose as an affinity reagent to purify the GyrB fusion protein, followed by sodium dodecyl sulfate-polyacrylamide gel electrophoresis (SDS–PAGE) fractionation and immunoblotting with antibodies to either the fusion partner or the GyrB domain. The basic method is as follows.

[22] C. J. Marshall, *Nature (London)* **383,** 127 (1996).
[23] Z. Luo, G. Tzivion, P. J. Belshaw, D. Vavvas, M. Marshall, and J. Avruch, *Nature (London)* **383,** 181 (1996).

1. Wash cells twice with phosphate-buffered saline (PBS). Lyse the cells in lysis buffer consisting of 50 mM Tris-HCl (pH 7.5), 100 mM NaCl, 50 mM β-glycerophosphate, 1 mM EDTA, 1 mM EGTA, 1 mM dithiothreitol (DTT), 1% Triton X-100, aprotinin (1 μg/ml), leupeptin (1 μg/ml), and phenylmethylsulfonyl fluoride (PMSF, 100 μg/ml).

2. Remove nuclei by centrifugation at 10,000 rpm for 5 min in a tabletop centrifuge. Incubate the postnuclear supernatant for 4 hr with novobiocin–agarose. [Novobiocin–agarose can be purchased from ICN (Irvine, CA) or produced according to a coupling procedure outlined by Maxwell and colleagues.[24,25]]

3. Precipitate agarose beads by brief centrifugation (5000g for 1 min). Remove the supernatant and wash three times with lysis buffer and once with PBS. Resuspend the beads in Laemmli buffer and carry out SDS–PAGE fractionation and immunoblotting. For detecting chimeric proteins, we recommend using an antibody that recognizes the fusion partner domain of the GyrB fusion protein. Alternatively, a monoclonal antibody (7D3; Lucent LTD, contact at *gyrase@le.ac.uk*), which recognizes the GyrB domain,[26] can be used (1 μg/ml). This antibody has not been used successfully for immunoprecipitations.

In addition, it is frequently helpful to quantitate the amount of GyrB fusion protein within a cell. This is particularly easy to do when both the endogenous (unmodified) and GyrB-tagged proteins are present within the same cell. For example, we can compare the relative amounts of Raf–GyrB with those of endogenous Raf by carrying out immunoblots of whole-cell lysates with an antibody directed against Raf. Because the GyrB domain contributes an additional 24 kDa to the overall size of the protein, the endogenous Raf can be readily distinguished from Raf–GyrB.

Finally, it may be desirable to confirm that coumermycin not only binds to the GyrB fusion protein but also promotes dimer formation. Procedures developed by O'Farrell *et al.*[18] and Gormley *et al.*[10] permit chemical cross-linking and analysis of coumermycin-induced dimers. In general, this involves preparing hypotonic lysates of transfected cells via Dounce homogenization in lysis buffer [50 mM sodium phosphate (pH 8.1), 5 mM EGTA, 5 mM magnesium chloride] followed by shearing cells through a 26-gauge needle. The lysates are then treated for 15 min at room temperature with

[24] W. L. Staudenbauer and E. Orr, *Nucleic Acids Res.* **9**, 3589 (1981).

[25] A. H. Maxwell and A. J. Howells, *in* "Protocols for DNA topoisomerases 1 DNA topology and enzyme purification" (M. A. Bjornsti and N. Osherhoff, eds.), pp. 135–144, Humana Press, Totowa, New Jersey, 1999.

[26] M. Thornton, M. Armitage, A. Maxwell, B. Dosanjh, A. J. Howells, V. Norris, and D. C. Sigee, *Microbiology* **140**, 2371 (1994).

coumermycin or novobiocin as a control, after which dimerized proteins are then cross-linked with a chemical cross-linker such as BS3 or dimethyl suberimidate. The cross-linked products are then fractionated by SDS–PAGE and analyzed by immunoblotting with an antibody to either the GyrB domain or the fusion partner.

An alternative approach, using a BIAcore system (Biacore, Uppsala, Sweden), was developed by O'Farrell et al.[18] to demonstrate dimerization of STAT3–GyrB chimeras. This method employs double-stranded oligonucleotides (oligos) comprising multiple copies of the STAT3-binding site, which have been biotinylated at the 5' end [oligos annealed in 10 mM Tris (pH 7.5), 0.1 mM EDTA]. These STAT3-binding oligonucleotides are then affixed to a streptavidin sensor chip by flowing 10μl of a 1 μM double-stranded oligonucleotide solution (diluted in HBS [10 mM HEPES (pH 7.4), 0.15 M NaCl, 3.4 mM EDTA, and 0.05% surfactant P20; Pharmacia Biosensor, San Diego, CA] over the surface of the chip. The surface is washed once with 0.5 M NaCl–HBS (1 min) and once with 0.5% (w/v) SDS–water (1 min), and then reequilibrated in HBS. Cell lysates containing STAT3–GyrB are then prepared and poly(dI · dC) (100 μg/ml) added to block nonspecific binding. Dimerization is induced by addition of coumermycin for 15 min at 25°. Lysates are then applied to the sensor chip at a flow rate of 5 μl/min. Incubation in the presence of novobiocin instead of coumermycin is used as a negative control.

Evaluating Coumermycin-Induced Dimerization and
  Biological Responses

A characteristic feature of biological responses elicited by chemically induced dimerization is revealed through simple dose–response studies. Because activation by this method depends on steady state conditions favoring dimer formation, a bell-shaped curve should be obtained when plotting coumermycin concentration versus biological response. An example of this is shown in Fig. 3, wherein coumermycin-induced activation of Raf–GyrB is examined by assaying the phosphorylation of Erk1 by Mek in intact cells. Maximal activation occurred at 900 nM coumermycin, while higher doses inhibited activation. This result is consistent with the model that activation occurs by dimerization, because at high coumermycin concentrations excess drug should inhibit dimer formation. In all systems examined so far, maximal activation occurs at coumermycin concentrations between 100 nM and 1 μM. Because this dose range is fairly broad, a dose–response analysis should be performed to determine the concentration of coumermycin required for optimal activation. Novobiocin, the monomeric form of coumermycin, is a useful control for specificity because

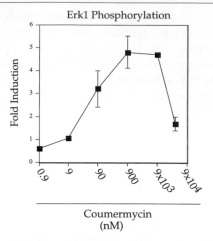

Fig. 3. Coumermycin-induced Raf–GyrB activation exhibits a biphasic dose response. NIH 3T3 fibroblasts were transfected with a construct encoding a myristoylated form of Raf–GyrB. Cells were serum starved overnight and subsequently stimulated with increasing amounts of coumermycin. Activation was assessed by monitoring the increase in Erk1 phosphorylation relative to vehicle (DMSO) alone.

it binds to the same site on GyrB as coumermycin but cannot drive dimerization. Furthermore, novobiocin almost completely blocks coumermycin-induced effects when present in 10- to 20- fold molar excess. Thus novobiocin is a useful reagent for terminating coumermycin-induced dimerization and possibly coumermycin-induced biological responses as well.

When preparing coumermycin and novobiocin for subsequent experiments a few practical factors need to be considered. Coumermycin itself is poorly soluble in water, and therefore stock solutions should be made in dimethyl sulfoxide (DMSO, soluble up to 40 mg/ml). Novobiocin, in contrast, is soluble in water at up to 100 mg/ml. Stock solutions of coumermycin can be diluted directly into cell culture media at concentrations of up to 10 μg/ml (9 μM). At higher concentrations coumermycin may precipitate out of solution or begin to inhibit eukaryotic topoisomerase II enzymes. Coumermycin has also been reported to bind to serum proteins.[14] This could affect the dose response obtained when stimulating cells in the presence or absence of fetal bovine serum, although in our hands we have not seen any dramatic influence of serum in altering the effective coumermycin concentration. Finally, although coumermycin readily penetrates cell membranes, experiments done with cell extracts require 10-fold lower coumermycin concentrations than those used to stimulate intact cells, most likely reflecting the improved access of coumermycin to its target.[15]

Conclusions

The coumermycin–GyrB system is a useful strategy for regulating protein homodimerization and activation both in cell extracts and in intact cell culture systems. Coumermycin-based dimerization strategies are currently being used to identify biological responses (such as changes in gene transcription) induced by the Raf and Jak/STAT signal transduction pathways. As the utility of dimerization-based strategies becomes evident we expect that they will be applied successfully to an increasing number of signal transduction pathways. An important development in this process will be the generation of heterodimeric reagents incorporating GyrB-binding groups. The relative ease by which novobiocin can be synthetically modified to incorporate novel chemical entities should permit facile development of such reagents. Finally, it will be particularly interesting to investigate how specific pathways interact by using chemically distinct dimerization methods to regulate the activity of each independently. Such approaches will provide important insights into the overall regulation of cell signaling.

Acknowledgments

We thank Petra Doerfler and Karsten Sauer for comments on the manuscript, and Anthony Maxwell for advice on GyrB–coumermycin interactions.

[32] Use of Glutathionc S-Transferase and Break Point Cluster Region Protein as Artificial Dimerization Domains to Activate Tyrosine Kinases

By Yoshiro Maru

Introduction

From retroviral activation of cellular genes we know that oncogenes are generated by genetic alterations including deletion, point mutation, fusion with another gene, and rearrangement. Naturally occurring oncogenes in human cancer also contain these types of genetic changes.[1,2] The

[1] T. Hunter and J. A. Cooper, "The Enzymes," Vol. XVII, p. 191. Academic Press, San Diego, California, 1986.
[2] J. M. Bishop, *Science* 235, 305 (1987).

BCR-ABL oncogene found in human chronic myelogenous leukemia is generated by a reciprocal translocation between chromosome 9 and 22 t(9;22), which juxtaposes the BCR gene on chromosome 22 with the ABL gene on 9.[3] The ABL gene encodes a tyrosine kinase whose activity is highly regulated *in vivo*. The fusion to BCR constitutively activates the ABL tyrosine kinase activity, and the amino-terminal BCR sequence 1–60 with its oligomerizing property makes an essential contribution to activation.[4,5] Multimerization-mediated activation of tyrosine kinases is well known in receptor tyrosine kinases, in which ligand-induced dimerization of the receptor increases its activity as a tyrosine kinase enzyme under physiological circumstances.[6] Why does the oligomerization activate receptor tyrosine kinase enzymes? Ligand-dependent oligomerization induces autophosphorylation of the receptor. Autophosphorylation of specific tyrosine residues has been reported to enhance enzymatic activities of receptors for insulin[7] and hepatocyte growth factor (Met).[8] Positive regulation of enzymatic activity by autophosphorylated tyrosine residues that may serve as allosteric sites is also observed in cytoplasmic tyrosine kinases such as c-Src[9] and v-Fps.[10] However, this is not the case for the epidermal growth factor receptor[11] and BCR-ABL[12]; both of these need oligomerization for activation. Almost all receptor tyrosine kinase oncogenes, including Tpr-Met, Tpr-Trk (nerve growth factor receptor), Rfp-Ret, and TEL-PDGFβR (platelet-derived growth factor receptor), that are activated by DNA rearrangement are fused to sequences with motifs that are predicted to form a coiled coil.[13,14] A question concerning whether ABL activation in physiological settings requires oligomerization remains to be answered. On the basis of the crystal structure of c-Src and sequence homology between c-

[3] Y. Maru and O. N. Witte, "Application of Basic Science to Hematopoiesis and Treatment of Disease," p. 123. Raven Press, New York, 1993.

[4] J. R. WcWhirter, D. L. Galasso, and J. Y. J. Wang, *Mol. Cell. Biol.* **13,** 7587 (1993).

[5] Y. Maru, D. E. Afar, O. N. Witte, and M. Shibuya, *J. Biol. Chem.* **271,** 15353 (1996).

[6] C. H. Heldin, *Cell* **80,** 213 (1995).

[7] P. A. Wilden, J. M. Backer, C. R. Kahn, D. A. Cahill, G. J. Schroeder, and M. F. White, *Proc. Natl. Acad. Sci. U.S.A.* **87,** 3358 (1990).

[8] L. Naldini, E. Vagna, R. Ferracini, P. Longati, L. Gandino, M. Prat, and P. M. Comoglio, *Mol. Cell. Biol.* **11,** 1793 (1991).

[9] H. Piwnica-Worms, K. B. Saunders, T. M. Roberts, A. E. Smith, and S. H. Cheng, *Cell* **49,** 75 (1987).

[10] K. Meckling-Hansen, R. Nelson, P. Branton, and T. Pawson, *EMBO J.* **6,** 659 (1987).

[11] A. Honegger, T. J. Dull, D. Szapary, A. Komoriya, R. Kris, A. Ullrich, and J. Schlessinger, *EMBO J.* **7,** 3053 (1988).

[12] A. M. Pendergast, M. L. Gishizky, M. H. Havlik, and O. N. Witte, *Mol. Cell. Biol.* **13,** 1728 (1993).

[13] G. A. Rodrigues and M. Park, *Curr. Opin. Genet. Dev.* **4,** 15 (1994).

[14] T. R. Golub, G. F. Barker, M. Lovett, and D. G. Gilliland, *Cell* **77,** 307 (1994).

Src and ABL, it is presumed that the SH3 domain of ABL binds to the catalytic domain (CD) directly or with the SH2-CD linker sequence in between and thus autoinhibits the kinase activity.[15] The fact that mutations of amino acids in the ABL SH3 domain, SH2-CD linker, or CD that are thought to be important to maintain this structure upregulate the kinase activity supports this idea. Not only BCR but also the Ets-related transcription factor TEL, with its oligomerizing property, have been shown to activate ABL (TEL-ABL) in human acute myeloid leukemia with a t(9;12;14) translocation.[16] It is possible that oligomerization may also abolish this structure to release the autoinhibition. Here we describe artificial activation of tyrosine kinase enzymes by tagging sequences such as glutathione *S*-transferase (GST) or the break point cluster region (BCR) sequence, which can confer oligomerizing abilities on them.

Glutathione *S*-Transferase and Break Point Cluster Region Protein

Glutathione *S*-transferases appear to play an essential role in the biotransformation of xenobiotics. They catalyze reactions in which the sulfur atom of glutathione provides electrons for reactive electrophiles, resulting in the formation of glutathione conjugates.[17] In the rat liver at least 10 different isoenzymes have been found in the cytosol fraction. The enzymes are ubiquitous and abundant in mammalian tissues. X-Ray crystallography of GST isoenzyme 3-3 has revealed that it is globular in shape and forms an asymmetric dimer.[18,19] The two subunits contact one another through hydrogen bonds and electrostatic and hydrophobic interactions. Purification of GST is efficiently performed with glutathione-immobilized Sepharose, in which the ligand is attached by coupling with the oxirane group of epoxy-activated Sepharose (glutathione-Sepharose). GST encoded by *Schistosoma japonicum* has been cloned and widely utilized as a useful tool in the expression and purification of recombinant proteins in the form of GST fusions in *Escherichia coli*.[20,21] GST-tagged proteins expressed in *E. coli* in large quantities can be easily purified by single-step column chromatography with glutathione-Sepharose.

[15] D. Barila and G. Superti-Furga, *Nature Genet.* **18**, 280 (1998).
[16] T. R. Golub, A. Goga, G. F. Barker, D. E. H. Afar, J. McLaughlin, S. K. Bohlander, J. D. Rowley, O. N. Witte, and D. G. Gilliland, *Mol. Cell. Biol.* **16**, 4107 (1996).
[17] B. Mannervik, *Adv. Enzymol. Relat. Areas Mol. Biol.* **57**, 357 (1985).
[18] X. Ji, P. Zhang, R. N. Armstrong, and G. L. Gilliland, *Biochemistry* **31**, 10169 (1992).
[19] M. W. Parker, M. L. Bello, and G. Federici, *J. Mol. Biol.* **213**, 221 (1990).
[20] D. B. Smith, M. R. Rubira, R. J. Simpson, K. M. Davern, W. U. Tiu, P. G. Board, and G. F. Mitchell, *Mol. Biochem. Parasitol.* **27**, 249 (1988).
[21] D. B. Smith and K. S. Johnson, *Gene* **67**, 31 (1988).

BCR was initially identified as a fusion partner of ABL in the formation of the BCR-ABL oncogene found in human chronic myelogenous leukemias. Computer analysis of the BCR sequence 1–70 revealed the presence of a heptad repeat of hydrophobic residues that might contain an amphipathic α helix; and this could be a coiled coil domain that mediates oligomerization.[4,13] Baculovirus-expressed BCR was partially purified by Superose 6 gel-filtration column chromatography as a large molecule of 650 kDa, which is the size of a homotetramer.[22] P210 BCR-ABL expressed in baculovirus was also purified as a molecule whose size was greater than 800 kDa.[23] Cross-linking studies have shown that the amino-terminal 60 amino acids have the ability to form a tetrameric structure when expressed in *E. coli.*[5] These data support the idea that BCR-ABL forms a multimeric structure that is mediated by the BCR amino-terminal sequence. The significance of this structure is evidenced by the fact that deletion of the amino-terminal 40 amino acids results in a complete loss of biochemical as well as biological activity *in vivo.*[24] Once activated, BCR-ABL is autophosphorylated as in the case of ligand-stimulated receptor tyrosine kinases, and Tyr-177 in the BCR sequence provides a site for phosphotyrosine/SH2 domain-mediated binding to the Ras activator Grb-2/SOS complex.[25]

Thus fusion with either a GST or BCR sequence can be useful to multimerize proteins, and if they require multimerization for activation the artificial fusion proteins could be constitutively active.

Plasmid Construction

cDNA fragments were expressed in the form of GST fusion proteins in mammalian cells as follows. The *Aat*II site (position 1224) of the pGEX-3X vector was converted to an *Xba*I site by blunting with Klenow followed by *Xba*I linker addition. The *Hinc*II–*Xba*I fragment containing the GST cDNA and the stop codons downstream was cloned into the *Sma*I and *Xba*I sites of pGEM4 (Promega, Madison, WI), whose original Asp-718 site was converted to *Xba*I, and whose *Eco*RI site was destroyed by blunting and religation [pGEM4-GST : Asp718 to *Xba*I : *Eco*RI(–)]. We named it

[22] Y. Maru and O. N. Witte, *Cell* **67**, 459 (1991).

[23] A. M. Pendergast, R. Clark, E. S. Kawasaki, F. P. McCormick, and O. N. Witte, *Oncogene* **4**, 759 (1989).

[24] A. J. Muller, J. C. Young, A. M. Pendergast, M. Pondel, N. R. Landau, D. R. Littman, and O. N. Witte, *Mol. Cell. Biol.* **11**, 1785 (1991).

[25] A. M. Pendergast, L. A. Quilliam, L. D. Cripe, C. H. Bassing, Z. Dai, N. Li, A. Batzer, K. M. Rabun, C. J. Der, J. Schlessinger, and M. L. Gishizky, *Cell* **75**, 175 (1993).

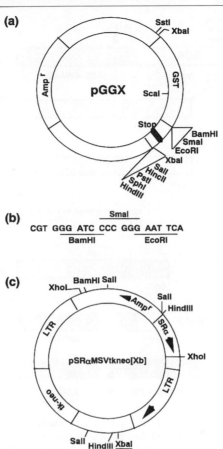

FIG. 1. Plasmids for GST fusion constructs. (a) Structure of pGGX. It contains a multiple cloning site (*Bam*HI/*Sma*I/*Eco*RI) and multiple stop codons downstream (stop). The frame of the cloning site is shown in (b). GST fusion cDNA fragments were cut out as *Xba*I fragments by using the *Xba*I sites upstream of the GST gene and downstream of the stop codons and can be cloned into the *Xba*I site of the retroviral expression vector pSRalphaMSVtkneo[Xb] (c) as described previously.[5,26]

pGGX (Fig. 1). The unique *Sma*I, *Bam*HI, and *Eco*RI sites were retained for cloning cDNA fragments. Into these sites we ligated four cDNAs: (1) a *Bam*HI-to-*Eco*RI fragment encoding P185 BCR-ABL with the amino-terminal amino acids 1–160 of BCR deleted[5] (Fig. 2), (2) a *Bgl*II-to-*Eco*RI fragment encoding amino acids 414–426 of BCR (TrpProAsnAspAsp-GluGlyAlaPheHisGlyAspAla) followed by an ABL sequence starting from exon II, and (3) an *Eco*RI-to-*Eco*RI fragment encoding the full-length

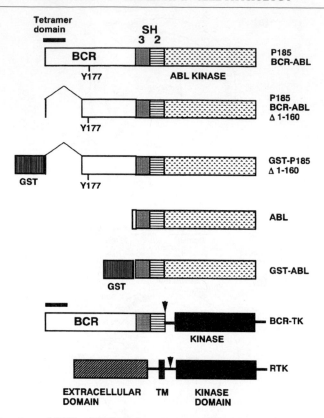

FIG. 2. Structure of GST and BCR fusion constructs. The tetramer domain, Tyr-177 (Grb-2-binding site), and the SH3 and SH2 domains are shown. The amino-terminal fusion of the GST sequence to P185 BCR-ABL delta 1–160 (GST-P185delta1–160) and ABL (GST-ABL) is shown. The extracellular domain, the transmembrane domain (TM), and the kinase domain of receptor tyrosine kinase (RTK) such as Trk and KDR (see text) are shown. The tyrosine kinase domain (TK) of the RTK was fused to the BCR sequence (BCR-TK).

xeroderma pigmentosum group B protein (XPB).[26] A *Xho*I linker was added to the *Sma*I site in pGGX. (4) A *Bam*HI-to-*Xho*I fragment carrying XPB 203–782, which does not contain the nuclear localization signal, was cloned into pGGX : *Sma*I to *Xho*I.[26] The resulting in-frame fusion constructs (1) GST-P185Δ1–160, (2) GST-ABL (this construct contains 13 amino acids of BCR origin), (3) GST-XPB 1–782 (full-length), and (4) GST-XPB 203–782 were cut out as *Xba*I fragments and cloned into the pSRalphaMSVtkneo

[26] N. Takeda, M. Shibuya, and Y. Maru, *Proc. Natl. Acad. Sci. U.S.A.* **96,** 203 (1999).

retrovirus vector, whose unique *Eco*RI site was converted to *Xba*I by blunting and *Xba*I linker addition (pSRalphaMSVtkneo[Xb])[5,24] (Fig. 1).

We have also tried to activate receptor tyrosine kinases Trk and KDR, receptors for nerve growth factor and for vascular endothelial growth factor, respectively, by fusion to the BCR sequence containing the oligomerization domain (Fig. 2: BCR-TK). The 5' BCR-ABL *Xba*I–*Kpn*I fragment from pGEM4/P185 BCR-ABL and the *Kpn*I–*Eco*RI fragment from pGEX-3X(1021–1120)P210[SH2] were ligated.[27] This 5' *Xba*I–*Eco*RI fragment containing the first BCR exon, ABL SH3, and SH2 domains was used for fusion to the following kinases. Construction of BCR-TRK was started with pLM6, kindly provided by M. Barbacid.[27]. *Eco*RI linkers (12 bp, CCGGAATCCGG) were added to the *Bal*I site (position 1487) and *Eco*RI sites in pUC118 were converted to *Hin*dII sites. The *Eco*RI–*Hin*dIII fragment containing the Trk kinase domain was fused to the *Xba*I–*Eco*RI fragment described above. The resulting BCR-TRK (*Xba*I–*Hin*dIII) sequences were cloned into pSRalphaMSVtkneo[Xb]. BCR-KDR was constructed as follows: The unique *Bam*HI site (position 2412) in KDR,[28] which is localized in the sequence corresponding to the submembranous region of the receptor, was converted to a *Not*I site by Klenow treatment and *Not*I linker ligation. The 3' *Not*I–*Xba*I fragment containing the tyrosine kinase domain to the carboxy terminus was fused to the 5' *Xba*I–*Eco*RI fragment whose 3' end was converted to *Not*I to make an in-frame fusion. The resulting BCR-KDR was cloned into the retroviral expression vector pSRalphaMSVtkneo[Xb].

Expression and Purification of Glutathione S-Transferase-Tagged Proteins from Mammalian Cells

The retroviral expression vector pSRalphaMSVtkneo can be used for establishing replication-incompetent retroviruses as well as for transient transfection in Cos cells or 293T cells. Intracellular localization is an important factor for activation of oncogenes. For example, there are two different forms of mouse ABL proteins, type I and type IV, with different amino termini that are generated by alternative splicing of the first exon. The type IV ABL protein is nuclear. Deletion of the SH3 domain from type IV ABL not only activates the tyrosine kinase activity but also relocalizes the protein from the nucleus to the cytoplasm.[29] Because GST is basically a cytosolic

[27] Y. Maru, S. Yamaguchi, and M. Shibuya, *Oncogene* **16,** 2585 (1998).
[28] B. I. Terman, M. Dougher-Vermazen, M. E. Carrion, D. Dimitrov, D. C. Armellino, D. Gospodarowicz, and P. Bohlen, *Biochem. Biophys. Res. Commun.* **187,** 1579 (1992).
[29] R. A. van Etten, P. Jackson, and D. Baltimore, *Cell* **58,** 669 (1989).

enzyme, we examined whether the well-known nuclear protein XPB (xeroderma pigmentosum group B), a component of the basal transcription factor TFIIH, was relocalized to the cytoplasm by fusion to GST. Immunostaining showed that almost all GST-XPB proteins are in the nucleus [Fig. 3 (2)]. When the amino-terminal sequence that contains the nuclear localization signal was deleted (XPB 203–782), the GST-XPB 203–782 proteins were localized both in the cytoplasm and the nucleus [Fig. 3 (1)]. However, this does not rule out the possibility that intracellular localization of some other nuclear proteins fused to GST may be governed by GST. The endogenous ABL proteins are found both in the cytoplasm (e.g., mouse type I) and the nucleus (type IV). Immunostaining of the GST-ABL proteins whose ABL sequence starts from exon II showed that they were in the cytoplasm (data not shown).

Cells were lysed in protein extraction buffer containing 50 mM HEPES (pH 7.4)–150 mM NaC1–1% (v/v) Triton X-100–2 mM sodium vanadate–10 mM NaF–10 mM sodium pyrophosphate–1 mM EDTA–1 mM phenylmethylsulfonyl fluoride (PMSF)–leupeptin (20 μg/ml). Glutathione-Sepharose was washed twice in 50 mM HEPES, pH 7.4, and was incubated with the extracted protein for 90 min in a rotator. Beads were washed twice in the protein extraction buffer and twice in 50 mM HEPES, pH 7.4. Bound proteins were released by boiling for 5 min in Laemmli sample buffer. Alternatively bound proteins were eluted in 5–10 mM glutathione (reduced

(1) XPB 203-782   (2) XPB 1-782

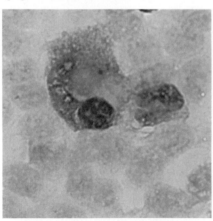 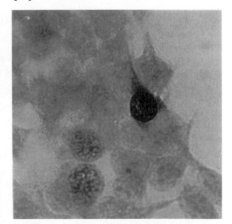

FIG. 3. Immunostaining of the nuclear protein XPB fused to GST. XPB 203–782 (1) and XPB 1–782 (2) were transiently expressed in 293T cells, and anti-XPB staining was performed with an immunoperoxidase system (Vectastain ABC kit).

form) in HEPES (pH 7.4). Sodium dodecyl sulfate (SDS; 0.02%, w/v) can be included in the protein extraction buffer but it reduces the efficiency of protein elution in glutathione.

Activation of Tyrosine Kinases by Fusion with Glutathione
  S-Transferase or Break Point Cluster Region Protein

Deletion of the oligomerizing amino-terminal BCR sequence (e.g., positions 1–40) results in a complete loss of activation of BCR-ABL. GST was tagged to P185 BCR-ABLΔ1–160 (GST-P185Δ1–160) and we wondered whether the tyrosine kinase activity as well as the transforming potentials could be restored to P185Δ1–160. The autophosphorylation activity *in vivo* of GST-P185Δ1–160 expressed in susceptible Rat1 fibroblasts was detected but was less than 50% of the wild-type P185BCR-ABL.[5] The recruitment of the Ras activator Grb-2 adaptor protein to GST-P185Δ1–160 implies that the Tyr-177 retained in this GST fusion construct was phosphorylated. Tyrosine phosphorylations of Shc and p62Dok in GST-P185Δ1–160-expressing Rat1 cells were also detected, but represented only 30% of that in wild-type cells.[5] Reactivation of these biochemical activities was correlated with biological potential. The transforming ability of GST-P185Δ1–160 as judged by the Rat1 cell transformation was 30–40% of that of wild type. This indicates that GST with its dimerizing ability can at least partially substitute for the BCR tetramerization domain-containing sequence in BCR-ABL in terms of both biochemical and biological potential. As described above, the reactivated and autophosphorylated GST-P185Δ1–160 binds to the Ras activator Grb-2/SOS complex, possibly through the phosphotyrosine–SH2 domain interaction. The tyrosine-to-phenylalanine mutation at position 177 in BCR-ABL (P185BCR-ABL177F) has been shown to dramatically abolish the transforming potential toward Rat1 fibroblasts.[25] However, P185BCR-ABL177F is still capable of transforming hematopoietic cells, possibly because it can phosphorylate cellular substrates, Shc and Stat5, for example, that are essential for transformation.[30,31] This information raises a simple question concerning whether oligomerization by itself can increase the enzymatic activity of ABL. The nonspecific aggregation of proteins, which is observed in overexpression systems, mimicks oligomerization. The ABL proteins overexpressed in baculovirus, Cos cells, or 293T cells are autophosphorylated.[32] However, the autophosphorylation of the

[30] A. Goga, J. McLaughlin, D. E. H. Afar, D. C. Saffran, and O. N. Witte, *Cell* **82**, 981 (1995).
[31] K. Shuai, J. Halpern, J. ten Hoeve, X. Rao, and C. L. Sawyers, *Oncogene* **13**, 247 (1996).
[32] A. M. Pendergast, A. J. Muller, M. H. Havlik, R. Clark, F. McCormick, and O. N. Witte, *Proc Natl. Acad. Sci. U.S.A.* **88**, 5927 (1991).

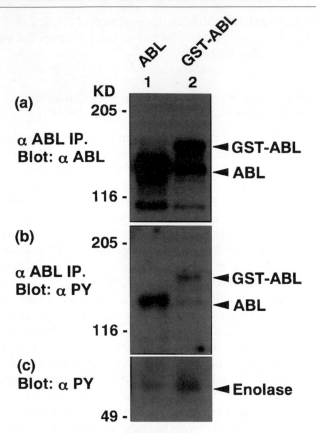

Fig. 4. GST-ABL expressed in 293T cells has enhanced tyrosine kinase activity. Anti-ABL immunoprecipitates prepared from 293T cells transiently transfected with ABL (lane 1) or GST-ABL (lane 2) were subjected to anti-ABL (a) or anti-phosphotyrosine (PY20) (b) Western blotting. In (c) anti-ABL immunoprecipitates were subjected to *in vitro* kinase reactions with enolase as a substrate. Supernatants containing enolases after reactions were probed with anti-PY antibody (PY20).

ABL protein does not augment its enzymatic activity, i.e., the ability to phosphorylate substrates.[12] To answer the question, we transiently overexpressed GST-ABL in 293T cells. As shown in Fig. 4b, the overexpressed GST-ABL proteins were autophosphorylated *in vivo* as reported in the case of the overproduced ABL protein.[32] When anti-ABL immunoprecipitates were subjected to *in vitro* kinase reactions with enolase as a substrate, GST-ABL was found to be more potent as a tyrosine kinase enzyme than ABL overexpressed in a similar fashion (Fig. 4c). Thus the dimeric structure rather than the overexpression appears to contribute to the enhancement of the enzymatic activity, at least *in vitro*.

We have also tried to activate receptor tyrosine kinases in a ligand-independent fashion by fusion to BCR. The ABL tyrosine kinase domain and downstream sequences in BCR-ABL were replaced with the kinase domain of receptor tyrosine kinases Trk or KDR. The resulting artificial chimeric proteins BCR-TRK and BCR-KDR, respectively, were expressed in either 293T cells or Rat1 fibroblasts, and were found to be activated[27] (Fig. 5). Although those constitutively activated tyrosine kinases of receptors are no longer localized in the membrane, they exert biological effects that resemble those mediated by the original receptors. For example, nerve growth factor causes neurite extension in neuronal PC12 cells. Expression of BCR-TRK in PC12 cells also induced neurite extensions.[27] Biological functions of receptor tyrosine kinases whose ligands are unknown could be investigated by utilizing constitutively active forms of the kinase that are engineered by tagging multimerization elements such as GST or BCR. However, it should be noted that this type of BCR fusion protein could

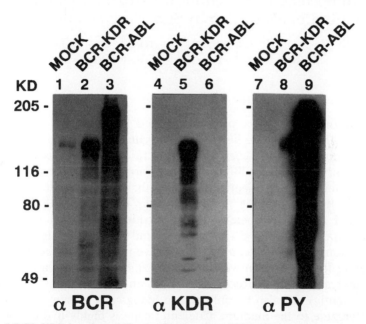

FIG. 5. BCR-KDR is activated. Total cell lysates from 293T cells that were mock transfected (lanes 1, 4, and 7) or transfected with BCR-KDR (lanes 2, 5, and 8) or BCR-ABL (lanes 3, 6, and 9) were subjected to anti-BCR (lanes 1–3), anti-KDR (lanes 4–6), and anti-PY (lanes 7–9) Western blotting. Note that BCR-KDR is tyrosine phosphorylated (lane 8).

have a phosphotyrosine at BCR position 177 that mediates the interaction with the Ras activator Grb-2/SOS.

Concluding Remarks

Artificial oligomerization of tyrosine kinases by fusion to GST or BCR could result in their activation. Molecular mechanisms may involve disruption of negative regulation or mimic physiological activation as in the case of receptor tyrosine kinases. In the BCR-ABL oncogene, oligomerization of ABL is necessary but not sufficient for its full activation. The artificial activation of tyrosine kinases of as yet unknown biological function may potentially give a clue to uncover their functional roles.

[33] Recombinant Aequorin as Tool for Monitoring Calcium Concentration in Subcellular Compartments

By Valerie Robert, Paolo Pinton, Valeria Tosello, Rosario Rizzuto, and Tullio Pozzan

Aequorin as Ca^{2+} Indicator

Aequorin is a 21-kDa protein, isolated from jellyfish of the genus *Aequorea*, that emits blue light in the presence of calcium. The aequorin originally purified from the jellyfish is a mixture of different isoforms called "heterogeneous aequorin."[1] In its active form the photoprotein includes an apoprotein and a covalently bound prosthetic group, coelenterazine. As schematically shown in Fig. 1, when calcium ions bind to the three high-affinity E-F hand sites, coelenterazine is oxidized to coelenteramide, with a concomitant release of carbon dioxide and emission of light. Although this reaction is irreversible, *in vitro* an active aequorin can be obtained by incubating the apoprotein with coelenterazine in the presence of oxygen and 2-mercaptoethanol. Reconstitution of an active aequorin (expressed recombinantly) can also be obtained in living cells by simple addition of coelenterazine to the medium. Coelenterazine is highly hydrophobic and has been shown to permeate cell membranes of various cell types, ranging

---

[1] O. Shimomura, *Biochem. J.* **234**, 271 (1986).

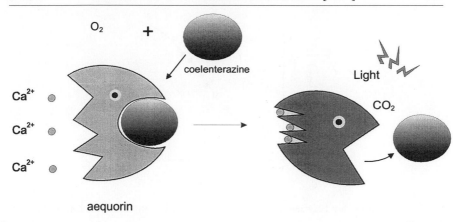

FIG. 1. Scheme of the Ca^{2+}-induced photon emission process.

from the slime mold *Dictyostelium discoideum* to mammalian cells and plants.[2]

Different coelenterazine analogs have been synthesized that confer to the reconstituted protein specific luminescence properties.[3] A few synthetic analogs of coelenterazine are now commercially available from Molecular Probes (Eugene, OR).

The possibility of using aequorin as a calcium indicator is based on the existence of a well-characterized relationship between the rate of photon emission and the free Ca^{2+} concentration. For physiological conditions of pH, temperature, and ionic strength, this relationship is more than quadratic in the $[Ca^{2+}]$ range of $10^{-5}-10^{-7}$ M. The presence of three Ca^{2+}-binding sites in aequorin is responsible for the high degree of cooperativity, and thus for the steep relationship between photon emission rate and $[Ca^{2+}]$ (Fig. 2). $[Ca^{2+}]$ can be calculated from the formula L/L_{max}, where L is the rate of photon emission at any instant during the experiment and L_{max} is the maximal rate of photon emission at saturating $[Ca^{2+}]$. The rate of aequorin luminescence is independent of $[Ca^{2+}]$ at high ($>10^{-4}$ M) and low $[Ca^{2+}]$ ($<10^{-7}$ M). However, as described below in more detail, it is possible to expand the range of $[Ca^{2+}]$ that can be monitored with aequorin. Although aequorin luminescence is not influenced by either K^+ or Mg^{2+} (which are the most abundant cations in the intracellular environment and thus the most likely source of interference in physiological experiments), both ions are competitive inhibitors of Ca^{2+}-activated luminescence. Aequorin

[2] T. Pozzan, R. Rizzuto, P. Volpe, and J. Meldolesi, *Physiol. Rev.* **74**, 595 (1994).
[3] O. Shimomura, B. Musicki, Y. Kishi, and S. Inouye, *Cell Calcium* **14**, 373 (1993).

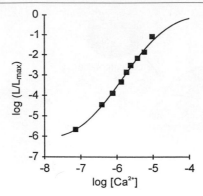

FIG. 2. Relationship between the free Ca^{2+} concentration and the rate of aequorin photon emission.

photon emission can also be triggered by Sr^{2+}, but its affinity is about 100-fold lower than that of Ca^{2+}, while lanthanides have high affinity for the photoprotein (e.g., are a potential source of artifacts when they are used to block Ca^{2+} channels). pH was also shown to affect aequorin luminescence but at values below pH 7. Because of the characteristics described above, experiments with aequorin need to be done under well-controlled conditions of pH and ionic concentrations, notably of Mg^{2+}.

Recombinant Aequorins

For a long time the only reliable way to introduce aequorin into living cells was to microinject the purified protein. This procedure is time consuming and laborious, and requires special care in handling of the purified photoprotein. Alternative approaches (scrape loading, reversible permeabilization, etc.) have been rather unsuccessful. These procedures are not described in this chapter. The cloning of the aequorin gene[4] has opened the way to recombinant expression and thus has largely expanded the applications of this tool for investigating Ca^{2+} handling in living cells. In particular, recombinant aequorin can be expressed not only in the cytoplasm, but also in specific cellular locations by including specific targeting sequencing in the engineered cDNAs.

Extensive manipulations of the N terminal of aequorin have been shown not to alter the chemiluminescence properties of the photoprotein and its Ca^{2+} affinity. On the other hand, even marginal alterations of the C terminal

[4] S. Inouye, N. Masato, Y. Sakaki, Y. Takagi, T. Miyaka, S. Iwanaga, T. Miyata, and F. I. Tsugi, *Proc. Natl. Acad. Sci. U.S.A.* **82,** 3154 (1985).

either abolish luminescence altogether or drastically increase Ca^{2+}-independent photon emission.[5] As demonstrated by Watkins and Campbell,[6] the C-terminal proline residue of aequorin is essential for the long-term stability of the bound coelenterazine. For these reasons, all targeted aequorins synthesized in our laboratory include modifications of the photoprotein N terminal. Three targeting strategies have been adopted.

1. Inclusion of a minimal targeting signal sequence with the photoprotein cDNA. This strategy was initially used to design the mitochondrial aequorin and was also followed to synthesize an aequorin localized in the nucleus and in the lumen of the Golgi apparatus.

2. Fusion of the cDNA encoding aequorin to that of a resident protein of the compartments of interest. This approach has been used to engineer aequorins localized in the sarcoplasmic reticulum (SR), in the nucleoplasm and cytoplasm (shuttling between the two compartments depending on the concentration of steroid hormones), on the cytoplasmic surface of the endoplasmic reticulum (ER) and Golgi, and in the subplasmalemma cytoplasmic rim.

3. Addition to the aequorin cDNA of sequences that encode polypeptides that bind to endogenous proteins. This strategy was adopted to localize aequorin in the ER lumen.

We routinely included in all the recombinant aequorins the HA1 epitope tag, which facilitates the immunocytochemical localization of the recombinant protein in the cell.

Chimeric Aequorin cDNAs

Below we briefly describe the constructs produced in our laboratory (Fig. 3). A few other constructs have been produced in other laboratories and are not dealt with in detail here.

Cytoplasm: cytAEQ

An unmodified aequorin cDNA encodes a protein that in mammalian cells is located in the cytoplasm and, given its small size, also diffuses into the nucleus.[7] An alternative construct is also available that is located on the outer surface of the ER and of the Golgi apparatus.[8] This construct

---

[5] M. Nomura, S. Inouye, Y. Ohmiya, and F. I. Tsuji, *FEBS Lett.* **295,** 63 (1991).
[6] N. J. Watkins and A. K. Campbell, *Biochem. J.* **293,** 181 (1993).
[7] M. Brini, F. De Giorgi, M. Murgia, R. Marsault, M. L. Massimino, M. Cantini, R. Rizzuto, and T. Pozzan, *Mol. Biol. Cell.* **8,** 129 (1997).
[8] R. Rizzuto, P. Pinton, W. Carrington, F. S. Fay, K. E. Fogarty, L. M. Lifshitz, R. A. Tuft, and T. Pozzan, *Science* **280,** 1763 (1998).

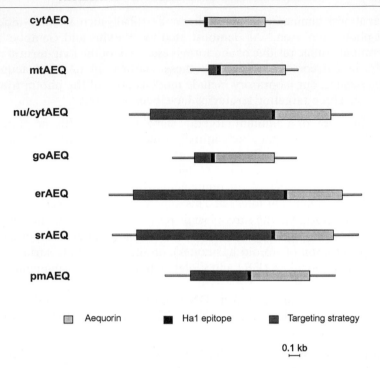

FIG. 3. Schematic representation of aequorin chimeras.

was intended to drive the localization of aequorin to the inner surface of the plasma membrane, given that it derives from the fusion of the aequorin cDNA with that encoding a truncated metabotropic glutamate receptor (mgluR1). The encoded chimeric protein, however, remains trapped on the surface of the ER and Golgi apparatus, with the aequorin polypeptide facing the cytoplasmic surface of these organelles. The cytoplasmic signal revealed by this chimeric aequorin is indistinguishable from that of a cytoplasmic aequorin, but it has the advantage of being membrane bound and excluded from the nucleus.

Mitochondria (mtAEQ): mtAEQ was the first targeted aequorin generated in the laboratory, and it has been successfully employed to measure the $[Ca^{2+}]$ of the mitochondrial matrix of various cell types. This construct includes the targeting presequence of subunit VIII of human cytochrome *c* oxidase fused to the aequorin cDNA.[9]

[9] R. Rizzuto, A. W. Simpson, M. Brini, and T. Pozzan, *Nature (London)* **358,** 325 (1992).

Nucleus (nuAEQ): Two nuAEQ constructs are presently available. The first consists of a hybrid cDNA encoding aequorin and the nuclear localization signal of the glucocorticoid receptor (excluding the hormone-binding domain). The expressed protein is constitutively located in the nucleus.[10] The second construct contains a much larger portion of the same receptor, including the hormone-binding domain. In the presence of glucocorticoids, the chimeric protein is translocated to the nucleus, while in its absence it is predominantly cytoplasmic. The advantage of this construct is that it allows measurement of the cytoplasmic and nucleoplasmic [Ca^{2+}] in the same transfected cell population, depending on the addition of the hormone. Obviously, careful controls must be carried out to verify that glucocorticoids do not interfere with the Ca^{2+} response.[7]

Golgi (goAEQ): To drive the expression of aequorin in the Golgi lumen, the aequorin cDNA has been fused to the cDNA encoding the transmembrane portion of sialyltransferase, a resident protein of the lumen of the medium-*trans*-Golgi.[11]

Endoplasmic Reticulum (erAEQ): The erAEQ includes the leader (L) and the VDJ and Ch1 domains of an IgG$_{2b}$ heavy chain fused at the N terminus of aequorin. Retention in the ER depends on the presence of the Ch1 domain, which is known to interact with high affinity with the lumenal ER protein BiP.[12]

Sarcoplasmic Reticulum (srAEQ): The srAEQ chimera results from the fusion of aequorin with calsequestrin, a protein confined in the terminal cisternae of striated muscle SR.[7]

Subplasma Membrane Region (pmAEQ): The pmAEQ construct derives from the fusion of the aequorin cDNA with that of SNAP25. The latter protein is part of the neurosecretory machinery and is recruited to the inner surface of the plasma membrane after palmitoylation of specific cysteine residues.[13]

To expand the range of Ca^{2+} sensitivity that can be monitored with the various targeted aequorins we have also employed in many of our constructs a mutated form of the photoprotein (Asp119 → Ala). This point mutation affects specifically the second EF hand motif of wild-type aequorin.[14] The

---

[10] M. Brini, M. Murgia, L. Pasti, D. Picard, T. Pozzan, and R. Rizzuto, *EMBO J.* **12,** 4813 (1993).

[11] P. Pinton, T. Pozzan, and R. Rizzuto, *EMBO J.* **18,** 5298 (1998).

[12] M. Montero, M. Brini, R. Marsault, J. Alvarez, R. Sitia, T. Pozzan, and R. Rizzuto, *EMBO J.* **14,** 5467 (1995).

[13] R. Marsault, M. Murgia, T. Pozzan, and R. Rizzuto, *EMBO J.* **16,** 1575 (1997).

[14] J. M. Kendall, G. Sala-Newby, V. Ghalaut, R. L. Dormer, and A. K. Campbell, *Biochem. Biophys. Res. Comm.* **187,** 1091 (1992).

affinity for Ca^{2+} of this mutated aequorin is about 20-fold lower than that of the wild-type photoprotein. Chimeric aequorins with the mutated isoform are presently available for the cytoplasm, the mitochondrial matrix, the ER and SR, the Golgi apparatus, and the subplasma membrane region.

Cell Preparation and Transfection

Although in a few cases the aequorin cDNA has been microinjected, the most commonly employed method to obtain expression of the recombinant protein is transfection. Various expression plasmids have been employed, some commercially available (pMT2, pcDNA1 and -3), others kindly provided by colleagues. It is not the purpose of this chapter to describe all the transfection procedures, the choice of which mainly depends on the cell type employed. The calcium phosphate procedure is by far the simplest and least expensive and it has been used successfully to transfect a number of cell lines, including HeLa, L929, L, Cos 7, A7r5, and PC12 cells, as well as primary cultures of neurons and skeletal muscle myotubes. Other transfection procedures have also been employed, such as liposomes, the "gene gun," and electroporation. Viral constructs for some aequorins are also available.[15,16]

In this section we briefly describe the calcium phosphate procedure, a simple and convenient transfection method for HeLa cells and rat myotubes. Whenever a new cell type is investigated, we always start with this procedure and only if problems arise do we adopt more sophisticated approaches.

HeLa Cells

One day before the transfection step, HeLa cells in Dulbecco's modified Eagle's medium (DMEM) supplemented with 10% (v/v) fetal calf serum (FCS) are plated on a 13-mm round coverslip at 30–50% confluence. Just before the transfection procedure, cells are washed with 1 ml of fresh medium.

Skeletal Muscle Myotubes

Primary cultures of skeletal muscle are prepared from the posterior limb muscles of newborn rats. Primary cultures are initiated from satellite cells obtained after four successive treatments with 0.125% (w/v) trypsin in phosphate-buffered saline. Cells are then plated in DMEM supplemented

[15] C. M. Rembold, J. M. Kendall, and A. K. Campbell, *Cell Calcium* **21,** 69 (1997).
[16] M. T. Alonso, M. J. Barrero, E. Carnicero, M. Montero, J. Garcia Sancho, and J. Alvarez, *Cell Calcium* **24,** 87 (1998).

with 10% FCS in 10-cm petri dishes at a density of 10^6 cells/ml. After 1 hr of incubation at 37°, nonadherent cells are collected and seeded at a density of 2×10^5 cells onto 13-mm coverslips coated with gelatin [2% (w/v) in phosphate-buffered saline (PBS)]. The myoblasts are then transfected during the second day of culture, i.e., before fusion occurs.

Calcium Phosphate Transfection Procedure

The following stock solutions need to be prepared and kept at $-20°$ until use.

CaCl$_2$, 2.5 M

HEPES-buffered solution (HBS): 280 mM NaCl, 50 mM HEPES, Na$_2$HPO$_4$ (pH 7.12), 1.5 mM

Tris–EDTA (TE): 10 mM Trizma base, EDTA (pH 8), 1 mM

All solutions are sterilized by filtration, using 0.22-μm pore size filters.

For one coverslip, 5 μl of 2.5 M CaCl$_2$ is added to the DNA dissolved in 45 μl of TE. Routinely, 4 μg of DNA is used to transfect one coverslip. The solution is then mixed by vortexing with 50 μl of HBS and incubated for 20 to 30 min at room temperature. The cloudy solution is then added directly to the cell monolayer. Eighteen to 24 hr after addition of the DNA, the cells are washed with PBS (two or three times until the excess precipitate is completely removed). Using this protocol the transfected cells are usually between 30 and 50%. Although an optimal transfection is obtained after an overnight incubation, we found that substantial aequorin expression, sufficient for most experimental conditions, is also obtained by incubation for only 6 hr with the calcium phosphate–DNA complex.

Stable clones can also be obtained by cotransfecting with the aequorin cDNA another plasmid encoding resistance to neomycin and then selecting the cells with neomycin at 0.8 mg/ml.[17]

After removing the calcium phosphate precipitate, myoblasts are cultured in DMEM supplemented with 2% (v/v) horse serum (HS) to stimulate the fusion of myoblasts into myotubes. The cell medium is then changed every 2 days. The myotubes can be maintained under these conditions for more than 10 days. The number of myotubes expressing the recombinant aequorin is usually about 50% and the cells continue to express the polypeptide during the whole period of culture.

Localization of Expressed Proteins

An essential aspect of targeted aequorin methodology is the accuracy of the subcellular localization. Although mistargeting has been rarely observed

---

[17] R. Rizzuto, M. Brini, and T. Pozzan, *Methods Cell. Biol.* **40**, 339 (1994).

with the constructs described above, it cannot be taken for granted that this will not occur in some cell model. We thus suggest that any time a new cell type is employed the subcellular localization be checked by immunocytochemistry (Fig. 4). The inclusion of the HA1 epitope tag in all the constructs described above is thus extremely convenient. Briefly, coverslips of transfected cells are fixed with 4% (v/v) formaldehyde in PBS for 20 min, washed two or three times with PBS and then incubated for 10 min in 50 mM NH$_4$Cl in PBS. Permeabilization of cells is obtained with 0.5% (v/v) Triton X-100 in PBS for 5 min followed by two or three washes with PBS. After a step of saturation of unspecific sites with 1% (w/v) gelatin for at least 30 min, cells are incubated with the anti-HA1 monoclonal antibody 12CA5 at 1:100 dilution for 1 hr in a wet chamber. The cells are

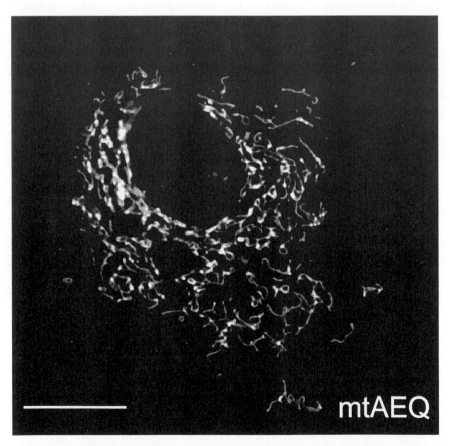

mtAEQ

FIG. 4. Immunochemical analysis of mtAEQ localization in HeLa cells. Bar: 10 μm.

then washed with PBS and incubated with fluorescein- or rhodamine-labeled anti-mouse IgG antibodies.

A simple, indirect alternative method to verify the subcellular distribution of the recombinant aequorin is to use plasmids in which the cDNA encoding aequorin is substituted with cDNA encoding green fluorescent protein (GFP). The assumption is that aequorin and GFP behave as passive cargoes and the subcellular localization depends exclusively on the targeting strategy. So far this assumption has been verified for all the constructs described above.

Reconstitution of Functional Aequorin

Once expressed the recombinant aequorin must be reconstituted into the functional photoprotein. This is accomplished by incubating cells with the synthetic coelenterazine for variable periods of time (usually 1–3 hr) and under conditions of temperature and $[Ca^{2+}]$ that depend on the compartment investigated. Practically, coelenterazine is dissolved at 0.5 mM in pure methanol as a 100× stock solution kept at −80°. This solution tolerates several freeze–thaw cycles. However, we recommend that the supply of coelenterazine solution be split into small aliquots (50 μl). Coelenterazine must be protected from light.

For compartments with low $[Ca^{2+}]$ under resting conditions (cytosol, mitochondria, and nucleus) the cells transfected with the appropriate recombinant aequorins are simply incubated at 37° in fresh DMEM supplemented with 1% (v/v) FCS and 5 μM coelenterazine. Higher or lower coelenterazine concentrations can also be used, if necessary. Good reconstitution is achieved with 1 hr of incubation, but an optimal reconstitution requires 2–3 hr.

For compartments endowed with high $[Ca^{2+}]$ under resting conditions (ER, SR, and the Golgi apparatus), to obtain good reconstitution and interpretable data it is first necessary to reduce the $[Ca^{2+}]$ in the organelle, otherwise aequorin is immediately consumed after reconstitution and in steady state little functional photoprotein is present in cells. Depletion of Ca^{2+} from the organelles has been (and can be) achieved in different ways. Below we describe a few simple protocols.

Depletion protocol 1: Cells are incubated at 37° for 5 min in KRB solution (Krebs–Ringer modified buffer: 125 mM NaCl, 5 mM KCl, 1 mM Na₃PO₄, 1 mM MgSO₄, 5.5 mM glucose, 20 mM HEPES, pH 7.4) supplemented with 3 mM EGTA, 10 μM ionomycin, and 10 μM tBuBHQ (an inhibitor of the endosarcoplasmic reticulum Ca^{2+} ATPases). After washing with KRB containing 100 μM EGTA, 5% (w/v) bovine serum albumin, and 10 μM tBuBHQ, cells are further incubated in the same medium supplemented with 5 μM coelenterazine for 1 hr, but at 4°.

Depletion protocol 2: All conditions are as described above, but iono-mycin is omitted and cells are treated instead with an agent (histamine for HeLa cells) that induces the release of Ca^{2+} from the ER through opening of the inositol triphosphate (InsP3) receptors.

Depletion protocol 3: For skeletal muscle myotubes transfected with srAEQ we have designed a slightly different protocol that better preserves the functionality of the cells. Briefly, the depletion of the SR is carried out for 2 min at room temperature in a KRB solution containing 10 mM caffeine (to open the ryanodine receptor) and 30 μM tBuBHQ to prevent refilling of the SR. The cells are then incubated in a KRB solution containing 100 μM EGTA and 5 μM coelenterazine for 1 hr at 4°.

Slight variations in these depletion protocols have been used both by our group and other investigators. Here it is necessary to stress a few general aspects of the procedure: (1) The more efficient the Ca^{2+} depletion, the better the reconstitution; (2) some compartments (e.g., the Golgi and in part of the ER) can be grossly altered morphologically by the Ca^{2+} depletion protocol. The incubation at 4° largely prevents these morphologi-cal changes, without altering the efficacy of the reconstitution; and (3) if ionophores or sarcoplasmic-endoplasmic-reticulum Ca^{2+}-ATPase (SERCA) inhibitors are employed for depletion they must be removed completely before starting the experiment. For this reason extensive washing of the cell monolayer with bovine serum albumin (BSA) is recommended at the end of the reconstitution procedure.

In the case of cells transfected with the aequorin targeted to the subplas-malemmal space, the protocol of reconstitution may require an incubation in Ca^{2+}-free medium. In A7r5 we have demonstrated that the efficiency of reconstitution is considerably augmented in media without extracellular Ca^{2+}.[13] This was not the case in rat myotubes.

Luminescence Detection

The aequorin detection system is derived as described by Cobbold and Lee[18] and is based on the use of a low-noise photomultiplier placed in close proximity (2–3 mm) to aequorin-expressing cells. The cell chamber, which is on the top of a hollow cylinder, is adapted to fit a 13-mm-diameter coverslip. The volume of the perfusing chamber is kept to a minimum (about 200 μl). The chamber is sealed on the top with a coverslip, held in

---

[18] P. H. Cobbold and J. A. C. Lee. J. C. McCormack and P. H. Cobbold, eds., p. 55. Oxford University Press, Oxford, 1991.

place with a thin layer of silicon. Cells are continuously perfused via a peristaltic pump with medium thermostatted via a water jacket at 37°. The photomultiplier (EMI 9789 with amplifier-discriminator) is kept in a dark box and cooled at 4°. During manipulations on the cell chamber, the photomultiplier is protected from light by a shutter. During aequorin experiments, the shutter is opened and the chamber with cells is placed in close proximity to the photomultiplier. The output of the amplifier-discriminator is captured by an EMI C600 photon-counting board in an IBM- compatible microcomputer and stored for further analysis.

Ca^{2+} Measurement

For the cells transfected with cytosolic, mitochondrial, or nuclear aequorins, the coverslip with the transfected cells is transferred to the luminometer chamber and perfused with KRB saline solution in the presence of 1 mM CaCl$_2$ to remove the excess coelenterazine. The stimuli or drugs to test are added to the perfusing medium and reach the cells with a lag time that depends on the rate of the flux and the length of the tubes. To make the stimulation more rapid and homogeneous the rate of the peristaltic pump is set to its maximum speed. Under these conditions we calculated that the whole monolayer is homogeneously exposed to the stimuli in 2 sec. At the end of the experiments, all the aequorin is discharged by permeabilizing the cells with a hypotonic solution containing digitonin (100 μM) and CaCl$_2$ (10 mM). A typical experiment in HeLa cells is presented in Fig. 5.

For erAEQ-, srAEQ-, or GoAEQ-transfected cells, unreacted coelenterazine and drugs are removed by prolonged perfusion (3–6 min) with a saline solution containing 600 μM EGTA and 2% (w/v) BSA. BSA is then removed from the perfusion buffer and the refilling of the compartments is started by perfusing the medium containing either 1 mM CaCl$_2$ or SrCl$_2$ (Fig. 6). Note that BSA increases the luminescence background level.

We found that, despite the depletion protocol and the use of a low Ca^{2+} affinity aequorin mutant, the rate of aequorin consumption on Ca^{2+} refilling is so rapid that most aequorin is consumed in 30 sec and the calibration of the signal in terms of [Ca^{2+}] becomes unreliable.[12] Two alternative solutions to this problem have been developed: (1) the use of Sr^{2+} as a Ca^{2+} surrogate and (2) the reconstitution not with the wild-type coelenterazine, but with the analog coelenterazine n, which reduces the rate of aequorin photon emission at high [Ca^{2+}]. In the latter case [Ca^{2+}] between 10^{-4} and 10^{-3} M can be reliably calibrated.[11,19] Finally it should be stressed that

[19] V. Robert, F. De Giorgi, M. L. Massimino, M. Cantini, and T. Pozzan, *J. Biol. Chem.* **273**, 30372 (1998).

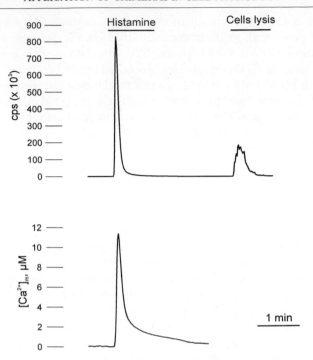

FIG. 5. Changes in mitochondrial $[Ca^{2+}]$ in HeLa cells. *Top:* Luminescence data. *Bottom:* Calibrated $[Ca^{2+}]$. Where indicated histamine (100 μM) was added. At the end of the experiments, cells were lysed by digitonin plus 10 mM $CaCl_2$. Cps, counts per second.

even a minor missorted fraction of aequorin in a low-$[Ca^{2+}]$ compartment, undetectable by morphological methods, can lead to a substantial artifactual underestimation of the Ca^{2+} levels in the high-$[Ca^{2+}]$ compartments.[20] A simple empirical solution to this problem has been described by Maechler et al.[21]

Conversion of Luminescent Signal into Ca^{2+} Concentration

To transform luminescence values into $[Ca^{2+}]$ values, we have used the method described by Allen and Blink.[22] The method relies on the relationship between $[Ca^{2+}]$ and the ratio between the light intensity re-

[20] M. Montero, J. Alvarez, W. J. J. Scheenen, R. Rizzuto, J. Meldolesi, and T. Pozzan, *J. Cell. Biol.* **3,** 601 (1997).
[21] P. Maechler, D. Kennedy, E. Sebo, T. Pozzan, and C. B. Wollheim. *J. Biol. Chem.* **18,** 12583 (1999).
[22] D. G. Allen and J. R. Blinks, *Nature (London)* **273,** 509 (1978).

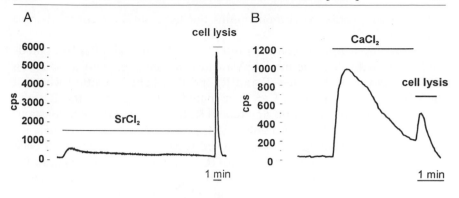

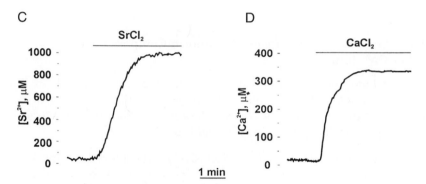

FIG. 6. Monitoring of $[Ca^{2+}]$ and $[Sr^{2+}]$ in rat myotubes. (A and B) Crude luminescence data. (C and D) Calibrated $[Sr^{2+}]$ and $[Ca^{2+}]$. The SR was depleted as described. Functional srAEQ was reconstituted with wild-type coelenterazine (A–C) or coelenterazine n (B–D). Where indicated, the perfusion medium was supplemented with 1 mM SrCl₂ or CaCl₂.

corded in physiological conditions (L, counts per second) and that which would have been reported if all the aequorin were instantaneously exposed to saturating $[Ca^{2+}]$ (L_{max}). Given that the rate constant of aequorin consumption at saturating $[Ca^{2+}]$ is 1.0 sec⁻¹, a good estimate of L_{max} can be obtained from the total aequorin light output recorded from the cells after discharging all the aequorin. This usually requires the addition of excess Ca^{2+} and detergents as shown in the preceding section. As aequorin is being consumed continuously, it must be stressed that, for calibration purposes, the value of L_{max} is not constant and decreases steadily during the experiment. The value of L_{max} to be used for $[Ca^{2+}]$ calculations at every time point along the experiment should be calculated as the total light output of the whole experiment minus the light output recorded before that point.

The relationship between the ratio (L/L_{max}) and [Ca^{2+}] has been modeled mathematically. The model postulates that each of the Ca^{2+}-binding sites has two possible states, T and R, and that light is emitted when all the sites are in the R state. Ca^{2+} is assumed to bind only in the R state. This model contains three parameters: KR, the Ca^{2+} association constant, KTR = [T]/[R], and n, the number of Ca^{2+}-binding sites. The values we obtained for the recombinantly expressed recombinant aequorin for each parameter are KR = 7.23 10^6 M^{-1}, KTR = 120, n = 3. The equation for the model reported by Allen et al.[23] provides the algorithm we used to calculate the [Ca^{2+}] values at each point where the ratio = (L/L_{max})$1/n$.

$$Ca^{2+} (M) = \text{ratio} + (\text{ratio KTR}) - 1/KR - (\text{ratio KR})$$

Advantages and Disadvantages of Aequorin Compared with Other Ca^{2+} Indicators

Today numerous indicators are available to measure [Ca^{2+}]. In this section, we briefly describe some advantages and disadvantages of the photoprotein over the most widely used fluorescent indicators, the tetracarboxilate dyes such as Fura-2, Indo-1, and Fluo-3.

1. The photoprotein aequorin is exclusively localized in the cytosol and is completely excluded from organelles. The fluorescent dyes, if loaded using their intracellularly trappable AM-esters, are usually found also in intracellular organelles such as the mitochondria and the ER. This intracellular trapping can complicate the interpretation of the data obtained with the dyes. Furthermore, the dyes are usually slowly released into the medium, while aequorin loss requires the death of the cell.

2. Because of the low luminescence background of cells and the steepness of the Ca^{2+} dependence of aequorin luminescence, minor changes in [Ca^{2+}] can be easily appreciated.

3. Obviously both the photoproteins and the fluorescent Ca^{2+} indicators increase the Ca^{2+}-buffering capacity of the cells and thus potentially may interfere with cellular Ca^{2+} homeostasis. The high signal-to-noise ratio characteristic of aequorin allows, however, the use of the photoprotein at concentrations that are two to three orders of magnitude lower than those necessary for the fluorescent indicators. The immediate advantage of using

[23] D. G. Allen, J. R. Blinks, and F. G. Prendergast, Science 195, 996 (1971).

aequorin is thus a reduced Ca^{2+} buffering that minimizes artifactual quantitative and qualitative changes in Ca^{2+} dynamics.[24]

4. For many years, a major limitation of the use of aequorin has been the need to introduce the polypeptide into living cells by traumatic methods such as microinjection. Because of this restriction, the use of aequorin as a Ca^{2+} indicator has been limited to large and robust cells (eggs, muscle fibers, giant synapses of squid, etc.). The cloning of the aequorin cDNA[4] had offered the possibility of expressing the recombinant protein into virtually all cell types. In addition, aequorin cDNA can be cotransfected with a gene of interest and allows a simple and effective way to monitor in a large number of cells the effect of that gene on Ca^{2+} homeostasis. Such studies are far more complex and time consuming when using the Ca^{2+} indicators.[24]

5. The major impact of aequorins in the field of signaling mechanisms depends on the possibility of obtaining accurate targeting of the protein in subcellular regions. Such a selective localization has never been achieved with fluorescent dyes.

6. The most obvious disadvantage of aequorins with respect to the fluorescent Ca^{2+} indicators is the low number of photons (<1 photon per molecule compared to $>10^4$ photons in the case of fluorescent dyes) that can be emitted by aequorin and the irreversible nature of the photon emission reaction. For these reasons, single-cell analysis of Ca^{2+} transients with cytosolic aequorin (or with any of the targeted chimeras) requires expensive equipment and has intrinsically low time and spatial resolution. This type of analysis is far simpler and more accurate with the fluorescent dyes.

7. Calibration of aequorin signals in terms of [Ca^{2+}] most often requires cell lysis (but see Ref. 25 for an alternative approach). Many fluorescent Ca^{2+} indicators can be calibrated in intact cells with the "ratio mode." Furthermore, because of the steep relationship between [Ca^{2+}] and the rate of photon emission, the signal of aequorin is biased toward cells (or compartments) with higher [Ca^{2+}]. This disadvantage can be, in some cases, turned into a bonus.[13]

8. Aequorin is consumed while measuring Ca^{2+}, while this phenomenon is negligible with fluorescent dyes.

In summary, aequorins, and in particular targeted recombinant aequorins, may represent for some applications a useful, and sometime superior, tool (compared with fluorescent Ca^{2+} indicators) to investigate Ca^{2+} signal-

[24] M. Brini, R. Marsault, C. Bastianutto, J. Alvarez, T. Pozzan, and R. Rizzuto, *J. Biol. Chem.* **17,** 9896 (1995).
[25] M. R. Knight, N. D. Read, A. K. Campbell, and A. J. Trewavas, *J. Cell. Biol.* **1,** 83 (1993).

ing in living cells. Quite recently the group of Tsien[26] has introduced the Ca^{2+} indicators named "cameleons," molecularly engineered proteins capable of coupling the advantages of aequorins in terms of selective targeting to the high signal characteristics of fluorescent molecules. This technique is presently in its infancy, but its potential in this field is enormous.

[26] A. Miyawaki, J. Llopis, R. Heim, J. M. McCaffery, J. A. Adams, M. Ikura, and R. Y. Tsien, *Nature* (*London*) **388**, 882 (1997).

[34] Recombinant Aequorin as Reporter of Changes in Intracellular Calcium in Mammalian Cells

By JENNY STABLES, LARRY C. MATTHEAKIS, RAY CHANG, and STEPHEN REES

Introduction

The photoprotein aequorin, from the coelenterate jellyfish *Aequorea victoria,* is a bioluminescent complex formed from the 21-kDa apoaequorin, the luminophore cofactor coelenterazine, and molecular oxygen.[1] Apoaequorin is a 189-amino acid polypeptide chain containing three calcium-binding sites (EF-hand structures).[2] After the binding of calcium to these sites, aequorin undergoes a conformational change, converting into an oxygenase. The subsequent oxidation of coelenterazine by the bound molecular oxygen results in the production of apoaequorin, coelenteramide, CO_2, and light with a λ_{max} for emission of 469 nm, which can be detected by conventional luminometry. The kinetics of light emission after the binding of calcium to the aequorin complex is rapid and is typically complete in less than 10 sec.

Aequorin has been used for many years as a reporter of changes in intracellular calcium concentration in mammalian cells or *Xenopus* oocytes.[3] In these experiments loading of cells with aequorin has involved microinjection of purified protein. However, more recently the cloning of the aequorin cDNA has allowed expression of this protein, both transiently

[1] M. Brini, R. Marsault, C. Bastianutto, J. Alvarez, T. Pozzan, and R. Rizzuto, *J. Biol. Chem.* **270**, 9896 (1995).
[2] S. Inouye, M. Noguchi, Y. Sakaki, Y. Tagaki, T. Miyata, S. Iwanaga, T. Miyata, and F. Tsuji, *Proc. Natl. Acad. Sci. U.S.A.* **82**, 3154 (1985).
[3] C. Ashley and A. Campbell (eds.), "The Detection and Measurement of Free Calcium." Elsevier/North Holland, Amsterdam, 1979.

and stably, in a range of cell types and has greatly expanded the utility of aequorin as a calcium reporter.[2,4,5] Aequorin offers a highly sensitive method with which to assess changes in the level of intracellular calcium, which offers significant time savings over the use of calcium-sensitive fluorescent dyes such as Fura-2 or Fluo-3. The detection of changes in intracellular calcium concentration using expressed aequorin does not require loading of the cells with dye, the extensive washing steps involved in the dye-loading procedure, the leaching of dye from the cells, or fluorescence quenching associated with the use of these dyes.

To expand the use of aequorin a number of modified aequorins have been constructed in which expression of the protein is targeted to particular cellular compartments in order to measure calcium changes within those compartments. This includes aequorin targeted to the mitochondria,[6] nucleus,[7] and endoplasmic reticulum.[8] Furthermore, the construction of fusion proteins with aequorin has facilitated the analysis of local calcium changes. A calcium-sensitive adenylyl cyclase/aequorin fusion protein has been used to report the changes in intracellular calcium concentration that regulate this enzyme.[9] In a further application aequorin has been used as a reporter to monitor changes in gene expression.[10] In these experiments a secreted aequorin was generated following the fusion of the signal peptide of human follistatin to the N terminus of apoaequorin. Changes in gene expression were thus monitored according to the degree of aequorin luminescence in the cell culture medium.

The $G\alpha_{q/11}$ family of heterotrimeric G proteins has five members: $G\alpha_q$, $G\alpha_{11}$, $G\alpha_{14}$, $G\alpha_{15}$, and $G\alpha_{16}$. Agonist activity at G protein-coupled receptors (GPCRs) that couple to G-protein α subunits of this class results in the activation of the phosphoinositidases of the phospholipase $C\beta$ (PLCβ) class to generate the second-messenger metabolites sn-1–2 diacylglycerol (sn-1–2 DAG) and inositol 1,4,5-triphosphate (IP$_3$). This is followed by an increase in intracellular calcium concentration as a consequence of the release of calcium from intracellular stores (Fig. 1). The measurement

[4] Y. Sheu, L. Kricka, and D. Pritchett, *Anal. Biochem.* **209,** 343 (1993).

[5] D. Button and M. Brownstein, *Cell Calcium* **14,** 663 (1993).

[6] R. Rizzuto, M. Brini, C. Bastianutto, R. Marsault, and T. Pozzan, *Methods Enzymol.* **260,** 417 (1995).

[7] M. Badminton, J. Kendall, G. Sala-Newby, and A. Campbell, *Exp. Cell Res.* **216,** 236 (1995).

[8] J. Kendall, M. Badminton, R. Dormer, and A. Campbell, *Anal. Biochem.* **221,** 173 (1994).

[9] Y. Nakahashi, E. Nelson, K. Fagan, E. Gonzales, J.-L. Guillou, and D. Cooper, *J. Biol. Chem.* **272,** 18093 (1997).

[10] S. Inouye and F. Tsuji, *Anal. Biochem.* **201,** 114 (1992).

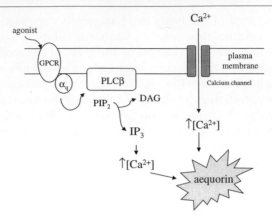

FIG. 1. Use of cell lines expressing the calcium-sensitive photoprotein aequorin as a screening system for agonist activity at GPCRs and ion channels. Agonist-mediated activation of either a 7-TM receptor or an ion channel, which results in calcium mobilization, will generate an increase in luminescence.

of calcium release from intracellular stores has principally involved the preloading of cells with calcium-sensitive fluorescent dyes followed by the measurement of an increase in fluorescence after agonist activation of the receptor. However, more recently the apoaequorin cDNA has been expressed in mammalian cells and used to detect agonist activation of many GPCRs. This includes a histamine receptor,[1] a 5-HT$_2$ serotonin receptor,[4] the V$_{1A}$ vasopressin receptor, and the α_1-adrenoceptor.[5] Furthermore, the Edg 2 and Edg 4 lysophosphatidic acid receptors,[11] the GAL R2 galanin receptor,[12] the prostanoid receptors EP1 and EP3,[13] and the B1 bradykinin receptor[14] have also been adapted to the aequorin reporter system.

To further the use of aequorin as a reporter of GPCR signal transduction, aequorin has been expressed in mammalian cells together with the so-called promiscuous G-protein α subunit Gα_{16} to develop "universal" detection systems for agonist activity at GPCRs.[15] Heterologous expression

[11] S. An, T. Bleu, Y. Zheng, and E. J. Goetzl, *Mol. Pharmacol.* **54,** 881 (1988).
[12] L. F. Kolakowski, G. P. O'Neill, A. D. Howard, S. R. Broussard, K. A. Sullivan, S. D. Feighner, M. Sawzdargo, T. Nguyen, S. Kargman, L. L. Shiao, D. L. Hreniuk, C. P. Tan, J. Evans, M. Abramovitz, A. Chateauneuf, N. Coulombe, G. Ng, M. P. Johnson, A. Tharian, H. Khoshbouei, S. R. George, R. G. Smith, and B. F. O'Dowd, *J. Neurochem.* **71,** 2239 (1998).
[13] Y. Boie, R. Stocco, N. Sawyer, D. M. Slipetz, M. D. Ungrin, F. Neuschafer-Rube, G. P. Puschel, K. M. Metters, and M. Abramowitz, *Eur. J. Pharmacol.* **340,** 227 (1997).
[14] T. MacNeil, S. Feighner, D. L. Hreniuk, J. F. Hess, and L. H. Van der Ploeg, *Can. J. Physiol. Pharmacol.* **75,** 735 (1997).
[15] X. Zhu and L. Birnbaumer, *Proc. Natl. Acad. Sci. U.S.A.* **93,** 2827 (1996).

in mammalian cells of $G\alpha_{16}$, or its murine homolog $G\alpha_{15}$, allows the coupling of a range of GPCRs that do not normally couple to this G protein to stimulate PLCβ activity to result in an increase in intracellular calcium concentration.[16,17] This includes the β_2-adrenoreceptor, the dopamine D_1 receptor, the vasopressin V_2 receptor and the adenosine A_{2A} receptor, all of which normally couple through $G\alpha_S$ to generate an activation of adenylyl cyclase. Similarly, the normally $G\alpha_i$-coupled muscarinic acetylcholine M_2 receptor, the serotonin 5-HT$_{1A}$ receptor, and the δ-opioid receptor will couple to $G\alpha_{16}$ in a recombinant cell line expressing this protein. In this assay system ligand activity at potentially any GPCR can be detected as an increase in aequorin luminescence as a consequence of receptor coupling to $G\alpha_{16}$.

Another mechanism for increasing the intracellular concentration of calcium is by entry of extracellular calcium through ion channels (Fig. 1). Aequorin can be used to measure opening of a calcium-selective ion channel, provided that the rise in intracellular calcium concentration is sufficient to stimulate the aequorin complex before the channel inactivates. Examples of such ion channels include the voltage-gated calcium channels, the purinergic P2X receptors, and the vanilloid receptors. A single report demonstrates the use of aequorin to study calcium entry through the L-type calcium channel.[18]

The P2X receptors are a unique family of ligand-gated cation channels expressed in a variety of cells.[19] They are equally permeable to Na^+ and K^+, but have a fourfold higher permeability to Ca^{2+}. The P2X receptors are transmembrane proteins, and a variety of experiments suggests that they exist as both homomeric and heteromeric complexes. All P2X channels are activated by extracellular ATP, but they can have distinct phenotypes when expressed individually or in various combinations. For example, homomeric P2X$_2$ receptors are activated by ATP, but not by the ATP derivative $\alpha\beta$meATP.[19] P2X$_3$ receptors open in response to $\alpha\beta$meATP, but the current desensitizes within hundreds of milliseconds.[20] The coexpression of P2X$_2$ and P2X$_3$ results in a heteromeric complex that is activated by $\alpha\beta$meATP, but exhibits the long desensitization time of P2X$_2$ (>10 sec).[20,21]

[16] S. Offermans and M. Simon, *J. Biol. Chem.* **270**, 15175 (1995).

[17] D. Wu, Y. Kuang, Y. Wu, and H. Jiang, *J. Biol. Chem.* **270**, 16008 (1995).

[18] A. Maeda, S. Nishimura, K. Kameda, T. Imagawa, M. Shigekawa, and E. L. Barsoumian, *Anal. Biochem.* **242**, 31 (1996).

[19] R. A. North and E. A. Barnard, *Curr. Opin. Neurobiol.* **7**, 346 (1997).

[20] C. Lewis, S. Neidhart, C. Holy, R. A. North, G. Buell, and A. Surprenant, *Nature (London)* **377**, 432 (1995).

[21] E. Kawashima, D. Estoppey, C. Virginio, D. Fahmi, S. Rees, A. Surprenant, and R. A. North, *Receptors Channels* **5**, 53 (1998).

Thus, the expression of single P2X subtypes or combinations of subtypes can result in a significant increase in the intracellular calcium concentration after stimulation with the appropriate ligand.

In this chapter we describe the use of recombinant aequorin to detect changes in intracellular calcium concentration in mammalian cells after agonist activation of $G\alpha_{q/11}$-coupled G protein-coupled receptors and after calcium entry through P2X family ion channels. In contrast to the use of single-tube luminometers for the detection of aequorin luminescence we describe the detection of aequorin luminescence in a 96-well microplate format, using either an injector luminometer or the fluorescence imaging plate reader (FLIPR).

Methods

Aequorin Expression Plasmids

All DNA manipulations are performed by standard methods.[22] The vectors pMAQ2, containing cytoplasmic expressed apoaequorin (cytAEQ[1]), and pcDNA1/mtAEQ, containing mitochondria-targeted apoaequorin (mtAEQ[23]), are obtained from Molecular Probes (Eugene, OR). For the analysis of GPCR signaling in transient transfection assays, the cDNAs are excised from these vectors on digestion with EcoRI and transferred into the EcoRI site of pcDNA3 (InVitrogen, San Diego, CA) to generate pcDNA3/cytAEQ and pcDNA3/mtAEQ. For the generation of stable mammalian cell lines for GPCR studies both aequorin cDNAs are also transferred into the EcoRI site of pCIN.[24] For P2X ion channel studies, mitochondria-targeted apoaequorin is transferred on an EcoRI restriction fragment from pcDNA1/mtAEQ (Molecular Probes) into the same site of pBluescript-II-KS (Stratagene, La Jolla, CA). This resulting plasmid is subsequently digested with XhoI and XbaI, and the apoaequorin fragment is transferred into the same sites of pcDNA3.1/Hyg (InVitrogen) to generate mtAEQ/pcDNA3.1-Hyg.

Cloning of Receptor and Ion Channel cDNAs

The following GPCR and P2X ion channel cDNAs have been cloned at Glaxo Wellcome (Research Triangle Park, NC): the human C5a receptor, human neurokinin NK_1 receptor, human adenosine A_1 receptor, human adenosine A_{2A} receptor, human β_2-adrenoceptor, human angiotensin AT_1

[22] J. Sambrook, E. Fritsch, and T. Maniatis, "Molecular Cloning: A Laboratory Manual," 2nd ed. Cold Spring Harbor Laboratory Press, Cold Spring Harbor, New York, 1989.
[23] R. Rizzuto, A. Simpson, M. Brini, and T. Pozzan, *Nature (London)* **358,** 325 (1994).

receptor, human endothelin ET_A receptor, human oxytocin (OT) receptor, human vasopressin V1a receptor, human vasopressin V1b receptor, human melatonin ML_{1A} receptor, and rat $P2X_2$ and $P2X_3$ ion channel subunits. In each case the cDNA sequence agrees with the published sequences available in the GenBank database.

For stable expression in Chinese hamster ovary (CHO) cells the adenosine A_1, adenosine A_{2A}, and melatonin ML_{1A} receptors, and the β_2-adrenoceptors, have been cloned into the expression vector pCIN.[24] The vasopressin V1a, vasopressin V1b, and oxytocin receptors have been cloned into the expression vector pCIH, a derivative of pCIN in which the neomycin selection marker is replaced with hygromycin. The C5a receptor cDNA has been cloned into pcDNA3 (InVitrogen) for transient expression. The cloning of human $G\alpha_{16}$ and the construction of the plasmid pCIH/G16 is described in Stables *et al.*[25]

Stable Expression Studies

Chinese hamster ovary cells stably expressing the human NK_1 receptor (CHO/NK1), ET_A endothelin receptor, and AT_1 angiotensin receptor have been derived at Glaxo Wellcome. All cells are maintained at 37° with 5% CO_2 and 95% humidity. Untransfected CHO cells and CHO/NK1 cells are grown in CHO medium [Dulbecco's modified Eagle's medium (DMEM)–F12, supplemented with 5% (v/v) fetal calf serum (FCS) and 2 mM glutamine; all reagents from Life Technologies, Gaithersburg, MD]. CHO cells expressing the ET_A endothelin receptor, and the AT_1 angiotensin receptor, are grown in CHO medium supplemented with G418 (1 mg/ml; Life Technologies). CHO cells stably expressing the G-protein α subunit $G\alpha_{16}$ (CHO/G16), derived after transfection of CHO cells with the expression vector pCIH/G16, are maintained in CHO medium supplemented with hygromycin B (0.4 mg/ml; Roche Biosciences, Mannheim, Germany).[25]

CHO cells stably expressing cytoplasmic and mitochondrial aequorin (CHOQ and CHOMQ, respectively) are derived after transfection of CHO cells with the expression vectors pCIN/cytAEQ or pCIN/mtAEQ, using a calcium phosphate transfection kit (InVitrogen).[25] After antibiotic selection in G418 (1 mg/ml), individual clones are expanded and analyzed for expression of apoaequorin by stimulating the endogenous purinergic $P2Y_2$ receptor found on CHO cells with ATP (data not shown).

[24] S. Rees, J. Coote, J. Stables, S. Goodson, S. Harris, and M. Lee. *BioTechniques* **20,** 102 (1996).
[25] J. Stables, A. Green, F. Marshall, N. Fraser, E. Knight, M. Sautel, G. Milligan, M. Lee, and S. Rees, *Anal. Biochem.* **252,** 115 (1997).

CHO cells coexpressing cytoplasmic aequorin and either the oxytocin receptor, the vasopressin V1a receptor, or the vasopressin V1b receptor are derived after transfection of CHOQ cells with the expression vectors pCIH/OT, pCIH/V1a, and pCIH/V1b, respectively. After antibiotic selection in G418 (1 mg/ml) and hygromycin B (0.4 mg/ml), clones are isolated and analyzed for coexpression of aequorin and receptors by stimulation with vasopressin (data not shown). To prepare cells for aequorin assay, cell lines expressing both aequorin and a GPCR are plated at 40% confluence in opaque 96-well microtiter plates (Packard Biosciences, Meriden, CT).

Human embryonic kidney (HEK) 293 cells stably expressing the rat P2X$_2$ ion channel, and CHO cells coexpressing the rat P2X$_2$ and rat P2X$_3$ ion channels, are isolated as described.[21] For transfection with mtAEQ/pcDNA3.1-Hyg, cells are electroporated with 20 μg of plasmid in a Bio-Rad (Hercules, CA) GenePulser set at 400 V and 250 μF, and the transfected cells are grown in CHO medium. The HEK293 and CHO transfected populations are selected in the presence of 100- and 700-μg/ml concentrations of hygromycin B (Life Technologies), respectively. Individual clones are expanded and tested for ligand-induced luminescence on an FLIPR.

Transient Expression Studies

Transient expression studies have been performed after the transfection of CHO cells stably expressing a GPCR with either pcDNA3/cytAEQ or pcDNA3/mtAEQ, or after the transfection of CHOQ or CHOMQ cells with the appropriate GPCR expression vector. In either case cells are grown to 40% confluency in a 96-well opaque microtiter plate and then transfected with the appropriate expression plasmid (100 ng/well) or, in the case of cotransfection studies, a 50-ng/well concentration of each plasmid and the LipofectAMINE reagent according to the manufacturer instructions (Life Technologies).

Aequorin Assay

All aequorin assays described in this chapter are conducted within 96-well microtiter plates. The use of 96-well plate-based aequorin assays allows the rapid screening of many samples, and is particularly amenable to compound screening within the pharmaceutical industry. As aequorin luminescence has a signal half-life of less than 10 sec, signal detection requires the use of a luminometer equipped with reagent injectors in order to permit rapid recording of light emission after compound addition. In all GPCR-signaling experiments aequorin luminescence is measured at room temperature in the 10 sec after compound addition, using the Integrate Flash mode on a Dynatech (McLean, VA) ML3000 injector luminometer. These assays are performed in opaque 96-well microtiter plates. In transient expression studies aequorin assays are performed 48 hr posttransfection. For experiments using stable cell lines, cells are plated 18 hr prior to the assay. In either case cells are maintained in the absence of G418 after plating into 96-well microtiter plates, as some reports have indicated that G418 may interfere with the process of calcium mobilization.[26] G418 is a known inhibitor of phospholipase C, and is also thought to have an effect on plasma membrane calcium channels.

For all aequorin assays performed to characterize GPCR signal transduction, cells are incubated at 37° in CHO medium containing 5 μM coelenterazine for 3 hr, in order to reconstitute the holoenzyme apoaequorin. Test compounds are diluted in extracellular medium [125 mM NaCl, 5 mM KCl, 2 mM MgCl$_2$ · 6H$_2$O, 0.5 mM NaH$_2$PO$_4$ · H$_2$O, 5 mM NaHCO$_3$, 10 mM HEPES, 10 mM glucose, 0.1% (w/v) bovine serum albumin (BSA), pH 7.4], and added at 2× final concentration in a volume equal to that already in the assay plate (to give a final volume of 200 μl). After compound addition aequorin luminescence is detected as described. In the studies described in this chapter calcium signaling is simply measured as an increase in light units. However, for the accurate determination of the molar change in intracellular calcium concentration it is necessary to calibrate the luminometer.[1]

For P2X ion channel assays, luminescence is measured on an FLIPR instrument (Molecular Devices, Merlo Park, CA). The FLIPR instrument enables the simultaneous addition of test compound to, and the rapid recording of either fluorescence or luminescence from, every well within a 96-well microtiter plate. Use of the FLIPR can greatly enhance the throughput of calcium assays when using the generation of aequorin luminescence as the assay readout. For ion channel assays, cells expressing P2X receptors are suspended in CHO medium containing 5 μM coelenterazine, and approximately 1 × 10^5 cells (contained in a volume of 120 μl) are added to each well of a black 96-well clear-bottom plate (Corning Costar, Acton, MA). The plate is incubated at 37° for 5 hr to allow reconstitution of the aequorin complex. To measure luminescence on the FLIPR, the camera lens aperture is adjusted to F/1.4, and the camera exposure and update times are set to 10 and 10.6 sec, respectively. The P2X ligands are dissolved in DMEM–F12 medium and serial dilutions are added to the wells of a 96-well plate. The assay plate containing cells and the ligand plate are placed on the FLIPR instrument, and the software is programmed

[26] H. Sipma, L. Van der Zee, A. Den Hertog, and A. Nelemans, *Eur. J. Pharmacol.* **305,** 207 (1996).

to deliver 30 μl of ligand to each well. Luminescence is measured at room temperature in a dark room.

Results

Types of Coelenterazine

The luminescent characteristics of the aequorin complex are, in part, dependent on the type of coelenterazine used to reconstitute the holoenzyme. A number of analogs of coelenterazine have been derived that confer different calcium affinities and spectral properties to the aequorin complex. For example, when comparing the high-sensitivity coelenterazine (hcp-) with the low-sensitivity type (n-), the intensity of the aequorin luminescence can vary by up to 10,000-fold. Another group of coelenterazines (e-) emits luminescence at two separate peaks (400–405 nm, 440–475 nm); the intensity ratio of these two peaks varies with calcium concentration.[27] In these studies we have used native coelenterazine purchased from Molecular Probes. We find that the intensity of aequorin luminescence obtained when using native coelenterazine is sufficiently bright to enable detection by either the Dynatech ML3000 injection luminometer or the FLIPR apparatus.

Comparison of Mitochondrial and Cytoplasmic Aequorin as Reporters of Intracellular Calcium Changes

A number of reports have indicated that when expressed in mammalian cells as reporters of changes in cytoplasmic calcium concentration, significantly greater luminescence is obtained when using mitochondrially targeted aequorin in contrast to that obtained with cytoplasmically expressed aequorin.[25,28] To demonstrate that mtAEQ is able to accurately report the increase in cytoplasmic calcium concentration that results from agonist activation of a $G\alpha_q$-coupled GPCR, both mtAEQ and cytAEQ were stably expressed in CHO cells. CHO cells endogenously express a P2Y$_2$ purinergic receptor. A concentration–response curve to the purinergic agonist ATP was constructed, using aequorin luminescence as readout. Application of agonist resulted in a large, rapid, and transient concentration-related increase in luminescence with both reporters. At near-maximal agonist concentrations mtAEQ yielded a luminescent signal 10-fold brighter than

---

[27] O. Shimomura, B. Musicki, Y. Kishi, and S. Inouye, Cell Calcium 14, 373 (1993).
[28] R. Rizzuto, C. Bastianutto, M. Murgia, and T. Pozzan, J. Cell Biol. 126, 1183 (1994).

cytAEQ (Fig. 2). The calculated median effective concentration (EC_{50}) of ATP was 8.5 ± 1.8 μM with cytAEQ as the reporter and 5.8 ± 1.6 μM with mtAEQ as the reporter. This value is in agreement with previous studies, demonstrating that the agonist-mediated increase in intracellular

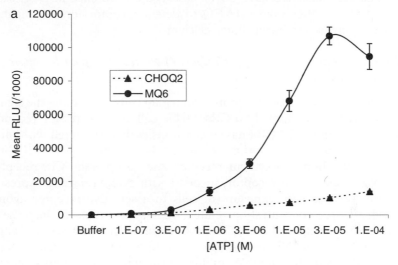

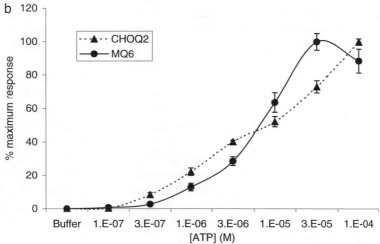

FIG. 2. ATP stimulation of aequorin luminescence in CHO cells stably transfected with cytAEQ (CHOQ2) or mtAEQ (MQ6). Data are expressed as mean relative light units (RLU). (a) Concentration–response curves were constructed in triplicate with the agonist ATP. (b) The data are expressed as a percentage of the maximum luminescent response to ATP. The EC_{50} values are 8.5 ± 1.8 μM (cytAEQ, ▲, broken line) and 5.8 ± 1.6 μM (mtAEQ, ●, solid line).

calcium concentration is accurately reported with aequorin regardless of the subcellular location of the reporter.[25] Because of the larger signal obtained when using mtAEQ, this reporter was used in all subsequent experiments in which aequorin was expressed transiently, and in the majority of the stable cell lines. We have never observed a difference in EC_{50} when using mtAEQ rather than cytAEQ to detect receptor- or ion channel-mediated increases in intracellular calcium.

Aequorin as Reporter of $G\alpha_{q/11}$-Coupled G Protein-Coupled Receptor Signal Transduction

We have expressed aequorin in mammalian cells to examine the signaling of many $G\alpha_{q/11}$-coupled GPCRs. Stable cell lines expressing recombinant or endogenous GPCRs have been transiently transfected with mitochondrially targeted aequorin. We have also created stable cell lines expressing the human oxytocin receptor, the vasopressin V1a receptor, or the vasopressin V1b receptor, together with cytoplasmically expressed aequorin. In each case exposure of cells to a selective receptor agonist induced a large, concentration-related, rapid, and transient increase in aequorin liminescence (Fig. 3). The fold increase in luminescence ranges from 63-fold for the AT_1 receptor to 768-fold for the V1a receptor, making the latter reporter suitable for high-throughput compound screening purposes (Fig. 3a). The range in luminescence generated reflects a number of factors, including the inherent ability of the receptor to couple to calcium mobilization, the level of expression of both the GPCR and aequorin, and the cellular location of the recombinant aequorin.

Aequorin has also been widely used in our laboratories and elsewhere for the quantitative characterization of receptor agonists.[4,5,11–13] Full concentration–response curves for the agonists oxytocin and vasopressin were constructed in the CHO cell line coexpressing the human oxytocin receptor and cytoplasmic aequorin. EC_{50} values were 22 ± 5 nM for oxytocin and 16 ± 4 nM for vasopressin. We find that the use of aequorin to assess agonist efficacy at GPCRs generates the same rank order of potency as obtained when using other traditional measures of $G\alpha_{q/11}$-coupled GPCR singaling, including the measurement of changes in the level of intracellular calcium using fluorescent dyes, or the measurement of the accumulation of radiolabeled inositol phosphates.

Aequorin as General Reporter of G Protein-Coupled Receptor Signal Transduction

Following the coexpression of aequorin with the so-called "promiscuous" G-protein α subunit, $G\alpha_{16}$, a generic screening system has been devel-

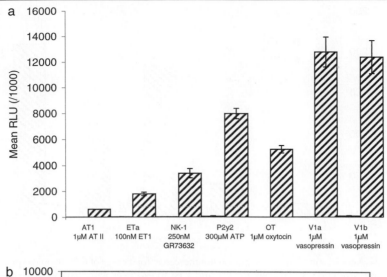

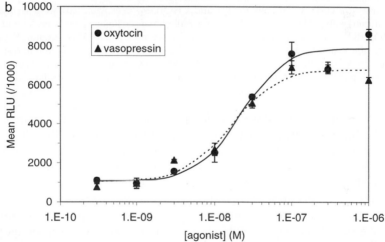

FIG. 3. Agonist stimulation of aequorin luminescence for a range of $G\alpha_{q/11}$-coupled GPCRs. (a) pcDNA3/mtAEQ was transiently transfected into CHO cells stably expressing the human endothelin ET_A, human angiotensin AT_1, human neurokinin NK-1, and hamster P2y2 purinergic receptors. The human V1a and V1b vasopressin receptors, and the human OT oxytocin receptor, were stably coexpressed with cytAEQ. Data are from one experiment representative of three, performed in triplicate, and are expressed as mean relative light units (RLU). Solid bars show the level of luminescence generated in response to assay medium alone, and hatched bars show the response to a near-maximal concentration of relevant agonist. (b) Concentration–response curves for oxytocin (●, solid line) and vasopressin (▲, dotted line) were constructed in a cell line stably coexpressing cytoplasmic aequorin and the human oxytocin receptor.

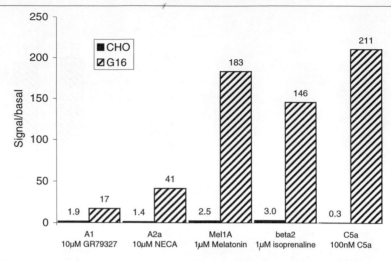

FIG. 4. Agonist stimulation of aequorin luminescence for a range of GPCRs in the presence of coexpressed $G\alpha_{16}$. CHO cells (solid) and CHO/$G\alpha_{16}$ cells (hatched) were transiently cotransfected with pcDNA3/mtAEQ together with one of the human adenosine A1 and A2a receptors, the β_2-adrenoceptor, the ML_{1A} melatonin receptor, and the C5a receptor. For each receptor, the luminescence generated is expressed as the fold increase over basal in response to a near-maximal concentration of agonist. Data are from one experiment representative of three, each performed in triplicate. The data values are included above each bar.

oped that allows the assessment of agonist activity at a large number of GPCRs that normally are not able to couple to members of the $G\alpha_{q/11}$ G-protein family.[25,29] The $G\alpha_{q/11}$ family G-protein $G\alpha_{16}$ is capable of functional interaction with a wide range of GPCRs. When transiently expressed in CHO cells with a mitochondrially targeted aequorin reporter, the $G\alpha_i$-coupled human melatonin ML_{1A} and adenosine A_1 receptors, and the $G\alpha_s$-coupled human β_2-adrenoceptor and adenosine A_{2a} receptors, were shown to have no effect on calcium mobilization after exposure to receptor agonist. In addition, we were unable to elicit an aequorin readout after stimulation of the C5a receptor, even though this receptor has been classified as coupling to $G\alpha_q$ (Fig. 4). However, transient coexpression of all these receptors, together with a mitochondrially targeted aequorin reporter in CHO cells stably expressing the G-protein $G\alpha_{16}$, resulted in agonist-mediated mobilization of calcium at each of these receptors, as detected by an increase in aequorin luminescence. As seen with the $G\alpha_q$-coupled receptors, the fold increase in luminescence varies, in this case from 17- to 211-fold. In addition to the explanations mentioned previously, it is likely that different receptors

[29] G. Milligan, F. Marshall, and S. Rees, *Trends Pharmacol. Sci.* **17,** 235 (1996).

will couple to $G\alpha_{16}$ with a range of affinities. These data demonstrate the utility of $G\alpha_{16}$ and an aequorin reporter as tools to develop generic assay systems for the majority of GPCRs, regardless of their natural coupling mechanism. In these studies we have observed that the efficacy values observed for agonist activation of $G\alpha_{16}$ can vary substantially from those obtained when measuring receptor signaling through the normal G-protein partner. However, in our hands the rank order of potency of a series of agonists is maintained.

Aequorin as Reporter of Ion Channel Activity

We have also investigated the use of recombinantly expressed aequorin to measure ion entry through calcium-selective ion channels. The increase in intracellular calcium concentration seen after ion channel opening can be transient, hence aequorin can be used to report on ion channel opening only if the channel remains open long enough for enough calcium to enter the cell and activate the aequorin complex. As with the detection of changes in intracellular calcium concentration after GPCR signal transduction, the intensity of light emission from aequorin is proportional to the intracellular calcium concentration, thus enabling dose–response curves to be determined for ligand-gated or voltage-gated ion channels.[18] We have used the FLIPR to measure aequorin luminescence in stable cell lines expressing apoaequorin and the homomeric $P2X_2$ or heteromeric $P2X_2/P2X_3$ purinergic channels (Fig. 5). In HEK 293 cells stably expressing the $P2X_2$ channel and mtAEQ, the agonist ATP was able to cause an increase in aequorin luminescence with an EC_{50} of 2.8 ± 0.3 μM (Fig. 5a). Similarly, in CHO cells stably expressing both $P2X_2$ and $P2X_3$ together with mtAEQ the channel agonist $\alpha\beta$meATP caused an increase in aequorin luminescence with an EC_{50} of 1.2 ± 0.2 μM (Fig. 5b). These values are similar to the EC_{50} values of approximately 10 μM (for ATP at the $P2X_2$ channel) and 1 μM (for $\alpha\beta$meATP at the $P2X_3$ channel) previously determined by fluorescent measurements and electrophysiological analysis[19, 20] (also R. Chang and L. C. Mattheakis, unpublished experiments, 1998). While we have not investigated the use of aequorin for the study of other calcium-selective ion channels it is likely that this reporter could be used to characterize calcium entry through voltage-gated calcium channels and the vanilloid receptors, among others.

Concluding Remarks

In this chapter we have described the use of recombinant aequorin for the detection of changes in intracellular calcium concentration as a

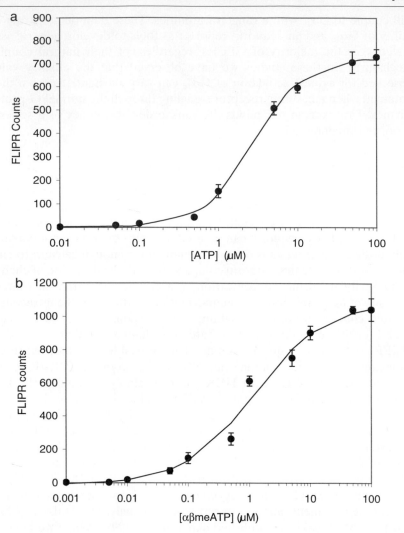

FIG. 5. Agonist-mediated activation of calcium-selective ion channels to stimulate aequorin luminescence. Changes in luminescence were quantified on a FLIPR. (a) Concentration–response curves for ATP were constructed in HEK 293 cells stably expressing both aequorin (mtAEQ) and the homomeric P2X$_2$ purinergic channel. (b) Concentration–response curves for $\alpha\beta$meATP were constructed in CHO cells stably expressing both aequorin (mtAEQ) and the heteromeric P2X$_2$/P2X$_3$ purinergic channel.

consequence of GPCR-mediated release of calcium from intracellular stores, and as a result of calcium entry through purinergic P2X family calcium channels. Aequorin assays are fast, simple to perform, and highly sensitive, and are amenable to assay in microplate formats. As such, aequorin has become the reporter of choice in our laboratories for the study of drug-induced changes in intracellular calcium concentration. In contrast to the use of the calcium-sensitive dyes, or the direct injection into cells of aequorin protein, the ability to target recombinant aequorin to specific intracellular compartments, or to express aequorin as a fusion protein with the protein of interest, allows the study of calcium changes in discrete cellular compartments. For the detection of aequorin luminescence we use either the Dynatech ML3000 luminometer equipped with reagent injectors or the FLIPR instrument. The use of these instruments enables the rapid pharmacological characterization of both GPCRs and calcium-selective ion channels according to the generation of aequorin luminescence. Furthermore, as these assay systems enable the rapid characterization of many hundreds or thousands of new chemicals for activity at the GPCR or ion channel under study, aequorin assays have become widely used within the pharmaceutical industry for compound screening. The two major disadvantages of aequorin assays have been the cost of the cofactor coelenterazine and the half-life of aequorin luminescence. The cost of coelenterazine has decreased dramatically as a number of new manufacturers have entered the market. Aequorin luminescence has a signal half-life of less than 10 sec, hence a luminometer equipped with reagent injectors is required for the detection of the luminescent signal. A number of detection systems are now available that are suitable for the detection of aequorin luminescence. However, the availability of a mutant aequorin, or an aequorin assay, in which the half-life of aequorin luminescence is lengthened will further increase the usefulness of this reporter.

Acknowledgments

We thank the following members of the Molecular Pharmacology Unit (Glaxo Wellcome Research and Development) for their assistance in the cloning of receptors and G proteins, and the construction of cell lines: Neil Fraser, Andy Green, Elaine Murrison, Jason Brown, Emma Knight, and Liz Eggleston.

[35] Monitoring Protein Conformations and Interactions by Fluorescence Resonance Energy Transfer between Mutants of Green Fluorescent Protein

By Atsushi Miyawaki *and* Roger Y. Tsien

Introduction

Mutants of Green Fluorescent Protein with Altered Colors

Green fluorescent protein (GFP) is a spontaneously fluorescent protein from the jellyfish *Aequorea victoria.*[1] It can be genetically concatenated to many other proteins, and the resulting fusion proteins are usually fluorescent and often preserve the biochemical functions and cellular localization of the partner proteins. GFP fusions have major advantages over previous techniques for fluorescent labeling of proteins by covalent reaction with small molecule dyes. The chimeric fluorescent proteins are generated *in situ* by gene transfer into cells or organisms, obviating high-level heterologous expression, purification, *in vitro* labeling, and microinjection of recombinant proteins. Targeting signals can be used to direct localization of the chimeras to particular tissues, cells, organelles, or subcellular sites. The sites of labeling are defined exactly, giving a molecularly homogeneous product without the use of elaborate protein chemistry.

Mutagenesis has produced GFP mutants with shifted wavelengths of excitation and emission that can serve as donors and acceptors for fluorescence resonance energy transfer (FRET). FRET is a nondestructive spectroscopic method[2-10] that can be used to monitor the proximity and relative angular orientation of fluophores in single living cells. The donor and acceptor fluorophores can be on separate proteins to see intermolecular

[1] R. Y. Tsien, *Annu. Rev. Biochem.* **67,** 509 (1998).
[2] L. Stryer, *Annu. Rev. Biochem.* **47,** 819 (1978).
[3] B. Herman, *Methods Cell Biol.* **30,** 219 (1989).
[4] T. M. Jovin and D. J. Arndt-Jovin, *Annu. Rev. Biophys. Biophys. Chem.* **18,** 271 (1989).
[5] P. S. Uster and R. E. Pagano, *J. Cell Biol.* **103,** 1221 (1986).
[6] R. Y. Tsien, B. J. Bacskai, and S. R. Adams, *Trends Cell Biol.* **3,** 242 (1993).
[7] G. W. Gordon, G. Berry, X. H. Liang, B. Levine, and B. Herman, *Biophys. J.* **74,** 2702 (1998).
[8] P. I. Bastiaens and A. Squire, *Trends Cell Biol.* **9,** 48 (1999).
[9] T. W. J. Gadella, G. N. M. van der Krogt, and T. Bisseling, *Trends Plant Sci.* **4,** 287 (1999).
[10] J. R. Lakowicz, "Principles of Fluorescence Spectroscopy," 2nd Ed. Plenum Press, New York, 1999.

association, or attached to the same macromolecule to detect its conformational changes. Two pairs of GFP mutants have been used for FRET: BFP (blue)–GFP (green) and CFP (cyan)–YFP (yellow).[1,11,12] Further improvements of GFP mutants for FRET have been made as follows. (1) For better folding of GFPs at 37° in mammalian cells, mammalian codon bias and some amino acid substitutions have been introduced to make enhanced fluorescent proteins: EBFP, ECFP, EGFP, and EYFP[1,13]; (2) (E)YFP (S65G/S72A/T203Y) proved to be highly sensitive around neutral pH (pK_a 7). Two amino acid substitutions (V68L/Q69K) were found to lower its pK_a to about 6.[14] EYFP-V68L/Q69K or EYFP.1 (S65G/V68L/Q69K/S72A/T203Y) is less pH sensitive and therefore a more reliable acceptor than the original EYFP for most applications. In the remainder of this chapter we use the terms CFP and YFP to mean the generic classes of cyan and yellow mutants, whereas EYFP and EYFP. 1 refer to the specific variants listed above.

Monitoring Fluorescence Resonance Energy Transfer via Emission Ratioing

Steady state FRET is most conveniently observed by exciting the sample at the donor excitation wavelengths while measuring the ratio of fluorescence intensities emitted at wavelengths corresponding to the emission peaks of the donor (donor channel) versus those of the acceptor (FRET channel). The occurrence of FRET diminishes the signal in the donor channel while increasing the amplitude of the FRET channel. However, simple measurements of the intensities in the two channels and their ratio are perturbed by several factors other than the FRET efficiency. These interfering factors include uncertainties in the relative concentrations of donor and acceptor and spectral cross-talk of two main kinds: (1) direct excitation of the acceptor at the donor excitation wavelengths, and (2) leakage of donor emission into the FRET channel. Leakage of acceptor fluorescence into the donor channel is hardly ever a problem because emission spectra usually cut off quite abruptly at their short-wavelength borders.

These uncertainties are minimized if both the donor and acceptor are

[11] R. Y. Tsien and D. C. Prasher, "GFP: Green Fluorescent Protein Strategies and Applications" (M. Chalfie and S. Kain, eds.). John Wiley & Sons, New York, 1998.
[12] R. Heim, *Methods Enzymol.* **302,** 408 (1999).
[13] A. Miyawaki, J. Llopis, R. Heim, J. M. McCaffery, J. A. Adams, M. Ikura, and R. Y. Tsien, *Nature (London)* **388,** 882 (1997).
[14] A. Miyawaki, O. Griesbeck, R. Heim, and R. Y. Tsien, *Proc. Natl. Acad. Sci. U.S.A.* **96,** 2135 (1999).

fused to the same partner protein to form a three (or more)-component chimera whose conformation is sensitive to the biochemical environment. Obviously the stoichiometry of donor and acceptor is then fixed. Although spectral cross-talk reduces the dynamic range over which the emission ratio can vary, that emission ratio for any given construct can still be empirically calibrated in terms of the extent of biochemical reaction. The first examples of such chimeras were simple fusions of BFPs and GFPs to protease-sensitive linkers. Proteolysis of such linkers *in vitro* disrupted FRET.[15,16] More recently, fusions with linkers cleavable by important intracellular proteases have been introduced to assay the activity of those proteases in living cells.[17] However, these assays still monitor irreversible hydrolysis of covalent bonds. Could the same principle of FRET between different colors of GFP be used to monitor reversible conformational equilibria and fluctuating intracellular signals? Our first successful demonstration of the feasibility of this principle was a family of Ca^{2+} indicators, as explained below.

Construction of Cameleons, Indicators for Ca^{2+} Based on Calmodulin

Initial Design and Screening

Many effects of Ca^{2+} in cells are mediated by the binding of Ca^{2+} to calmodulin, which causes calmodulin to bind and activate target proteins. Ikura and collaborators solved the nuclear magnetic resonance solution structure of calmodulin bound to M13,[18] the 26-residue calmodulin-binding peptide of myosin light-chain kinase. They then fused the C terminus of calmodulin to M13 by means of a Gly-Gly spacer and verified that this hybrid protein changed from a dumbbell-like extended form to a compact globular structure on binding Ca^{2+}. This hybrid protein, calmodulin–M13, gave us a potential starting point for a Ca^{2+} indicator. We initially sandwiched calmodulin–M13 between BFP and S65T.[13] The chimeric proteins (see Fig. 1 for generic structure) were expressed in bacteria and analyzed for FRET. The amino acid sequences of the boundary regions between the calmodulin–M13 hybrid and the GFPs proved critical to the optimization of protein folding (formation of chromophore) and Ca^{2+}-dependent changes in FRET. Numerous deletions, insertions, and amino acid substitutions were tested, as shown in Table I.

[15] R. Heim and R. Y. Tsien, *Curr. Biol.* **6,** 178 (1996).

[16] R. D. Mitra, C. M. Silva, and D. C. Youvan, *Gene* **173,** 13 (1996).

[17] X. Xu, A. L. V. Gerard, B. C. B. Huang, D. C. Anderson, D. G. Payan, and Y. Luo, *Nucleic Acids Res.* **26,** 2034 (1998).

[18] M. Ikura, G. M. Clore, A. M. Gronenborn, G. Zhu, C. B. Klee, and A. Bax, *Science* **256,** 632 (1992).

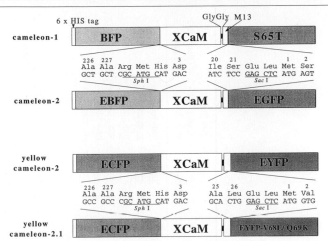

FIG. 1. Domain structures of cameleons (cameleon-1, cameleon-2, YC2, and YC2.1) expressed in bacteria for *in vitro* characterization, showing sequences of the boundaries between the donor GFP and *Xenopus* calmodulin (XCaM) and between M13 and the acceptor GFP.

Folding and Fluorescence of Green Fluorescent Protein Mutants. Intramolecular FRET obviously requires that both the donor and acceptor GFPs remain fluorescent after fusion. However, in many variants of our chimeras, either the BFP or the S65T was not fluorescent, probably because of misfolding. Also, some of the chimeras were easily proteolyzed in bacteria or during purification. Resistance to proteolysis of the chimeras was finally checked in mammalian cells; in HeLa cells transfected with the cDNAs, the intact chimera was uniformly distributed in the cytosolic compartment but excluded from the nucleus, as expected for a 74-kDa protein without targeting signals. On the other hand, the chimeras that were susceptible to proteolysis entered the nucleus. Although we earlier referred to the molecular homogeneity of GFP-fused proteins, trial and error is often necessary to obtain fusions that express well without precipitation or proteolysis.

Responsivity of Fluorescence to Ca^{2+}. Only intact chimeras containing both functional BFP and S65T were screened for responsivity to Ca^{2+}. Sometimes only one amino acid substitution changed the FRET efficiency dramatically; we found it impossible to predict in advance what sorts of amino acid sequences should be put in the boundaries (Table I). Empirically, we found it better to fuse the two GFPs rigidly to the calmodulin–M13 hybrid protein without extra spacers at the boundaries, leaving the Gly-Gly spacer between the calmodulin and M13 as the only obvious hinge in the whole chimera. Thus removal of the 11 C-terminal amino acids of the BFP, slightly more than the 9 residues too disordered to be seen in the

TABLE I
VARIOUS CHIMERIC PROTEINS FOR Ca^{2+} INDICATORS CONTAINING BFP AND S65T[a]

Chimera	BFP (donor) (-AAGITHGMDELYK)	Xenopus calmodulin (MHDQLTEEQIAEFKE---TAK)	Linker (GGK)	M13 peptide (RRW---FKKISSSGAL)	Linker (EL)	GFP:S65T (acceptor) (MSKGEELF-)	Qualitative grade
1	-AAGITHMDELYK	GAGMHDQLTEEQIAEFKE---TAK	GGK	RRW---FKKISSSGAL	EL	MSKGEELF-	C[b]
2	-AGIATHG	MHDQLTEEQIAEFKE---TAK	GGK	RRW---FKKISSSGAL	EL	MSKGEELF-	B
3	-AAGITHG	MREEQIAEFKE---TAK	GGK	RRW---FKKISSSGAL	EL	MSKGEELF-	B
4	-AAGITHG	MQEEQIAEFKE---TAK	GGK	RRW---FKKISSSGAL	EL	MSKGEELF-	C
5	-AAGITHG	MHEEQIAEFKE---TAK	GGK	RRW---FKKISSSGAL	EL	MSKGEELF-	C
6	-AAGITHG	MLEEQIAEFKE---TAK	GGK	RRW---FKKISSSGAL	EL	MSKGEELF-	C
7	-AAGITHG	MPEEQIAEFKE---TAK	GGK	RRW---FKKISSSGAL	EL	MSKGEELF-	C
8	-AAGITHG	MLQIAEFKE---TAK	GGK	RRW---FKKISSSGAL	EL	MSKGEELF-	Not folded
9	-AGIT#	MHDQLTEEQIAEFKE---TAK	GGK	RRW---FKKISSSGAL	EL	MSKGEELF-	D
10	-AAC	MHDQLTEEQIAEFKE---TAK	GGK	RRW---FKKISSSGAL	EL	MSKGEELF-	Not folded
11	-AAR	MHDQLTEEQIAEFKE---TAK	GGK	RRW---FKKISSSGAL	EL	MSKGEELF-	B
12	-AAG	MHDQLTEEQIAEFKE---TAK	GGK	RRW---FKKISSSGAL	EL	MSKGEELF-	C
13	-AAS	MHDQLTEEQIAEFKE---TAK	GGK	RRW---FKKISSSGAL	EL	MSKGEELF-	C
14	-AAR	MHDQLTEEQIAEFKE---TAK	GGK	RRW---FKKISSSG	EL	MSKGEELF-	B
15	-AAR	MHDQLTEEQIAEFKE---TAK	GGK	RRW---FKKIS	EL	MSKGEELF-	A

16	-AAR	MH QLTEEQIAEFKE---TAK	GGK	RRW---FKKIS	EL	MSKGELF-	D
17	-AAR	M QLTEEQIAEFKE---TAK	GGK	RRW---FKKIS	EL	MSKGELF-	Not folded
18	-AAR	MHDQLTEEQIAEFKE---TAK	GGK	RRW---FKKIA	EL	MSKGELF-	B
19	-AAR	MHDQLTEEQIAEFKE---TAK	GGK	RRW---FKKIC	EL	MSKGELF-	B
20	-AAR	MHDQLTEEQIAEFKE---TAK	GGK	RRW---FKKI	EL	MSKGELF-	B
21	-AAR	M #EEQIAEFKE---TAK	GGK	RRW---FKKISSSGAL	EL	MSKGELF-	C
22	-AAR	M #EQIAEFKE---TAK	GGK	RRW---FKKISSSGAL	EL	MSKGELF-	B
23	-AAR	M LQIAEFKE---TAK	GGK	RRW---FKKISSSGAL	EL	MSKGELF-	C
24	-AAR	M QIAEFKE---TAK	GGK	RRW---FKKISSSGAL	EL	MSKGELF-	A
25	-AAR	M #EEQIAEFKE---TAK	GGK	RRW---FKKIS	EL	MSKGELF-	Not folded
26	-AAR	M LQIAEFKE---TAK	GGK	RRW---FKKIS	EL	MSKGELF-	Not folded
27	-AAR	MHDQLTEEQIAEFKE---TAK	GGGGS	RRW---FKKIS	EL	MSKGELF-	Proteolyzed
28	-AAR	MHDQLTEEQIAEFKE---TAK	(GGS)$_4$	RRW---FKKIS	EL	MSKGELF-	Proteolyzed

a The genes for BFP and *Xenopus* calmodulin were linked using an *Sph*I site; the amino acid sequences encoded by the *Sph*I site are underlined. The underlined amino acid sequences (EL) between the M13 peptide and S65T are encoded by an *Sac*I site. #, Introduced amino acid site #. The quality of the Ca^{2+}-dependent change in FRET was qualitatively graded as A–D, with A being the best. Chimera 15 gave above 80% change in FRET, and was termed cameleon-1.

b The change in FRET was not reversible. "Not folded" indicates either the BFP or the S65T was not fluorescent, or the whole protein was not synthesized. "Proteolyzed" indicates proteolysis occurred between calmodulin and M13 peptide.

BFP or GFP crystal structures,[19,20] improved the Ca^{2+}-dependent change in FRET significantly. The capriciousness of these permutations made us call these Ca^{2+} indicators "cameleons." Like real chameleons, they readily change color, and retract and extend a long tongue (M13) into and out of the mouth of the calmodulin (CaM).

We tried to select bacterial clones producing good Ca^{2+} indicator proteins by measuring FRET directly from colonies on a plate, when they were soaked with Ca^{2+} and then EGTA in the presence of Ca^{2+} ionophore. But the signals obtained depended on many factors other than Ca^{2+}: size of colony, degree of misfolding and/or proteolysis inside bacteria, efficiency of exposure of the proteins to the reagents, and so on. Because quantitation of FRET in colonies was unreliable, we had to purify each protein, check its integrity, and then analyze FRET. An alternative screening system would be fluorescence-activated cell sorting (FACS), which measures fluorescence from every bacterium. Unfortunately, FACS is not suited to comparing the fluorescence of given clones before and after a manipulation such as changing the Ca^{2+}.

The efficiency of FRET theoretically should not be affected by exchanging the positions of the donor and acceptor fluorophores, because the distance between the chromophores and the orientation factor remains unchanged. However, the quality of chimeric proteins is not necessarily the same in the two cases. In our experience, when two GFPs are fused to the N and C termini of a host protein, the N-terminal GFP is generally better folded than the C-terminal GFP. In our initial screens we put the BFP at the N terminus and S65T at the C terminus, because the fluorescence of the latter is more reliably seen by eye and would imply that both GFPs must have folded correctly. In such complete constructs, we tried exchanging the BFP and S65T, but these chimeras (S65T–calmodulin–M13–BFP) empirically showed a somewhat smaller Ca^{2+}-dependent FRET response, about 1.65-fold maximal ratio change, compared with 1.8 to 1.9-fold for BFP–CaM–M13–S65T. Therefore we settled on the latter ordering.

Importance of Relative Orientations of Green Fluorescent Protein Mutants. Multidimensional nuclear magnetic resonance (NMR) had shown that the Ca^{2+}-saturated calmodulin–M13 hybrid protein shows a compact globular structure similar to that of the Ca^{2+}–calmodulin–M13 intermolecular complex.[18] The three-dimensional structure of the Ca^{2+}-free, extended

[19] R. M. Wachter, B. A. King, R. Heim, K. Kallio, R. Y. Tsien, S. G. Boxer, and S. J. Remington, *Biochemistry* **36,** 9759 (1997).
[20] M. Ormö, A. B. Cubitt, K. Kallio, L. A. Gross, R. Y. Tsien, and S. J. Remington, *Science* **273,** 1392 (1996).

form of the hybrid protein has not been solved, although Porumb et al.[21] showed that its conformation differed from that of the Ca^{2+} complex. Furthermore, we do not know how the two GFPs fused to the hybrid protein are positioned. Current indications are that the Ca^{2+}-dependent increase in FRET is due more to a change in the relative orientations of the two GFP chromophores rather than the distance between them. The belief is based on observations that small structural alterations such as adding or subtracting single amino acids from the linker regions, or substituting a circularly permuted CFP for native CFP,[22] can cause profound changes in the effect of Ca^{2+} on FRET. Such alterations should not greatly affect the distance between the chromophores but could well change their relative orientations drastically.

Expression of Cameleon in Mammalian Cells

The prototype cameleon, cameleon-1 (Fig. 1), was efficiently expressed and folded in bacteria and increased its ratio of ultraviolet-excited 510:445 nm emissions by 70% on binding Ca^{2+}. The decrease in blue and increase in green emission indicated that Ca^{2+} increased the efficiency of FRET from BFP to S65T, consistent with the expected decrease in distance between the two ends of the protein. Cameleon-1 displayed a biphasic Ca^{2+} dependency with apparent dissociation constants of 70 nM and 11 μM, and therefore could report a wide range of Ca^{2+} concentration from 10^{-9} to 10^{-4} M. Despite the promising Ca^{2+} sensitivity observed in *in vitro* experiments, the fluorescence of cameleon-1 was not bright enough for Ca^{2+} imaging in mammalian cells. For adequate expression and brightness of the mutant GFPs in mammalian cells, enhanced GFPs (EBFP and EGFP) encoded by sequences containing mammalian codon usage and including amino acid mutations for improved folding at 37° were developed.[1] The substitution of EBFP and EGFP for BFP and S65T improved the expression of cameleon in mammalian cells. The resulting cameleon, cameleon-2 (Fig. 1), was able to report agonist-evoked changes in cytosolic Ca^{2+} concentration ($[Ca^{2+}]_c$) in HeLa cells. Cameleon-2 also appeared to be expressed better than cameleon-1 in bacteria.

Longer-Wavelength Cameleons

Another concern was that blue mutants such as EBFP are the dimmest and most bleachable of the GFPs. Their excitation peaks are in the ultravio-

[21] T. Porumb, P. Yau, T. S. Harvey, and M. Ikura, *Protein Eng.* **7,** 109 (1994).
[22] G. S. Baird, D. A. Zacharias, and R. Y. Tsien, *Proc. Natl. Acad. Sci. U.S.A.* **96,** 11241 (1999).

let at 382 nm, which is potentially injurious, excites the most cellular autofluorescence, and could interfere with the use of caged compounds. Therefore, enhanced cyan and yellow fluorescent proteins (ECFP and EYFP) were substituted for EBFP and EGFP, respectively, to make "yellow cameleons" (YCs). ECFP has two excitation peaks of nearly equal amplitude at 434 and 452 nm. To minimize direct excitation of the acceptor, the ECFP should be excited at the 434-nm peak or even shorter wavelengths, and the acceptor should be EYFP rather than EGFP. Another theoretical advantage of the ECFP–EYFP pair is that it gives stronger FRET than EBFP–EGFP, assuming other factors are equal. The calculated distance R_0 at which FRET is 50% efficient between randomly oriented chromophores is 5 nm for ECFP–EYFP versus 4 nm for EBFP–EGFP.[12] Accordingly, the boundary regions were reoptimized for maximal Ca^{2+} dependence of FRET; the C terminus of the M13 peptide in YCs was extended by five amino acids. The resulting YCs showed 1.8- to 1.9-fold changes in emission ratio from zero to saturating Ca^{2+}.

pH Sensitivity of Green Fluorescent Protein Based Fluorescence Resonance Energy Transfer

Intracellular pH varies among organelles and under conditions such as mitogen stimulation and metabolic stress. If FRET is used to measure protein conformations or interactions in intact cells, the donor and acceptor should be indifferent to physiological changes in pH. However, every GFP mutant is at least somewhat pH sensitive, in that all can be quenched by sufficiently acidic pH. The most important effects of pH are on the quantum efficiency of the donors EBFP and ECFP (Fig. 2) and the absorbance spectra of the acceptors EGFP and EYFP (Fig. 3). Decreasing pH quenches the emissions of EBFP and ECFP with little effect on their absorbance spectra, indicating that their quantum yields are depressed by acid, but the effects are not serious until the pH falls below 6. However, the original EYFP has quite a high pK_a of 6.9 for both its absorbance and emission spectra[23] (Fig. 3), indicating that its absorbance, not its quantum yield, is pH sensitive. This pK_a is uncomfortably close to cytosolic pH, rendering the first generation of cameleons quite pH sensitive. The obvious solution was to find a less pH-sensitive version of EYFP. Adding the mutation Q69K to 10C (S65G, V68L, S72A, T203Y) dropped the apparent pK_a to 6.1,[14] decreasing the sensitivity to pH changes between 6 and 8 (Fig. 3). 10C Q69K (EYFP-V68L/Q69K or EYFP.1) could be substituted for the original EYFP without altering the Ca^{2+}-dependent FRET changes of YCs, because

[23] J. Llopis, J. M. McCaffery, A. Miyawaki, M. G. Farquhar, and R. Y. Tsien, *Proc. Natl. Acad. Sci. U.S.A.* **95,** 6803 (1998).

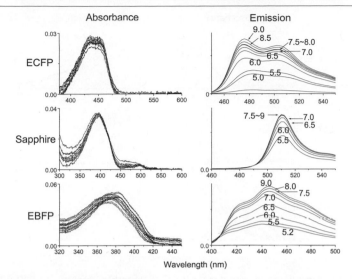

Fig. 2. pH dependency of donor GFPs. *Left:* Absorbance spectra of EBFP (F64L/Y66H/Y145F), Sapphire (T203I/S72A/Y145F), and ECFP (F64L/S65T/Y66W/N146I/M153T/V163A/N164H) in buffers of different pH. The individual spectra are not labeled because they are so nearly overlapping. "Sapphire" (synonymous with "H9-40") is a UV-excited, green-emitting mutant[1,12] included here for completeness, although it is of little utility as an FRET donor to EGFP or EYFP because its emission overlaps too greatly with theirs. *Right:* Emission spectra measured in buffers of the indicated pH. Excitation was at 380 nm for EBFP, 400 nm for Sapphire, and 432 nm for ECFP.

the two EYFPs have the same fluorescence properties other than pH sensitivity. YC2 incorporating EYFP.1 was constructed and termed YC2.1 (Fig. 1). The best samples of YC2.1 now show emission ratio changes of 2.0- to 2.1-fold between zero and saturating Ca^{2+}; the slight improvement over YC2 is probably due to better protein purification rather than to the V68L/Q69K mutations.

Despite the improvement in pK_a values, pH sensitivity should not be forgotten. Checking or clamping of ambient pH is still desirable to prevent artifacts. It is unfortunate that acidity causes the donors to lose quantum yield and the acceptors to lose absorbance, because the two negative effects reinforce each other. Had the losses been in donor absorbance or acceptor quantum yield, they would not have affected the efficiency of FRET.

Future Prospects for Improvement of Cameleons

Aside from their pH sensitivity, the YFPs have other drawbacks as FRET acceptors, such as their small Stokes shift and their photochemical

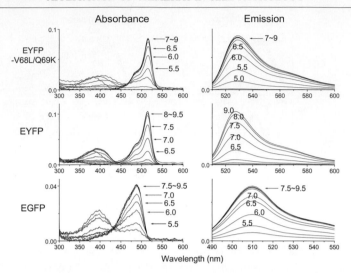

FIG. 3. pH dependency of acceptor GFPs. *Left:* Absorbance spectra of EYFP.1 (S65G/V68L/Q69K/S72A/T203Y), EYFP (S65G/S72A/T203Y), and EGFP (F64L/S65T) at the indicated pH. *Right:* pH dependency of emission spectra of the acceptor GFPs. Excitation was at 500 nm for EYFP.1 and EYFP, and at 480 nm for EGFP.

instability (photochromism) at high light intensities,[24] although as will be seen, the photochromism can be useful for quantifying FRET. More recently, novel yellow and red fluorescent proteins have been cloned from corals.[25] Although these new proteins have not yet been fully characterized or optimized for mammalian expression, they offer the exciting prospect of accepting FRET from EGFP. Such pairs would use a donor whose quantum yield is pH independent and allow us to make "red cameleons."

Advances in single-molecule detection and single-molecule spectroscopy by laser-induced fluorescence offer new tools for the study of individual macromolecules.[13] It has become possible to observe the fluorescence from single molecules of YC2.1, trapped in agarose gels to prevent lateral diffusion.[25a]

Procedures

Here we present a series of *in vitro* experiments for YC2.1.

[24] R. M. Dickson, A. B. Cubitt, R. Y. Tsien, and W. E. Moerner, *Nature (London)* **388,** 355 (1997).
[25] M. V. Matz, A. F. Fradkov, Y. A. Labas, A. P. Savitsky, A. G. Zaraisky, M. L. Markelov, and S. A. Lukyanov, *Nature Biotechnol.* **17,** 969 (1999).
[25a] S. Brasselet, A. Miyawaki, and W. E. Moerner, Submitted (2000).

Expression and Purification of Recombinant YC2.1 Protein. The YC2.1 coding sequence is subcloned into *Bam*HI and *Eco*RI restriction sites of expression plasmid pRSETB (InVitrogen, San Diego, CA), which encodes a fusion protein containing a six-histidine (His$_6$) tag and an enterokinase cleavage site upstream from the insert. For bacterial expression, *Escherichia coli* JM109(DE3) is transformed with the plasmid and grown on LB plates containing ampicillin (0.1 mg/ml). A 2-ml overnight culture is used to inoculate 100–500 ml of medium. Overgrowth of the colonies and cultures before inoculation should be avoided. The bacteria are grown at 25° to an optical density of 0.4–0.8 at 600 nm and induced with 0.1–1 m*M* isopropyl-thiogalactoside for 12–24 hr. Cells are harvested by centrifugation. The bacterial pellet looks greenish-yellow, primarily due to the absorption of ECFP and EYFP.1. The pellet is resuspended in 30 ml of phosphate-buffered saline (PBS), pH 7.4, and lysed with a French press in the presence of protease inhibitors (leupeptin, pepstatin A, and phenylmethylsulfonyl fluoride). The lysate is then clarified by centrifuging at 12,000*g* for 30 min at 4°. Efficient extraction by a French press gives a greenish-yellow supernatant and a whitish pellet of cellular debris. Binding of the His$_6$ tag to Ni-NTA agarose (Qiagen, Chatsworth, CA) is carried out in a batch mode; the supernatant is transferred to a new tube, to which 0.5–1 ml of the 50% (v/v) Ni-NTA slurry is added, and mixed gently on a rotary shaker at 4° for 1 hr. The lysate–Ni-NTA mixture is loaded into a column, which is washed with 10 volumes of PBS, and then with 10 volumes of TN300 buffer [Tris-HCl (pH 7.4), 300 m*M* NaCl]. The recombinant YC2.1 protein is eluted with 1–3 ml of elution buffer (100 m*M* imidazole in TN300). Imidazole is removed by passing the eluate through a Sephadex G-25 column, eluting with buffer A [50 m*M* HEPES–KOH (pH 7.4), 100 m*M* NaCl]. The protein sample is concentrated with a Centricon 30 filter (Amicon, Danvers, MA), and then used for size-exclusion or ion-exchange chromatography.

Monitoring the Integrity of YC2.1. Figure 4 shows elution profiles of the concentrated eluates obtained by size-exclusion chromatography. A size-exclusion column (Biosep SEC-S3000, 300 × 7.5 mm; Phenomenex, Torrance, CA) is equilibrated with buffer A, and is linked to absorbance (at 280 nm) and fluorescence (excitation at 432 nm; emission at 502 or 528 nm) detectors. Thus the effluent is monitored on line for total protein concentration and for concentration of the desired protein at the same time. YC2.1 recombinant protein is eluted in almost a single peak (Fig. 4A). A slight shoulder on the right side may represent minor proteolysis. These results indicate that nearly all the proteins in the Ni-NTA eluate are full-length YC2.1. In contrast, Fig. 4B shows a chimera containing the linker GGSGGSGGS instead of the GG used in YC2.1. This protein sample is

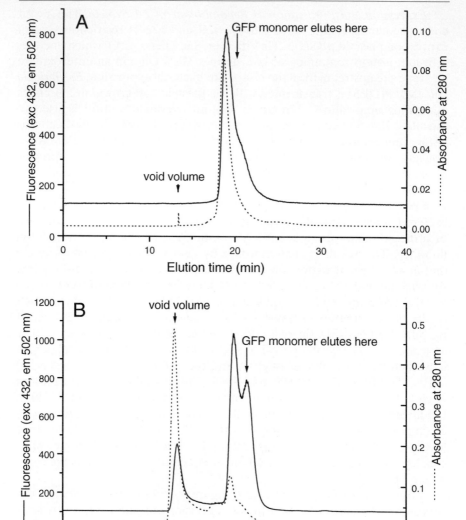

FIG. 4. Size-exclusion chromatography of fluorescent proteins. Total protein and the fluorescence of ECFP were monitored by absorbance at 280 nm (dashed lines) and by fluorescence (solid lines; excitation at 432 nm, emission at 502 nm). The void volume and the elution position for GFP monomer are indicated. The absorbance peaks appear slightly earlier than the matching fluorescence peaks because the effluent passed first through the absorbance detector, and then through the fluorescence detector. (A) Elution profiles for recombinant YC2.1 protein purified by Ni-NTA chromatography. (B) Elution profiles for an unsatisfactory chimera with GGSGGSGGS instead of GG as the linker between the CaM and M13.

prepared in the same way as mentioned above. Most of the protein aggregates and emerges in the void volume with little fluorescence. The peak of the desired protein is low and not well separated in the absorbance profile. Also, there clearly appears a fluorescence peak representing proteolyzed products, the elution time of which is almost the same as that of GFP monomer in this chromatographic system.

Spectroscopic Analysis of YC2.1 Protein; Quantification of Fluorescence Resonance Energy Transfer by Proteolysis. The YC2.1 protein sample is then spectroscopically characterized. Fluorescence measurements are performed with a fluorometer (Fluorolog; SPEX Industries, Edison, NJ). An excitation wavelength of 432 nm is used. The protein is diluted to a final concentration of about 10 μM with buffer A containing 100 μM $CaCl_2$. Compared with the reference spectra for separate samples of ECFP and EYFP.1 (Fig. 5), the emission spectrum of Ca^{2+}-saturated YC2.1 excited at ECFP wavelengths (Fig. 6, dashed line) consists mostly of EYFP.1 emission, showing considerable FRET from ECFP to EYFP.1. EGTA is then added to a final concentration of 500 μM, which should reduce the free Ca^{2+} to 14 nM. A reciprocal change in the two emission intensities is observed (solid line in Fig. 6), indicating reduction in FRET. Even in the absence of Ca^{2+} ion, however, significant FRET takes place on YC2.1, because the two GFPs cannot more infinitely apart from each other. We have tried to

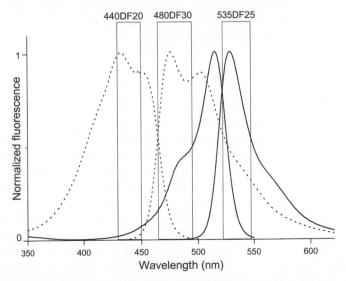

FIG. 5. Normalized excitation and emission spectra of ECFP and EYFP.1. The passbands of the excitation filter (440DF20) and two emission filters (donor channel, 480DF30; FRET channel, 535DF25) are indicated by boxes.

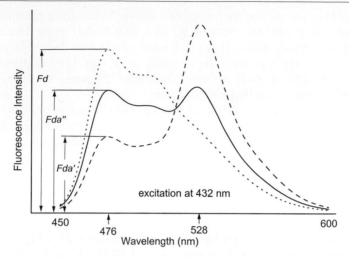

FIG. 6. Emission spectra of YC2.1 (excited at 432 nm) of the Ca^{2+}-saturated (dashed line) and Ca^{2+}-unsaturated (solid line) forms, and after proteolysis by trypsin (dotted line). The wavelengths giving the emission peaks of ECFP (476 nm) and EYFP.1 (528 nm) are indicated.

quantitate the FRET of the Ca^{2+}-saturated and Ca^{2+}-unsaturated forms, by measuring the fluorescence intensity of donor both in the absence (F_d) and presence (F_{da}) of acceptor. Generally, the efficiency of FRET (E) is given by Eq. (1)[10]:

$$E = 1 - (F_{da}/F_d) \tag{1}$$

At a wavelength of 476 nm, ECFP (donor) exhibits a major emission peak, whereas EYFP.1 (acceptor) does not emit light at all; the fluorescence emission at 476 nm comes only from the donor (Fig. 5). Therefore F'_{da} and F''_{da} in Fig. 6 are the F_{da} values of the Ca^{2+}-saturated and Ca^{2+}-unsaturated forms, respectively. To obtain F_d, 10 μg of trypsin is added into the cuvette. Separate experiments have verified that the fluorescences of the donor and acceptor GFPs are totally resistant to trypsin. This procedure causes complete dissociation of ECFP and EYFP.1 at room temperature within 10 min. As the fluorescence of EYFP.1 decreases, that of ECFP is de-quenched to reach F_d. The obtained spectrum (Fig. 6, dotted line) is almost identical to that of ECFP only (Fig. 5), except for a tiny hump around 528 nm from EYFP.1 that is slightly cross-excited at 432 nm. Utilizing the F_d, F'_{da}, and F''_{da} values, the FRET efficiencies of YC2.1 at saturating versus 14 nM free Ca^{2+} are calculated to be 54 and 25%, respectively.

Quantifying Fluorescence Resonance Energy Transfer Efficiency

In the preceding discussion, the ratio of acceptor to donor emissions has been used as an index of the extent of FRET. This emission ratio is advantageous because it is the fluorescence parameter that is easiest to measure frequently and nondestructively while canceling out variations in cell thickness, excitation intensity, emission sensitivity, and overall indicator concentration. On the other hand, the observed emission ratio is still a function not only of the FRET efficiency but also the stoichiometry of acceptors relative to donors, the ratio of their quantum efficiencies, the relative efficiencies of the filters and detector(s) used to collect the emissions, the extent to which the excitation wavelength adventitiously excites the acceptor, and the extent to which the donor emission spills over into the acceptor emission band. Figure 5 shows that CFP can be selectively excited at 400–440 nm with little but not completely negligible direct excitation of YFP, but there is no wavelength at which YFP emission can be collected without including some direct emission from CFP. How do we disentangle the FRET efficiency or the extent of biochemical interaction from all these complicating factors? Several approaches have been employed in the literature, all of which have different advantages and restrictions.

Mathematical Correction

In principle the perturbing parameters listed above could be carefully measured with pure samples of donor and acceptor, preferably using the same instrument as used for the actual biological measurement, if possible in cells similar to those in which the FRET is to be measured. With three fluorescence measurements (excitation of donor band, measurement at donor and acceptor emission bands; excitation of acceptor, measurement of acceptor emission), the stoichiometry of acceptors to donors can in principle be deduced and then corrected for, allowing extraction of the FRET efficiency.[7] In our view, the problems are that the mathematical formalism is complex and not intuitive, that the result often requires subtraction of nearly equal quantities and may therefore be sensitive to small errors in the correction parameters, and that the separate donor and acceptor samples must have the same optical properties (except for FRET) as the interacting partners.

Internal Calibration

If the biological sample could be forced to two known states of high versus low protein association, faster than expression levels can change, the emissions under those reference conditions can be used to calibrate previous or subsequent measurements on that sample. For example, if Ca^{2+} can be clamped *in situ* to high and low levels, using ionophores and chelators, emission ratios measured from the cells before Ca^{2+} clamping can be retrospectively calibrated in terms of the Ca^{2+} saturation of the indicator. Furthermore, if the titration curve for the indicator has been characterized *in vitro,* and it is assumed that it is the same in intact cells, the absolute $[Ca^{2+}]$ concentrations may be calculated.[13,14] The problem is that for most novel protein interactions, it is not known how to turn the interaction on and off rapidly and completely *in situ.*

External Calibration

In the external calibration approach, the emission ratio associated with the condition in which FRET is suspected (e.g., BFP-X and GFP-Y, where X and Y may form a heterodimer) is compared with that from separate cells in which FRET is unlikely to occur. Examples of such negative controls include transfections with unfused donor and acceptor GFP mutants (e.g., BFP and GFP), or fusions to noninteracting partners (e.g., BFP-X and GFP-$Z,$ where X and Z do not interact), or fusions to the interacting partners but in the presence of a dominant-negative competitor such as excess unfused X or Y. A positive control such as BFP fused to GFP through an appropriate spacer may also be included for comparison.[26]

For example, to demonstrate a direct interaction between Bax and Bcl-2 proteins in mammalian cells, Mahajan *et al.*[27] coexpressed BFP–Bcl-2 and GFP–Bax fusion proteins in the same cells, and confirmed their colocalization within mitochondria. They then observed significant FRET from BFP to GFP when BFP–Bcl-2 was excited, fairly similar to that seen from a BFP–GFP fusion. In contrast, the green-to-blue emission ratio was much lower from a noninteracting pair of fusion proteins, cytochrome c–GFP and BFP–Bcl-2, coexpressed and observed in separate experiments. Similarly, Day[26] showed FRET between BFP fused to the transcription factor Pit-1 and the analogous GFP–Pit-1. Cotransfection of BFP–Pit-1 and GFP fused to a nuclear localization sequence was used as a negative control, while BFP fused to GFP through a three-amino acid linker was taken as a positive control.

The main problem with this general approach is that separate cells and transfections are being compared so that any variations in relative expression levels of the donor and acceptor constructs would produce artifactual variations in emission ratio, obscuring the effects of FRET. To go

[26] R. N. Day, *Mol. Endocrinol.* **12,** 1410 (1998).
[27] N. P. Mahajan, K. Linder, G. Berry, G. W. Gordon, R. Heim, and B. Herman, *Nature Biotechnol.* **16,** 547 (1998).

beyond a qualitative statement that FRET is occuring requires detailed mathematical corrections (see Mathematical Correction, above).

Donor Lifetime Measurement

One of the most elegant methods for measuring FRET is to measure the excited state lifetime of the donor. For a homogeneous population, the energy transfer efficiency E is given not only by Eq. (1) but also by Eq. (2):

$$E = 1 - (\tau_{DA}/\tau_{D}) \tag{2}$$

where τ_{DA} and τ_{D} are the excited state lifetimes of the donor in the presence and absence of the acceptor, respectively. τ_{DA} can be measured dynamically and nondestructively at every pixel of the image and compared with τ_{D} measured from reference cells containing only donor. Only donor emission is monitored; emissions from the acceptor are disregarded, and large excesses of unbound acceptor have no effect. Because these lifetimes are independent of protein expression levels per se and most other perturbations, they should be quantitatively comparable from sample to sample. Several successful examples of FRET involving GFP mutants have been published.[8,9,28] The main disadvantage of lifetime imaging is that the equipment is expensive and not yet available in a ready-made commercial system. A subsidiary problem is that the usual and most accurate ways to determine lifetimes are wasteful of photons, for two reasons. Photons coming from the acceptor are discarded, even though they are the dominant emission if FRET is efficient, because the quantum yield of the acceptor GFP or YFP considerably exceeds that of the donor BFP or CFP. Also, the classic and most accurate means to measure lifetimes require that only emissions passing through narrow temporal or frequency windows are collected; these windows are systematically scanned to map the decay characteristics, but cause considerable photon losses requiring longer observation to compensate.

Acceptor Bleaching

Equation (1) indicates that the efficiency of FRET is given by the extent of dequenching of the donor when the acceptor is removed. For example, FRET from Cy3 to Cy5 has been quantified by photobleaching the Cy5.[29] The same principle is useful for quantifying the efficiency of FRET from

[28] T. Ng, A. Squire, G. Hansra, F. Bornancin, C. Prevostel, A. Hanby, et al., Science 283, 2085 (1999).
[29] P. I. Bastiaens, I. V. Majoul, P. J. Verveer, H. D. Soling, and T. M. Jovin, EMBO J. 15, 4246 (1996).

CFP to YFP, because strong illumination at long wavelength (>500 nm) bleaches YFP without affecting CFP, providing an optical means to eliminate YFP *in situ* whenever desired. Such bleaching is applicable to any CFP and YFP fusions inside cells under the microscope, whereas the trypsin cleavage described above (see Spectroscopic Analysis of YC2.1 Protein) is only feasible *in vitro* and depends on the linker being much more vulnerable to cleavage than the GFP mutants. No specialized equipment is required, unlike excited state lifetime imaging. Although the same principle might be tried with BFP and GFP, the differential bleaching with that pair would be more difficult because BFP is more photolabile than CFP, whereas GFP is more photostable than YFP. A unique feature of YFP, not shared by the other mutants or organic dyes, is that the bleaching is at least partly reversible by illumination at UV wavelengths.[14,24] Individual molecules of YFP can be seen to undergo many cycles of bleaching and regeneration, but in a large population of molecules the reversibility is not complete (see results below), so photobleaching is best done at the end of the experiment. It then provides a retrospective absolute calibration of FRET efficiency for the more sensitive and nondestructive dual-emission ratioing performed during cell dynamics. The following sections describe experiments using YFP photobleaching to quantify FRET in several systems.

Split Yellow Cameleon-2.1

Split yellow cameleon-2.1 undergoes FRET between CFP and YFP, the efficiency of which is controllable by Ca^{2+}. It consists of an equimolar mixture of the two constituents of yellow cameleon-2.1: ECFP–CaM and M13–EYFP.1 (Fig. 7[29a]).[13] We first discuss how to measure the FRET efficiency when the components are saturated with Ca^{2+} to make a complex in a cuvette and in a solution under a microscope.

Preparation of Purified Proteins

Recombinant proteins (ECFP–CaM and M13–EYFP.1) were expressed using the T7 expression system [pRSETB (InVitrogen)/JM109(DE3)]. *Escherichia coli* cells are transformed with the plasmids and selected at 37° on LB plates containing ampicillin (100 μg/ml). A single colony is picked into 2 ml of LB medium containing ampicillin (100 μg/ml), and grown overnight at 37°. A 100-fold dilution is made, and the culture is grown at room tempcrature until it reaches a density of approximately OD_{600} 0.5, and then protein expression is induced with isopropy1-β-D-thiogalactopyra-

[29a] M. Kozak, *J. Cell Biol.* **108,** 229 (1989).

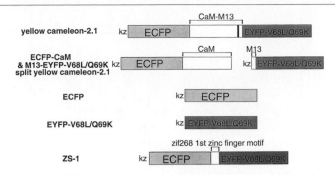

FIG. 7. Schematic structures of fusion proteins containing ECFP and/or EYFP.1 for mammalian expression. kz, Kozak sequence for optimal translational initiation in mammalian cells.[29a]

noside at a final concentration of 0.2 mM. Growth is continued for a further 12–24 hr. The cells are lysed with a French press, and the polyhistidine-tagged proteins are purified from the cleared lysates on nickel-chelate columns (Qiagen). The protein samples in the eluates are concentrated with a Centricon-30 (Amicon), and are further purified by gel-filtration or ion-exchange chromatography (Mono Q).

In Vitro Spectra of Split Yellow Cameleon-2.1

Emission spectra of the purified proteins in a cuvette are measured with a fluorometer (SPEX Industries). ECFP–CaM and M13–EYFP.1 are mixed at 1 : 1 stoichiometry (5 μM each) in a buffer [100 mM HEPES–KOH (pH 7.4), 50 mM KCl, 0.1 mM EGTA], and the two proteins dissociate completely (our unpublished results), resulting in no FRET (Fig. 8, left). The emission spectrum (excited at 432 nm) is almost identical to that of ECFP only, except for a small hump around 528 nm, which is from EYFP.1 cross-excited slightly at 432 nm. When Ca^{2+} is added to the solution at a final concentration of 1 mM, the 528 nm : 476 nm emission ratio increases by about fourfold (Fig. 8, left). From Eq. (1) and the donor intensities at 476 nm with and without Ca^{2+}, an energy transfer efficiency in the Ca^{2+}-saturated state is calculated to be 0.43, assuming that the Ca^{2+}-free form has zero interaction and energy transfer efficiency.

Imaging Split Yellow Cameleon-2.1 in Solution under Microscope

Bleaching of YFP is readily achieved under a microscope with a standard 150-W xenon lamp and a ×40 objective lens with high numerical aperture, such as 1.2 or greater. Proteins in solutions containing 50 mM HEPES–

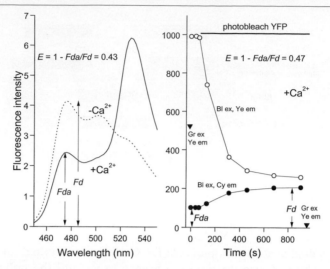

FIG. 8. FRET efficiency of split cameleon-2.1 saturated with Ca^{2+} *in vitro. Left:* Emission spectra of split yellow cameleon-2.1 in a cuvette, excited at 432 nm, at zero (dotted line) and saturating (solid line) Ca^{2+}. Fluorescence intensities at a wavelength of 476 nm are designated as F_d and F_{da}, respectively. *Right:* Fluorescence intensities through the FRET channel (535DF25, open circles labeled "Bl ex, Ye em"), donor channel (480DF30, closed circles labeled "Bl ex, Cy em"), and direct excitation of the acceptor (closed triangles labeled "Gr ex, Ye em") from a microscopic droplet containing split yellow cameleon-2.1 saturated with Ca^{2+} under light mineral oil. Intense illumination with a 540DF23 filter is indicated by the bar "photobleach YFP." The signals of the donor channel before and after the illumination are designated as F_{da} and F_d, respectively.

KOH (pH 7.4) and 50 mM KCl are injected through glass pipettes into light mineral oil on a coverslip. The droplets of the solutions are 50–150 μm in diameter, so that they are completely confined within a field. Imaging is performed on a Zeiss Axiovert microscope with a cooled CCD camera (Photometrics, Tucson, AZ), controlled by MetaFluor 2.75 software (Universal Imaging, West Chester, PA).

First, ECFP–CaM and M13–EYFP.1 are separately injected into light mineral oil. They are imaged with blue and green excitations attenuated through neutral density filters of 1–4% transmission. Photobleaching is achieved by irradiating without neutral density filters through 535DF25 or 540DF23 interference filters from Omega Optical or Chroma Technologies (both Brattleboro, VT); the first number gives the center of the passband while the second number is the full width at half-maximal transmission, e.g., 540DF23 should pass 528.5–551.5 nm. These wavelengths are chosen to be on the long-wavelength edge of the YFP absorbance band in order to minimize any chance of photobleaching CFP. After the completion of the work described below, we found that the above described passbands are unnecessarily conservative, in that a 525DF40 filter passing 505–545 nm bleaches YFP much faster but still avoids any damage to CFP (unpublished results of C. Y. Cho and R. Y. Tsien). YFP is regenerated with UV (330WB80: nominally 290–370 nm, but wavelengths below 330 nm are blocked by the glass objective). Although it is difficult to measure the power emerging from the objective in a microscopic beam of > 1.0 NA, we did measure the power entering the objective as 16 and 3 mW for the 540DF23 and 330WB80 filters, respectively. Figure 9 (top) shows that EYFP.1, for example within M13–EYFP.1, can be completely bleached with ~540 nm under conditions that have no effect on ECFP, for example within ECFP–CaM. The complete loss of long-wavelength EYFP.1 absorbance at this stage is also confirmed in another experiment, in which the protein in a cuvette is bleached by a laser at 532 nm, followed by measurement of absorbance. UV illumination causes partial recovery of the photobleached EYFP.1 ($21 \pm 2\%$ of the original fluorescence intensity, $n = 3$).

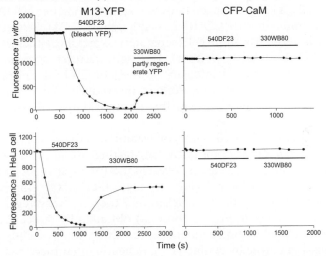

FIG. 9. Selective photobleaching of YFP *in vitro* (*top*) and in cells (*bottom*). M13–EYFP.1 in a droplet under mineral oil (*top left*) or expressed in HeLa cells (*bottom left*) was exposed to intense irradiation through 540DF23 and 330WB80 filters. The fluorescence intensity was monitored with a 495DF10 excitation filter, a 505DCLP dichroic mirror, and a 535DF25 emission filter. Likewise, ECFP–CaM under mineral oil (*top right*) and in HeLa cells (*bottom right*) was exposed to the same irradiation while monitoring its fluorescence with a 440DF20 excitation filter, a 455DRLP dichroic mirror, and a 480DF30 emission filter.

Figure 8 (right) shows a first test of the photobleach method to quantify FRET. A droplet of a 1 : 1 mixture of ECFP–CaM and M13–EYFP.1 in saturating^{2+}, under mineral oil, is monitored by conventional emission ratioing using a 440DF20 excitation filter, a 455 DRLP dichroic mirror, and two emission filters (donor channel, 480DF30 for ECFP; FRET channel, 535DF25 for EYFP.1) alternated by a filter changer (Lambda 10-2; Sutter Instruments, San Rafael, CA). Photobleaching is performed as before; its completeness is verified by directly exciting the EYFP.1 via a 495DF10 excitation filter and a 505DCLP dichroic mirror, and monitoring through a 535DF25 emission filter. Bleaching reduces but does not eliminate the signal in the FRET channel; the residual emission is the cross-talk of the long-wavelength tail of ECFP into the yellow channel. Meanwhile the EYFP.1 bleaching dequenches the ECFP emission from F_{da} to F_d. The increase was 47% of F_d, corresponding to an energy transfer efficiency adequately close to that obtained in Fig. 8 (left) by comparing high versus low Ca^{2+}. But only the photobleach method is generalizable to live cells and to any pair of fusions to CFP and YFP.

Imaging Split Yellow Cameleon-2.1 within Cells under Microscope

HeLa cells at 22° are imaged between 2 and 5 days after cDNA transfection with Lipofectin (GIBCO-BRL, Gaithersburg, MD). Selective bleaching of YFP is verified in HeLa cells separately expressing either ECFP–CaM or M13–EYFP.1 (Fig. 9, bottom). UV light restores a greater fraction ($42 \pm 6\%$ of the original intensity, $n = 8$) of EYFP.1 intensity in live cells than was possible *in vitro*. Perhaps the reducing environment of the cytoplasm partly inhibited irreversible photooxidative destruction of the EYFP.1.

A supramaximal dose (0.1 mM) of histamine in the presence of extracellular Ca^{2+} (1.3 mM) evokes a fairly long-lasting rise in cytosolic [Ca^{2+}] in HeLa cells, leading to as usual increase in the FRET channel and decrease in the donor ECFP channel (Fig. 10, top). During the stimulation, the EYFP.1 is deliberately photobleached and then partially regenerated, as shown by the images in Fig. 10 (bottom) obtained by direct excitation of the EYFP.1. The bleaching dequenches the ECFP to F_d, slightly above the donor emission before histamine (F_{da}), indicating that there has been only 5.5% energy transfer efficiency in the unstimulated cells, consistent with a low but nonzero resting [Ca^{2+}]. The minimum ECFP emission at high [Ca^{2+}], F'_{da} corresponds to 42% energy transfer. Thus a single photobleach at the end of the experiment can retrospectively calibrate the energy transfer efficiency at all previous time points, assuming no bleaching or redistribution of the ECFP.

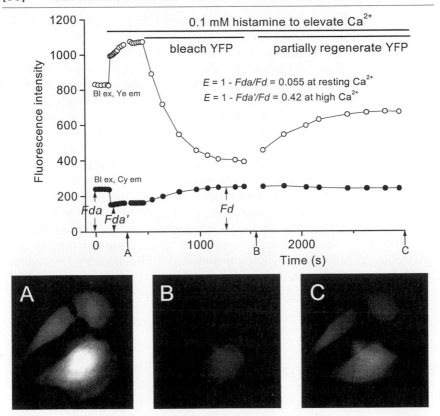

FIG. 10. Quantitation of FRET efficiency of split yellow cameleon-2.1 in a HeLa cell by the photobleaching method. *Top:* Emission intensities through the FRET channel (open circles) and donor channel (closed circles), both excited with a 440DF20 filter. The application of histamine and intense irradiation with 540DF23 and 330WB80 filters are indicated by horizontal bars. Fluorescence intensities of donor channel before (F_{da}) and after (F'_{da}) the addition of histamine, and after the 540DF23 illumination (F_d) are shown by arrows. *Bottom:* Fluorescence images of the M13-EYFP.1 taken with a 495DF10 excitation filter, a 505DCLP dichroic mirror, and a 535DF25 emission filter at times A, B, and C indicated in the top graph. The pictures were printed with a fairly low-contrast printer that exaggerates the residual brightness at time B.

Photobleach Measurements of Fluorescence Resonance Energy Transfer in Yellow Cameleon-2.1 and between Unfused Cyan Fluorescent Protein and Yellow Fluorescent Protein

In yellow cameleon-2.1, the donor (ECFP) and the acceptor (EYFP.1) are joined by CaM and M13. Although the donor and acceptor can be separated *in vitro* by trypsin (Fig. 6), trypsin cannot be applied within cells. However, photobleaching works in intact cells (Fig. 11) just as with the split cameleon. Application of 0.1 mM histamine followed by a 10 μM concentration of an antagonist, cyproheptadine, produces a Ca^{2+} transient indicated by the reciprocal changes in the two emission intensities. After recovery of basal $[Ca^{2+}]$, the EYFP.1 is photobleached. The efficiency of FRET at the peak of Ca^{2+} transient (Fig. 11, F_{da}) is 43%, similar to that for split yellow cameleon-2.1 (42%). The efficiency of FRET after recovery from histamine (Fig. 6, F'_{da}) is 30%, indicating significant FRET even at basal Ca^{2+}, limited by the length of the CaM–M13 tether between the ECFP and EYFP.1 and any tendency for the two GFP mutants to dimerize. The 30 and 43% values for energy transfer efficiency are reasonably consistent with the trypsin-based measurements (Fig. 6) of 25 and 54% *in vitro*, considering that cytosolic Ca^{2+} does not traverse the full range from zero to saturating. Also in reasonable agreement, excited state lifetime imaging has yielded an independent measurement of 25% FRET for YC2 in unstimulated cowpea protoplasts.[9]

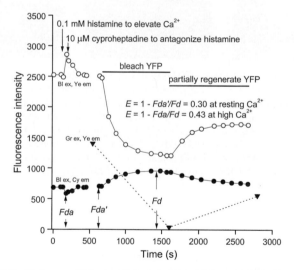

FIG. 11. FRET efficiency of intact yellow cameleon-2.1 in a HeLa cell measured by the photobleaching method as in Fig. 10. The fluorescence intensities of the donor channel during the Ca^{2+} transient (F_{da}), after recovery from histamine (F'_{da}), and after the 540DF23 illumination (F_d) are shown by arrows. Fluorescence of directly excited EYFP.1 is indicated by filled triangles.

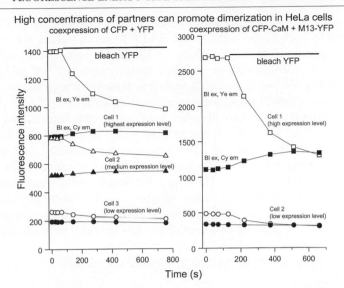

FIG. 12. Protein concentration-dependent FRET in unstimulated HeLa cells expressing ECFP and EYFP.1. *Left:* Fluorescence intensities through FRET and donor channels are shown by open and closed symbols, respectively, for three cells expressing estimated concentrations of 300 μM (squares), 200 μM (triangles), and 80 μM (circles) of each of the two proteins. *Right:* Similar results for two cells coexpressing ECFP–CaM and EYFP.1 at high and low levels.

Detection of Green Fluorescent Protein Dimerization inside Cells

Wild-type GFP undergoes changes in its absorption spectrum as a function of protein concentration,[30] implying some form of aggregation via hydrophilic and hydrophobic interactions, which are visible in some but not all GFP crystal forms.[1,31] Any intrinsic affinity of CFP and YFP for each other might also promote interactions between the fusion proteins containing the donor and acceptor GFPs, leading to artifactual interactions. To test whether ECFP and EYFP.1 interact at high concentrations *in vivo,* we expressed the two proteins in HeLa cells with identical promoters (Fig. 7) to achieve roughly equivalent expression levels. Concentrations of proteins in cells were estimated by knowing the thickness of cells and comparing their fluorescence brightness with those of known concentrations of fluorescent proteins in a wedge-shaped microchamber as previously described.[14] Figure 12 (left) shows data from three neighboring cells expressing different

[30] W. W. Ward, H. J. Prentice, A. F. Roth, C. W. Cody, and S. C. Reeves, *Photochem. Photobiol.* **35,** 803 (1982).
[31] F. Yang, L. G. Moss, and G. N. Phillips, Jr., *Nature Biotechnol.* **14,** 1246 (1996).

levels of each of the two proteins; the molar ratio of ECFP to EYFP.1 was almost one in each of the cells. When EYFP.1 was bleached, we observed slight but reproducible dequenching of ECFP fluorescence in cells containing high concentrations of proteins (cell a, squares, 300 μM; cell b, triangles, 200 μM), but not in cells with low concentrations (cell c, circles, 80 μM). Such concentration-dependent dimerization of GFP was also detected in unstimulated HeLa cells when split yellow cameleon-2.1 was highly concentrated (Fig. 12, right). It seems reasonable to assume that FRET in an CFP–YFP dimer, where the two proteins are touching one another, would be as strong as in Ca^{2+}-saturated cameleons or other covalent fusions incorporating CFP and YFP. Therefore the small FRET values obtained imply that dimerization is detectable but far from complete at the highest expression levels, as if the effective dissociation constant were in the millimolar range. A dissociation constant of 100 μM for GFP homodimerization *in vitro* was estimated by analytical ultracentrifugation,[32] but large amounts of cytosolic proteins might well be somewhat competitive.

Zinc Sensor Protein Analogous to Cameleons

Can the principle of the cameleons be adapted to measure analytes other than Ca^{2+}? If the donor and acceptor are linked with just a CaM-binding peptide, then intermolecular binding of unlabeled $(Ca^{2+})_4$–CaM to the central peptide disrupts FRET.[33] This system has been used to determine free $(Ca^{2+})_4$–CaM concentrations in living cells.[34] To explore ions separate from Ca^{2+} signaling, we made a zinc sensor protein (ZS-1) by fusing ECFP and EYFP.1 with a linker of 39 amino acids corresponding to a zinc finger motif from a mouse transcription factor, zif268 (Fig. 7). The cDNA for ZS-1 was created by replacing the *Sph*1–*Sac*1 fragment for CaM–M13 in yellow cameleon-2.1 with a fragment encoding the first zinc finger motif of mouse zif268 (from D. Saffen, University of Tokyo, Japan). The amino acid sequence of the introduced motif was MHERPYACPVESCDRRFSRSDELTRHIRIHTGQKEL, in which amino acids conserved in this family of zinc fingers have been underlined. Figure 13 (left) shows the spectra of ZS-1 in a cuvette. FRET was considerable even in the absence of Zn^{2+}; the emission ratio (535 : 480 nm) was 3.5, comparable to the ratio when the two GFP mutants were linked with floppy spacers.[12] In the presence of Zn^{2+} (0.1 mM), the ratio increased further to reach 6.7. The emission ratio change of the zinc sensor protein (ZS-1) was

[32] G. N. Phillips, *Curr. Opin. Struct. Biol.* **7,** 821 (1999).
[33] V. A. Romoser, P. M. Hinkle, and A. Persechini, *J. Biol. Chem.* **272,** 13270 (1997).
[34] A. Persechini and B. Cronk, *J. Biol. Chem.* **274,** 6827 (1999).

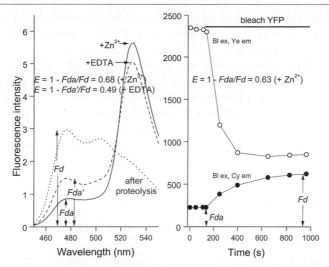

FIG. 13. Emission spectra and FRET efficiencies of a zinc sensor protein ZS-1 *in vitro*. *Left:* Emission spectra of ZS-1 in a cuvette, excited at 432 nm in 0.1 mM ZnCl$_2$ (solid line), 0.3 mM EDTA (dashed line), and after proteolysis (dotted line). Fluorescence intensities at 476 nm when the protein was saturated with Zn^{2+} (F_{da}) and proteolyzed (F_d) are indicated. *Right:* Fluorescence intensities through FRET channel (open circles) and donor channel (closed circles) from a solution containing ZS-1 saturated with 0.2 mM ZnCl$_2$ in mineral oil. The signals of the donor channel before and after photobleaching the acceptor are F_{da} and F_d, respectively.

Zn^{2+} specific and reversible. After proteolysis by trypsin, the ratio went down to 0.65. Assuming no FRET after proteolysis, the efficiency of FRET was 49% in Zn^{2+}-free EDTA versus 68% when Zn^{2+} saturated. The latter value was checked by photobleaching the EYFP in a droplet under mineral oil on the microscope stage. The donor emission from ZS-1 with 0.1 mM ZnCl$_2$ was dequenched by 63% (Fig. 13, right), in adequate agreement with the estimate obtained by cleaving the linker. As far as we know, ZS-1 with Zn^{2+} has the highest efficiency of FRET for any CFP–YFP chimera so far. The transition dipoles of ECFP and EYFP.1 are presumably positioned to be even more favorable for FRET when the Zn^{2+} binding increases the ordering of the zinc finger motif.

Conclusions

FRET between GFP mutants fused to host proteins has great potential as a general method for imaging dynamic changes in ligand concentrations and the conformations and interactions of those host proteins.[35] Because a great deal of trial-and-error may be needed to optimize the spectral response, FRET readouts are not yet well suited to screening random libraries to find hitherto unknown partners. Instead, FRET can provide noninvasive monitoring with high spatial and temporal resolution of protein–protein interactions and ligand-induced conformational changes that are already known to be important and reasonably stoichiometric. Such measurements can be made much more quantitative by photobleaching the FRET acceptor, preferably a YFP, at the end of the experiment, because this easy step provides a reliable internal calibration of the FRET efficiency.

Acknowledgments

This work was supported by the Howard Hughes Medical Institute, National Institutes of Health (NS27177 to R.Y.T.), and a Human Frontiers Science Program long-term fellowship to A.M.

[35] R. Y. Tsien and A. Miyawaki, *Science* **280,** 1954 (1998).

[36] Studies of Signal Transduction Events Using Chimeras To Green Fluorescent Protein

By TOBIAS MEYER and ELENA OANCEA

Introduction

Specificity in signal transduction often requires that signaling events occur at a precise time and place. While second messengers can transduce signals from one cellular site to another by passive diffusion or self-propagation,[1] many signaling steps are now thought to be mediated by the translocation of signaling proteins to and from local sites of action. This model suggests that specificity often results from the binding of signaling proteins to adapter proteins,[2,3] cytoskeletal[4] and membrane components,[5] as well as other targets. In most cases, translocation is thought to require a diffusion

[1] T. Meyer, *Cell* **64,** 675 (1991).
[2] C. S. Zuker and R. Ranganathan, *Science* **283,** 650 (1999)
[3] M. Colledge and J. D. Scott, *Trends Cell Biol.* **9,** 216 (1999).
[4] K. Shen, M. F. Teruel, K. Subramanian, and T. Meyer, *Neuron* **21,** 593 (1998).
[5] S. J. Leevers, H. F. Paterson, and C. J. Marshall, *Nature (London)* **369,** 411 (1994).

step and an increase in the local binding affinity. Such changes in local binding affinity can be mediated by the phosphorylation of signaling proteins,[6-8] the production of soluble and lipid second messengers,[9-16] or changes in the availability of binding partners. Regulated localization and translocation processes cannot readily be studied *in vitro* or by using immunolocalization methods. This makes it necessary to develop methods for monitoring the localization and translocation of signaling proteins in the intact cellular environment.

This chapter assumes that the reader is familiar with the expression of green fluorescent protein (GFP) constructs and with different fluorescence imaging methods. Those methods have been described in several previous reviews (*Methods in Cell Biology* volume 58 focused on green fluorescent protein and fluorescence imaging techniques[17]). Here we focus on the more specific question of how GFP-tagged signaling proteins and GFP-tagged signaling domains can be used to understand signal transduction processes. First, we compare the usefulness of GFP fusion proteins versus immunofluorescence approaches for investigating protein localization. We then discuss how GFP-tagged signaling proteins can be used to explore the translocation of signaling proteins and evaluate when GFP-tagged individual signaling domains are useful to understand specific signaling events such as phosphorylation or the production of lipid second messengers. Furthermore, we discuss how GFP-tagged mutant signaling proteins and domains can be used to understand the activation steps of particular signaling proteins. Finally, we discuss the molecular construction of GFP-tagged full-length signaling proteins and GFP-tagged signaling domains.

[6] T. Stauffer and T. Meyer, *J. Cell Biol.* **139,** 1447 (1997).

[7] K. Shen and T. Meyer, *Science* **284,** 162 (1999).

[8] L. S. Barak, S. S. Ferguson, J. Zhang, and M. G. Caron, *J. Biol. Chem.* **272,** 27497 (1997)

[9] E. Oancea, M. N. Teruel, A. Quest, and T. Meyer, *J. Cell Biol.* **140,** 485 (1998).

[10] T. Stauffer and T. Meyer, *Curr. Biol.* **8,** 343 (1998).

[11] C. D. Kontos, T. Stauffer, W. P. Yang, L. Haung, M. A. Blanar, T. Meyer, and K. G. Peters, *Mol. Cell Biol.* **18,** 4131 (1998).

[12] K. Hirose, S. Kadowaki, M. Tanabe, H. Takeshima, and M. Iino, *Science* **284,** 1527 (1999).

[13] A. Schurmann, M. Schmidt, M. Asmus, S. Bayer, F. Fliegert, S. Koling, S. Massmann, C. Schilf, M. C. Subauste, M. Voss, K. H. Jakobs, and H. G. Joost, *J. Biol. Chem.* **274,** 9744 (1999).

[14] K. Venkateswarlu, F. Gunn-Moore, J. M. Tavare, and P. J. Cullen, *J. Cell Sci.* **112,** 1957 (1999).

[15] P. J. Lockyer, S. Wennstrom, S. Kupzig, K. Venkateswarlu, J. Downward, and P. J. Cullen, *Curr. Biol.* **9,** 265 (1999).

[16] E. Oancea and T. Meyer, *Cell* **95,** 307 (1998).

[17] K. F. Sullivan and S. A. Kay (eds.), *Methods Cell Biol.* **58,** 1 (1998).

Measurement Procedures

Comparison between Green Fluorescent Protein and Immunofluorescence
Approaches for Investigating Localization of Signaling Proteins

Although many important findings about the intracellular distribution of signaling proteins can be obtained by using immunofluorescence studies, GFP-tagged signaling proteins offer an opportunity to visualize the localization and translocation of proteins in living cells in a dynamic manner. The main problem for both imaging approaches is that fluorescence microscopy is limited by the resolution of light, which allows the separation of only those cellular structures that are approximately 0.2 to 0.5 μm apart. Thus, structures that are 200 nm (or less) apart cannot be resolved either by immunofluorescence or GFP fusion protein methods.

Several important differences exist between the use of immunofluorescence or GFP fusion proteins for understanding the localization of signaling proteins.

1. The GFP approach can be used to visualize the localization of proteins in living cells while immunofluorescence requires that cells be fixed before fluorescence images are taken. Although fixation can introduce artifacts in the distribution of signaling proteins, different fixation procedures have the advantage of "washing out" soluble or weakly bound signaling proteins as well as signaling proteins that are bound to membranes and other structures. This "washout" often allows visualization of the distribution of tightly bound signaling proteins much more clearly in immunofluorescence studies compared with GFP distribution studies in living cells. Comparison between the cellular distribution of a particular protein by a GFP versus immunofluorescence approach can be useful to determine the fraction of tightly bound protein. Because GFP can be measured in fixed cells, GFP fusion proteins can also be used to measure directly the fraction of protein that binds to structures that are retained intact in the fixation protocol.

2. The use of GFP-tagged proteins eliminates the need for antibodies against a protein of interest and at the same time eliminates potential problems with antibody specificity. However, the GFP tag, which is a 27-kDa protein, may also interfere with the localization of signaling when used as a tag. Both C- and N-terminal GFP tags should be tested to reduce the possibility that the GFP tag interferes with the correct localization of the protein of interest.

3. Signaling proteins are often expressed at low copy numbers and their intracellular docking sites might also be limited. If the expression of a GFP-tagged protein exceeds the available docking sites, a distribution different

from that of the endogenous protein will be observed. This is not a problem in immunofluorescence studies, which typically measures the distribution of the native protein.

4. While individual GFP molecules can be monitored *in vitro,* autofluorescence of cells imposes a severe limit on single-molecule GFP measurements in living cells. For example, the background from mitochondria and other structures can correspond to an estimated 10 to 100 local GFP molecules (see Ref. 17, p. 31 for the calibration method used for this estimate). Thus, in order to become visible in a living cell, more than 10 GFP molecules must typically be present at an individual spot (the size of this spot is defined by the light resolution of the microscope). This limits the GFP approach to signaling proteins that have high local concentrations of docking sites and bound GFP fusion proteins. In contrast, when using immunofluorescence, this problem is circumvented by the use of secondary antibodies, which makes it possible to have multiple fluorophores per antigen. Also, the background fluorescence can be significantly reduced by the fixation process.

In summary, this suggests that immunofluorescence remains a better suited method to determine the subcellular localization(s) of signaling protein if high specificity antibodies are available, if no dynamic measurements are needed, and if the binding sites or structures are preserved during fixation. An advantage of immunofluorescence is that the distribution of native proteins can be measured, the fluorescence signal per signaling protein is higher, and subcellular structures are more clearly apparent due to a washout in the fixation process and lower autofluorescence. In contrast, GFP-tagged signaling proteins have the advantage of showing the distribution of signaling proteins in living cells. While the GFP approach is less sensitive for proteins present at low concentrations, it also eliminates possible artifacts due to fixation and antibody specificity.

Using Green Fluorescent Protein-Tagged Signaling Proteins or Protein Domains as Functional Fluorescent Indicators for Signaling Processes

The main advantage of using GFP-tagged signaling proteins is probably not to study their intracellular localization but to understand dynamic translocation events and to obtain cellular readouts for intermediate signal transduction steps. Translocation of signaling proteins to or from the plasma membrane,[18–22] nucleus,[23–25] and specialized structures such as caveolae[6]

[18] N. Sakai, K. Sasaki, N. Ikegaki, Y. Shirai, Y. Ono, and N. J. Saito, *J. Cell Biol.* **139,** 1465 (1997).

[19] J. Sloan-Lancaster, J. Presley, J. Ellenberg, T. Yamazaki, J. Lippincott-Schwartz, and L. E. Samelson, *J. Cell Biol.* **143,** 613 (1998).

or postsynaptic densities[7] have been studied by GFP approaches in our and other laboratories. By visualizing such translocation events and by measuring time courses of translocation, it is possible to understand some of the basic principles that govern cellular signal transduction pathways. In many cases, signaling protein translocation can be used as a cellular readout for the particular signaling step and pharmacological and other perturbations can be effectively studied.

What are the limitations for monitoring the translocation of GFP-tagged signaling proteins? Some of the same problems apply for this type of translocation measurements as discussed above for the localization of signaling proteins. Some of the more specific concerns and solutions are as follows.

1. In terms of microscopy methods, most fluorescence measurements of GFP-tagged intracellular signaling proteins require that transfected cells be grown on coated coverslips. Alternatively, cells can be transfected after being plated onto the coverslips. Depending on the cell type and particular cellular distribution of the signaling protein, conventional or confocal fluorescence imaging can be the appropriate methods to analyze the distribution of signaling proteins. Confocal microscopy is typically of minimal help for flat cells or thin cell processes. For practical considerations, both imaging methods should be tried (also, consider that many available microscopes are not optimally aligned or equipped with optimal GFP filters). Because the signal-to-noise ratio is always a problem for uses of GFP in signal transduction, light intensities of the laser or lamp and the duration of exposure should be chosen to match as closely as possible the bleach limit. We found that a final bleached fraction of ~10–30% of the cellular fluorescence during the time course of an experiment is a good compromise for optimal image quality.

2. In many cases, the fraction of translocating versus stationary signaling proteins is relatively small and quantitative image analysis tools must be used to measure translocation. For example, ratios of cytosolic versus nu-

[20] F. D. Brown, N. Thompson, K. M. Saqib, J. M. Clark, D. Powner, N. T. Thompson, R. Solari, and M. J. Wakelam, *Curr. Biol.* **8,** 835 (1998).

[21] H. Yokoe and T. Meyer, *Nature Biotechnol.* **14,** 1252 (1996).

[22] C. A. Parent, B. J. Blacklock, W. M. Froehlich, D. B. Murphy, and P. N. Devreotes, *Cell* **95,** 81 (1998).

[23] M. Koster and H. Hauser, *Eur. J. Biochem.* **260,** 137 (1999).

[24] H. Htun, J. Barsony, I. Renyi, D. L. Gould, and G. L. Hager, *Proc. Natl. Acad. Sci. U.S.A.* **93,** 4845 (1996).

[25] V. Georget, J. M. Lobaccaro, B. Terouanne, P. Mangeat, J. C. Nicolas, and C. Sultan, *Mol. Cell. Endocrinol.* **129,** 17 (1997).

clear or plasma membrane versus cytosolic fluorescence intensities can be graphed as a function of time. Such measurements of distribution ratios as a function of time can be made with series of images and software that measure the fluorescence intensity in both regions of interest as a function of time (using free software such as NIH Image or commercial software such as Metamorph, Universal Imaging Corp.). How large an amplitude is needed to reliably study translocation? As a guideline, a greater than 10% enrichment in the fluorescence in the compartment should occur for a maximal receptor stimulus.

For the example of plasma membrane translocation, a direct way to determine the time course of translocation is to take a series of images through the midsection of cells and to simultaneously measure the relative fluorescence intensity in a region of the plasma membrane versus the cytosol. The relative change in the plasma membrane fluorescence can then be calculated either as $(I_{PM} - I_{cyt})/I_0$, as $(I_{PM} - I_{cyt})/(I_{PM} + I_{cyt})$, or as $(I_{PM} - I_{cyt})/I_{cyt}$, where I_{PM} is the average fluorescence intensity of the plasma membrane region in which the fluorescence is measured, I_{cyt} is the average cytosolic fluorescence, and I_0 is the initial total average fluorescence intensity. Such an approach assumes that both the cytosolic and plasma membrane fluorescence are uniform and remain uniform during the translocation process. The ratio calculated above for each of the sequential images can then be represented as a function of time. An alternative analysis method can be used if the confocal microscope is focused onto the cell surface at the coverslip. The fluorescence intensity change in the surface can then be used to measure plasma membrane translocation or dissociation.

A problem for both methods is cell movement, which is often observed after receptor stimulation. Many cell types flatten or ruffle on receptor activation, in which case the contour or membrane density of the cell changes. In such cases, an analysis can be used in which line profiles across a cell are plotted as a function of time and the amplitude difference between plasma membrane and cytosol can be used for a translocation graph. As an example of such a line profile analysis, Fig. 1 shows images and line plots for the plasma membrane translocation of the C1 domain of the γ isoform of protein kinase C (PKCγ) at different time points. The time course shown is an average of 12 cells. Having a series of line profiles of the cell, the same ratio as described above can be calculated. The calculated ratio for each time point can then be represented as a function of time. Three different parameters can be determined from such a graph: the amplitude, the delay, and the half-maximal time for translocation.

If the time course of translocation is monitored for a long enough period of time, it can also give information about the reversibility of the process. Most of the proteins that translocate from cytosol to the plasma membrane

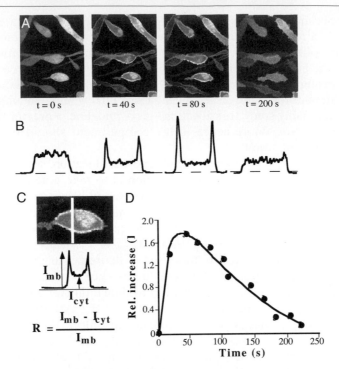

FIG. 1. Line profile analysis of plasma membrane translocation of C1–GFP in response to receptor activation. (A) Sequential images of RBL cells expressing C1–GFP taken immediately before, and 40, 80, and 200 sec after the stimulation of PAF receptors. The images shown were not corrected for photobleaching. (B) For each cell in a given image, a line intensity profile across the cell was obtained. Typical intensity profiles are shown at each of the four time points. (C) Schematic representation of the method used to calculate a relative increase in plasma membrane staining. A relative increase in plasma membrane localization was calculated from the plasma membrane (I_{PM}) and the average cytosolic fluorescence intensity (I_{cyt}), respectively. (D) The plasma membrane translocation was represented as a relative increase in plasma membrane localization (R) and plotted as a function of time. The PAF ligand was added at $t = 0$ sec. The resulting curve represents the time course of plasma membrane translocation of C1–GFP in response to PAF receptor activation. [Modified from E. Oancea and T. Meyer, *Cell* **95,** 307 (1998).]

do so reversibly. The reverse translocation in the presence of the receptor stimulus in the extracellular medium usually reflects the desensitization of the receptor or a negative feedback on the translocation of the protein of interest.

3. Cell-to-cell differences in translocation events are often significant and are an expression of the variability in the composition or state of cultured cells (e.g., the number of days in culture, the composition of the culture medium, number of passages, and the density of cells can introduce

marked differences). Experiments must therefore be repeated multiple times under defined culture conditions in order to obtain the appropriate signal-to-noise ratio.

In many cases, docking partners of the GFP-tagged protein must be overexpressed in the same cell in order to resolve the translocation event. This enables the GFP-tagged protein to have a similar concentration as its binding partners. Coexpression also improves the signal-to-noise ratio because a larger fraction of the fluorescent protein translocates to the docking site.

4. Depending on the cell type and particular cellular distribution of the signaling protein, conventional or confocal fluorescence imaging approaches should be used for an optimal analysis of the protein distribution. Because the signal-to-noise ratio is always a limitation for uses of GFP in translocation studies, light intensities from the laser or lamp should be kept high (~10–30% bleaching by the end of a translocation experiment is again a good compromise). For dynamic measurements, the signal in an image can in many cases be optimized by reducing the total number of images taken and by increasing the size of the pinhole (in a confocal microscope).

5. Signaling proteins are often modular, being composed of independent functional domains. The translocation and activation process for such proteins will therefore consist of a number of sequential steps in which each particular domain will perform its function. GFP has been used to measure the translocation of signaling proteins in response to receptor activation. Besides different isoforms of PKC, which have been shown to translocate to the plasma membrane,[16] PLD1, β-arrestin, PLA2, ZAP-70, RAF, Akt, and others were shown to undergo similar translocation. We found that in many cases, the signal-to-noise ratio and sensitivity of translocation events are much higher for minimal protein domains when compared with the full-length protein. This is in part due to competing alternative binding interactions by other parts of a protein and also to self-inhibition of binding domains by internal protein-binding interactions (i.e., SH2, C1, or PH domains).

In the case of the C2 domain, the translocation to the plasma membrane is triggered by the binding of calcium and negatively charged lipids to the C2 domain. For SH2 domains, the recruitment occurs in response to plasma membrane tyrosine phosphorylation events, and for the C1 domain from PKCγ, diacylglycerol (DAG) generated in the plasma and nuclear membrane can be detected by the induced translocation of the C1 domain. Similar to the C1 domain, PH domains tagged with GFP can be used as fluorescent indicators for the phosphoinositol lipids they bind to (e.g., phosphatidylinositol 4,5-bisphosphate or 3,4,5-trisphosphate).

For the protein domains or proteins that bind particular second messengers or lipids, GFP tagging of these domains can generate fluorescent translocation indicators for the particular messengers or signaling event. Such an indicator has the advantage that it will detect physiologically relevant signals in living cells. The accumulation of second messengers or phosphorylated residues can be determined by measuring the time course of translocation for the protein domain. One disadvantage of such indicators is the fact that they can often act as dominant negative inhibitors, and that a fluorescent indicator consisting of a GFP-labeled protein or domain will occupy some of its interaction partner. This is a problem if the downstream effects of a particular event are to be investigated. However, if the respective messenger is generated at high enough concentration, the concentration of messenger buffered by the fluorescent indicator can be relatively small. To address this problem, the indicator can be made more fluorescent so that fewer indicator molecules are needed to detect the signal. Tandem or tetrameric GFP tags can be made for that purpose.

6. Fluorescence recovery after photobleaching (FRAP) measurements are often useful to (1) determine which fraction of signaling proteins is mobile and which fraction is immobile, (2) measure the dissociation time constant of signaling proteins from an immobilized structure such as post synaptic density (PSD) or plasma membrane microdomains, or (3) measure the average diffusion coefficient of a signaling protein. In these measurements, a small region is bleached by focusing a laser onto a particular site or by scanning a small region of the cell with high laser power (using a confocal laser). If the mobility or dissociation time is fast, laser power above 10 mW and rapid imaging must be used. A variation of this method is to increase the visible fluorescence of some GFP variants by UV-laser pulses (up to threefold) and to track the fluorescence within the cell as a function of time.[21]

In summary, different problems in signal transduction can be investigated by using GFP-tagged full-length constructs of signaling proteins and/or GFP-tagged domains of signaling proteins. The main limitation for this application of GFP-tagged signaling proteins is the often low concentration of signaling proteins, which makes many translocation events not readily observable. Depending on the investigated signaling process, conventional or confocal microscopy, FRAP measurements and quantitative image analysis can be used to gain insights into mechanisms of signal transduction. Each translocating signaling protein or domain can also be used as a single-cell readout to explore the upstream players in a particular signal transduction pathway.

Structure–Function Analysis of Signaling Proteins

A particularly powerful application of GFP-tagged signaling proteins is the *in vivo* understanding of activation and regulatory mechanisms as well as the study of sequential activation steps for a signaling protein. This application of GFP technology uses deletion mutants, individual GFP-tagged domains, and site-directed mutants of a particular signaling protein to understand its regulation in the cellular context. This type of application can be used effectively if a signaling protein exhibits a regulated change in its localization or distribution. For example, the effect of phosphorylation, second-messenger binding, or protein–protein binding interactions can be readily explored in a live cell context, which often leads to insights concerning the interplay of their structure and function.

What are the advantages of such an *in vivo* structure–function analysis?

1. This approach works only if a signaling protein translocates as part of a regulatory event. Furthermore, the translocation event must be measurable by light microscopy, which requires sufficiently high concentrations of docking sites as well as intracellular target structures that can be readily resolved by light microscopy.

2. Selective mutant signaling proteins can be made. For example, phosphorylation of serine and threonine residues can be explored by making mutant GFP-tagged proteins with alanine or glutamate (or aspartate) residues, respectively, to simulate unphosphorylated and phosphorylated forms of the protein. Deletion mutants and individual domains can be tested, binding domains can be disabled, and isoforms and splice variants can be explored. Nevertheless, all mutations should be made with some knowledge of the three-dimensional structure of a protein to prevent misslocalizations due to a missfolding of proteins.

3. The accessibility of subdomains of signaling proteins that bind second messengers, membranes, or proteins can be explored by comparing the translocation and localization of the full-length protein with that of the individual domains. Such comparative studies can be made by titrating a receptor ligand (or drug) and by quantifying the amplitude of the translocation response or by analyzing the time course of translocation for each of the constructs.

In many signaling proteins, internal binding interactions within the signaling protein are used to generate sequential or delayed activation processes. For example, conventional PKCs contain tandem C1 and C2 domains as well as a catalytic domain. While the C1 domain binds to phorbol ester and DAG, the C2 domain binds to plasma membrane in the presence of calcium signals. When the time course of translocation to the plasma membrane in response to an increase in the intracellular calcium concentration was compared for PKC and for its C2 domain, similar translocation time constants were measured (Fig. 2A and C). This observation can be used to conclude that the C2 domain of PKC is in a conformation that allows it to readily bind calcium and translocate the full protein to the plasma membrane. In contrast, the C1 domain translocates on a time scale

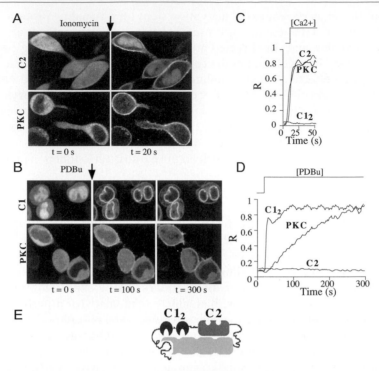

FIG. 2. Experiments that test the orientation of the C1 and C2 domains for inactive PKCγ. (A) Confocal fluorescence images of RBL cells expressing C2–GFP (*top*) or PKCγ–GFP (*bottom*). When ionomycin (1 μM) was added to cells expressing either C2–GFP or PKCγ–GFP, maximal translocation of PKCγ–GFP as well as C2–GFP occurred rapidly. Twenty-second time points are shown on the right. (B) When PDBu (1 μM) was added to cells expressing either C1–GFP (*top*) or PKCγ–GFP (*bottom*), a translocation of the two GFP fusion proteins occurred with markedly different time courses. *Middle:* Taken 100 sec after stimulation; C1–GFP is shown to be maximally translocated while PKCγ–GFP is only partially translocated. PKCγ–GFP reaches maximal translocation only after 300 sec (*bottom right*). It should be noted that a significant fraction of C1–GFP is nuclear localized in the unstimulated cell and rapidly translocated to the nuclear membrane after PDBu addition. (C) Time course of the plasma membrane translocation of C1–GFP, C2–GFP, and PKCγ–GFP after addition of the calcium ionophore ionomycin. The relative plasma membrane translocation in a series of sequential images was calculated as shown in Fig. 1D. The rapid translocation time course observed for C2–GFP and PKCγ–GFP suggests an "outside" orientation for the calcium-binding sites of the C2 domain in PKCγ. (D) Time course of the plasma membrane translocation of C1–GFP, C2–GFP, and PKCγ–GFP after addition of the phorbol ester PDBu. The rapid translocation time course observed for C1–GFP is contrasted by a much slower time course of PKCγ–GFP translocation, suggesting an "inside" orientation of the diacylglycerol-binding sites of the C1 domain in PKCγ. (E) Model for the initial orientations of the diacylglycerol-binding sites in the C1 domain and the calcium-binding site in the C2 domain of PKCγ. The C1 domain is sterically hindered while the C2 domain is facing the surface of the protein and can readily bind calcium. This model supports a sequential activation process by calcium and diacylglycerol. [Modified from E. Oancea and T. Meyer, *Cell* **95**, 307 (1998).]

more than 10 times faster than PKC itself and is much more sensitive to phorbol ester (Fig. 2B and D). This suggests that the C1 domain is not readily accessible for binding DAG in the inactive protein and that the binding of the C1 to DAG must occur subsequent to the calcium-triggered plasma membrane translocation of PKC (Fig. 2E).

In summary, this type of approach is a powerful tool for understanding many signaling proteins but requires that a significant amount of biochemical characterization of the signaling protein of interest be done beforehand or in parallel with the *in vivo* measurements.

Practical Considerations for Labeling Signaling Proteins

The first practical aspect in making a signaling protein chimera is to choose its C- or N-terminal end for the GFP fusion (or in some cases an internal GFP tag might be suitable). When the crystal structure of the signaling protein is known, the best choice is the terminus that points away from likely protein–protein interaction sites or from catalytically relevant regions. Especially when structural data are not available, biochemical characterization is important to define whether an end is suitable for GFP fusion. For this purpose, cell extracts from transfected cells or purified fusion constructs can be used (a fusion construct may also contain an additional histidine or other tag for facilitated purification). Enzymatic as well as ligand- or protein-binding assays are useful for such characterizations. For example, for PKC, both ends can be fused with PKC without affecting its *in vitro* activity. In contrast, the C-terminal end of calmodulin kinase II (CaMKII) is part of the oligomer and CaMKII oligomerization is disrupted if a GFP tag is linked at the C terminus, while oligomerization and functional activity is preserved with an N-terminal GFP tag.

If no obvious reason exists why a specific end should not be chosen, it is useful to make C- and N-terminal constructs and compare the localization and translocation of both fusion proteins. If the two constructs show the same translocation behavior, it is likely that the GFP tag is inert for the localization and translocation events studied inside the cell.

It is not uncommon that a transfected GFP fusion construct is nonfluorescent or enzymatically inactive. In addition, when fusion proteins are made by ligation of a protein of interest into existing cloning sites of a GFP expression vector, the typically introduced additional linker amino acids can act as an additional targeting sequence (GFP itself does not seem to have a significant amount of binding interactions). Incorrect expression or folding is in some cases correlated with the presence of bright intracellular spots that are apparent in some cells. These spots likely correspond to

protein aggregates and usually occur on the background of correctly localized proteins (the localization of such aggregates does not usually change
with stimulation). After sequencing the linker regions and preferably the
entire fusion construct, the linker must be evaluated more closely. It might
be necessary to design a specific linker instead of utilizing available cloning sites.

If no or weak fluorescence is observed in cells, if aggregates are observed,
or if the observed localizations is likely an artifact, the linker between the
GFP and the signaling protein should be increased (or decreased) and
replaced with inert linker sequences such as glycine-alanine repeats (e.g.,
a GAGAGA linker with three repeats). In our experience, such linkers
with glycine-alanine repeats work in many cases. For most of the proteins,
two or three glycine-alanine repeats were sufficient to eliminate the problems (in some cases we had to use five repeats).

If a fusion construct expresses with low efficiency or if the expression
level should be kept low so that its concentration does not exceed a limited
number of intracellular docking sites, it is necessary to increase the fluorescence per signaling protein. In such cases, multi-GFP tags can be used in
the fusion protein. Dimeric and tetrameric GFP tags have already been
successfully tested for this purpose.

*Practical Consideration for Making Green Fluorescent Protein Chimera
with Minimal Protein Domains*

When using minimal protein domains, rather than the entire signaling
protein, most of the practical considerations that were listed above still
apply. One advantage of using domains is that crystal structures are available for many domains important in signal transduction. Because signaling
domains are usually compact structures with accessible ends, either N- or
C-terminus fusion proteins with GFP can typically be made. Nevertheless,
in order to allow for proper folding of the domains, which are usually much
smaller in size than the GFP, the linker sequence is particularly critical. In
some cases, we have successfully used multiple glycine-alanine repeats for
C-terminal as well as N-terminal fusions for such constructs (two to five
repeats). Alternatively, we had in some cases better results when using
approximately 5 to 10 flanking amino acids of the signaling protein as
linkers. A caution to this approach is that native hydrophobic flanking
sequences may not be suitable as linkers.

One concern (or advantage) with overexpression of signaling domains
is that they may act as dominant negative constructs that prevent the
activation of the native protein. This is different from the expression of
functional GFP-tagged signaling proteins. For example, expressed GFP-

tagged SH2 domains or PH domains can function as dominant negative constructs while the expression of Syk or Akt, which contain the particular domain, may enhance the particular signaling pathway.

Acknowledgments

We thank M. F. Teruel, W. Chen, and K. Shen for stimulating discussions. This work was supported by NIH Grants GM-48113 and GM-51457.

[37] Use of Fusions to Green Fluorescent Protein in the Detection of Apoptosis

By EVE SHINBROT, COLLIN SPENCER, VALERIE NATALE, and STEVEN R. KAIN

Introduction

Apoptosis, or programmed cell death, is one of the most widely studied areas in biology. Apoptosis describes a highly regulated and specialized form of cell death that is important in physiological and pathological cellular processes. The morphological and physiological changes associated with apoptosis have been well characterized.[1,2] Cellular changes in response to apoptotic stimuli include changes in plasma membrane characteristics, nuclear condensation, and cleavage of DNA into nucleosomal ladders.[3,4] Because it is significant in many disparate processes, apoptosis research traverses scientific fields and disciplines. The availability of widely applicable, easy to use detection methods for identifying apoptotic cells in different systems is key to the advancement of the field; fluorescent protein technologies are an example of such methodology.

Using fluorescent proteins in apoptotic detection methods allows visualization of cellular changes in real time and offers an easy to use method to detect and identify apoptotic cell populations. One method fuses enhanced green fluorescent protein (EGFP)[5–8] to another protein that binds to apop-

[1] J. F. Kerr, *J. Pathol.* **105,** 13 (1971).

[2] J. F. Kerr, A. H. Wyllie, and A. R. Currie, *Br. J. Cancer* **26,** 239 (1972).

[3] A. H. Wyllie, J. F. Kerr, and A. R. Currie, *Int. Rev. Cytol.* **68,** 251 (1980).

[4] A. H. Wyllie, *Nature (London)* **284,** 555 (1980).

[5] M. Chalfie, Y. Tu, G. Euskirchen, W. W. Ward, and D. C. Prasher, *Science* **263,** 802 (1994).

[6] B. P. Cormack, R. H. Valdivia, and S. Falkow, *Gene* **173,** 33 (1996).

[7] S. Inouye and F. I. Tsuji, *FEBS Lett.* **351,** 211 (1994).

totic cells. Annexin V–EGFP is a novel reagent that can detect apoptotic cells by identifying changes that occur in their plasma membranes.[9] A second reagent, EGFP-F, uses EGFP to illuminate membrane morphological changes that occur in response to apoptotic stimuli. EGFP-F can be used to highlight the successive morphological changes that occur during apoptosis.[10]

Annexin-V–EGFP

Apoptosis-induced plasma membrane changes are commonly observed in many systems. One such change is the externalization of phosphatidylserine (PS). PS is usually found on the inner plasma membrane but it becomes externalized during apoptosis.[11–14] Detection of externalized PS is a widely applied technique.[15–20] Annexin V is a 35- to 36-kDa protein consisting of functionally distinguishable calcium and phospholipid-binding domains that bind externalized PS in a calcium-dependent manner.[15,21] Annexin V can be fused to EGFP, and this fusion protein can be used for fluorescent detection of apoptotic cells.[9] Annexin-V–EGFP retains the PS-binding properties of annexin V, while maintaining the photostable fluorescent properties of EGFP, thus providing a specific, photostable reagent for detecting apoptotic cells.[9,21]

We have evaluated the ability of annexin V–EGFP to detect a variety

[8] D. C. Prasher, V. K. Eckenrode, W. W. Ward, F. G. Prendergast, and M. J. Cormier, *Gene* **111**, 229 (1992).

[9] J. D. Ernst, L. Yang, J. L. Rosales, and V. C. Broaddus, *Anal. Biochem.* **260**, 18 (1998).

[10] E. S. Shinbrot, C. M. Spencer, and S. R. Kain, *BioTechniques* **26**, 1064 (1999).

[11] S. J. Martin, D. M. Finucane, G. P. Amarante-Mendes, G. A. O'Brien, and D. R. Green, *J. Biol. Chem.* **271**, 28753 (1996).

[12] S. J. Martin, C. P. Reutelingsperger, A. J. McGahon, J. A. Rader, R. C. van Schie, D. M. LaFace, and D. R. Green, *J. Exp. Med.* **182**, 1545 (1995).

[13] C. P. Reutelingsperger, G. Hornstra, and H. C. Hemker, *Eur. J. Biochem.* **151**, 625 (1985).

[14] G. Rimon, C. E. Bazenet, K. L. Philpott, and L. L. Rubin, *J. Neurosci. Res.* **48**, 563 (1997).

[15] V. A. Fadok, D. R. Voelker, P. A. Campbell, J. J. Cohen, D. L. Bratton, and P. M. Henson, *J. Immunol.* **148**, 2207 (1992).

[16] G. Koopman, C. P. Reutelingsperger, G. A. Kuijten, R. M. Keehnen, S. T. Pals, and M. H. van Oers, *Blood* **84**, 1415 (1994).

[17] M. van Engeland, F. C. Ramaekers, B. Schutte, and C. P. Reutelingsperger, *Cytometry* **24**, 131 (1996).

[18] M. van Engeland, L. J. Nieland, F. C. Ramaekers, B. Schutte, and C. P. Reutelingsperger, *Cytometry* **31**, 1 (1998).

[19] J. F. Tait, D. Gibson, and K. Fujikawa, *J. Biol. Chem.* **264**, 7944 (1989).

[20] G. Zhang, V. Gurtu, S. R. Kain, and G. Yan, *BioTechniques* **23**, 525 (1997).

[21] J. D. Ernst, A. Mall, and G. Chew, *Biochem. Biophys. Res. Commun.* **200**, 867 (1994).

of apoptotic events. Annexin V–EGFP can identify apoptotic cells via microscopic examination, flow cytometric analysis, and microplate fluorimetry. Annexin V–EGFP can detect apoptosis in a variety of cell types that was induced by numerous stimuli, and its binding is specific to apoptotic cells.

EGFP-F

One of the most reliable methods for identifying apoptotic cells is by observing morphological changes that occur as apoptosis progresses.[22] The formation of protrusions in the plasma membrane, usually referred to as *membrane blebbing,* is a hallmark of late-stage apoptosis. Membrane blebbing is most often assessed via light and electron microscopy.[22] However, fluorescent protein technology allows easy observation of subtle membrane changes via fluorescence microscopy. pEGFP-F (Clontech, Palo Alto, CA) targets EGFP to the plasma membrane of transfected cells[23] and therefore can track the formation of membrane protrusions occurring in apoptosis. Expression of EGFP-F results in a fluorescent green plasma membrane, allowing clear visualization of morphological changes, and has the added benefit of marking transfected cells.

Experimental Protocol

General Apoptosis Assay with Annexin V–EGFP

1. Induce apoptosis in cells. In this example, we incubate Jurkat cells for 8 hr in the presence of anti-Fas antibody (clone CH-11, 200 ng/ml; PanVera, Madison, WI).
2. Collect cells by centrifuging them at 900 rpm for 5 min.
3. Rinse the cell pellet in binding buffer (10 mM HEPES, 150 mM NaCl, 5 mM KCl, 1 mM MgCl$_2$, 1.8 mM CaCl$_2$).
4. Centrifuge at 900 rpm for 5 min, discard the supernatant, and resuspend the pellet in binding buffer.
5. Add annexin V–EGFP (Clontech) for a final concentration of 0.5 μg/ml.
6. Incubate for 5 min at room temperature in the dark.
7. Rinse the cells in binding buffer and analyze for positive annexin V staining (cell death positive) by flow cytometric analysis, microscopic examination, or a fluorescent plate reader.

---

[22] Z. Darzynkiewicz, E. Bedner, F. Traganos, and T. Murakami, *Hum. Cell* **11,** 3 (1998).
[23] W. Jiang and T. Hunter, *BioTechniques* **24,** 349, 352, 354 (1998).

Microscopic Analysis

Cells are placed on a glass slide and covered with a coverslip. They are observed under a Zeiss (Thornwood, NY) fluorescence microscope, using a fluorescein isothiocyanate (FITC) filter set. Microscopic examination (Fig. 1A; see color plate) shows bright EGFP fluorescence around the plasma membrane, indicating positive annexin V–EGFP binding to apoptotic cells. Uninduced cells do not display this fluorescence (data not shown).

Flow Cytometric Analysis

The cells are analyzed with a FACSCalibur (Becton Dickinson, San Jose, CA) equipped with an argon single laser excitation light at 488 nm. The signal generated by EGFP is detected with a green (FITC) signal detector (Fig. 1B; see color plate).

Plate Reader Analysis

Cells are grown in Dynatech (Chantilly, VA) plates with black opaque wells and treated as described above. Annexin V-positive cells are detected with a PerSeptive Biosystems (Framingham, MA) Cytofluor II multiwell plate reader, using an excitation wavelength of 485/20 nm and emission of 530/30 nm with a gain of 70. Apoptotic populations of cells are easily detected (Fig. 1C; see color plate).

Attached Cells

To analyze annexin V–EGFP binding to adherent cells without removing them from their plates, we use HeLa cells (1×10^5/well) grown in two-well chamber slides (Falcon culture slides; Becton Dickinson, Franklin Lakes, NJ). After induction of apoptosis with staurosporine ($1 \mu M$), early apoptotic cells start to become rounded. We gently rinse the wells with annexin V–EGFP binding buffer, and then incubate the cells with annexin V–EGFP for 10 min. On examining them by fluorescence microscopy, the cells that were annexin V–EGFP-positive cells also have a rounded or blebbed morphology (Fig. 1D; see color plate).

Annexin V–EGFP Binding Assay

We performed a dose–response study. To determine the sensitivity of annexin V–EGFP, Jurkat cells are induced to undergo apoptosis with anti-Fas antibody (200 μg/ml). The concentration of annexin V–EGFP is titered

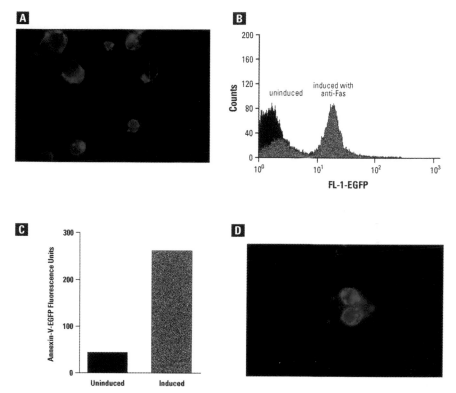

FIG. 1. Apoptotic Jurkat cells identified using Annexin-V-EGFP. Jurkat cells were incubated with 200 ng/ml anti-Fas monoclonal antibody (CH-11) for 8 hours. Cells were stained with 0.5 μg/ml Annexin V-EGFP. (A) Apoptotic cells visualized using fluorescent microscopy (100× magnification). (B) Apoptotic cells were detected using Flow Cytometric analysis. (C) Apoptotic cells identified using a plate reader. Jurkat cells were plated in 96-well plates, then induced to undergo apoptosis using anti-Fas antibody. The cells were rinsed, then incubated with Annexin V-EGFP. Apoptotic cells were detected using a fluorescent plate reader. (D) 1×10^5 HeLa cells were grown on two chamber slides, induced to undergo apoptosis with $1~\mu M$ staurosporine, and visualized by fluorescent microscopy using a FITC filter.

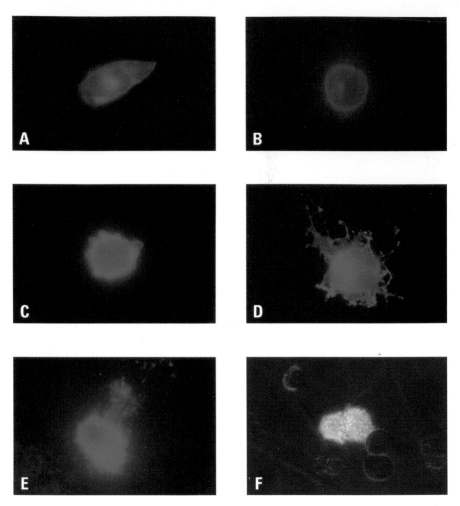

FIG. 5. This figure was previously published in Ref. 10. 1×10^5 HeLa cells were plated in 2-well chamber slides (Nunc, Corning). The cells were transfected with 4 μg/ml pEGFP-F using CLONFECTIN as a transfection reagent according to the manufacturer's protocol. Cells were examined for expression of the EGFP-F transgene 24 hours after transfection. Transfected cells displayed a green fluorescent plasma membrane, the cell was elongated and attached (5A). Apoptosis was induced by the addition of 1 μM staurosporine. The cell started to appear round, and was less attached to the plate surface (5B). The characteristic apoptotic membrane blebbing was clearly visible, EGFP-F fluorescence outlined various bubble-like projections from the plasma membrane (5C). Membrane convolutions include extensive surface projections that start to form membrane-bound cytoplasmic fragments (5D). Apoptotic bodies form from membrane-bound cellular fragments (5E). Positively transfected HeLa cells with EGFP-F are clearly distinguishable from nontransfected neighbors (5F). The effect of 1 μM staurosporine is clearly visible in EGFP-F transfected cells; membrane blebbing is clear and detectable.

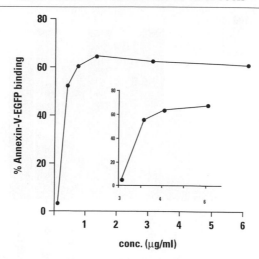

Fig. 2. Dose titration of annexin V–EGFP binding. Jurkat cells were induced to undergo apoptosis by UV irradiation. Annexin V–EGFP binding was titered at various concentrations to determine assay sensitivity. Apoptotic cells were easily separated from nonapoptotic populations by using annexin V–EGFP concentrations ranging from 5 to 0.25 μg/ml.

from 5 to 0.05 μg/ml (Fig. 2). Apoptotic populations are clearly detected by flow cytometry with annexin V–EGFP at 5 to 0.5 μg/ml. At 0.25 μg/ml it becomes difficult to separate positive and negative populations, although doing so is possible with effort.

Apoptosis Stimuli

We tested the ability of annexin V–EGFP to bind to cells by a variety of stimuli (Fig. 3). Apoptotic stimuli are added directly to the cell culture medium.

1. *Fas receptor activation:* A Fas receptor antibody (CH-11, 200 ng/ml) that mimics the action of Fas ligand binding is added directly to Jurkat cells for 8 hr.

2. *UV treatment:* Jurkat cells are exposed to UV treatment[24] for 5 min, and allowed to continue in culture for 3 hr. At this time the annexin V–EGFP assay is performed.

3. *RNA synthesis inhibitor:* 50 μM actinomycin D (Clontech), which induces apoptosis by blocking RNA synthesis,[25] is added directly to the culture medium for 6 hr.

[24] S. J. Martin and T. G. Cotter, *Int. J. Radiat. Biol.* **59,** 1001 (1991).
[25] S. J. Martin, *Immunol. Lett.* **35,** 125 (1993).

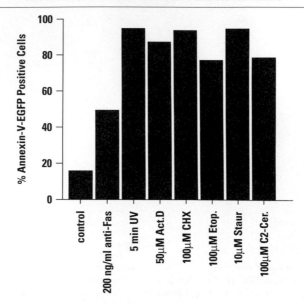

Fig. 3. Annexin V–EGFP detection of apoptotic Jurkat cells induced by various stimuli. Jurkat cells were induced to undergo apoptosis by anti-Fas antibody, UV irradiation (5 min), 100 μM cyclohexamide, 100 μM etoposide, 10 μM staurosporine, or 100 μM C2-ceramide, for 8 hr. Cells were stained with annexin V–EGFP (0.5 μg/ml). Apoptotic cells were detected by flow cytometry.

4. *Protein synthesis inhibitor:* Cycloheximide (100 μM; Clontech), which induces apoptosis by blocking protein synthesis,[25] is added directly to the culture medium overnight.

5. *Staurosporine:* Staurosporine (10 μM/ml; Clontech), a microbial alkaloid, acts as a protein kinase inhibitor and induces apoptosis in mammalian cells.[26] We incubate cells in this reagent for 6 hr.

6. *Etoposide:* Etoposide (100 μM; Clontech) induces cell death by inhibiting topoisomerase activity, which is important for DNA synthesis.[27] Cells are incubated in this reagent for 6 hr.

Annexin V–EGFP identifies apoptotic cells in response to each stimuli tested. These results show that annexin V–EGFP can detect apoptotic Jurkat cells induced by different mechanisms.

---

[26] T. Tamaoki, H. Nomoto, I. Takahashi, Y. Kato, M. Morimoto, and F. Tomita, *Biochem. Biophys. Res. Commun.* **135,** 397 (1986).
[27] P. D'Arpa and L. F. Liu, *Biochim. Biophys. Acta* **989,** 163 (1989).

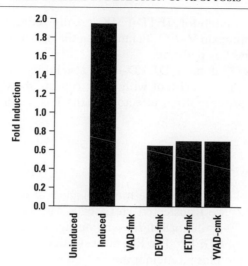

FIG. 4. Inhibitor profile. Jurkat cells were induced with anti-Fas antibody (CH-11), and incubated with various caspase inhibitors. Cells were analyzed by flow cytometry.

Inhibitors

We have examined annexin V–EGFP binding in the presence of caspase inhibitors (Fig. 4). The involvement of caspases in apoptosis has been the subject of recent reviews.[28–32] Jurkat cells are induced with anti-Fas antibody (CH-11, 200 ng/ml). Cell-permeable, noncleavable peptide analogs of caspase substrates are used to inhibit caspase activity. The caspase inhibitors (all supplied by Clontech) are added to the cells immediately before induction.

1. We add a broad-spectrum caspase inhibitor, VAD-FMK, which inhibits all caspase family proteases. Annexin V–EGFP binding is completely inhibited.

2. When an irreversible inhibitor of caspase 1, YVAD-CMK, is added to the cells, annexin V–EGFP binding is reduced; however, apoptotic populations are still detectable.

[28] D. Green and G. Kroemer, *Trends Cell Biol.* **8,** 267 (1998).
[29] M. Garcia-Calvo, E. P. Peterson, B. Leiting, R. Ruel, D. W. Nicholson, and N. A. Thornberry, *J. Biol. Chem.* **273,** 32608 (1998).
[30] S. J. Martin and D. R. Green, *Cell* **82,** 349 (1995).
[31] D. W. Nicholson and N. A. Thornberry, *Trends Biochem. Sci.* **22,** 299 (1997).
[32] N. A. Thronberry and Y. Lazebnik, *Science* **281,** 1312 (1998).

3. A caspase 8 inhibitor, IETD-FMK, greatly reduces the population of cells that bind annexin V–EGFP, indicating that it is an efficient inhibitor of apoptosis in the Fas pathway.

4. The caspase 3 inhibitor, DEVD-FMK, also inhibits annexin V–EGFP binding. This result is consistent with other reports showing that caspase 8 and caspase 3 are important effectors of the Fas death receptor apoptotic cascade.

EGFP-F Apoptosis Assay

To visualize plasma membrane morphological changes in response to apoptotic stimuli,[10] we transfected HeLa cells with pEGFP-F, a plasmid that contains the 20-amino acid farnesylation signal from c-Ha-Ras added to the C terminus of EGFP.[23] The Ras farnesylation signal targets heterologous proteins to the plasma membrane.[33,34]

1. HeLa cells are grown in two-well chamber slides to a density of 1×10^5 cells/well.

2. EGFP-F (4 μg/ml; Clontech) is transfected into each well by liposome-mediated transfection[35] (CLONfectin transfection reagent).

3. Transfected cells are identified by their green fluorescent plasma membranes. These cells appear flat and remain attached to their wells (Fig. 5A; see color plate).

4. Staurosporine (1 μM) is added to each well to induce apoptosis. In response, the cells begin to appear rounded (Fig. 5B; see color plate).

5. Next, the borders of the membranes display bubble-like projections, which are accentuated due to EGFP fluorescence (Fig. 5C; see color plate).

6. When the cells dissociate into apoptotic bodies (Fig. 5D and E; see color plate), small membrane fragments are still visible.

Tracking plasma membrane changes offers many advantages. EGFP-F can be used in cotransfection experiments to determine the ability of transfected genes to inhibit or induce apoptosis. Figure 5F (see color plate) illustrates this application. This application may be useful for apoptosis

[33] A. Aronheim, D. Engelberg, N. Li, N. al-Alawi, J. Schlessinger, and M. Karin, *Cell* **78,** 949 (1994).

[34] J. F. Hancock, K. Cadwallader, H. Paterson, and C. J. Marshall, *EMBO J.* **10,** 4033 (1991).

[35] G. Zhang, V. Gurtu, T. H. Smith, P. Nelson, and S. R. Kain, *Biochem. Biophys. Res. Commun.* **236,** 126 (1997).

studies, along with any study involving monitoring changes in plasma membrane morphology.

Discussion

Fluorescent protein technologies add many advantages to apoptotic detection methods. The use of EGFP apoptotic indicators allows identification and detection of apoptotic cells in kinetic studies monitoring the progression of apoptosis. Two such reagents are annexin V–EGFP and EGFP-F.

Annexin V–EGFP retains the binding properties of annexin V, with the added benefit of EGFP, making it an excellent reagent for detecting apoptotic cells.[9,21] The EGFP fusion provides an easily detectable, photostable addition to annexin V detection. Annexin V–EGFP binding of externalized phosphatidylserine is easily detectable by flow cytometry, fluorescence microscopy, and with a fluorescence multiwell plate reader. We observed annexin V–EGFP binding on Jurkat suspension cells and adherent HeLa cells. This reagent would work well for double-staining procedures, making it especially useful for flow cytometry and fluorescence microscopy. Because annexin V–EGFP can be detected in a multiwell plate reader, it lends itself to use in high-throughput analysis. The annexin V–EGFP binding procedure is simple, quick, and highly specific, a requirement for high-throughput sample analysis. Annexin V–EGFP is a sensitive reagent, which is able to clearly detect apoptotic populations at concentrations as low as 0.5 μg/ml, making it economical for use in high-throughput analysis.

Annexin V–EGFP specifically detected apoptotic cells in response to various stimuli. We tested many types of stimuli ranging from receptor-mediated events (anti-Fas antibody) to RNA synthesis, protein synthesis, and protein kinase inhibitors. In each case, annexin V–EGFP detected apoptotic cells. In addition, the reagent was able to detect apoptosis in many different cell types. These properties of annexin V–EGFP indicate that it is useful for detecting apoptosis in many different systems, under different stimuli.

Addition of the 20-amino acid farnesylation signal to EGFP results in targeting of this fusion protein to the plasma membrane.[23] EGFP-F creates a bright green fluorescent membrane, allowing easy detection of subtle changes in morphology.[10] The plasma membrane changes that occur during apoptosis can be followed in real time, using cells transfected with EGFP-F.

Other EGFP fusion proteins may be used to follow apoptotic changes and mechanisms involved in various apoptotic pathways. EGFP may be added as a C-terminal or N-terminal fusion to key proteins in the apoptotic cascade. These may include cytochrome c,[36,37] Bid,[38] Aif, Bax,[37] or any other apoptotic proteins of interest. Determining the locations of these genes before apoptosis and after applying various stimuli would be interesting. The addition of EGFP to these proteins will allow observation of the movement of these proteins within the cell in response to various conditions. The addition of EGFP would also allow the dynamic tracing of the movement of these apoptotic proteins within the cell in real time, and may help to determine the mechanisms involved in their actions.

[36] J. Llopis, J. M. McCaffery, A. Miyawaki, M. G. Farquhar, and R. Y. Tsien, *Proc. Natl. Acad. Sci. U.S.A.* **95,** 6803 (1998).
[37] N. P. Mahajan, K. Linder, G. Berry, G. W. Gordon, R. Heim, and B. Herman, *Nature Biotechnol.* **16,** 547 (1998).
[38] H. Li, H. Zhu, C. J. Xu, and J. Yuan, *Cell* **94,** 491 (1998).

[38] Synapto-pHluorins: Chimeras between pH-Sensitive Mutants of Green Fluorescent Protein and Synaptic Vesicle Membrane Proteins as Reporters of Neurotransmitter Release

By Rafael Yuste, Rebecca B. Miller, Knut Holthoff, Shifang Zhang, and Gero Miesenböck

Introduction

A Primer on Presynaptic Physiology

What has been termed the "standard model"[1] of neurotransmitter release is based on classic experiments performed on the frog neuromuscular junction by B. Katz and his school.[2,3] According to the standard model, transmitter is stored in synaptic vesicles and released in a quantal all-or-none fashion. The probability of release is a function of the presynaptic membrane potential; it is low (but not zero) at resting potential and rises dramatically when an action potential invades the terminal.

At the neuromuscular junction, several hundred quanta are released

[1] C. F. Stevens, *Cell* **72**(Suppl.), 55 (1993).
[2] B. Katz, "Nerve, Muscle, and Synapse." McGraw-Hill, New York, 1966.
[3] B. Katz, "The Release of Neural Transmitter Substances." Liverpool University Press, Liverpool, UK, 1969.

METHODS IN ENZYMOLOGY, VOL. 327
0076-6879/00 $30.00

per action potential[3]; at central synapses, the number is estimated to be either zero or one,[4] and the percentage of transmission failures is significant.[5-8] Release appears to occur from a site morphologically identified as the active zone.[9] Neuromuscular junctions have many release sites, central synapses usually one.[10] Released quanta are thought to be drawn from a docked or "readily releasable" pool of vesicles.[11] At central synapses, estimates of the size of this pool range from 2 to 27 vesicles.[10,12]

Once a vesicle is released, its membrane is reinternalized[13-15] after a variable span of time ($\sim$1–20 sec),[16-20] along a route that may or may not include the endosomal system,[14,21-24] and its interior is refilled with transmitter. Uptake of classic neurotransmitters is mediated by transporters powered by a proton electrochemical gradient across the vesicle membrane; this gradient is established and maintained by vacuolar H^+-ATPases.[25-27] After refilling with transmitter, the vesicle becomes again available for release. It is unclear whether this occurs after transition through a "reserve" pool, or whether a newly refilled vesicle can immediately enter the readily releasable pool. The number of distinct functional vesicle pools at central synapses (in addition to the readily releasable pool) is unknown, and so,

[4] C. F. Stevens and Y. Wang, *Neuron,* **14,** 795 (1995).
[5] N. A. Hessler, A. M. Shirke, and R. Malinow, *Nature (London)* **366,** 569 (1993).
[6] C. Rosenmund, J. D. Clements, and G. L. Westbrook, *Science* **262,** 754 (1993).
[7] C. F. Stevens and Y. Wang, *Nature (London)* **371,** 704 (1994).
[8] V. N. Murthy, T. J. Sejnowski, and C. F. Stevens, *Neuron* **18,** 599 (1997).
[9] J. E. Heuser, T. S. Reese, M. J. Dennis, Y. Jan, L. Jan, and L. Evans, *J. Cell Biol.* **81,** 275 (1979).
[10] T. Schikorski and C. F. Stevens, *J. Neurosci.* **17,** 5858 (1997).
[11] C. Rosenmund and C. F. Stevens, *Neuron,* **16,** 1197 (1996).
[12] C. F. Stevens and T. Tsujimoto, *Proc. Natl. Acad. Sci. U.S.A.* **92,** 846 (1995).
[13] B. Ceccarelli, W. P. Hurlbut, and A. Mauro, *J. Cell Biol.* **57,** 499 (1973).
[14] J. E. Heuser and T. S. Reese, *J. Cell Biol.* **57,** 315 (1973).
[15] W. J. Betz and G. S. Bewick, *Science* **255,** 200 (1992).
[16] W. J. Betz and G. S. Bewick, *J. Physiol. (London)* **460,** 287 (1993).
[17] H. von Gersdorff and G. Matthews, *Nature (London)* **367,** 735 (1994).
[18] T. A. Ryan, S. J. Smith, and H. Reuter, *Proc. Natl. Acad. Sci. U.S.A.* **93,** 5567 (1996).
[19] T. A. Ryan, *Neuron* **17,** 1035 (1996).
[20] H. von Gersdorff and G. Matthews, *Annu. Rev. Physiol.* **61,** 725 (1999).
[21] L. Clift-O'Grady, A. D. Linstedt, A. W. Lowe, E. Grote, and R. B. Kelly, *J. Cell Biol.* **110,** 1693 (1990).
[22] E. Grote, J. C. Hao, M. K. Bennett, and R. B. Kelly, *Cell,* **81,** 581 (1995).
[23] J. H. Koenig and K. Ikeda, *J. Cell Biol.* **135,** 797 (1996).
[24] V. N. Murthy and C. F. Stevens, *Nature (London)* **392,** 497 (1998).
[25] Q. Al-Awqati, *Annu. Rev. Cell Biol.* **2,** 179 (1986).
[26] P. R. Maycox, J. W. Hell, and R. Jahn, *Trends Neurosci.* **13,** 83 (1990).
[27] Y. Liu and R. H. Edwards, *Annu. Rev. Neurosci.* **20,** 125 (1997).

as a consequence, is the functional status of the majority of the ~200 vesicles that are discerned morphologically[10,28] at each synapse.

Imaging Presynaptic Function

Early efforts to image neurotransmitter release focused on the neuromuscular junction and were inspired by previous work with horseradish peroxidase (HRP). HRP, when added to the extracellular medium, becomes trapped in recycling synaptic vesicles and can, after fixation and sectioning, be visualized by electron microscopy.[13,14,29,30] To adapt this principle to living preparations, fluorescent dyes were substituted for HRP. Initially, soluble dyes such as sulforhodamine 101 or fluorescein-5,6-sulfonic acid were used[31]; later, they were replaced by a family of amphipathic compounds that are virtually nonfluorescent in aqueous solution but become brightly fluorescent after they insert into a lipid bilayer. These compounds, FM 1-43 and its analogs, are internalized with recycling synaptic vesicle membranes and viewed after residual dye present in the plasma membrane has been washed off.[15,16,32,33]

FM 1-43 has been employed successfully not only at neuromuscular junctions of vertebrates[15,16,32] and invertebrates,[34] but also at "central" synapses formed by dissociated neurons in culture.[8,18,24,35–37] Bulk-applied synthetic dyes, however, often face problems of specificity and access in intact or semiintact preparations such as brain slices, where many synapses are crowded in small volumes and access of staining and wash solutions is limited.

A possible alternative is to encode a protein-based optical indicator in DNA.[38] Such indicators not only solve the problem of probe access (they are generated biosynthetically and can thus be used even in intact animals), they also provide genetic control over which synapses are labeled: indicator expression can for instance be restricted to certain cell types (by cell type-

---

[28] K. M. Harris and P. Sultan, *Neuropharmacology* **34,** 1387 (1995).

[29] E. Holtzman, A. R. Freeman, and L. A. Kashner, *Science* **173,** 733 (1971).

[30] B. Ceccarelli, W. P. Hurlbut, and A. Mauro, *J. Cell Biol.* **54,** 30 (1972).

[31] J. W. Lichtman and R. S. Wilkinson, *J. Physiol. (London)* **393,** 355 (1987).

[32] W. J. Betz, F. Mao, and G. S. Bewick, *J. Neurosci.* **12,** 363 (1992).

[33] A. J. Cochilla, J. K. Angleson, and W. J. Betz, *Annu. Rev. Neurosci.* **22,** 1 (1999).

[34] M. Ramaswami, K. S. Krishnan, and R. B. Kelly, *Neuron* **13,** 363 (1994).

[35] T. A. Ryan and S. J. Smith, *Neuron* **14,** 983 (1995).

[36] T. A. Ryan, H. Reuter, B. Wendland, F. E. Schweizer, R. W. Tsien, and S. J. Smith, *Neuron* **11,** 713 (1993).

[37] T. A. Ryan, H. Reuter, and S. J. Smith, *Nature (London)* **388,** 478 (1997).

[38] G. Miesenböck and J. E. Rothman, *Proc. Natl. Acad. Sci. U.S.A.* **94,** 3402 (1997).

specific promoters and enhancers) or the elements of neural circuits (by using transneuronal tracers as the expression vector[39,40]). At present, however, there is a price to be paid for these benefits: even with the preferred current incarnation of genetically encoded indicator, termed synapto-pHluorin,[41] the optical signal associated with transmitter release is smaller than that obtained with FM 1-43 or its analogs.

Synapto-pHluorins

Mode of Action

Because the ~200 vesicles at a typical central synapse[10,28] cannot be resolved optically, vesicles that have undergone exocytosis must be distinguished spectrally from vesicles that have not. In the case of FM 1-43, this distinction is made through dye content: vesicles undergoing exocytosis release the dye that has been trapped in their lumen, while resting vesicles remain stained.[15,16,32,33] Each quantal release thus decreases the overall fluorescence intensity of the terminal.[8,24,37]

Synapto-pHluorins use pH-sensitive mutants of green fluorescent protein (GFP), termed "pHluorins," to afford the distinction between resting and exocytosed vesicles.[41] As mentioned above, resting vesicles maintain a proton electrochemical gradient across their membrane to keep neurotransmitter concentrated in their interior.[27] As a consequence, the vesicle interior is acidified to a pH of ~5.7.[41-43] pHluorins located in these vesicles are in the "off" state, with a characteristic fluorescence excitation spectrum (see below). On exocytosis, the vesicle interior becomes continuous with the extracellular space, the pH rises to ~7.4, and the pHluorin excitation spectrum switches to the "on" state. Because total fluorescence is the linear sum of emissions from all synapto-pHluorins in a terminal, increases in the "on" content of this composite spectrum indicate neurotransmitter release.

Structure

Synapto-pHluorins are chimeric membrane proteins composed of two modules[41]: (1) a pHluorin module that reports local pH, and (2) the synaptic

---

[39] H. G. Kuypers and G. Ugolini, *Trends Neurosci.* **13,** 71 (1990).

[40] P. R. Lowenstein and L. W. Enquist (eds.), "Protocols for Gene Transfer in Neuroscience." John Wiley & Sons, New York, 1996.

[41] G. Miesenböck, D. A. De Angelis, and J. E. Rothman, *Nature (London)* **394,** 192 (1998).

[42] R. G. Johnson and A. Scarpa, *J. Cell Biol.* **251,** 2189 (1976).

[43] H. H. Füldner and H. Stadler, *Eur. J. Biochem.* **121,** 519 (1982).

vesicle membrane protein VAMP-2,[44,45] which attaches the pHluorin to the inner vesicle surface. The pHluorin amino terminus is fused to the carboxy terminus of VAMP-2, which is located in the vesicle lumen. This results in type II topology, with a membrane anchor segment that also serves as a noncleavable signal peptide. The two modules are joined via the flexible linker[46] sequence (Ser-Gly-Gly)$_2$-Thr-Gly-Gly.

Synapto-pHluorin cDNAs are available as MluI–XbaI fragments in the mammalian expression vector pCI (GenBank accession number U47119; Promega, Madison, WI). Their generic structure is mouse VAMP-2 (GenBank accession number AF007168)—AgeI site-containing linker sequence (AGC GGC GGA AGC GGC GGG ACC GGT GGA)—pHluorin. At the 5′ junction between vector and insert, the MluI site is followed by a Kozak sequence and the initiating ATG (ACG CGT GCC ACC ATG. . .). At the 3′ junction, a terminator codon (TAA) lies between AgeI and XbaI sites. Initiator- and terminator-less pHluorin-coding sequences are thus released and can be swapped after AgeI digestion.

Ratiometric and Ecliptic pHluorins

pHluorins come in two flavors: "ratiometric" and "ecliptic."[41] Ratiometric pHluorin (GenBank accession number AF058694) shows a reversible excitation ratio change between pH 8.0 and 5.0, with a response time of less than 20 msec. As pH is lowered, the dominant fluorescence excitation peak is shifted from 395 to 475 nm (Fig. 1A). Ratioing the fluorescence intensities at the two excitation wavelengths provides a measure of pH that, like all ratiometric indices, is insensitive to variations in excitation

---

[44] M. Baumert, P. R. Maycox, F. Navone, P. De Camilli, and R. Jahn, *EMBO J.* **8,** 379 (1989).

[45] L. A. Elferink, W. S. Trimble, and R. H. Scheller, *J. Biol. Chem.* **264,** 11061 (1989).

[46] J. S. Weissman and P. S. Kim, *Cell* **71,** 841 (1992).

Fig. 1. Fluorescence excitation spectra of (A) ratiometric and (B) ecliptic pHluorin. Spectra were recorded in a Perkin-Elmer LS-50B spectrofluorimeter at a constant emission wavelength of 508 nm; bandwidths for excitation and emission were 8 nm. Samples consisted of purified recombinant pHluorin, at chromophore concentrations of 27.5 μM, in 50 mM sodium acetate plus 50 mM sodium cacodylate, adjusted to the indicated pH values, and 100 mM NaCl, 1 mM MgCl$_2$, and 1 mM CaCl$_2$. Chromophore concentrations were estimated on the basis of the 447-nm absorbance of acid-denatured pHluorin at pH 11.8 and an extinction coefficient of 44,100 M^{-1} cm^{-1} [W. W. Ward, *in* "Bioluminescence and Chemiluminescence" (M. A. DeLuca and W. D. McElroy, eds.), p. 235. Academic Press, New York, 1981.] The ordinate scales are normalized to the 395-nm peak of ratiometric pHluorin at pH 7.5.

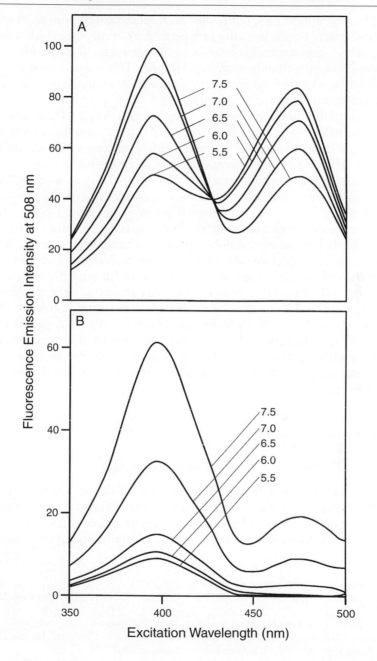

light intensity, optical path legth, and fluorophore concentration. Because the ratiometric pHluorin does not possess a "dark" state, exocytosis must be distinguished against a background of resting vesicles and other subcellular structures that may contain synapto-pHluorin. This requires that the percentage of vesicles undergoing exocytosis be sufficiently large to alter the composite spectrum of the terminal.[41]

Ecliptic pHluorin (GenBank accession number AF058695) differs from ratiometric pHluorin in that its 475-nm band attains a nonfluorescent state (is "eclipsed") at a pH below 6.0; the 395-nm band remains weakly fluorescent (Fig. 1B). Again, the spectral change is reversible within less than 20 msec after returning to neutral pH. The availability of a "dark" state should essentially eliminate background fluorescence due to resting vesicles: Vesicles whose internal pH is below 6.0 are invisible under 475-nm excitation and become fluorescent only as they exocytose and their "internal" pH rises to that of the extracellular space. In a mast cell line, this allowed detection of individual vesicle fusion events as sudden appearances of fluorescence excited at 475 nm, in locations known (through previous illumination with 395-nm light) to contain secretory granules.[41] The diameter of these granules ($\sim$500 nm)[47-49] is $\sim$10-fold larger than that of synaptic vesicles ($\sim$50 nm),[9,10,28] which facilitates detection $\sim$100-fold. As discussed below, attempts to image single synaptic vesicles would require ecliptic pHluorins with improved properties, as well as additional measures, such as a confocal aperture or multiphoton excitation, to eliminate background fluorescence.

Preparations for Study

Dissociated Cultures of Rat Hippocampal Neurons

Our procedure is a modification of that of Goslin, Asmussen, and Banker.[50] Their chapter (as well as much of the remainder of the book of which it is part) is an encyclopedic source on neural cultures.[51]

Two days before cultures are prepared, 25-mm round glass coverslips,

[47] S. J. Burwen and B. H. Satir, *J. Cell Biol.* **73,** 660 (1977).
[48] D. Lawson, M. C. Raff, B. Gomperts, C. Fewtrell, and N. B. Gilula, *J. Cell Biol.* **72,** 242 (1977).
[49] D. E. Chandler and J. E. Heuser, *J. Cell Biol.* **86,** 666 (1980).
[50] K. Goslin, H. Asmussen, and G. Banker, *in* "Culturing Nerve Cells" (G. Banker and K. Goslin, eds.), p. 339. MIT Press, Cambridge, Massachusetts, 1998.
[51] G. Banker and K. Goslin (eds.), "Culturing Nerve Cells" 2nd Ed. MIT Press, Cambridge, Massachusetts, 1998.

No. 1 thickness, are placed in porcelain staining racks (Thomas Scientific, Swedesboro, NJ), briefly rinsed with distilled water, and soaked overnight in nitric acid. The next day, the coverslips are washed extensively with distilled water, autoclaved, air dried, and dispensed into tissue culture-grade six-well plates. Glass and plastic surfaces are coated with 0.5 mg/ml poly-D-lysine (Sigma, St. Louis, MO) in 100 mM sodium borate, pH 8.5. After application of 1–2 ml of coating solution per well, the plates are left in a tissue culture incubator overnight until shortly before use (see below). Eight six-well plates, providing a plating surface equivalent to that of forty-eight 35-mm dishes, are usually sufficient to accommodate the cells derived from one litter.

Cultures are prepared from rat fetuses at embryonic day 19 (E19), obtained from timed-pregnant females (Taconic, Germantown, NY). The rat is killed with carbon dioxide and its uterus exposed through a midline abdominal incision. Fetuses are removed and placed on a gauze pad on a bed of crushed ice and tap water. The consistency of the bed of ice must be firm enough to support the weight of a dozen or so rat pups (the typical number of animals per litter), but should contain sufficient liquid to partially cover the fetuses and protect them from drying out.

The next steps are performed under a dissection microscope in a tissue culture hood, using sterile tools. Tools (one pair of Dewecker iris scissors, one forceps, two microspatulas) are sealed in Vis-U-All II sterilization pouches (Surgicot, Research Triangle Park, NC) and autoclaved. Fetuses are pinned to a plastic mat with three 20-gauge hypodermic needles, one through each front paw and one through the snout. Working from the back of the head with Dewecker iris scissors (Miltex; obtained from SSR Fine Surgical Tools, Oyster Bay, NY) and Dumont No. 5 forceps, skin and skull (which at this age is paper-thin) are cut in the midline and removed. After cutting brainstem and optic nerves, the cerebral hemispheres and diencephalon are carefully mobilized with two Handi-Hold microspatulas (Fisher Scientific, Pittsburgh, PA) and placed in a 60-mm tissue culture dish containing ice-cold EBSS (Earle's balanced salt solution) with 10 mM HEPES–NaOH, pH 7.0.

The two hemispheres are separated with a sagittal cut and rotated such that their medial sides point up. Using the tips of the forceps to pin the hemisphere to the bottom of the tissue culture dish, diencephalon and basal ganglia are removed in a single, shallow scooping movement of the microspatula. This exposes the temporal horn of the lateral ventricle and brings the crescent-shaped hippocampus into view. Starting with the easily discernable chorioid plexus, the meninges are peeled off the tissue, if possible in a single movement; the task is complete once blood vessels are no longer visible on the surface of the brain. The hippocampus is removed with small cuts and gentle pulls and tears, placed in a conical 50-ml Falcon tube containing 10 ml of EBSS with 10 mM HEPES–NaOH, pH 7.0, and stored on ice. With some practice, the dissection will take less than 5 min per fetus, and a whole litter can be processed in less than 1 hr.

With about 15 min of dissection work left, the procedure is briefly interrupted to prepare a tissue dissociation solution containing two enzymes: papain (20 U/ml) to disaggregate the tissue, and RNase-free DNase I (10 U/ml) (Boehringer Mannheim, Indianapolis, IN) to counter viscosity increases following accidental cell lysis. Papain, a sulfhydryl protease, is supplied as the inactive mercurypapain and needs to be activated by mild reduction (1 mM cysteine) and chelation of mercury (0.5 mM EDTA). A good source of enzyme and both activators is Vial 2 of the Worthington (Freehold, NJ) papain dissociation system. The lyophilized contents of two vials are reconstituted in 10 ml of EBSS with 10 mM HEPES–NaOH (pH 7.0) and DNase I and incubated for 15 min at 36° to ensure full activity.

Once all hippocampi are collected in cold HEPES–EBSS, the solution is aspirated and replaced with prewarmed, filtered (0.22-μm pore size) dissociation solution. Incubation for 90 min at 36° will sufficiently disaggregate the tissue to allow mechanical dissociation into a single-cell suspension (see below) but will not produce any conspicuous changes: the tissue will stay intact and settle to the bottom of the tube. At the end of the 90-min incubation, the dissociation solution is aspirated and the tissue washed three times with 25 ml of growth medium. Growth medium consists of basal medium Eagle with Earle's salts, without L-glutamine, supplemented with 25 mM HEPES–NaOH (pH 7.4), 10% (v/v) heat-inactivated donor horse serum (JRH Biosciences, Lenexa, KS), 20 mM glucose, 1 mM sodium pyruvate, 0.2% (v/v) Mito+ Serum Extender (Collaborative Biomedical Products, Bedford, MA), and penicillin (100 U/ml) plus streptomycin (0.1 mg/ml).

After the final wash, the tissue is resuspended in 2 ml of growth medium and triturated with a Pasteur pipette for 10–20 passes. The pipette is gently held, at a slight angle, against the bottom of the 50-ml Falcon tube, to reduce the opening through which the tissue is forced to a narrow slit. To avoid frothing, the pipette is never emptied completely. Once the cell suspension looks homogeneous, its volume is brought to 10 ml with growth medium and trituration resumed for five passes with a 10-ml serological pipette. After 1 min of settling, density and viability of cells in the supernatant are evaluated in a hemocytometer. A 20-μl aliquot of cell suspension is mixed with an equal volume of 0.4% (w/v) trypan blue solution (Sigma) and allowed to stand for 5 min. About 90% of cells should exclude the dye, no cell clumps should be visible, and cell densities will typically range from 200,000 to 500,000/ml.

In preparation for plating, the poly-D-lysine solution is removed from the six-well plates containing coverslips while the enzymatic dissociation is in progress, and the wells are rinsed twice with filtered (0.22-μm pore size) water. The desired number of cells (or, simply, ~2% of the total yield) is added per well, in a total of 3 ml of growth medium. The surplus medium added during plating (about half the amount would be sufficient) is a convenient way to generate conditioned medium: As the neurons differentiate in culture, they become extremely sensitive to medium changes; often, replacement of their growth medium with fresh medium, even with the batch in which they were originally plated, causes near-instant cell death. To avoid such accidents, we keep our cultures in the same medium for periods of more than 4 weeks, and take care to produce sufficient conditioned medium to be able to perform all necessary manipulations (see below). The only modification of the plating medium is the addition of 10 μM cytosine arabinoside 3 days after plating, to arrest the proliferation of nonneuronal cells. Cultures are maintained in a humidified incubator at 37°, in an atmosphere containing 5% CO_2.

Cortical Slice Cultures

Cortical brain slices are harvested from C57 mice at postnatal days 0 to 4 (P0 to P4), using autoclaved dissection tools and filtered (0.22-μm pore size) solutions. The mouse is cryoanesthetized on a bed of ice for 10–15 min, scrubbed with 70% (v/v) ethanol, and decapitated with scissors. The head is transferred to a 35-mm tissue culture dish filled with cold HEPES–artificial cerebrospinal fluid [HEPES–ACSF: 10 mM HEPES–NaOH (pH 7.3), 140 mM NaCl, 5 mM KCl, 1 mM CaCl$_2$, 1 mM MgCl$_2$, 24 mM glucose]. Working in a tissue culture hood, skin and skull are carefully cut along the dorsal cranial fissure with scissors and forceps and parted sideways. The brain is removed and placed into a fresh tissue culture dish filled with cold HEPES–ACSF. Under a dissection microscope, the cerebral hemispheres are separated with a No. 15 surgical blade and oriented with the diencephalon facing down. Cerebellum and mesencephalon are carefully dissected; then, after the hemisphere is rotated such that its medial side faces up, the diencephalon is removed with a flat-ended spatula. The remaining piece of tissue, representing cortex and hippocampus, is trimmed into a rectangular block with one edge parallel to the dorsal cranial fissure.

Immediately before use, the tissue chopper (TC-2 tissue sectioner; Sorvall, Norwalk, CT) is sterilized with 70% (v/v) ethanol inside the tissue culture hood and the rectangular block of tissue positioned such that the chopping orientation is perpendicular to the dorsal cranial fissure. The tissue is cut into ten 300-μm thick slices at an intermediate speed setting.

Using a flat-ended spatula, the slices are transferred to a fresh culture dish containing cold HEPES–ACSF. They are separated from each other with a surgical blade as soon as possible and rested on ice for 20–30 min.

Slices are maintained in medium containing, per 100 ml, 50 ml of basal medium Eagle, 25 ml of Hanks' balanced salt solution, and 25 ml of horse serum (HyClone, Logan, UT). We find a consistently higher transfection rate with HyClone serum than with material from other vendors. The medium is supplemented with 10 mM HEPES–NaOH (pH 7.3), 36 mM glucose, and penicillin (100 U/ml) plus streptomycin (0.1 mg/ml). The slices are kept at an air–medium interface,[52] supported on 30-mm Millicell-CM insert chambers with 0.4-μm membranes (Millipore, Bedford, MA). Working in a tissue culture hood, the membrane of the insert is first saturated with medium. To this end, the insert chamber is briefly lifted out of its container, to allow addition of 1 ml of culture medium, and then returned to its original position. An additional 0.5 ml of culture medium is applied on top of the insert. Individual slices, typically three to six per chamber, are arranged with a flat-ended spatula on the membrane such that they lie flat and are separated by at least 2–3 mm, because they will spread somewhat during culture. After the slices have settled, the medium is aspirated and the insert chamber transferred to a six-well culture plate containing exactly 1 ml of medium per well. Cultures are maintained in a humidified incubator at 37°, in an atmosphere containing 5% CO_2. During medium changes, which are performed every other day, the insert chambers are simply transferred to a fresh six-well plate prefilled with 1 ml of medium per well. Slices can be kept in culture for at least 3 weeks.

Acute Cortical Slices

Slices of visual cortex are obtained from C57 mice (Taconic) at P15 to P20. Mice are anesthetized with an intraperitoneal injection of a mixture of ketamine (120 mg/kg; Sigma) and xylazine (10 mg/kg; Sigma) in saline. When the animals no longer respond to pain (toe pinch), they are decapitated with scissors. The skull is exposed through a midline incision with a scalpel. Using medium-size scissors, the skull is carefully cut in the midline and the two flaps folded sideways, baring the cortex. After cranial nerves and brainstem are gently severed with a bent spatula, the brain is transferred to a beaker filled with ice-cold bicarbonate–ACSF that is bubbled with 95% O_2 and 5% CO_2; the solution contains 124 mM NaCl, 3 mM KCl, 1 mM $CaCl_2$, 3 mM $MgSO_4$, 1.25 mM NaH_2PO_4, 26 mM $NaHCO_3$, and

---

[52] L. Stoppini, P. A. Buchs, and D. Muller, *J. Neurosci. Methods* **37**, 173 (1991).

10 mM glucose. The brain is allowed to cool for at least 2 min and then trimmed with a razor blade. Two coronal cuts remove cerebellum and frontal lobe; one horizontal cut removes ventral structures, including midbrain and most of the diencephalon.

To mount the trimmed brain for slicing, an agar cube [5% (w/v) in ACSF] is glued to the plastic base of the Vibratome. The tissue block is then set in place in a vertical orientation (frontal surface pointing up) and the posterior (occipital) plane glued to the plastic base with cyanoacrylate (Superglue). The ventral (horizontal) plane faces the agar block. The chamber is flooded with ice-cold ACSF, and the brain is sliced into 400-μm-thick coronal slices, using a Vibratome (TPI, St. Louis, MO) or a Vibroslicer (TPI). Individual slices are gently removed with the blunt end of a Pasteur pipette and deposited in the incubation chamber, which consists of supported nylon netting inside a plastic beaker filled with bicarbonate–ACSF that is bubbled with 95% O_2 and 5% CO_2 and kept at room temperature. Experiments are performed within 1–11 hr after slicing.

Expression of Synapto-pHluorins

Viral Vectors

Of many alternatives,[40,53,54] adenovirus, adeno-associated virus, vaccinia virus, Semliki Forest virus, and Sindbis virus among them, we use herpes simplex virus type I (HSV-I) to direct expression of synapto-pHluorins in neurons. HSV-I is a DNA virus with a genome of ~150 kb that can accept foreign DNA sequences in one of two ways.[55,56] A heterologous expression cassette can be inserted into a full-length viral genome by homologous recombination, to generate a recombinant virus vector.[57–59] Or, the expression cassette can be carried on an "amplicon" plasmid that contains an HSV origin of replication and a cleavage/packaging sequence. When

[53] M. G. Kaplitt and A. D. Loewy (eds.), "Viral Vectors: Gene Therapy and Neuroscience Applications." Academic Press, San Diego, California, 1995.

[54] R. S. Slack and F. D. Miller, *Curr. Opin. Neurobiol.* **6**, 576 (1996).

[55] X. O. Breakefield and N. A. DeLuca, *N. Biol.* **3**, 203 (1991).

[56] D. J. Fink, N. A. DeLuca, W. F. Goins, and J. C. Glorioso, *Annu. Rev. Neurosci.* **19**, 265 (1996).

[57] M. F. Shih, M. Arsenakis, P. Tiollais, and B. Roizman, *Proc. Natl. Acad. Sci. U.S.A.* **81**, 5867 (1984).

[58] P. Marconi, D. Krisky, T. Oligino, P. L. Poliani, R. Ramakrishnan, W. F. Goins, D. J. Fink, and J. C. Glorioso, *Proc. Natl. Acad. Sci. U.S.A.* **93**, 11319 (1996).

[59] D. M. Krisky, D. Wolfe, W. F. Goins, P. C. Marconi, R. Ramakrishnan, M. Mata, R. J. Rouse, D. J. Fink, and J. C. Glorioso, *Gene Ther.* **5**, 1593 (1998).

transfected into a mammalian cell, the HSV sequences of the plasmid allow the amplicon plasmid to be replicated and packaged into virions, provided the necessary machinery is supplied by a copropagated helper virus.[60]

Our preferred HSV-I strain is THZ.3, the kind gift of D. Krisky and J. Glorioso (University of Pittsburgh, Pittsburgh, PA).[58,59] THZ.3 is a deletion mutant deficient in three of the five immediate-early (α) genes ($\alpha4$, $\alpha22$, and $\alpha27$); its replication cycle is therefore blocked at an early stage and its pathogenic potential is minimal. In addition, THZ.3 lacks VHS (encoded by the U_L41 gene), a structural component of the virion responsible for the early shutoff of host protein synthesis during infection.[61,62] The virus is propagated on a cell line termed 7B,[59] which provides the two essential immediate-early gene products ($\alpha4$ and $\alpha22$) in *trans* to complement functions deleted from THZ.3.

7B cells are grown in 75-cm^2 flasks in Dulbecco's modified Eagle's medium (DMEM) with 10% (v/v) fetal bovine serum, penicillin (100 U/ml), and streptomycin (0.1 mg/ml). Infections are performed at near confluency with 0.05 PFU/cell, in growth medium with 10 mM HEPES–NaOH, pH 7.1. When generalized cytopathic effect develops (3–4 days after infection), we let an additional 24 hr pass before collecting the medium (which will contain large floating sheets of detached cells) and scraping residual debris off the flask. Medium and debris are pooled and subjected to three freeze–thaw cycles at $-80°$. The viral stock is titrated for plaques on 7B cells overlaid with 0.5% (w/v) methylcellulose, aliquoted and stored at $-80°$. Titers of 10^7–10^8 PFU/ml are typically obtained.

To generate a recombinant virus vector,[58,59,63] 7B cells are similarly infected and THZ.3 virions harvested, with the exception that cellular debris is collected by sequential low-speed (4000g for 10 min) and high-speed centrifugation (100,000g for 1 hr) and the supernatants are discarded. Pellets from both spins are combined and resuspended in 0.3 ml of Tris–EDTA (TE), pH 8.0, per 75-cm^2 flask. Total genomic DNA (which is mostly viral) is extracted with the help of a QIAamp blood kit (Qiagen, Chatsworth, CA), using two spin columns per 75-cm^2 flask. Digestion with *Pac*I releases a *lacZ* insert from the $\alpha22$ locus of THZ.3 and creates two large genomic fragments with $\alpha22$ flanking sequences. The gap in the viral genome is repaired by recombination with a *Not*I-linearized pB5 vector,[63] in which a synapto-pHluorin transciptional unit is flanked by homologous $\alpha22$

[60] R. R. Spaete and N. Frenkel, *Cell* **30**, 295 (1982).
[61] N. Schek and S. L. Bachenheimer, *J. Virol.* **55**, 601 (1985).
[62] A. D. Kwong, J. A. Kruper, and N. Frenkel, *J. Virol.* **62**, 912 (1988).
[63] D. M. Krisky, P. C. Marconi, T. Oligino, R. J. Rouse, D. J. Fink, and J. C. Glorioso, *Gene Ther.* **4**, 1120 (1997).

sequences. Plasmid and viral DNA are mixed at a 1:1 mass ratio and transfected into 7B cells at 80–90% confluency, using 8 μl of LipofectAMINE reagent (Life Technologies, Gaithersburg, MD) per microgram of DNA, and 0.15–0.2 μg of DNA per cm^2 of surface area. Once viral plaques appear (after 4–6 days), recombinants are purified to homogeneity by limiting dilution on 7B cells and examined for synapto-pHluorin expression by fluorescence microscopy.

Amplicon vectors are based on the pα4"a" backbone.[64] The plasmid is purified on a Qiagen column and transfected into 7B cells at 80–90% confluency, using 8 μl of LipofectAMINE reagent per microgram of DNA, and 0.15–0.2 μg of DNA per cm^2 of surface area. Twenty-four hours after transfection, the cells are infected with 0.1 PFU of THZ.3 (which in this case serves as the helper virus) per cell, in a minimal volume of growth medium containing 10 mM HEPES–NaOH, pH 7.1. To achieve high vector-to-helper ratios, the viral inoculum (0.5–1 ml per 75-cm^2 flask) is removed after a 90-min adsorption period and replaced with 6 ml of growth medium. The flasks are kept on a gently rocking platform until significant cytopathic effect develops; virions are harvested 24 hr later and stored after three freeze–thaw cycles at −80°. To estimate the vector-to-helper ratio of the stock, two assays are performed: (1) a plaque assay on 7B cells, which allow helper virions to replicate and form plaques while vector virions (whose genome consists of concatenated copies of the amplicon plasmid) are replication deficient, and (2) a test for synapto-pHluorin expression on a cell line that does not support THZ.3 replication, e.g., Vero. If the ratio appears low (i.e., less than 1 vector particle per 20 helper particles), it may be improved by serially passaging the stock on 7B cells. The outcome of such efforts is, however, often uncertain because the replication dynamics of two interacting viral populations are complex.[65]

Dissociated hippocampal neurons are infected after 2–4 weeks in culture. The virus stock is diluted to the desired multiplicity in conditioned medium, which is produced by maintaining the cultures in an excess of medium; of the 3 ml of growth medium initially added per well, at least one-half is dispensible. We typically draw 1–1.5 ml of medium from each well to be infected to prepare the viral inoculum. The remainder is set aside and reapplied when the inoculum is withdrawn after a 90-min infection period. Neurons are usually imaged 24–72 hr after infection (Fig. 2).

---

[64] M. S. Lawrence, D. Y. Ho, R. Dash, and R. M. Sapolsky, *Proc. Natl. Acad. Sci. U.S.A.* **92**, 7247 (1995).
[65] A. D. Kwong and N. Frenkel, *in* "Viral Vectors: Gene Therapy and Neuroscience Applications" (M. G. Kaplitt and A. D. Loewy, eds.), p. 25. Academic Press, San Diego, California, 1995.

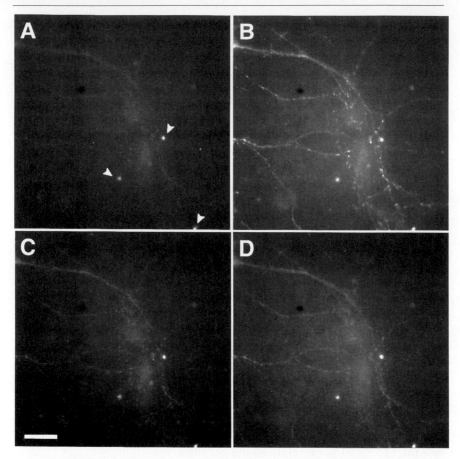

FIG. 2. Wide-field images of hippocampal neurons in dissociated culture, expressing ratiometric synapto-pHluorin after infection with an HSV amplicon vector. Numerous *en passant* synapses formed by infected neurons are visible. Fluorescence was excited with 12-nm-wide bands of the xenon arc spectrum, centered at 410 nm (*left*) and 470 nm (*right*). Images were collected before (*top*) and after (*bottom*) the addition of 50 mM NH$_4$Cl to neutralize the intravesicular pH. After addition of NH$_4$Cl, nerve terminals display fluorescence changes in accordance with movement of ratiometric pHluorin from an acidic to a neutral environment (see Fig. 1A): Fluorescence excited at 410 nm increases [compare (A) and (C)], while fluorescence excited at 470 nm decreases [compare (B) and (D)]. Arrowheads denote dirt particles, which do not undergo pH-dependent fluorescence changes. Scale bar: 25 μm.

Particle-Mediated Gene Transfer

Cortical slice cultures are transfected with the help of the Helios gene gun system (Bio-Rad, Hercules, CA). Synapto-pHluorin in pCI (Promega) is purified on a Qiagen column and precipitated onto the gold microcarrier particles according to the Helios gene gun instructions, with the exception that polyvinylpyrrolidone is omitted. Tubing is prepared with a special cutter (Helios system; Bio-Rad) and filled with an ethanolic solution containing 100 μg of DNA and 25 mg of gold particles (1.0-μm diameter; Bio-Rad). After 2–3 min, the gold particles precipitate and adhere to the walls of the tubing, which is then dried under nitrogen. Tubing sets are stable with desiccant at 4°, and can be used for transfections for up to 1 month.

Slices are transfected after a minimum of 48 hr in culture.[66] The gene gun system is operated according to the instruction manual, except where noted. To reduce slice damage due to high-pressure helium flow, the spacer located in front of the gun is covered with a nylon mesh (90-μm diameter; Small Parts, Miami Lakes, FL). In addition, the diffuser is removed from the barrel because in our experience it does not increase the number of transfected cells. To avoid cross-contamination of DNA samples, one barrel and one cartridge holder are used per sample, and empty cartridge holders and barrels are kept under UV for 20 min before loading. After adjusting the helium pressure to 150 psi, two or three "preshots" are fired with an empty cartridge to clean the helium pathway and the mesh and to stabilize the pressure. The cartridge is then locked in place and the gun positioned manually on top of the culture plate. Held perpendicular to the plate, with the spacer almost touching the insert, the gun is fired once per well.

While high transfection efficiencies are easily achieved, we often find it difficult to control the copy number of transfected genes adequately. In most cases, transfected cells express excessively high synapto-pHluorin levels, with the adverse consequence that nerve terminals appear no longer stained exclusively. Rather, synapto-pHluorin is expressed throughout the cell, including its dendritic arbor and soma (Fig. 3). Protocols to optimize the gene dosage for particle-mediated transfection are currently under investigation.

Transgenesis

Perhaps their most powerful advantage, synapto-pHluorins offer a natural way to translate the functional organization of an organism into a pattern

---

[66] D. C. Lo, A. K. McAllister, and L. C. Katz, *Neuron* **13,** 1263 (1994).

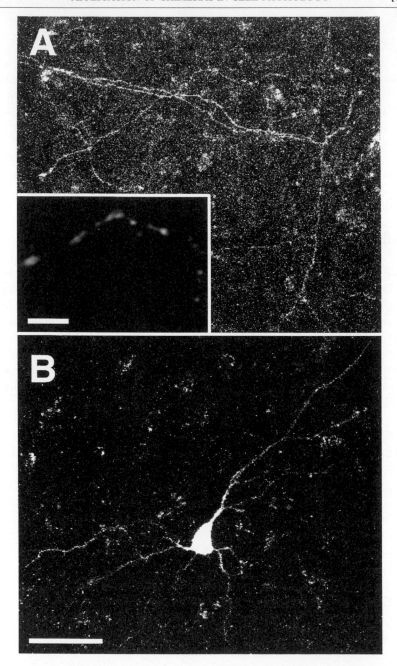

of probe expression. This is an extremely valuable property for tissues composed of many different cell types, because only one or a few of them can be singled out for analysis, based on a functional characteristic[38,41]: expression of a certain receptor, synthesis of a certain neurotransmitter, common developmental lineage, or location in a certain anatomical region. Because such characteristics are almost invariably linked to expression of a particular gene (or set of genes), the expression pattern of a genetically encoded indicator can be tailored to match them. This usually requires techniques of transgenesis in genetically tractable organisms such as *Caenorhabditis elegans*,[67] *Drosophila*,[68] zebrafish,[69] or mouse[70]. Expression of the indicator is controlled by the appropriate *cis*-acting sequences, using transgenes employing characterized promoters, homologous recombination,[71] or "enhancer detection" lines providing *trans*-activators in defined spatial patterns.[72,73]

Optical Imaging

A wealth of information on many aspects of optical imaging and image processing can be found in the books by Inoué and Spring,[74] Bracewell,[75]

[67] C. C. Mello, J. M. Kramer, D. Stinchcomb, and V. Ambros, *EMBO J.* **10,** 3959 (1991).
[68] G. M. Rubin and A. C. Spradling, *Science* **218,** 348 (1982).
[69] A. Meng, J. R. Jessen, and S. Lin, *Methods Cell Biol.* **60,** 133 (1999).
[70] B. Hogan, R. Beddington, F. Constantini, and E. Lacy, "Manipulating the Mouse Embryo: A Laboratory Manual." Cold Spring Harbor Laboratory Press, Cold Spring Harbor, New York, 1994.
[71] M. R. Capecchi, *Science* **244,** 1288 (1989).
[72] C. J. O'Kane and W. J. Gehring, *Proc. Natl. Acad. Sci. U.S.A.* **84,** 9123 (1987).
[73] E. Bier, H. Vaessin, S. Shepherd, K. Lee, K. McCall, S. Barbel, L. Ackerman, R. Carretto, T. Uemura, E. Grell, L. Y. Jan, and Y. N. Jan, *Genes Dev.* **3,** 1273 (1989).
[74] S. Inoué and K. R. Spring, "Video Microscopy," 2nd Ed. Plenum, New York, 1997.
[75] R. N. Bracewell, "Two-Dimensional Imaging." Prentice-Hall, Englewood Cliffs, New Jersey, 1995.

FIG. 3. Two-photon images of neurons in a cortical slice culture, transfected with ecliptic synapto-pHluorin by particle-mediated gene transfer. (A and B) Projections of individual neurons reconstructed from 24 optical sections spaced 1 μm apart. Staining in (A) appears to be axon specific; at higher magnification, individual presynaptic terminals are clearly visible (*inset*). Staining in (B) is diffusely somatodendritic. In all likelihood, differences in synapto-pHluorin expression levels account for the differences in cellular localization. Scale bars: (A) 25 μm; (B) 50 μm; *inset,* 5 μm.

and Castleman,[76] as well as in the volumes edited by Pawley[77] and Yuste, Lanni, and Konnerth.[78]

Optical Components for Fluorescence Microscopy

As measured on recombinant protein in solution, the fluorescence excitation maxima of ratiometric and ecliptic pHluorin are 395 and 475 nm (Fig. 1). Although useful information can often be obtained by recording changes in fluorescence emitted on excitation of a single-wavelength band ($\Delta F/F$), an ability to address both excitation bands is generally desirable and easily achieved with a suitable combination of light source, interference filters or monochromators, and dichroic beamsplitters.

In our initial development of the technique,[41] we used a monochromator equipped with a 75-W xenon arc lamp and 12-nm diffraction grating (Polychrome II; Till Photonics, Planegg, Germany) as the light source. This allowed us to record excitation spectra through the microscope and optimize our particular configuration. For the short-wavelength feature, the optimal excitation band was centered at 410 rather than 395 nm, probably because of a drop in the transmittance of our objective at about 400 nm. For the long-wavelength feature, the signal-to-noise ratio was maximal if the center of the excitation band was moved from 475 to 470 nm, to minimize bleedthrough of excitation light into the emission channel.

While a xenon arc lamp and monochromator make an ideal illuminator for development and troubleshooting, they may be a less ideal combination once the optimal conditions have been established. The mean luminous densities of xenon arc lamps are by a factor of 2–5 smaller than those of mercury arc lamps,[74] and monochromators typically pass a significantly narrower spectral band than interference filters. Together, these factors can noticeably limit the amount of excitation light delivered to the specimen. The short-wavelength band of ratiometric pHluorin, for example, although a prominent excitation peak *in vitro* (Fig. 1A), fluoresces only dimly under excitation by the 404- to 416-nm band of the xenon spectrum (Fig. 2A and C). The 400- to 440-nm band of the mercury spectrum, which contains two intense lines at 405 and 436 nm, might be a better choice for microscopy.

[76] K. R. Castleman, "Digital Image Processing." Prentice-Hall, Englewood Cliffs, New Jersey, 1996.
[77] J. B. Pawley (ed.), "Handbook of Biological Confocal Microscopy," 2nd Ed. Plenum, New York, 1995.
[78] R. Yuste, F. Lanni, and A. Konnerth (eds.), "Imaging Living Cells: A Laboratory Manual." Cold Spring Harbor Laboratory Press, Cold Spring Harbor, New York, 1999.

Based on these considerations (and cautioning that we have not yet had practical experience with interference filters for excitation of the short-wavelength feature), the following excitation bandpass filters would appear suitable: for the short-wavelength band, Chroma (Brattleboro, VT) HQ 415/30, HQ 420/40, or D 410/60; for the long-wavelength band, any of a number of enhanced GFP (EGFP) or fluorescein filters, such as Chroma HQ 470/40 or HQ 480/40, or Zeiss (Thornwood, NY) 450–490. We feel that the bandwidths of the two excitation filters (D 410/30 and HQ 470/20) in the dedicated pHluorin set from Chroma may be unnecessarily narrow.

The pHluorin emission band, with a peak at 508 nm, overlaps that of GFP and fluorescein, and can be captured by any of a number of GFP or fluorescein bandpass filters (Chroma HQ 525/50 or HQ 535/50, Zeiss BP 515–565). It is wise to match a particular emission filter to a particular dichroic. If both pHluorin peaks are to be excited, an "extended-reflection" dichroic such as Chroma 495 DCXR or 500 DCXR, or Zeiss FT 510, is imperative; dichroics included in the popular Chroma High Q filter sets, such as Q 495 LP or Q 505 LP, are unsuitable because they transmit up to 70% of the 400- to 450-nm band. Extended-reflection dichroics characteristically possess a somewhat shallower slope than dichroics of the Q series, i.e., the transition from reflection to transmission occurs with an optical density (OD) of 2 in ~26 nm, as opposed to an OD of 2 in ~17 nm. To avoid bleedthrough with extended-reflection dichroics, the cutoff wavelength of the "reddest" excitation bandpass filter and the cuton wavelength of the emission bandpass filter should be kept at least 25 nm apart.

Two-photon excitation spectra can differ considerably from their single-photon counterparts.[79] At the time of writing, systematic tests have been performed only with ecliptic pHluorin. Its two-photon excitation spectrum is similar to the single-photon case, except that the wavelength axis is scaled by a factor of 1.95 and the sharp single-photon peaks are broadened. Importantly, wavelengths above 900 nm excite a state with ecliptic properties: its fluorescence is completely quenched at acidic pH.

Microscope Layout

For thin samples such as dissociated cultures grown on coverslips, an inverted microscope (Zeiss Axiovert 100 or 135 TV), equipped with a high-NA oil immersion objective (Zeiss Plan-Neofluar ×40/1.3, ×63/1.25, or ×100/1.3) and a bottom camera port ("Keller hole") that places specimen

[79] C. Xu and W. W. Webb, *J. Opt. Soc. Am.* **13,** 481 (1996).

and detector on the same optical axis provides unsurpassed light collection efficiency. The glass coverslip is inserted into an MS-502 clamp chamber (ALA Scientific Instruments, Westbury, NY) that fits into a thermostatted PDMI-2 microincubator (Medical Systems, Greenvale, NY) mounted on the microscope stage. Experiments are performed in 25 mM HEPES–NaOH (pH 7.4), 119 mM NaCl, 2.5 mM KCl, 2 mM CaCl$_2$, 2 mM MgCl$_2$, 30 mM glucose.

For thick samples such as slice cultures or acute brain slices, an upright microscope with a fixed stage and objective focusing [Zeiss Axioskop FS or FS2, Olympus (Melville, NY) BX50WI], equipped with water immersion objectives with long working distances (Zeiss Achroplan ×40/0.8 or ×63/0.9; Olympus ×40/0.8 or ×60/0.9), performs best. The slice is held in a custom-built Plexiglas chamber using a small "harp," which consists of a U-shaped flat platinum wire with small strands of nylon thread glued onto it.

Image Collection

Wide-field images are collected with either a cooled CCD (PentaMax-512EFT frame-transfer camera with 12-bit EEV CCD-37 chip and fiberoptically coupled Gen IV image intensifier; Princeton Instruments, Trenton, NJ) or SIT camera (C2400-08; Hamamatsu Photonics, Bridgewater, NJ). The CCD is controlled through MetaMorph/MetaFluor software (Universal Imaging, West Chester, PA) running on a Pentium-based PC. The SIT camera is connected to a frame grabber (LG-3; Scion Corporation, Frederick, MD) in a Power MacIntosh running NIH Image (http://rsb.info.nih.gov/nih-image/). To minimize photodamage, the mercury arc lamp (Olympus) is shuttered (Uniblitz; Vincent Associates, Rochester, NY) and the Polychrome II monochromator synchronized with image acquisition. Macros for shutter control in NIH Image were written by C.-B. Chien (University of California at San Diego, San Diego, CA) and customized.

Using the monochromator as the light source, typical exposure times for the PentaMax CCD range from 20 to 200 msec/frame, depending on synapto-pHluorin expression level and intensifier gain. When the CCD is operated at high frame rates with accordingly short exposure times, the intensifier is electronically shuttered ("gated") to prevent smearing during frame transfer. The SIT camera is used either in time-lapse mode, with images taken every 5–30 sec and averaged over 16 frames, or at video rates (33 msec/frame) without averaging.

Two-photon images[80] are collected on a custom-built instrument employing scanners of the Fluoview confocal microscope (Olympus). A Ti:sapphire laser producing 130-fsec pulses at 75 MHz (Mira; Coherent) is pumped by either an 8-W argon ion laser (Innova; Coherent) or a 5-W solid-state laser (Verdi; Coherent). The microscope is equipped with a ×60/0.9 water immersion objective with improved infrared transmission (Olympus) and optical components from Olympus and Spindler-Hoyer (Göttingen, Germany). External PMTs (Hamamatsu, Tokyo, Japan) are used in whole-area detection mode[80]; they are mounted either on top of the objective or on top of the trinocular tube. Images are reconstructed with Fluoview software (Olympus).

Two-photon images of fluorescent structures (putative nerve terminals) are acquired at the highest digital zoom (×10), resulting in a spatial resolution of 20 pixels/μm. For reconstructions of neurons, serial images in the z plane are taken every 1 μm in order to obtain a complete picture of the imaged field. For time-lapse sequences, images are collected every 2–45 sec. In some cases, two to five focal planes 0.5–1 μm apart are scanned at each time point; these are later projected into a single plane using custom-written macros in NIH Image. Images are aligned visually to correct for drift in the xy plane.

Calibration

To relate ratiometric or fluorescence intensity ($\Delta F/F$) measurements to changes in pH, a series of calibration images needs to be recorded under conditions in which the pHluorin is exposed to solutions of defined pH. This can be accomplished in a number of ways.

The first option is glutathione–agarose beads decorated with recombinant pHluorin, which is produced as a glutathione-*S*-transferase (GST) fusion protein. *Escherichia coli* strain BL-21 is transformed with pHluorin in pGEX-2T (Amersham Pharmacia Biotech, Piscataway, NJ) and grown to an OD_{600} of 1.0. Protein synthesis is induced with 0.1 mM isopropyl-β-D-thiogalactopyranoside (IPTG) at 25° and induction continued overnight. Bacteria are lysed by lysozyme treatment, sonication, or passage through a French pressure cell or cell disruptor, and the lysate is cleared by centrifugation at 100,000g for 1 hr. The supernatant is incubated with glutathione–agarose beads (glutathione–Sepharose 4B; Amersham Pharmacia Biotech, or Sigma) for ~30 min at room temperature. Beads and bound GST–pHluorin are collected by low-speed centrifugation (500g for 1 min),

[80] W. Denk, J. H. Strickler, and W. W. Webb, *Science* **248,** 73 (1990).

washed, equilibrated in buffers spanning the desired pH range, and spotted onto glass slides or coverslips for imaging. The beads are large enough to settle quickly by gravity and do not undergo Brownian motion.

Alternatively, cells expressing a glycosylphosphatidylinositol (GPI)-anchored pHluorin at their surface[41] can be imaged in buffers of defined pH. HeLa cells (or any other easily transfected cell type available in the laboratory) are grown on coverslips and transiently transfected with pCI (Promega) carrying a pHluorin cDNA sandwiched between sequences encoding a 30-amino acid preprolactin signal peptide on the one hand and the 37 carboxy-terminal amino acids of decay accelerating factor, which contain a signal for GPI anchor addition,[81] on the other. Cells are transfected using 5 μl of SuperFect reagent (Qiagen) per microgram of plasmid DNA, and 0.2 μg of DNA per cm^2 of surface area.

A cautionary remark is appropriate with respect to either of these calibration methods. GST fusion proteins form dimers, and GPI-anchored proteins can cluster in glycolipid "rafts" at the cell surface.[82] When wild-type GFP is dimerized via GST or clustered via GPI anchors, the forced proximity of two or more GFP molecules causes some distortion of the excitation spectrum of the protein.[83] It is conceivable that pHluorins, which as direct descendants of wild-type GFP retain many of its spectral characteristics,[41,84–87] suffer similar distortions. A way to avoid the issue is to express soluble pHluorin in the cytoplasm, and to set up a "proton clamp" across the plasma membrane. HeLa cells are transfected with pHluorin in pCI, using the conditions for transfection of GPI-anchored pHluorin, and incubated in buffers of defined pH. The buffers are supplemented with 10 μM nigericin ionophore (Sigma) and contain K$^+$ instead of Na$^+$ as the major cation. This collapses the K$^+$ gradient across the membrane and allows H$^+$-driven K$^+$/H$^+$ exchange between cytoplasm and extracellular medium, to equalize the proton concentration in both compartments.[88,89] It is advisable

[81] I. W. Caras, G. N. Weddell, M. A. Davitz, V. Nussenzweig, and D. W. Martin, Jr., *Science* **238**, 1280 (1987).
[82] K. Simons and E. Ikonen, *Nature (London)* **387**, 569 (1997).
[83] D. A. De Angelis, B. Miesenböck, B. V. Zemelman, and J. E. Rothman, *Proc. Natl. Acad. Sci. U.S.A.* **95**, 12312 (1998).
[84] H. Morise, O. Shimomura, F. H. Johnson, and J. Winant, *Biochemistry* **13**, 2656 (1974).
[85] R. Y. Tsien, *Annu. Rev. Biochem.* **67**, 509 (1998).
[86] D. C. Prasher, V. K. Eckenrode, W. W. Ward, F. G. Prendergast, and M. J. Cormier, *Gene* **111**, 229 (1992).
[87] M. Chalfie, Y. Tu, G. Euskirchen, W. W. Ward, and D. C. Prasher, *Science* **263**, 802 (1994).
[88] G. R. Bright, G. W. Fisher, J. Rogowska, and D. L. Taylor, *J. Cell Biol.* **104**, 1019 (1987).
[89] M. Kneen, J. Farinas, Y. Li, and A. S. Verkman, *Biophys. J.* **74**, 1591 (1998).

to ascertain the proper operation of the clamp with the help of a fluorescent pH indicator that is spectrally distinct from pHluorin and can be imaged simultaneously. Choices include carboxy SNARF-1 and carboxy SNAFL-2, which are available as cell-permeant esters from Molecular Probes (Eugene, OR); pertinent filter sets are available from Chroma.

A quick method to elicit a pHluorin response, and to roughly estimate the acidity of the environment of a synapto-pHluorin, is to add 25–50 mM NH$_4$Cl to the medium[41] (Fig. 2). NH$_4$Cl releases ammonia, which diffuses across cellular membranes and quenches free protons in acidified organelles, thus neutralizing their internal pH.[90] The effect is fully reversible after washout of NH$_4$Cl.[90]

Outlook

All methods in synaptic physiology ultimately strive to resolve the quantal nature of neurotransmission. For decades the exclusive domain of electrophysiology, quantal synaptic events have only recently entered the realm of optical imaging. In presynaptic terminals, individual vesicles have been stained with FM 1-43, and quantal dye release has been monitored.[8,24,37] In postsynaptic spines, calcium transients activated by single quanta of transmitter have been visualized with fluorescent calcium indicators.[91,92]

It is hoped that continued development will refine synapto-pHluorins to such a point that they will be able to perform at a comparable level. From our current vantage point, this presents a formidable technical challenge. At the vesicular level, geometry and functionality constrain the number of synapto-pHluorin molecules that can be packed onto the surface of a single vesicle. pHluorins with enhanced brightness, synapto-pHluorins carrying multiple fluorescent modules, and methods to fully substitute endogenous vesicle proteins with pHluorin-tagged derivatives would help to maximize the fluorescent signal emitted per released quantum of neurotransmitter.

At the synaptic level, the signal recorded from a single terminal is composed of fluorescence emitted by all synapto-pHluorins located within the limits of optical resolution, including molecules at the cell surface (where they appear after vesicle release), in resting vesicles, and in the endosomal pathway. pHluorins undergoing sharp transitions between nonfluorescent and brightly fluorescent states would help to enhance the signal due to molecules in states of activation. Targeting modules with particularly clean

[90] F. R. Maxfield, J. Cell Biol. 95, 676 (1982).
[91] R. Yuste and W. Denk, Nature (London) 375, 682 (1995).
[92] Z. F. Mainen, R. Malinow, and K. Svoboda, Nature (London) 399, 151 (1999).

subcellular distributions, with high selectivity for synaptic vesicles[22] or the relevant functional pools of synaptic vesicles, would help to further reduce background fluorescence due to nonsynaptic organelles or nonfunctional vesicle pools.

[39] Studying Organelle Physiology with Fusion Protein-Targeted Avidin and Fluorescent Biotin Conjugates

By Minnie M. Wu, Juan Llopis, Stephen R. Adams,
J. Michael McCaffery, Ken Teter, Markku S. Kulomaa,
Terry E. Machen, Hsiao-Ping H. Moore, and Roger Y. Tsien

Introduction

We have developed and used a novel method for studying the lumenal pH of specific cellular organelles: a membrane-permeable, pH-sensitive fluorescein-biotin derivative is targeted to specific organelles expressing avidin chimera proteins. Until recently, the major hurdle to studying organelle pH in live cells had been the lack of appropriate methods for targeting pH probes to specific organelles. Several groups have now targeted pH dyes specifically to the Golgi, *trans*-Golgi network (TGN), and endoplasmic reticulum (ER) of live, intact cells by using the following methods: microinjection of cells with pH dye-filled liposomes, which appeared to fuse preferentially with the Golgi,[1] targeting a fluorescently labeled bacterial toxin to the Golgi or ER via retrograde transport,[2-4] and expressing pH-sensitive green fluorescent protein (GFP) mutants in specific organelles.[5-7] All these methods have provided important contributions to the understanding of the regulation of organelle pH; however, each approach has certain drawbacks. Microinjection of dye-containing liposomes is technically difficult, and fusion of the liposomes to Golgi membranes may result in perturbation of

[1] O. Seksek, J. Biwersi, and A. S. Verkman, *J. Biol. Chem.* **270,** 4967 (1995).

[2] J. H. Kim, C. A. Lingwood, D. B. Williams, W. Furuya, M. F. Manolson, and S. Grinstein, *J. Cell Biol.* **134,** 1387 (1996).

[3] N. Demaurex, W. Furuya, S. D'Souza, J. S. Bonifacino, and S. Grinstein, *J. Biol. Chem.* **273,** 2044 (1998).

[4] J. H. Kim, L. Johannes, B. Goud, C. Antony, C. A. Lingwood, R. Daneman, and S. Grinstein, *Proc. Natl. Acad. Sci. U.S.A.* **95,** 2997 (1998).

[5] M. Kneen, J. Farinas, Y. Li, and A. S. Verkman, *Biophys. J.* **74,** 1591 (1998).

[6] J. Llopis, J. M. McCaffery, A. Miyawaki, M. G. Farquhar, and R. Y. Tsien, *Proc. Natl. Acad. Sci. U.S.A.* **95,** 6803 (1998).

[7] G. Miesenböck, D. A. De Angelis, and J. E. Rothman, *Nature (London)* **394,** 192 (1998).

the ionic composition and function of the Golgi. Using fluorescently labeled bacterial toxins to measure pH is more advantageous, but globotriaosyl ceramide, the cell surface glycolipid necessary for toxin internalization, is not expressed on all cell types. Also, the Golgi localization of the pH probe relies on the kinetics of retrograde transport of the toxin, and signal from other, adjacent organelles in the secretory pathway cannot be eliminated. Furthermore, the toxin-derived pH probe labels only compartments that can be reached by retrograde transport (e.g., endosomes, Golgi, and ER). Organelles such as secretory granules and mitochondria remain inaccessible. Targeting pH-sensitive GFP mutants to cellular compartments is the least invasive and most specific method for studying organelle pH, but most GFPs are nonratiometric, and thus prone to artifacts due to changes in focus or dye concentration, and they do not exhibit optimal pK_a values for all organelle environments.

We developed a new method for measuring organelle pH because we wished to use ratiometric dyes with pK_a values close to the pH found in the Golgi and ER (and other organelles). A fluorescein–biotin-based ratiometric pH dye (Flubi-2) with pK_a 6.7 was targeted as a nonfluorescent membrane-permeable form (Flubida-2) to the ER and Golgi, using avidin chimera proteins. This chapter describes how the avidin chimera proteins were expressed in specific organelles of mammalian cells, the synthesis and characteristics of the Flubida dyes, and the methods used for labeling avidin-containing compartments with Flubi dyes and then studying pH regulation in mammalian organelles.

Experimental Strategy and Construction of Avidin–KDEL and Sialyltransferase–Avidin Plasmids

The strategy for targeting Flubi-2 to organelles is shown in Fig. 1A and C. We constructed avidin chimera plasmids that would express avidin chimera proteins localized to specific intracellular compartments on the basis of their targeting sequences. Cells expressing these avidin chimera proteins were loaded with Flubida-2 (Fluorescein biotin diacetate-2, the membrane-permeable, nonfluorescent ester form of Flubi-2), which accumulates in the cell after esterase hydrolysis of the charge-masking acetate esters. The high affinity of biotin for avidin retains the dye in the cellular compartment expressing the chimera while unbound dye leaks out of other cellular compartments during a subsequent washing step. All the chimera proteins were constructed so that the avidin portion would be in the organelle lumen. The targeting sequence for sialyltransferase (ST), a *trans*-Golgi resident enzyme, was fused to avidin (AV) to make the single-pass trans-membrane ST–avidin fusion protein (ST–AV), which was used to target

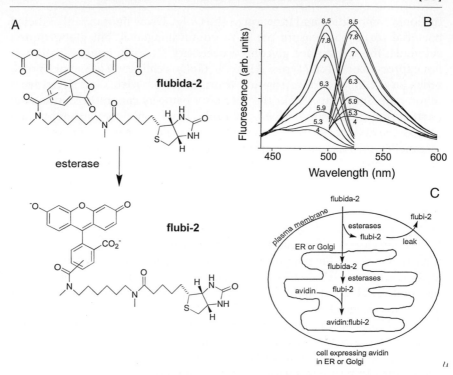

FIG. 1. A strategy for targeting pH-sensitive dyes to organelles. (A) Chemical structures of Flubi-2 and its membrane-permeable, nonfluorescent ester form, Flubida-2. (B) *In vitro* pH dependence of the excitation spectrum (emission, 530 nm) and emission spectrum (excitation, 480 nm) of Flubi-2. (C) Organelle-specific avidin chimera proteins were transiently expressed in HeLa cells. Transfected cells were loaded with 2–4 μM Flubi-2 overnight (at least 10 hr), before chasing with normal growth medium (at least 2 hr) to minimize background fluorescence. Labeled cells are then ready for ratio imaging experiments.

Flubi dyes to the Golgi lumen. Soluble avidin (AV) was appended with the C-terminal tetrapeptide KDEL, an ER retrieval motif, to give soluble AV–KDEL, which was used to target Flubi dyes to the ER lumen.

Sialyltransferase was polymerase chain reaction (PCR) amplified from human α-2,6-sialyltransferase–avidin (B. Seed, Massachusetts General Hospital, Boston, MA), using primers (5'-cgcgggaagcttgccaccatgattcacac-caacctg-3' and 5'-cgcgggcggatcctgggtgctgcttgagga-3') that introduced 5' *Hin*dIII and 3' *Bam*HI restriction sites, allowing isolation of a fragment containing the cytosolic, transmembrane, and truncated lumenal domains of ST. This truncated ST contains the 17-amino acid sequence sufficient for Golgi retention.[8] Avidin was PCR amplified from chicken avidin (courtesy of M. Kulomaa, University of Tampere, Finland), using primers (5'-cgcggggatcccgccagaaagtgctcgctg-3' and 5'-cgcgggggcggccgct cactccttctgtgtgcg-3') that allowed isolation of full-length avidin with 5' *Bam*HI and 3' *Not*I restriction sites. ST and AV were triple-ligated into pCDM8 plasmid (courtesy of B. Seed), using *Hind*III, *Bam*HI, and *Not*I restriction sites. Avidin–KDEL (AV–KDEL) was PCR amplified from chicken avidin with primers (5'-atccaagcttgctgcagagatggtgc-3' and 5'-taatg-gatcctcacagctcgtctttctccttctgtgtgcgcaggcg-3') that appended the KDEL tet-rapeptide sequence to the C terminus of avidin and allowed cloning of AV–KDEL by 5' *Hind*III and 3' *Bam*HI restriction sites into the pcDNA3 vector (InVitrogen, San Diego, CA).

Expression of Avidin–KDEL and Sialyltransferase–Avidin in
  Mammalian Cells

AV–KDEL and ST–AV were expressed in HeLa cells, which were cultured at 37° in Dulbecco's modified Eagle's medium (DMEM; Biowhittaker, Walkersville, MD) supplemented with 10% (v/v) fetal calf serum (Sigma, St. Louis, MO), penicillin (20 U/ml), and streptomycin (20 mg/ml) (Biowhittaker) and grown in the presence of air–6% CO_2. HeLa cells were transiently transfected with either AV–KDEL or ST–AV DNA by electroporation. Cells from a 50–75% confluent 10-cm dish were trypsinized, washed twice, and resuspended in 0.8 ml of ice-cold sucrose buffer [270 mM sucrose, 7 mM sodium phosphate buffer (pH 7.4), 1 mM MgCl$_2$]. Ten micrograms of DNA (AV–KDEL or ST–AV) was added to the cell suspension just prior to electroporation (GenePulser; Bio-Rad, Hercules, CA) in ice-cold 0.4-mm gap cuvettes (Bio-Rad) at 500 V, 200 Ω, 25 μF. After electroporation, cells recovered on ice for 10 min before being replated in normal growth medium. Peak expression of AV–KDEL and ST–AV for transiently transfected HeLa cells was 48 to 72 hr posttransfection. We also generated a cell line stably expressing AV–KDEL by selection with G418. Neither transient nor stable expression of avidin chimera proteins appeared to affect cell health or viability, because transfected cells were similar in morphology and growth rate to untransfected cells. In addition to the AV–KDEL and ST–AV chimera proteins, we also constructed avidin chi-mera proteins that localize to the ER–Golgi intermediate compartment (ERGIC–avidin), medial-Golgi (MG160–avidin), and secretory granules (proopiomelanocortin–avidin).

The intracellular localization of the ER- and Golgi-localized avidin

---

[8] S. H. Wong, S. H. Low, and W. Hong, *J. Cell Biol.* **117,** 245 (1992).

chimera proteins was confirmed by fluorescence staining [with fluorescein isothiocyanate (FITC)–biotin or fluorescent antibodies] and by electron microscopy. For fluorescence staining, 48 to 60 hr posttransfection, cells plated on 22-mm coverslips were washed twice with phosphate-buffered saline (PBS), fixed in 4% (w/v) paraformaldehyde for 20 min at room temperature, and permeabilized in ice-cold 100% methanol for 15 sec. Cells were incubated for 30 min in 1% (w/v) bovine serum albumin (BSA)–PBS before incubation in either FITC–biotin (Molecular Probes, Eugene, OR) or in goat anti-avidin D antibody (Vector Laboratories, Burlingame, CA) for 1 hr at room temperature. Cells stained with FITC–biotin were then rinsed once with PBS and mounted on slides with a nonbleach reagent (KPL mounting medium; Kirkegaard & Perry, Gaithersburg, MD) and viewed with a Zeiss (Oberkochen, Germany) Axiophot fluorescence microscope with ×63 oil objective. For electron microscopy, 48 to 60 hr posttransfection, cells were washed with PBS, fixed in 4% (w/v) paraformaldehyde–0.05% (v/v) glutaraldehyde for 1 hr at room temperature, permeabilized with 0.1% (w/v) saponin in 1% (w/v) BSA–PBS, and sequentially incubated in primary goat anti-avidin antibody followed by horseradish peroxidase-conjugated secondary antibody diluted in 1% (w/v) BSA–PBS containing 0.1% (w/v) saponin for 1 hr each. The cells were washed in 0.1 M cacodylate-HCl, pH 7.4, followed by fixation in 2% (v/v) glutaraldehyde contained in 0.1 M cacodylate-HCl, pH 7.4, for 1 hr, incubation in diaminobenzidine, postfixation in 0.1 M cacodylate containing 1% (w/v) OsO_4 and 1% (w/v) KFeCN, and then processing as previously described.[9] Figure 2 shows the intracellular localization of AV–KDEL by immunoperoxidase staining, visualized by electron microscopy.

Rescue of Mislocalized Avidin Chimera Proteins

Oligomerization of avidin due to the well-known homotetrameric interactions appeared not to affect targeting of AV–KDEL or ST–AV.[10] However, avidin oligomerization may become problematic with some constructs. For example, we have found that avidin–furin and avidin–P-selectin were not targeted, as expected, to the TGN and secretory granules, respectively, in neuroendocrine (AtT-20) cells, but to the nuclear envelope and large, membrane-bound vesicles, which may have been ER derived. One possibility is that the mistargeting was due to aggregate formation when an oligomeric targeting signal (P-selectin can form hexamers, and furin forms dimers) was combined with tetrameric avidin. If so, overexpression of native

[9] J. M. McCaffery and M. G. Farquhar, *Methods Enzymol.* **257,** 259 (1995).
[10] N. M. Green, *Adv. Protein Chem.* **29,** 85 (1975).

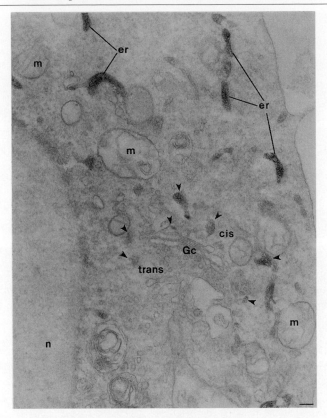

FIG. 2. Immunoperoxidase staining of AV–KDEL. NRK cells expressing AV–KDEL were immunoperoxidase labeled with a polyclonal antibody against avidin and then visualized by electron microscopy. Extensive labeling is seen confined to the ER (solid lines) and dilated rims of the *cis*-Golgi network (arrowheads) but is not detected on the *trans*-most Golgi cisternae, consistent with previously reported KDEL localizations. Bar: 0.2 μm.

soluble avidin (not fused to any membrane-bound targeting sequence) should compete with the avidin chimera proteins for tetramerization and prevent the formation of mistargeted aggregates. To test this, we made a recombinant adenovirus carrying a tetracycline-repressible soluble avidin construct. Expression of this virus in AtT-20 cells rescued the mislocalization of avidin–furin (and avidin–P-selectin, not shown). Cells expressing avidin–furin alone showed aberrant localization of FITC–biotin staining to the nuclear envelope, ER, and vesicular structures (Fig. 3B), which differed from the TGN localization of native furin (Fig. 3A). Coexpression of avidin–furin with soluble avidin resulted in correct localization to the TGN (Fig. 3C); the observed FITC–biotin staining was due to membrane-

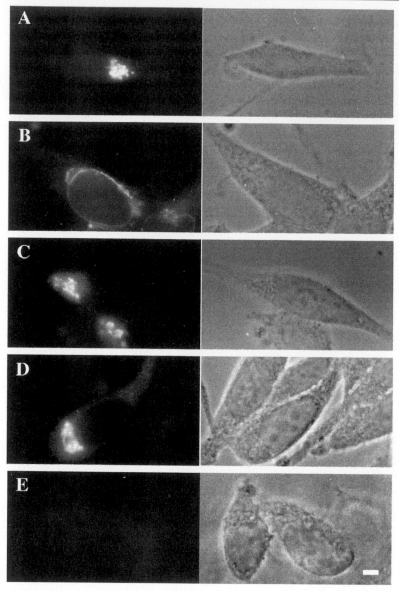

Fig. 3. Rescue of mislocalized avidin–furin. (A) The normal TGN localization of furin is shown in fixed, permeabilized, and stained (with anti-FLAG antibody) AtT-20 cells expressing an FLAG-tagged, full-length Furin protein. (B) Avidin–furin in AtT-20 cells is mislocalized to the nuclear envelope, ER, and vesicular structures. Cells were fixed, permeabilized, and stained with FITC–biotin. (C) When avidin–furin was coexpressed with soluble avidin, using a recombinant adenovirus, and treated with cycloheximide (100 μg/ml) for 3 hr, avidin–furin

bound avidin–furin and not soluble avidin, because in this experiment the cells were pretreated with cycloheximide (100 μg/ml) for 3 hr to clear soluble avidin from the secretory pathway (see below).

We performed control experiments (Fig. 3D and E) to ensure that the staining in the rescued AtT-20 cells was not due to soluble avidin alone. Untransfected cells were infected with avidin–adenovirus and then fixed, permeabilized, and stained or, alternatively, treated with cycloheximide (100 μg/ml) to inhibit new protein synthesis before being fixed, permeabilized, and stained. Control cells that had been infected with avidin–adenovirus showed FITC–biotin staining in a perinuclear structure, probably the Golgi, on its way to being secreted (Fig. 3D). After a 3-hr cycloheximide treatment, all the soluble avidin was chased out of the cells (Fig. 3E).

Thus, as is the case with all chimera proteins, careful controls should be performed to assure proper organelle localization of the avidin chimera proteins. In some cases in which avidin chimera proteins become mistargeted due to oligomerization, correct targeting can be achieved by the rescue strategy described here or by constructing avidin chimera proteins with a mutant, monomeric avidin (M. Kulomaa, unpublished results, 1999).

Synthesis and *in Vitro* Characterization of Flubi Dyes

The chemical structures of the Flubi dyes are shown in Fig. 4. All the dyes are membrane permeable and can be used in the ratiometric mode. Four Flubi dyes, each with a different pK_a, allow measurements of pH in organelles with different pH values by matching the pK_a of the indicator dye with the pH environment of the organelle lumen. The range of pK_a values of the Flubi dyes allow measurements in organelles that range in lumenal pH from 7.2 (ER) to 4.5 (lysosomes). The Flubi-2 dye was advantageous for studies of pH regulation in the ER (~pH 7.2) and Golgi (~pH 6.4), because it had a pK_a of about 6.5 (see below, pK_a 6.8 for ER *in situ*, pK_a 6.6 for Golgi *in situ*). Another advantage of using the ratiometric Flubi dyes compared with nonratiometric pH indicators such as most pH-sensitive GFP mutants is that changes in fluorescence intensity due to variation in

---

was correctly localized to the TGN, as detected by FITC–biotin staining. (D) Control cells that had been infected with the avidin adenovirus alone showed FITC–biotin staining in a perinuclear structure, probably the Golgi, on its way to being secreted. (E) After avidin-infected control cells [e.g., cells in (D)] were treated with cycloheximide (100 μg/ml) for 3 hr, all the soluble avidin was chased out of the cells, resulting in no specific FITC–biotin staining. Bar: 5 μm.

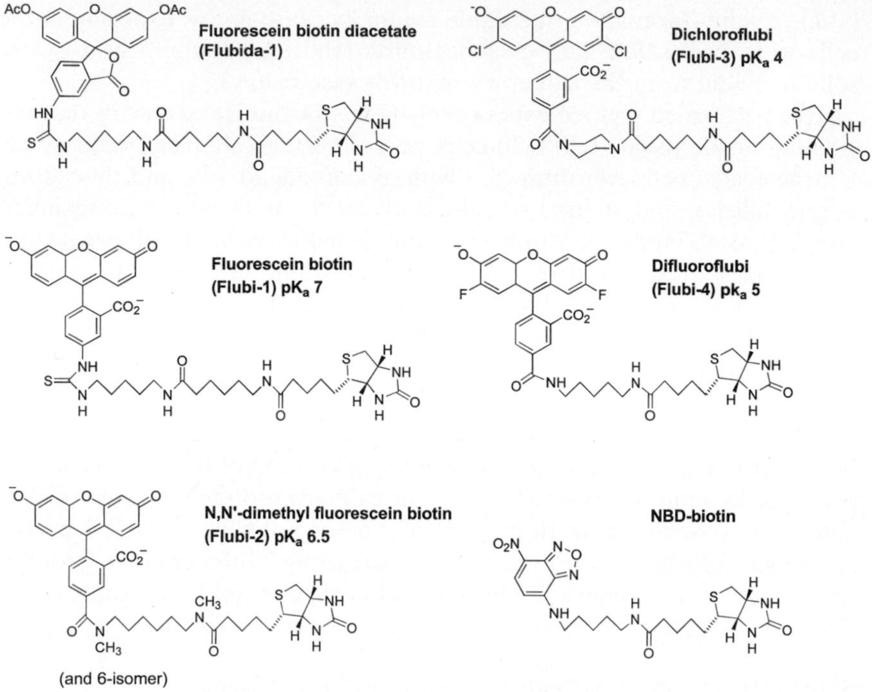

FIG. 4. Chemical structures of the Flubi dyes. Flubida-1 is the nonfluorescent and membrane-permeable form of Flubi-1. Corresponding structures for Flubida-2, -3, and -4 are not illustrated but contain the fluorescein moiety in the same acetylated and lactonized form. The pK_a values refer to the apparent H^+ dissociation constants for the avidin-bound dyes measured *in vitro*.

path length or dye concentration are eliminated, because these factors affect both the pH-sensitive and pH-insensitive wavelengths equally. The Flubi dyes have an excitation spectrum with a pH-sensitive wavelength at 490 nm and pH-insensitive wavelength at 440 nm (see Fig. 1B for the pH dependencies of the excitation and emission spectra of Flubi-2 *in vitro*). Hydrolysis of Flubida-2 to Flubi-2 inside the cell occurs in the cytosol and organelles, as demonstrated by the appearance of the fluorescent species (Flubi-2) on cleavage of the acetate groups.

Synthesis of Flubida-1

Fluorescein biotin (1.8 mg, 2.2 μmol) was suspended in acetic anhydride (2 ml), refluxed until a colorless solution was formed (about 2 hr), and evaporated. Thin-layer chromatography (TLC) [SiO$_2$–10% (v/v) methanol–

CHCl$_3$] revealed two major products that turned fluorescent on heating the plate. Only the least polar product was positive when sprayed with a biotin-specific stain.[11] Reactions with model compounds indicated that these reaction conditions resulted in partial acylation of the biotin ureido group, giving a product that did not test positive for biotin and did not bind avidin. The mixture was used without further purification.

Synthesis of Flubida-2

N,N′-Dimethyl-(6-aminohexyl)biotinamide. N,N′-Dimethylhexane-2,6-diamine (65 μl, 0.37 mmol; Aldrich, Milwaukee, WI) was added with rapid stirring to a solution of succinimidyl D-biotin (25 mg, 0.073 μmol) in dry dimethyl fluoride (DMF, 0.5 ml) under argon. After 2 hr at room temperature, the white solid was removed by centrifugation, the supernatant was evaporated to dryness and redissolved in absolute ethanol, and the product was separated by preparative layer chromatography [Analtech (Newark, DE) SiO$_2$ G plate eluted with butanol–hydroxyacetate–H$_2$O (12:3:5, by volume)] and visualized at one edge with biotin spray[11] (R_f 0.4). The product was extracted twice with hot ethanol and evaporated to dryness. Silica was removed by centrifugation through a 0.2-μm pore size filter after dissolution in a few milliliters of 20% (v/v) methanol–CHCl$_3$. Evaporation of the filtrate gave an off-white solid. Yield, 20 mg (74%). ^{1}H-NMR (200 MHz, δ ppm, CDCl$_3$-CD$_3$OD) 1.0–1.6 (m's, 14H, -CH_2-), 2.16 (t, 2H, -CH_2-CO), 2.47 (s, 3H, *Me*HN-), 2.55 (d, 2H, biotin S-CH_2), 2.71 (s, MeN-CO), 2.75 (s, mm, CH_2NH, biotin S-CH_2), 2.97 (m, 1H, biotin S-CH-), 3.13 (t, partially obscured by CD$_3$OD peaks, CH_2-N-CO), 4.1-4.4 (multiplets partially obscured by water peak, biotin N-CH). Electrospray mass spectrum: 371.3 (M+1); calc. 371.5.

Flubida-2. N,N′-Dimethyl-(6-aminohexyl) biotinamide (7.5 mg, 20 μmol) was dissolved in 20% (v/v) methanol–CHCl$_3$ (1 ml) and 5 (and 6)-carboxyfluorescein; succinimidyl ester (15 mg, 33 μmol) dissolved in dry DMF (1 ml) was added, followed by N,N Diisopropyl-ethylamine (DIEA) (100 μl). The red solution was stirred overnight, evaporated, and resuspended in acetic anhydride (2 ml) and pyridine (5 μl) and heated at 60° with stirring for 20 min. The colorless reaction mixture was evaporated and the product separated by SiO$_2$ chromatography [5–10% (v/v) methanol–CHCl$_3$] as a gummy white solid. Yield, 4 mg (25%). ^{1}H-NMR (200 MHz, δ ppm, CDCl$_3$-CD$_3$OD) was complex, containing peaks characteristic for a 5 (and 6)-substituted fluorescein diacetate, a N,N′-dimethylhexanediamide, and biotin. Electrospray mass spectrum: 813.3 (M + 1); calc. 813.9.

---

[11] D. B. McCormick and J. A. Roth, *Methods Enzymol.* **18A,** 383 (1970).

Flubi-2. Flubida-2 [10 μl of 3.5 m*M* solution in dimethyl sulfoxide (DMSO)] was diluted in methanol (50 μl), and aqueous 1 *M* KOH (10 μl) was added at room temperature. After 30 min, the brightly fluorescent solution was neutralized with 1 *M* HEPES (20 μl), evaporated, and redissolved in H_2O (35 μl).

Synthesis of Flubida-3

t-Butyl O,O-Diacetyl 2',7'-dichlorofluorescein-(5,6)-carboxamidopiperazine-4-carboxylate. 5-Carboxy-2',7'-dichlorofluorescein diacetate (132 mg, 0.25 mmol) was suspended in dry CH_2Cl_2 (0.5 ml) under argon. After adding 4-methyl morpholine (33 μl, 0.3 mmol), the resulting solution was cooled to 0° and isobutyl chloroformate (36 μl, 0.28 mmol) was added with stirring. After 15 min, *t*-butyl 1-piperazinecarboxylate (51 mg, 0.28 mmol) was added, and the reaction mixture was slowly allowed to warm up to room temperature (over 2 hr) and then evaporated. The crude product was dissolved in ethylacetate (25 ml), washed with 1 *M* HCl (20 ml) and brine (20 ml), dried, and evaporated to a yellow oil. After triturating with ethanol, 45 mg of starting material was recovered by filtration as a white solid. The evaporated filtrate was separated by SiO_2 chromatography by elution with ethyl acetate–hexane (30–66%, v/v) to yield the product as a white solid. Yield, 43.7 mg (25%).

Flubi 3. *t*-Butyl *O,O*-diacetyl 2',7'-dichlorofluorescein-(5,6)-carboxamidopiperazine-4-carboxylate (10 mg, 14 μmol) was dissolved in dry CH_2Cl_2 (1 ml) under argon and treated with trifluoroacetic acid (100 μl) at room temperature for 2 hr. After evaporation and redissolution in methanol (1 ml) and H_2O (1 ml), the acetyl esters were removed by the addition of sufficient 1 *M* KOH to maintain a pH of 10–11 for about 5 min. The dark red solution was neutralized to about pH 7.5 with 1 *M* HCl, evaporated, and redissolved in water to make a 10 m*M* solution. This solution (1.25 ml, 10 m*M*) was mixed with a DMF solution of biotin-succinimidyl ester (1.25 ml, 10 m*M*) and an aqueous solution of 4-methyl morpholine (0.5 *M*, 125 μl). After 72 hr at room temperature, capillary electrophoresis (20 kV, 50 m*M* borate, pH 8.5) indicated reaction completion, and the reaction mixture was evaporated. The resulting crude product was redissolved in aqueous sodium bicarbonate (<1 ml), filtered (0.2-μm pore size centrifugal filter), and reprecipitated with aqueous HCl. The resulting orange solid was collected by centrifugation, washed with cold H_2O, and desiccated *in vacuo* over phosphorus pentoxide.

Flubida-3. Crude Flubi-3 free acid (5 mg, 5.9 μmol) was suspended in acetic anhydride (2 ml) and pyridine (5 μl) and heated at 60° for a few minutes until colorless. After evaporation, the product was dissolved in 5%

(v/v) methanol–CHCl$_3$ (5 ml), washed with 1 M HCl (5 ml), dried, and evaporated. The resulting material was essentially pure by TLC [10% (v/v) methanol–CHCl$_3$). ^{1}NMR (CDCl$_3$-CD$_3$OD; δ ppm) 0.9–1.5 (m's, 16H), 1.9 (m, 4H) 2.17 (s, 6H, OCOCH$_3$), 2.52 (d, 1H), 2.72 (dd, 1H), 2.96 (t, 2H), 3.2 (m, 1-2H), 3.4 (broad m, 8H, piperazine), 4.2 (m, 2H), 6.71 (s, 1H), 7.01 (s, 1H), 7.10 (s, 1H), 7.26 (s, 1H), 7.56 (dd, 1H), 7.97 (d, 1H).

Synthesis of Flubida-4

Biotin cadaverine (1 mg, 3 μmol) and Oregon Green 5-carboxylate, succinimidyl ester (1 mg, 1.9 μmol) were suspended in dry DMF (200 μl). After brief heating to dissolve the biotin derivative, diisopropylethylamine (10 μl) was added to give a red-orange solution that was kept at room temperature for 2 hr. The reaction mixture was evaporated, suspended in acetic anhydride (1 ml) and pyridine (5 μl), and stirred until colorless. After evaporation, the product was separated by silica gel chromatography, eluting with 10% (v/v) methanol–CHCl$_3$ to give a colorless gum. Electrospray mass spectrum: 807.2, calculated for $(M + 1)^+$ 807.8.

Synthesis of NBD Biotin

Biotin cadaverine (2.5 mg, 7.6 μmol) was dissolved in DMF (0.2 ml) and H$_2$O (0.2 ml) with gentle heating; aqueous 1 M KPO$_4$ (pH 7, 10 μl) and NBD chloride (4-chloro-7-nitrobenzofurazan; 2 mg, 10 μmol; predissolved in 10 μl of dry DMF) were then added. The solution turned yellow and after 4 hr was evaporated, dissolved in 5% (v/v) methanol–CHCl$_3$ (25 ml), washed with aqueous 1 M KPO$_4$ (pH 7, 10 ml), dried, and evaporated. The product was separated by silica gel chromatography by elution with 20% (v/v) methanol–CHCl$_3$.

pH Titration of Flubi-2 in the Endoplasmic Reticulum and Golgi of Intact Cells

An *in situ* calibration was performed at the end of each experiment to generate a calibration curve of 490 nm/440 nm excitation ratio (emission, 520–560 nm) versus pH, which was used to convert ratio values to pH. At the end of every experiment, cells and organelles were clamped to different pHs by perfusing calibration solutions of at least four different pHs. The calibration solutions contained (in mM): 70 NaCl, 70 KCl, 1.5 K$_2$HPO$_4$, 1 MgSO$_4$, 10 HEPES [or morpholineethanesulfonic acid (MES)], 2 CaCl$_2$, and 10 glucose, were titrated to different pH values (8.2, 7.0, 6.5, 6.0, 5.5), and supplemented with 10 μM each of nigericin (Sigma, St. Louis, MO), the

K$^+$/H$^+$ exchange ionophore, and monensin (Sigma), the Na$^+$/H$^+$ exchange ionophore. We used these calibration solutions rather than the more traditional high [K$^+$]–nigericin method because it was possible that the organelles contained different [Na$^+$] and [K$^+$] than the cytosol. Thus, this calibration method made no assumptions about cytosolic or organelle [Na$^+$] and [K$^+$] and allowed equilibration of both Na$^+$ and K$^+$ to drive the equilibration of H$^+$. The 490 nm/440 nm fluorescence ratio was plotted versus pH of the calibration solution; the data were fit to a sigmoidal curve (InPlot; Graph Pad, Irvine, CA), and the resulting fit was used to convert the ratio values to pH values using Eq. (1):

$$pH = pK + \log[(R - R_{min})/(R_{max} - R)] \qquad (1)$$

where R is the ratio of Flubi-2 fluorescence intensity excited at 490 nm/440 nm wavelength light; R_{min} and R_{max} are the minimum and maximum values, respectively, determined from the curve fit, and the pK is the pK_a determined from the fit. We generated average calibration curves for Flubi-2 localized in the ER (Fig. 5A) and Golgi (Fig. 5B). Flubi-2 had a similar pH dependence whether expressed in the lumen of the ER bound to AV–KDEL (Fig. 5A; pK_a 6.8 ± 0.03 SEM, n = 13), in the Golgi lumen bound to ST–AV (Fig. 5B, pK_a 6.6 ± 0.04 SEM, n = 43), or *in vitro* (Fig. 1B, the pK_a of the complex avidin–Flubi-2 separated by gel filtration using a Sephadex G25 column was 6.53).

Labeling Endoplasmic Reticulum and Golgi of Live Cells with Flubi-2

Flubida-2 was dissolved in DMSO (approximately 2 mM), mixed 1 : 1 with 20% (w/v in DMSO) Pluronic F-127 (Molecular Probes), and then diluted to the desired final concentration (2–4 μM) with serum-free DMEM. Thirty to 48 hr posttransfection with either AV–KDEL or ST–AV DNA, HeLa cells were rinsed once with serum-free DMEM and loaded with 2–4 μM Flubida-2 for 3–5 hr (or overnight for 10–15 hr). Background fluorescence was minimized by chasing the labeled cells with normal growth medium for at least 2 hr, to allow excess dye–biotin to exit from the cytosol. The strong avidin–biotin interaction (K_a 10^{15} M^{-1}) ensured stable, specific avidin–Flubi-2 binding that resisted washing. Biotin starvation of the cells was not necessary before Flubida-2 loading, as staining was bright and stable.

Studying Endoplasmic Reticulum and Golgi pH with Targeted Avidin–Flubi-2 System and Fluorescence Ratio Imaging

We measured the pH of Flubi-2-loaded ER and Golgi compartments in live, intact cells by digitally processed fluorescence ratio imaging. AV–

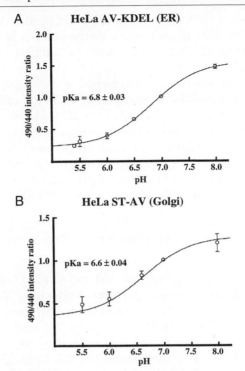

FIG. 5. *In situ* characterization of Flubi-2: 490 nm/440 nm excitation ratio (emission, 520–560 nm) versus pH_{ER} and pH_G. pH values of the ER and Golgi were monitored by the targeted avidin–Flubi method. Calibration solutions containing nigericin and monensin (10 μM each) were perfused onto cells at the end of each experiment. At least four solutions of different pH values (pH 8.2, 7.0, 6.5, 6.0, 5.5, and/or 5.0) were used. Calibration curves for both ER (A) and Golgi (B) were used to convert ratio values to pH and to determine the pK_a of the dye. The pK_a of Flubi-2 bound to AV–KDEL in the ER was 6.8 ± 0.03 SEM ($n = 13$), while Flubi-2 in the Golgi lumen bound to ST–AV had a pK_a of 6.6 ± 0.04 SEM ($n = 43$). To compare separate experiments, 490 nm/440 nm intensity ratios were normalized (pH 7 = 1.0 ratio).

KDEL- or ST–AV-expressing cells were plated on coverslips, labeled with Flubi-2, and mounted in an open perfusion chamber (at room temperature or heated to 32–37°) on an inverted IM35 Zeiss microscope with a ×63 oil immersion objective. Images were transmitted to a Dage (Michigan City, IN) 68 SIT camera, which collected emission (520–560 nm; Omega Optical, Brattleboro, VT) images of the cells during alternate excitation at 490 and 440 ± 5 nm (using a Lambda-10 filter wheel; Sutter Instruments, Novato, CA). Individual images for each wavelength were averaged over eight frames by a digital image processor (Axon Imaging Workbench; Axon

Instruments, Foster City, CA) run on a 133-MHz Pentium computer (Gateway, N. Sioux City, SD). Since single experiments can require as much as 200 MB of memory, all data were written to compact disks using a CD recorder (Hi Val, Santa Ana, CA).

The avidin–Flubi method was developed with the goal of determining how pH is regulated in different compartments in the cell. We have measured the average ER pH (pH_{ER}) and Golgi pH (pH_G) in intact HeLa cells bathed in pH 7.4 Ringer solution (containing in mM: 141 NaCl, 2 KCl, 1.5 K_2HPO_4, 1 $MgSO_4$, 10 HEPES, 2 $CaCl_2$, 10 glucose). Average pH_{ER} was 7.2 ± 0.2 SD ($n = 26$ cells), close to the average cytosolic pH (pH_c), measured with the widely used cytosolic pH dye BCECF-AM (7.4 ± 0.2 SD, $n = 119$); steady state pH_G averaged 6.4 ± 0.3 SD ($n = 46$).

Measuring Buffer Capacity and Rates of H^+ Transport across Organelle Membranes

In this section, we discuss experimental and methodological approaches required to make quantitative comparisons of rates of H^+ flux across organelle membranes. During investigations of the mechanisms that give rise to the different pH values in the ER and Golgi,[12] it became clear that the different pH values were generated in large part by having different rates of pumps and leaks for H^+. Specifically, the ER has the same pH as the cytosol because it expresses no active H^+-pumping ability, and there is a large leak for H^+. The Golgi is more acidic than the ER because the Golgi has an active H^+ v-ATPase that is able to acidify the Golgi lumen despite the presence of a H^+ leak that is only three times smaller than that of the ER.

Determining H^+ leak rates from measurements of pH requires knowledge of the buffering capacity, β, as shown by Eq. (2):

$$J^H \text{ (mmol } H^+\text{/liter/min)} = \beta \text{ (mmol } H^+\text{/liter/pH)} \times \Delta pH/min \quad (2)$$

where J^H is the H^+ flux and ΔpH is the change in pH. We determined β for the cytosol, ER, and Golgi by measuring magnitudes of rapid increases in pH of these compartments in response to application of known concentrations (30 or 40 mM) of NH_4Cl Ringer (same as Ringer solution except 30 or 40 mM NaCl is replaced with 30 or 40 mM NH_4Cl). Because buffer capacities may be pH dependent,[13] it is important to determine β of the Golgi and ER with bafilomycin-pretreated cells (500 nM, 2 hr) so that the baseline pH values, prior to NH_4Cl treatment, were the same for both

[12] M. M. Wu, J. Llopis, S. Adams, J. M. McCaffery, M. G. Farquhar, M. S. Kulomaa, T. E. Machen, H.-P. H. Moore, and R. Y. Tsien, *Chem. Biol.* **7**, 197 (2000).
[13] W. H. Weintraub and T. E. Machen, *Am. J. Physiol.* **257**, G317 (1989).

compartments. Bafilomycin pretreatment also blocked H^+-pumping ability of the organelles, which could alter the magnitude of the pH responses elicited by NH_3/NH_4^+ treatment. ER and Golgi pH both increased immediately due to the rapid entry of NH_3, which is in equilibrium with NH_4^+ in solution.[14] Using pK 9.0 for the $NH_4^+ \rightarrow H^+ + NH_3$ reaction and knowing $[NH_4^+]$ from the amount added to the Ringer solution and assuming that NH_3 equilibrates equally across all cell membranes, β was calculated from the rapid increase in pH (extrapolated to time zero) during the switch from Ringer to NH_3/NH_4^+-containing Ringer according to Eq. (3)[13,14]:

$$\beta = [NH_4^+]/\Delta pH \tag{3}$$

We have used two approaches to measure the H^+ leak rates across organelle (ER and Golgi) membranes. One approach measures the leak out of ER and Golgi membranes in intact cells, while the second method can be used to quantitate both the leak into and out of organelles of plasma membrane-permeabilized cells. For experiments using intact cells, the rates of H^+ leak out of the cytosol and organelles were determined by measuring the pH recovery of these compartments from an acid load. Prior to each experiment, cells were pretreated for 2 hr with 500 nM bafilomycin (H^+ v-ATPase inhibitor) to ensure we eliminated the H^+ pump contribution and only measured H^+ permeabilities. In separate experiments, cytosolic pH was monitored with the pH-sensitive dye BCECF-AM, while ER and Golgi pH were monitored by the avidin-Flubi-2 method. The compartments were acid loaded by an extended 20-min treatment with 40 mM NH$_4$Cl Ringer. The long treatment was required because the Golgi appeared to have a low permeability to NH_4^+.[12] After washout of the NH$_4$Cl-containing Ringer with Na-free Ringer [Na$^+$ was replaced with N-methyl-D-glucamine (NMG)], the pH values of all three compartments typically acidified to pH <6.4 and remained acidic, probably because the lack of Na$^+$ prevented the Na$^+$/H$^+$ exchanger in the plasma membrane from recovering the cytosolic pH. The H^+ leak rates out of the cytosol, ER, and Golgi were then measured when the Na$^+$-free Ringer was replaced with Na$^+$-containing Ringer. By comparing rates of H^+ leak out of the ER and Golgi, insights can be gained into the relative "leakiness" or permeability of these organelles to H^+. This approach, which should be applicable to other organelles in which steady state pH is determined by countering activities of H^+ v-ATPase(s) and H^+ leak(s), is advantageous because cells and organelles remain in their intact state, and rates of H^+ leak are performed under conditions in which the transorganelle membrane pH (i.e., from organelle to cytosol) is always the same. A drawback to this method is that it allows conclusions only about

---

[14] A. Roos and W. F. Boron, *Physiol. Rev.* **61**, 296 (1981).

relative H^+ leaks because the rate of pH recovery will always be limited by the rate of pH recovery due to plasma membrane mechanisms. In addition, the method requires controls to assure that the leak and pump are not affected by the presence or absence of Na^+.

The second approach we used to study the H^+ leak across organelles required selective permeabilization of the plasma membrane, using strepto-lysin-O (SL-O), and then establishment of pH gradients across the ER and Golgi membranes. This method allowed us to monitor both the H^+ leak into and leak out of the organelles. SL-O is a bacterial toxin that binds cholesterol and forms pores at 37°. We used SL-O to selectively permeabil-ize the plasma membrane of cells by using the following protocol. One hundred microliters of reconstitution buffer [10 mM HEPES (pH 7.2), 10 mM dithiothreitol (DTT), 0.1% (w/v) BSA] was added to lyophilized SL-O (from S. Bhakdi, University of Mainz, Germany) to give a 1-mg/ml stock solution. HeLa cells expressing AV–KDEL or ST–AV were loaded with Flubida-2, chased in normal growth medium, and pretreated with bafilo-mycin (500 nM, 2 hr). The cells were rinsed twice with ice-cold Ringer before a 15-min incubation (on ice) in cold SL-O (5 mg/ml)–Ringer solution. The cold incubation allowed SL-O to bind to plasma membrane cholesterol without pore formation. Excess, unbound SL-O was removed by washing cells twice with ice-cold Ringer, and permeabilization was induced by incu-bating cells in a pH 7.4 intracellular buffer (IB consisted of, in mM: 110 potassium gluconate, 20 KCl, 0.1 CaCl$_2$, 2.7 K$_2$HPO$_4$, 10 HEPES, 10 MES, 10 glucose, 5 Mg-ATP, 2 Na-GTP, 2 magnesium acetate, titrated to different pH values with NMG base) for 15 min at 37° to allow pore formation. We confirmed plasma membrane permeabilization of cells at the end of each experiment by treating the cells with 1 μM propidium iodide, which bound to DNA and turned nuclei red only in permeabilized cells. Because the efficiency of SL-O permeabilization is cell type dependent, the amount of toxin should be titrated for each different cell type.

After the Flubi-labeled, bafilomycin-pretreated cells were permeabil-ized with SL-O, we measured the H^+ leak into the ER and Golgi by establishing transmembrane pH gradients by changing the outside solution from pH 7.4 IB to pH 6.0 IB. The H^+ leak out of the membranes was determined by changing the outside solution from pH 6.0 IB to pH 7.4 IB. Using this protocol, we could also test whether counterions (such as K^+ and Cl^-) were required for the H^+ leak into and out of the organelles. In the ion replacement experiments, all chloride was replaced with gluconate and all potassium replaced with NMG to make Cl^--free and K^+-free IB solutions, respectively. For both the intact and SL-O permeabilized experi-ments, we were able to quantitate the H^+ leak into or out of the different compartments by fitting the data to exponential equations and obtaining rate constants and half-times for the pH recoveries.

The advantages of the permeabilized cell approach are that it allows more quantitative measurement of H^+ leak (independently of plasma membrane pH regulation) and better control of ionic conditions. The disadvantage is that despite the fact that SL-O permeabilization was performed to minimize damage, this procedure could have induced subtle changes in H^+ permeability or resulted in the loss of important cytosolic regulatory factors. However, because Golgi membranes of permeabilized cells still responded to NH_3 and still maintained an acidic pH of <6.5 when not pretreated with bafilomycin (data not shown), SL-O did not seem to grossly alter organelle membranes.

Conclusion

We developed a novel method for measuring pH in specific intracellular organelles. Avidin chimera proteins were localized to specific compartments by appending organelle-specific targeting sequences to avidin. Cells expressing the organelle-specific avidin chimera proteins were loaded with Flubida-2, a membrance-permeable, pH-sensitive fluorescein–biotin derivative. Flubi-2 accumulated specifically and stably in avidin-containing organelles, allowing us to monitor steady state pH. Although there were several potential drawbacks of this method, none became problematic in our experiments. The presence of avidin in the cells (which might have been expected to bind all cellular stores of biotin and kill the cells) did not affect either viability or survival of the cells, even in stably transfected cells (M. Wu, K. Teter, and H. P. Moore, unpublished observations, 1999). Thus, it was not necessary to add extra biotin to the medium to counter the presence of avidin in the cells. In some cases, avidin chimera proteins were mislocalized, probably due to extended oligomerization of both the targeting domains and the avidin domains of the chimera proteins, but we were able to rescue the mistargeted proteins by coexpression of soluble avidin. Finally, control experiments indicated that pretreatment of cells with biotin-free medium did not improve labeling of cells with Flubi-2 (J. Llopis, unpublished observations, 1999).

The targeted avidin–Flubi approach was advantageous because it allowed specific targeting of a ratiometric pH-sensitive dye to the ER and Golgi of mammalian cells (as well as other compartments; M. Wu, T. E. Machen, and H.-P. Moore, unpublished observations, 1999). The method provides flexibility in that the Flubi dyes could be modified chemically to provide different pH sensitivities that would be suitable for making measurements in both neutral organelles (e.g., ER) as well as in organelles that may vary in acidity from pH 6.4 (e.g., Golgi) to pH 4.5 (lysosomes). Using this method, we have performed a comparative study of pH regulation between the ER and Golgi of HeLa cells.[12] Fluorescence ratio imaging experiments can be performed on Flubi-loaded, avidin chimera protein-expressing cells under conditions in which the cells are SL-O permeabilized and/or pretreated with pharmacological agents (bafilomycin, brefeldin A). Fluorescence measurements can be made in any compartment that can retain an avidin chimera protein, and in any cell type that can express the proteins.

Additional Applications

The targeted avidin–biotin indicator method we have described in this chapter can be modified to study physiological events or processes other than pH regulation. To demonstrate this versatility we labeled cells expressing AV–KDEL with an NBD–biotin conjugate (Fig. 4), which proved to diffuse across cell membranes. The staining correlated with ER markers and was indistinguishable from that obtained with Flubi-2, except that the fluorescence was blue. Thus, biotin conjugates of fluorophores or fluorescent indicators other than fluorescein can potentially be made in order to label organelles with other colors (for dual-staining procedures) or to study their function in living cells. Cell lines developed expressing targeted avidin could thus be used for other purposes beyond monitoring organelle pH.

Acknowledgments

We thank Marilyn Farquhar and the ICC/EM Core Facility at UCSD (NIH Grant CA58689). We also thank George Oster, Michael Grabe, Hongyun Wang, and members of the Machen, Moore, and Tsien laboratories for helpful discussions.

[40] Fluorescent Labeling of Recombinant Proteins in Living Cells with FlAsH

By B. Albert Griffin, Stephen R. Adams, Jay Jones, and Roger Y. Tsien

Introduction

Chemical labeling of specific sites in proteins is usually achieved by reaction of single cysteine residues (either native or introduced by site-targeted mutagenesis) with appropriate thiol-reactive derivatives.[1] Other reactive amino acids such as lysines or glutamates are generally too abundant in proteins to allow specific reaction with appropriately reactive probe. This general approach has, however, been limited to *in vitro* modification of purified proteins. Biological studies of such labeled proteins in cells require their reintroduction by disruptive techniques such as microinjection or electroporation, which often greatly limits the scope of such experiments. Labeling of single cysteine residues in specific proteins in living cells is precluded by the millimolar concentrations of competing thiol from glutathione and other proteins.

Alternative approaches generally require fusion of the desired protein (at the DNA level, followed by transfection of cells) with intrinsically fluorescent proteins such as green fluorescent protein (GFP) or proteins to which specific small molecules can be targeted (reviewed in Tsien and Miyawaki[2]). However, such proteins are large (e.g., 30 kDa for GFP), often larger than the target protein, and are restricted in their fusion sites and in the functionalities that can be targeted (e.g., GFP is limited to fluorescence). Incorporation of unnatural amino acids has been achieved by *in vitro* translation methods[3] or in ion channels expressed in oocytes[4] but cannot yet be applied more generally. *In vitro* protein ligation[5] of fluorescently labeled peptides or polypeptides to construct full-length proteins is a promising new approach that may be applicable to living cells.

[1] G. T. Hermanson, "Bioconjugate Techniques." Academic Press, San Diego, California, 1996.

[2] R. Y. Tsien and A. Miyawaki, *Science* **280,** 1954 (1998).

[3] V. W. Cornish, D. Mendel, and P. G. Schultz, *Angew. Chem. Int. Ed. Eng.* **34,** 621 (1995).

[4] M. W. Nowak, J. P. Gallivan, S. K. Silverman, C. G. Labarca, D. A. Dougherty, and H. A. Lester, *Methods Enzymol.* **293,** 504 (1998).

[5] T. W. Muir, D. Sondhi, and P. A. Cole, *Proc. Natl. Acad. Sci. U.S.A.* **95,** 6705 (1998).

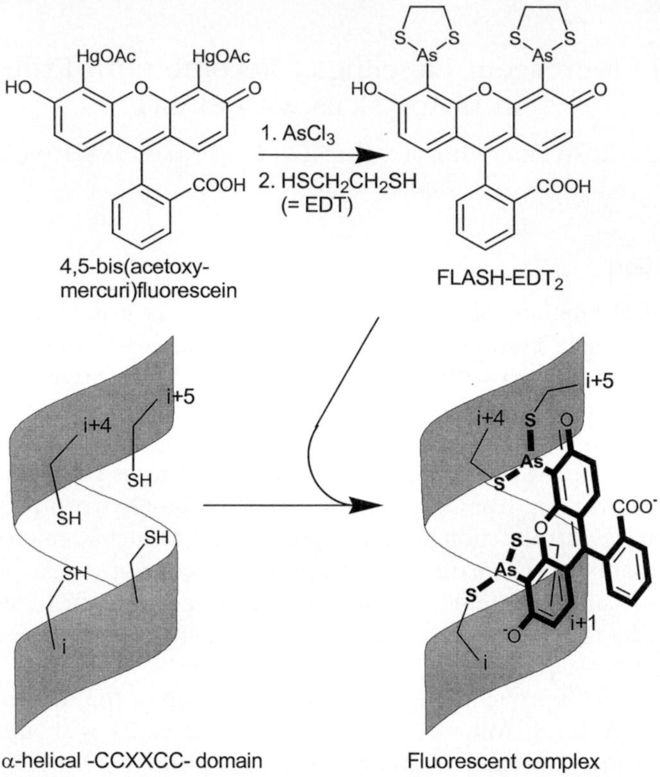

FIG. 1. The synthesis of FlAsH–EDT$_2$ from fluorescein mercuric acetate and arsenic trichloride, followed by reaction with 1,2-ethanedithiol (EDT). The fluorescent complex of FlAsH with an α-helical peptide containing the CC$X$$X$CC FlAsH site is shown schematically as the two arsenic atoms bridging the i, $i+1$ and $i+4$, $i+5$ thiols, respectively. An alternative conformation involving i, $i+4$ and $i+1$, $i+5$ binding of the arsenics is also possible.

Our approach[6] to site-specific labeling of proteins in living cells has been to utilize the well-known affinity of arsenoxides $(R—As=O)$ for a pair of closely spaced cysteines. To prevent labeling of such endogenous cellular sites (and the associated toxicity), a fluorescein containing two arsenoxides (FlAsH) was designed that has a much higher affinity for four appropriately spaced cysteines (CC$X$$X$CC, where X is any amino acid other than cysteine) in an α-helical conformation (Fig. 1). Such motifs are sufficiently uncommon in naturally occurring proteins to permit specific modification of the target protein incorporating the introduced FlAsH site

[6] B. A. Griffin, S. R. Adams, and R. Y. Tsien, *Science* **281,** 269 (1998).

in living cells. By labeling in the presence of the arsenoxide antidote 1,2-ethanedithiol (EDT), nonspecific labeling and toxicity can be minimized because EDT forms more stable complexes with arsenic than do pairs of cysteines. Furthermore, FlAsH complexed with two EDT molecules, (FlAsH–EDT$_2$) is membrane permeable and nonfluorescent yet becomes brightly fluorescent on binding the CCXXCC site, thereby decreasing background signal from unbound dye during labeling. The tetracysteine site can be attached as an N- or C-terminal tag or incorporated into a known α-helical structure. Addition of a high concentration (millimolar) of EDT reverses the binding of FlAsH to the tetracysteines, permitting reversible labeling. Chemical modification of the fluorescein moiety allows incorporation of different photochemical properties (e.g., different colors for multicolor analysis or fluorescence resonance energy transfer) or use as a handle to target other small molecules to proteins modified with the FlAsH site.

FlAsH may also be used to label purified proteins *in vitro* as an alternative to fluorescein iodoacetamide or maleimide reagents, with the advantage that the tetracysteine-binding site can be labeled without affecting single cysteines in the molecule. In addition, the restricted rotational mobility and fixed orientation of FlAsH bound to the FlAsH site (L. Gross and S. R. Adams, unpublished results, 1999), compared with conventional fluorescein labeling reagents, may be advantageous for studies of protein mobility.

Synthesis of FlAsH and Its Derivatives

FlAsH–EDT$_2$ can be synthesized by a one-pot two-step synthesis from commercially available fluorescein mercuric acetate (Fig. 1). A similar method has been used to make analogs of FlAsH containing additional functionalities (e.g., $-CO_2H$, $-NH_2$, and $-Cl$) with similar yields (our unpublished results, 1999).

Synthesis of FlAsH–EDT$_2$

Fluorescein mercuric acetate (85 mg, 0.1 mmol; Aldrich, Milwaukee, WI) was suspended in dry N-methyl pyrrolidinone (NMP; 1.5 ml) under argon. Arsenic trichloride (167 μl, 2 mmol) (*CAUTION:* Highly toxic! Use fumehood!) was added followed by palladium acetate (a few milligrams) and dry N,N-diisopropylethylamine (140 μl, 0.8 mmol). The resulting pale yellow solution was stirred at room temperature for 3 hr. The reaction mixture was poured into 50 ml of a stirred 1:1 (v/v) mixture of acetone and 0.25 M pH 7 phosphate buffer (to give a final pH of about 4–5), and

1,2-ethanedithiol (EDT, 99%; 285 μl, 3.4 mmol; Fluka, Buchs, Switzerland) was added immediately. The mixture rapidly turned cloudy. CHCl$_3$ (25 ml) was added with continual stirring. After 15 min, the mixture was diluted with water (50 ml) and separated, the aqueous layer was further extracted with CHCl$_3$ (two 25-ml volumes), and the combined extracts were dried over anhydrous sodium sulfate and evaporated to near dryness, using a water aspirator only. Any precipitates formed during extraction were ignored and removed during filtration of the drying agent. The oily orange residue was dissolved in toluene (50 ml) and washed with brine (three times, 50 ml each), dried, and evaporated to near dryness. This step removed NMP and could be omitted if care was taken not to overload the column during the subsequent chromatography. Alternatively, residual NMP (and EDT) can be removed from the toluene extract by a final (careful!) evaporation under oil vacuum-pump pressure (<1 mmHg). The product was purified by column chromatography on silica gel 60 (230–400 mesh, 20 g packed in toluene; E. Merck, Darmstadt, Germany), as the first orange band eluted with 1:9 (v/v) ethyl acetate–toluene. Fractions containing FlAsH–EDT$_2$ should not be completely concentrated to dryness, because the solid does not redissolve if left under vacuum for even a short time. (A possible explanation could be that vacuum removal of EDT might leave behind the free arsenoxide; arsenoxides are known to be prone to polymerization when concentrated.) Trituration with 95% ethanol overnight at 4° gave an off-white solid, melting point 155° with decomposition. Yield, 24 mg (36%). Keep at −20° protected from light. Solutions of FlAsH–EDT$_2$ in dimethyl sulfoxide (DMSO) or ethanol have been kept frozen for several months without significant deterioration, although some precipitation can occur with samples containing additional EDT.

^{1}H NMR (200 MHz, CDCl$_3$ with a trace of CD$_3$OD): 2.3 (broad singlet, OH), 3.57 (multiplet, 8 protons, -SCH$_2$CH$_2$S-), 6.60 (doublet, J = 8.8Hz, 2 protons, H-2' and H-7'), 6.69 (d, J = 8.8 Hz, 2 protons, H-1' and H-8'), 7.19 (d, 1 proton, H-7), 7.66 (m, 2 protons, H-5, 6), 8.03 (d, 1 proton, H-4). Electrospray mass spectroscopy in negative ion mode indicated a monoisotopic mass for the −1 ion of 663.0 Da (theoretical 662.85). The extinction coefficient was 4.1×10^4 M^{-1} cm^{-1} at the absorbance maximum of 507.5 nm in pH 7 buffer.

In Vitro Peptide and Protein Labeling with FlAsH–EDT$_2$

Labeling of tetracysteine peptides by FlAsH *in vitro* is easily accomplished and can be conveniently monitored by the concomitant increase in fluorescence. Successful labeling of proteins has been achieved at the N or C termini by addition of the sequence of a model peptide that binds FlAsH

(EAA ARE ACC REC CAR A). (The N-terminal tryptophan included in the original model peptide for convenient quantification is not necessary for FlAsH binding.) To date only a few examples of different proteins have been tried, so a minimal FlAsH site has yet to be defined. An internal FlAsH site in an existing α helix of calmodulin was generated by mutation of four amino acids to cysteine and successfully labeled with FlAsH *in vitro* and in living cells.[6]

For labeling *in vitro* it is important that the cysteines that bind to the probe be completely reduced because FlAsH will not react with disulfides. If reduction of cysteines is required, we usually treat a concentrated stock solution of the peptide (millimolar) with either dithiothreitol (DTT) or triscarboxyethylphosphine (TCEP), using standard methodologies. The reduced solution is then diluted to micromolar concentrations for labeling and study. High concentrations of DTT (tens of millimolar) may compete with the tetracysteines for FlAsH and decrease labeling and so should be avoided in the final labeling solution. The rate of labeling with FlAsH is pH sensitive, as reaction requires the cysteine to be in the thiolate form. Adequate reaction rates occur at pH 7 for labeling at micromolar concentrations but lower pH values may be used with higher concentrations of reactants and longer reaction times.

Efficient labeling generally requires the presence of small monothiols (1 mM) such as mercaptoethanol (2-ME) or 2-mercaptoethanesulfonic acid (MES). Monothiols have a weak affinity (millimolar) for arsenoxides and may aid in shuttling the arsenics into the correct position for good binding. In the absence of a small monothiol the labeling proceeds more slowly. Addition of the nonthiol reductant TCEP does not increase the rate of labeling, indicating that it is not the reductive power of 2-ME that is responsible for the enhanced reactivity. Addition of 2-ME to a FlAsH solution can result in the development of some fluorescence even in the absence of a target peptide and the presence of excess dithiols. The source of this fluorescence has not been determined but may be due to formation of 2-ME adducts with the arsenics, which quench the fluorescein less effectively than EDT. The fluorescence intensity of these adducts is low compared with that produced when FlAsH binds to the target peptide. A moderate fluorescent enhancement on labeling with FlAsH (particularly when using an excess), may be indicative of decomposition or of significant fluorescent impurities left over from synthesis. Incubating the FlAsH–EDT$_2$ with a slight excess of EDT prior to reaction can often decrease such background, suggesting that slow hydrolysis of the EDT group from the FlAsH occurs with time. Excess FlAsH may be removed after completion of labeling by standard techniques such as reversed-phase high-performance liquid chromatography (HPLC) or gel filtration.

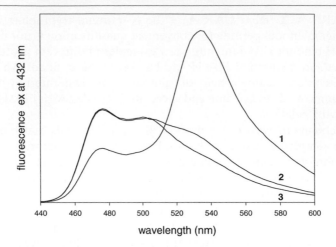

FIG. 2. Fluorescence emission spectra of cyan fluorescent protein (CFP) linked to a FlAsH target sequence by a protease recognition site. (1) Intact fusion protein labeled with FlAsH–EDT$_2$, showing FRET from CFP to FlAsH. (2) FRET is disrupted when fusion protein is cleaved by trypsin. (3) Addition of 5 mM BAL (British Anti-Lewisite, 2,3-dimercaptopropanol) removes FlAsH from the tetracysteine target sequence, making it nonfluorescent. The fusion of CFP (GFP residues 1–227; F64L, S65T, Y66W, N146I, M153T, V163A) to the amino acid sequence   RMRPPGPGDEVDGVDEVAKKSKEPGELAEAAAREACCRECCAREAA-AREAAAR was accomplished by standard molecular biology techniques. The recombinant protein was expressed in *E. coli* and purified by means of a polyhistidine tag. It was desalted with a Millipore Ultrafree-4 filter with a 10-kDa cutoff, and then resuspended in 100 mM HEPES. The fusion protein at a concentration of 100 μM (determined with a Bio-Rad DC protein assay kit) was reduced overnight in PBS, pH 7.3, containing 1 mM DTT. Labeling was accomplished by adding 1 μM FlAsH–EDT$_2$ and 10 μM EDT to 1 μM reduced protein in 2 ml of PBS containing 1 mM 2-ME. The labeling took about 2 hr for completion. Trypsin (GIBCO-BRL; final concentration, 0.4 mg/ml) quickly cleaved the FlAsH peptide adduct from CFP. Addition of 1 μl of BAL dissociated the FlAsH probe from its target sequence.

Cyan Fluorescent Protein–FlAsH Construct

FlAsH can be used to label recombinant proteins containing the FlAsH target sequence *in vitro*. Again it is necessary that the four cysteines in the FlAsH target sequence be in the reduced state so that they are available to bind the probe. Purification of recombinant proteins containing reduced tetracysteine-binding sites can be facilitated by including reducing agents (such as 2-ME, DTT, and TCEP) at all stages in the work-up. Figure 2 shows the spectra of a recombinant protein in which a FlAsH target sequence is linked via a peptidase cleavage site to CFP.[7] Cyan fluorescent protein (CFP) is a mutant GFP with blue-shifted excitation and emission spectra that

[7] R. Y. Tsien, *Annu. Rev. Biochem.* **67,** 509 (1998).

undergoes fluorescence resonance energy transfer (FRET) to FlAsH if sufficiently close in space (<5 nm) and orientated appropriately. After labeling with FlAsH but before the addition of trypsin, there is efficient energy transfer between CFP and the FlAsH fluorophore as evidenced by reduced fluorescence at 475 nm compared with the fluorescence at 528 nm (spectrum 1, Fig. 2). On cleavage with trypsin, the 528-nm signal is greatly reduced while that at 475 nm is increased, indicating direct emission from CFP and lack of FRET (spectrum 2, Fig. 2). After cleavage was complete, a 5 mM concentration of the dithiol (British Anti-Lewisite BAL, or 2,3-dimercaptopropanol) was added to dissociate the FlAsH fluorophore from the target sequence. Spectrum 3 in Fig. 2 shows that this reagent reduced the fluorescent signal of the FlAsH–peptide conjugate but had no effect on the CFP emission at 475 nm. This demonstrates that the peptide cleavage was indeed complete and that the residual fluorescence of the FlAsH–peptide conjugate was due to direct excitation of the fluorophore and not to FRET caused by incomplete cleavage by the peptidase. This is a control that cannot be easily performed when using other methods of protein labeling.

FlAsH Labeling in Cells

One of the most useful features of the FlAsH labeling system is the ability of the probe to label recombinant proteins in living cells. The non-fluorescent reagent is applied to the outside of cells, crosses the plasma membrane, finds its target within the cell, binds, and becomes fluorescent. The specificity of FlAsH binding is improved by the addition of EDT to the loading solution. The empirically determined concentration ratio of 10 μM EDT to 1 μM FlAsH–EDT$_2$ decreases staining of endogenous site while still allowing FlAsH to bind to the designed motif.

Typically the loading solution for labeling cells is 1 μM FlAsH–EDT$_2$ and 10 μM EDT in HEPES-buffered saline (HBS; containing either glucose or, for reduction of background fluorescence, 1 mM sodium pyruvate). The empirically found effect of pyruvate may result from a change in the redox state of the cells. The FlAsH–EDT$_2$ and EDT, both in DMSO (1 μl each of stock solutions of 1 mM FlAsH–EDT$_2$ and 10 mM EDT), are mixed first and then diluted with the buffer (1 ml). It is important to use freshly made EDT solutions because oxidation can occur readily. The solution can be incubated at room temperature for 15 min to ensure that any FlAsH–EDT$_2$ that may have become unprotected during storage rebinds EDT before application to cells. The cells to be stained are rinsed with HBS to remove serum proteins that may slow labeling by binding FlAsH–EDT$_2$ (see below). The labeling solution is then added and the cells incubated for about 1 hr at room temperature. Some labelings may take longer and can be conveniently monitored by fluorescence microscopy. It is useful to label a control of mock-transfected cells to be sure that the fluorescence seen is indeed due to labeling of the desired protein containing the tetracysteine-binding site. Nontransfected cells are not an ideal control, because transfection can introduce artifactual fluorescence staining by increasing cellular debris.

Suppression of Background Staining in Cells

The ratio of reagents outlined above may not sufficiently lower background staining to allow straightforward use of the FlAsH labeling technique. This is especially evident when cells other than HeLa are used (e.g., ECV 304, HEK 293, CHO, 3T3, 3T6) or when the protein to be labeled is expressed at low levels. Increasing the ratio of EDT to FlAsH–EDT$_2$ may further reduce the background staining, but at the expense of desired binding. The appropriate ratio may have to be determined empirically for each protein construct and cell type.

The addition of two nonfluorescent compounds, one membrane permeant and one membrane impermeant, to the loading solution greatly reduces the undesired background fluorescence in cells. *In vitro,* low millimolar concentrations of EDT (or the more water-soluble BAL) are sufficient to completely reverse the binding of FlAsH to a tetracysteine model peptide. In cells, however, some FlAsH staining persists even at high concentrations of EDT (up to 30% of the background staining in wild-type HeLa). The idea that an uncharged molecule might help reduce this non-dithiol-responsive background arose from the observation of the fluorescent interaction of FlAsH–EDT$_2$ with bovine serum albumin (BSA) *in vitro.* A solution of FlAsH–EDT$_2$ becomes significantly more fluorescent when BSA is added (with an estimated quantum yield of 0.1 compared with 0.5 for the FlAsH–peptide complex). This fluorescence is not reduced by the addition of millimolar EDT. An extremely high affinity of the FlAsH arsenics for BSA is unlikely because all cysteines in BSA with the exception of one are oxidized to disulfides, and thus BSA presents no consensus site for tight binding of FlAsH. The most likely explanation is that FlAsH–EDT$_2$ remains intact when binding to a hydrophobic pocket on BSA, but becomes more fluorescent because of a change in environment. BSA is well known for binding hydrophobic dyes and boosting their fluorescence. Quenching of fluorescence in free FlAsH–EDT$_2$ is believed to result from deactivation of the excited state by electron transfer from the arsenic atom to the xanthene fluorophore and/or by vibrational rotation of the As–EDT groups. Hydrophobic binding of FlAsH–EDT$_2$ to BSA probably partially prevents these modes of quenching. Binding is reversible, addition of a tetracysteine-containing peptide to the complex results in formation of peptide–FlAsH complex, although at a slower rate than in the absence of BSA.

Similar hydrophobic sites in cells may bind FlAsH–EDT$_2$ and give fluorescence that is unresponsive to high concentrations of dithiol. Such unwanted fluorescence might be reduced if molecules could be found that would preferentially occupy such greasy cellular binding sites, either by higher affinity or by mass action, thus displacing FlAsH–EDT$_2$. We screened 30 FlAsH–EDT$_2$ analogs and nonfluorescent dyes (including fluorescein derivatives, EDT adducts of phenylarsenoxides, and hydrophobic dyes) for their ability to reduce FlAsH background staining in untransfected HeLa cells. Of the compounds screened, the commercially available dye Disperse Blue 3 (Fig. 3; Aldrich sample further purified by recrystallization from toluene) was the most effective at reducing background fluorescence (Fig. 4). Typically, 20 μM dye added to the loading buffer is more than sufficient to remove virtually all of the background nonresponsive to dithiols. Other mechanisms may also be operating when uncharged compounds reduce background staining by FlAsH–EDT$_2$. For example, the background reducing agent may bind close to the FlAsH–EDT$_2$ molecule and thereby reduce fluorescence by quenching rather than by displacement.

Dead or dying cells are brightly stained (even untransfected cells) by the standard FlAsH loading solution (1 μM FlAsH–EDT$_2$ and 10 μM EDT), probably by exposure of hydrophobic sites that bind the dye. In the study of single cells by fluorescence microscopy, these bright cells can often be ignored. However, when populations of cells are to be studied these bright cells may overwhelm the desired signal.

Part of the fluorescence in the bright, rounded up cells is removed by the addition of an uncharged dye such as Disperse Blue 3. Suppression of the bulk of the remaining background in these dead cells can be accom-

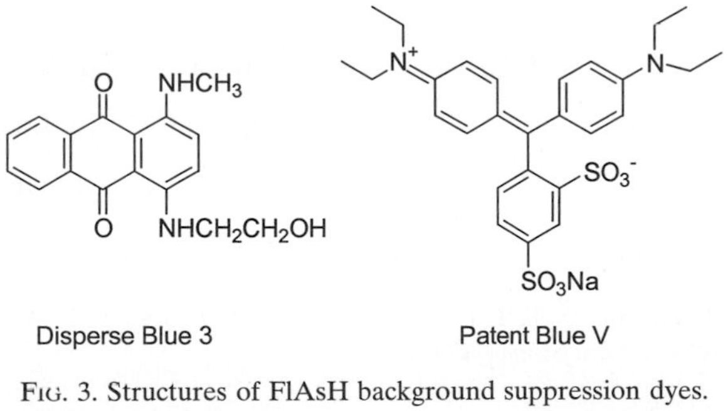

FIG. 3. Structures of FlAsH background suppression dyes.

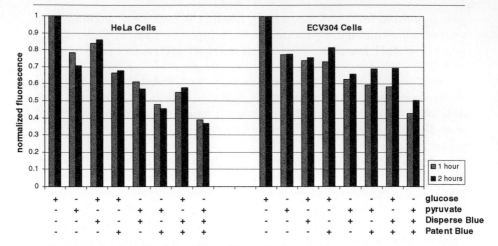

FIG. 4. Demonstration of the additive effect of background suppression techniques. Cells, either HeLa or EVC304, were split into 96-well plates and allowed to grow to confluence overnight in DMEM supplemented with 10% (v/v) FBS. Medium was removed and the cells rinsed once with 200 μl of HBS per well. The indicated components, along with 1 μM FlAsH–EDT$_2$ and 10 μM EDT, in 100 μl of HBS were applied to cells, which were then incubated at room temperature. Fluorescence was measured with a CytoFluor multiwell plate reader 1 and 2 hr after adding the FlAsH solutions. Filters: excitation, 485DF20; emission, 530DF25. Concentrations: glucose, 10 mM (1.8 g/liter); sodium pyruvate, 1 mM; Disperse Blue 3, 20 μM; Patent Blue V, 1 mM.

plished by flooding the cells with quenchers at high concentration. Membrane-impermeant dyes have access to the interior of cells whose cytoplasmic membrane has been compromised, but not to the interior of living cells. We found that the commercially available dye Patent Blue V (Fluka; Fig. 3) is effective, while some dyes such as trypan blue are not. Other charged dyes or combinations may reduce this source of background further.

Demonstration of Background Suppression in Transfected Cells

The additives that were developed with untransfected cells were demonstrated to be effective in transfected cells. Figure 5[6,8,9] (top) illustrates the

[8] M. J. B. van den Hoff, A. F. M. Moorman, and W. H. Lamers, *Nucleic Acids Res.* **20,** 2902 (1992).

[9] G. Zlokarnik, P. A. Negulescu, T. E. Knapp, L. Mere, N. Burres, L. Feng, M. Whitney, K. Roemer, and R. Y. Tsien, *Science* **279,** 84 (1998).

effect in HeLa cells as an example of a cell line that shows low background staining. Figure 5 (bottom) demonstrates the greater effect achieved in 3T6 cells, which have higher background staining. In these experiments the background suppression agents were added with the FlAsH–EDT$_2$ loading solution and remained on the cells for the duration of labeling. The uncharged compound and the membrane-impermeant dyes may be added after FlAsH–EDT$_2$ labeling if it is found desirable to limit the exposure of cells to these agents, although no toxic effects were apparent. After labeling is complete, the FlAsH–EDT$_2$ solution is removed from the cells and replaced with a rinse solution. For best results, the rinse solution should contain Disperse Blue, Patent Blue, and 10 μM EDT to minimize retention of the unbound and less membrane-permeable FlAsH–EDT$_2$ by cellular hydrophobic sites during rinsing.

FlAsH staining of intact bacteria expressing proteins with FlAsH target sites requires higher concentrations of FlAsH–EDT$_2$ (10–20 μM) in the presence of 2-ME (1–5 mM) for several hours. Bacterial lysis (by freeze–thawing) increases the rate of labeling, suggesting decreased permeability of bacteria cell walls to FlAsH compared with mammalian cells (S. R. Adams and A. Miyawaki, unpublished results, 1999). Similarly, preliminary experiments with yeast indicate that FlAsH does not have access to the interior of the cell until yeast spheroblasts are formed by removal of the cell wall (B. A. Griffin and G. Odorizzi, unpublished results, 1999).

Summary and Outlook

FlAsH labeling of recombinant proteins for cellular localization studies can be considered an alternative to the popular method using GFP fusions,[7] with the FlAsH method having some advantages. The size of the fluorescent tag is considerably smaller: Bound FlAsH has a molecular weight of less than 600 and the addition of a FlAsH target site can be as small as four introduced cysteines (with negligible change in molecular weight) with an appended peptide adding less than 2 kDa. This compares with a molecular weight of 30,000 for GFP, which is therefore more likely to perturb the native structure and function of the tagged protein. Both FlAsH and GFP tagging generate a fluorescent protein with similar brightness (the product of the extinction coefficient and fluorescence quantum yield). However, multiple FlAsH sites could be introduced into a protein so that considerably brighter labeling would aid in detecting low-abundance proteins.

Fusions with GFP are generally limited to the C and N termini of proteins, although insertion between some domains may be tolerated (e.g.,

Loading conditions: 1   2   3

HeLa with
FLASH site

HeLa mock
transfected

Loading conditions: 1   2   3

3T6 with
FLASH site

3T6 mock
transfected

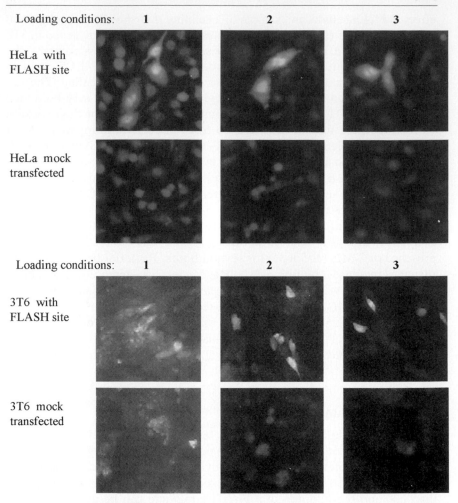

Fig. 5. Comparison of FlAsH labeling in cells with and without background suppression techniques. A construct encoding the FlAsH target peptide AEAAAREACCRECCARA appended to the C terminus of *Xenopus* calmodulin was inserted into pcDNA3 by standard molecular biology techniques.[6] Cells, either HeLa or 3T6, were dissociated from cell culture dishes, using a low calcium buffer, spun down, and then resuspended in a buffer[8] that resembles intracellular ionic concentrations. Ten micrograms of plasmid was added to 1 ml of the above, containing about 10^7 cells. The mixture was electroporated at 0.324 kV with a Bio-Rad GenePulser II equipped with a Bio-Rad Capacitance Extender Plus. Immediately after electroporation the mixture was diluted into medium [DMEM–10% (v/v) FBS without antibiotics] and split into several culture flasks. A control plasmid encoding β-lactamase[9], which does not contain a FlAsH target sequence, was similarly transfected into cells. The next day the cells were stained in HBS using: conditions 1 (1 μM FlAsH–EDT$_2$, 10 μM EDT and 10 mM glucose), or conditions 2 (1 μM FlAsH–EDT$_2$, 10 μM EDT, 1 mM pyruvate, 20 μM Disperse Blue 3) or conditions 3 (conditions 1 with 1 mM Patent Blue V added). In the HeLa

in the Shaker potassium channel[10]) and insertion of proteins into GFP[11] (at specific tolerant sites) without loss of fluorescence has been achieved. Internal FlAsH sites may be tolerated in numerous surface α-helical regions of a protein, more readily permitting location of the fluorophore at nonperturbing sites.

Color mutants of GFP span the blue to yellow range of the spectrum; FlAsH currently is limited to green emission although a red variant has been developed that is a good FRET acceptor for GFP and YFP (S. R. Adams and J. L. Llopis, unpublished results, 1999). In contrast, FRET between color mutants of GFP has been limited to BFP–GFP and CFP–YFP pairs, necessitating excitation with ultraviolet or violet light, respectively, with its inherent drawbacks of higher autofluorescence and less convenient laser lines. The description of red fluorescent proteins[12] from coral may eventually allow use of the more favorable GFP–RFP or YFP–RFP pairs.

A major advantage of the FlAsH system is the comparative ease of chemical modification of FlAsH. Coupled with the synthetic versatility of organic chemistry, this enables the incorporation of functionalities other than fluorescence into targeted proteins or peptides. Some modifications already investigated include addition of photosensitizing groups, magnetic resonance imaging agents, membrane-impermeant groups, cross-linking groups, fluorescent Ca^{2+} sensors, or reactive groups for immobilization (our unpublished results, 1999).

Disadvantages of FlAsH include the requirement of FlAsH binding for reduced cysteines. Labeling of proteins in oxidizing environments (such as the secretory pathway or extracellular) requires *in situ* reduction prior to labeling (S. R. Adams and Y. Yao, unpublished results, 1999). More importantly, higher background staining is generally seen in the FlAsH labeling technique compared with GFP chimeras. This can limit the usefulness of the technique for protein localization (especially of proteins

---

[10] M. S. Siegel and E. Y. Isacoff, *Neuron* **19,** 735 (1997).
[11] G. S. Baird, D. A. Zacharias, and R. Y. Tsien, *Proc. Natl. Acad. Sci. U.S.A.* **96,** 11241 (1999).
[12] M. V. Matz, A. F. Fradkov, Y. A. Labas, A. P. Savitsky, A. G. Zaraisky, M. L. Markelov, and S. A. Lukyanov, *Nature Biotechnol* **17,** 969 (1999).

---

experiment, the staining solution was replaced with solutions containing all components except FlAsH–EDT$_2$, before images were collected from a cooled CCD with Axon Imaging Workbench software. Filters: excitation, 495DF10, 505 dichroic; emission, 535DF50, In the 3T6 experiment, the staining solution was not removed before images were recorded on a Pixera digital camera. Filters: excitation, 450DF50, 480 dichroic; emission, 485 long pass.

expressed in low abundance) although the suppression techniques described above help decrease this problem significantly. The use of FlAsH as an FRET acceptor from CFP (or the red version from GFP) suffers much less from this problem as only specifically bound FlAsH is excited through FRET. Background FlAsH staining is not significantly excited by the wavelengths used for CFP. Further optimization of the FlAsH target site, perhaps through screening of peptide libraries, should still allow strong binding at concentrations of dithiol that minimize background staining.

[41] Ubiquitin Fusion Technique and Its Descendants

By ALEXANDER VARSHAVSKY

The ubiquitin (Ub) fusion technique was developed in 1985–1986, through experiments in which a segment of DNA encoding the 76-residue Ub was joined, in frame, to DNA encoding *Escherichia coli* β-galactosidase (βgal).[1,2] When the resulting protein fusion was expressed in the yeast *Saccharomyces cerevisiae* and detected by radiolabeling and immunoprecipitation with an anti-βgal antibody, only the moiety of βgal was observed, even if the labeling time was short enough to be comparable to the time (1–2 min) required for translation of the Ub-βgal open reading frame (ORF). It was found that in eukaryotic cells the Ub moiety of the fusion was rapidly cleaved off after the last residue of Ub (Fig. 1).[1] The proteases involved are called deubiquitylating[3] enzymes (DUBs) or Ub-specific processing proteases (UBPs).[4–7] A eukaryotic cell contains more than 10 distinct DUBs, all of which are highly specific for the Ub moiety. The *in vivo*

---

[1] A. Bachmair, D. Finley, and A. Varshavsky, *Science* **234**, 179 (1986).

[2] A. Varshavsky, *Proc. Natl. Acad. Sci. U.S.A.* **93**, 12142 (1996).

[3] Ubiquitin whose C-terminal (Gly-76) carboxyl group is covalently linked to another compound is called the *ubiquityl* moiety, the derivative terms being *ubiquitylation* and *ubiquitylated*. The term *Ub* refers to both free ubiquitin and the ubiquityl moiety. This nomenclature, which is also recommended by the Nomenclature Committee of the International Union of Biochemistry and Molecular Biology,[19] brings Ub-related terms in line with the standard chemical terminology.

[4] K. Wilkinson and M. Hochstrasser, *in* "Ubiquitin and the Biology of the Cell" (J.-M. Peters, J. R. Harris, and D. Finley, eds.). Plenum Press, New York, 1998.

[5] J. W. Tobias and A. Varshavsky, *J. Biol. Chem.* **266**, 12021 (1991).

[6] R. T. Baker, J. W. Tobias, and A. Varshavsky, *J. Biol. Chem.* **267**, 23364 (1992).

[7] C. A. Gilchrist, D. A. Gray, and R. T. Baker, *J. Biol. Chem.* **272**, 32280 (1997).

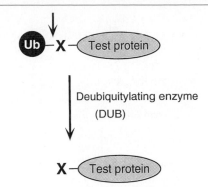

FIG. 1. The ubiquitin fusion technique. Linear fusions of Ub to other proteins are cleaved after the last residue of Ub by deubiquitylating enzymes (DUBs) (see text).[1, 2]

cleavage at the Ub–polypeptide junction of a Ub fusion has been shown to be largely cotranslational.[8,9]

One physiological function of the cleavage reaction (Fig. 1) is to mediate the excision of Ub from its natural DNA-encoded fusions either to itself (poly-Ub)[10] or to specific ribosomal proteins.[11,12] Many of the DUB proteases that catalyze the cleavage of linear Ub fusions can also cleave Ub off its branched, posttranslationally formed conjugates, in which Ub is joined either to itself, as in a multi-Ub chain, or to other proteins.[4,13] A branched Ub–protein conjugate usually comprises a multi-Ub chain covalently linked to an internal lysine residue of a substrate protein. The ubiquitylated substrate is processively degraded by the 26S proteasome, an ATP-dependent multisubunit protease.[14–17] For reviews of the Ub system, see Refs. 18–22.

Another finding about the DUB-mediated cleavage reaction (Fig. 1)

[8] N. Johnsson and A. Varshavsky, *EMBO J.* **13**, 2686 (1994).
[9] G. C. Turner and A. Varshavsky, submitted (2000).
[10] D. Finley, E. Özkaynak, and A. Varshavsky, *Cell* **48,** 1035 (1987).
[11] K. L. Redman and M. Rechsteiner, *Nature (London)* **338,** 438 (1989).
[12] D. Finley, B. Bartel, and A. Varshavsky, *Nature (London)* **338**, 394 (1989).
[13] C. M. Pickart, *FASEB J.* **11**, 1055 (1997).
[14] O. Coux, K. Tanaka, and A. L. Goldberg, *Annu. Rev. Biochem.* **65**, 801 (1996).
[15] W. Baumeister, J. Walz, F. Zühl, and E. Seemüller, *Cell* **92**, 367 (1998).
[16] M. Rechsteiner, *in* "Ubiquitin and the Biology of the Cell" (J. M. Peters, J. R. Harris, and D. Finley, eds.), pp. 147–189. Plenum Press, New York, 1998.
[17] G. N. DeMartino and C. A. Slaughter, *J. Biol. Chem.* **274**, 22123 (1999).
[18] M. Hochstrasser, *Annu. Rev. Genet.* **30**, 405 (1996).
[19] A. Varshavsky, *Trends Biochem. Sci.* **22**, 383 (1997).
[20] A. Hershko and A. Ciechanover, *Annu. Rev. Biochem.* **76**, 425 (1998).
[21] T. Maniatis, *Genes Dev.* **13**, 505 (1999).
[22] L. Hicke, *Trends Cell Biol.* **9**, 107 (1999).

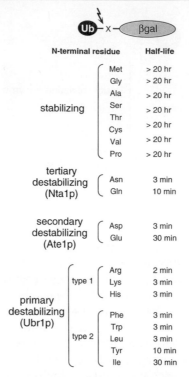

FIG. 2. The N-end rule of the yeast *S. cerevisiae*.[2] Specific residues at the N terminus of a test protein such as βgal are produced by the Ub fusion technique (Fig. 1 and text). The *in vivo* half-lives of the corresponding *X*-βgal proteins are indicated on the right. Stabilizing N-terminal residues (Met, Gly, Ala, Ser, Thr, Cys, Val, and Pro) are not recognized by Ubr1p (N-recognin), the E3 component of the N-end rule pathway. Primary destabilizing N-terminal residues (Arg, Lys, His, Phe, Trp, Leu, Tyr, and Ile) are directly bound by either type 1 or type 2 substrate-binding sites of Ubr1p. Secondary destabilizing N-terminal residues are arginylated by the *ATE1*-encoded Arg-tRNA-protein transferase (R-transferase), yielding the N-terminal Arg, a primary destabilizing residue. Tertiary destabilizing N-terminal residues Asn and Gln are deamidated by the *NTA1*-encoded N-terminal amidohydrolase (Nt-amidase), yielding the secondary destabilizing residues Asp and Glu, respectively. The N-end rule of mammalian cells is similar but contains fewer stabilizing residues.[2]

led to the discovery of the N-end rule, a relation between the *in vivo* half-life of a protein and the identity of its N-terminal residue (Fig. 2).[1] First, it was shown that the cleavage of a Ub–*X*–polypeptide fusion after the last residue of Ub takes place regardless of the identity of a residue *X* at the C-terminal side of the cleavage site, proline being the single exception. By allowing a bypass of the "normal" N-terminal processing of a newly formed protein, this result yielded an *in vivo* method for placing different residues at the N termini of otherwise identical proteins. Second, it was found that the *in vivo* half-lives of the resulting test proteins were determined by the identities of their N-terminal residues, a relation referred to as the N-end rule (Fig. 2).[1] The N-end rule pathway, which targets the N terminus-specific degradation signals, called the N-degrons, is one pathway of the Ub system. For a review and work on the N-end rule pathway, see Refs. 2 and 23–31.

The Ub fusion technique (Figs. 1 and 2) remains the method of choice for producing, *in vivo*, the desired N-terminal residue in a protein of interest. Owing to the constraints of the genetic code, nascent proteins bear N-terminal methionine (formyl-Met in prokaryotes). The known methionine aminopeptidases (MAPs), which remove N-terminal Met, do so only if the residue to be exposed is stabilizing according to the yeast-type N-end rule.[2,32] In other words, MAPs do not cleave off N-terminal methionine if it is followed by any of the 12 destabilizing residues (Fig. 2). The Ub-specific DUB proteases are free of this constraint, except when the residue X of a Ub–X–polypeptide is proline, in which case the cleavage still takes place but at a much lower rate.[1,33] More recently, a specific DUB was identified that can efficiently cleave at the Ub–proline junction.[7]

The Ub fusions can be deubiquitylated *in vitro* as well.[25,34,35] The high activity and specificity of DUBs should make them the reagents of choice for applications that involve, for example, the removal of affinity tags from overexpressed and purified proteins. Unfortunately, there are no commercially available DUBs at present, in part because of difficulties encountered in purifying and stabilizing large DUBs such as *S. cerevisiae* Ubp1p, and also because Proteinix (Rockville, MD), a company that has held the licenses for Ub fusion patents over the last decade, has not commercialized this technology.

Another major application of the Ub fusion technique resulted from the observations that expression of a protein as a Ub fusion can dramatically

[23] C. Byrd, G. C. Turner, and A. Varshavsky, *EMBO J.* **17**, 269 (1998).
[24] Y. T. Kwon, Y. Reiss, V. A. Fried, A. Hershko, J. K. Yoon, D. K. Gonda, P. Sangan, N. G. Copeland, N. A. Jenkins, and A. Varshavsky, *Proc. Natl. Acad. Sci. U.S.A.* **95**, 7898 (1998).
[25] I. V. Davydov, D. Patra, and A. Varshavsky, *Arch. Biochem. Biophys.* **357**, 317 (1998).
[26] Y. T. Kwon, A. S. Kashina, and A. Varshavsky, *Mol. Cell. Biol.* **19**, 182 (1999).
[27] Y. T. Kwon, F. Lévy, and A. Varshavsky, *J. Biol. Chem.* **274**, 18135 (1999).
[28] F. Lévy, J. A. Johnston, and A. Varshavsky, *Eur. J. Biochem.* **259**, 244 (1999).
[29] T. Suzuki and A. Varshavsky, *EMBO J.* **18**, 101 (1999).
[30] Y. Xie and A. Varshavsky, *Curr. Genet.* **36**, 113 (1999).
[31] P. O. Falnes and S. Olsnes, *EMBO J.* **17**, 615 (1999).
[32] R. A. Bradshaw, W. W. Brickey, and K. W. Walker, *Trends Biochem. Sci.* **23**, 263 (1998).
[33] E. S. Johnson, B. W. Bartel, and A. Varshavsky, *EMBO J.* **11**, 497 (1992).
[34] D. K. Gonda, A. Bachmair, I. Wünning, J. W. Tobias, W. S. Lane, and A. Varshavsky, *J. Biol. Chem.* **264**, 16700 (1989).
[35] R. T. Baker, *Curr. Opin. Biotechnol.* **7**, 541 (1996).

augment the yield of the protein.[36–39] The yield enhancement effect of Ub was observed with short peptides as well.[40,41] This and other applications of Ub fusions are described below, with references to the original articles and specific constructs.

Production and Uses of N-Degrons

An N-degron comprises the destabilizing N-terminal residue of a protein and an internal lysine residue.[2,29,42,43] A set of N-degrons containing different N-terminal residues that are destabilizing in a given cell defines the N-end rule of the cell.[2] The lysine determinant of an N-degron is the site of formation of a substrate-linked multi-Ub chain.[13,18,44] A way to produce an N-degron in a protein of interest is to express the protein as a Ub fusion in which the junctional residue (which becomes N-terminal on removal of the Ub moiety) is destabilizing (Fig. 2). An appropriately positioned internal lysine residue (or residues) is the second essential determinant of N-degron. Many natural proteins lack such "targetable" lysines, and therefore would remain long-lived even if their N-terminal residue were replaced by a destabilizing residue. One way to bypass this difficulty is to link a protein of interest to a relatively short (<50 residues) portable N-degron that contains both an N-terminal destabilizing residue (produced through a Ub fusion) and a requisite lysine residue(s). The earliest portable N-degron of this kind is still among the strongest known (Fig. 3B).[1,29,42] It was found, using the new strategy of a screen in the sequence space of just two amino acids, lysine and asparagine, that certain sequences containing exclusively lysines and asparagines can function *in vivo* as highly effective N-degrons.[29] The portability and modular organization of N-degrons make possible a variety of applications whose common feature is the conferring of a constitutive or conditional metabolic instability on a protein of interest.

[36] T. R. Butt, S. Jonnalagadda, B. P. Monia, E. J. Sternberg, J. A. Marsh, J. M. Stadel, D. J. Ecker, and S. T. Crooke, *Proc. Natl. Acad. Sci. U.S.A.* **86**, 2540 (1989).

[37] D. J. Ecker, J. M. Stadel, T. R. Butt, J. A. Marsh, B. P. Monia, D. A. Powers, J. A. Gorman, P. E. Clark, F. Warren, and A. Shatzman, *J. Biol. Chem.* **264**, 7715 (1989).

[38] P. Mak, D. P. McDonnell, N. L. Weigel, W. T. Schrader, and B. W. O'Malley, *J. Biol. Chem.* **264**, 21613 (1989).

[39] R. T. Baker, S. A. Smith, R. Marano, J. McKee, and P. G. Board, *J. Biol. Chem.* **269**, 25381 (1994).

[40] Y. Yoo, K. Rote, and M. Rechsteiner, *J. Biol. Chem.* **264**, 17078 (1989).

[41] A. Pilon, P. Yost, T. E. Chase, G. Lohnas, T. Burkett, S. Roberts, and W. E. Bentley, *Biotechnol. Prog.* **13**, 374 (1997).

[42] A. Bachmair and A. Varshavsky, *Cell* **56**, 1019 (1989).

[43] C. P. Hill, N. L. Johnston, and R. E. Cohen, *Proc. Natl. Acad. Sci. U.S.A.* **90**, 4136 (1993).

[44] V. Chau, J. W. Tobias, A. Bachmair, D. Marriott, D. J. Ecker, D. K. Gonda, and A. Varshavsky, *Science* **243**, 1576 (1989).

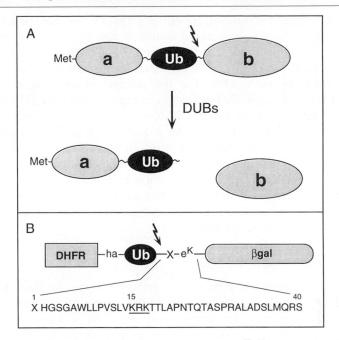

FIG. 3. The UPR (ubiquitin/protein/reference) technique.[79] (A) A tripartite fusion containing **a**, the reference protein moiety whose C terminus is linked, via a spacer peptide, to the Ub moiety. The C terminus of Ub is linked to **b**, a protein of interest. *In vivo*, this tripartite fusion is cotranslationally cleaved[9] by deubiquitylating enzymes (DUBs) at the Ub–**b** junction, yielding equimolar amounts of the unmodified protein **b** and **a**–Ub, the reference protein **a** bearing a C-terminal Ub moiety. If **a**–Ub is long-lived, a measurement of the ratio of **a**-Ub to **b** as a function of time or at steady state yields, respectively, the *in vivo* decay curve or the relative metabolic stability of protein **b**.[29,79] (B) Example of a specific UPR-type Ub fusion.[29] This fusion contains the following elements: DHFRha, a mouse dihydrofolate reductase (DHFR) moiety extended at the C terminus by a sequence containing the hemagglutinin-derived ha epitope; the Ub moiety (more specifically, the Ub^{R48} moiety bearing the Lys → Arg alteration at position 48); a 40-residue, *E. coli* Lac repressor-derived sequence, termed e^K [extension (*e*) containing lysines (*K*)] and shown below in single-letter abbreviations for amino acids; a variable residue *X* between Ub and e^K; the *E. coli* βgal moiety lacking the first 24 residues of wild-type β-Gal. The lightning arrow indicates the site of *in vivo* cleavage by DUBs.[29]

N-Degron and Reporter Proteins

A change in the physiological state of a cell that is preceded or followed by the induction or repression of specific genes can be monitored through the use of promoter fusions to a variety of protein reporters, such as, for example, βgal, β-glucuronidase, luciferase, and green fluorescent protein (GFP). A long-lived reporter is useful for detecting the induction of genes, but is less suitable for monitoring either a rapid repression or a temporal pattern that involves an up- and downregulation of a gene of interest. A sufficiently short-lived reporter is required in such settings. The metabolically unstable $X–\beta$gal proteins of the initial N-end rule study[1] (Fig. 2) were the first such reporters. Over the last decade, other protein reporters, including those described above, were metabolically destabilized by extending them with either a portable N-degron or a "nonremovable" Ub moiety.[45–47] The latter is targeted by a distinct Ub-dependent proteolytic pathway called the UFD pathway (Ub/fusion/degradation).[1,48] These metabolically unstable proteins, expressed as Ub fusions, should be particularly useful in settings where the concentration of the reporter must reflect a recent level of gene activity. Portable N-degrons were also used to destabilize specific protein antigens, thereby enhancing the presentation of their peptides to the immune system.[49,50]

N-Degron and Conditional Mutants

A frequent problem with conditional phenotypes is their leakiness, i. e., unacceptably high residual activity of either a temperature-sensitive (*ts*) protein at nonpermissive temperature or a gene of interest in the "off" state of its promoter. Another problem is "phenotypic lag," which often occurs between the imposition of nonpermissive conditions and the emergence of a relevant null phenotype. Phenotypic lag tends to be longer with proteins that are required in catalytic rather than stoichiometric amounts.

In one application of Ub fusions and the N-end rule pathway to the problem of phenotypic lag, a constitutive N-degron (produced as a Ub fusion) was linked to a protein expressed from an inducible promoter.[51] This method is constrained by the necessity of using a heterologous promoter and by the constitutively short half-life of a target protein, whose levels may therefore be suboptimal under permissive conditions. An alternative approach is to link the N-degron to a normally long-lived protein in a strain in which the N-end rule pathway can be induced or repressed. Such strains have been constructed with *S. cerevisiae*,[52,53] but can also be designed in

[45] C. K. Worley, R. Ling, and J. Callis, *Plant Mol. Biol.* **37**, 337 (1998).
[46] H. Deichsel, S. Friedel, A. Detterbeck, C. Coyne, U. Hamker, and H. K. MacWilliams, *Dev. Genes Evol.* **209**, 63 (1999).
[47] I. Paz, J.-R. Meunier, and M. Choder, *Gene* **236**, 33 (1999).
[48] E. S. Johnson, P. C. Ma, I. M. Ota, and A. Varshavsky, *J. Biol. Chem.* **270**, 17442 (1995).
[49] A. Townsend, J. Bastin, K. Gould, G. Brownlee, M. Andrew, B. Coupar, D. Boyle, S. Chan, and G. Smith, *J. Exp. Med.* **168**, 1211 (1988).
[50] T. Tobery and R. F. Siliciano, *J. Immunol.* **162**, 639 (1999).
[51] E. C. Park, D. Finley, and J. W. Szostak, *Proc. Natl. Acad. Sci. U.S.A.* **89**, 1249 (1992).
[52] Z. Moqtaderi, Y. Bai, D. Poon, P. A. Weil, and K. Struhl, *Nature (London)* **383**, 188 (1996).
[53] M. Ghislain, R. J. Dohmen, F. Levy, and A. Varshavsky, *EMBO J.* **15**, 4884 (1996).

other species, including mammalian cells. The metabolic stabilities, and hence also the levels of N-degron-bearing proteins, in a cell with an inducible N-end rule pathway are either normal or low, depending on whether Ubr1p, the recognition (E3) component of the N-end rule pathway, is absent or present.[52,53] These conditional mutants can be constructed with any cytosolic or nuclear protein whose function tolerates an N-terminal extension.

Yet another design is a portable N-degron that is inactive at a low (permissive) temperature but becomes active at a high (nonpermissive) temperature. Such an N-degron was constructed, using the Ub fusion technique, from a specific *ts* allele of the 20-kDa mouse dihydrofolate reductase (DHFR) bearing the N-terminal arginine, a strongly destabilizing residue.[54] Linking this DHFR-based, heat-inducible N-degron to proteins of interest yielded a new class of *ts* mutants, called *td* (temperature-activated degron). The *td* method does not require an often unsuccessful search for a *ts* mutation in a gene of interest. If the corresponding protein can tolerate N-terminal extensions, the corresponding *td* fusion is functionally unperturbed at permissive temperature. In contrast, low activity of a *ts* protein at permissive temperature is a frequent problem with conventional *ts* mutants. The *td* method eliminates or reduces the phenotypic lag, because the activation of N-degron results in rapid disappearance of a *td* protein. Another advantage of the *td* technique is the possibility of employing two sets of conditions: a *td* protein-expressing strain at permissive versus nonpermissive temperature or, alternatively, the same strain versus a congenic strain lacking the N-end rule pathway, with both strains at nonpermissive temperature.[54] This powerful internal control, provided in the *td* technique by two alternative sets of permissive/nonpermissive conditions, is unavailable with conventional *ts* mutants. Since 1994, a few laboratories described successful uses of the *td* method to construct *ts* alleles of specific proteins (e.g., Refs. 55 and 56). A recent modification of the *td* technique combines the galactose-inducible overexulsion of Ubr1p and the temperature-sensitive (*td*) N-degron.[56a]

N-Degron and Conditional Toxins

A major limitation of the current pharmacological strategies stems from the absence of drugs that are specific for two or more independent molecular

[54] R. J. Dohmen, P. Wu, and A. Varshavsky, *Science* **263**, 1273 (1994).
[55] G. Caponigro and R. Parker, *Genes Dev.* **9**, 2421 (1995).
[56] J. Wolf, M. Nicks, S. Deitz, E. van Tuinen, and A. Franzusoff, *Biochem. Biophys. Res. Commun.* **243**, 191 (1998).
[56a] K. Labib, J. A. Tercero, and J. F. X. Diffley, *Science* **288,** 1643 (2000).

targets. For the reasons discussed in detail elsewhere,[57,58] it is desirable to have a therapeutic agent that possesses a multitarget, combinatorial selectivity, which requires the presence of two or more predetermined targets in a cell and simultaneously the absence of one or more targets for the drug to exert its effect. Note that simply combining two or more "conventional" drugs against different targets in a multidrug regimen would not yield the multitarget selectivity, because the two drugs together would perturb not only cells containing both targets but also cells containing either one of the targets.

A strategy for designing protein-based reagents that are sensitive to the presence or absence of more than one target at the same time was proposed in 1995.[57] A key feature of these reagents is their ability to utilize codominance, the property characteristic of many signals in proteins, including degrons and nuclear localization signals (NLSs). Codominance, in this context, refers to the ability of two or more signals in the same molecule to function independently and not to interfere with each other. The critical property of a degron-based multitarget reagent is that its intrinsic toxicity is the same in all cells, whereas its half-life (and, consequently, its steady state level and overall toxicity) in a cell depends on the protein composition of the cell, specifically on the presence of "target" proteins that have been chosen to define the profile of a cell to be eliminated.[57] A related but different design involves a toxic protein made short-lived (and therefore relatively nontoxic) by the presence of a degradation signal such as an N-degron. (The latter is produced by the Ub fusion technique.) If a cleavage site for a specific viral processing protease is placed between the toxic moiety of the fusion and the N-degron, the fusion would be cleaved in virus-infected cells but not in uninfected cells. As a result, the toxic moiety of the fusion would become long-lived (and therefore more toxic) only in virus-infected cells.[59] The codominance concept and the ideas about protein-size multitarget reagents have been extended to small ($<$1-kDa) multitarget compounds.[58]

Overproduction of Proteins as Ubiquitin Fusions

A major application of the Ub fusion technique is its use to augment the yields of recombinant proteins.[36–39] This approach increases the yield of short peptides as well.[40,41,60] The yield-enhancing effect of Ub was ob-

[57] A. Varshavsky, *Proc. Natl. Acad. Sci. U.S.A.* **92**, 3663 (1995).
[58] A. Varshavsky, *Proc. Natl. Acad. Sci. U.S.A.* **95**, 2094 (1998).
[59] A. Varshavsky, *Cold Spring Harbor Symp. Quant. Biol.* **60**, 461 (1996).
[60] T. H. LaBean, S. A. Kauffman, and T. R. Butt, *Mol. Divers.* **1**, 29 (1995).

served not only with eukaryotic cells (where the Ub moiety is present in a nascent fusion but not in its mature counterpart) but also in prokaryotes, which lack the Ub system, including DUBs, and therefore retain the Ub moiety in a translated fusion.[35-37] (*Escherichia coli* transformed with a plasmid expressing the *S. cerevisiae* DUB Ubp1p acquires the ability to deubiquitylate Ub fusions.[61])

The yield-enhancing effect of Ub stems at least in part from rapid folding of the nascent Ub moiety, whose presence at the N terminus of an emerging polypeptide chain may thereby partially protect the still unfolded chain from attacks by proteolytic pathways of the cytosol (most of these pathways are a part of the Ub system). The remarkably strong increases in protein yield even in eukaryotic cells, where the Ub moiety of the fusion is retained transiently (for it is rapidly removed by DUBs), suggest that this protection by Ub is particularly critical during translation, when an emerging, partially unfolded polypeptide chain may present degrons that are buried in the folded version of the same polypeptide. The chaperone role of Ub in this setting reflects one of its physiological functions. Specifically, the experiments with natural Ub fusions containing ribosomal proteins have shown that the transient presence of Ub in front of a ribosomal protein moiety is required for the efficient incorporation of that moiety into the nascent ribosomes,[12] most likely because of the transient protection effect described above.

The Ub-mediated increase in total yield is often accompanied by an even greater increase in the solubility of overexpressed protein. In this regard, the effect of Ub is analogous to that of several other proteins, such as thioredoxin[62] and maltose-binding protein (MBP).[63] When these moieties are cotranslationally linked to a protein of interest, they often increase its yield and solubility. A model of the underlying mechanism suggested for MBP[63] may also be relevant to the effect of Ub moiety. Specifically, a partially unfolded nascent protein is presumed to weakly interact with the nearby (upstream) MBP moiety, thereby transiently precluding intermolecular self-interactions that could result in irreversible aggregation before the protein has had the time to attain its mature conformation.[63]

The first engineered Ub fusions utilized pUB23-X, a family of high copy plasmids that expressed Ub–X–βgal proteins containing different junctional residues (X) in *S. cerevisiae* from a galactose-inducible, glucose-repressible promoter.[1,42] Subsequent designs facilitated the construction of ORFs encoding Ub–X–polypeptide fusions by introducing a *Sac*II (*Sst*II)

---

[61] J. W. Tobias, T. E. Shrader, G. Rocap, and A. Varshavsky, *Science* **254**, 1374 (1991).

[62] E. R. LaVallie, E. A. DiBlasio, S. Kovacic, K. L. Grant, P. F. Schendel, and J. M. McCoy, *BioTechnology* **11**, 187 (1993).

[63] R. B. Kapust and D. S. Waugh, *Protein Sci.* **8**, 1668 (1999).

site within the codons for the last three residues of the Ub moiety.[39] In this cloning scheme, an ORF of interest is amplified by polymerase chain reaction (PCR) and a primer in which the 5' extension encodes the last three residues of Ub. Another cloning route employs double-stranded oligonucleotides with *Sac*II cohesive ends that are used to join the DNA fragments.[39] The expression of a resulting Ub–*X*–polypeptide fusion in a eukaryotic cell (or in a prokaryotic cell that contains the *S. cerevisiae* Ubp1p DUB) yields an *X*–polypeptide bearing a predetermined N-terminal residue *X* (Figs. 1 and 2).

In their natural milieu, proteins of biotechnological or pharmacological interest are often products of the secretory pathway, and therefore are cleaved by signal peptidase on their entrance into the endoplasmic reticulum (ER). This cleavage frequently yields destabilizing residues at the N-termini of these proteins. When the same proteins are overexpressed in the cytosol of a heterologous bacterial or eukaryotic host, their N-terminal methionine tends to be retained, because MAPs cannot cleave off N-terminal methionine if it is followed by a destabilizing residue (see above). It is in these, quite frequent, cases that the expression of a protein as a Ub–*X*–protein fusion attains two aims at once: producing a protein of interest bearing the desired N-terminal residue (Fig. 2) and also, quite often, increasing the yield of the protein, in comparison with an otherwise identical expression of the Ub-lacking protein.[35]

There are numerous examples of Ub-mediated increases in the yield and solubility of overexpressed proteins. For instance, a conventional heterologous expression of the *Streptomyces* tyrosinase in *E. coli* yielded inactive enzyme, whereas expression of tyrosinase as a Ub fusion resulted in an abundant and active enzyme.[64] Another example of the use of Ub fusions in *E. coli* was an abundant expression of the soluble human collagenase catalytic domain. In contrast, the expression of the same protein in the absence of N-terminal Ub moiety resulted in low yield and insoluble product.[65] A 60-fold increase in the yield of the human pi class glutathione transferase GSTP1 was observed on the addition of a Ub-coding sequence to the *GSTP1* ORF.[39] A strong increase in protein yield in *E. coli* was reported with a combination of the T7 RNA polymerase promoter system and Ub fusions.[66] Several other examples of the Ub fusion approach to

[64] K. Han, J. Hong, H. C. Lim, C. H. Kim, Y. Park, and J. M. Cho, *Ann. N.Y. Acad. Sci.* **721**, 30 (1994).

[65] M. R. Gehring, B. Condon, S. A. Margosiak, and C. C. Kan, *J. Biol. Chem.* **270**, 22507 (1995).

[66] M. H. Koken, H. H. Odijk, M. Van Duin, M. Fornerod, and J. H. Hoeijmakers, *Biochem. Biophys. Res. Commun.* **195**, 643 (1993).

protein overexpression[38,67-69] are described in an earlier review by Baker.[35] More recently, Hondred and colleagues applied the Ub fusion technique to augment protein expression in transgenic plants.[70]

Ubiquitin-Assisted Dissection of Protein Translocation across Membranes

A 1994 method called UTA (ubiquitin translocation assay) employs Ub as a kinetic probe in the context of signal sequence-bearing Ub fusions.[8] After emerging from ribosomes in the cytosol, a protein may remain in the cytosol, or may be transferred to compartments separated from the cytosolic space by membranes. With a few exceptions, noncytosolic proteins begin journeys to their respective compartments by crossing membranes that enclose intracellular organelles such as the ER and mitochondria in eukaryotes or the periplasmic space in bacteria. Amino acid sequences that enable a protein to cross the membrane of a compartment are often located at the protein's N terminus. These "signal" sequences[71] are targeted by translocation pathways specific for each compartment. The translocation of a protein across a compartment membrane can start before the synthesis of the protein is completed, resulting in docking of the still translating ribosome at the transmembrane channel. The UTA technique takes advantage of the rapid (cotranslational) cleavage of a Ub fusion to examine temporal aspects of protein transport across the ER membrane in living cells.[8] Specifically, if a Ub fusion that has been engineered to bear an N-terminal signal sequence (SS) upstream of the Ub moiety is cleaved in the cytosol by DUBs, the fusion's reporter moiety would fail to be translocated into the ER. Conversely, if a nascent SS mediates the docking of a translating ribosome at the transmembrane channel rapidly enough, or if the fusion Ub moiety is located sufficiently far downstream of the SS, then by the time the Ub moiety emerges from the ribosome the latter is already docked, and the nascent Ub moiety enters the ER before it can fold and/or be targeted by DUBs. Thus, the cleavage at the Ub moiety of an SS-bearing Ub fusion in the cytosol can serve as an *in vivo* kinetic marker and a tool for analyzing targeting in protein translocation.[8] The temporal sensitivity of the UTA technique stems from rapid folding of the nascent Ub moiety

[67] E. A. Sabin, C. T. Lee-Ng, J. R. Shuster, and P. J. Barr, *Bio Technology* **7**, 705 (1989).
[68] E. Rian, R. Jemtland, O. K. Olstad, J. O. Gordeladze, and K. M. Gautvik, *Eur. J. Biochem.* **213**, 641 (1993).
[69] M. Coggan, R. Baker, K. Miloszewski, G. Woodfield, and P. Board, *Blood* **9**, 2455 (1995).
[70] D. Hondred, J. M. Walker, D. E. Mathews, and R. D. Vierstra, *Plant Physiol.* **119**, 713 (1999).
[71] G. Blobel, *Proc. Natl. Acad. Sci. U.S.A.* **77**, 1496 (1980).

that precludes its translocation and makes it a substrate of DUBs in the cytosol shortly after the emergence of the fusion Ub moiety from the ribosome.

Split-Ubiquitin Sensor for Detection of Protein–Protein Interactions

Another Ub-based method, termed the split-Ub sensor or USPS (Ub/split/protein/sensor), makes it possible to detect and monitor a protein–protein interaction as a function of time, at the natural sites of this interaction in a living cell.[72] These capabilities of the split-Ub technique distinguish it from the two-hybrid assay.[73] The design of a split-Ub sensor is based on the following observations: when a C-terminal fragment of the 76-residue Ub (C_{ub}) was expressed as a fusion to a reporter protein, the fusion was cleaved by DUBs only if an N-terminal fragment of Ub (N_{ub}) was also expressed in the same cell. This reconstitution of native Ub from its fragments, detectable by the *in vivo* cleavage assay, was not observed with a mutationally altered N_{ub}. However, if C_{ub} and the altered N_{ub} were each linked to polypeptides that interact *in vivo*, the cleavage of the fusion containing C_{ub} was restored, yielding a generally applicable assay for kinetic and equilibrium aspects of the *in vivo* protein interactions.[72]

Enhancement of Ub reconstitution by interacting polypeptides linked to fragments of Ub stems from a local increase in concentration of one Ub fragment in the vicinity of the other. This in turn increases the probability that the two Ub fragments coalesce to form a quasinative Ub moiety, whose (at least) transient formation results in the irreversible cleavage of the fusion by DUBs. This cleavage can be detected readily, and can be followed as a function of time or at steady state.[72,74] Unlike the two-hybrid method, which is based on the apposition of two structurally independent protein domains whose folding and functions do not require direct interactions between the domains, the split-Ub assay involves reconstituting the conformation of a small, single-domain protein. Applications of the split-Ub sensor have shown that this assay is capable of detecting transient *in vivo* interactions such as the binding of a signal sequence of a translocated protein to Sec62p, a component of the ER channel.[74] Different reporter readouts and selection-based screens have been devised for the split-Ub assay, making it possible to use this method for identifying the *in vivo* ligands of a protein

[72] N. Johnsson and A. Varshavsky, *Proc. Natl. Acad. Sci. U.S.A.* **91**, 10340 (1994).
[73] S. Fields and O. Song, *Nature (London)* **340**, 245 (1989).
[74] M. Dünnwald, A. Varshavsky, and N. Johnsson, *Mol. Biol. Cell* **10**, 329 (1999).

of interest, similar to the main application of the two-hybrid assay.[75,76] Split-protein sensors analogous to split-Ub but employing other proteins, such as DHFR, have been developed as well.[77,78]

UPR Technique

Direct measurements of the *in vivo* degradation of intracellular proteins require a pulse–chase assay. It involves the labeling of nascent proteins for a short time with a radioactive precursor ("pulse"), the termination of labeling through the removal of radiolabel and/or the addition of a translation inhibitor, and the analysis of a labeled protein of interest at various times afterward ("chase"), using immunoprecipitation and sodium dodecyl sulfate–polyacrylamide gel electrophoresis (SDS–PAGE), or analogous techniques. Its advantage of being direct notwithstanding, a conventional pulse–chase assay is fraught with sources of error. For example, the immunoprecipitation yields may vary from sample to sample; the volumes of samples loaded on a gel may vary as well. If the labeling for specific chase times is done with separate batches of cells (as is the case, e.g., with anchorage-dependent mammalian cell cultures), the efficiency of labeling is yet another unstable parameter of the assay. As a result, pulse–chase data tend to be semiquantitative at best, lacking the means to correct for these errors.

A robust and convenient "internal reference" strategy was described in 1996. This strategy, an extension of the original Ub fusion method, was termed the UPR (ubiquitin/protein/reference) technique.[79] UPR can compensate for several sources of data scatter in a pulse–chase assay (Fig. 3). UPR employs a linear fusion in which Ub is located between a protein of interest and a reference protein moiety (Fig. 3A). The fusion is cotranslationally cleaved by DUBs after the last residue of Ub, producing equimolar amounts of the protein of interest and the reference protein bearing the C-terminal Ub moiety. If both the reference protein and the protein of interest are immunoprecipitated in a pulse–chase assay, the relative amounts of the protein of interest can be normalized against the reference protein in the same sample.[28,29,79] The UPR technique (Fig. 3)

[75] I. Stagljar, C. Korostensky, N. Johnsson, and S. te Heesen, *Proc. Natl. Acad. Sci. U.S.A.* **95**, 5187 (1998).

[76] S. Wittke, N. Lewke, S. Müller, and N. Johnsson, *Mol. Biol. Cell* **10**, 2519 (1999).

[77] I. Remy and S. W. Michnick, *Proc. Natl. Acad. Sci. U.S.A.* **96**, 5394 (1999).

[78] J. N. Pelletier, F. X. Campbell-Valois, and S. W. Michnick, *Proc. Natl. Acad. Sci. U.S.A.* **95**, 12141 (1998).

[79] F. Lévy, N. Johnsson, T. Rumenapf, and A. Varshavsky, *Proc. Natl. Acad. Sci. U.S.A.* **93**, 4907 (1996).

can thus compensate for the scatter of immunoprecipitation yields, sample volumes, and other sources of sample-to-sample variation. The increased accuracy afforded by UPR underscored the insufficiency of the current "half-life" terminology, because the *in vivo* degradation of many proteins deviates from first-order kinetics. For a discussion of this problem and the terminology for describing nonexponential decay, see Refs. 29 and 79.

Ubiquitin Sandwich Technique

Nascent polypeptides emerging from the ribosome may, in the process of folding, present degradation signals similar to those recognized by the Ub system in misfolded or otherwise damaged proteins. It has been a long-standing question whether a significant fraction of nascent polypeptides is cotranslationally degraded. Determining whether nascent polypeptides are actually degraded *in vivo* has been difficult because at any given time the nascent chains of a particular protein species are of different sizes, and therefore would not form a band on electrophoresis in a conventional pulse–chase assay. The Ub sandwich technique[9] makes it possible to detect cotranslational protein degradation by measuring the steady state ratio of two reporter proteins whose relative abundance is established cotranslationally.

Operationally, the Ub sandwich technique[9] is a three-protein version of the UPR assay.[79] A polypeptide to be examined for cotranslational degradation, termed **B**, is sandwiched between two stable reporter domains **A** and **C** in a linear fusion protein. The three polypeptides are connected via Ub moieties to create a fusion protein of the form **AUb–BUb–CUb**. The independent polypeptides **AUb**, **BUb**, and **CUb** that result from the cotranslational cleavage of **AUb–BUb–CUb** by DUBs are called modules. The DUB-mediated cleavage establishes a kinetic competition between two mutually exclusive events during the synthesis of the **AUb–BUb–CUb** fusion: cotranslational UBP cleavage at the **BUb–CUb** junction to release the long-lived **CUb** module or, alternatively, cotranslational degradation of the entire **BUb–CUb** nascent chain by the 26S proteasome. In the latter case, the processivity of proteasome-mediated degradation results in the destruction of the Ub moiety between **B** and **C** before it can be recognized by UBPs. The resulting drop in levels of the **CUb** module relative to levels of **AUb**, referred to as the C/A ratio, reflects the cotranslational degradation of domain **B**. This measurement provides a minimal estimate of the total amount of cotranslational degradation, because nonprocessive cotranslational degradation events that do not extend into the **C** domain are not detected. The Ub sandwich method was used to demonstrate that more than

50% of nascent protein molecules bearing an N-degron can be degraded cotranslationally in *S. cerevisiae*, never reaching their mature size before their destruction by processive proteolysis.[9]

If cotranslational protein degradation by the Ub system is found to be extensive for at least some wild-type proteins (surveys of natural proteins remain to be carried out by this new technique), it could be accounted for as an evolutionary trade-off between the necessity of identifying and destroying degron-bearing mature proteins and the mechanistic difficulty of distinguishing between posttranslationally and cotranslationally presented degrons. Cotranslational protein degradation may also represent a previously unrecognized form of protein quality control, which destroys nascent chains that fail to fold correctly. These and other questions about physiological aspects of the cotranslational protein degradation can now be addressed directly in living cells through the Ub sandwich technique.[9]

Concluding Remarks

The Ub fusion technique is made possible by the ability of DUBs to cleave a Ub fusion *in vivo* or *in vitro* after the last residue of Ub irrespective of the flanking sequence context. Since its development, the Ub fusion technique has given rise to a number of applications whose common feature is utilization of the rapid and highly specific cleavage of a Ub-containing fusion by DUBs. Among these applications is the UPR technique, which increases the accuracy of pulse–chase and analogous measurements. I hope that the use of UPR will spread, supplanting the conventional, far less accurate pulse–chase protocols that lack a reference protein. The Ub sandwich technique, a descendant of UPR, has made it possible to determine the extent of cotranslational protein degradation *in vivo* for any protein of interest. One important feature of the Ub moiety is its ability, as a part of linear fusions, to increase the yields and solubility of overexpressed proteins or short peptides in either eukaryotic or bacterial hosts. In yet another class of Ub-based applications, the demonstrated coalescence of peptide-size Ub fragments into a quasinative Ub fold has yielded the split-Ub sensor for detecting protein interactions *in vivo*. Ub fusions continue to be useful in a remarkable variety of ways.

Acknowledgments

I am most grateful to the former and current members of my laboratory, whose work made possible some of the advances described in this review. I thank Daniel Finley (Harvard Medical School) and Rohan Baker (Australian National University) for their comments on the manuscript. Our studies are supported by grants from the National Institutes of Health (GM31530 and DK39520).

[42] Use of Phosphorylation Site Tags in Proteins

By Sidney Pestka, Lei Lin, Wei Wu, and Lara Izotova

Labeled proteins are used in a wide variety of applications. These include ligand–receptor interactions, pharmacokinetics, diagnostic imaging, and therapy. In particular, monoclonal antibodies labeled with many different isotopes have been used in diagnostic imaging as well as therapy. Labeling of proteins to high radiospecific activity in the laboratory has commonly been carried out by various iodination procedures that result in the incorporation of ^{125}I into proteins. Although the procedure is reasonably convenient and rapid, the risks associated with the use of ^{125}I are significant, so that monitoring of the thyroid gland is required when using radioactive iodine. In addition, when monoclonal antibodies are labeled with iodine or other methods, the monoclonal antibodies are significantly altered by the chemical procedures used. Chemical modification of proteins is a random process, with the modifications occurring throughout the protein chain. Accordingly, many of the chemical modifications inactivate the molecules labeled. In confronting this problem at the time it was necessary for us to obtain labeled ligands, we designed a new procedure that provided for a gentle method of labeling proteins enzymically with the use of isotopes substantially safer to employ than radioactive iodine. The procedure involves the introduction of a phosphokinase recognition site into the protein by constructing an expression vector encoding the protein with the new protein kinase recognition site. The procedure is convenient and rapid. It can be applied generally to virtually all proteins.

As noted above, the procedure we used incorporates a protein kinase recognition site into the protein of interest by recombinant DNA procedures. The coding sequence for the protein kinase recognition site is incorporated into the sequence encoding the protein in an appropriate expression vector. Because we were studying interferon binding to receptors, we first modified a number of interferons by these procedures.[1-3] Since assays for biological activity of interferons are quite sensitive, being able to detect picograms of these molecules, we assessed the effect of incorporation of the phosphorylation sites into the interferons. If, indeed, these sites could

---

[1] B. L. Li, J. A. Langer, B. Schwartz, and S. Pestka, *Proc. Natl. Acad. Sci. U.S.A.* **86,** 558 (1989).

[2] P. Wang, L. Izotova, T. M. Mariano, R. J. Donnelly, and S. Pestka, *J. Interferon Res.* **14,** 41 (1994).

[3] X. X. Zhao, B. L. Li, J. A. Langer, G. Van Riper, and S. Pestka, *Anal. Biochem.* **178,** 342 (1989).

be introduced without affecting the biological activity significantly, they could be used for ligand-binding studies, pharmacokinetics, and other research areas such as in the development of radioimmunoassays (RIAs) that would be much safer than those based on [125]I. It was surprising to us that the modifications of the interferons did not alter the profile of activity or the specific activity of the interferons either before or after phosphorylation.[1–3] After modifying a number of interferons in this manner, we went on to modify monoclonal antibodies to begin to develop these for diagnostic and therapeutic applications.

Because the cAMP-dependent protein kinase (PKA) recognition site was well characterized and the enzyme was commercially available, we first used this recognition site.[1–5] Later we showed that the same principle could be used for other protein kinase recognition sites: casein kinase I,[6] casein kinase II,[7] and Src tyrosine kinase.[8,8a] All these recognition sites could be used effectively in these applications and the sites could be introduced into proteins by site-specific mutation of the coding sequence. The casein kinase I and casein kinase II recognition sites contain acidic rather than basic residues found in the cAMP-dependent protein kinase recognition site. The Src tyrosine kinase recognition site is a neutral site. Having the various types of sites available for phosphorylation increases the choice of the amino acid residues of the phosphorylation site. A wider choice of residues is important to increase the range of protein charge and structure that could be introduced to minimize altering the activity, the pharmacokinetics, and other properties of the protein. In addition, the use of radiolabeled phosphate (^{33}P or ^{32}P) in proteins would have a greater margin of safety than many isotopes such as ^{125}I and many of the heavy metals such as ^{111}In, ^{99m}Tc, ^{90}Y, and [186]Re used for labeling of monoclonal antibodies.

The introduction of a kinase recognition site into proteins keeps their essential structure intact, in contrast to chemical conjugation of chelating agents necessary to bind heavy metal radioisotopes[9,10] or chemical linking of peptides with a cAMP-phosphorylation site.[11] This is especially significant

---

[4] L. Lin, B. Daugherty, J. Schlom, and S. Pestka, *Cancer Res.* **56,** 4250 (1996).

[5] L. Lin, S. D. Gillies, Y. Lan, L. Izotova, W. Wu, J. Schlom, and S. Pestka, *Int. J. Oncol.* **13,** 115 (1998).

[6] L. Lin, S. D. Gillies, J. Schlom, and S. Pestka, *Protein Expr. Purif.* **15,** 83 (1999).

[7] L. Lin, S. D. Gillies, J. Schlom, and S. Pestka, *Anticancer Res.* **18,** 3971 (1998).

[8] L. Lin, S. D. Gillies, J. Schlom, and S. Pestka, *Int. J. Oncol.* **13,** 725 (1998).

[8a] H. C. Cheng, H. Nishio, O. Hatase, S. Ralph, and J. H. Wang, *J. Biol. Chem.* **267,** 9248 (1992).

[9] D. M. Goldenberg, *Am. J. Med.* **94,** 297 (1993).

[10] J. Schlom, "Biological Therapy of Cancer" (V. T. Devita, S. Hellman, and S. A. Rosenberg, eds.), p. 507. J. B. Lippincott, Philadelphia, 1995.

[11] B. M. Foxwell, H. A. Band, J. Long, W. A. Jeffery, D. Snook, P. E. Thorpe, G. Watson, P. J. Parker, A. A. Epenetos, and A. M. Creighton, *Br. J. Cancer* **57,** 489 (1988).

as antibodies are being used as human therapeutics, where it is important to minimize antigenicity. The attachment of peptides chemically to a protein creates new epitopes not present in any natural proteins of the body, as both the chemical linkers used and the peptides attached form multiple new epitopes. In fact, the conjugation of a peptide to larger proteins has been the standard procedure for many decades to prepare antibodies to peptide antigens. In the process, when peptides are coupled to proteins there are at least two new epitopes added to the protein: the peptide itself and the combination of the peptide region and protein to which it is attached. In addition, the attachment of peptides to proteins increases the immunogenicity of proteins to which they are attached.[12] Therefore, attachment of peptides to proteins that are immunogenic or to those that are not immunogenic provides a new entity with increased immunogenicity. Because antibody production in response to administration of mouse monoclonal antibodies to humans results in production of human anti-mouse antibodies (HAMAs), profoundly limiting multiple use of the antibodies as radiolabeled diagnostics or therapeutics, investigators have developed chimeric and humanized monoclonal antibodies to minimize immunogenicity.[10] The procedures we have developed and describe in this chapter minimize these problems and provide a convenient method to prepare radiolabeled proteins.

Enzyme Reactions

Protein kinases are phosphotransferases that transfer a phosphate from the γ-phosphate of ATP (or another nucleoside triphosphate such as GTP) to the acceptor amino acid of a protein substrate. Serine, threonine, and tyrosine are the most common acceptors in proteins that are phosphorylated. However, other amino acids that have been found to be phosphorylated include histidine, arginine, lysine, cysteine, aspartic acid, and glutamic acid.[13,14] The protein kinases have been classified according to the amino acid acceptor into protein-serine/threonine, protein-tyrosine, protein-histidine (histidine, arginine, and lysine as acceptors), protein-cysteine, and protein-aspartyl/glutamyl kinases. Many protein kinases have been well characterized and can be used to phosphorylate proteins *in vitro*. The best characterized of the protein kinases are the protein-serine/threonine and protein-tyrosine kinases that most frequently use ATP as the high-energy phosphate donor.

---

[12] S. Dagan, E. Tzehoval, M. Fridkin, and M. Feldman, *J. Biol. Response Modif.* **6,** 625 (1987).
[13] T. Hunter and B. M. Sefton, *Methods Enzymol.* 200 (1991).
[14] T. Hunter and B. M. Sefton, *Methods Enzymol.* 201 (1991).

Radioactivity Options

The protein kinases discussed in this chapter utilize ATP transferring the γ-phosphate to the hydroxyl group of the amino acids serine and threonine or to the phenolic hydroxyl group of tyrosine in the protein. Therefore, the radioactive labels that can be used for such transfers are ^{32}P, ^{33}P, and ^{35}S. The sulfate can be transferred as a thiosulfate on the terminal phosphate of ATP. When reagents with a long half-life are desired, ^{35}S with a half-life of 87 days would be suitable. However, the energy of the β emission from ^{35}S is low, so that it cannot be used for imaging or therapy in animals. The high-energy β emission of ^{32}P is suitable for both imaging and therapeutic applications, making it useful for labeling monoclonal antibodies and other reagents for a wide variety of uses. Radioimmunoassays with ^{32}P, for example, would be several times more sensitive than radioimmunoassays with ^{125}I. In the case of monoclonal antibodies, the use of ^{32}P can substitute for many isotopes that require covalent linkage to the antibodies through chelating agents that are covalently linked to the protein.

Both ^{32}P and ^{33}P can be used conveniently for labeling, as both isotopes are available commercially as [γ-^{32}P]ATP and [γ-^{33}P]ATP. In addition, as noted above the γ-^{35}S analog of ATP, ATP γ-thiophosphate, provides the ability to incorporate ^{35}S into proteins with a substantially longer half-life (87 days) than ^{32}P (14.2 days) or ^{33}P (24.4 days). The half-lives of ^{125}I and ^{131}I are 60 and 8.1 days, respectively. Because of the shorter half-life of ^{32}P than ^{125}I, the specific radioactivity of ^{32}P-labeled proteins is more than fourfold that of ^{125}I-labeled proteins per incorporated radioisotope moiety. Furthermore, multiple kinase recognition sites can be incorporated into proteins without altering the activity of the molecule,[5] whereas this is difficult with iodination because the greater the number of random sites of covalent modification, the greater the inactivation of the proteins. Although radioactive derivatives provide sensitive levels of detection, it should be noted that the nonradioactive phosphate can also be incorporated into these proteins. The proteins can be tagged with nonradioactive ^{31}P that permits the detection of the phosphoserine and phosphotyrosine derivatives sensitively by Western blotting with antibodies against phosphoserine and phosphotyrosine residues, respectively, in the intact proteins or by techniques such as mass spectroscopy. The use of nonradioactive phosphate is attractive in studies such as pharmacokinetics where the phosphorylated proteins could be injected into patients without exposure to ionizing radiation.

Genetic Engineering

Two general methods have been used for introduction of phosphorylation sites into proteins: site-specific mutations and fusions. Phosphorylation sites can be introduced into proteins as fusion segments to the N terminus or the C terminus, or even as internal segments. In addition, it is possible to introduce phosphokinase recognition sites within the protein itself. This can often be accomplished by changing one amino acid with little or no modification of the function or biological activity of the protein. It is desirable to have some useful information about the protein before choosing a site to modify. Modification of the proteins can be made by a wide variety of genetic engineering techniques that are not described in this chapter.

Site-specific mutations were used to introduce new phosphorylation sites into the context of the primary sequence of interferons[1,2,4] [also R. Donnelly and S. Pestka, unpublished data for interleukin 2 (IL-2), 1999; W. Wu and S. Pestka, unpublished data for monoclonal antibody chCC49-WW5P, 1999]. Fusions were constructed with a variety of proteins and various phosphorylation sites[3,5–8,15–28] [interferon,[3] monoclonal antibodies,[5–8] retinoblastoma protein,[15,19] osteopontin,[16] c-Fos,[17] calmodulin,[18] diphtheria toxin,[20] endothelial growth factors,[20] enterotoxins,[20] lyphokines,[20] ricin,[20,21] antibody fragments,[23] microtubule-associated protein[25]; *Escherichia coli* RNA polymerase β' and σ subunits[27,28]; IL-3, C. Miyamoto, personal communication, 1999; interferon β (IFN-β), X.-X. Zhou, J. Langer, and S. Pestka, unpublished data, 1999; IL-10 and vIL-10, L. Izotova, S. Kotenko, S. Saccani and S. Pestka, unpublished observations, 1999]. In these cases, the activities of the modified and of the phosphorylated proteins remained intact. The details of the various constructions are given in the specific reports cited. Foxwell *et al.*[11] used a nongenetic engineering proce-

[15] P. D. Adams, X. Li, W. R. Sellers, K. B. Baker, X. Leng, J. W. Harper, Y. Taya, and W. G. Kaelin, Jr., *Mol. Cell. Biol.* **19,** 1068 (1999).

[16] S. Ashkar, D. B. Teplow, M. J. Glimcher, and R. A. Saavedra, *Biochem. Biophys. Res. Commun.* **191,** 126 (1993).

[17] M. A. Blanar and W. J. Rutter, *Science* **256,** 1014 (1992).

[18] R. Fischer, Y. Wei, and M. Berchtold, *BioTechniques* **21,** 292 (1996).

[19] W. G. J. Kaelin, Jr., W. Krek, W. R. Sellers, J. A. DeCaprio, F. Ajchenbaum, C. S. Fuchs, T. Chittenden, Y. Li, P. J. Farnham, and M. A. Blanar, *Cell* **70,** 351 (1992).

[20] D. Mohanraj, J. L. Wahlsten, and S. Ramakrishnan, *Protein Expr. Purif.* **8,** 175 (1996).

[21] D. Fryxell, B. Y. Li, D. Mohanraj, B. Johnson, and S. Ramakrishnan, *Biochem. Biophys. Res. Commun.* **210,** 253 (1995).

[22] S. J. Moss, C. A. Doherty, and R. L. Huganir, *J. Biol. Chem.* **267,** 14470 (1992).

[23] D. Neri, H. Petrul, G. Winter, Y. Light, R. Marais, K. E. Britton, and A. M. Creighton, *Nature Biotechnol.* **14,** 485 (1996).

[24] D. Ron and H. Dressler, *BioTechniques* **13,** 866 (1992).

[25] R. E. Stofko-Hahn, D. W. Carr, and J. D. Scott, *FEBS Lett.* **302,** 274 (1992).

[26] D. Zamanillo, E. Casanova, A. Alonso-Llamazares, S. Ovalle, M. A. Chinchetru, and P. Calvo, *Neurosci. Lett.* **188,** 183 (1995).

[27] T. M. Arthur and R. R. Burgess, *J. Biol. Chem.* **273,** 31381 (1998).

[28] R. R. Burgess, T. M. Arthur, and B. C. Pietz, *Methods Enzymol.* **328,** Chap. 11, in press (2000).

dure to introduce the cAMP-dependent protein kinase recognition site into monoclonal antibodies by using peptides containing the PKA sequence and then covalently linking these chemically to a monoclonal antibody. The phosphorylated monoclonal antibody was stable in serum, as were the attached phosphates. However, these peptides were randomly attached to the protein, generated many new epitopes that would enhance the antigenicity of the modified monoclonal antibody, and yielded a heterogeneous product in contrast to the homogeneous product obtained by the introduction of phosphorylation sites at defined sites in the protein through genetic engineering.

Protein Kinase Recognition Sites

Protein kinase recognition sites have been identified for many kinases. A summary of useful amino acid recognition motifs is given in Table I based on data reported.[29-39] Although the recognition sites can be optimized for each protein or peptide for a given enzyme, the recognition sites listed in Table I show amino acid sequences of sites we have found to be effective for four kinases when introduced into proteins.[1-8] Although both serine and threonine can be used for the protein-serine/threonine kinases, it was reported that the apparent K_m for threonine was 37-fold higher than the comparable sequence with serine[40] and that, in general, serine is a much better acceptor than threonine.[37] Thus, the transfer of the phosphate to serine often occurs at a higher efficiency than the transfer to threonine with the protein-serine/threonine kinases. This results from the substantial differences in K_m of the reactions.

[29] A. M. Edelman, D. K. Blumenthal, and E. G. Krebs, *Annu. Rev. Biochem.* **56,** 567 (1987).
[30] D. B. Glass and E. G. Krebs, *Annu. Rev. Pharmacol. Toxicol.* **20,** 363 (1980).
[31] T. Hunter and J. A. Cooper, *Annu. Rev. Biochem.* **54,** 897 (1985).
[32] B. E. Kemp and R. B. Pearson, *Trends Biochem. Sci.* **15,** 342 (1990).
[33] E. G. Krebs and J. A. Beavo, *Annu. Rev. Biochem.* **48,** 923 (1979).
[34] R. B. Pearson and B. E. Kemp, *Methods Enzymol.* **200,** 62 (1991).
[35] O. Zetterqvist and U. Ragnarsson, *FEBS Lett.* **139,** 287 (1982).
[36] F. Marchiori, F. Meggio, O. Marin, G. Borin, A. Calderan, P. Ruzza, and L. A. Pinna, *Biochim. Biophys. Acta* **971,** 332 (1988).
[37] O. Marin, F. Meggio, F. Marchiori, G. Borin, and L. A. Pinna, *Eur. J. Biochem.* **160,** 239 (1986).
[38] O. Marin, A. Calderan, P. Ruzza, G. Borin, F. Meggio, N. Grankowski, and F. Marchiori, *Int. J. Pept. Protein Res.* **36,** 374 (1990).
[39] F. Meggio, J. W. Perich, E. C. Reynolds, and L. A. Pinna, *FEBS Lett.* **283,** 303 (1991).
[40] B. E. Kemp, D. J. Graves, E. Benjamini, and E. G. Krebs, *J. Biol. Chem.* **252,** 4888 (1977).

TABLE I
RECOGNITION SITES FOR VARIOUS PROTEIN KINASES[a]

Protein kinase	Recognition site
cAMP-dependent protein kinase (PKA)[b–j]	ArgArgXaaSerXab
Casein kinase[k,l]	AspAspAspAspSer**IleAsp**AspAspAspAspSer
Casein kinase II[f,i,m–p]	AspAspAspSerGluGluAsp
Src tyrosine kinase[q,r]	LysValGluLysIleGlyGluGlyThrTyrGlyValValTyrLys

[a] For introduction of the cAMP-dependent protein kinase recognition site, we used Arg-Arg-Xaa-Ser-Xab, where Xaa is usually Ala, and Xab has been Val, Leu, Met, or Gln.[1,2,4,5] However, as shown previously a wide variety of amino acids can substitute for Xaa and Xab effectively.[29,32–35] For the casein kinase I site, we used two AspAspAspAspSer sequences in tandem connected with **IleAsp**, shown in boldface,[6] based on sequences reported by Meggio *et al.*[39] For the casein kinase II site, we[7] used the sequence shown here.[29,34,36–38] For the Src tyrosine kinase site we used the sequence shown above,[8] based on the sequence reported.[8,8a]

[b] B. L. Li, J. A. Langer, B. Schwartz, and S. Pestka, *Proc. Natl. Acad. Sci. U.S.A.* **86,** 558 (1989).

[c] P. Wang, L. Izotova, T. M. Mariano, R. J. Donnelly, and S. Pestka, *J. Interferon Res.* **14,** 41 (1994).

[d] L. Lin, B. Daugherty, J. Schlom, and S. Pestka, *Cancer Res.* **56,** 4250 (1996).

[e] L. Lin, S. D. Gillies, Y. Lan, L. Izotova, W. Wu, J. Schlom, and S. Pestka, *Int. J. Oncol.* **13,** 115 (1998).

[f] A. M. Edelman, D. K. Blumenthal, and E. G. Krebs, *Annu. Rev. Biochem.* **56,** 567 (1987).

[g] B. E. Kemp and R. B. Pearson, *Trends Biochem. Sci.* **15,** 342 (1990).

[h] E. G. Krebs and J. A. Beavo, *Annu. Rev. Biochem.* **48,** 923 (1979).

[i] R. B. Pearson and B. E. Kemp, *Methods Enzymol.* **200,** 62 (1991).

[j] O. Zetterqvist and U. Ragnarsson, *FEBS Lett.* **139,** 287 (1982).

[k] L. Lin, S. D. Gillies, J. Schlom, and S. Pestka, *Protein Expr. Purif.* **15,** 83 (1999).

[l] F. Meggio, J. W. Perich, E. C. Reynolds, and L. A. Pinna, *FEBS Lett.* **283,** 303 (1991).

[m] L. Lin, S. D. Gillies, J. Schlom, and S. Pestka, *Anticancer Res.* **18,** 3971 (1998).

[n] F. Marchiori, F. Meggio, O. Marin, G. Borin, A. Calderan, P. Ruzza, and L. A. Pinna, *Biochim. Biophys. Acta* **971,** 332 (1988).

[o] O. Marin, F. Meggio, F. Marchiori, G. Borin, and L. A. Pinna, *Eur. J. Biochem.* **160,** 239 (1986).

[p] O. Marin, A. Calderan, P. Ruzza, G. Borin, F. Meggio, N. Grankowski, and F. Marchiori, *Int. J. Pept. Protein Res.* **36,** 374 (1990).

[q] L. Lin, S. D. Gillies, J. Schlom, and S. Pestka, *Int. J. Oncol.* **13,** 725 (1998).

[r] H. C. Chen, H. Nishio, O. Hatase, S. Ralph, and J. H. Wang, *J. Biol. Chem.* **267,** 9248 (1992).

Modified Proteins with Phosphorylation Sites Introduced

Since our publication describing the introduction of phosphorylation sites into proteins to generate proteins that can be labeled conveniently and gently with little or no effect on the protein structure and function,[1,3] a number of reports have appeared taking advantage of this concept. We used the procedure first with interferons.[1-3] The interferons could be gently labeled so that the products retained structure and function. The ^{32}P-labeled interferons were used for binding and cross-linking studies with the receptor components.[1-3,41-45] These studies used phosphorylatable human interferon α (IFN-α) species: Hu-IFN-αA, Hu-IFN-αB2, and Hu-IFN-αA/D.

Several reports described the construction and use of similar vectors for incorporation of protein kinase recognition sites into proteins. Blanar and Rutter[17] designed an expression vector to encode proteins with an amino-terminal extension containing a PKA recognition site. They fused the region to the leucine zipper dimerization motif of c-Fos, which was then used to screen a cDNA library. The phosphorylated protein was used to study protein–protein interactions. Kaelin et al.[19] used the vector prepared by Blanar and Rutter[17] to prepare a ^{32}P-labeled, glutathione-S-transferase (GST)-retinoblastoma (pRB) fusion protein to screen expression libraries efficiently. Continuing these studies with the same fusions proteins, Adams et al.[15] determined the region of pRB phosphorylated by cyclin A-cdk2 and cyclin E-cdk2. Ron and Dressler[24] constructed an expression vector similar to that of Blanar and Rutter[17] and Kaelin et al.[19] The vector encoded a GST tag and a PKA site that could be fused to protein-coding regions. The ^{32}P-labeled protein could be used identify interacting proteins. Stofko-Hahn et al.[25] constructed a vector with a PKA site, a factor Xa cleavage site, and a calmodulin-binding peptide. Fusion proteins could easily be purified on a calmodulin affinity matrix and the fused protein phosphorylated with [γ-^{32}P]ATP and PKA. Ashkar et al.[16] prepared a fusion protein with osteopontin containing a protein kinase recognition site and GST to study the function and posttranslational modifications of osteopontin. Moss et al.[22] prepared fusion proteins with the intracellular domains of the γ-aminobutyric acid type A (GABAA) receptor subunits to determine the recognition sites for PKA and PKC.

Zamanillo et al.[26] used fusion proteins of the dopamine receptor subtypes to study their phosphorylation by PKA and PKC. Xiao et al.[46] constructed fusion proteins with recognition sites for PKA, casein kinase II,

[41] J. R. Cook, C. M. Cleary, T. M. Mariano, L. Izotova, and S. Pestka, J. Biol. Chem. 271, 13448 (1996).

[42] I. Flores, T. M. Mariano, and S. Pestka, J. Biol. Chem. 266, 19875 (1991).

[43] S. Pestka, Semin. Oncol. 24, S9 (1997).

[44] J. Soh, T. M. Mariano, J. K. Lim, L. Izotova, O. Mirochnitchenko, B. Schwartz, J. A. Langer, and S. Pestka, J. Biol. Chem. 269, 18102 (1994).

[45] A. P. Alexenko, J. Li, N. Mathialagan, L. Izotova, T. M. Mariano, S. Pestka, and R. M. Roberts, J. Interferon Cytokine Res. 15, S97 (1995).

[46] C. Y. Xiao, S. Hubner, R. M. Elliot, A. Caon, and D. A. Jans, J. Biol. Chem. 271, 6451 (1996).

and the cyclin-dependent kinase cdc2. They demonstrated that nuclear import of simian virus 40 (SV40) large tumor antigen (T-ag) fusion proteins, normally regulated by the motif comprising phosphorylation sites for casein kinase II and the cyclin-dependent kinase cdc2, could be controlled by PKA if the PKA recognition site is substituted for the casein kinase II site. Fischer et al.[18] used a vector that contained sequences encoding a FLAG epitope, GST, and a PKA recognition site. They introduced the coding sequence of calmodulin to construct a calmodulin fusion protein. The ^{32}P-labeled calmodulin fusion protein was used to identify calmodulin-binding proteins in cellular extracts to replace more laborious techniques that employed ^{125}I-labeled calmodulin or nonradioactive biotinylated calmodulin. Arthur and Burgess[27] and Burgess et al.[28] used N-terminal hexahistidine (His$_6$)–PKA fusions of E. coli RNA polymerase β' and σ subunits to map the interaction site between one of the σ subunits and the β' subunit of the core enzyme.

Neri et al.[23] developed protein kinase tags for antibody fragments secreted from bacteria to identify immunoreactivity. After labeling the fragments with [γ-^{32}P]ATP and casein kinase II, the antibody fragments were used to determine affinity to specific antigens. They concluded that "In contrast to non site-specific methods such as radioiodination, antibodies labeled with casein kinase II retain full immunoreactivity."

Mohanraj et al.[20] prepared an expression vector for introduction of a PKA site at the amino terminus of proteins. They prepared seven different phosphorylatable proteins, demonstrating the generality of the methodology. After labeling with ^{32}P, they concluded that "All seven proteins used in this study could be expressed with the phosphorylation sequence at their amino terminus and specifically labeled without loss of biological activity."

Protein Purification

Before phosphorylation the proteins are purified by standard procedures. Because the introduction of the phosphorylation sites does not materially change the structure of the proteins, the usual procedures used for the unmodified proteins has been found to be effective.[1-8] When fusion proteins have tags such as GST, FLAG, or calmodulin, then affinity chromatography can be used.

Phosphorylation Procedures

Materials and Reagents

Adenosine 5'-triphosphate, γ-^{32}P-labeled ([Iγ-^{32}P]ATP), >5000 Ci/mmol, > 185 TBq/mmol (NEN Life Science Products, http://

www.nenlifesci.com; Amersham Pharmacia Biotech, *http://www.
nycomed-amersham.com*): The amount required for the reaction is
dried in a Savant (*http://www.savec.com*) Speed-Vac concentrator
prior to use

Adenosine 5'-triphosphate, γ-^{33}P-labeled ([γ-^{33}P]ATP), >1000 Ci/
mmol, >37 TBq/mmol (Amersham Pharmacia Biotech): The
amount required for the reaction is dried in a Savant Speed-Vac
concentrator prior to use

Adenosine 5'-(γ-thio)triphosphate, γ-^{35}S-labeled ([γ-^{32}S]ATP), >1000
Ci/mmol, >37 TBq/mmol (New England Nuclear, *http://www.
nenlifesci.com;* Amersham Pharmacia Biotech): The amount required for the reaction is dried in a Savant Speed-Vac concentrator
prior to use

Catalytic subunit of bovine heart cAMP-dependent protein kinase,
$\geq$20,000 units/mg (Sigma-Aldrich, *https://www.sigma-aldrich.com*):
This is prepared in dithiothreitol (DTT, 6 mg/ml, 39 mM) at 12.5
units/μl and stored at $-70°$ or below in small aliquots

Recombinant Phosphorylatable Interferons and Cytokines

Human IFN-αA-P1 (PBL BioMedical Laboratories, New Brunswick,
NJ; *http://www.interferonsource.com*)
Human IFN-αB2-P (PBL BioMedical Laboratories)
Human IFN-αA/D, Universal type I interferon (PBL BioMedical Lab-
oratories)
Ovine IFN-τ-P (PBL BioMedical Laboratories)
Bovine IFN-τ-P (PBL BioMedical Laboratories)
Human interleukin 2 (PBL BioMedical Laboratories)

Recombinant Phosphorylatable Monoclonal Antibodies

Monoclonal antibody IgG1-PKA (PBL BioMedical Laboratories):
This monoclonal antibody contains a cassette of two cAMP-depen-
dent protein kinase recognition sites, PKA sites, per heavy chain.
Monoclonal antibody IgG1-PKA-6 (PBL BioMedical Laboratories):
This monoclonal antibody contains a cassette of six cAMP-depen-
dent protein kinase recognition sites, PKA sites, per heavy chain.
Monoclonal antibody IgG1-CKI (PBL BioMedical Laboratories): This
monoclonal antibody contains a casein kinase I protein kinase recog-
nition site in each heavy chain.
Monoclonal antibody IgG1-CKII (PBL BioMedical Laboratories):
This monoclonal antibody contains a casein kinase II protein kinase
recognition site in each heavy chain.

Monoclonal antibody IgG1-Tyr (PBL BioMedical Laboratories): This monoclonal antibody contains a Src protein tyrosine kinase recognition site in each heavy chain.

Radiolabeling of Interferon

The following procedure usually yields high levels of labeling.

1. About 0.5–1 μg of protein at 30° is incubated for 60 min with 1 mCi of [γ-^{32}P]ATP and 15–60 units of the catalytic subunit of the cAMP-dependent protein kinase. The reaction volume of 25 μl should also contain 20 mM Tris-HCl (pH 7.4), 100 mM NaCl, 12 mM MgCl$_2$, and 5–20 mM dithiothreitol (depending on how much kinase is used).

2. After the reaction, the reaction mixture is diluted with 0.25–0.5 ml of a cold solution of bovine serum albumin (5 mg/ml) in 10 mM sodium pyrophosphate, pH 6.7. The sodium pyrophosphate is used to inhibit dephosphorylation.

3. The solution is then dialyzed twice against 1 liter or three times against 250 ml of 10 mM sodium pyrophosphate, pH 6.7, at 4° to remove the unincorporated [γ-^{32}P]ATP. Dialysis is performed for at least 6 hr before changing each solution. After dialysis, the labeled protein is stored in small aliquots in liquid nitrogen.

Typically, the degree of phosphorylation has ranged from about 890 to 2600 Ci/mmol (45–135 μCi/μg) for Hu-IFN-αA-P1,[1] 1111 Ci/mmol ($\sim$57 μCi/μg) for Hu-IFN-αB2-P, and 1028 Ci/mmol ($\sim$53 μCi/μg) for Hu-IFN-αA/D(Bgl).[2] The same procedure can be used to label the proteins with [γ-^{33}P]ATP or [γ-^{35}S]adenosine 5'-(γ-thio)triphosphate.

Phosphorylation of Proteins with cAMP-Dependent Phosphorylation Site

The following conditions are used for phosphorylation of proteins containing the cAMP-dependent phosphorylation site. For the highest labeling, [^{32}P]ATP should be used at the highest specific activity. This radiolabeled compound comes as an aqueous solution in 0.01 M Tricine. In general the best way to handle the material is to dry the entire contents in a Speed-Vac and dissolve the residue in a small amount of water at the concentration as noted below. The catalytic subunit of the cAMP-dependent protein kinase from bovine heart is used in the reaction below. The reaction conditions are as follows.

The MAb chCC49-6P is labeled with [γ-^{32}P]ATP and the cAMP-dependent protein kinase as described previously.[5] Approximately 10 μg of MAb is incubated at 30° for 60 min with 0.5 mCi of [γ-^{32}P]ATP (specific activity,

6000 Ci/mmol; Du Pont-New England Nuclear) and 25 units of the catalytic subunit of cAMP-dependent protein kinase from bovine heart muscle [specific activity, $\geq$20,000 units/mg (Sigma), dissolved in DTT at 6 mg/ml] in 25 μl of 20 mM Tris-HCl (pH 7.4), 100 mM NaCl, and 12 mM MgCl$_2$, and then cooled on ice to stop the reaction. After addition of 300 μl of bovine serum albumin (5 mg/ml) in 10 mM sodium pyrophosphate, pH 6.7, at 4°, the 0.325-ml reaction mixture is dialyzed against 1000 volumes of phosphate-buffered saline (PBS) overnight at 4°. Dialysis buffer is changed twice. Incorporation of radioactivity into the monoclonal antibodies is measured with a liquid scintillation spectrometer after precipitation of the protein with trichloroacetic acid (TCA).[47] To remove the labile ^{32}P, the final product in 0.325 ml is adjusted to pH 7.4 with 1 M Tris base, and then incubated at 37° overnight. The labeled material is then stored in a liquid nitrogen freezer in small aliquots until use.

After the reaction, the samples are dialyzed as noted above. However, alternative methods for removal of the free phosphate and ATP can be used, such as centrifugal filter separators or small disposable gel-filtration columns. Although all these procedures work to remove the phosphate and residual ATP, we have found that dialysis is the simplest and most effective method. It should be noted that we have found in evaluating the stability of the attached phosphate that two rates of phosphate release are observed.[4–8] One is a rapid release reflecting loosely bound phosphate with the major attached phosphate being relatively stable. When this is observed, we simply carry out the dialysis at 37° overnight instead of at 4° to remove the loosely attached phosphate from the final phosphorylated protein. As expected, on radioautography the final phosphorylated product comigrates with the corresponding Coomassie blue-stained protein after polyacrylamide gel electrophoresis. When fusion proteins containing epitope tags such as FLAG, Myc, HA (influenza hemagglutin), or GST are used, the protein can be isolated on beads containing antibody or a ligand for the tag. After washing the beads, the labeling reaction can be performed with the proteins to be labeled on the beads. After the reaction, the radioactivity and other reaction ingredients can be removed by washing the beads. The phosphorylated protein can then be removed from the affinity column with the affinity ligand or by acidification.

It is common to have high specific activity protein products, at the level of 5000 Ci/mmol. The specific activity can be raised substantially by using a cassette of one or more phosphorylation sites.[5] The following are the

---

[47] S. Pestka, *J. Biol. Chem.* **247**, 4669 (1972).

conditions for phosphorylation of proteins with the casein kinase I, casein kinase II, and Src tyrosine kinases.

Phosphorylation of Proteins with Casein Kinase I Phosphorylation Site

MAb chCC49CKI is labeled with [γ-^{32}P]ATP and the casein kinase I protein kinase as described.[6] Approximately 10 μg of MAb is incubated at 37° for 60 min with 0.5 mCi of [γ-^{32}P]ATP and 15 units of casein kinase I (Upstate Biotechnology, *http://www.upstatebiotech.com*) in 25 μl of 50 mM Tris-HCl (pH 7.5), 100 mM NaCl, and 12 mM MgCl$_2$, and then cooled on ice to stop the reaction. After addition of 300 μl of bovine serum albumin (5 mg/ml) in 10 mM sodium pyrophosphate, pH 6.7, at 4°, the 325-μl reaction mixture is dialyzed against 1000 volumes of PBS overnight at 4°. Dialysis buffer is changed twice. Incorporation of radioactivity into MAb chCC49CKI is measured with a liquid scintillation spectrometer after precipitation of the protein with TCA.[47] Details of the dialysis procedure are given above, in the section describing phosphorylation of the PKA site. To remove the labile phosphate, the phosphorylated protein ([^{32}P]MAb chCC49CKI) is preincubated at 37° for 24 hr in the medium in which it is to be used or in pH 7.4 buffer such as PBS. As noted above, this procedure was established because it was apparent that a percentage of the [^{32}P]phosphate attached to the protein was labile and could be eliminated by 24 hr preincubation.[4–8]

Phosphorylation of Proteins with Casein Kinase II Phosphorylation Site

MAb chCC49CKII is labeled with [γ-^{32}P]ATP and the casein kinase II as described.[7] Approximately 10 μg of MAb is incubated at 30° for 60 min with 0.5 mCi of [γ-^{32}P]ATP and 1 μl (0.23 unit; Upstate Biotechnology) of the casein kinase II in 25 μl of 20 mM Tris-HCl (pH 7.4), 100 mM NaCl, and 12 mM MgCl$_2$, and then cooled on ice to stop the reaction. After addition of 300 μl of bovine serum albumin (5 mg/ml) in 10 mM sodium pyrophosphate, pH 6.7, the 0.325-ml reaction mixture is dialyzed against PBS overnight at 4°. Dialysis buffer is changed twice. Incorporation of radioactivity into MAb chCC49CKII is measured with a liquid scintillation spectrometer after precipitation of the protein with trichloroacetic acid.[47] Details of the dialysis procedure are given above, in the section describing phosphorylation of the PKA site. To remove the labile phosphate, the phosphorylated protein [^{32}P]MAb chCC49CKII is preincubated at 37° for 24 hr in the medium in which it is to be used or in pH 7.4 buffer such as PBS.

Phosphorylation of Proteins with Src Tyrosine Kinase
  Phosphorylation Site

MAb chCC49Tyr is labeled with $[\gamma\text{-}^{32}P]ATP$ and the Src tyrosine kinase as described.[8] Approximately 10 μg of MAb is incubated at 30° for 60 min with 0.5 mCi of $[\gamma\text{-}^{32}P]ATP$ and 15 units of tyrosine protein kinase (Src kinase, $p60^{c\text{-}src}$; Upstate Biotechnology) in 25 μl of 50 mM Tris-HCl (pH 7.0) and 50 mM MgCl$_2$, and then cooled on ice to stop the reaction. After addition of 300 μl of bovine serum albumin (5 mg/ml) in 10 mM sodium pyrophosphate, pH 6.7, at 4°, the 325-μl reaction mixture is dialyzed against PBS overnight at 4°. Dialysis buffer is changed twice. Incorporation of radioactivity into MAb chCC49Tyr is measured with a liquid scintillation spectrometer after precipitation of the protein with TCA.[47] Details of the dialysis procedure are given above, in the section describing phosphorylation of the PKA site. To remove the labile phosphate, the phosphorylated protein $[^{32}P]$MAb chCC49CKII is preincubated at 37° for 24 hr in the medium in which it is to be used or in pH 7.4 buffer such as PBS.

Binding of Labeled Monoclonal Antibody to Surface Antigens

Binding of $[^{32}P]$MAb chCC49K1 and $[^{32}P]$MAb chCC49-6P to MCF-7 4C10 human breast carcinoma cells is performed as described.[5] MCF-7 4C10 cells used for binding studies are grown to confluence in six-well tissue culture plates in medium [Dulbecco's modified Eagle's medium (DMEM), GIBCO, Grand Island, NY] supplemented with 10% (v/v) fetal bovine serum, sodium pyruvate (0.05 mg/ml), insulin (0.005 mg/ml), and 0.5× nonessential amino acids, and treated with Hu-IFN-γ (1000 units/ml) for 48 hr before assay. Twenty minutes before assay, the medium is removed, and 1 ml of fresh medium is added to each well. $[^{32}P]$MAb chCC49K1 is then added with or without a 100-fold excess of unlabeled MAb chCC49K1 or MAb chCC49. After incubation at room temperature for 1 hr, the plates are placed on ice and each well is washed four times with 2 ml of ice-cold PBS to remove the unbound radioactive ligand. After washing, 1 ml of 1% (w/v) sodium dodecyl sulfate (SDS) is added to each well to remove cells and the ^{32}P-labeled MAb. Radioactivity is determined in a liquid scintillation counter by placing the 1-ml sample in 5 ml of a scintillation fluor. The binding with $[^{32}P]$MAb chCC49-6P is determined similarly, except that cells are cultured in a 96-well microtiter plate because the specific radioactivity of the labeled antibody is higher.

The binding of monoclonal antibody to TAG-72 antigen on breast

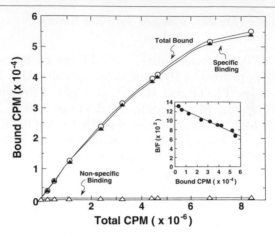

FIG. 1. Binding of monoclonal antibody chCC49K1 to MCF-7 4C10 breast carcinoma cells. Labeling of chCC49K1 monoclonal antibody and its binding to MCF-7 4C10 cells was performed as described in text.[5] Nonspecific binding values were subtracted from total bound counts to obtain the quantity of specific counts bound shown on the ordinate. [Reproduced from L. Lin, S. D. Gillies, Y. Lan, L. Izotova, W. Wu, J. Schlom, and S. Pestka, *Int. J. Oncol.* **13**, 115 (1998).]

cancer cell line MCF7 is shown in the Fig. 1. The [^{32}P]MAb chCC49K1 is bound to human breast carcinoma cells MCF-7 4C10 treated with Hu-IFN-γ (1000 units/ml) with the specific binding approaching saturation at the higher concentrations tested (Fig. 1). Scatchard analysis[48] of the data indicated these MCF-7 4C10 human breast carcinoma cells have 1900 receptors per cell, calculated from the binding of [^{32}P]MAb chCC49K1. The binding of [^{32}P]MAb chCC49K1 to the cells is blocked by unlabeled MAb chCC49K1. The binding of [^{32}P]MAb chCC49-6P to cells (data not shown) is similar to that shown for binding of [^{32}P]MAb chCC49K1 (Fig. 1). It is noteworthy that the binding of the ^{32}P-labeled monoclonal antibody to the cells is almost the same in the absence or presence of competitor. This is commonly seen with the ^{32}P-labeled proteins. It was initially a surprising result when compared with ^{125}I-labeled proteins. Once this is seen, it is not necessary to add competitor routinely to carry out binding studies. Thus the ^{32}P-labeled proteins can be used in ligand-binding studies without competitors, saving large amounts of the often difficult to obtain and expensive proteins. However, there always should be controls performed to ascertain that there is little or no difference in the presence of competitor.

[48] G. Scatchard, *Ann. N.Y. Acad. Sci.* **51**, 660 (1949).

Binding of Interferon to Its Cell Surface Receptor

The binding of ^{32}P-labeled human IFN-αB2 and IFN-αA/D to bovine MDBK cells and human Daudi cells is shown in Fig. 2. Here also it can be seen that the binding of the labeled interferon in the presence or absence of competitor is almost identical. To explain why ^{32}P-labeled proteins seem to have low nonspecific binding we hypothesize that the gentle labeling procedures with the protein kinases as well as the labeling of the protein at a specific site account for the low nonspecific binding. Chemical labeling of proteins occurs randomly at targeted residues such as lysine or tyrosine, depending on the chemical labeling procedure. As a result some of the protein is likely to be inactivated by attachment and modification of the amino acids in certain positions. Accordingly, some of the randomly labeled molecules in the population labeled by chemical means are likely to be inactive. In addition, often chemical labeling conditions can denature some of the protein. Another contribution to the higher nonspecific binding with ^{125}I-labeled proteins is that the iodine atom is a large, effectively nonpolar group, known to enhance nonspecific hydrophobic iterations. Whereas chemical labeling provides a population of different molecules with the label attached at different places along the chain, the protein kinases label

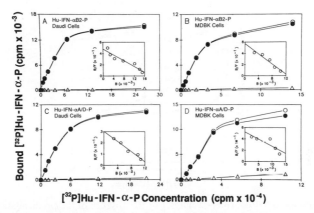

[32P]Hu - IFN - α - P Concentration (cpm x 10[-4])

FIG. 2. Binding of ^{32}P-labeled Hu-IFN-αB2-P and Hu-IFN-αA/D-P to cells. [32P]Hu-IFN-αB2-P or [32P]Hu-IFN-αA/D-P was added to cells at the indicated concentrations in the absence (total binding, O) or presence (nonspecific binding, $\triangle$) of excess nonradioactive Hu-IFN-αB2-P or Hu-IFN-αA/D-P, respectively, and incubated for 1 hr at 24°. Specific binding ($\bullet$) represents the difference between the total and nonspecific binding. (A and B) Binding of [32P]Hu-IFN-αB2-P to Daudi and MDBK cells, respectively; (C and D) binding of [32P]Hu-IFN-αA/D-P to Daudi and MDBK cells, respectively. *Inset:* Scatchard analysis of the specific binding data (B, radioactivity of ligands bound to cells; F, radioactivity of free or unbound ligands). [Reproduced from P. Wang, L. Izotova, T. M. Mariano, R. J. Donnelly, and S. Pestka, *J. Interferon Res.* **14**, 41 (1994).]

the protein at one or more specific residues along the protein chain. We have used various protein kinases to label proteins. Where it is necessary to conserve a negative, positive, or neutral charge on the segment of the molecule to be labeled, it is possible to do so by the appropriate choice of protein kinase recognition site. This also contributes to retaining the activity of the protein.

Stability of Phosphorylated Proteins

As noted above, some of the phosphorylated proteins exhibit two rates of decay; a rapid loss of the phosphate and a relatively slow stable phase. This is shown in Fig. 3. When this is seen, incubating the phosphorylated product after the reaction at 37° during the dialysis procedure for a period of about 24 hr at 37° eliminates the rapid decay phase, as shown in Fig. 4. The observation of a rapid decay phase was surprising. Nevertheless, the loosely bound phosphate can easily be eliminated by incubating a phosphorylated product at 37° as described above. The percentage of [^{32}P]phosphate retained on the [^{32}P]MAb chCC49K1 was determined by comparing the radioactivity at different time points with that of the initial value in buffer

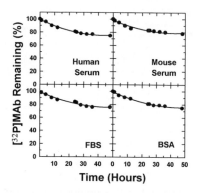

Fig. 3. Stability of [^{32}P]MAb-chCC49CK1 in the absence of the preincubation step. Each reaction contained 0.5 ml of human serum, mouse serum, fetal bovine serum, or a solution of bovine serum albumin (5 mg/ml in PBS), 125 μl of 1 M Tris-HCl (pH 7.4), and 15 μl of the [^{32}P]MAb-chCC49CKI (1.1 × 10^7 cpm) for a total volume of 0.64 ml and incubated at 37°. Portions of 20 μl were taken in duplicate over a 48-hr period to determine the stability of [^{32}P]phosphate attached to the MAb. The percentage of ^{32}P remaining on the [^{32}P]MAb chCC49K1 in sera and buffer over a 48-hr period at 37° is shown. The experiments were performed as described in text, except that no preincubation step was used to remove loosely attached phosphate after the phosphorylation reaction. [Data from L. Lin, S. D. Gillies, J. Schlom, and S. Pestka, *Protein Expr. Purif.* **15**, 83 (1999).]

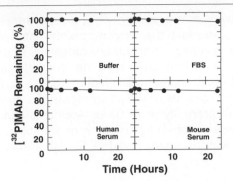

Fig. 4. Stability of [^{32}P]MAb chCC49K1 in various sera. Labile phosphate was removed before use of the labeled monoclonal antibody as described in text. The percentage of ^{32}P remaining on the [^{32}P]MAb chCC49K1 in sera and buffer over a 24-hr period at 37° is shown. The experiments were performed as described in text. [Data from L. Lin, S. D. Gillies, Y. Lan, L. Izotova, W. Wu, J. Schlom, and S. Pestka, *Int. J. Oncol.* **13**, 115 (1998).]

and various sera (Fig. 4). It can be seen that about 93–96% of the phosphate remained stably attached to the MAb after a 24-hr incubation in buffer, fetal bovine, human, or mouse serum.

Molecular Modeling to Introduce Phosphorylation Sites

Because the crystal structures of many proteins are known, it is possible to use the coordinates of the crystal structure to evaluate the effects of the insertion of the phosphorylation site and the phosphate itself. We have been able to utilize this effectively to predict useful sites within a protein in which to introduce phosphorylation sites with minimal modification of the protein structure (W. Wu and S. Pestka, unpublished studies, 2000).

Concluding Remarks

The preceding summary provides a description of the use of this new procedure for the modification and labeling of proteins that ordinarily do not contain phosphorylation sites. However, some proteins contain phosphorylation sites and thus can be phosphorylated directly. We have used phosphorylated IFN-γ, which contains a cAMP-dependent phosphorylation site, quite effectively as a ligand.[42,49–71] The procedures described in this chapter provide a product that is conveniently labeled to high specific activity and that is as active as the parental molecule. The binding specificity of these proteins is so high that often the competition portions of the experiment to eliminate nonspecific binding are not required. Depending on the specific requirements of the experiments, the phosphorylation sites can be introduced internally or as fusion segments added to the protein. In addition, the use of labeled phosphate or sulfate instead of iodine to label proteins provides a safer procedure for laboratory personnel.

[49] R. Fields, T. M. Mariano, S. Stein, and S. Pestka, *J. Interferon Res.* **8,** 549 (1987).
[50] Y. Hibino, T. M. Mariano, C. S. Kumar, C. A. Kozak, and S. Pestka, *J. Biol. Chem.* **266,** 6948 (1991).
[51] Y. Hibino, C. S. Kumar, T. M. Mariano, D. Lai, and S. Pestka, *J. Biol. Chem.* **267,** 3741 (1992).
[52] V. Jung, A. Rashidbaigi, C. Jones, J. A. Tischfield, T. B. Shows, and S. Pestka, *Proc. Natl. Acad. Sci. U.S.A.* **84,** 4151 (1987).
[53] V. Jung, C. Jones, A. Rashidbaigi, D. D. Geyer, H. G. Morse, R. B. Wright, and S. Pestka, *Somatic Cell Mol. Genet.* **14,** 583 (1988).
[54] V. Jung, C. Jones, C. S. Kumar, S. Stefanos, S. O'Connell, and S. Pestka, *J. Biol. Chem.* **265,** 1827 (1990).
[55] S. V. Kotenko, L. S. Izotova, B. P. Pollack, T. M. Mariano, R. J. Donnelly, G. Muthukumaran, J. R. Cook, G. Garotta, O. Silvennoinen, J. N. Ihle, and S. Pestka, *J. Biol. Chem.* **270,** 20915 (1995).
[56] S. V. Kotenko, L. S. Izotova, B. P. Pollack, G. Muthukumaran, K. Paukku, O. Silvennoinen, J. N. Ihle, and S. Pestka, *J. Biol. Chem.* **271,** 17174 (1996).
[57] C. S. Kumar, T. M. Mariano, M. Noe, A. K. Deshpande, P. M. Rose, and S. Pestka, *J. Biol. Chem.* **263,** 13493 (1988).
[58] C. S. Kumar, G. Muthukumaran, L. J. Frost, M. Noe, Y.-H. Ahn, T. M. Mariano, and S. Pestka, *J. Biol. Chem.* **264,** 17939 (1989).
[59] J. A. Langer, A. Rashidbaigi, V. Jung, H.-F. Kung, and S. Pestka, "Identification of the Receptor for Immune Interferon on Human Cell Lines Using ^{32}P-Labeled Human Immune Interferon." International Society for Interferon Research, Clearwater Beach, Florida, 1985.
[60] J. A. Langer and S. Pestka, *J. Interferon Res.* **5,** 637 (1985).
[61] J. A. Langer and S. Pestka, *Methods Enzymol.* **119,** 305 (1986).
[62] J. A. Langer, A. Rashidbaigi, and S. Pestka, *J. Biol. Chem.* **21,** 9801 (1986).
[63] T. M. Mariano, C. A. Kozak, J. A. Langer, and S. Pestka, *J. Biol. Chem.* **262,** 5812 (1987).
[64] T. M. Mariano, L. Izotova, and S. Pestka, "Cytokines—A Practical Approach" (F. R. Balkwill, ed.), p. 77. IRL Press-Oxford University Press, New York, 1994.
[65] A. Rashidbaigi, H.-F. Kung, and S. Pestka, *J. Biol. Chem.* **260,** 8514 (1985).
[66] A. Rashidbaigi, J. A. Langer, V. Jung, C. Jones, H. G. Morse, J. Tischfield, J. J. Trill, H.-F. Kung, and S. Pestka, *Proc. Natl. Acad. Sci. U.S.A.* **83,** 384 (1986).
[67] A. Rashidbaigi, H.-F. Kung, and S. Pestka, *Methods Enzymol.* **119,** 315 (1986).
[68] A. Rashidbaigi, S. Stefanos, V. Jung, and J. A. Langer, *J. Interferon Res.* **8,** 641 (1988).
[69] S. Stefanos, Y. H. Ahn, and S. Pestka, *J. Interferon Res.* **9,** 719 (1989).
[70] S. Stefanos and S. Pestka, *J. Biol. Regul. Homeost. Agents* **4,** 57 (1990).
[71] H.-F. Kung and E. Bekesi, *Methods Enzymol.* **119,** 296 (1986).

Acknowledgments

We thank Stephen Gillies, Jeffery Schlom, and Bruce Daugherty for input and advice in many parts of these projects and Eleanor Kells for assistance in preparing the manuscript. This study was supported in part by United States Public Health Services Grants RO1-CA46465 from the National Cancer Institute and RO1 AI36450 from the National Institute of Allergy and Infectious Diseases. A special award from the Milstein Family Foundation to S.P. provided additional support for a variety of efforts in this project.

Author Index

Numbers in parentheses are footnote reference numbers and indicate that an author's work is referred to although the name is not cited in the text.

A

Aasland, R., 61
Abbas, A. K., 357
Abbas-Terki, T., 386(57), 388(57), 391, 398(57)
Abe, K., 342
Abelson, J., 188
Abramowitz, M., 458, 466(12, 13)
Abrams, C. S., 331, 338(10)
Achacoso, J. A., 318(7), 319
Acheson, A., 34, 198(8), 199
Ackerman, L., 537
Ackermann, A., 252
Acland, P., 370, 373(13)
Acosta, M., 374
Adair, B. D., 165
Adam, A., 314
Adam, E. J. H., 6
Adam, S. A., 6
Adamczewski, M., 227(52), 228
Adams, J. A., 456, 473, 474(13), 482(13), 488(13), 490(13)
Adams, P. D., 598, 601(15)
Adams, S. P., 333, 334(18)
Adams, S. R., 472, 546, 560, 561(12), 564(12), 565, 566, 569(6), 574(6), 576(6)
Adelman, J. P., 255
Adelstein, S., 229, 230(3), 231, 234, 235(3), 237, 238
Aderem, A., 318(14, 15), 319(14, 15), 320
Aebersold, R., 188
Aebi, M., 95, 96(5), 97(5), 99(5), 188, 191, 192, 281
Aeder, S. E., 47
Afar, D. E., 430, 431, 432(5), 433(5), 435(5)
Ahmad, M., 177
Ahmed, N. N., 340, 341

Ahmed, S. N., 365, 366, 366(98), 367(98)
Ahn, Y.-H., 612
Ainsworth, J. R., 370
Aitchison, J. D., 294
Ajchenbaum, F., 598, 602(19)
Ajioka, R. S., 56
Akazawa, C., 123
Akimov, S., 374
Al-Alawi, N., 318, 319(2), 338, 520
Alarcon, B., 212, 282
Al-Awqati, Q., 523
Albagli, O., 386, 387(13), 397
Albalat, R., 386(38), 387(38), 390
Alber, G., 213, 229
Alberola-Ila, J., 424, 425(15), 428(15)
Alberti-Segui, C., 64, 65(5)
Aldea, M., 415
Aldrich, C. J., 362, 363(82)
Aldrich, T. H., 34, 198(7, 8), 199
Alessi, D. R., 332, 338(13), 340(13)
Alexanko, A. P., 601
Ali, J. A., 422
Allchin, E. H., 42
Allen, D. G., 452, 454
Allen, L. A. H., 318(14, 15), 319(14, 15), 320
Allende, M. L., 146, 147, 150(14), 152(14)
Alonso, M. T., 446
Alonso-Llamazares, A., 598, 601(26)
Altman, A., 6
Alvarez, J., 364, 445, 446, 451(12), 452, 455, 456, 458(1)
Amalric, F., 370, 373
Amaral, M. C., 227
Amarante-Mendes, G. P., 514
Amberg, D. C., 5
Ambros, V., 537
Ambrosini, E., 166
Amiot, M., 213, 422

C

Q

Subject Index

A

ABL
  breakpoint cluster region fusion protein
    activation of kinase, 439–440
    plasmid construction, 433–435
    dimerization and activity, 430–431
    translocation in cancer, 429–430, 432
N-Acetyl-β-D-glucosaminidase, lysosomal
  marker assay, 54
Aequorin calcium indicator
  advantages and limitations, 454–457, 471
  calcium affinity mutant, 445–446
  calcium concentration determination
    calculations, 441, 452–454
    data acquisition, 451–452
  coelenterazine analogs, 441, 464
  G protein-coupled receptor signal trans-
    duction analysis
    G_α subunit coexpression, 458–459
    microtiter plate assays, 462–464
    overview of signal transduction,
      457–458
    performance as calcium indicator, 466,
      468–469, 471
    plasmids, 460–461
    stable expression studies, 461–462
    transient expression studies, 462
  ion specificity, 442
  luminescence detection, 450–451, 460,
    462–464, 471
  photon emission process, 440–441, 456
  purinergic P2X receptor calcium flux
    analysis
    channel activity overview, 459–460
    microtiter plate assays, 462–464
    performance as calcium indicator, 469,
      471
    plasmids, 460–461
    stable expression studies, 461–462
    transient expression studies, 462
  reconstitution into functional photo-
    protein
    calcium depletion, 449–450

coelenterazine addition, 449
stoichiometry of binding, 440, 456
subcellular targeting
  comparison of mitochondrial and cyto-
    plasmic reporters, 464–466
  cytoplasm, 443–444
  endoplasmic reticulum, 445
  Golgi, 445
  immunofluorescence microscopy,
    447–449
  mitochondria, 444
  nucleus, 445
  sarcoplasmic reticulum, 445
  strategies, 443
  subplasma membrane region, 445
terminus deletion studies, 442–443
transfection
  calcium phosphate transfection,
    446–447
  HeLa cell preparation, 446
  skeletal muscle myotube preparation,
    446–447
AKT, fatty acylated chimeric protein,
  340–341
Alkaline phosphatase fusion proteins
  activity assays, 19–20, 26–27
  APtag vectors
    overview, 21
    clone insertion, 21–24
    transfection, 24–25
  dimerization, 19, 200
  immunoprecipitation, 27–29
  placental isoform for fusion, 19
  purification, 25–26
  receptor study applications
    binding assays
      caveats, 200
      cell surface binding, 201, 203–204
      cell-free system, 204–205
    binding *in situ*
      cultured cells, 34–35
      tissue sections, 32–33
      whole-mount embryos, 29–32
    cloning of novel receptors and ligands

657

ISBN 0-12-182228-1